Tutorial:

Modern Design and Analysis of DISCRETE-EVENT COMPUTER SIMULATIONS

Edward J. Dudewicz and Zaven A. Karian

IEEE CATALOG NUMBER EHO227-9
LIBRARY OF CONGRESS NUMBER 85-60381
IEEE COMPUTER SOCIETY ORDER NUMBER 597
ISBN 0-8186-0597-9

COVER DESIGNED BY JACK I. BALLESTERO

IEEE COMPUTER SOCIETY

THE INSTITUTE OF ELECTRICAL AND ELECTRONICS ENGINEERS, INC.
IEEE

COMPUTER SOCIETY PRESS

Published by IEEE Computer Society Press
1730 Massachusetts Avenue, N.W.
Washington, D.C. 20036-1903

COVER DESIGNED BY JACK I. BALLESTERO

IEEE Catalog Number EH0227-9
Library of Congress Number 85-60381
IEEE Computer Society Order Number 597
ISBN 0-8186-0597-9 (Paper)
ISBN 0-8186-4597-0 (Microfiche)

Order from: IEEE Computer Society
Post Office Box 80452
Worldway Postal Center
Los Angeles, CA 90080

IEEE Service Center
445 Hoes Lane
Piscataway, NJ 08854

THE INSTITUTE OF ELECTRICAL AND ELECTRONICS ENGINEERS, INC.

Preface

Simulation has become **the most widely used management science/operations research technique**, replacing mathematical programming applications in many instances. Managers, researchers, and professionals in nearly all endeavors, and especially in computers and engineering, today recognize that they need to understand and be able to apply this modern technique for their research and decision making.

Simulation is particularly useful in situations where the following constraints apply: a complete mathematical problem formulation is not available or possible, the available analytical methods require simplified assumptions that distort the true nature of the problem, the available analytical methods are so complex that they are impractical to use, it is too complex or too expensive to conduct real-world experiments, or it is necessary to compress time so that a system or process is observable over an extended period.

The **objective of this tutorial** is **to provide a working understanding of the design, implementation, and analysis of computer simulations.** The **emphasis** is thus **on methods for application rather than on theory,** and **our goal is for the reader to be able to properly apply and interpret the design and analysis aspects covered in his/her own simulation studies of systems.**

The authors gratefully acknowledge support provided by a sabbatical leave from Syracuse University, and by the Robert C. Good Fellowship of Denison University, which contributed to the development of this work. They also wish to express their appreciation to C. G. Stockton, Bill Carroll, and Margaret Brown of the IEEE Computer Society for their assistance in the editorial process of this project.

Edward J. Dudewicz
Syracuse, New York

Zaven A. Karian
Granville, Ohio

Introduction

Experimental sampling to seek out answers to statistical questions (e.g., W.S. Gosset's experimental sampling in 1908, under the pseudonym Student, in which he sought to determine the distribution function of the sample correlation coefficient in normal samples, and also to bolster his faith in what is now known as Students' t-distribution which then rested on a rather ad hoc theoretical analysis) **was the precursor of modern digital simulation.** With the advent of the modest electronic digital computer, simulations of more complex questions became possible, however great programming effort was required as was relatively great cost. With the relative ease of programming major simulations in modern software, and with the relative economy of computations **today, simulations of great size and complexity are being attempted.** In order **to obtain valid results** in these settings, as well as to obtain them in more modest settings, **considerations of a statistical nature are essential. These considerations go under the name of "modern design and analysis of experiments" and are the subject of this text as they relate to discrete-event computer simulations.**

Table of Contents

Part I. Random Number Generation and Testing of Random Number Generators

Today, simulation and Monte Carlo studies play a crucially important role in many fields. Such studies depend, for their validity, on a source of numbers on the interval 0.00000000 to 1.00000000 which appear to be random and independent of one another. Sources of such numbers are called **random number generators**, and their **history** (as well as **the need for voluminous quantities of such numbers in modern simulation and Monte Carlo studies**) is covered in the first article by Dudewicz (1975a). That article also gives a brief summary of some of the methods used to provide such numbers on a digital computer.

Testing of the random number generator one will use for goodness (as well as for speed) is crucial to the validity of any simulation study. A description of **how such testing is properly done**, with some results for certain widely-available generators is given in the next two articles by Dudewicz (1976) and Dudewicz (1980), while the fourth article (Dudewicz (1975b)) considers **how to obtain chi-square percentage points needed in such testing.**

The fifth article by Borgwaldt and Brandl (1981) is typical of much of what appears in the applied literature on random number generators: good scientists (re)discover the need to test random number generators, and retest one of the widely available generators (such as RANDU), finding it bad. Often such articles also present a "new" generator which passes some cursory testing, but which may otherwise be quite bad.

The sixth article by Dudewicz and Ralley (1982) notes **which popular generators have been extensively tested** (and so users need not re-test them, as the process of testing is time-consuming and costly if done well, and should thus be restricted to generators at one's facility which have not been the subjects of such testing in the past). Some details of a "new" bad generator from the literature are also given.

Full details of testing of random number generators are available in *The Handbook of Random Number Generation and Testing* cited in the sixth article. The TESTRAND code described in the seventh article by Dudewicz and Ralley (1983) is available for a nominal cost, and will allow many users to easily test their own computer center's generators. That full code is also available in the *Handbook* just-cited.

RANDOM NUMBERS: THE NEED, THE HISTORY, THE GENERATORS

Edward J. Dudewicz

The Onio State University, Columbus, Ohio 43210, U.S.A.

SUMMARY

Today simulation and Monte Carlo studies play an important and ever more significant role in virtually every field of human endeavor, and such studies often consume large amounts of computer time. Nearly every such study requires, for its execution, a source of random numbers (i.e. numbers which appear to be independent uniform random variables on the range 0.0 to 1.0). Historically statisticians have attempted to provide quality random numbers in quantity in various ways, the most common today being via numeric algorithms executed within a digital computer. Statistical testing can (although it has not yet) rank these algorithms on speed and goodness.

Keywords: Uniform Random Numbers, Pseudo-Random Numbers, Random Number Generators, Observational, Internal Physical, Internal Numeric, Mid-Square, Congruential, Feedback Shift Register, Simulation, Monte Carlo.

1. INTRODUCTION

Today simulation and Monte Carlo studies play an important and ever more significant role in virtually every field of human endeavor. Such studies arise in business [including such areas as job-shop scheduling (Ashour and Vaswani, 1972) and marketing (Browne, 1972)]; the humanities [psychoanalysis (von Zeppelin and Moser, 1973)]; science [air pollution (Katz, 1973); chemistry (Manock, 1972); genetics (Madalena and Hill, 1972)], and social science [social conflict (Chesser, 1972); housing policies (Rider, 1973)]. An amusing example is given in Schmidt (1974) where PK (psychokinesis) was tested by seeing how well a subject (thought to have some PK ability) could affect the output of a binary random number generator. According to Schmidt, there was "... significant evidence of PK ..." shown by the experiment. (Of course a perceptive subject could perhaps capitalize on correlation in the generators; one might speculate that PK shown here is a result of such correlation.) An example which gives food for thought is given by Matsuda (1973), where ability to generate random numbers is used as a measure of creative thinking [the more random the numbers, the more creative the thinking!]. Such studies often consume large amounts of computer time. Nearly every such study requires, for its execution, a source (called a generator) of numbers which appear to be independent uniform random variables on the range 0.0 to 1.0; such numbers

Reprinted with permission from *Statistical Distributions in Scientific Work*, G.P. Patil et al. (Editors), Volume 2, pages 25-36.

are called pseudo-random numbers or more commonly and somewhat imprecisely simply random numbers (Knuth, 1969, p.3). Due to the importance of such studies, whose results may directly and significantly influence the lives of millions [for example, see Federal Highway Administration (1973) and Milstein (1973)] and the fact that non-independence and/or non-uniformity in the random number generator may utterly vitiate a study, access to "good" random number generators is of prime importance to investigators using simulation methodology. Due to the cost (in computer time) of such studies, access to "fast" random number generators is of co-prime importance.

Despite these needs for access to "good" and "fast" random number generators, and despite the existence of a large body of literature in the area (the bibliography of Nance and Overstreet (1972) lists 491 papers and books, and many others have appeared since that bibliography was compiled), the typical situation is that a computer center or a computer language will offer one (rarely two or more) generators which are at best weak and poorly documented. Often, the only generator available is RANDU [see IBM (1970,p.77) for details of the generator and Learmonth and Lewis (1973, p.167) for some of its defects]. We omit details of a package developed at Ohio State to alleviate this situation; a copy of this package, whose acronym is IRCCRAND, is available on request from the author.

Below we cover the need for and history of random number generators, random number generation methods, and (allude to) the general testing of random number generators. As von Neumann (1951, p.37) said, " ... I think nobody who is practically concerned will want to use a sequence produced by any method without testing it statistically ...".

2. THE NEED, THE HISTORY, THE GENERATORS

It has been said (Learmonth and Lewis, 1973, p.167) that "Many of today's statistical questions are being answered through large-scale simulation. The generation of good pseudo-random deviates for simulation and Monte Carlo experiments is of prime importance.". In this light, and in light of the fact (loc. cit.) that "... many generators are being used whose statistical properties make them a hindrance rather than an aid in such experiments," we see both the need for random numbers and the need to take the generation of random numbers for one's simulation seriously (and not, for example, to use a generator because it is the only easily accessible one).

The importance of simulation and Monte Carlo studies, where the random numbers find their use, has been outlined in the Introduction. As Dean C. Jackson Grayson, Jr., of Tulane University has stated (Naylor, Balintfy, Burdick, and Chu, 1966, p.v), "With computer simulation, one can gain insight into complex systems, build and test theories, and peer dimly, but explicitly into the future. The final choice of an action or a theory to fit reality still rests with the human involved, but computer simulation is capable of providing powerful assistance as an analytical tool.". This importance of simulation and Monte Carlo and its widespread use has led to the development of numerous special simulation languages [see Tocher (1965) and Teichroew and Lubin (1966); e.g. SIMSCRIPT, GPSS, DYNAMØ, SIMULA, etc.] and to the vital necessity of incorporating randomness into such studies. Randomness occurs in reality in such instances as those where: items are neither completely predictable from, nor perfectly correlated with, available knowledge and accepted causes; items are unexpected; results of an experiment differ from the norm; and systems are not controllable in every detail. Hence the provision of randomness (via random numbers) is essential if a simulation model is to adequately mirror reality. How the need to produce randomness in attributes (cash requirements, priorities, routings through machines, rainfall amounts, etc.) and in event times (service times, inter-arrival times, times between rainfalls, etc.) has been filled historically is the subject of the remainder of this section. Note that while the need in practice is seldom a need for uniform random numbers on the interval 0 to 1, essentially all random number needs can be met via use of such uniform random numbers on the interval 0 to 1.

While the decade 1940-1949 saw the development of the first modern computers, which led to extensive developments in the principles, techniques and capabilities of simulation and Monte Carlo, the area was in existence in embryonic form even before the advent of modern computing machinery. For example, we may recall the Buffon Needle problem (Kuo, 1965, pp.269-272), which motivated experiments in the second half of the 19th century (Hall, 1873) in which experimenters threw needles haphazardly onto a ruled board and used the observed number of needle-line intersections to estimate π. On rare occasion the emphasis of such embryonic uses was even on original discovery rather than on comforting verification (the latter is about all one can claim for a simulation estimate of π). For example, in 1908 W. S. Gosset (writing under the pseudonym Student) used experimental sampling to seek out the distribution function of the sample correlation coefficient in normal samples, and also (Hammersley and Handscomb, 1964, p.7) to "bolster his faith in his so-called t-distribution, which he had devised by a somewhat shaky and incomplete theoretical analysis". At this stage random numbers were generated by <u>observational</u> methods [also called "physical devices" (Hammersley and Handscomb, 1964, pp.26, 159-160)] such as (Hull and Dobell, 1962, p.245) rolling dice, drawing from a supposedly well-stirred urn, spinning a roulette wheel, etc.. For example if one spins a spinner which has the circumference marked off from 0.00 through 1.00, then it might be reasonable in some instances to assume the result is a random number between 0 to 1. Rather than go through the (significant) effort of such a process each time random numbers are needed, it makes sense to construct (and test) such numbers and tabulate them for others to use when the need arises. Thus, at Karl Pearson's suggestion, L. H. C. Tippett prepared such a table (Tippett, 1927) consisting of digits which he had collected from census reports. This appears to be the first such table, and the only one as late as 1938 (Kendall and Babington-Smith, 1938, p.156). These digits, of which Tippett gave 41,600, have been tested for randomness by a variety of methods and seem satisfactory (Gage, 1943). However Yule (1938) thought the numbers "patchy". Since 41,600 digits were an insufficient number for lengthy sampling investigations even before the advent of the modern digital computer (one can form only 4160 ten-digit random numbers from such a collection) a table of 100,000 digits was published in 1939; this is the table of Kendall and Babington-Smith (1951). These digits were produced mechanically by a method described in Kendall and Babington-Smith (1939), who also tested (and found acceptable) the digits produced. A number of other tables (see Hull and Dobell, 1962, p.246) were also produced at about this point in time, including tables derived from telephone directory numbers (Kendall and Babington-Smith, 1938, pp.156-157, 164) and from logarithm tables [15,000 random digits tabled by Fisher and Yates (1938), compiled from among the 15th to 19th digits in certain sections of A. J. Thompson's Logarithmica Britannica; for some tests see Kendall and Babington-Smith, 1939, pp.59-61]. Tabulating efforts seem to have ended with publication of the monumental million random digits of the Rand Corporation [Rand (1955); these digits are also available on punched cards], which were produced (Brown, 1951) by an electronic roulette wheel constructed explicitly for this purpose. However, special-purpose tables are still being constructed, often on a digital computer, and published. For example Clark (1966) gave 501,120 random digits and 100,224 normal deviates along with statistics of their subsets which allow for efficient stratified sampling; see also Clark and Holz (1960). However, as Vickery (1939, p.62) noted, "These numbers [those of Kendall and Babington-Smith] ... do not begin to supply the need for such numbers in contemporary statistical investigations. A further difficulty lies in the vast amount of labour involved in drawing, by the use of these numbers, a sample such as one of 10,000 from a population of 500,000.". The disfavour into which such tables and direct observational methods fell is felt in the prophetic statement of Brown (1951) that "My own personal hope for the future is that we won't have to build any more random digit generators. It was an interesting experiemnt, it fulfilled a useful purpose, and one can do it again that way, if necessary, but it may not be asking too much to hope that ... some ... numerical process will permit us to compute our random numbers as we need them. The advantages of such a method are fairly obvious in large-scale computation where extensive tabling operations are relatively clumsy.".

Since observational methods involve a number of dubious assumptions (e.g. if one derives one's numbers from telephone book numbers, one may need to assume an assignment of numbers independent of names as well as ignoring the exchange, while if one uses license plate numbers on (e.g.) cars similar assumptions will be needed) tables developed from them may often fail statistical tests. However, even if they pass, one will have a large volume to store (on tape, cards, or disc), which will therefore be expensive (since, as we noted previously, a need for 2, 3, or 4 million random digits is not unusual). A third fault of such tables is that the process of reading them from storage into the computer memory is a relatively slow process and hence can impede execution of the simulation program ... its execution may have to be suspended while numbers are read in. One alternative considered briefly was the use of an internal physical source, such as particles given off by decay of a radioactive source [the first mention of use of a radioactive source in connection with random numbers seems to have been due to Vickery (1939) in a slightly different context], or noise in an electronic circuit (for references see Hull and Dobell, 1962, p.246). Some major faults (for others see Hull and Dobell, 1962, p.247) of this method are that debugging becomes difficult (since a different random number stream is obtained each time the program is run), and that if one's simulation situation involves comparing various alternatives then each alternative is simulated with a different sequence of random numbers (which is undesirable since there is sampling efficiency in using the same sequence of random numbers for the different alternatives). The above considerations led to the development of internal numeric sources of random numbers, which use a deterministic numerical process to produce sequences of (pseudo) random numbers which can pass statistical tests for randomness (i.e. which cannot be distinguished from "true randomness").

The first internal numeric source was the mid-square method due to von Neuman and Metropolis in about 1946 (see Hull and Dobell, 1962, p.247, and Hammer, 1951, p.33). With this method one starts with some 2n digit number, squares it, and takes the middle 2n digits of the 4n digit product as the next number. The length of such a sequence of numbers (before it gets into a cycle) depends on the starting number (and is thus "indeterminate"), and may be small (Forsythe, 1951). For example 165_{10} on an 8-bit binary machine is represented as $x_0 = 10100101$, whose square is 0110101001011001, which yields as its middle 8 digits 10100101; this is the starting number, hence the sequence yields a length of only 1 before repeating. Similarly with $x_0 = 10_{10}$, whose square is 0100.

Another method frequently discussed is the use of successive digits of the infinite expansion of some transcendental number (such as π, e, etc.). Some variants are based on ergodic theorems of Weyl (1916) and use such numbers as $\pi n^2 - [\pi n^2]$ for $n = 1, 2, \ldots$; see Hull and Dobell (1962, pp.247-248) for details and recent references. This, like the mid-square method, is of little use (though for different reasons): the generation process is long and (since digits are stored) we come back to table storage problems.

The internal numeric source which is used most frequently at present is the multiplicative congruential (or power residue) method where one takes residues of successive powers of a number x to be the successive numbers. In this method, as in the mid-square method, the numbers involved are treated as integers (and integer arithmetic operations are used in the computer) until, just before use, a decimal point is placed before the digits in the random sequence:

$$(n^{th} \text{ number in the sequence}) \equiv x_n = x^n \bmod m. \qquad (1)$$

Since formula (1) involves raising a number x to very large powers it would be difficult to implement on most computers (with any accuracy after reduction modulo m); however it can be shown [recalling that $x^n \bmod m$ is the remainder when x^n is divided by m (e.g. $17 \bmod 5 = 2$ since $17 \div 5 = 3 + 2 / 5$), one finds $x^n \bmod m = (x^n + xkm) \bmod m = (x(x^{n-1} + km)) \bmod m = (x(x^{n-1} \bmod m)) \bmod m = (x \cdot x^{n-1}) \bmod m$; justification of the steps of this deviation is an exercise in elementary number theory] that an equivalent expression is

$$x_n = (x \cdot x_{n-1}) \bmod m, \qquad (2)$$

and this is easily implemented. The first generator of this type was given by D. H. Lehmer (see Hull and Dobell, 1962, p.247) in 1949 for the ENIAC, an 8-digit decimal machine. His choices of x and m were $x = 23$ and $m = 10^8 + 1$ for reasons of computational efficiency: one need simply multiply the current 8-digit number by 23, remove the top 2 digits and subtract them from the remaining number (and, if this number is negative, add $10^8 + 1$) to obtain the next 8-digit number in the sequence. For example, if $x_{n-1} = 10741101$ then this process is

$$\begin{array}{rl} 10741101 & x_{n-1} \\ 23 & \quad \text{multiply by 23} \\ \hline 32223303 & \\ 21482202 & \\ \hline 0247045323 & 23x_{n-1} \\ -02 & \\ \hline 47045321 & x_n \text{ (if } \geq 0\text{; if } < 0 \text{ add } 10^8 + 1). \end{array} \qquad (3)$$

To find a mod b one subtracts "b" from "a" as many times as possible without having the result go negative (so 0247045323 mod 10^6 = 045323). Hence, this process, which avoids the fairly slow operation of division on the computer in finding $x \cdot x_{n-1}$ mod m, yields precisely x_n as given in (2).

When a multiplicative congruential random number generator as specified in (2) is to be used, one must specify x (called the multiplier, and often denoted by ρ in the literature), x_0 (the starting value, called the seed), and m (the modulus). At present the most common choice of m is $m = r^s$, where r is the base of the number system of the computer being used (usually 2, 10, or 16) and s is the word length on the computer; this results in a fast mod operation. [When two integers of word length s each are multiplied on the computer, a product 2s places long results, usually stored in two registers, say the A register for the top s places, and Q register for the low s places.

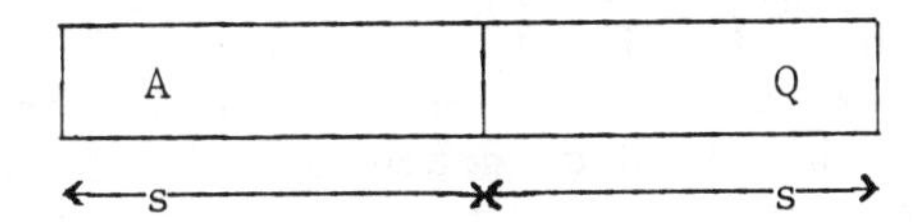

To perform the mod r^s operation on this product, one simply drops the contents of the A register. Finally one places a decimal point before the digits in the Q register in order to obtain random numbers on (0, 1) instead of on (0, $r^s - 1$). From one generator one can obtain several streams of random numbers simply by choosing different values for the seed x for each.] Some standard recommendations for a binary machine (r = 2) are: choose x_0 to be any odd integer, and choose x to be of the form $8t \pm 3$ (t = 1, 2, 3, ...). This will guarantee a cycle length of 2^{s-2} terms. For example, if s = 35 (as for the IBM 7040, 7044, 7090, and 7094) one obtains approximately 8.5 billion terms before repeating, which is desirable; considerably smaller cycles can be obtained if one chooses $x \neq 8t \pm 3$ for any t. It also guarantees (see IBM, 1959, pp.5-6, 11) that the bits of x_n are periodic; in particular if $x_n = b_s b_{s-1} \ldots b_4 b_3 b_2 b_1$ (where each of $b_s, \ldots, b_1$ is either 0 or 1) then $b_1 = 1$ always, b_2 and b_3 either do not change or else alternate as n changes, and b_i has period 2^{i-2} (i = 4, ..., s). Hence one should not use the bits of x_n as random bits (rather, if random bits are needed then one should, e.g., let the nth bit be 0 if $x_n \leq 0.5$ and 1 if $x_n > 0.5$). It is standard (see IBM, 1959, p.5) to note: that choosing a multiplier x with few 1's yields a faster multiplication [this is true on machines (such as the IBM 1130) which have software multiplication, but not on machines (such as the IBM 7090, 370, etc.) which have hardware multiplication]; that x close to $\sqrt{r^s}$ is a good choice for a multiplier; and that some seeds x_0 are better than others (in particular seeds with many leading zeros should be avoided). The analogous standard recommendations for a decimal machine (r = 10) are: choose x_0 to be any number not divisible by 2 or 5, and choose x to be of the form ±(3, 11, 13, 19, 21, 27, 29, 37, 53, 59, 61, 67, 69, 77, 83, 91) mod 200.

This will guarantee a cycle length of $5 \cdot 10^{s-2}$ terms (e.g. 500 million terms on an $s = 10$ digit machine such as the B220). It also guarantees periodicity in the digits as for a binary machine. Again it is standard to recommend an x with a small sum-of-digits and close to $\sqrt{10^s}$ (e.g. if $s = 10$, $x = 0000100011$: has sum of digits 3, which is small; is close to $\sqrt{10^{10}} = 100000$; and has $0000100011 \bmod 200 = 11$, which is acceptable). These recommendations generalize in the analogous way to a machine of base r; e.g., the seed x_0 should not be divisible by factors of r. It should be noted that some of these standard recommendations have been shown in recent literature to lead to serious problems. For example, the recommendation of x close to $\sqrt{r^s}$ given by IBM (1959, p.5) and still being issued by IBM currently can lead to serious difficulties in the distribution of triplets of random numbers from a generator, thus contradicting a "standard" recommendation (Coveyou and MacPherson, 1967, p.119). Hence extreme care is needed in the selection of a generator and one should probably not use one's own [as one might be led to do after reading IBM (1959) ... which is still being issued in 1974].

Among other methods one may find noted in the literature, but which are not widely used today, are: the Fibonacci series method, which uses $x_n = (x_{n-1} + x_{n-2}) \bmod r^s$, which has a large period, but which also has such disadvantages as runs up and down and a loss of speed compared to (2) due to the memory access needed to keep two old values; the additive congruential method, which uses $x_n = (x_{n-1} + x_{n-k}) \bmod r^s$, where the run property of $k = 2$ does not exist for sufficiently large k ($k > 10$), but which requires storing k values and circulating on them (perhaps by a wrap-around file); and the mixed congruential method which uses $x_n = (x \cdot x_{n-1} + c) \bmod r^s$ with $c \neq 0$ [see Greenberger (1961) for some notes on such generators and Greenberger (1965) for some cases of poor performance]. The Fibonacci series method has been shown to be bad (see Rotenberg, 1960, p.75), while Coveyou and MacPherson (1967, p.101) state that the claim that the mixed congruential generators ($c \neq 0$) differ significantly from the multiplicative congruential generators in their statistical properties is "superstition".

The most promising internal numeric source under development at present is the generalized feedback shift register (GFSR) algorithm, which utilizes a linear congruence modulo 2 to produce its sequence; this will not be discussed at length in the present paper. Its advantages (arbitrarily long computer-independent period, speed and good multidimensional properties) will make it the generator of the future unless serious non-random properties are discovered for its sequences in the near future.

Given one (or several) proposed methods of generating a sequence $x_1, x_2, x_3, \ldots$ of random numbers, the question arises "How random are these numbers?" Some results of testing (with generation times) will be given in Dudewicz (1975), which will also consider the ranking, on speed and goodness, of algorithms for providing random numbers.

3. ACKNOWLEDGEMENTS

The author wishes to acknowledge the support of the Instruction & Research Computer Center at The Ohio State University (Dr. Roy F. Reeves, Director), and the programming assistance of Mr. Dale J. Schroeder of the IRCC. This work was supported in part by the U.S. Army Research Office - Durham.

REFERENCES

Ashour, S. and Vaswani, S.D. (1972). Simulation, 18, 1-10.

Bates, C.B. and Zirkle, J.A. (1971). Analysis of random numbers from four random number generators. Technical Report 4-71, Systems Analysis Group, U.S. Army Combat Developments Command, Fort Belvoir, Virginia 22060, August 1971.

Brown, G.W. (1951). In Monte Carlo Method, A.S. Householder (ed.). National Bureau of Standards, Washington, D.C., 31-32.

Browne, W.G. (1972). Interfaces, Bulletin of the Institute of Management Sciences, Meeting Issue,March 1972,121-22.(Abstract).

Chesser, R.J. (1972). A computer simulation model of conflict-cooperative behavior between social units. 42nd National Meeting of the Operations Research Society of America, Nov. 8-10, 1972, Atlantic City, New Jersey.

Clark, C.E. (1966). Random Numbers in Uniform and Normal Distribution with Indices for Subsets. Chandler Publishing Company, San Francisco.

Clark, C.E. and Holz, B.W. (1960). Exponentially Distributed Random Numbers. Johns Hopkins Press, Baltimore.

Coveyou, R.R. and MacPherson, R.D. (1967). Journal of the Association for Computing Machinery, 14, 100-119.

Dudewicz, E.J. (1975). Speed and quality of random numbers for simulation. In preparation.

Dudewicz, E.J. and Ramberg, J.S. (1975). Random Variable Generation, Digital Computers and Simulation. In Preparation.

Dudewicz, E.J., Ramberg, J.S. and Tadikamalla, P.R. (1974). Annual Technical Conference Transactions of the American Society for Quality Control, 28, 407-418.

Federal Highway Administration (1973). Prospectus "Adaptation of a freeway simulation model for studying incident detection and control". RFP-72, Office of Research, U.S. Department of Transportation, Washington, D.C..

Fisher, R.A. and Yates, F. (1938). Statistical Tables for Biological, Agricultural and Medical Research. Oliver & Boyd, Edinburgh.

Forsythe, G.E. (1951). In Monte Carlo Method, A.S. Householder (ed.). National Bureau of Standards, Washington, D.C., 34-35.

Gage, R. (1943). Journal of the American Statistical Association, 38, 223-227.

Good, I.J. (1969). The American Statistician, 23, 42-45.

Gorenstein, S. (1967). Communications of the Association for Computing Machinery, 10, 111-118.

Greenberger, M. (1961). Journal of the Association for Computing Machinery, 8, 163-167.

Greenberger, M. (1965). Communications of the Association for Computing Machinery, 8, 177-179.

Hall, A. (1873). The Messenger of Mathematics, 2, 113-114.

Hammer, P.C. (1951). In Monte Carlo Method, A.S. Householder (ed.). National Bureau of Standards, Washington, D.C., 33.

Hammersley, J.M. and Handscomb, D.C. (1964). Monte Carlo Methods. Wiley, New York.

Hull, T.E. and Dobell, A.R. (1962). SIAM Review, 4, 230-54.

IBM (1959; First Edition Reprinted December 1969). Random Number Generation and Testing. Reference Manual GC20-8011-0, International Business Machines Corporation, White Plains, New York.

IBM (1970). System/360 scientific subroutine package, Version III, programmer's manual, program number 360A-CM-03X, Manual GH20-0205-4(Fifth Ed.), IBM Corporation, White Plains, New York.

Katz, P.L. (1973). A generalized computer simulation of urban air pollution as related to energy use. 6th Hawaii International Conference on System Sciences, Jan. 9-11, 1973, Honolulu, Hawaii.

Kendall, M.G. and Babington-Smith, B. (1938). Journal of the Royal Statistical Society, 101, 147-166.
Kendall, M.G. and Babington-Smith, B. (1939). Supplement to the Journal of the Royal Statistical Society, 6, 51-61.
Kendall, M.G. and Babington-Smith, B. (1951). Tables of Random Sampling Numbers. Cambridge University Press, Cambridge.
Knuth, D.E. (1969). The Art of Computer Programming, Volume 2: Seminumerical Algorithms. Addison-Wesley, Reading, Mass.
Kuo, S.S. (1965). Numerical Methods and Computers. Addison-Wesley, Reading, Mass.
Learmonth, G.P. and Lewis, P.A.W. (1973). Proceedings of the Computer Science and Statistics Seventh Annual Symposium on the Interface, W.J. Kennedy (ed.). Statistical Laboratory, Iowa State University, 163-171.
Madalena, F.E. and Hill, W.G. (1972). Genetical Research, Cambridge, 20, 75-99.
Manock, J.J. (1972). Simulation of chemical reactions with a digital computer. 24th Annual American Chemical Society Southeastern Regional Meeting, Nov. 2-4, 1972, Birmingham, Alabama.
Marsaglia, G., Ananthanarayanan, K. and Paul, N. (1973). How to use the McGill random number package "SUPER-DUPER". School of Computer Science, McGill University, Montreal.
Matsuda, K. (1973). Japanese Psychological Research, 15, 101-108.
Milstein, J.S. (1973). Dynamics of the Vietnam War, A Quantitative Analysis and Predictive Computer Simulation. Ohio State University Press, Columbus, Ohio.
Nance, R.E. and Overstreet, C.,Jr. (1972). Computing Reviews, 13, 495-508.
Naylor, T.H., Balintfy, J.L., Burdick, D.S. and Chu, K. (1966). Computer Simulation Techniques. Wiley, New York.
Neave, H.R. (1973). Applied Statistics, 22, 92-97.
Rand Corporation (1955). A Million Random Digits with 100,000 Normal Deviates. The Free Press, Glencoe, Illinois.
Rider, K.L. (1973). A simulation study of four proposed housing policies for New York City. Winter Simulation Conference, Jan. 17-19, 1973, San Francisco, California.
Rotenberg, A. (1960). Journal of the Association for Computing Machinery, 7, 75-77.
Schmidt, H. (1974). Journal of Parapsychology, 38, 47-55.
Teichroew, D. and Lubin, J.F. (1966). Communications of the Association for Computing Machinery, 9, 723-741.
Tippett, L.H.C. (1927). Random Sampling Numbers. Cambridge University Press, Cambridge.
Tocher, K.D. (1965). Operational Research Quarterly, 16, 189-217.
Vickery, C.W. (1939). Supplement to the Journal of the Royal Statistical Society, 6, 62-66.
von Neumann, J. (1951). In Monte Carlo Method, A.S. Householder (ed.). National Bureau of Standards, Washington, D.C., 36-38.
von Zeppelin, I. and Moser, U. (1973). International Journal of Psycho-Analysis, 54, 79-84.
Weyl, H. (1916). Mathematische Annalen, 77, 313-52.
Yule, G.U. (1938). Journal of the Royal Statistical Society, 101, 167-72.

Speed and Quality of Random Numbers for Simulation

EDWARD J. DUDEWICZ

The Ohio State University, Columbus, Ohio 43210

SIMULATION and Monte Carlo studies play an important and ever more significant role in business, and such studies often consume large amounts of computer time. Hence, quality and speed are considerations of prime importance with regard to random numbers used in such studies. A number of random number sources have been tested for quality and speed, and the implications of the results for users of simulation methodology are given.

Introduction

Today simulation and Monte Carlo studies play an important and ever more significant role in virtually every field of business (for example, job-shop scheduling as described by Ashour and Vaswani [1], marketing as described by Browne [2], and process capability as described by Schafer [29]) and many fields of science (see Halton [5]). Such studies often consume large amounts of computer time. Nearly every such study requires, for its execution, a source (called a generator) of numbers which appear to be independent uniform random variables on the range 0.0 to 1.0; such numbers are called pseudo-random numbers or [20] random numbers. Due to the importance of such studies and the fact that bad random numbers may utterly vitiate a study, access to "good" random number generators is of prime importance to investigators using simulation methodology. Due to the cost (in computer time) of such studies, access to "fast" random number generators is of co-prime importance.

Despite these needs for access to "good" and "fast" random number generators, and despite the existence of a large body of literature in the area, the typical situation is that a computer center or a computer language will offer one (rarely two or more) generator which is at best weak and poorly documented. The only generator is often RANDU, which is described on p. 77 of International Business Machines Corporation [18] and is known to have substantial defects (see p. 167 of Learmonth and Lewis [28].) For this reason the current author developed a documented package of random number generators at The Ohio State University's Instruction aud Research Computer Center (IRCC). The package is called IRCCRAND [15] and a report [15] (including a program listing) as well as a tape are available upon request for nominal fees. (Requests for [15] should be addressed to: Publications Secretary, The Statistics Laboratory, The Ohio State University, 1958 Neil Avenue, Columbus, Ohio 43210, enclosing $3.50. Information on ordering the tape will be enclosed with the report.)

In the development of this package of random number generators we sought to include a spread of generators from good-costly to bad-cheap, and generators using a variety of methods to generate their sequences of random numbers. The reason for the good-costly to bad-cheap spread is that in some pilot studies a rough approximation to randomness is sufficient (in which case there is no need to pay for the high quality random numbers which arise from a

Dr. Dudewicz is Associate Professor (Department of Statistics) and Mathematical Analyst (Instruction & Research Computer Center) at The Ohio State University, Columbus, Ohio. His research was supported in part by the U. S. Army Research Office, Durham. A previous version of this paper was presented at the 1975 ATC in San Diego and appeared in the ATC Transactions.

KEY WORDS: Random Numbers, Pseudorandom Numbers, Monte Carlo, Congruential, Shift-Register, Timing, Uniformity Testing.

good generator when less costly random numbers will do as well for the purpose at hand). The reasons for including generators using a variety of methods to generate their sequences of random numbers are several. First, in order to insure that a study has not been vitiated by non-random properties of the generator used it is often desired to re-run at least part of a simulation study using a quite different generator (after which, if the two sets of results agree, one has more confidence that generator deviance from randomness is not of such a serious nature as to vitiate the simulation study). (Such vitiation may occur even in cases thought well-justified theoretically, through interactions of two or more good approximations to produce a bad overall approximation. For a surprising example see Neave [12].) Second, with the present state of knowledge about random number generators it seems ill-advised to provide one generator (while perhaps damning all others; e.g. Marsaglia, Ananthanarayanan, and Paul [23] state that in SUPER-DUPER (which they call a "random number package") the congruential part of their generator "··· uses the multiplier 69069 ··· much better than any of the highly touted but poorly justified multipliers used for the past 20 years."); this would disallow easy facility for runs with different generators, and might be found to be folly when new properties of generators are uncovered in the years ahead. (Indeed, results of Marsaglia [9], [10] on regularities in congruential random number generators led him to state [10] that they "... must be considered unsatisfactory...".) Thus IRCCRAND [15] is designed to be a true package, not just a one generator package.

The need for and history of random number generators has been covered in more detail elsewhere [16]. In the first section the random number sources tested for speed and quality with examples of their use will be detailed, the testing procedures will be detailed in the next section, and results of the tests, with their implications for persons using simulation methodology, will be detailed in the third section.

The Random Number Generators

The random number generators which were chosen for inclusion in IRCCRAND were chosen with several criteria in mind. We wanted (for reasons discussed in the Introduction) to include a spread of generators from good-costly to bad-cheap, to include generators using a variety of methods to generate their sequences of random numbers, and (so that we might provide guidance on their speed and goodness relative to other generators to users) to include widely used or widely available generators. We will now briefly describe the generators included and note the sources from which they were obtained (as well as sources from which more detail can be obtained). (Note that it is sometimes the case that the code used by various authors does not clearly correspond to their intended algorithms due to subtleties of computer handling of various operations; some brief notes are provied by Koehl [21]. The practical effect, if any, of these subtleties may be pursued in a later paper; the present study evaluates the code actually used and currently available to practitioners.)

Generator *RN1* was given by Lurie and Mason [8], and is a modified version of a generator given by Marsaglia and Bray [11]. The method used (called a *composite congruential* generator by Lurie and Mason [8], p. 364) is to let

$$x_n = (v_n + w_n) \bmod m_1 \qquad (1)$$

where v_n and w_n each come from multiplicative congruential generators, in particular from

$$\begin{cases} v_n = 65539\ v_{n-1} \bmod m_2 \\ w_n = 262147\ w_{n-1} \bmod m_2 , \end{cases} \qquad (2)$$

and $m_1 = r^s = 2^{31}$ while $m_2 = 2^{32}$. (The reason for this situation is as follows (see Marsaglia and Bray [11], pp. 757–758): On the IBM 360 and 370, FORTRAN integers are stored as 32 binary digits, and multiplication of two integers produces a 32-bit integer (the ordinary product mod 2^{32}), thus the simple multiplication corresponding to $65539v_{n-1}$ in RN1 yields a mod 2^{32} as in (2). However, in algebraic expressions a stored integer I is considered positive if $0 \leq I < 2^{31}$, and is considered $I - 2^{32}$ (negative) if $2^{31} \leq I \leq 2^{32} - 1$, hence the algebraic expression corresponding to $v_n + w_n$ in RN1 yields a mod 2^{31} as in (1).) (Note that $65539 = 2^{16} + 3$ and was noted as producing "excellent test results" by Marsaglia and Bray [11], p. 759, while $262147 = 2^{18} + 3$ was not mentioned in Marsaglia and Bray [11].) (In reading the RN1 program, note that $.46566 \times 10^{-9} = 2^{-31}$ is used to map an integer between 0 and 2^{31} onto the range 0 to 1.)

Generator *RN2* was given by International Business Machines Corporation [18], and is widely known by the name RANDU. The method used is the multiplicative congruential generator

$$x_n = 65539x_{n-1} \bmod 2^{31}. \qquad (3)$$

While this generator is at present widely used (indeed it may easily be the single most widely used generator) it has also (as discussed in the Introduction) been widely and severely criticized. (In reading the RN2 program, note that $2147483647 + 1 = 2^{31}$.)

Generator *KERAND* is identical in function to generator RN2. However while RN2 is written in FORTRAN, KERAND is written in assembler and thus has faster execution (more random numbers generated per unit of time). The program was provided by Keane [19].

Generator *IRANU* maps the random numbers provided by KERAND onto a user-specified set of integers ILOW, ILOW + 1, $\cdots$, IHIGH − 1, IHIGH (in order to provide random integers between ILOW and IHIGH). The program was provided by Scott [26].

Generator *RN4* is identical in function to generator RN2. However, the coding of RN4 is that used by the Statistical Package for the Social Sciences (SPSS) of Nie, Bent, and Hull [24] (also see Nie and Hull [25]), pp. 78ff, when it samples from a file. (The random number generator itself cannot be accessed by a user of SPSS; rather, it is implicitly used when one uses the instructions which are provided in SPSS for sampling a file in order to provide a $100p$ % sample of the entire file for a user-specified value of $p (0 < p < 1)$. Note that the SAMPLE instruction provided for this purpose in SPSS includes each item in the sample with probability p. Thus in sampling a file of (e.g.) 42 items at $100p = 20$ %, one may obtain a sample of size 7 at one time, a sample of size 8 at another time, and later perhaps a sample of size 11, even though 20 % of 42 is 8.4. This is because sampling is done item by item rather than file-wise (which is more complex to implement, but usually more reasonable).)

Generator *RN3* was given by Marsaglia and Bray [11] for the IBM 360. They also considered the IBM 7094 and the SRU (Univac) 1108. For further notes on the latter see Grosenbaugh [4]. (Recall that generator RN1 given by Lurie and Mason [8] is a modified version of that of Marsaglia and Bray [11].) The method used is a *composite* generator, which relies on three basic multiplicative congruential generators

$$\begin{cases} l_n = 65539\, l_{n-1} \bmod m_2 \\ m_n = 33554433\, m_{n-1} \bmod m_2 \\ k_n = 362436069\, k_{n-1} \bmod m_2 \end{cases} \quad (4)$$

where $m_2 = 2^{32}$. (Note that $65539 = 2^{16} + 3$, $33554433 = 2^{25} + 1$, and 362436069 were noted as producing "excellent test results" by Marsaglia and Bray[11]. In reading the RN3 program note that $.23283064 \times 10^{-9} = 2^{-32}$ while $16777216 = 2^{24}$.) One starts with 128 initial odd integer values stored in the locations $N(1), N(2), \cdots, N(128)$, and uses sequences l_n and m_n to choose one out of these 128 locations. The random number generated is based on the number in that location, say $N(J)$, as well as l_n and m_n. The location $N(J)$ is then refilled with a number from generator k_n. (Thus, except for the initial 128 values, the vector $N(1), \cdots, N(128)$ contains numbers from the sequence k_n. These are "shuffled" by using generators l_n and m_n to choose a random element of the vector $N(1), \cdots, N(128)$. The random number generated is not simply $N(J)$ if J is the location chosen, but also depends on l_n and m_n.) A subroutine is provided which initializes $N(1), \cdots, N(128), l_0, m_0, k_0$, etc. (RN3INT), and should be called once before RN3 is used. (However, a sophisticated user may provide his own initialization if he so desires.)

Generator *RN5* was given by Kruskal [6] and is now the generator used in OMNITAB II (see Hogben, Peavy, and Varner [17]). (Note that the original OMNITAB as used at Iowa State University (see Chamberlain and Jowett [14]) used a multiplicative congruential (non-composite) generator suggested by Marsaglia and Bray [11], p. 758, and hence did not suffer from the small period of the generator of (5).) The generator is a multiplicative congruential one,

$$x_n = 5^3 x_{n-1} \bmod 2^{13}, \quad (5)$$

and can be properly used by computers with relatively small word size (and hence is "extremely portable"). This generator has a period of only $2^{11} = 2048$, however, and will thus repeat itself rapidly in a simulation of any size (and hence may be appropriate mainly for either classroom illustration or small preliminary studies). (In reading the RN5 program note that $2^{13} = 8192$. This generator showed up well in a Coveyou-MacPherson analysis (see Kruskal [6]) indicating a weakness of that analysis as a *sole* criterion of goodness of a generator.)

Generator *RNCG* is a portable mixed congruential generator, which allows the user to specify the constants for any mixed congruential generator within his machine's limits. This is included mainly for use in studying random number generators, and it is in general not recommended that a simulation user choose his own generator constants (at least not without substantial reason, such as a solid treatment of those constants in the literature). One widely-used specification (e.g. this was used in the CUPL language (see Walker [27]) and was studied by Coveyou and MacPherson) is

$$x_n = 452807053 x_{n-1} \bmod 2^{31}. \quad (6)$$

(Note that $5^{15} \bmod 2^{31} = 30517578125 \bmod 2^{31} = 452807053$.)

The six generation routines *IUNI*, *IVNI*, *UNI*, *VNI*, *RNOR*, and *REXP* are drawn from the random number generator package SUPER-DUPER issued by McGill University School of Computer Science (see Marsaglia, Ananthanarayanan, and Paul [23]). Initialization routines used with these routines are *MCGILL* and *RSTART*. (Routines called EXPTH and TNORTH are also in the package, but are used internally, and should not be called by a user.) The basic random number generator used combines the multiplicative congruential generator

$$x_n = 69069\, x_{n-1} \bmod 2^{32} \qquad (7)$$

and an FSR generator. (Note that $69069 = 3 \cdot 7 \cdot 11 \cdot 13 \cdot 23$.) UNI yields random numbers on (0, 1), VNI yields random numbers on $(-1, 1)$, IUNI yields random integers on $(0, 2^{31})$, and IVNI yields random integers on $(-2^{31}, 2^{31})$. RNOR yields standard normal variables, while REXP yields exponential variables with mean 1.

The basic multiplicative congruential generator (7) of SUPER-DUPER yields x_n as 32 bits (each a 0 or a 1). The FSR generator (this generator uses a right shift 15 and a left shift 17, that is it is based on the trinomial $x^{17} + x^{15} + 1$) yields an nth number y_n (say) which also has 32 bits. These are then added by *exclusive or* addition (usually denoted by $\oplus$; recall that $1 \oplus 1 = 0$, $1 \oplus 0 = 0 \oplus 1 = 1$, and $0 \oplus 0 = 0$) to produce z_n, the nth random number. For example, let us calculate z_n for two arbitrary values of x_n and y_n:

$$\begin{cases} x_n = 01001 \cdots 01010 \\ y_n = 00101 \cdots 10110 \\ z_n = x_n \oplus y_n = 01100 \cdots 11100. \end{cases} \qquad (8)$$

This combination yields a generator with a period of $2^{46} - 2^{29} \approx 7 \times 10^{13}$ (see Learmonth and Lewis [28], p. 164); this period had been stated by Marsaglia, Ananthanarayanan, and Paul [23] as being approximately 5×10^{18}).

The eight generation routines *INT*, *RANDOM*, *NORMAL*, *EXPON*, *SINT*, *SRAND*, *SNORM*, and *SEXPON* are drawn from the random number generator package LLRANDOM issued by the Naval Postgraduate School (see Learmonth and Lewis [22]). Initialization routine *LLRAND* is used with these routines. (Routines called REXPTH and RNORTH are also in the package, but are used internally, and should not be called by a user.) The basic random number generator used is a multiplicative congruential one studied by Lewis, Goodman, and Miller [7]:

$$x_n = 16807\, x_{n-1} \bmod (2^{31} - 1). \qquad (9)$$

(Note that $16807 = 7^5$.) INT yields random integers, RANDOM yields random numbers on (0.0, 1.0), NORMAL yields standard normal variables,

TABLE 1. IRCCRAND'S [15] Generators, A Summary

Generator	Source	Method	Provides
RN1	Lurie and Mason [8]	Composite congruential	Random numbers
RN2	International Business Machines Corporation [18]	Multiplicative congruential (RANDU)	" "
KERAND	Keane [19]	RN2 in assembler	" "
IRANU	Scott [26]	Maps RN2 onto specified integers	Random integers
RN4	Nie, Bent, and Hull [24] (SPSS)	RN2	Random numbers
RN3	Marsaglia and Bray [11]	Composite	" "
RN5	Kruskal [6] (OMNITAB II)	Multiplicative congruential	" "
RNCG	IRCCRAND [15]	General mixed congruential	" "
IUNI	Marsaglia, Ananthanarayanan, and Paul[23]*	Congruential and FSR	Random integers $[0, 2^{31})$
IVNI	" " " "	" " "	Random integers on $[-2^{31}, 2^{31})$
UNI	" " " "	" " "	Random numbers on [0, 1)
VNI	" " " "	" " "	Random numbers on (-1, 1)
RNOR	" " " "	Transformation	Standard normals
REXP	" " " "	Transformation	Exponentials, mean 1
INT (SINT)	Learmonth and Lewis [22] (LLRANDOM)	Multiplicative congruential	Random integers (shuffled)
RANDOM (SRAND)	" " " "	" "	Random numbers (shuffled)
NORMAL (SNORM)	" " " "	Transformation	Standard normals (shuffled)
EXPON (SEXPON)	" " " "	Transformation	Exponentials, mean 1 (shuffled)

*SUPER-DUPER

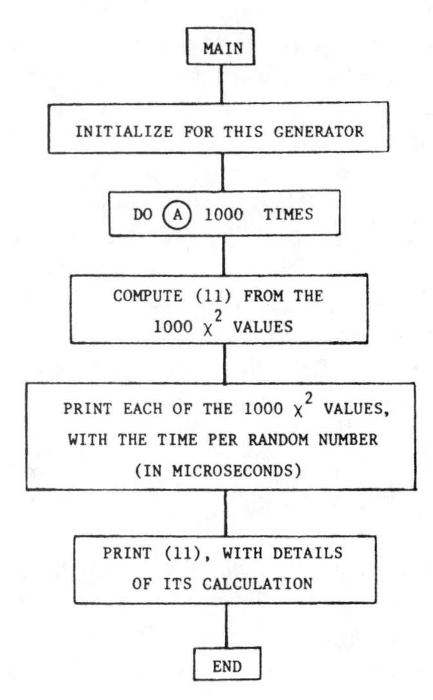

FIGURE 1. Flowchart of the Testing of Each Generator.

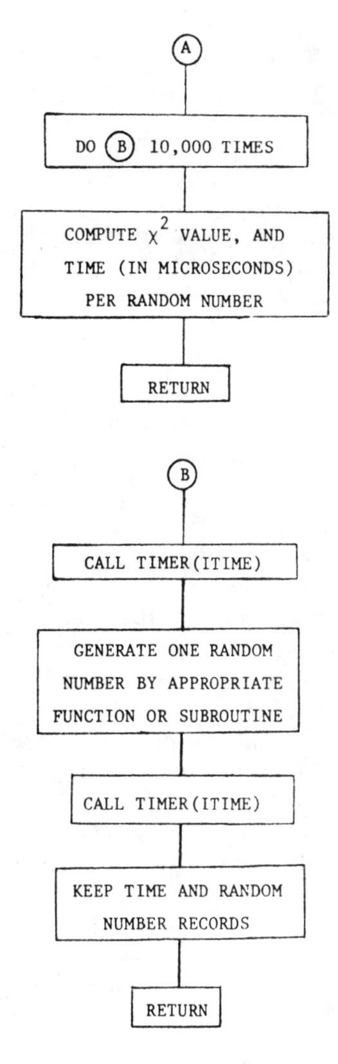

FIGURE 1. Flowchart of the Testing of Each Generator (Continued).

and EXPON yields exponential variables with mean 1. Routines SINT, SRAND, SNORM, and SEXPON perform similarly, but shuffle the sequence of random numbers or deviates so as to obtain (hopefully) better randomness properties.

The generators available in IRCCRAND [15] (with their sources, methods, and functions) are summarized in Table 1. (Detail on how to use these generators, with examples and program listings, is given in IRCCRAND [15].)

The Testing Procedures

Given one (or several) proposed methods of generating a sequence $x_1, x_2, x_3, \cdots$ of random numbers, the question arises "How random are these numbers?" Remember that our uses of this sequence of numbers $x_1, x_2, x_3, \cdots$ are predicated on the proposition that $x_1, x_2, x_3, \cdots$ are (or are indistinguishable from) independent uniform random variables on the interval 0 to 1. Hence, all tests of uniformity on the interval 0 to 1 and of independence are possible candidates for use in this context. Our testing is directed toward applying one of these tests to a substantial part of the sequence, for each method, using the results to rank generators on both quality and time needed for generation.

The test used is a sensitive chi-square test for uniformity of distribution, and is described by Gorenstein [3] and used by Bates and Zirkle [13]. Briefly, one proceeds as follows. Some number of random numbers (in our case 10,000) are generated. These are classified according to where they fall in the range of the uniform distribution (namely 0 to 1), and a chi-square statistic is calculated to measure the deviation of the actual observed frequencies from what one would expect to find if the distribution were truly uniform on the interval 0 to 1. More specifically, we took the range (0 to 1) of the uniform distribution and partitioned it into $K = 100$ intervals (0.00, 0.01), (0.01, 0.02), $\cdots$, (0.98, 0.99), (0.99, 1.00). If the distribution is truly uniform one expects on the average to find $E_j = (10{,}000)\cdot(1/100) = 100$ of the 10,000 random numbers in the jth interval ($j = 1, 2, \cdots, 100$), but actually finds some (usually different) number 0_j of the 10,000 random numbers in the jth interval ($j = 1, 2, \cdots, 100$). One then calculates the chi-square statistic which measures the discrepancy between $E_1, E_2, \cdots, E_{100}$ and $0_1, 0_2, \cdots, 0_{100}$, namely

$$\chi^2 = \sum_{j=1}^{K} (0_j - E_j)^2/E_j = \sum_{j=1}^{100} \frac{(0_j - 100)^2}{100}. \quad (10)$$

(Note that under actual uniformity on the interval 0 to 1, the statistic χ^2 has an approximate chi-square distribution with $K - 1 = 100 - 1 = 99$ degrees of freedom, say $\chi^2(99)$.) At this point, usually one performs a hypothesis test to see if the calculated χ^2 value in (10) is unusually larger or smaller than one would expect under true randomness. For example let $\chi^2(99, \gamma)$ be the value such that a $\chi^2(99)$ variable exceeds $\chi^2(99, \gamma)$ with probability $1 - \gamma$. Then to test for uniformity at level $\alpha = 0.01$ one rejects if either $\chi^2 > \chi^2(99, .995)$ or $\chi^2 < \chi^2(99, .005)$. However, this test applies only to the 10,000 random numbers used in calculating (10); if one uses another 10,000 random numbers one can obtain a different result. In particular if one obtains 10,000 random numbers a large number of times then one will accept a number of times and (usually) also reject a number of times. However, it is possible to combine all of this information into one test, which also tests whether the deviations as measured by χ^2 occur in proper proportions for true uniformity.

To perform the "overall" test referred to above, we proceed as follows. From the first 10,000 random numbers calculate χ^2 as in (10) and call the result χ_1^2. Similarly obtain χ_2^2 from the second 10,000 random numbers, $\cdots$, χ_{1000}^2 from the thousandth 10,000 random numbers. Then (under uniformity) χ_1^2, χ_2^2 $\cdots$, χ_{1000}^2 should behave like a random sample from a $\chi^2(99)$ distribution; this can also be tested using a chi-square test. To perform this test with 100 equally probable intervals, we need to find the percentage points $\chi^2(99, .01)$, $\chi^2(99, .02)$, $\cdots$, $\chi^2(99, .99)$ of a $\chi^2(99)$ distribution. Once these are found we know that each of the 100 intervals $(0, \chi^2(99, .01))$, $(\chi^2(99, .01), \chi^2(99, .02))$, $\cdots$, $(\chi^2(99, .98), \chi^2(99, .99))$, $(\chi^2(99, .99), +\infty)$ is expected to contain $E_1 = \cdots = E_{100} = 10$ of the 1000 values $\chi_1^2, \cdots, \chi_{1000}^2$. We find the actual numbers in each interval $(0_1, \cdots, 0_{100})$ and calculate

$$\chi_T^2 = \sum_{j=1}^{100} \frac{(0_j - 10)^2}{10}, \quad (11)$$

which has an approximate chi-square distribution $\chi^2(99)$ under uniformity. We then test χ_T^2 at a desired level α. Using the values of (11) for several generators we can also rank the generators on quality of their uniformity. The results of this testing (with generation times) are given in the next section.

TABLE 2. Summary χ_T^2's and Timings

Generator	Source	χ_T^2	γ such that $P[\chi^2(99) \le \chi_T^2] = \gamma$	Time* per Random Number (microseconds)
RANDOM	Learmonth and Lewis [22] (LLRANDOM)	80.20	.08+	14.6
RN3	Marsaglia and Bray [11]	80.00	.08+	38.2
UNI	Marsaglia, Ananthanarayanan, and Paul [23] (SUPER-DUPER)	106.00	.70+	7.4
SRAND	Learmonth and Lewis [22] (LLRANDOM)	108.60	.76+	16.8
RN1	Lurie and Mason [8]	119.80	.92+	21.8
KERAND	Keane [19]	125.00	.96+	15.0
RN2	International Business Machines Corporation [18] (RANDU)	----**	----**	29.8***
RN4	Nie, Bent, and Hull [24] (SPSS)	----**	----**	27.0***
RN5	Kruskal [6] (OMNITAB II)	----****	.99+	39.6***

* This is the smallest of the 1000 times found in our runs, less 2 microseconds (clock-reading time). The smallest is taken as our best estimate since it is the one with the least interrupt time included.

** RN2 and RN4 produced the same random numbers, and these were the same as those given in KERAND (except that occasionally those of KERAND were 6×10^{-8} larger). Hence these entries are essentially the same as those for KERAND and a run of 1000 was not made; a run of 100 was made to check on timing.

*** As in footnote*, but based on 100 times since a long run was not needed to test goodness.

**** Based on a run of 100 χ^2's, find a χ_T^2 of 9900; all of the 100 "observations" fell in the first cell (χ^2's .01 point).

The Speed and Goodness of the Random Number Generators

In this section we describe our results on the speed and the goodness of the generators in IRCCRAND [15]. For each generator, the testing was performed as indicated in the flowchart of Figure 1. Namely, we wish to perform the test based on (11) described previously, and at the same time to keep track of the generation times per random variable in microseconds. The results χ_T^2 (with corresponding percentage points, as well as generation times per random number) are given in Table 2.

We now wish to interpret Table 2 and make some recommendations for users. We can rank the generators on speed, on goodness, and on a combination (this latter requires a judgment which is difficult to make) and from Table 2 we see that: on goodness alone RN3 and RANDOM stand out; on time alone UNI stands out; and on a weighted combination RANDOM seems preferable. Our overall recommendation: *use either RANDOM or UNI.*

Acknowledgments

The support of the Instruction & Research Computer Center at The Ohio State University (Dr. Roy F. Reeves, Director) is gratefully acknowledged. Thanks are due to numerous IRCC personnel for aid with this project, in particular Mr. Dale J. Schroeder (programming), Mr. Ernest W. Leggett, Jr. (helpful comments), and Mr. Mark J. Ebersole (timing routines).

References

Periodicals:

1. Ashour, S., and Vaswani, S. D., "A GASP Simulation Study of Job-Shop Scheduling," *Simulation*, Vol. 18, 1972, pp. 1–10.
2. Browne, W. G., "Trends of Simulation in Marketing Management," Abstract, *Interfaces, Bulletin of The Institute of Management Sciences*, Meeting Issue, March 1972, pp. 121–122.
3. Gorenstein, S., "Testing a Random Number Generator," *Communications of the Association for Computing Machinery*, Vol. 10, 1967, pp. 111–118.
4. Grosenbaugh, L. R., "More on Fortran Random Number Generators," *Communications of the Association for Computing Machinery*, Vol. 12, 1969, p. 639.
5. Halton, J. H., "A Retrospective and Prospective Survey of the Monte Carlo Method," *SIAM Review*, Vol. 12, 1970, pp. 1–63.
6. Kruskal, J. B., "Extremely Portable Random Number Generator," *Communications of the Association for Computing Machinery*, Vol. 12, 1969, pp. 93–94.
7. Lewis, P. A. W.; Goodman, A. S.; and Miller, J. M., "A Pseudo-Random Number Generator for the System/360," *IBM Systems Journal*, Vol. 8, 1969, pp. 136–146.
8. Lurie, D., and Mason, R. L., "Empirical Investigation of Several Techniques for Computer Generation of Order Statistics," *Communications in Statistics*, Vol. 2, 1973, pp. 363–371.
9. Marsaglia, G., "Random Numbers Fall Mainly in the Planes," *Proceedings of the National Academy of Sciences*, Vol. 61, 1968, pp. 25–28.
10. Marsaglia, G., "Regularities in Congruential Random Number Generators," *Numerische Mathematik*, Vol. 16, 1970–71, pp. 8–10.
11. Marsaglia, G., and Bray, T. A., "One-line Random Number Generators and their Use in Combinations," *Communications of the Association for Computing Machinery*, Vol. 11, 1968, pp. 757–759.
12. Neave, H. R., "On Using the Box-Muller Transformation with Multiplicative Congruential Pseudo-Random Number Generators," *Applied Statistics*, Vol. 22, 1973, pp. 92–97.

Books:

13. Bates, C. B., and Zirkle, J. A., "Analysis of Random Numbers from Four Random Number Generators," *Technical Report 4-71*, Systems Analysis Group, U.S. Army Combat Development Command, Fort Belvoir, Virginia 22060, August 1971.
14. Chamberlain, R. L., and Jewett, D., *The OMNITAB Programming System—A Guide for Users*, Statistical Laboratory, Iowa State University, Ames, Iowa, 1968.
15. Dudewicz, E. J., *IRCCRAND—The Ohio State University Random Number Generator Package*, Technical Report No. 104, Department of Statistics, The Ohio State University, Columbus, Ohio 43210, October 1974.
16. Dudewicz, E. J., "Random Numbers: The Need, the History, the Generators," *Statistical Distributions in Scientific Work, Vol. 2: Model Building and Model Selection*, edited by G. P. Patil, S. Kotz, and J. K. Ord, D. Reidel Publishing Company, Dordrecht, Holland, 1975, pp. 25–36.
17. Hogben, D.; Peavy, S. T.; and Varner, R. N., *OMNITAB II User's Reference Manual*, Technical Note 552, National Bureau of Standards, Washington, D.C. 20234, October 1971.
18. International Business Machines Corporation, "System/360 Scientific Subroutine Package, Version III, Programmer's Manual, Program Number 360A-CM-03X," *Manual GH20-0205-4* (Fifth Edition), IBM Corporation, White Plains, New York, August 1970.
19. Keane, C. G., "Uniformly Distributed Random Numbers—Real, Integer," one page typed description and two pages of computer source code listing, Instruction & Research Computer Center, The Ohio State University, Columbus, Ohio, 1969.
20. Knuth, D. E., *The Art of Computer Programming, Volume 2/Seminumerical Algorithms*, Addison-Wesley Publishing Company, Inc., Reading, Massachusetts, 1969.
21. Koehl, F. S., *The IRCCRAND Random-Number Generator Package*, FORTLIB writeup, Instruction & Research Computer Center, The Ohio State University, Columbus, Ohio, 1975.
22. Learmonth, G. P., and Lewis, P. A. W., "Naval Postgraduate School Random Number Generator Package

LLRANDOM," *Technical Report*, Naval Postgraduate School, Monterey, California, June 1973.

23. Marsaglia, G.; Ananthanarayanan, K.; and Paul, N., "How to Use the McGill Random Number Package "SUPER-DUPER"," four page typed description, School of Computer Science, McGill University, Montreal, Quebec, Canada, 1973.
24. NIE, N.; BENT, D. H.; AND HULL, C. H., *SPSS—Statistical Package for the Social Sciences*, McGraw-Hill, Inc., New York, 1970.
25. NIE, N. H., AND HULL, C. H., *SPSS—Statistical Package for the Social Sciences: Update Manual*, National Opinion Research Center, University of Chicago, Chicago, 1973.
26. SCOTT, T. J., "Uniformly Distributed Random Numbers Within Limits—Integer," one page typed description and one page of computer source code listing, Instruction & Research Computer Center, The Ohio State University, Columbus, Ohio, 1968.
27. WALKER, R. J., *An Instruction Manual for CUPL—The Cornell University Programming Language*, Cornell University, Ithaca, New York, 1967.

Transactions and Proceedings

28. LEARMONTH, G. P., AND LEWIS, P. A. W., "Statistical Tests of Some Widely Used and Recently Proposed Uniform Random Number Generators," *Proceedings of the Computer Science and Statistics Seventh Annual Symposium on the Interface* (W. J. Kennedy, editor), Statistical Laboratory, Iowa State University, Ames, Iowa, 1973, pp. 163–171.
29. SCHAFER, D., "Process Integrity and Capability Through Statistical Methods—A Report," *Annual Technical Conference Transactions of the American Society for Quality Control*, Vol. 27, 1973, pp. 374–380.

SPEED AND QUALITY OF RANDOM NUMBERS FOR SIMULATION, II*

Edward J. Dudewicz, Professor
The Ohio State University, Columbus, Ohio 43210

ABSTRACT

In a previous paper results of testing of random number sources (important in simulation and Monte Carlo studies) for quality and speed were given, with their implications for users. Extensive further work on testing these (and additional) random number generators now allows new (surprising) conclusions. Code for a recommended generator is included.

INTRODUCTION

Simulation and Monte Carlo studies today play a greater role than ever before in all fields, including quality control. For example, Gutt and Gruska (1977) use it to predict quality problems which may result from variation in manufacturing and assembly operations.

As noted in Dudewicz (1976), "Such studies often consume large amounts of computer time. Nearly every such study requires, for its execution, a source (called a generator) of numbers which appear to be independent uniform random variables on the range 0.0 to 1.0; such numbers are called pseudo-random numbers or... random numbers. Due to the importance of such studies and the fact that bad random numbers may utterly vitiate a study, access to "good" random number generators is of prime importance to investigators using simulation methodology. Due to the cost (in computer time) of such studies, access to "fast" random number generators is of co-prime importance."

To help fill these needs, the current author developed a documented package of random number generators at The Ohio State University's Instruction and Research Computer Center (IRCC), and a report (including a program listing and information on ordering a tape) detailed this package (Dudewicz (1974)).

While that work included 10 random number generators and evaluated them extensively with one major test (see Dudewicz (1976) for details of the generators and their testing), much remained to be done. Namely, the generators considered were to be given expanded capabilities (e.g.: returning a vector of random numbers at a time, thus reducing the overhead of subroutine calls substantially in many applications; returning quantities needed to restart a generator from an intermediate finishing point; and adding double and extended precision random number generation), new generators proposed elsewhere since 1974 were to be included (e.g., those which showed up well in work of Hoaglin (1976)), and a battery of extensive testing on numerous types of departure from uniformity was to be accomplished. This has now been done, and is detailed in Dudewicz and Ralley (1979).

TESTS INCLUDED

The tests utilized in our evaluation of random number generators number 15 (11 of which are applicable to all generators and include the chi-square on chi-square test of Dudewicz (1976)), and are briefly described in Table I. These tests include (and go far beyond) all those suggested by Knuth (1969) for testing of random number generators.

* This investigation was supported in part by Grant Number 1 R01 CA26254-01, awarded by the National Cancer Institute, DHEW.

TABLE I. Tests included in TESTRAND

TESTRAND Name	Generic Name	Generates values for χ^2 on χ^2?	Generates values for K-S on K-S?	Language
TST01	Chi-square on chi-square, and K-S on K-S	-	-	FORTRAN
TST02	Uniform Distribution (simple chi-square)	Yes	-	FORTRAN
TST03	Lagged Correlation	-	-	FORTRAN
TST04	Gap	Yes	-	FORTRAN
TST05	Coupon Collector's	Yes	-	FORTRAN
TST06	K-S Goodness-of-Fit	-	Yes	FORTRAN
TST07	Permutation	Yes	-	FORTRAN
TST08	Poker	Yes	-	FORTRAN
TST09	Runs Up	Yes	-	FORTRAN
TST10	Serial Pairs	Yes	-	FORTRAN
TST11	Maximum-of-t	-	Yes	FORTRAN
Special Tests for Linear Congruential Generators				
TSTM1	Serial Correlation			FORTRAN
TSTM2	Potency and Period			FORTRAN (extended precision)
TSTM3	Prime Factorization			FORTRAN
TSTM4	Spectral			PL/1*

* Only the Spectral Test, TSTM4, is written in a language other than FORTRAN.

RESULTS OF TESTING

Of the random number generators tested (called URN01, URN02, ..., URN20), seven passed the goodness tests: URN01, URN02, URN12, URN13, URN14, URN15, URN20. Of these, URN02, URN12, URN14 are relatively slow; hence we would recommend most users choose from among URN01, URN13, URN15, URN20.

URN01 is equivalent to the unshuffled basic random number generator GGU3 of IMSL (1977); it is very surprising that its "shuffled" version URN04 (called GGUS3 by IMSL) failed in our testing (due to the gap test results). Users of IMSL software should therefore beware of this putatively superior generator (but can rely on URN01 with confidence insofar as its results in this testing are concerned). URN13 is due to Ahrens and Dieter (1974), while URN15 and URN20 were suggested by Hoaglin (1976).

The code for URN20 is given below; the testing results reported suggest it should be very suitable for use by most experimenters.

```
C                                                                         ØØØØØØ1Ø
C     SUBROUTINE URN2Ø(IX,X,NBATCH)                                       ØØØØØØ2Ø
C                                                                         ØØØØØØ3Ø
C************************************************************************ ØØØØØØ4Ø
C                                                                         ØØØØØØ5Ø
C USAGE                                                                   ØØØØØØ6Ø
C     CALL URN2Ø(IX,X,NBATCH)                                             ØØØØØ17Ø
C                                                                         ØØØØØ18Ø
C DESCRIPTION OF PARAMETERS                                               ØØØØØ19Ø
C     IX...FOR THE FIRST ENTRY THIS MAY CONTAIN ANY ODD INTEGER NUMBER    ØØØØØ2ØØ
C        BETWEEN 1 AND 999999999.                                         ØØØØØ21Ø
C     X...IS THE RESULTING ARRAY OF SINGLE PRECISION FLOATING POINT       ØØØØØ23Ø
C        RANDOM NUMBERS ON (Ø.Ø,1.Ø)                                      ØØØØØ24Ø
C     NBATCH...IS THE NUMBER OF RANDOM NUMBERS DESIRED TO BE GENERATED    ØØØØØ25Ø
C           PER SUBROUTINE CALL.                                          ØØØØØ26Ø
C                                                                         ØØØØØ27Ø
C REMARKS                                                                 ØØØØØ28Ø
C     THIS SUBROUTINE IS SPECIFIC TO SYSTEM/36Ø AND SYSTEM/37Ø COMPUTERS  ØØØØØ29Ø
C                                                                         ØØØØØ3ØØ
C SUBROUTINES AND FUNCTION SUBPROGRAMS REQUIRED                           ØØØØØ33Ø
C     NONE                                                                ØØØØØ34Ø
C                                                                         ØØØØØ35Ø

C************************************************************************ ØØØØØ51Ø

URN2Ø    CSECT                                                            ØØØØØØ1Ø
         USING URN2Ø,15                                                   ØØØØØØ2Ø
         STM   14,12,12(13)                                               ØØØØØØ3Ø
         ST    13,SAVE+4                                                  ØØØØØØ4Ø
         LR    11,13          COPY ADDR OF CALLING SAVE INTO REG 11       ØØØØØØ5Ø
         LA    13,SAVE        ADDR OF SAVE IN REG13                       ØØØØØØ6Ø
         ST    13,8(11)       STORE ADDR OF SAVE IN CALLING PROG SAVE     ØØØØØØ7Ø
*                                                                         ØØØØØØ8Ø
         LM    2,4,Ø(1)       REG(2)=ADDR(IX) REG(3)=ADDR(X)              ØØØØØØ9Ø
*                             REG(4)=ADDR(NBATCH)                         ØØØØØØ95
         L     11,Ø(4)        VALUE OF NBATCH IN REG11                    ØØØØØ1ØØ
         L     1Ø,=F'Ø'       INIT INDEX                                  ØØØØØ11Ø
         L     9,=X'4ØØØØØØØ'  EXPONENT FOR FLOATING POINT CONVERSION     ØØØØØ12Ø
LOOP     L     7,Ø(2)         VALUE OF IX IN REG7                         ØØØØØ13Ø
         M     6,MLT          REG6=QUOT, REG7=REM MOD 2**32               ØØØØØ14Ø
         SLDL  6,1            PUTS QUOT MOD 2**31 IN REG6                 ØØØØØ15Ø
         SRL   7,1            PUTS REM MOD 2**31  IN REG7                 ØØØØØ16Ø
         AR    6,7            REG6 = REM MOD 2**31-1=                     ØØØØØ17Ø
*                                REM MOD 2**31 + QUOT MOD 2**31           ØØØØØ171
         C     6,=F'Ø'        CHECK FOR OVERFLOW                          ØØØØØ172
         BNL   UP1            BRANCH TO UP1 IF SUM IS NOT NEG             ØØØØØ174
         SLL   6,1            DIVIDE REG6 BY 2**31                        ØØØØØ176
         SRL   6,1            REM MOD 2**31 IN REG6                       ØØØØØ177
         A     6,=F'1'        ADD QUOT FROM DIVISION TO REM TO            ØØØØØ178
*                              OBTAIN REM MOD 2**31-1 IN REG6             ØØØØØ179
UP1      ST    6,Ø(2)         STORE VALUE OF IX FOR NEXT RND NOS          ØØØØØ19Ø
         LPR   6,6            ABSOL VALUE OF IX IN REG6                   ØØØØØ2ØØ
         L     5,=X'4EØØØØØØ'  EXPONENT IN REG5 (16**14)                  ØØØØØ21Ø
         STM   5,6,DOUBLE     STORE FLT PT NOS IN REGS5 AND 6             ØØØØØ22Ø
         LD    2,DOUBLE       LOAD FLT PT REG2 WITH FLT PT NOS            ØØØØØ23Ø
         AD    2,=D'Ø.Ø'      NORMALIZE BY ADDING Ø.                      ØØØØØ24Ø
         DD    2,MODUL           DIVIDE BY 2**31-1                        ØØØØØ25Ø
         STD   2,Ø(1Ø,3)      STORE NORMALIZED FLT PT NOS IN X(INDEX)     ØØØØØ26Ø
         LA    1Ø,4(1Ø)       INCREMENT INDEX                             ØØØØØ27Ø
         BCT   11,LOOP        BOTTOM OF LOOP                              ØØØØØ28Ø
         L     13,SAVE+4      RESTORE REGISTERS                           ØØØØØ29Ø
         LM    14,12,12(13)                                               ØØØØØ3ØØ
         BCR   15,14          RETURN TO CALLING PROGRAM                   ØØØØØ31Ø
MLT      DC    F'2Ø278128Ø8   MULTIPLIER FOR URN2Ø
DOUBLE   DS    D                                                          ØØØØØ34Ø
MODUL    DC    X'487FFFFFFFØØØØØØ'                                        ØØØØØ35Ø
SAVE     DS    18F                                                        ØØØØØ36Ø
         END                                                              ØØØØØ37Ø
```

REFERENCES

1. J. H. Ahrens and V. Dieter: "Non-Uniform Random-Numbers," Institut fur Math. Statistik, Technische Hochschule in Graz, A 8010 Graz, Hamerlingg. 6, VI, Austria, 1974, P.1-10.

2. Dudewicz, E.J.: "IRCCRAND-The Ohio State University Random Number Generator Pacakge," Technical Report No. 104 (Statistics Laboratory Publication No. 1), Department of Statistics, The Ohio State University, Columbus, Ohio, 1974.

   (Requests should be addressed to: Publications Secretary, The Statistics Laboratory, The Ohio State University, 1958 Neil Avenue, Columbus, Ohio 43210, enclosing $3.50. Information on ordering the tape will be enclosed with the report.)

3. Dudewicz, E.J.: "Speed and Quality of Random Numbers for Simulation," Journal of Quality Technology, Vol. 8, 1976, pp. 171-178, 234.

4. Dudewicz, E.J. and Ralley, T.: "TESTRAND-Random Number Generation and Testing Pacakge (Including IRCCRAND)," Technical Report No. 173, Department of Statistics, The Ohio State University, Columbus, Ohio, 1979. (To order, write to author.)

5. Gutt, J.D. and Gruska, G.F.: "Variation Simulation," Annual Technical Conference Transactions of the American Society for Quality Control, Vol. 31, 1977, pp. 557-563.

6. Hoaglin, D.C.: "Theoretical Properties of Congruential Random-Number Generators: An Empirical View," Memorandum NS-340, Department of Statistics, Harvard University, Cambridge, Massachusetts, November 1976.

7. IMSL: "IMSL Library 1 Reference Manual," International Mathematical and Statistical Libraries, Inc., Sixth Floor, GNB Building, 7500 Bellaire Blvd., Houston, Texas 77036, 1977.

8. Knuth, D.E.: The Art of Computer Programming, Volume 2/Seminumerical Algorithms, Addison-Wesley Publishing Company, Inc., Reading,Massachusetts, 1969.

LCS 500:70:000

Reprinted with permission from *Proceedings of the Computer Science and Statistics Eighth Annual Symposium on the Interface*, J.W. Frane (Editor), Health Sciences Computing Facility, University of California, Los Angeles, California, 1975, pages 217-221.

CHI-SQUARE PERCENTAGE POINTS FOR CHI-SQUARE DISTRIBUTION TESTING OF CHI-SQUARE VALUES, WITH AN APPLICATION TO RANDOM NUMBER GENERATORS

Edward J. Dudewicz
The Ohio State University
Columbus, Ohio

Abstract

In order to test the hypothesis that random variables $Z_1, Z_2, \ldots, Z_{K^*}$ each have a chi-square distribution with n-1 degrees of freedom (say $\chi^2(n-1)$), one may desire to use a chi-square test with 100 equally probable intervals. This means one will need to know the percentage points $\chi^2(n-1,.01)$, $\chi^2(n-1,.02), \ldots, \chi^2(n-1,.99)$ of a $\chi^2(n-1)$ distribution. However the most extensive tables of $\chi^2(n-1,\gamma)$ available in the literature (Harter (1964)) do not suffice for this purpose. Several methods are available for the calculation and/or approximation of $\chi^2(n-1,\gamma)$, including: linear interpolation based on existing tables; normal approximation (two types); Cornish-Fisher approximation; and Aitken interpolation based on existing tables. Calculations for the case n = 100 yielded some surprising difficulties and some surprising conclusions about the "best" method. The table resulting from these calculations is given. An application is given to the testing of random number generators (with n = 100, K* = 1000), and the ranking of generators on speed and goodness is discussed.

1. INTRODUCTION

Given one (or several) proposed methods of generating a sequence $x_1, x_2, x_3, \ldots$ of random numbers, the question arises (see Good (1969) for a brief and lucid discussion) "How random are these numbers?" Remember that uses of this sequence of numbers $x_1, x_2, x_3, \ldots$ are predicated on the proposition that $x_1, x_2, x_3, \ldots$ are (or are indistinguishable from) independent uniform random variables on the interval 0 to 1. Hence, all tests of uniformity on the interval 0 to 1 and of independence are possible candidates for use in this context, and hence, as has been noted by Knuth (1969), p. 35, "There is literally no end to the number of tests that can be conceived ..." to quantitatively measure randomness. Some of the most-used tests (see, e.g., Knuth (1969), pp. 34-100) include: frequency chi-square tests; Kolmogorov-Smirnov tests; serial tests; gap tests; poker (or partition) tests; coupon collector's tests; permutation tests; runs tests; maximum tests; lagged-product tests; tests on subsequences; and spectral tests. There are many tests because there are many ways in which a sequence $x_1, x_2, x_3, \ldots$ can fail to be random, and different tests are good at detecting various of these departures.

In the literature, one commonly finds random number generators tested by applying 5 to 10 of the above tests (or variations to them) to a limited part of the sequence. (E.g. Bates and Zirkle (1971) applied various tests to 50,000 numbers from various sequences. Our testing below is based on 10,000,000 numbers from various sequences. However, Learmonth and Lewis (1973) ran tests based on over 6,500,000 numbers from various sequences, and Mosimann (1974) has recently called our attention to the fact that testing along these lines was performed at the National Institutes of Health; the results were never published and are no longer available.) Elsewhere (see Dudewicz (1974)) we have given details of a documented package of random number generators developed at The Ohio State University's Instruction and Research Computer Center (IRCC), and (see Dudewicz (1975)) of its testing. The package, intended to make a variety of generators (with documentation) available, is called IRCCRAND and a report (including a program listing) as well as a tape are available upon request for nominal fees. Since the testing has been described elsewhere (Dudewicz (1975)) we will not repeat the details here. The test is a chi-square test for uniformity of distribution (while good performance on this test is necessary for a random number generator to be of high-quality, it is not sufficient; and it might often be appropriate to do additional testing) and the calculation of $\chi^2(99, .01),\ldots,$ $\chi^2(99, .99)$ is essential to this test. These are not available in easily accessible form in the literature, and will undoubtedly be desired for use by others who desire to perform extensive tests on random number generators in the future.

2. EXISTING TABLES

The most extensive table of values of χ^2 percentage points available in the literature seems to be that of Harter (1964a), which contains $\chi^2(99, \gamma)$ to 6 significant figures for $\gamma = .0001, .0005, .001, .005, .01, .025, .05, .1(.1).9, .95, .975, .99, .995, .999, .9995, .9999$. However, we need these values $\chi^2(99, \gamma)$ for $\gamma = .01(.01).99$; hence Harter's table (while useful) is not nearly sufficient. From this information we have the values of Figure 1, a subset of those actually needed.

3. CALCULATION/APPROXIMATION

In order to "fill out" Figure 1 to $\gamma = .01(.01).99$, several methods are available. First, one can perform linear interpolation on the values of Figure 1. This will yield approximations to the true values for which we will have no error bound and no reason to believe the error is suitably small. Hence, values generated in this manner would be appropriate only for a crude analysis, and not for the sort of analysis proposed and done in Dudewicz (1975) (which could be vitiated by this introduction of noise).

A second method would use the fact that a $\chi^2(99)$ variable is "approximately" normal with mean 99 and variance 198 to derive an approximation to the chi-square percentage point $\chi^2(99, \gamma)$ from the corresponding standard normal percentage point $z(\gamma)$ (this can be calculated to any accuracy which might conceivably be needed for use in (1), via subroutine CDFNI given by Milton and Hotchkiss (1969)) via

$$\chi^2(99, \gamma) \approx z(\gamma)\sqrt{198} + 99. \tag{1}$$

Again we have no error bound, hence such value would be appropriate only for a crude analysi (Approximation (1) follows from use of the ce limit theorem. A more accurate approximation (see Lancaster (1969), p. 21) supposed to be

$$\chi^2(99, \gamma) \approx 0.5(z(\gamma) + \sqrt{197})^2. \tag{2}$$

However, this still allows us no error bound.)

A third method is to use an approximation based on the Cornish-Fisher expansion, namely (see Rao, Mitra, and Matthai (1966), p. 64)

$$\begin{aligned}\chi^2(99, \gamma) \approx\ & 99 + \sqrt{99}(z(\gamma)\sqrt{2}) + \tfrac{2}{3}(z^2(\gamma) - 1) \\ & + \frac{z^3(\gamma)-7z(\gamma)}{(\sqrt{99})(9\sqrt{2})} - \frac{6z^4(\gamma)+14z^2(\gamma)-32}{(99)(405)} \\ & + \frac{9z^5(\gamma)+256z^3(\gamma)-433z(\gamma)}{(99\sqrt{99})(4860\sqrt{2})} \\ & + \frac{12z^6(\gamma)-243z^4(\gamma)-923z^2(\gamma)+1472}{(99^2)(25515)} \\ & - \frac{3753z^7(\gamma)+4353z^5(\gamma)-289517z^3(\gamma)-289717z(\gamma)}{(99^2\sqrt{99})(9185400\sqrt{2})}.\end{aligned} \tag{3}$$

Again we have no error bound. However, we know there is a sound basis for the underlying expan-

Figure 1: Some chi-square percentage points.

γ	$\chi^2(99, \gamma)$	γ	$\chi^2(99, \gamma)$
0.01	69.2299	0.60	101.928
0.05	77.0463	0.70	105.868
0.10	81.4492	0.80	110.607
0.20	87.0052	0.90	117.407
0.30	91.1663	0.95	123.225
0.40	94.8259	0.99	134.642
0.50	98.3341	1.00	∞

sion and hence expect the approximation may be adequate.

A fourth method, and the final one which we will consider, is to use Aitken interpolation on a table of the incomplete gamma function $I(u,p)$, given by

$$I(u,p) \equiv \frac{1}{\Gamma(p+1)} \int_0^{u/\sqrt{p+1}} v^p e^{-v} dv, \qquad (4)$$

since

$$\chi^2(99, \gamma) = \sqrt{198}\, u(\gamma) \qquad (5)$$

where $u(\gamma)$ solves $\gamma = I(u, 48.5)$. This method is suggested by Harter (1964a) for use in conjunction with the tables of Harter (1964b), and is expected to be adequate because of the results reported by Harter. (Harter (1964a) also gives two other methods of computing $\chi^2(99, \gamma)$, which he used to check this method in constructing his tables.)

The results of computations (in double-precision with FØRTRAN on the IBM 370/165 at the Instruction & Research Computer Center at The Ohio State University) using these five methods (namely: linear interpolation on Figure 1; a normal approximation via (1); an improved normal approximation via (2); approximation through the Cornish-Fisher expansion via (3); and Aitken interpolation via (5)) are given in Figure 2. It can be seen (largely by reference to Figure 1) that: linear interpolation is unacceptably rough; the normal approximation, along with its improvement, is often worse than crude linear interpolation; the Cornish-Fisher based approximation is unbelievably accurate; the Aitken interpolation is identical to the Cornish-Fisher based approximation for $\gamma \geq 0.23$ (but experiences difficulties in its last decimal place for $\gamma \leq 0.22$); and for all γ for which Harter (1964a) tables $\chi^2(99, \gamma)$ and for which we also calculated, there is precise agreement with the Cornish-Fisher based approximation (except at $\gamma = 0.10$, where Harter gave 81.4492 while we find 81.4493). Due to these results we believe the Cornish-Fisher based approximation has given values exact up to one unit in the last digit reported (and used these values in calculations for testing in Dudewicz (1975)).

4. ACKNOWLEDGMENTS

The support of the Instruction & Research Computer Center at The Ohio State University (Dr. Roy F. Reeves, Director) is gratefully acknowledged. Thanks are due to numerous IRCC personnel for aid with this project, in particular Mr. Dale J. Schroeder. This work was supported in part by the U.S. Army Research Office - Durham.

5. REFERENCES

1. Bates, C.B. and Zirkle, J.A., "Analysis of random numbers from four random number generators," Technical Report 4-71, Systems Analysis Group, U.S. Army Combat Developments Command, Fort Belvoir, Virginia 22060, August 1971.

2. Dudewicz, E.J., IRCCRAND - The Ohio State University Random Number Generator Package, Technical Report No. 104, Department of Statistics, The Ohio State University, Columbus, Ohio 43210, October 1974.

3. Dudewicz, E.J. (1975), "Speed and quality of random numbers for simulation," Annual Technical Conference Transactions of the American Society for Quality Control, Vol. 29 (1975), to appear.

4. Good, I.J., "How random are random numbers?," The American Statistician, Vol. 23 (1969), pp. 42-45.

5. Harter, H.L., "A new table of percentage points of the chi-square distribution," Biometrika, Vol. 51 (1964a), pp. 231-239.

6. Harter, H.L., New Tables of the Incomplete Gamma-Function Ratio and of Percentage Points of the Chi-Square and Beta Distributions, U.S. Government Printing Office, Washington, D.C., 1964b.

7. Knuth, D.E., The Art of Computer Programming, Volume 2/Seminumerical Algorithms, Addison-Wesley Publishing Company, Inc., Reading, Massachusetts, 1969.

8. Lancaster, H.O., The Chi-squared Distribution, John Wiley & Sons, Inc., New York, 1969.

9. Learmonth, G.P. and Lewis, P.A.W., "Statistical tests of some widely used and recently proposed uniform random number generators," Proceedings of the Computer Science and Statistics Seventh Annual Symposium on the Interface (W.J. Kennedy, editor), Statistical Laboratory, Iowa State University, Ames, Iowa, 1973, pp. 163-171.

10. Milton, R.C. and Hotchkiss, R., "Computer evaluation of the normal and inverse normal distribution functions," Technometrics, Vol. 11 (1969), pp. 817-822.

11. Mosimann, J., Personal communication, 1974.

12. Rao, C.R., Mitra, S.K., and Matthai, A., Formulae and Tables for Statistical Work, Statistical Publishing Society, Calcutta, India, 1966.

Figure 2: Chi-square percentage points via five methods.

P	LINEAR INTERP	NORMAL APPROX	NORMAL IMPR	CORNISH-FISHER	AITKEN-LAGRANGE	NO. BELOW	NO. ABOVE
0.01	69.2299	66.2654	68.5541	69.2299	69.2301	2	18
0.02	71.1184	70.1012	71.7832	72.2880	72.2881	2	18
0.03	73.1381	72.5349	73.8705	74.2754	74.2755	2	18
0.04	75.0922	74.3657	75.4604	75.7949	75.7948	3	17
0.05	77.0463	75.8549	76.7661	77.0463	77.0464	4	16
0.06	77.9269	77.1224	77.8864	78.1226	78.1225	5	15
0.07	78.8075	78.2338	78.8753	79.0746	79.0745	5	15
0.08	79.6880	79.2289	79.7660	79.9336	79.9338	5	15
0.09	80.5686	80.1339	80.5804	80.7204	80.7204	5	15
0.10	81.4492	80.9670	81.3338	81.4493	81.4493	5	15
0.11	82.0048	81.7412	82.0370	82.1306	82.1307	5	15
0.12	82.5604	82.4665	82.6986	82.7724	82.7724	5	15
0.13	83.1160	83.1503	83.3247	83.3805	83.3806	5	15
0.14	83.6716	83.7986	83.9205	83.9599	83.9599	5	15
0.15	84.2272	84.4161	84.4901	84.5143	84.5143	10	10
0.16	84.7828	85.0067	85.0366	85.0469	85.0469	10	10
0.17	85.3384	85.5737	85.5629	85.5603	85.5603	10	10
0.18	85.8940	86.1197	86.0712	86.0566	86.0566	10	10
0.19	86.4496	86.6469	86.5635	86.5377	86.5377	10	10
0.20	87.0052	87.1573	87.0414	87.0052	87.0052	10	10
0.21	87.4213	87.6526	87.5065	87.4605	87.4605	10	10
0.22	87.8374	88.1343	87.9599	87.9048	87.9047	10	10
0.23	88.2535	88.6035	88.4027	88.3390	88.3390	10	10
0.24	88.6696	89.0614	88.8360	88.7642	88.7642	10	10
0.25	89.0858	89.5091	89.2606	89.1812	89.1812	10	10
0.26	89.5019	89.9473	89.6772	89.5907	89.5907	10	10
0.27	89.9180	90.3770	90.0865	89.9934	89.9934	10	10
0.28	90.3341	90.7987	90.4893	90.3898	90.3898	10	10
0.29	90.7502	91.2132	90.8860	90.7806	90.7806	10	10
0.30	91.1663	91.6210	91.2772	91.1663	91.1663	10	10
0.31	91.5323	92.0228	91.6633	91.5472	91.5472	10	10
0.32	91.8982	92.4189	92.0449	91.9238	91.9238	10	10
0.33	92.2642	92.8099	92.4223	92.2966	92.2966	10	10
0.34	92.6301	93.1961	92.7959	92.6659	92.6659	10	10
0.35	92.9961	93.5781	93.1660	93.0320	93.0320	10	10
0.36	93.3621	93.9560	93.5330	93.3953	93.3953	10	10
0.37	93.7280	94.3304	93.8973	93.7560	93.7560	10	10
0.38	94.0940	94.7015	94.2590	94.1145	94.1145	10	10
0.39	94.4599	95.0696	94.6186	94.4710	94.4710	10	10
0.40	94.8259	95.4351	94.9762	94.8259	94.8259	10	10
0.41	95.1767	95.7982	95.3321	95.1793	95.1793	10	10
0.42	95.5257	96.1591	95.6867	95.5315	95.5315	10	10
0.43	95.8784	96.5182	96.0400	95.8828	95.8828	10	10
0.44	96.2292	96.8757	96.3924	96.2333	96.2333	10	10
0.45	96.5800	97.2318	96.7442	96.5834	96.5834	10	10
0.46	96.9308	97.5868	97.0954	96.9332	96.9332	10	10
0.47	97.2816	97.9409	97.4464	97.2829	97.2829	10	10
0.48	97.6325	98.2943	97.7973	97.6329	97.6329	10	10
0.49	97.9833	98.6472	98.1485	97.9832	97.9832	10	10
0.50	98.3341	99.0000	98.5000	98.3341	98.3341	10	10

Figure 2: Chi-square percentage points via five methods.

P	LINEAR INTERP	NORMAL APPROX	NORMAL IMPR	CORNISH-FISHER	AITKEN-LAGRANGE	NO. BELOW	NO. ABOVE
0.51	98.693	99.353	98.852	98.686	98.686	10	10
0.52	99.053	99.706	99.205	99.039	99.039	10	10
0.53	99.412	100.059	99.559	99.393	99.393	10	10
0.54	99.772	100.413	99.915	99.749	99.749	10	10
0.55	100.131	100.768	100.272	100.106	100.106	10	10
0.56	100.490	101.124	100.630	100.465	100.465	10	10
0.57	100.850	101.482	100.991	100.827	100.827	10	10
0.58	101.209	101.841	101.354	101.191	101.191	10	10
0.59	101.569	102.202	101.720	101.558	101.558	10	10
0.60	101.928	102.565	102.088	101.928	101.928	10	10
0.61	102.322	102.930	102.459	102.301	102.301	10	10
0.62	102.716	103.298	102.834	102.678	102.678	10	10
0.63	103.110	103.670	103.213	103.059	103.059	10	10
0.64	103.504	104.044	103.595	103.444	103.444	10	10
0.65	103.898	104.422	103.982	103.834	103.834	10	10
0.66	104.292	104.804	104.374	104.229	104.229	10	10
0.67	104.686	105.190	104.771	104.630	104.630	10	10
0.68	105.080	105.581	105.174	105.036	105.036	10	10
0.69	105.474	105.977	105.583	105.449	105.449	10	10
0.70	105.868	106.379	105.998	105.868	105.868	10	10
0.71	106.342	106.787	106.420	106.296	106.296	10	10
0.72	106.852	107.201	106.850	106.731	106.731	10	10
0.73	107.290	107.623	107.289	107.175	107.175	10	10
0.74	107.764	108.053	107.737	107.629	107.629	10	10
0.75	108.238	108.491	108.194	108.093	108.093	10	10
0.76	108.711	108.939	108.663	108.569	108.569	10	10
0.77	109.185	109.396	109.143	109.057	109.057	10	10
0.78	109.659	109.866	109.636	109.558	109.558	10	10
0.79	110.133	110.347	110.144	110.074	110.074	10	10
0.80	110.607	110.843	110.667	110.607	110.607	10	10
0.81	111.287	111.353	111.207	111.157	111.157	10	10
0.82	111.967	111.880	111.767	111.728	111.728	10	10
0.83	112.647	112.426	112.348	112.321	112.321	10	10
0.84	113.327	112.993	112.952	112.939	112.939	10	10
0.85	114.007	113.584	113.584	113.585	113.585	10	10
0.86	114.687	114.201	114.247	114.263	114.263	15	5
0.87	115.367	114.850	114.944	114.978	114.978	15	5
0.88	116.047	115.534	115.682	115.735	115.735	15	5
0.89	116.727	116.259	116.467	116.542	116.542	15	5
0.90	117.407	117.033	117.309	117.407	117.407	15	5
0.91	118.571	117.866	118.217	118.342	118.342	15	5
0.92	119.734	118.771	119.208	119.364	119.364	15	5
0.93	120.898	119.766	120.303	120.495	120.495	15	5
0.94	122.061	120.878	121.531	121.765	121.765	15	5
0.95	123.225	122.145	122.939	123.225	123.225	15	5
0.96	126.079	123.634	124.605	124.955	124.955	15	5
0.97	128.933	125.465	126.667	127.103	127.103	15	5
0.98	131.788	127.899	129.435	129.996	129.996	15	5
0.99	134.642	131.735	133.858	134.642	134.642	15	5

Evidence of Significant Bias in an Elementary Random Number Generator.

H. Borgwaldt, V. Brandl

Kernforschungszentrum Karlsruhe,
Postfach 3640,
D-7500 Karlsruhe 1

An elementary pseudo random number generator for isotropically distributed unit vectors in 3-dimensional space has been tested for bias. This generator uses the IBM-supplied routine RANDU and a transparent rejection technique. The tests show clearly that non-randomness in the pseudo random numbers generated by the primary IBM generator leads to bias in the order of 1 percent in estimates obtained from the secondary random number generator. FORTRAN listings of 4 variants of the random number generator called by a simple test programme and output listings are included for direct reference.

1. Introduction.

Monte Carlo techniques have been popular in reactor neutron physics calculations in situations where diffusion theory is not accurate enough and the geometry too complicated for transport codes. Meanwhile the performance of 2-dimensional neutronic transport codes has been improved considerably, reducing the field of Monte Carlo applications.

But, in reactor technology neutron transport is not the only area of Monte Carlo application. The response surface method, used in reactor safety research, takes an approximate response surface equation as a fast-running substitute for the accurate response of a complex safety code to input parameter vectors. Conventional Monte Carlo techniques are, then, used to sample repeatedly input vectors from an assumed probability distribution, evaluate the approximate responses, and obtain finally an estimate of the probability distribution of the response in form of a histogram (or a set of moments) /1/. The results thus obtained depend on several factors: the goodness of fit of the response function approximation in the region of concern, the sample size, the use of special sampling techniques (e.g. Latin hypercube sampling), and finally on the properties of the random number generator (RNG).

Another typical Monte Carlo application, also in reactor safety research , is fault tree evaluation by simulation, in cases where analytical methods are not available. Here too one must rely on a reasonable behaviour of the RNGs used. Therefore, it seems appropriate to communicate, as a general warning, adverse experience originating from a neutron transport application.

RNGs for sampling from arbitrary distributions can be realized by several means, e.g. transformation, rejection and special techniques. The common feature of all these techniques is that they use an input stream of values from a primary RNG, usually supplied with the computer software. This RNG yields uniformly distributed values in the open interval (0., 1.). They must be sufficiently random for all practical applications. If this cannot be assured then all derived results may be questioned.

It has been recognized long ago that RNGs can be demonstrated to be far from perfect /2/. On the other hand, the authors have, as many other practitioners, believed that, at least for the established RNGs, their imperfections can be demonstrated only by sophisticated mathematical methods, based on the theory of numbers or similar tools. Therefore, we expected that straight-forward applications should not show any effects comparable to the inevitable statistical errors, known to decrease with the square root of the sample size. This conviction got lost, when one of us (V. B.) investigated neutron transport in an anisotropic medium using a modified version of the Monte Carlo neutron transport code KAMCCO /3/.

In addition to the expected anisotropy of the z-direction versus the transversal directions the results showed also a marked anisotropy in the (x,y)-plane not explainable by any feature of the physical model. After some search, in which coding errors, especially in the ASSEMBLER versions of RNGs and truncation effects were suspected, we recognized that a secondary, derived RNG used for generating isotropically distributed unit vectors in 3-dimensional space was very sensitive to the inherent weakness of the IBM-supplied primary RNG RANDU /4/.

2. Specification of the RNG tested.

In 3-dimensional space the marginal distribution for each component of isotropically distributed vectors (normalized to unit length) is uniform in the interval (-1.0, 1.0). The projection of such vectors into any 2-dimensional plane has an isotropic distribution of directions in this plane. This leads to a simple recipe for the pertinent RNG of pseudo random vectors (X,Y,Z) :

Step 1: Sample one component, e.g. Z, from the uniform distribution in the interval (-1.0, 1.0), using RANDU.

Step 2: Sample similarly the remaining components X, Y.

Step 3: If the point (X,Y) is inside a circular disk of unit radius, then continue; else return to step 2 (This rejection technique has an efficiency of 78.5 percent).

Step 4: Normalize the projection (X,Y) such that the complete vector (X,Y,Z) gets unit normalization.

With an ideal primary RNG this secondary RNG should perform very well. For one 3-component vector an average of only 3.55 calls of the primary RNG and one call of the SQRT function are needed.

3. Test procedure, including variants of the RNG.

The marginal distributions of the absolute values of each vector component are uniform on the interval (0., 1.) with a mean of 0.50 .

Table 1. Results for reference case A.

DEMONSTRATION OF BIASSED R.N.G.   CASE A

SAMPLE SIZE: 100000,   RUNS: 10

RUN	BIAS (PCT.) FOR X	BIAS (PCT.) FOR Y	BIAS (PCT.) FOR Z	BIAS (ST. DEV.) FOR X	BIAS (ST. DEV.) FOR Y	BIAS (ST. DEV.) FOR Z
1	0.548	-0.139	-0.480	3.00	-0.76	-2.63
2	0.665	0.100	-0.839	3.64	0.55	-4.61
3	0.328	0.218	-0.440	1.79	1.19	-2.42
4	0.616	0.153	-0.587	3.38	0.84	-3.23
5	0.501	0.098	-0.604	2.74	0.54	-3.31
6	0.319	0.380	-0.600	1.75	2.08	-3.29
7	0.352	0.011	-0.448	1.93	0.06	-2.46
8	0.808	0.022	-0.809	4.42	0.12	-4.45
9	0.349	0.291	-0.706	1.90	1.59	-3.87
10	0.413	0.314	-0.667	2.26	1.72	-3.66

Table 2. Results of test series B.

DEMONSTRATION OF BIASSED R.N.G.   CASE B

SAMPLE SIZE: 100000,   RUNS: 10

RUN	BIAS (PCT.) FOR X	BIAS (PCT.) FOR Y	BIAS (PCT.) FOR Z	BIAS (ST. DEV.) FOR X	BIAS (ST. DEV.) FOR Y	BIAS (ST. DEV.) FOR Z
1	0.638	-0.087	-0.483	3.50	-0.48	-2.65
2	0.838	-0.003	-0.838	4.58	-0.02	-4.60
3	0.519	0.183	-0.440	2.85	1.00	-2.42
4	0.707	0.144	-0.588	3.88	0.79	-3.23
5	0.526	0.212	-0.604	2.88	1.16	-3.32
6	0.477	0.345	-0.599	2.62	1.89	-3.29
7	0.490	-0.051	-0.449	2.68	-0.28	-2.46
8	0.906	0.015	-0.808	4.96	0.08	-4.44
9	0.432	0.306	-0.706	2.36	1.68	-3.87
10	0.619	0.220	-0.668	3.38	1.21	-3.67

This was taken as a criterion for the test programmes reproduced in the Appendix. The results given are deviations in percents for the estimated mean absolute value of all 3 vector components. In addition, these errors have been converted to standard deviations to show their significance. Although the IBM-supplied RNG RANDU yields single precision values only, we have employed double precision throughout the test programmes to eliminate any possible truncation effect. For each test case an adequate sample size of 100,000 realisations and a sequence of 10 runs was chosen to obtain significant results.

Case A of our test programme is the reference case, coded as explained above. The cases B, C, and D each contain one modification versus the reference case. Case A (cf. Table 1) shows over 10 runs an average bias for the x-component of .49 percent, and a bias of -.62 percent for the z-component. These values are quite high and look significant, corresponding to 2.7 and -3.4 (single run) standard deviations, respectively. Throughout the series of 10 runs there is no change of sign in the errors for these 2 components. As to the estimates for the y-component, the registered average deviation of .15 percent corresponding to .79 standard deviations is significantly smaller. The sign of the error is positive in 9 out of 10 runs, indicating bias also for this component. But here a more careful analysis would be necessary to exclude pure coincidence.

For Case B the order, in which vector components are determined, has been changed. The z-component is selected after the (x,y)-direction has been determined. Note, that under these circumstances the random numbers used to determine X, Y, Z are always in sequence, whereas in the reference Case A the rejection technique for X, Y sometimes breaks up this triplet into one isolated random number (for Z) and a doublet (for X, Y), with an even number of rejected random numbers in between. The results (cf. Table 2) seem to indicate that the behaviour of the modified RNG becomes worse for the x-component. In terms of standard deviations the average errors of the x-, y-, and z-components become 3.4, .70, and -3.4, respectively. For reasons unexplained, the y-component shows the best behaviour of all 4 cases, considering not only the magnitude of errors but also the higher number of sign changes.

Following a suggestion by E. Gelbard /5/, we have next attempted to decouple somewhat the selection of random numbers used for generating the z-component and the (x,y)-pair, respectively. For Case C this is done by inserting one blind call to the primary RNG RANDU after determining the z-component. Table 3 shows no qualitative changes, in comparison with the reference Case A. The mean deviation in the x-component, .50 percent or 2.8 standard deviations, and the corresponding value for the z-component, -.61 percent or -3.4 standard deviations, stay practically unchanged. Note that through the loop the pairs of random numbers used for the (x,y)-combination of one vector and the random numbers used for the z-component of the following vector still form triplet sequences.

Only the last Case D shows a significant improvement. For this case a blind call of the primary RNG RANDU has been inserted before determining the z-component of the random vectors. Now (cf. Table 4) the mean errors for the x- and z-components are reduced to .19 percent or 1.1 standard deviations and .03 percent or .14 standard deviations, respectively. The corresponding value of -.19 percent (or -1.1 standard deviations) for the y-component seems to indicate that the bias has been partially shifted to this component. But to corroborate this evidence a much more detailed analysis would be necessary.

4. Conclusions.

The sample calculations done for this communication demonstrate very clearly that non-randomness in the pseudo random numbers generated by a standard primary RNG like IBM's RANDU can easily lead to bias in the order of 1 percent in estimates from secondary RNGs. This is much more than can be tolerated. We also see, from the Case D data, which direction to take in order to overcome such an effect, at least for this special application. But we are left with a very uneasy feeling, what tricks any RNG may play in situations which are less transparent.

What we really need, is not a RNG which has passed certain statistical tests for randomness; it may fail in the very next one. Instead we would need a RNG, which could be proved to show approximate randomness by some

Table 3. Results of test series C.

DEMONSTRATION OF BIASSED R.N.G. CASE C

SAMPLE SIZE: 100000, RUNS: 10

	BIAS (PCT.) FOR			BIAS (ST. DEV.) FOR		
RUN	X	Y	Z	X	Y	Z
1	0.418	0.212	-0.574	2.29	1.16	-3.15
2	0.251	0.183	-0.352	1.38	1.00	-1.93
3	0.534	0.037	-0.409	2.93	0.20	-2.25
4	0.302	0.416	-0.517	1.66	2.28	-2.85
5	0.540	0.419	-0.892	2.95	2.29	-4.90
6	0.418	0.132	-0.455	2.29	0.72	-2.50
7	0.613	0.029	-0.664	3.36	0.16	-3.64
8	0.701	0.048	-0.670	3.83	0.26	-3.69
9	0.553	0.240	-0.881	3.03	1.31	-4.83
10	0.704	-0.110	-0.681	3.85	-0.60	-3.73

Table 4. Results of test series D.

DEMONSTRATION OF BIASSED R.N.G. CASE D

SAMPLE SIZE: 100000, RUNS: 10

	BIAS (PCT.) FOR			BIAS (ST. DEV.) FOR		
RUN	X	Y	Z	X	Y	Z
1	0.139	-0.328	0.200	0.76	-1.80	1.10
2	0.072	-0.058	-0.006	0.40	-0.32	-0.04
3	0.177	-0.346	0.166	0.97	-1.90	0.91
4	0.019	-0.047	0.132	0.11	-0.26	0.73
5	0.068	0.135	-0.202	0.37	0.74	-1.11
6	0.322	-0.005	-0.293	1.77	-0.03	-1.61
7	0.328	-0.378	0.052	1.79	-2.07	0.29
8	0.152	-0.328	0.195	0.84	-1.79	1.07
9	0.221	-0.225	-0.016	1.21	-1.24	-0.09
10	0.438	-0.342	0.031	2.40	-1.87	0.17

practical standard. By now, sufficient mathematical tools should be available, e.g. in the theory of numbers, the theory of programme complexity, and in tools usually employed to develop cryptographic algorithms.

We want to close with one short remark. It seems to have been widely accepted that primary RNGs should be extremely fast-running. RANDU, like many standard generators, is of the congruential type. Starting with an arbitrary odd integer N(0) for initialisation, a sequence of pseudo random odd integers N(i) is generated by the recursive relation

N(i+1) = A * N(i) modulo (2**31),   for i=0,1,2,...

with A = 65539 = 2**16 + 3.

These pseudo random integers are then normalized. Such procedures are extremely simple, which also means that their programme complexity is very low. Therefore, we may suspect that the generated sequence is far from random /6/. Yet, the use of such extremely simple, fast-running procedures seems to be completely unnecessary. For most realistic Monte Carlo applications a break-down of the computer times shows that only a very minor part is used in calls of the primary RNG. This means that introducing more complex primary RNGs will in most cases not affect adversely the performance of Monte Carlo programmes.

5. References:

/1/ W. Sengpiel, Report KfK 2965 (1980)

/2/ R.R. Coveyou, R.D. MacPherson, Journ. of the ACM, Vol. 14, pp. 100-119 (1967)

/3/ G. Arnecke, H. Borgwaldt, V. Brandl, M. Lalovic, Report KFK 2190 (1976)

/4/ "System/360 Scientific Subroutine Package", Fifth Edition, IBM Corporation (August 1970)

/5/ E. Gelbard, private communication (1980)

/6/ C.P. Schnorr, "Zufälligkeit und Wahrscheinlichkeit", Lecture Notes in Mathematics 218, Springer-Verlag, Berlin-Heidelberg-New York (1971)

6. Appendix: Listings of FORTRAN test programmes.

```
C     TEST-PROGRAMME FOR CHECKING BIAS IN THE ABSOLUTE VALUES
C     OF THE COMPONENTS OF VECTORS INTENDED TO BE UNIFORMLY
C     DISTRIBUTED ON THE 3-DIMENSIONAL UNIT SPHERE.
C
C     REFERENCE CASE A.
C
      REAL*8 UNIT/1./,TWO/2./,HALF/.5/,X,Y,Z,SUMX,SUMY,SUMZ,VARX,VARY,
     +       VARZ,DEVX,DEVY,DEVZ,RAND,TERM
      IA = 1
      RAND = 0.
C     RANDU-ROUTINE INITIALIZED
      ITOT = 100000
      NRUN = 10
      WRITE (6,1000) ITOT,NRUN
      ICONT = 1
  100 SUMX = 0.
      SUMY = 0.
      SUMZ = 0.
      VARX = 0.
      VARY = 0.
      VARZ = 0.
      DO 300 I=1,ITOT
        CALL RANDU (IA,IB,RAND)
        IA = IB
        Z = TWO*RAND-UNIT
C     Z UNIFORM IN (-1.,1.)
  200   CALL RANDU (IA,IB,RAND)
        IA = IB
        X = TWO*RAND-UNIT
        CALL RANDU (IA,IB,RAND)
        IA = IB
        Y = TWO*RAND-UNIT
        TERM = X**2+Y**2
        IF (TERM.GT.UNIT) GOTO 200
C     X,Y UNIFORM IN UNIT DISK
        TERM = DSQRT((UNIT-Z**2)/TERM)
        X = X*TERM
        Y = Y*TERM
C     X,Y NORMALIZED
        SUMX = SUMX+DABS(X)
        VARX = VARX+X**2
        SUMY = SUMY+DABS(Y)
        VARY = VARY+Y**2
        SUMZ = SUMZ+DABS(Z)
  300   VARZ = VARZ+Z**2
      TERM = DFLOAT(ITOT)
      SUMX = SUMX/TERM
      SUMY = SUMY/TERM
      SUMZ = SUMZ/TERM
C     MEAN ABSOLUTE VALUES OF VECTOR COMPONENTS
```

```
      VARX = DSQRT((VARX/TERM-SUMX**2)/TERM)
      VARY = DSQRT((VARY/TERM-SUMY**2)/TERM)
      VARZ = DSQRT((VARZ/TERM-SUMZ**2)/TERM)
C     STANDARD DEVIATIONS OF ESTIMATES
      SUMX = (SUMX-HALF)/HALF
      SUMY = (SUMY-HALF)/HALF
      SUMZ = (SUMZ-HALF)/HALF
C     RELATIVE DEVIATIONS OF ESTIMATES
      DEVX = (SUMX*HALF)/VARX
      DEVY = (SUMY*HALF)/VARY
      DEVZ = (SUMZ*HALF)/VARZ
C     NORMALIZED DEVIATIONS OF ESTIMATES
      WRITE (6,2000) ICONT,SUMX,SUMY,SUMZ,DEVX,DEVY,DEVZ
      ICONT = ICONT+1
      IF (ICONT.LE.NRUN) GOTO 100
      STOP
 1000 FORMAT ('0'/'0'/'0'/'0'/'0'20X,'DEMONSTRATION OF BIASSED R.N.G. '
     +        ,'     CASE A'/'0',20X,'SAMPLE SIZE: ',I8,',      RUNS: ',
     +        I4/1X/'0',26X,'BIAS (PCT.) FOR',17X,'BIAS (ST. DEV.) FOR'
     +        /'0',13X,'RUN',7X,'X',9X,'Y',9X,'Z',13X,'X',9X,'Y',9X,
     +        'Z'/1X)
 2000 FORMAT ('0',I15,1X,2P3F10.3,3X,0P3F10.2)
      END
```

******

```
C     TEST-PROGRAMME FOR CHECKING BIAS IN THE ABSOLUTE VALUES
C     OF THE COMPONENTS OF VECTORS INTENDED TO BE UNIFORMLY
C     DISTRIBUTED ON THE 3-DIMENSIONAL UNIT SPHERE.
C
C     CASE B, Z-COMPONENT AFTER X, Y.
C
      REAL*8 UNIT/1./,TWO/2./,HALF/.5/,X,Y,Z,SUMX,SUMY,SUMZ,VARX,VARY,
     +       VARZ,DEVX,DEVY,DEVZ,RAND,TERM
      IA = 1
      RAND = 0.
C     RANDU-ROUTINE INITIALIZED
      ITOT = 100000
      NRUN = 10
      WRITE (6,1000) ITOT,NRUN
      ICONT = 1
  100 SUMX = 0.
      SUMY = 0.
      SUMZ = 0.
      VARX = 0.
      VARY = 0.
      VARZ = 0.
      DO 300 I=1,ITOT
  200   CALL RANDU (IA,IB,RAND)
        IA = IB
        X = TWO*RAND-UNIT
        CALL RANDU (IA,IB,RAND)
        IA = IB
        Y = TWO*RAND-UNIT
```

```
         TERM = X**2+Y**2
         IF (TERM.GT.UNIT) GOTO 200
C     X,Y UNIFORM IN UNIT DISK
         CALL RANDU (IA,IB,RAND)
         IA = IB
         Z = TWO*RAND-UNIT
C     Z UNIFORM IN (-1.,1.)
         TERM = DSQRT((UNIT-Z**2)/TERM)
         X = X*TERM
         Y = Y*TERM
C     X,Y NORMALIZED
         SUMX = SUMX+DABS(X)
         VARX = VARX+X**2
         SUMY = SUMY+DABS(Y)
         VARY = VARY+Y**2
         SUMZ = SUMZ+DABS(Z)
  300    VARZ = VARZ+Z**2
      TERM = DFLOAT(ITOT)
      SUMX = SUMX/TERM
      SUMY = SUMY/TERM
      SUMZ = SUMZ/TERM
C     MEAN ABSOLUTE VALUES OF VECTOR COMPONENTS
      VARX = DSQRT((VARX/TERM-SUMX**2)/TERM)
      VARY = DSQRT((VARY/TERM-SUMY**2)/TERM)
      VARZ = DSQRT((VARZ/TERM-SUMZ**2)/TERM)
C     STANDARD DEVIATIONS OF ESTIMATES
      SUMX = (SUMX-HALF)/HALF
      SUMY = (SUMY-HALF)/HALF
      SUMZ = (SUMZ-HALF)/HALF
C     RELATIVE DEVIATIONS OF ESTIMATES
      DEVX = (SUMX*HALF)/VARX
      DEVY = (SUMY*HALF)/VARY
      DEVZ = (SUMZ*HALF)/VARZ
C     NORMALIZED DEVIATIONS OF ESTIMATES
      WRITE (6,2000) ICONT,SUMX,SUMY,SUMZ,DEVX,DEVY,DEVZ
      ICONT = ICONT+1
      IF (ICONT.LE.NRUN) GOTO 100
      STOP
 1000 FORMAT ('0'/'0'/'0'/'0'/'0'20X,'DEMONSTRATION OF BIASSED R.N.G. '
    +        ,'      CASE B'/'0',20X,'SAMPLE SIZE: ',I8,',      RUNS: ',
    +        I4/1X/'0',26X,'BIAS (PCT.) FOR',17X,'BIAS (ST. DEV.) FOR'
    +        /'0',13X,'RUN',7X,'X',9X,'Y',9X,'Z',13X,'X',9X,'Y',9X,
    +        'Z'/1X)
 2000 FORMAT ('0',I15,1X,2P3F10.3,3X,0P3F10.2)
      END
```

```
C     TEST-PROGRAMME FOR CHECKING BIAS IN THE ABSOLUTE VALUES
C     OF THE COMPONENTS OF VECTORS INTENDED TO BE UNIFORMLY
C     DISTRIBUTED ON THE 3-DIMENSIONAL UNIT SPHERE.
C
C     CASE C, SELECTION OF (X,Y) MADE MORE INDEPENDENT.
C
      REAL*8 UNIT/1./,TWO/2./,HALF/.5/,X,Y,Z,SUMX,SUMY,SUMZ,VARX,VARY,
     +       VARZ,DEVX,DEVY,DEVZ,RAND,TERM
      IA = 1
      RAND = 0.
C     RANDU-ROUTINE INITIALIZED
      ITOT = 100000
      NRUN = 10
      WRITE (6,1000) ITOT,NRUN
      ICONT = 1
  100 SUMX = 0.
      SUMY = 0.
      SUMZ = 0.
      VARX = 0.
      VARY = 0.
      VARZ = 0.
      DO 300 I=1,ITOT
        CALL RANDU (IA,IB,RAND)
        IA = IB
        Z = TWO*RAND-UNIT
C     Z UNIFORM IN (-1.,1.)
        CALL RANDU (IA,IB,RAND)
        IA = IB
C     DECOUPLING OF (X,Y)-SELECTION
  200   CALL RANDU (IA,IB,RAND)
        IA = IB
        X = TWO*RAND-UNIT
        CALL RANDU (IA,IB,RAND)
        IA = IB
        Y = TWO*RAND-UNIT
        TERM = X**2+Y**2
        IF (TERM.GT.UNIT) GOTO 200
C     X,Y UNIFORM IN UNIT DISK
        TERM = DSQRT((UNIT-Z**2)/TERM)
        X = X*TERM
        Y = Y*TERM
C     X,Y NORMALIZED
        SUMX = SUMX+DABS(X)
        VARX = VARX+X**2
        SUMY = SUMY+DABS(Y)
        VARY = VARY+Y**2
        SUMZ = SUMZ+DABS(Z)
  300   VARZ = VARZ+Z**2
      TERM = DFLOAT(ITOT)
      SUMX = SUMX/TERM
      SUMY = SUMY/TERM
      SUMZ = SUMZ/TERM
C     MEAN ABSOLUTE VALUES OF VECTOR COMPONENTS
```

```
      VARX = DSQRT((VARX/TERM-SUMX**2)/TERM)
      VARY = DSQRT((VARY/TERM-SUMY**2)/TERM)
      VARZ = DSQRT((VARZ/TERM-SUMZ**2)/TERM)
C     STANDARD DEVIATIONS OF ESTIMATES
      SUMX = (SUMX-HALF)/HALF
      SUMY = (SUMY-HALF)/HALF
      SUMZ = (SUMZ-HALF)/HALF
C     RELATIVE DEVIATIONS OF ESTIMATES
      DEVX = (SUMX*HALF)/VARX
      DEVY = (SUMY*HALF)/VARY
      DEVZ = (SUMZ*HALF)/VARZ
C     NORMALIZED DEVIATIONS OF ESTIMATES
      WRITE (6,2000) ICONT,SUMX,SUMY,SUMZ,DEVX,DEVY,DEVZ
      ICONT = ICONT+1
      IF (ICONT.LE.NRUN) GOTO 100
      STOP
 1000 FORMAT ('0'/'0'/'0'/'0'/'0'20X,'DEMONSTRATION OF BIASSED R.N.G. '
     +         ,'     CASE C'/'0',20X,'SAMPLE SIZE: ',I8,',     RUNS: ',
     +         I4/1X/'0',26X,'BIAS (PCT.) FOR',17X,'BIAS (ST. DEV.) FOR'
     +         /'0',13X,'RUN',7X,'X',9X,'Y',9X,'Z',13X,'X',9X,'Y',9X,
     +         'Z'/1X)
 2000 FORMAT ('0',I15,1X,2P3F10.3,3X,0P3F10.2)
      END
```

******

```
C     TEST-PROGRAMME FOR CHECKING BIAS IN THE ABSOLUTE VALUES
C     OF THE COMPONENTS OF VECTORS INTENDED TO BE UNIFORMLY
C     DISTRIBUTED ON THE 3-DIMENSIONAL UNIT SPHERE.
C
C     CASE D, SELECTION OF Z MADE MORE INDEPENDENT.
C
      REAL*8 UNIT/1./,TWO/2./,HALF/.5/,X,Y,Z,SUMX,SUMY,SUMZ,VARX,VARY,
     +       VARZ,DEVX,DEVY,DEVZ,RAND,TERM
      IA = 1
      RAND = 0.
C     RANDU-ROUTINE INITIALIZED
      ITOT = 100000
      NRUN = 10
      WRITE (6,1000) ITOT,NRUN
      ICONT = 1
  100 SUMX = 0.
      SUMY = 0.
      SUMZ = 0.
      VARX = 0.
      VARY = 0.
      VARZ = 0.
      DO 300 I=1,ITOT
        CALL RANDU (IA,IB,RAND)
        IA = IB
C     DECOUPLING OF Z-SELECTION
        CALL RANDU (IA,IB,RAND)
        IA = IB
        Z = TWO*RAND-UNIT
C     Z UNIFORM IN (-1.,1.)
```

```
  200   CALL RANDU (IA,IB,RAND)
        IA = IB
        X = TWO*RAND-UNIT
        CALL RANDU (IA,IB,RAND)
        IA = IB
        Y = TWO*RAND-UNIT
        TERM = X**2+Y**2
        IF (TERM.GT.UNIT) GOTO 200
C     X,Y UNIFORM IN UNIT DISK
        TERM = DSQRT((UNIT-Z**2)/TERM)
        X = X*TERM
        Y = Y*TERM
C     X,Y NORMALIZED
        SUMX = SUMX+DABS(X)
        VARX = VARX+X**2
        SUMY = SUMY+DABS(Y)
        VARY = VARY+Y**2
        SUMZ = SUMZ+DABS(Z)
  300   VARZ = VARZ+Z**2
      TERM = DFLOAT(ITOT)
      SUMX = SUMX/TERM
      SUMY = SUMY/TERM
      SUMZ = SUMZ/TERM
C     MEAN ABSOLUTE VALUES OF VECTOR COMPONENTS
      VARX = DSQRT((VARX/TERM-SUMX**2)/TERM)
      VARY = DSQRT((VARY/TERM-SUMY**2)/TERM)
      VARZ = DSQRT((VARZ/TERM-SUMZ**2)/TERM)
C     STANDARD DEVIATIONS OF ESTIMATES
      SUMX = (SUMX-HALF)/HALF
      SUMY = (SUMY-HALF)/HALF
      SUMZ = (SUMZ-HALF)/HALF
C     RELATIVE DEVIATIONS OF ESTIMATES
      DEVX = (SUMX*HALF)/VARX
      DEVY = (SUMY*HALF)/VARY
      DEVZ = (SUMZ*HALF)/VARZ
C     NORMALIZED DEVIATIONS OF ESTIMATES
      WRITE (6,2000) ICONT,SUMX,SUMY,SUMZ,DEVX,DEVY,DEVZ
      ICONT = ICONT+1
      IF (ICONT.LE.NRUN) GOTO 100
      STOP
 1000 FORMAT ('0'/'0'/'0'/'0'/'0'20X,'DEMONSTRATION OF BIASSED R.N.G. '
     +        ,'    CASE D'/'0',20X,'SAMPLE SIZE: ',I8,',     RUNS: ',
     +        I4/1X/'0',26X,'BIAS (PCT.) FOR',17X,'BIAS (ST. DEV.) FOR'
     +        /'0',13X,'RUN',7X,'X',9X,'Y',9X,'Z',13X,'X',9X,'Y',9X,
     +        'Z'/1X)
 2000 FORMAT ('0',I15,1X,2P3F10.3,3X,0P3F10.2)
      END
```

MODERN AND EASY GENERATION OF RANDOM NUMBERS/ TESTING OF RANDOM NUMBER GENERATORS WITH TESTRAND

by

Edward J. Dudewicz
Department of Statistics

and

Thomas G. Ralley
Instruction and Research
Computer Center

The Ohio State University
Columbus, Ohio 43210

Summary

Simulation and Monte Carlo methods are finding increasing use in the study of applied problems and, accordingly, the importance of having available a random number generator of high quality has taken on greater importance. Vivid accounts have been given (e.g., in the Journal of Computational Physics) of the effects of using an inadequate generator in large scale simulations. In response to the need for assessing the quality of a generator and making available generators of assured quality, a system of computer programs, TESTRAND, has been developed at the Instruction and Research Computer Center of The Ohio State University. TESTRAND provides [4] uniformly distributed random numbers on the interval (0,1) and also provides a series of tests for assessing randomness of sequences of such random numbers.

The TESTRAND generator library [4] contains 20 generators selected because of their widespread use in the computing field, their advocacy by experts in the field, or investigations reporting particularly good testing characteristics.

The testing library contains 10 classical empirical tests which produce either chi-square or Kolmogorov-Smirnov statistics, three theoretical tests for linear multiplicative congruential generators, and two major tests which collect the chi-square or KS statistics from the classical tests and compare their empirical and theoretical distributions.

Using TESTRAND [4] the authors have been able to test a wide variety of different generators and to determine the extent to which philosophical arguments made for the underlying generating schemes depict the reality of the situation. The results of these investigations are described. An illustration of the ease and thoroughness of the possible testing is done using a recently highly touted generator from the chemical engineering literature; it fails miserably, and we see how and why it is nonrandom. All users of simulation and Monte Carlo techniques would be well advised to test their local generator(s) similarly. Details of how to obtain TESTRAND and its documentation are therefore given.

Testing the Quality of Random Number Generators

The sidespread use of simulation and Monte Carlo techniques in scientific investigations and the central role played by sequences of random numbers in the validity of studies using these techniques give emphasis to the importance of having random number generators of high and proven quality. In the past, a variety of historical accidents and subjective judgments have created myths concerning the construction of random number generator programs. Examining such generators, Knuth [8] comments on his own "super random number generator", Algorithm K:

> "The maching language program corresponding to the above algorithm was intended to be so complicated that a man reading a listing of it without explanatory comments would not know what the program was doing. ... [But], when this algorithm was first put onto a computer, it almost immediately converged to the 10 digit value 6065038420, which - by extraordinary coincidence - is transformed into itself by the algorithm."

TABLE OF TESTRAND GENERATORS

TESTRAND name	Generic name/Language	Reference
URN01	LLRANDOM (Assembler)	Learmouth and Lewis, Tech Report Naval Post Grad School (1973)
URN02	- (FORTRAN)	Marsaglia and Bray, CACM Vol. 11 (1968) pp. 757-759
URN03	SUPER-DUPER (Assembler)	Marsaglia, Paul and Ananthanarayanan, McGill School of Computer Science
URN04		See URN01: This is LLRANDOM with suffling
URN05	GFSR (FORTRAN)	Lewis and Payne, JACM Vol. 20 (1973) pp. 456-468
URN06	- (FORTRAN)	Lurie and Mason, Communications on Statistics Vol. 2 (1973) pp. 363-371
URN07	KERAND (Assembler)	IRCC Assembler version of RANDU
URN08	RANDU (FORTRAN)	IBM SSP Manual
URN09	RANDU as in SPSS (FORTRAN)	Nie, Bent and Hull, SPSS Update Manual (1973)
URN10	Generator, OMNITAB II (FORTRAN)	Kruskal, CACM Vol. 12 (1969) pp. 93-94
URN11	- (FORTRAN)	Coveyou and McPherson, JACM Vol. 14 (1967) pp. 100-119
URN12	-	See URN11; This is the double precision version of URN11
URN13	- (Assembler)	Ahrens and Dieter, Graz Inst. für Math Stat., Tech. Report (1974)
URN14	- (FORTRAN)	Zarling
URN15-URN20	- (Assembler)	Hoaglin, Technical Report Harvard Stat. Dept. NS-340

A striking example of a poor random number generator is RANDU, the generator supplied by IBM [7] as part of its original Scientific Subroutine Package. Coldwell [3] and Amadori [1] report the manner in which RANDU invalidated a particle simulation they conducted. Holmlid and Rynefors [6] report Monte

TABLE OF TESTRAND TESTS*

TESTRAND Name	Generic Description	Generates χ^2 on χ^2 Values	Generates K-S on K-S Values
TST01	Performs χ^2 on χ^2 and K-S on K-S		
TST02	Uniform Distribution (simple chi-square)	yes	
TST03	Lagged Correlation		
TST04	Gap	yes	
TST05	Coupon Collector	yes	
TST06	K-S Goodness of Fit		yes
TST07	Permutation	yes	
TST08	Poker	yes	
TST09	Runs Up	yes	
TST10	Serial Pairs	yes	
TST11	Maximum-of-t		yes
	Special Tests for Linear Congruential Generators		
TSTM1	Serial Correlation		
TSTM2	Potency and Period		
TSMT3	Prime Factorization		
TSTM4	Spectral		

*Source Language: FORTRAN, except for TSTM4 (PL/I)

Carlo trajectory studies of absolute complex formation cross sections in alkali-alkali halide relative scattering they reported earlier [5] were similarly invalidated. Fortunately, researchers are becoming more aware of the dangerous effects RANDU can have. Investigators at the Karlsruhe Kernforschungszentrum on the Project for Nuclear Safety [2] tested RANDU and found it defective before using it in Monte Carlo applications in nuclear reactor technology. However, applications to such a sensitive area as this demonstrate the need for the high quality random number generators available in [4]. The fact that RANDU was one of the first widely distributed random number generator programs has resulted in its being embedded in the system libraries of many computer installations and thus, by virtue of historical accident, one of the poorest generators has become one of the most widely used.

In other cases, people seem to have taken a particular point of view about the construction of random number generator programs without substantiation for their position. For example, if a random number generator is good, one would surely expect that the random selection of random numbers produced by this generator would be even better. Discussing this idea, Professor P.A.W. Lewis [9] wrote,

> "The first is the use of shuffling to improve the statistical properties of pseudo-random number generators. Shuffling is a fairly obvious idea which many people seem to have thought of... ."

He then describes the shuffling scheme. N random numbers from the generator are stored in a table. As a new random number is generated, the low order bits are used as an index into the table, the entry at this index is returned as output and the number just generated used to replace it in the table. Interestingly enough, tests [4] on the shuffled and non-shuffled versions show the non-shuffled generator to be superior. Thus, here as is true in science generally, a new idea needs to be subjected to careful testing before being accepted as having merit, and "common sense truths" turn out to be deceptively false "simple" solutions to complex problems.

The point to be made with these examples is that until such time as researchers have a deeper insight into the nature of random number generators, the quality of a generator is best determined by empirical testing with a collection [4] of accepted statistical tests.

A library of programs for the generation and testing of sequences of random numbers has been developed at Ohio State University. The system, TESTRAND, contains twenty generators producing sequences of uniformly distributed random numbers between 0 and 1, and fifteen tests to assess the quality of the generators. Facility for testing a generator supplied by the user is also provided.

Generators were chosen because of their wide usage or because of particularly good testing characteristics. Tests include all those suggested by Knuth [8] for testing random number generators. A special test, TST01, performing two powerful subtests, is also included. The first of these subtests performs a chi-square test on the chi-square values provided by other tests, while the second performs a Kolmogorov-Smirnov test on the K-S values generated by the Goodness-of-Fit (TST06) and the Maximum-of-t (TST11) tests. TST01 provides information about the statistical properties over long sequences of numbers (up to 10,000,000 numbers).

For the most part, the programs are in FORTRAN with some of the generators coded in Assembler language and one of the tests, the Spectral Test, written in PL/I. The package is intended for use on computers with IBM 370 architecture. The above tables summarize the generators and tests available.

Each generator in TESTRAND has been tested with the TESTRAND tests. A complete description of the procedures, testing results, and code is given in The Handbook of Random Number Generation and Testing with TESTRAND Computer Code [4].

Examples of Testing

To illustrate the use of TESTRAND to test generators, we present the following examples. Swain and Swain, in 1980, gave [10] a generator with some testing, which they recommended highly and is becoming widely used in the field of chemical engineering. This generator, ready for testing via TESTRAND [4], is

```
      SUBROUTINE URNWN(X,NBATCH)
      COMMON /OWNPRM/ INTIN(200),REALIN(200),IA,IR
      DIMENSION X(10000)
C
C  INITIAL VALUES ASSIGNED IN MAIN PROGRAM AS:
C     INTIN(1)=21468753
C     INTIN(2)=15796310
C     INTIN(3)=85896521
C
      M1=INTIN(1)
      M2=INTIN(2)
      M3=INTIN(3)
C
      DO 10 I=1,NBATCH
      M4=M1+M2+M3
      IF(M2.LT.50000000) M4=M4+1357
      IF(M4.GE.100000000) M4=M4-100000000
      IF(M4.GE.100000000) M4=M4-100000000
      M1=M2
      M2=M3
      M3=M4
      X(I)=M3*(1.0E-8)
   10 CONTINUE
C
      INTIN(1)=M1
      INTIN(2)=M2
      INTIN(3)=M3
      RETURN
      END
```

THE CHI-SQUARE ON CHI-SQUARES TEST WAS APPLIED (TESTRAND ROUTINE TST01), CENCODE = 1

TO 1000 GROUPS OF CHI-SQUARE VALUES GENERATED BY THE COUPON TEST
AND YIELDED THE FOLLOWING RESULTS:

COMPUTED CHI-SQUARE VALUES (EACH ON 15 DEGREES OF FREEDOM)
TESTS WITH COMPLETE SEGMENTS CONSISTING OF INTEGERS 1 TO 5 USING THE FRACTIONAL PART OF 100 * RANDOM NUMBER

J	PROB. P(J)	CHI-SQUARE (P, 15)	EXPECTED FREQUENCY	OBSERVED FREQUENCY	CHI-SQUARE CONTRIBUTION	J	PROB. P(J)	CHI-SQUARE (P, 15)	EXPECTED FREQUENCY	OBSERVED FREQUENCY	CHI-SQUARE CONTRIBUTION
1	0.01	5.2294	10.00	0	10.00	51	0.51	14.473	10.00	1	8.10
2	0.02	5.9849	10.00	0	10.00	52	0.52	14.608	10.00	2	6.40
3	0.03	6.5032	10.00	0	10.00	53	0.53	14.744	10.00	0	10.00
4	0.04	6.9137	10.00	0	10.00	54	0.54	14.881	10.00	0	10.00
5	0.05	7.2609	10.00	1	8.10	55	0.55	15.020	10.00	1	8.10
6	0.06	7.5661	10.00	0	10.00	56	0.56	15.159	10.00	1	8.10
7	0.07	7.8410	10.00	0	10.00	57	0.57	15.300	10.00	0	10.00
8	0.08	8.0930	10.00	0	10.00	58	0.58	15.443	10.00	0	10.00
9	0.09	8.3271	10.00	0	10.00	59	0.59	15.587	10.00	2	6.40
10	0.10	8.5468	10.00	1	8.10	60	0.60	15.733	10.00	1	8.10
11	0.11	8.7545	10.00	0	10.00	61	0.61	15.881	10.00	1	8.10
12	0.12	8.9523	10.00	0	10.00	62	0.62	16.031	10.00	2	6.40
13	0.13	9.1416	10.00	1	8.10	63	0.63	16.183	10.00	4	3.60
14	0.14	9.3236	10.00	0	10.00	64	0.64	16.337	10.00	1	8.10
15	0.15	9.4993	10.00	0	10.00	65	0.65	16.494	10.00	3	4.90
16	0.16	9.6695	10.00	0	10.00	66	0.66	16.653	10.00	2	6.40
17	0.17	9.8348	10.00	0	10.00	67	0.67	16.816	10.00	4	3.60
18	0.18	9.9959	10.00	0	10.00	68	0.68	16.981	10.00	4	3.60
19	0.19	10.1531	10.00	0	10.00	69	0.69	17.150	10.00	1	8.10
20	0.20	10.3070	10.00	1	8.10	70	0.70	17.322	10.00	1	8.10
21	0.21	10.4578	10.00	0	10.00	71	0.71	17.498	10.00	2	6.40
22	0.22	10.6059	10.00	0	10.00	72	0.72	17.677	10.00	4	3.60
23	0.23	10.7515	10.00	0	10.00	73	0.73	17.862	10.00	4	3.60
24	0.24	10.8950	10.00	0	10.00	74	0.74	18.051	10.00	0	10.00
25	0.25	11.0365	10.00	1	8.10	75	0.75	18.245	10.00	2	6.40
26	0.26	11.1763	10.00	0	10.00	76	0.76	18.445	10.00	10	0.0
27	0.27	11.3145	10.00	0	10.00	77	0.77	18.651	10.00	3	4.90
28	0.28	11.4512	10.00	0	10.00	78	0.78	18.863	10.00	4	3.60
29	0.29	11.5868	10.00	1	8.10	79	0.79	19.083	10.00	4	3.60
30	0.30	11.7212	10.00	0	10.00	80	0.80	19.311	10.00	4	3.60
31	0.31	11.8546	10.00	0	10.00	81	0.81	19.547	10.00	4	3.60
32	0.32	11.9872	10.00	1	8.10	82	0.82	19.793	10.00	4	3.60
33	0.33	12.1190	10.00	1	8.10	83	0.83	20.051	10.00	6	1.60
34	0.34	12.2502	10.00	1	8.10	84	0.84	20.320	10.00	7	0.90
35	0.35	12.3809	10.00	0	10.00	85	0.85	20.603	10.00	6	1.60
36	0.36	12.5111	10.00	2	6.40	86	0.86	20.902	10.00	6	1.60
37	0.37	12.6411	10.00	2	6.40	87	0.87	21.218	10.00	19	8.10
38	0.38	12.7708	10.00	2	6.40	88	0.88	21.555	10.00	12	0.40
39	0.39	12.9003	10.00	1	8.10	89	0.89	21.917	10.00	14	1.60
40	0.40	13.0298	10.00	0	10.00	90	0.90	22.307	10.00	9	0.10
41	0.41	13.1592	10.00	0	10.00	91	0.91	22.732	10.00	16	3.60
42	0.42	13.2888	10.00	0	10.00	92	0.92	23.199	10.00	13	0.90
43	0.43	13.4186	10.00	1	8.10	93	0.93	23.720	10.00	23	16.90
44	0.44	13.5486	10.00	0	10.00	94	0.94	24.311	10.00	14	1.60
45	0.45	13.6790	10.00	0	10.00	95	0.95	24.996	10.00	26	25.60
46	0.46	13.8098	10.00	0	10.00	96	0.96	25.816	10.00	38	78.40
47	0.47	13.9411	10.00	0	10.00	97	0.97	26.848	10.00	41	96.10
48	0.48	14.0730	10.00	0	10.00	98	0.98	28.260	10.00	53	184.90
49	0.49	14.2056	10.00	1	8.10	99	0.99	30.578	10.00	101	828.10
50	0.50	14.5389	10.00	2	6.40	100	1.00	********	10.00	500	******
						TOTAL			1000	1000	******

IF X IS A CHI-SQUARE RANDOM VARIABLE WITH 99 DEGREES OF FREEDOM, THEN P(X.LE. 134.64) = .99

Testing as performed by Swain and Swain was inadequate to detect it, but this generator failed the sensitive chi-square on chi-square tests based on the coupon test, the gap test, the permutation test, the poker test, and the runs up test due to its producing large chi-square values an extraordinarily large proportion of the time. For example, the above output is typical and illustrates this point.

Conclusions

We conclude that no simulation or Monte Carlo study of any importance should use any generator unless it is one of the generators described in [4] as having passed extensive testing, or the users subject it to similar extensive tests (see [4] for details and code which can easily be used to do this) and it passes.

References

1. Amadori, R. (1977). "Correlational defects in the standard IBM 360 random number generator and the classical ideal gas correlation function," Journal of Computational Physics, 24, 450-454.

2. Borgwaldt, H. and Brandl, V. (1981). "Evidence of a significant bias in an elementary random number generator," Technical Report, Kernforschungszentrum Karlsruhe, Institute für Neutronenphysik und Reaktortechnik.

3. Coldwell, R.L. (1974). "Correlational defects in the standard IBM 360 random number generator and the classical ideal gas correlation function," Journal of Computational Physics, 14, 223-226.

4. Dudewicz, E. J. and Ralley, T. G. (1981). The Handbook of Random Number Generation and Testing with TESTRAND Computer Code, American Sciences Press, Inc., Columbus, Ohio 43221-0161.

5. Holmlid, L. and Rynefors, K. (1976). Chem. Phys., 14, 403.

6. Holmlid, L. and Rynefors, K. (1978). "Uniformity of congruential pseudo-random number generators. Dependence on length of number sequence and resolution," Journal of Computational Physics, 26, 297-306.

7. IBM (1970). "System/360 scientific subroutine package, version III, programmer's manual, program number 360A-CM-03X," Manual GH20-0205-4 (Fifth Edition), IBM Corporation, White Plains, New York.

8. Knuth, D. E. (1969). The Art of Computer Programming, Volume 2/Seminumerical Algorithms, Addison-Wesley Publishing Company, Inc., Reading, Massachusetts.

9. Lewis, P. A. W. (1979). Numerical Computations Newsletter, IMSL, Issue 19, pp. 1-2.

10. Swain, C. G. and Swain, M. S. (1980). "A uniform random number generator that is reproducible, hardware-independent, and fast," Journal of Chemical Information and Computer Sciences, 20, 56-58.

11. Texas Instruments Master Library Manual (1977). Texas Instruments, Inc., Dallas, Texas.

TESTRAND: A Random Number Generation and Testing Library

Edward J. Dudewicz
Department of Mathematics
Syracuse University
Syracuse, NY 13210

Thomas G. Ralley
Instruction and Research Computer Center
The Ohio State University
Columbus, OH 43210

TESTRAND is a program library containing 20 routines for generating sequences of uniformly distributed random numbers on the interval [0, 1], and 14 tests for determining the extent to which a given sequence conforms to criteria for randomness. For the most part, programs are in FORTRAN, with some of the generators coded in IBM Assembler and one of the tests, the Spectral Test, written in PL/I. The package is intended for use on computers with IBM/360 or 370 architecture.

Generators in the library were selected for one of several reasons: because of their widespread use in the field; because of recommendations of experts; or because studies using the generator demonstrated particularly good testing characteristics. The test library is composed of 11 classical empirical tests, each of which produces either a chi-squared or Kolmogorov-Smirnov (K-S) statistic, and three theoretical tests for use with linear congruential generators. The test library includes all 12 of the tests recommended by Knuth (1969) for testing random number generators. Of the empirical tests, one performs two powerful subtests. The first subtest performs a chi-squared test on the chi-squared values produced by certain of the empirical tests, while the second performs a Kolmogorov-Smirnov test on the K-S values provided by the K-S goodness-of-fit and maximum-of-t tests. This permits, for example, a test run in which 1,000 batches of 10,000 random numbers each are generated and tested for uniformity of distribution and goodness of fit, producing a chi-squared and K-S statistic, respectively, for each batch. These 1,000 chi-squared and K-S values are stored and then tested to compare their empirical and theoretical distributions. In this way, information about the statistical properties of sequences of random numbers (up to 10,000,000 terms in length) is provided. Each generator in the package has been thoroughly tested. A description of the testing procedure and complete testing results can be found in Dudewicz and Ralley (1981).

TESTRAND provides other facilities as well. It can be used to test a generator supplied by the user, thus allowing evaluation of generators currently in use at the user's own installation. It provides a timing facility, giving the average time required to generate one term in a sequence of random numbers. It provides two easy-call generators for individuals requiring only a few random numbers, one with fixed initial seed and the other seeded from an internal clock. A TESTRAND User's Guide that provides an overview of the package and a detailed description of each of the generators and tests has been prepared. Also included in the guide are several examples with complete jobstreams, illustrating these different facilities.

The TESTRAND library consists of 99 routines and is large (approximately 15,000 card images when stored in 80 byte records). Each routine has been carefully documented with comments indicating how the routine is to be used and what resources are required. A complete source listing can be found in Dudewicz and Ralley (1981).

A copy of the TESTRAND library is available on magnetic tape from The Ohio State University Instruction and Research Computer Center for a charge, currently $90. The charge covers the cost of the tape, preparation, and mailing. An order form can be obtained from the Computer Center's Office of Customer Services and is also included in Dudewicz and Ralley (1981, pp. 379–380).

REFERENCES

DUDEWICZ, E.J., and RALLEY, T.G. (1981), *The Handbook of Random Number Generation and Testing With TESTRAND Computer Code,* 43221-0161, Columbus, Ohio: American Sciences Press.

KNUTH, D.E. (1969), *The Art of Computer Programming, Volume 2/Seminumerical Algorithms*, Reading, Mass.: Addison-Wesley.

Part II: Sampling from Univariate and Multivariate Distributions

Once a good source of random numbers on the interval (0.0, 1.0) is available and fully tested (see Part I for details), in typical use its output is transformed to **other statistical distributions** appropriate to the quantities which drive the simulation study. Most traditional is the attempt to obtain **the inverse function of the distribution function** of the random variable, since its evaluation at a uniform random number on the interval (0.0, 1.0) will yield a random variable with the desired distribution. However, most distributions important in practice cannot be inverted exactly, hence a variety of **approximate inverses** have been used. One of the most successful is described in the first article by Ramberg, Dudewicz, Tadikamalla, and Mykytka (1979), a four-parameter inverse which can fit a broad class of distributions from the normal and Student's t-distribution to skewed and truncated univariate distributions.

Another popular general method, especially useful where *exactly* the specified distribution (and not an approximation to it) is desired, is **the series method** described with applications and code for the Kolmogorov-Smirnov distribution and algorithms for the exponential and Raab-Green distributions, by Devroye (1981) in the second article.

Specific distributions have been the subject of extensive investigation. For example, **the gamma distribution** which is widely used as a model in applied studies (and **includes the exponential, Erlang, chi-square, F, and beta distributions** as either special cases or as simple combinations, and the normal as a limit) is the subject of numerous algorithms, which are of varying quality and accuracy. The third article by Tadikamalla and Johnson (1981) gives a comprehensive study of all algorithms, with code for the best algorithms.

Production of **multivariate random variables** which are appropriate for a given setting, is an area now seeing much activity. An overview is given in the fourth article by Johnson, Wang, and Ramberg (1984), with details not only for the traditional multivariate normal distribution, but also for other distributions which are important when the multi-normal model is not appropriate.

TECHNOMETRICS ©, VOL. 21, NO. 2, MAY 1979

A Probability Distribution and Its Uses in Fitting Data

John S. Ramberg
Systems and Industrial Engineering
The University of Arizona
Tucson, AZ 85721

Edward J. Dudewicz
Department of Statistics
The Ohio State University
Columbus, OH 43210

Pandu R. Tadikamalla
Graduate School of Business
The University of Pittsburgh
Pittsburgh, PA 15260

Edward F. Mykytka
Systems and Industrial Engineering
The University of Arizona
Tucson, AZ 85721

A four-parameter probability distribution, which includes a wide variety of curve shapes, is presented. Because of the flexibility, generality, and simplicity of the distribution, it is useful in the representation of data when the underlying model is unknown. A table based on the first four moments, which simplifies parameter estimation, is given. Further important applications of the distribution include the modeling and subsequent generation of random variates for simulation studies and Monte Carlo sampling studies of the robustness of statistical procedures.

KEY WORDS

Data fitting
Probability distribution
Systems of probability distributions
Moments
Random variate generation
Monte Carlo
Simulation

1. INTRODUCTION

Reasons for fitting a distribution to a set of data have been summarized by Hahn and Shapiro [6] (p. 195) as: the desire for objectivity, the need for automating the data analysis, and interest in the values of the distribution parameters. Although various empirical distributions already exist, e.g., the Pearson system and the Johnson system (see Chapter 7 of Hahn and Shapiro [6]) and the Burr distribution [1], we are presenting another distribution because of its simplicity, flexibility, and generality.

The new distribution is a generalization of Tukey's [16] lambda distribution. It was developed by Ramberg and Schmeiser [9, 10] for the purpose of generating random variates for Monte Carlo simulation studies because of the simple form of the resulting algorithm. (See (2).) A wide variety of curve shapes are possible with this distribution. Hence it is useful for the representation of data when the underlying model is unknown. Silver [14], for example, shows how the distribution can be used to approximate the safety factor in an inventory control model. It is also useful in Monte Carlo studies of the robustness of statistical procedures and for sensitivity analyses in simulation studies.

To illustrate the distribution, consider the histogram for 250 sample measurements of the coefficient of friction of a metal [6] (p. 219) given in Figure 1. The superimposed distribution was fitted by the methods described in this paper. This example is discussed further in Section 5.

In Section 2 some of the properties of the distribution are described. Section 3 contains a discussion of the use of the method of moments for fitting the distribution to data and a table to facilitate this procedure. Table construction and accuracy is described in Section 4.

2. THE PROPOSED DISTRIBUTION AND ITS PROPERTIES

A continuous probability distribution is usually defined by its distribution function or by its density function. Alternatively it can be defined by its per-

Received August 1976; revised May 1978

Winner of 1977 Shewell Award at ASQC Chemical Division Technical Conference

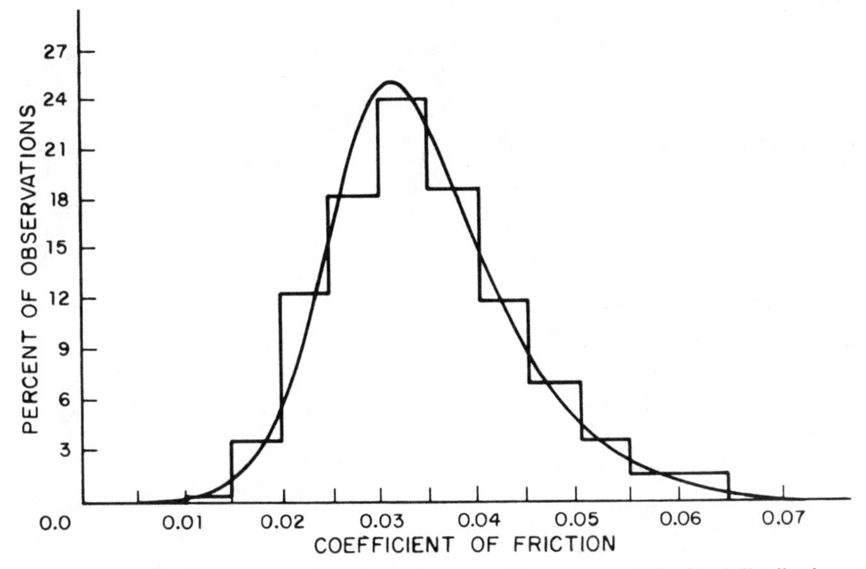

FIGURE 1. Coefficient of friction relative frequency histogram and the fitted distribution.

centile (or quantile) function, if the percentile function exists. The percentile function is simply the inverse of the distribution function. This concept is particularly useful in Monte Carlo simulation studies because of the following result: If X is a continuous random variable with percentile function R, and U is a uniform random variable on the interval zero to one, then the transformation $X = R(U)$ yields a random variable with the percentile function R.

A specific example is Tukey's [16] lambda function

$$R(p) = [p^{\lambda} - (1 - p)^{\lambda}]/\lambda \qquad (0 \leq p \leq 1), \quad (1)$$

which is defined for all nonzero lambda values. (As $\lambda \to 0$, the logistic distribution results.) Van Dyke [17] compared a normalized version of this function with Student's t distribution. Filliben [5] used this distribution to approximate symmetric distributions with a wide range of tail weights for studying location estimation problems of symmetric distributions. He also gave a very complete discussion of the properties of the percentile function. Joiner and Rosenblatt [7] studied the lambda distribution further and gave results on the sample range. Ramberg and Schmeiser [9] showed how this distribution could be used to approximate many of the well-known symmetric distributions and explored its application to Monte Carlo simulation studies.

Ramberg and Schmeiser [10] generalized (1) to a four-parameter distribution defined by the percentile function

$$R(p) = \lambda_1 + [p^{\lambda_3} - (1 - p)^{\lambda_4}]/\lambda_2 \quad (0 \leq p \leq 1), \quad (2)$$

where λ_1 is a location parameter, λ_2 is a scale parameter and λ_3 and λ_4 are shape parameters. This distribution, which includes the original lambda distribution, also permits skewed curves to be represented. Although the distribution function does not exist in "simple closed form," this should not be of concern to practitioners since the same is true of the normal distribution (whose percentiles are not nearly so easily computed). Another asymmetric generalization of (1) was considered by Ramberg [11].

The density function corresponding to (2) is given by:

$$\begin{aligned} f(x) &= f[R(p)] \\ &= \lambda_2[\lambda_3 p^{\lambda_3 - 1} + \lambda_4(1 - p)^{\lambda_4 - 1}]^{-1} \end{aligned} \qquad (0 \leq p \leq 1). \quad (3)$$

Plotting the density for given λ_1, λ_2, λ_3, λ_4 requires evaluation of (2) and (3) for values of p ranging from zero to one. Then $f[R(p)]$ is plotted on the y-axis versus $R(p)$ on the x-axis. Even though λ_1 does not explicitly appear in (3), the density is a function of λ_1 since it is defined in terms of $R(p)$, which depends upon λ_1, as can be seen from (2).

This four-parameter distribution includes a wide range of curve shapes as illustrated by the density plots in Figures 2a-2e. The densities are indexed by the values of their skewness ($\alpha_3 = E(x - \mu)^3/\sigma^3$) and kurtosis ($\alpha_4 = E(x - \mu)^4/\sigma^4$). Each has a mean of zero and a variance of one. The lambda values corresponding to these values of skewness and kurtosis are given in Table 1. In Figures 2a and 2b the skewness is fixed and three values of kurtosis are illustrated. In Figure 2a, $\alpha_3 = 0$ and $\alpha_4 = 3, 5, 9$. In Figure 2b, $\alpha_3 = 1$ and $\alpha_4 = 4, 6, 9$. In Figure 2c, the kurtosis is fixed ($\alpha_4 = 4$) and three values of skewness are illustrated ($\alpha_3 = 0$, 0.5 and 1.0).

A limited range of U-shaped curves are also possible and occur when $1 \leq \lambda_3, \lambda_4 \leq 2$. A few typical U-shaped densities are plotted in Figure 2d.

Figure 2e shows two densities which have approximately the same first four moments as the ex-

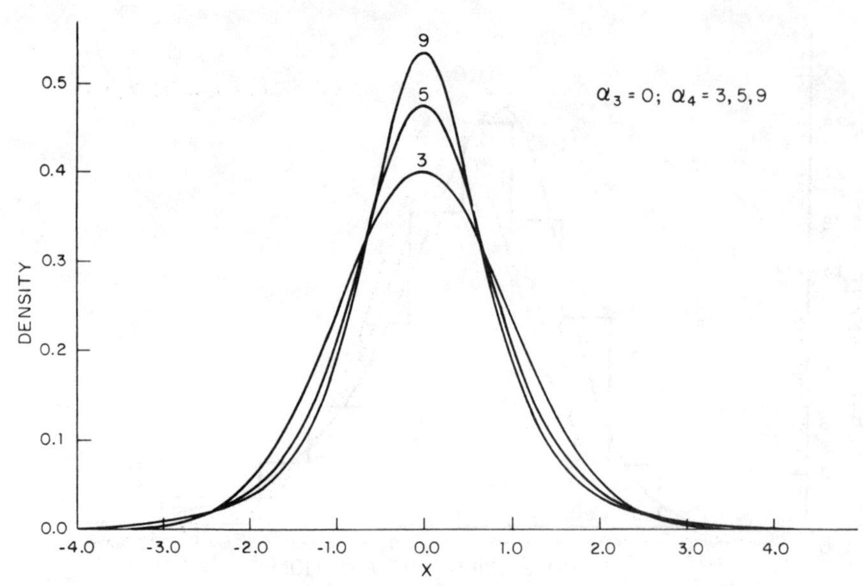

FIGURE 2a. Density plots for specified α_3 and α_4 values ($\alpha_3 = 0$; $\alpha_4 = 3, 5, 9$).

ponential distribution. The non-truncated density was obtained using the parameter values given in Table 4; the other density demonstrates that positively skewed, J-shaped curves result when $\lambda_3 = 0$. The latter curve provides a good approximation to the exponential density. Indeed, Schmeiser [13] has shown that the limiting distribution of this distribution is exponential with parameter θ as $\lambda_4 \to 0$ when $\lambda_1 = \lambda_3 = 0$ and $\lambda_2 = \lambda_4/\theta$.

The distribution can also provide good approximations to other well-known densities. For example, the distribution with $\lambda_1 = 0$, $\lambda_2 = 0.1975$, and $\lambda_3 = \lambda_4 = 0.1349$ results in an approximation to the normal distribution for which $\max_x |\Phi(x) - R^{-1}(x)| \approx .001$, where $\Phi(x)$ is the normal distribution function.

Although distributions are not necessarily determined by their moments, the moments often do provide useful information. In Figure 3 some distributions are characterized by their skewness and kurtosis. The normal, the rectangular, and the exponential distribution are each represented by a single point. The Student's t, the lognormal, the gamma, and the Weibull distributions are each represented by a line. The beta distribution is represented by a region of values. The proposed distribution covers the screened area; refer to Table 4 for some of the values included.

The Pearson and Johnson systems also cover large regions of this diagram (see, for example, Hahn and Shapiro [6]). Both of these systems incorporate a number of functional forms whereas the proposed distribution uses only one function and is computa-

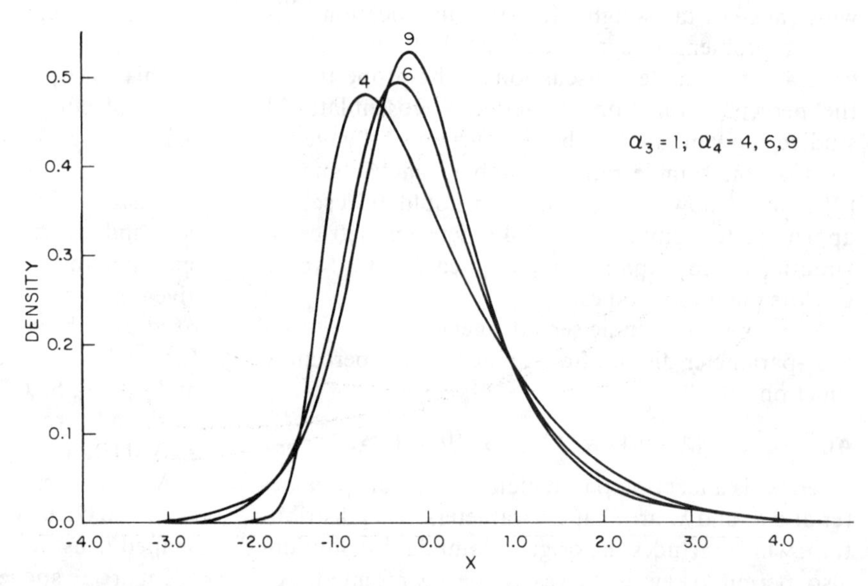

FIGURE 2b. Density plots for specified α_3 and α_4 values ($\alpha_3 = 1$; $\alpha_4 = 4, 6, 9$).

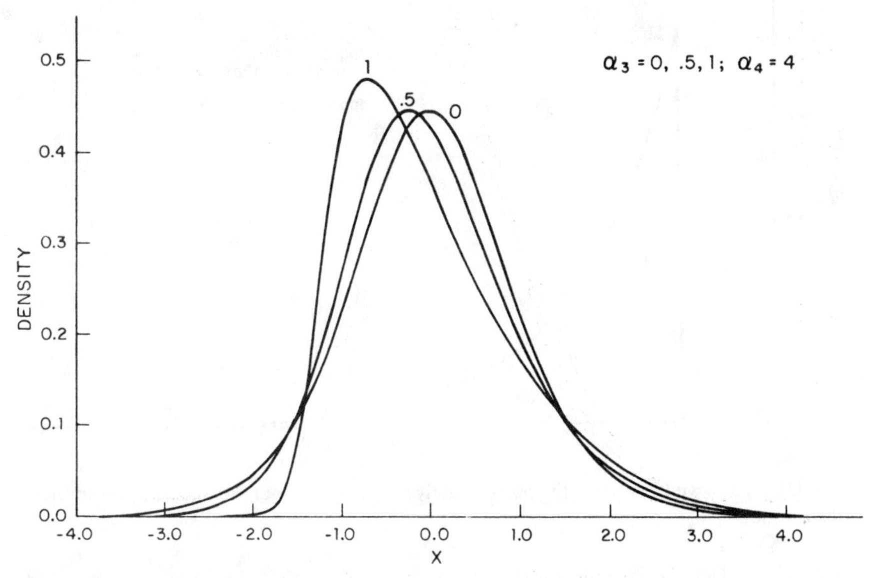

FIGURE 2c. Density plots for specified α_3 and α_4 values (α_3 = 0,.5,1; α_4 = 4).

tionally simpler. The Burr [1] distribution also covers a wide range of parameter values, but does not include symmetric distributions.

As originally indicated by Schmeiser [12], there are four regions of parameter values where the distribution is a legitimate one, i.e., the density function is nonnegative for all x and integrates to one. These regions, which are numbered arbitrarily 1, 2, 3 and 4 for reference purposes, are indicated in Figure 4. Our interest will center on Regions 3 and 4 since positive moments do not exist for the distributions in Regions 1 and 2. The values of X range from $R(0)$ to $R(1)$. These bounds, which depend on all of the lambda values, are given in Table 2. If $\lambda_3 \geq 1$, $f[R(0)] > 0$ and the distribution is truncated on the left. Similarly if $\lambda_4 \geq 1$, the distribution is truncated on the right.

Ramberg and Schmeiser [10] showed that the kth moment ($\lambda_1 = 0$), of the proposed distribution, when it exists, is given by

$$E(X^k) = \lambda_2^{-k} \sum_{i=0}^{k} \binom{k}{i} (-1)^i \beta(\lambda_3(k - i) + 1, \lambda_4 i + 1),$$

where β denotes the beta function, as defined, for example, in [2]. (The kth moment does not exist when any of the arguments of the beta function are negative. Thus, the kth moment exists if and only if $-1/k < \min(\lambda_3, \lambda_4)$.)

They also derived the following expressions for the mean, the variance, and the third ($\mu_3 = E(X - \mu)^3$) and fourth ($\mu_4 = E(X - \mu)^4$) moments about the mean for this distribution:

$$\mu = \lambda_1 + A/\lambda_2,$$

$$\sigma^2 = (B - A^2)/\lambda_2^2,$$

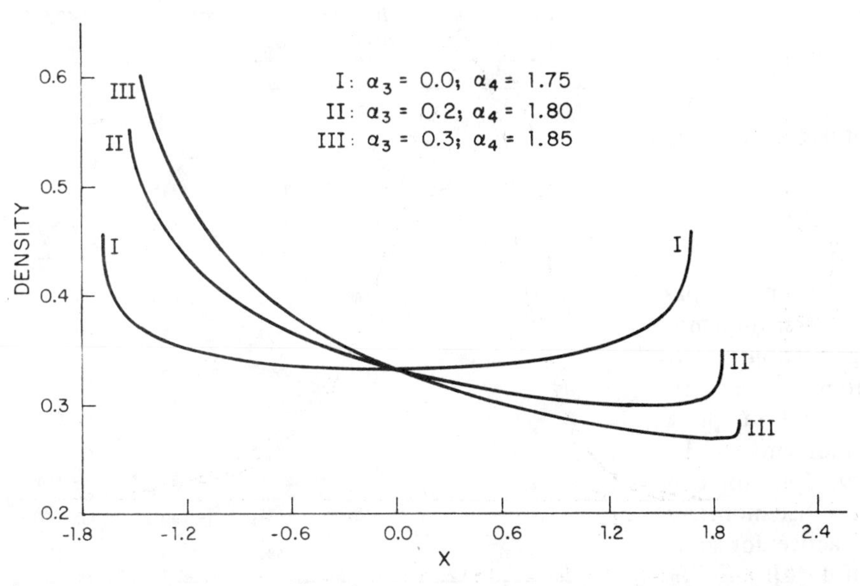

FIGURE 2d. Density plots for specified α_3 and α_4 values resulting in U-shaped densities.

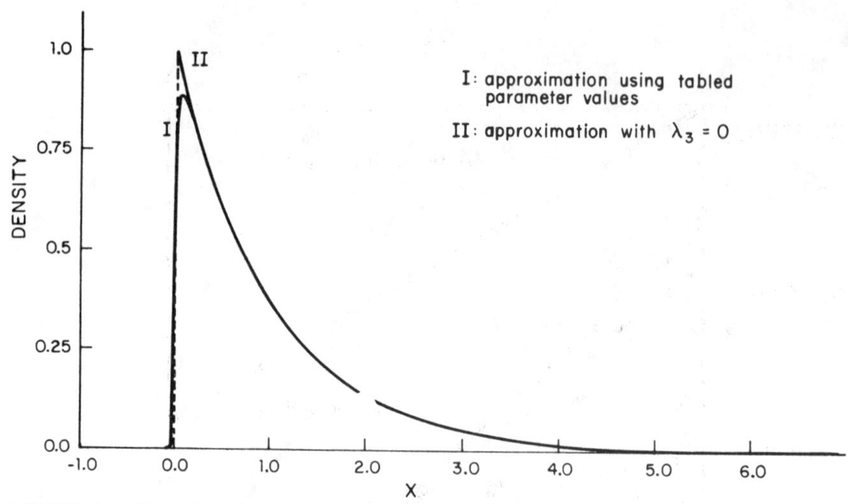

FIGURE 2e. Two densities with approximately the same first four moments as the exponential distribution.

$$\mu_3 = (C - 3AB + 2A^3)/\lambda_2^3,$$

$$\mu_4 = (D - 4AC + 6A^2B - 3A^4)/\lambda_2^4,$$

where

$$A = 1/(1 + \lambda_3) - 1/(1 + \lambda_4),$$

$$B = 1/(1 + 2\lambda_3) + 1/(1 + 2\lambda_4) - 2\beta(1 + \lambda_3, 1 + \lambda_4),$$

$$C = 1/(1 + 3\lambda_3) - 3\beta(1 + 2\lambda_3, 1 + \lambda_4) + 3\beta(1 + \lambda_3, 1 + 2\lambda_4) - 1/(1 + 3\lambda_4),$$

$$D = 1/(1 + 4\lambda_3) - 4\beta(1 + 3\lambda_3, 1 + \lambda_4) + 6\beta(1 + 2\lambda_3, 1 + 2\lambda_4) - 4\beta(1 + \lambda_3, 1 + 3\lambda_4) + 1/(1 + 4\lambda_4).$$

The skewness and kurtosis, as given by

$$\alpha_3 = \mu_3/\sigma^3 \tag{4}$$

and

$$\alpha_4 = \mu_4/\sigma^4, \tag{5}$$

are functions of λ_3 and λ_4, but do not depend upon λ_1 and λ_2.

3. PARAMETER ESTIMATION AND DISTRIBUTION FITTING

In this section we show how to determine the parameters of the distribution using the first four moments and how to fit the resulting distribution. It should be recognized that sample moments are sensitive to extreme observations and that the sampling variability of the third and fourth moments can be large. (See, for example, Dudewicz, Johnson and Ramberg [4].) However, we elected to use moments because of their widespread use in practice; for some properties of moments, see pp. 174ff of [2].

The values of λ_1, λ_2, λ_3 and λ_4 are given in Table 4 for selected values of α_3 and α_4 with $\mu = 0$ and $\sigma = 1$. (The construction and accuracy of this table is discussed in Section 4.) If the values of μ, σ, α_3 and α_4 are known, the lambda values are determined from Table 4 using as entry points the α_3 and α_4 values. One simply picks the values of λ_3 and λ_4 in Table 4 for which the α_3 and α_4 are closest to the desired values. If α_3 is negative, one uses its absolute value, and after finding the values of λ_3 and λ_4, interchanges their values and changes the sign of λ_1. (The density with a skewness of $-\alpha_3$ is the mirror image of the density with a skewness of α_3.)

Since the λ_1 and λ_2 values given in Table 4 are for a variate with a mean of zero and a variance of one, multiplying the resulting variate by σ and adding μ to it achieves the desired result. This reduces to computing

TABLE 1—*Lambda values for Figure 2 densities (see Table 4 for definition of + and $ symbols).*

α_3	α_4	λ_1	λ_2	λ_3	λ_4
0	1.75	0	.5943	1.4501	1.4501
0	3	0	.1974	.1349	.1349
0	4	0	.0262	.0148	.0148
0	5	0	-.0870	-.0443	-.0443
0	9	0	-.3203	-.1359	-.1359
0.2	1.80	.166	.5901	1.7680	1.1773
0.3	1.85	.246	.5852	1.934	1.062
0.5	4	-.290	.0604	.0259	.0447
1	4	-.886	.1333	.0193	.1588
1	6	-.379	-.0562	-.0187	-.0388
1	9	-.215	-.2356	-.0844	-.1249
2	9	.007	-.1081+	-.0407$	-.1076+
2	9	0	-.0580+	0	-.0580+

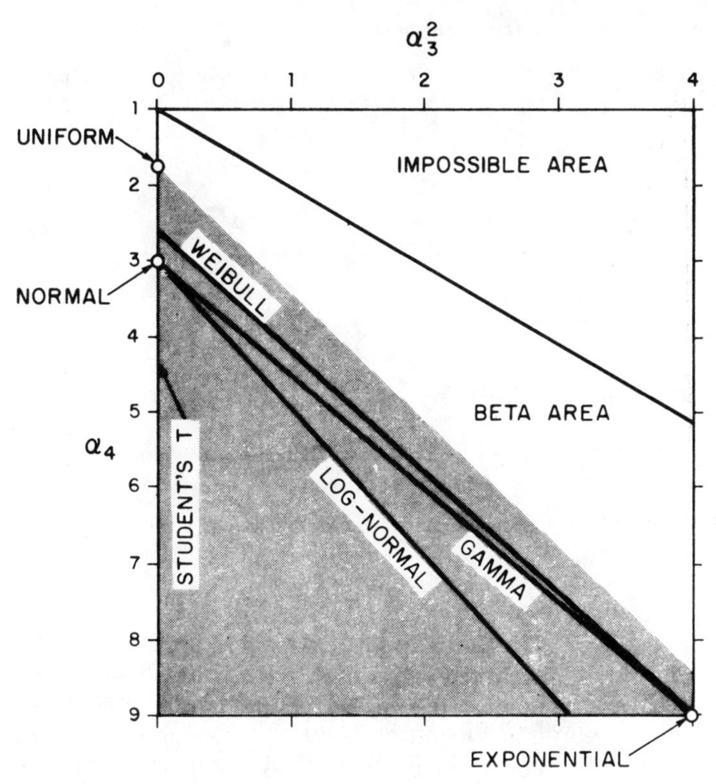

FIGURE 3. Characterization of distributions by their third and fourth moments (the proposed distribution covers the screened region).

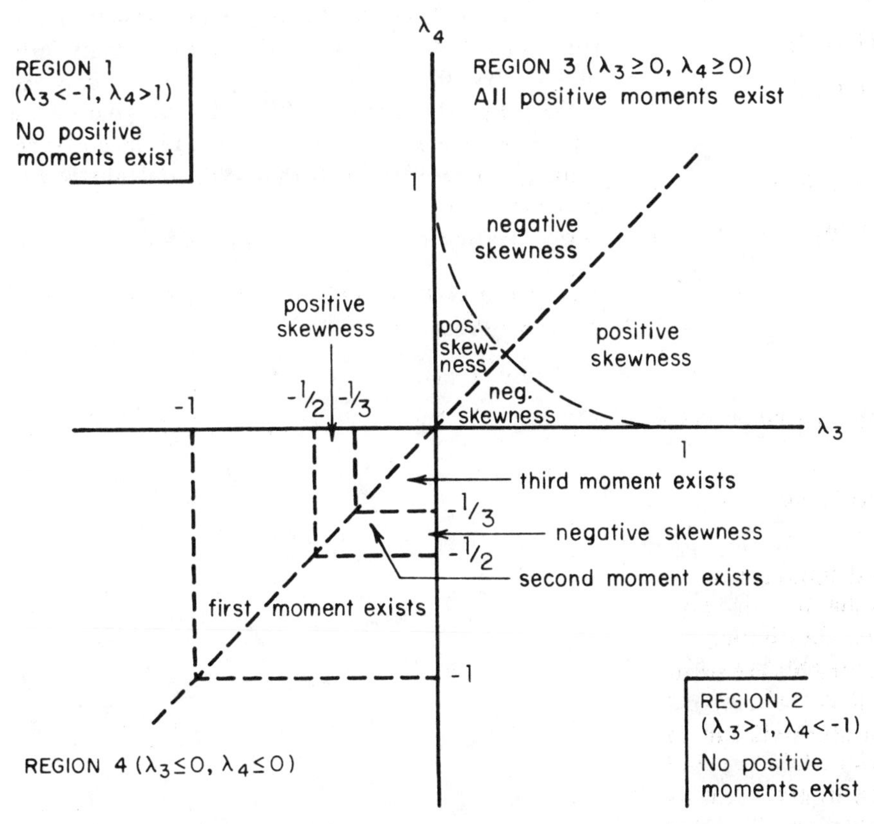

FIGURE 4. Some properties of the distribution which are dependent on the values of λ_3 and λ_4.

TABLE 2—*Lower and upper bounds of the distribution.*

Region	λ_2	λ_3	λ_4	Lower Bound	Upper Bound
		Value of			
	$\lambda_2 > 0$	$\lambda_3 > 0$	$\lambda_4 > 0$	$\lambda_1 - 1/\lambda_2$	$\lambda_1 + 1/\lambda_2$
3	$\lambda_2 > 0$	$\lambda_3 = 0$	$\lambda_4 > 0$	λ_1	$\lambda_1 + 1/\lambda_2$
	$\lambda_2 > 0$	$\lambda_3 > 0$	$\lambda_4 = 0$	$\lambda_1 - 1/\lambda_2$	λ_1
	$\lambda_2 < 0$	$\lambda_3 < 0$	$\lambda_4 < 0$	$-\infty$	∞
4	$\lambda_2 < 0$	$\lambda_3 = 0$	$\lambda_4 < 0$	λ_1	∞
	$\lambda_2 < 0$	$\lambda_3 < 0$	$\lambda_4 = 0$	$-\infty$	λ_1

$$\lambda_1(\mu, \sigma) = \lambda_1(0, 1)\sigma + \mu \tag{6}$$

and

$$\lambda_2(\mu, \sigma) = \lambda_2(0, 1)/\sigma. \tag{7}$$

Estimates of the lambda values from sample data are obtained in a similar manner using the third and fourth standardized sample moments

$$\hat{\alpha}_3 = m_3/(m_2)^{3/2}$$

$$\hat{\alpha}_4 = m_4/(m_2)^2,$$

where

$$m_2 = \sum_{i=1}^{n} (x_i - \bar{x})^2/n,$$

$$m_3 = \sum_{i=1}^{n} (x_i - \bar{x})^3/n,$$

and

$$m_4 = \sum_{i=1}^{n} (x_i - \bar{x})^4/n.$$

Computational forms for these sample moments are given on pp. 47–48 of [6].

The procedure outlined at the beginning of this section is followed with μ, σ, α_3 and α_4 replaced by $\bar{x}$, $\sqrt{m_2}$, $\hat{\alpha}_3$ and $\hat{\alpha}_4$, respectively. The resulting estimates of λ_i will be denoted by $\hat{\lambda}_i$ in subsequent sections.

To verify the goodness-of-fit one may plot (2) and compare it with the distribution function or the sample distribution function being approximated, and/or plot (3) and compare it with the density function or relative frequency histogram being approximated for verification. A FORTRAN program for the required computations is given in [3].

An approximate χ^2 goodness-of-fit statistic can

TABLE 3—*Actual versus expected frequencies based upon fitting the coefficient of friction data with the distribution.*

Coefficient of Friction	Actual Frequency	Expected Frequency
less than 0.015	1 } 10	9.725
0.015 - 0.020	9	
0.020 - 0.025	30	24.775
0.025 - 0.030	44	50.650
0.030 - 0.035	58	58.575
0.035 - 0.040	45	45.200
0.040 - 0.045	29	28.175
0.045 - 0.050	17	15.800
0.050 - 0.055	9	8.400
0.055 - 0.060	4 }	
0.060 or more	4	8.700

χ^2 value = 2.204, $P\{\chi^2_4 \leq 2.19\} = .30$ $P\{\chi^2_4 \leq 2.75\} = .40$

also be calculated. For a specified frequency histogram this requires the solution of (2) for p at each of the interval endpoints. This is straightforward numerically, since $R(p)$ is a well-behaved increasing function of p. Alternatively, intervals can be determined using the estimated percentile function with specified (perhaps equal) probabilities. Then one simply counts the number of observations falling in each interval and computes the approximate χ^2 statistic in the usual manner. (See, for example, p. 302 of [6].)

One might wish to use some criterion other than moments for estimating the parameters. For example, nonlinear least squares can be used to obtain the values of the parameters which minimize the squared distance between the percentile function (2) and the empirical percentile function.

4. TABLE CONSTRUCTION AND ACCURACY

Table 4 gives the values of λ_1, λ_2, λ_3 and λ_4 corresponding to values of α_3 and α_4 with zero mean and unit variance. (A shorter preliminary version of Table 4 appeared in [3].) The lambda values are given for α_3 ranging from 0.0 and 0.90 in steps of 0.05 and from 0.90 to 2.0 in steps of 0.1. For each α_3, lambda values are given for 39 values of α_4 in steps of 0.2, except where λ_3 and λ_4 are near zero and are given in steps of 0.1, starting from the lower limit indicated in Figure 3. The grid of α_3 and α_4 values is sufficiently fine so that interpolation is generally unnecessary. (The density plots of adjacent entries in the table will differ only slightly, even though the lambda values may differ substantially.)

The values of Table 4 have been rounded to four decimal digits, except where indicated by a plus sign (+) or a dollar sign ($). The plus sign (+) indicates that the parameter value has two leading zeroes and consequently has been rounded to six decimal digits and multiplied by 100. Similarly, those values indexed by a dollar sign ($) have four leading zeroes and have been rounded to eight digits and multiplied by 10,000. This was done to provide as many significant digits as possible in order to enhance the accuracy of the tabled value. Tabled values marked by a plus sign (+) should thus be multiplied by 10^{-2} and those marked by a dollar sign ($) should be multiplied by 10^{-4}.

The vast majority of the tabled values yield differences between the calculated and specified values of α_3 and α_4 of less than 0.01, i.e. $|\alpha_j(\lambda_3, \lambda_4) - \alpha_j| < 0.01, j = 3,4$. An asterisk (*) next to a tabled value of λ_j indicates that this difference is somewhat greater than 0.01. This error does not exceed 0.1 for α_4, 0.03 for $\alpha_3 < 1.0$ and 0.05 for $\alpha_3 \geq 1.0$. The values of λ_1 and λ_2 were checked in a similar fashion and yielded a mean within 0.01 of zero and a variance within 0.01 of one. These values result from closed form calculations and can be calculated by the user exactly.

For specified values of α_3 and α_4, the λ_3 and λ_4 values were originally computed by Tadikamalla [15] by solving the nonlinear equations

$$\alpha_3(\lambda_3, \lambda_4) = \alpha_3$$

$$\alpha_4(\lambda_3, \lambda_4) = \alpha_4,$$

using the subroutine ZSYSTM in the IMSL system (available from International Mathematical & Statistical Libraries, GNB Bldg., 7500 Bellaire, Houston, TX 77036). The expressions for $\alpha_3(\lambda_3, \lambda_4)$ and $\alpha_4(\lambda_3, \lambda_4)$ were obtained from equations (4) and (5).

The table given here is an extended and corrected version of Tadikamalla's. New values were obtained using the Nelder-Mead simplex procedure for function minimization where the objective function

$$f(\lambda_3, \lambda_4) = (\alpha_3(\lambda_3, \lambda_4) - \alpha_3)^2 + (\alpha_4(\lambda_3, \lambda_4) - \alpha_4)^2$$

was minimized over λ_3 and λ_4, subject to the constraint that $\lambda_3\lambda_4 > 0$, which insures that λ_3 and λ_4 are of the same sign. (See Olsson and Nelson [8] for a discussion and applications of the Nelder-Mead simplex procedure.)

Solutions other than those given in the table may exist for certain values of α_3 and α_4, usually with $\lambda_3 \geq 1$ and $\lambda_4 \geq 1$, for which the corresponding density functions are truncated.

Either of the two procedures just described can be used to obtain the lambda parameters corresponding to α_3 and α_4 values not in the table.

5. NUMERICAL EXAMPLE

We illustrate our method with the following data taken from Hahn and Shapiro [6]. Measurements of the coefficient of friction for a metal were obtained on 250 samples. The resulting values are summarized in the first two columns of Table 3. Hahn and Shapiro [6] give the following values for the moments which were calculated from the original data: $\bar{x} = 0.0345$, $\sqrt{m_2} = 0.0098$, $\hat{\alpha}_3 = 0.87$ and $\hat{\alpha}_4 = 4.92$. Rounding $\hat{\alpha}_3$ and $\hat{\alpha}_4$ to the nearest tabular values (0.85 and 4.9 respectively), the lambda values from Table 4 are $\hat{\lambda}_1 = -.413$, $\hat{\lambda}_2 = .0134$, $\hat{\lambda}_3 = .004581$, $\hat{\lambda}_4 = .0102$. The values for $\hat{\lambda}_1$ and $\hat{\lambda}_2$ are calculated as

$$\hat{\lambda}_1 = -.413\,(.0098) + .0345 = .0305$$

and

$$\hat{\lambda}_2 = .0134/.0098 = 1.3673.$$

Figure 1 shows the relative frequency histogram for this data and the probability density curve corresponding to the above lambda values. The observed and expected frequencies are given in Table 3. A comparison of the computed value of χ^2 (2.04) with the tabulated χ^2 values (see Table 3) indicates that the model fits the data quite well. However, since the parameters of the model are estimated by the method of moments rather than by the maximum likelihood method, the use of the χ^2 distribution is only approximate.

6. SUMMARY

A four-parameter distribution and a table facilitating parameter estimation using the first four sample moments have been presented. A wide variety of curve shapes are possible with this distribution as indicated by the figures in Section 2. Because of this flexibility and the inherent simplicity of this distribution it is useful in fitting data when, as is often the case, the underlying distribution is unknown. The definition of the distribution leads to a simple algorithm for generating random variates as is discussed in Section 2.

7. ACKNOWLEDGMENTS

The comments of the referees were helpful in improving the presentation and are acknowledged with thanks. Support for John S. Ramberg's research was provided by National Institutes of Health Grant No. GM22271-02. Support for Edward F. Mykytka's research was provided by the Graduate College of the University of Iowa.

REFERENCES

[1] BURR, I. W. (1973). Parameters for a general system of distributions to match a grid of α_3 and α_4. *Comm. Statist., 2*, 1–21.

[2] DUDEWICZ, E. J. (1976). *Introduction to Statistics and Probability*. New York: Holt, Rinehart and Winston.

[3] DUDEWICZ, E. J., RAMBERG, J. S., and TADIKAMALLA, P. R. (1974). A distribution for data fitting and simulation. *Annual Technical Conference Transactions of the American Society for Quality Control, 28*, 407–418.

[4] DUDEWICZ, E. J., JOHNSON, M. E. and RAMBERG, J. S. (1976). Fitting distributions to data with moments: sampling variability effects. *Annual Technical Conference Transactions of the American Society for Quality Control, 30*, 337–344.

[5] FILLIBEN, J. J. (1969). Simple and robust linear estimation of the location parameters of a symmetric distribution. Ph.D thesis, Princeton University.

[6] HAHN, G. J. and SHAPIRO, S. S. (1967). *Statistical Models in Engineering*. New York: John Wiley & Sons, Inc.

[7] JOINER, B. L. and ROSENBLATT, J. R. (1971). Some properties of the range in samples from Tukey's symmetric lambda distribution. *J. Amer. Statist. Assoc., 66*, 394–399.

[8] OLSSON, D. M. and NELSON, L. S. (1975). The Nelder-Mead simplex procedure for function minimization. *Technometrics, 17*, 45–51.

[9] RAMBERG, J. S. and SCHMEISER, B. W. (1972). An approximate method for generating symmetric random variables. *Comm. ACM, 15*, 987–990.

[10] RAMBERG, J. S. and SCHMEISER, B. W. (1974). An approximate method for generating asymmetric random variables. *Comm. ACM, 17*, 78–82.

[11] RAMBERG, J. S. (1975). A probability distribution with applications to Monte Carlo simulation studies. *Statistical Distributions in Scientific Work: Vol. 2—Model Building and Model Selection.* Edited by G. P. Patil, S. Kotz and J. K. Ord. Boston: D. Reidel Publishing Co.

[12] SCHMEISER, B. W. (1971). A general algorithm for generating random variables. Master's thesis, The University of Iowa.

[13] SCHMEISER, B. W. (1977). Methods for modelling and generating probabilistic components in digital computer simulation when the standard distributions are not adequate: a survey. *Proceedings of the Winter Simulation Conference*, 51–57.

[14] SILVER, E. A. (1977). A safety factor approximation based upon Tukey's lambda distribution. *Operational Research Quarterly, 28*, 743–46.

[15] TADIKAMALLA, P. R. (1975). Modeling and generating stochastic inputs for simulation studies. Ph.D. thesis, The University of Iowa.

[16] TUKEY, J. W. (1960). *The Practical Relationship Between the Common Transformations of Percentages of Counts and of Amounts.* Technical Report 36, Statistical Techniques Research Group, Princeton University.

[17] VAN DYKE, J. (1961). *Numerical Investigation of the Random Variable* $y = C(u^\lambda - (1 - u)^\lambda)$. Unpublished working paper, National Bureau of Standards Statistical Engineering Laboratory.

TABLE 4—*Lambda parameters for given values of skewness (α_3) and kurtosis (α_4) when μ = 0 and σ = 1.*

α_3 = 0.0

α_4	LAM 1	LAM 2	LAM 3	LAM 4
1.8	.0	.5774	1.0000	1.0000
2.0	.0	.4952	.5843	.5843
2.2	.0	.4197	.4092	.4092
2.4	.0	.3533	.3032	.3032
2.6	.0	.2949	.2303	.2303
2.8	.0	.2433	.1765	.1765
3.0	.0	.1974	.1349	.1349
3.2	.0	.1563	.1016	.1016
3.4	.0	.1191	.0742	.0742
3.6	.0	.0852	.0512	.0512
3.8	.0	.0545	.0317	.0317
4.0	.0	.0262	.0148	.0148
4.1	.0	.0128	.7140+	.7140+
4.2	.0	-.0659+	-.0363+	-.0363+
4.3	.0	-.0123	-.6706+	-.6706+
4.4	.0	-.0241	-.0130	-.0130
4.6	.0	-.0466	-.0246	-.0246
4.8	.0	-.0676	-.0350	-.0350
5.0	.0	-.0870	-.0443	-.0443
5.2	.0	-.1053	-.0528	-.0528
5.4	.0	-.1227	-.0606	-.0606
5.6	.0	-.1389	-.0677	-.0677
5.8	.0	-.1541	-.0742	-.0742
6.0	.0	-.1686	-.0802	-.0802
6.2	.0	-.1823	-.0858	-.0858
6.4	.0	-.1954	-.0910	-.0910
6.6	.0	-.2077	-.0958	-.0958
6.8	.0	-.2194	-.1003	-.1003
7.0	.0	-.2306	-.1045	-.1045
7.2	.0	-.2414	-.1085	-.1085
7.4	.0	-.2518	-.1123	-.1123
7.6	.0	-.2615	-.1158	-.1158
7.8	.0	-.2709	-.1191	-.1191
8.0	.0	-.2800	-.1223	-.1223
8.2	.0	-.2887	-.1253	-.1253
8.4	.0	-.2969	-.1281	-.1281
8.6	.0	-.3050	-.1308	-.1308
8.8	.0	-.3128	-.1334	-.1334
9.0	.0	-.3203	-.1359	-.1359

α_3 = 0.05

α_4	LAM 1	LAM 2	LAM 3	LAM 4
1.8	-1.703	.2861	.0000	.9502*
2.0	-1.229	.3122	.0505	.7603
2.2	-.802	.3314	.1128	.5802
2.4	-.375	.3328	.1876	.3941
2.6	-.143	.2924	.1973	.2605
2.8	-.083	.2429	.1625	.1903
3.0	-.059	.1975	.1276	.1425
3.2	-.046	.1565	.0974	.1061
3.4	-.038	.1194	.0718	.0770
3.6	-.033	.0856	.0499	.0530
3.8	-.027	.0548	.0311	.0327
4.0	-.026	.0264	.0146	.0153
4.1	-.024	.0132	.7184+	.7504+
4.2	-.024	.0704+	.0380+	.0397+
4.3	-.022	-.0120	-.6386+	-.6643+
4.4	-.022	-.0238	-.0126	-.0131
4.6	-.018	-.0462	-.0240	-.0248
4.8	-.019	-.0671	-.0342	-.0354
5.0	-.016	-.0867	-.0435	-.0448
5.2	-.016	-.1050	-.0519	-.0534
5.4	-.015	-.1222	-.0596	-.0612
5.6	-.014	-.1386	-.0667	-.0684
5.8	-.014	-.1538	-.0731	-.0750
6.0	-.013	-.1682	-.0791	-.0810
6.2	-.012	-.1820	-.0847	-.0866
6.4	-.012	-.1950	-.0899	-.0918
6.6	-.012	-.2074	-.0947	-.0967
6.8	-.011	-.2192	-.0992	-.1012
7.0	-.011	-.2303	-.1034	-.1054
7.2	-.010	-.2411	-.1074	-.1094
7.4	-.010	-.2515	-.1112	-.1132
7.6	-.979+	-.2613	-.1147	-.1167
7.8	-.999+	-.2707	-.1180	-.1201
8.0	-.928+	-.2797	-.1212	-.1232
8.2	-.906+	-.2884	-.1242	-.1262
8.4	-.931+	-.2968	-.1270	-.1291
8.6	-.912+	-.3048	-.1297	-.1318
8.8	-.852+	-.3125	-.1323	-.1343
9.0	-.837+	-.3201	-.1348	-.1368

α_3 = 0.10

α_4	LAM 1	LAM 2	LAM 3	LAM 4
1.8	-1.678	.2835	.0000*	.9071*
2.0	-1.271	.3028	.0412	.7373
2.2	-.872	.3177	.0941	.5700
2.4	-.515	.3164	.1477	.4116
2.6	-.269	.2863	.1678	.2831
2.8	-.164	.2417	.1486	.2033
3.0	-.117	.1977	.1205	.1503
3.2	-.092	.1572	.0936	.1111
3.4	-.076	.1203	.0698	.0803
3.6	-.065	.0866	.0490	.0552
3.8	-.057	.0558	.0308	.0342
4.0	-.049	.0276	.0149	.0163
4.1	-.048	.0142	.7606+	.8302+
4.2	-.046	.1440+	.0762+	.0828+
4.3	-.044	-.0109	-.5703+	-.6174+
4.4	-.041	-.0227	-.0118	-.0127
4.6	-.037	-.0452	-.0231	-.0247
4.8	-.036	-.0661	-.0332	-.0354
5.0	-.033	-.0857	-.0424	-.0450
5.2	-.032	-.1040	-.0507	-.0537
5.4	-.030	-.1213	-.0584	-.0616
5.6	-.028	-.1375	-.0654	-.0688
5.8	-.027	-.1530	-.0719	-.0755
6.0	-.027	-.1674	-.0778	-.0816
6.2	-.025	-.1811	-.0834	-.0872
6.4	-.024	-.1943	-.0886	-.0925
6.6	-.023	-.2066	-.0934	-.0973
6.8	-.023	-.2184	-.0979	-.1019
7.0	-.022	-.2297	-.1021	-.1062
7.2	-.021	-.2405	-.1061	-.1102
7.4	-.020	-.2507	-.1099	-.1139
7.6	-.020	-.2606	-.1134	-.1175
7.8	-.020	-.2699	-.1167	-.1208
8.0	-.019	-.2791	-.1199	-.1240
8.2	-.019	-.2878	-.1229	-.1270
8.4	-.018	-.2961	-.1258	-.1298
8.6	-.017	-.3041	-.1285	-.1325
8.8	-.017	-.3119	-.1311	-.1351
9.0	-.017	-.3193	-.1335	-.1376

α_3 = 0.15

α_4	LAM 1	LAM 2	LAM 3	LAM 4
1.8	-1.655	.2811	.0000*	.8700*
2.0	-1.323	.2934	.0314	.7204
2.2	-.940	.3056	.0782	.5623
2.4	-.617	.3031	.1215	.4194
2.6	-.376	.2791	.1435	.2994
2.8	-.244	.2397	.1350	.2156
3.0	-.177	.1980	.1135	.1586
3.2	-.138	.1584	.0901	.1167
3.4	-.114	.1219	.0682	.0843
3.6	-.098	.0884	.0485	.0581
3.8	-.086	.0577	.0310	.0363
4.0	-.076	.0294	.0155	.0178
4.1	-.073	.0160	.8378+	.9564+
4.2	-.069	.3217+	.1667+	.1890+
4.3	-.066	-.9113+	-.4680+	-.5278+
4.4	-.063	-.0210	-.0107	-.0120
4.6	-.058	-.0435	-.0218	-.0242
4.8	-.055	-.0644	-.0318	-.0351
5.0	-.051	-.0842	-.0410	-.0449
5.2	-.048	-.1025	-.0493	-.0537
5.4	-.045	-.1198	-.0569	-.0617
5.6	-.043	-.1361	-.0639	-.0690
5.8	-.042	-.1514	-.0703	-.0757
6.0	-.040	-.1660	-.0763	-.0819
6.2	-.038	-.1798	-.0819	-.0876
6.4	-.037	-.1928	-.0870	-.0929
6.6	-.035	-.2053	-.0919	-.0978
6.8	-.034	-.2172	-.0964	-.1024
7.0	-.033	-.2284	-.1006	-.1067
7.2	-.032	-.2392	-.1046	-.1107
7.4	-.031	-.2496	-.1084	-.1145
7.6	-.030	-.2593	-.1119	-.1180
7.8	-.029	-.2688	-.1153	-.1214
8.0	-.028	-.2780	-.1185	-.1246
8.2	-.028	-.2866	-.1215	-.1276
8.4	-.027	-.2948	-.1243	-.1304
8.6	-.027	-.3031	-.1271	-.1332
8.8	-.026	-.3108	-.1297	-.1357
9.0	-.025	-.3183	-.1322	-.1382

α_3 = 0.20

α_4	LAM 1	LAM 2	LAM 3	LAM 4
2.0	-1.387	.2841	.0212	.7090
2.2	-1.011	.2947	.0638	.5571
2.4	-.706	.2919	.1013	.4246
2.6	-.471	.2718	.1233	.3120
2.8	-.322	.2374	.1221	.2273
3.0	-.237	.1983	.1065	.1672
3.2	-.187	.1599	.0866	.1230
3.4	-.154	.1240	.0667	.0889
3.6	-.132	.0908	.0482	.0615
3.8	-.116	.0601	.0314	.0389
4.0	-.103	.0318	.0164	.0198
4.1	-.097	.0185	.9467+	.0113
4.2	-.093	.5707+	.2894+	.3429+
4.3	-.089	-.6641+	-.3342+	-.3929+
4.4	-.085	-.0185	-.9261+	-.0108
4.6	-.079	-.0410	-.0202	-.0233
4.8	-.074	-.0622	-.0302	-.0345
5.0	-.069	-.0818	-.0392	-.0444
5.2	-.065	-.1003	-.0475	-.0534
5.4	-.061	-.1176	-.0551	-.0615
5.6	-.058	-.1339	-.0621	-.0689
5.8	-.055	-.1494	-.0686	-.0757
6.0	-.053	-.1639	-.0745	-.0819
6.2	-.051	-.1778	-.0801	-.0877
6.4	-.049	-.1909	-.0853	-.0930
6.6	-.047	-.2034	-.0901	-.0980
6.8	-.045	-.2153	-.0947	-.1026
7.0	-.044	-.2265	-.0989	-.1069
7.2	-.043	-.2374	-.1029	-.1110
7.4	-.041	-.2477	-.1067	-.1148
7.6	-.040	-.2577	-.1103	-.1184
7.8	-.039	-.2671	-.1136	-.1218
8.0	-.038	-.2762	-.1168	-.1250
8.2	-.037	-.2850	-.1199	-.1280
8.4	-.036	-.2935	-.1228	-.1309
8.6	-.035	-.3014	-.1255	-.1336
8.8	-.035	-.3092	-.1281	-.1362
9.0	-.034	-.3168	-.1306	-.1387
9.2	-.034	-.3241	-.1330	-.1411

α_3 = 0.25

α_4	LAM 1	LAM 2	LAM 3	LAM 4
2.0	-1.465	.2748	.0105	.7034
2.2	-1.084	.2847	.0506	.5548
2.4	-.790	.2820	.0843	.4294
2.6	-.558	.2650	.1062	.3226
2.8	-.398	.2349	.1099	.2385
3.0	-.298	.1987	.0996	.1763
3.2	-.237	.1619	.0831	.1300
3.4	-.196	.1266	.0653	.0942
3.6	-.167	.0937	.0481	.0656
3.8	-.147	.0632	.0321	.0421
4.0	-.131	.0351	.0176	.0224
4.1	-.126	.0217	.0108	.0136
4.2	-.118	.8889+	.4408+	.5467+
4.3	-.113	-.3476+	-.1713+	-.2103+
4.4	-.108	-.0154	-.7540+	-.9175+
4.6	-.099	-.0380	-.0184	-.0220
4.8	-.094	-.0591	-.0282	-.0334
5.0	-.087	-.0790	-.0373	-.0436
5.2	-.082	-.0974	-.0455	-.0527
5.4	-.077	-.1149	-.0531	-.0610
5.6	-.073	-.1312	-.0601	-.0685
5.8	-.070	-.1467	-.0665	-.0754
6.0	-.067	-.1613	-.0725	-.0817
6.2	-.064	-.1753	-.0781	-.0876
6.4	-.062	-.1885	-.0833	-.0930
6.6	-.059	-.2010	-.0882	-.0980
6.8	-.058	-.2129	-.0927	-.1027
7.0	-.055	-.2242	-.0970	-.1070
7.2	-.054	-.2350	-.1010	-.1111
7.4	-.052	-.2455	-.1048	-.1150
7.6	-.051	-.2554	-.1084	-.1186
7.8	-.049	-.2649	-.1118	-.1220
8.0	-.048	-.2742	-.1151	-.1252
8.2	-.047	-.2829	-.1181	-.1283
8.4	-.046	-.2914	-.1210	-.1312
8.6	-.044	-.2995	-.1238	-.1339
8.8	-.044	-.3072	-.1264	-.1365
9.0	-.043	-.3147	-.1289	-.1390
9.2	-.042	-.3220	-.1313	-.1414

The parameter values given in this table are for a variate with zero mean and unit variance. The procedure for adjusting the parameters to reflect a different mean or variance is given in Section 3. A plus sign (+) next to a tabled value indicates that the value has two leading zeroes and should be multiplied by 10^{-2}. Similarly, a dollar sign ($) next to a tabled value indicates that the value should be multiplied by 10^{-4}. An asterisk (*) next to a tabled value of λ_j indicates that the difference between the calculated and specified values of α_j, i.e. $|\alpha_j(\lambda_3,\lambda_4) - \alpha_j|$, is somewhat greater than 0.01. See Section 4 for a discussion of the construction and accuracy of this table.

TABLE 4—*Continued*

$\alpha_3 = 0.30$

α_4	LAM 1	LAM 2	LAM 3	LAM 4
2.0	-1.550	.2660	.0000	.7020
2.2	-1.164	.2755	.0380	.5556
2.4	-.871	.2733	.0695	.4348
2.6	-.642	.2586	.0911	.3324
2.8	-.474	.2323	.0983	.2495
3.0	-.362	.1991	.0925	.1859
3.2	-.288	.1641	.0796	.1377
3.4	-.239	.1298	.0640	.1003
3.6	-.204	.0973	.0481	.0704
3.8	-.179	.0671	.0330	.0460
4.0	-.160	.0389	.0190	.0255
4.2	-.144	.0127	.6175+	.8035+
4.3	-.138	.0789+	.0380+	.0489+
4.4	-.131	-.0116	-.5554+	-.7057+
4.5	-.129	-.0231	-.0110	-.0139
4.6	-.121	-.0343	-.0163	-.0203
4.8	-.113	-.0554	-.0260	-.0319
5.0	-.105	-.0752	-.0350	-.0423
5.2	-.100	-.0939	-.0432	-.0517
5.4	-.094	-.1114	-.0508	-.0601
5.6	-.089	-.1279	-.0578	-.0678
5.8	-.085	-.1435	-.0643	-.0748
6.0	-.081	-.1582	-.0703	-.0812
6.2	-.078	-.1722	-.0759	-.0872
6.4	-.075	-.1854	-.0811	-.0927
6.6	-.072	-.1979	-.0860	-.0977
6.8	-.069	-.2100	-.0906	-.1025
7.0	-.067	-.2214	-.0949	-.1069
7.2	-.065	-.2325	-.0990	-.1111
7.4	-.063	-.2427	-.1028	-.1149
7.6	-.061	-.2528	-.1064	-.1186
7.8	-.060	-.2623	-.1098	-.1220
8.0	-.058	-.2716	-.1131	-.1253
8.2	-.056	-.2805	-.1162	-.1284
8.4	-.055	-.2889	-.1191	-.1313
8.6	-.054	-.2971	-.1219	-.1341
8.8	-.053	-.3050	-.1246	-.1367
9.0	-.052	-.3125	-.1271	-.1392
9.2	-.051	-.3197	-.1295	-.1416

$\alpha_3 = 0.35$

α_4	LAM 1	LAM 2	LAM 3	LAM 4
2.0	-1.539	.2639	.0000*	.6836*
2.2	-1.252	.2668	.0256	.5599
2.4	-.955	.2653	.0559	.4415
2.6	-.724	.2528	.0775	.3423
2.8	-.550	.2298	.0873	.2606
3.0	-.427	.1996	.0854	.1961
3.2	-.343	.1665	.0758	.1462
3.4	-.285	.1333	.0625	.1072
3.6	-.243	.1014	.0482	.0760
3.8	-.213	.0714	.0340	.0505
4.0	-.191	.0434	.0206	.0293
4.2	-.172	.0173	.8158+	.0112
4.3	-.163	.4870+	.2293+	.3090+
4.4	-.156	-.7105+	-.3332+	-.4431+
4.5	-.151	-.0187	-.8723+	-.0115
4.6	-.142	-.0298	-.0139	-.0180
4.8	-.132	-.0511	-.0236	-.0300
5.0	-.124	-.0710	-.0325	-.0407
5.2	-.117	-.0898	-.0407	-.0503
5.4	-.110	-.1074	-.0483	-.0589
5.6	-.105	-.1240	-.0553	-.0668
5.8	-.100	-.1396	-.0618	-.0739
6.0	-.096	-.1545	-.0678	-.0805
6.2	-.091	-.1685	-.0735	-.0865
6.4	-.088	-.1818	-.0787	-.0921
6.6	-.085	-.1945	-.0836	-.0973
6.8	-.082	-.2067	-.0883	-.1021
7.0	-.079	-.2181	-.0926	-.1066
7.2	-.077	-.2291	-.0967	-.1108
7.4	-.074	-.2396	-.1006	-.1147
7.6	-.072	-.2496	-.1042	-.1184
7.8	-.070	-.2593	-.1077	-.1219
8.0	-.068	-.2685	-.1109	-.1252
8.2	-.066	-.2775	-.1141	-.1283
8.4	-.065	-.2860	-.1170	-.1313
8.6	-.064	-.2942	-.1198	-.1341
8.8	-.062	-.3020	-.1225	-.1367
9.0	-.060	-.3096	-.1251	-.1392
9.2	-.059	-.3172	-.1276	-.1417

$\alpha_3 = 0.40$

α_4	LAM 1	LAM 2	LAM 3	LAM 4
2.2	-1.354	.2582	.0129	.5683
2.4	-1.043	.2580	.0430	.4500
2.6	-.808	.2473	.0648	.3527
2.8	-.627	.2273	.0767	.2720
3.0	-.494	.2000	.0782	.2069
3.2	-.400	.1690	.0718	.1555
3.4	-.333	.1371	.0609	.1149
3.6	-.284	.1060	.0482	.0824
3.8	-.248	.0764	.0351	.0558
4.0	-.222	.0485	.0223	.0337
4.2	-.200	.0224	.0103	.0149
4.3	-.190	.0100	.4597+	.6521+
4.4	-.182	-.0397+	-.0182+	-.0254+*
4.5	-.174	-.0136	-.6204+	-.8533+
4.6	-.166	-.0248	-.0113	-.0153
4.8	-.155	-.0462	-.0209	-.0277
5.0	-.146	-.0662	-.0297	-.0387
5.2	-.136	-.0850	-.0379	-.0485
5.4	-.129	-.1027	-.0455	-.0574
5.6	-.122	-.1194	-.0525	-.0654
5.8	-.115	-.1352	-.0591	-.0727
6.0	-.111	-.1501	-.0651	-.0794
6.2	-.106	-.1643	-.0708	-.0856
6.4	-.102	-.1778	-.0761	-.0913
6.6	-.098	-.1906	-.0811	-.0966
6.8	-.094	-.2026	-.0857	-.1014
7.0	-.091	-.2142	-.0901	-.1060
7.2	-.089	-.2253	-.0942	-.1103
7.4	-.086	-.2359	-.0981	-.1143
7.6	-.083	-.2459	-.1018	-.1180
7.8	-.081	-.2558	-.1053	-.1216
8.0	-.079	-.2650	-.1086	-.1249
8.2	-.077	-.2741	-.1118	-.1281
8.4	-.075	-.2827	-.1148	-.1311
8.6	-.073	-.2908	-.1176	-.1339
8.8	-.072	-.2988	-.1203	-.1366
9.0	-.070	-.3064	-.1229	-.1391
9.2	-.069	-.3139	-.1254	-.1416
9.4	-.067	-.3210	-.1278	-.1439

$\alpha_3 = 0.45$

α_4	LAM 1	LAM 2	LAM 3	LAM 4
2.2	-1.471	.2500	.0000	.5812
2.4	-1.138	.2511	.0305	.4608
2.6	-.894	.2424	.0528	.3641
2.8	-.707	.2248	.0663	.2840
3.0	-.565	.2003	.0707	.2184
3.2	-.460	.1716	.0674	.1657
3.4	-.384	.1412	.0590	.1236
3.6	-.329	.1110	.0480	.0897
3.8	-.287	.0818	.0361	.0619
4.0	-.255	.0542	.0241	.0388
4.2	-.230	.0282	.0126	.0193
4.3	-.221	.0158	.7045+	.0106
4.4	-.208	.4102+	.1833+	.2691+
4.5	-.200	-.7861+	-.3505+	-.5065+
4.6	-.192	-.0191	-.8511+	-.0121
4.8	-.178	-.0406	-.0180	-.0249
5.0	-.165	-.0607	-.0268	-.0362
5.2	-.157	-.0796	-.0349	-.0464
5.4	-.147	-.0975	-.0425	-.0555
5.6	-.140	-.1142	-.0495	-.0637
5.8	-.132	-.1302	-.0561	-.0712
6.0	-.127	-.1453	-.0622	-.0781
6.2	-.121	-.1595	-.0679	-.0844
6.4	-.116	-.1731	-.0733	-.0902
6.6	-.112	-.1860	-.0783	-.0956
6.8	-.108	-.1983	-.0830	-.1006
7.0	-.104	-.2098	-.0874	-.1052
7.2	-.101	-.2211	-.0916	-.1096
7.4	-.097	-.2316	-.0955	-.1136
7.6	-.095	-.2419	-.0992	-.1175
7.8	-.092	-.2518	-.1028	-.1211
8.0	-.090	-.2611	-.1061	-.1245
8.2	-.088	-.2702	-.1093	-.1277
8.4	-.085	-.2789	-.1124	-.1307
8.6	-.084	-.2871	-.1152	-.1336
8.8	-.081	-.2952	-.1180	-.1363
9.0	-.080	-.3029	-.1206	-.1389
9.2	-.078	-.3102	-.1231	-.1413
9.4	-.076	-.3176	-.1256	-.1437

$\alpha_3 = 0.50$

α_4	LAM 1	LAM 2	LAM 3	LAM 4
2.4	-1.245	.2445	.0178	.4748
2.6	-.987	.2376	.0410	.3770
2.8	-.790	.2225	.0561	.2969
3.0	-.639	.2006	.0630	.2307
3.2	-.525	.1742	.0625	.1768
3.4	-.440	.1454	.0566	.1332
3.6	-.376	.1163	.0476	.0979
3.8	-.329	.0877	.0369	.0689
4.0	-.290	.0604	.0259	.0447
4.2	-.262	.0345	.0149	.0243
4.3	-.248	.0221	.9582+	.0152
4.4	-.238	.0101	.4383+	.6815+
4.5	-.228	-.1612+	-.0700+	-.1066+
4.6	-.219	-.0128	-.5570+	-.8334+
4.8	-.202	-.0344	-.0149	-.0216
5.0	-.188	-.0546	-.0236	-.0333
5.2	-.177	-.0737	-.0317	-.0438
5.4	-.167	-.0917	-.0393	-.0532
5.6	-.157	-.1087	-.0464	-.0617
5.8	-.150	-.1246	-.0529	-.0694
6.0	-.142	-.1398	-.0591	-.0764
6.2	-.137	-.1542	-.0648	-.0829
6.4	-.131	-.1679	-.0702	-.0889
6.6	-.126	-.1809	-.0753	-.0944
6.8	-.122	-.1933	-.0800	-.0995
7.0	-.117	-.2050	-.0845	-.1042
7.2	-.114	-.2163	-.0887	-.1087
7.4	-.110	-.2270	-.0927	-.1128
7.6	-.107	-.2374	-.0965	-.1167
7.8	-.104	-.2473	-.1001	-.1204
8.0	-.101	-.2567	-.1035	-.1238
8.2	-.098	-.2659	-.1067	-.1271
8.4	-.095	-.2745	-.1098	-.1301
8.6	-.094	-.2830	-.1127	-.1331
8.8	-.091	-.2910	-.1155	-.1358
9.0	-.089	-.2986	-.1181	-.1384
9.2	-.088	-.3064	-.1207	-.1410
9.4	-.086	-.3134	-.1231	-.1433
9.6	-.084	-.3206	-.1255	-.1456

$\alpha_3 = 0.55$

α_4	LAM 1	LAM 2	LAM 3	LAM 4
2.4	-1.370	.2379	.4463+	.4931
2.6	-1.087	.2331	.0292	.3920
2.8	-.878	.2202	.0459	.3109
3.0	-.716	.2009	.0551	.2440
3.2	-.593	.1767	.0572	.1889
3.4	-.499	.1497	.0538	.1438
3.6	-.428	.1217	.0467	.1070
3.8	-.372	.0940	.0376	.0767
4.0	-.330	.0670	.0275	.0514
4.2	-.298	.0413	.0172	.0301
4.4	-.269	.0170	.7149+	.0118
4.5	-.257	.5355+	.2258+	.3644+
4.6	-.247	-.5954+	-.2515+	-.3975+
4.7	-.237	-.0169	-.7160+	-.0111
4.8	-.227	-.0276	-.0117	-.0178
5.0	-.213	-.0480	-.0203	-.0300
5.2	-.200	-.0671	-.0283	-.0408
5.4	-.187	-.0852	-.0359	-.0505
5.6	-.177	-.1024	-.0430	-.0593
5.8	-.169	-.1184	-.0495	-.0672
6.0	-.161	-.1338	-.0557	-.0745
6.2	-.153	-.1483	-.0615	-.0811
6.4	-.147	-.1620	-.0669	-.0872
6.6	-.141	-.1753	-.0721	-.0929
6.8	-.136	-.1878	-.0769	-.0981
7.0	-.131	-.1997	-.0814	-.1030
7.2	-.127	-.2111	-.0857	-.1075
7.4	-.123	-.2218	-.0897	-.1117
7.6	-.119	-.2322	-.0935	-.1157
7.8	-.115	-.2422	-.0972	-.1194
8.0	-.113	-.2519	-.1006	-.1230
8.2	-.110	-.2610	-.1039	-.1263
8.4	-.107	-.2698	-.1070	-.1294
8.6	-.104	-.2784	-.1100	-.1324
8.8	-.102	-.2864	-.1128	-.1352
9.0	-.100	-.2943	-.1155	-.1379
9.2	-.097	-.3019	-.1181	-.1404
9.4	-.095	-.3092	-.1206	-.1428
9.6	-.094	-.3164	-.1230	-.1452

TABLE 4—*Continued*

$\alpha_3 = 0.60$

α_4	LAM 1	LAM 2	LAM 3	LAM 4
2.4	-1.411	.2347	.0000*	.4951*
2.6	-1.198	.2286	.0171	.4098
2.8	-.972	.2180	.0355	.3265
3.0	-.800	.2009	.0467	.2583
3.2	-.665	.1791	.0514	.2020
3.4	-.562	.1539	.0504	.1554
3.6	-.482	.1273	.0454	.1171
3.8	-.420	.1005	.0379	.0854
4.0	-.372	.0740	.0289	.0589
4.2	-.335	.0486	.0194	.0366
4.4	-.302	.0244	.9911+	.0175
4.5	-.289	.0128	.5215+	.8965+
4.6	-.277	.1492+	.0611+	.1025+
4.7	-.266	-.9531+	-.3916+	-.6425+
4.8	-.256	-.0202	-.8326+	-.0134
5.0	-.238	-.0407	-.0168	-.0261
5.2	-.222	-.0600	-.0248	-.0373
5.4	-.209	-.0782	-.0323	-.0474
5.6	-.197	-.0956	-.0394	-.0565
5.8	-.187	-.1118	-.0460	-.0647
6.0	-.179	-.1273	-.0522	-.0722
6.2	-.171	-.1419	-.0580	-.0790
6.4	-.163	-.1559	-.0635	-.0853
6.6	-.157	-.1691	-.0686	-.0911
6.8	-.151	-.1818	-.0735	-.0965
7.0	-.146	-.1938	-.0781	-.1015
7.2	-.141	-.2052	-.0824	-.1061
7.4	-.137	-.2163	-.0865	-.1105
7.6	-.132	-.2267	-.0904	-.1145
7.8	-.128	-.2368	-.0941	-.1183
8.0	-.124	-.2465	-.0976	-.1219
8.2	-.121	-.2557	-.1009	-.1253
8.4	-.118	-.2647	-.1041	-.1285
8.6	-.115	-.2732	-.1071	-.1315
8.8	-.113	-.2815	-.1100	-.1344
9.0	-.110	-.2894	-.1127	-.1371
9.2	-.108	-.2970	-.1153	-.1397
9.4	-.105	-.3045	-.1179	-.1422
9.6	-.103	-.3116	-.1203	-.1445

$\alpha_3 = 0.65$

α_4	LAM 1	LAM 2	LAM 3	LAM 4
2.6	-1.329	.2240	.3908+	.4318
2.8	-1.076	.2157	.0246	.3443
3.0	-.889	.2010	.0380	.2742
3.2	-.744	.1812	.0449	.2162
3.4	-.630	.1582	.0464	.1682
3.6	-.542	.1330	.0435	.1283
3.8	-.472	.1072	.0377	.0952
4.0	-.418	.0813	.0300	.0674
4.2	-.374	.0564	.0215	.0440
4.4	-.338	.0323	.0126	.0239
4.5	-.324	.0207	.8137+	.0150
4.6	-.310	.9399+	.3719+	.6660+
4.7	-.297	-.1593+	-.0634+	-.1106+
4.8	-.285	-.0123	-.4921+	-.8391+
5.0	-.265	-.0328	-.0132	-.0216
5.2	-.248	-.0524	-.0211	-.0334
5.4	-.231	-.0707	-.0286	-.0438
5.6	-.219	-.0880	-.0356	-.0532
5.8	-.209	-.1046	-.0422	-.0618
6.0	-.198	-.1201	-.0484	-.0695
6.2	-.189	-.1350	-.0543	-.0766
6.4	-.181	-.1491	-.0598	-.0831
6.6	-.174	-.1625	-.0650	-.0891
6.8	-.167	-.1753	-.0700	-.0946
7.0	-.161	-.1874	-.0746	-.0997
7.2	-.155	-.1991	-.0790	-.1045
7.4	-.150	-.2100	-.0831	-.1089
7.6	-.145	-.2208	-.0871	-.1131
7.8	-.141	-.2309	-.0908	-.1170
8.0	-.137	-.2407	-.0944	-.1207
8.2	-.134	-.2501	-.0977	-.1242
8.4	-.130	-.2591	-.1010	-.1274
8.6	-.127	-.2677	-.1040	-.1305
8.8	-.124	-.2761	-.1069	-.1335
9.0	-.121	-.2840	-.1097	-.1362
9.2	-.119	-.2919	-.1124	-.1389
9.4	-.116	-.2994	-.1150	-.1414
9.6	-.114	-.3065	-.1174	-.1438
9.8	-.112	-.3136	-.1198	-.1461

$\alpha_3 = 0.70$

α_4	LAM 1	LAM 2	LAM 3	LAM 4
2.6	-1.368	.2217	.0000*	.4353*
2.8	-1.194	.2132	.0130	.3651
3.0	-.987	.2008	.0286	.2918
3.2	-.828	.1833	.0378	.2319
3.4	-.704	.1621	.0416	.1821
3.6	-.606	.1385	.0409	.1406
3.8	-.529	.1139	.0369	.1060
4.0	-.467	.0889	.0307	.0768
4.2	-.419	.0643	.0232	.0522
4.4	-.379	.0406	.0151	.0312
4.6	-.344	.0178	.6767+	.0130
4.7	-.331	.6799+	.2607+	.4872+
4.8	-.317	-.3917+	-.1512+	-.2750+
4.9	-.305	-.0144	-.5574+	-.9893+
5.0	-.294	-.0245	-.9565+	-.0166
5.2	-.276	-.0441	-.0173	-.0289
5.4	-.257	-.0626	-.0247	-.0398
5.6	-.243	-.0802	-.0317	-.0496
5.8	-.229	-.0967	-.0383	-.0584
6.0	-.219	-.1125	-.0445	-.0665
6.2	-.209	-.1275	-.0504	-.0738
6.4	-.199	-.1417	-.0560	-.0805
6.6	-.191	-.1554	-.0613	-.0867
6.8	-.184	-.1682	-.0662	-.0924
7.0	-.177	-.1805	-.0709	-.0977
7.2	-.170	-.1923	-.0754	-.1026
7.4	-.165	-.2036	-.0796	-.1072
7.6	-.160	-.2144	-.0836	-.1115
7.8	-.155	-.2246	-.0874	-.1155
8.0	-.151	-.2346	-.0910	-.1193
8.2	-.147	-.2439	-.0944	-.1228
8.4	-.143	-.2532	-.0977	-.1262
8.6	-.139	-.2618	-.1008	-.1293
8.8	-.136	-.2703	-.1038	-.1323
9.0	-.133	-.2784	-.1066	-.1352
9.2	-.130	-.2862	-.1093	-.1379
9.4	-.127	-.2937	-.1119	-.1404
9.6	-.125	-.3011	-.1144	-.1429
9.8	-.122	-.3081	-.1168	-.1452

$\alpha_3 = 0.75$

α_4	LAM 1	LAM 2	LAM 3	LAM 4
2.8	-1.334	.2104	.0000	.3903
3.0	-1.097	.2003	.0183	.3119
3.2	-.921	.1850	.0299	.2492
3.4	-.785	.1658	.0360	.1974
3.6	-.677	.1440	.0375	.1542
3.8	-.590	.1206	.0355	.1179
4.0	-.521	.0966	.0309	.0873
4.2	-.466	.0726	.0246	.0614
4.4	-.419	.0492	.0174	.0392
4.6	-.384	.0266	.9663+	.0202
4.7	-.367	.0156	.5749+	.0116
4.8	-.352	.4940+	.1833+	.3583+
4.9	-.339	-.5509+	-.2061+	-.3916+
5.0	-.324	-.0157	-.5915+	-.0109
5.2	-.306	-.0353	-.0134	-.0238
5.4	-.284	-.0539	-.0207	-.0352
5.6	-.268	-.0716	-.0276	-.0454
5.8	-.254	-.0884	-.0342	-.0547
6.0	-.240	-.1044	-.0405	-.0630
6.2	-.229	-.1195	-.0464	-.0706
6.4	-.219	-.1339	-.0520	-.0776
6.6	-.209	-.1476	-.0573	-.0840
6.8	-.201	-.1607	-.0623	-.0899
7.0	-.194	-.1731	-.0670	-.0954
7.2	-.188	-.1851	-.0715	-.1005
7.4	-.181	-.1964	-.0758	-.1052
7.6	-.175	-.2074	-.0799	-.1096
7.8	-.170	-.2177	-.0837	-.1137
8.0	-.165	-.2278	-.0874	-.1176
8.2	-.160	-.2375	-.0909	-.1213
8.4	-.156	-.2466	-.0942	-.1247
8.6	-.152	-.2554	-.0974	-.1279
8.8	-.148	-.2640	-.1004	-.1310
9.0	-.145	-.2722	-.1033	-.1339
9.2	-.142	-.2802	-.1061	-.1367
9.4	-.138	-.2879	-.1088	-.1393
9.6	-.135	-.2952	-.1113	-.1418
9.8	-.133	-.3023	-.1137	-.1442
10.0	-.130	-.3093	-.1161	-.1465

$\alpha_3 = 0.80$

α_4	LAM 1	LAM 2	LAM 3	LAM 4
3.0	-1.225	.1996	.6847+	.3356
3.2	-1.025	.1864	.0211	.2687
3.4	-.874	.1692	.0295	.2143
3.6	-.754	.1492	.0333	.1691
3.8	-.657	.1272	.0333	.1310
4.0	-.582	.1042	.0303	.0989
4.2	-.519	.0810	.0254	.0716
4.4	-.468	.0580	.0192	.0482
4.6	-.425	.0357	.0123	.0281
4.8	-.392	.0142	.5035+	.0107
4.9	-.375	.3770+	.1352+	.2770+
5.0	-.361	-.6291+	-.2278+	-.4531+
5.1	-.349	-.0164	-.5981+	-.0116
5.2	-.335	-.0261	-.9598+	-.0181
5.4	-.313	-.0449	-.0167	-.0301
5.6	-.295	-.0626	-.0235	-.0408
5.8	-.279	-.0795	-.0300	-.0504
6.0	-.264	-.0958	-.0363	-.0592
6.2	-.251	-.1110	-.0422	-.0671
6.4	-.240	-.1255	-.0478	-.0743
6.6	-.230	-.1394	-.0531	-.0810
6.8	-.220	-.1527	-.0582	-.0871
7.0	-.212	-.1653	-.0630	-.0928
7.2	-.204	-.1774	-.0676	-.0980
7.4	-.197	-.1889	-.0719	-.1029
7.6	-.191	-.2000	-.0760	-.1075
7.8	-.185	-.2104	-.0799	-.1117
8.0	-.180	-.2205	-.0836	-.1157
8.2	-.174	-.2304	-.0872	-.1195
8.4	-.169	-.2397	-.0906	-.1230
8.6	-.166	-.2488	-.0938	-.1264
8.8	-.161	-.2574	-.0969	-.1295
9.0	-.157	-.2658	-.0999	-.1325
9.2	-.154	-.2737	-.1027	-.1353
9.4	-.150	-.2815	-.1054	-.1380
9.6	-.147	-.2890	-.1080	-.1406
9.8	-.144	-.2962	-.1105	-.1430
10.0	-.141	-.3033	-.1129	-.1454
10.2	-.139	-.3100	-.1152	-.1476

$\alpha_3 = 0.85$

α_4	LAM 1	LAM 2	LAM 3	LAM 4
3.0	-1.303	.1985	.0000*	.3488
3.2	-1.145	.1875	.0110	.2912
3.4	-.973	.1723	.0220	.2332
3.6	-.838	.1541	.0281	.1855
3.8	-.732	.1336	.0301	.1455
4.0	-.645	.1119	.0291	.1117
4.2	-.577	.0895	.0256	.0829
4.4	-.519	.0671	.0206	.0582
4.6	-.472	.0451	.0146	.0370
4.8	-.430	.0238	.8001+	.0185
4.9	-.413	.0134	.4581+	.0102
5.0	-.398	.3503+	.1211+	.2612+
5.1	-.383	-.6701+	-.2345+	-.4896+
5.2	-.370	-.0165	-.5808+	-.0118
5.4	-.344	-.0353	-.0127	-.0244
5.6	-.324	-.0531	-.0193	-.0356
5.8	-.305	-.0703	-.0258	-.0457
6.0	-.290	-.0864	-.0319	-.0548
6.2	-.275	-.1019	-.0378	-.0631
6.4	-.262	-.1168	-.0435	-.0707
6.6	-.251	-.1307	-.0488	-.0776
6.8	-.241	-.1442	-.0539	-.0840
7.0	-.231	-.1570	-.0588	-.0899
7.2	-.223	-.1692	-.0634	-.0953
7.4	-.215	-.1809	-.0678	-.1004
7.6	-.207	-.1921	-.0720	-.1051
7.8	-.201	-.2028	-.0759	-.1095
8.0	-.195	-.2130	-.0797	-.1136
8.2	-.190	-.2229	-.0833	-.1175
8.4	-.184	-.2324	-.0868	-.1211
8.6	-.179	-.2416	-.0901	-.1246
8.8	-.175	-.2503	-.0932	-.1278
9.0	-.171	-.2587	-.0962	-.1309
9.2	-.167	-.2669	-.0991	-.1338
9.4	-.163	-.2748	-.1019	-.1366
9.6	-.159	-.2823	-.1045	-.1392
9.8	-.156	-.2897	-.1071	-.1417
10.0	-.153	-.2967	-.1095	-.1441
10.2	-.150	-.3037	-.1119	-.1464

TABLE 4—*Continued*

$\alpha_3 = 0.90$

α_4	LAM 1	LAM 2	LAM 3	LAM 4
3.2	-1.277	.1880	.0000	.3160
3.4	-1.085	.1751	.0133	.2548
3.6	-.933	.1586	.0218	.2039
3.8	-.814	.1397	.0260	.1615
4.0	-.717	.1193	.0269	.1258
4.2	-.639	.0979	.0251	.0953
4.4	-.575	.0762	.0214	.0693
4.6	-.522	.0547	.0164	.0468
4.8	-.478	.0337	.0106	.0273
5.0	-.439	.0132	.4328+	.0102
5.1	-.422	.3339+	.1111+	.2526+
5.2	-.407	-.6388+	-.2154+	-.4735+
5.3	-.394	-.0159	-.5428+	-.0116
5.4	-.379	-.0252	-.8694+	-.0180
5.6	-.353	-.0432	-.0152	-.0298*
5.8	-.334	-.0605	-.0215	-.0405
6.0	-.317	-.0768	-.0275	-.0500
6.2	-.301	-.0924	-.0334	-.0587
6.4	-.287	-.1073	-.0390	-.0666
6.6	-.273	-.1215	-.0444	-.0738
6.8	-.262	-.1352	-.0495	-.0805
7.0	-.252	-.1481	-.0544	-.0866
7.2	-.242	-.1606	-.0591	-.0923
7.4	-.233	-.1723	-.0635	-.0975
7.6	-.225	-.1838	-.0678	-.1024
7.8	-.218	-.1947	-.0718	-.1070
8.0	-.212	-.2051	-.0756	-.1113
8.2	-.205	-.2151	-.0793	-.1153
8.4	-.199	-.2246	-.0828	-.1190
8.6	-.194	-.2340	-.0862	-.1226
8.8	-.189	-.2428	-.0894	-.1259
9.0	-.185	-.2514	-.0924	-.1291
9.2	-.180	-.2597	-.0954	-.1321
9.4	-.176	-.2676	-.0982	-.1349
9.6	-.172	-.2753	-.1009	-.1376
9.8	-.168	-.2827	-.1035	-.1402
10.0	-.165	-.2900	-.1060	-.1427
10.2	-.162	-.2969	-.1084	-.1450
10.4	-.159	-.3035	-.1107	-.1472

$\alpha_3 = 1.00$

α_4	LAM 1	LAM 2	LAM 3	LAM 4
3.4	-1.253	.1772	.0000*	.2854*
3.6	-1.169	.1664	.4828+	.2490
3.8	-1.010	.1509	.0141	.1996
4.0	-.886	.1333	.0193	.1588
4.2	-.787	.1142	.0212	.1244
4.4	-.706	.0943	.0206	.0950
4.6	-.638	.0741	.0182	.0697
4.8	-.581	.0539	.0144	.0477
5.0	-.533	.0340	.9695+	.0285
5.2	-.492	.0146	.4383+	.0117
5.3	-.474	.5192+	.1584+	.4061+
5.4	-.445	-.0317+	-.0101+*	-.0242+*
5.5	-.442	-.0132	-.4176+	-.9946+
5.6	-.429	-.0222	-.7097+	-.0164
5.8	-.403	-.0395	-.0129	-.0282*
6.0	-.379	-.0562	-.0187	-.0388
6.2	-.358	-.0721	-.0244	-.0484
6.4	-.341	-.0873	-.0299	-.0571
6.6	-.325	-.1019	-.0352	-.0651
6.8	-.309	-.1158	-.0404	-.0723
7.0	-.297	-.1291	-.0453	-.0790
7.2	-.285	-.1419	-.0500	-.0852
7.4	-.275	-.1540	-.0545	-.0909
7.6	-.265	-.1658	-.0589	-.0962
7.8	-.256	-.1769	-.0630	-.1011
8.0	-.248	-.1878	-.0670	-.1058
8.2	-.241	-.1980	-.0707	-.1101
8.4	-.233	-.2079	-.0744	-.1141
8.6	-.227	-.2174	-.0778	-.1179
8.8	-.220	-.2267	-.0812	-.1215
9.0	-.215	-.2356	-.0844	-.1249
9.2	-.210	-.2440	-.0874	-.1281
9.4	-.204	-.2522	-.0904	-.1311
9.6	-.200	-.2602	-.0932	-.1340
9.8	-.195	-.2678	-.0959	-.1367
10.0	-.191	-.2752	-.0985	-.1393
10.2	-.187	-.2824	-.1010	-.1418
10.4	-.184	-.2893	-.1034	-.1442
10.6	-.180	-.2959	-.1057	-.1464

$\alpha_3 = 1.10$

α_4	LAM 1	LAM 2	LAM 3	LAM 4
3.8	-1.215	.1582	.0000*	.2379
4.0	-1.108	.1459	.6035+	.2013
4.2	-.974	.1294	.0125	.1607
4.4	-.869	.1117	.0157	.1267
4.6	-.781	.0932	.0165	.0977
4.8	-.708	.0743	.0154	.0727
5.0	-.647	.0552	.0128	.0508
5.2	-.596	.0365	.9168+	.0318
5.4	-.552	.0181	.4839+	.0150
5.5	-.532	.9038+	.2484+	.7342+
5.6	-.517	.0997+	.0279+	.0795+
5.7	-.497	-.8629+	-.2479+	-.6726+
5.8	-.481	-.0173	-.5046+	-.0132
6.0	-.451	-.0340	-.0103	-.0251
6.2	-.427	-.0501	-.0155	-.0358
6.4	-.403	-.0656	-.0208	-.0455
6.6	-.384	-.0805	-.0259	-.0544
6.8	-.366	-.0947	-.0309	-.0624
7.0	-.350	-.1084	-.0358	-.0698
7.2	-.335	-.1214	-.0405	-.0766
7.4	-.322	-.1341	-.0451	-.0829
7.6	-.311	-.1460	-.0494	-.0887
7.8	-.299	-.1577	-.0537	-.0941
8.0	-.289	-.1687	-.0577	-.0991
8.2	-.280	-.1794	-.0616	-.1038
8.4	-.271	-.1896	-.0653	-.1082
8.6	-.263	-.1994	-.0689	-.1123
8.8	-.256	-.2090	-.0724	-.1162
9.0	-.249	-.2180	-.0757	-.1198
9.2	-.242	-.2267	-.0788	-.1232
9.4	-.236	-.2353	-.0819	-.1265
9.6	-.231	-.2435	-.0848	-.1296
9.8	-.226	-.2513	-.0876	-.1325
10.0	-.221	-.2590	-.0903	-.1353
10.2	-.216	-.2664	-.0930	-.1379
10.4	-.211	-.2735	-.0955	-.1404
10.6	-.207	-.2804	-.0979	-.1428
10.8	-.203	-.2870	-.1002	-.1451
11.0	-.199	-.2936	-.1025	-.1473

$\alpha_3 = 1.20$

α_4	LAM 1	LAM 2	LAM 3	LAM 4
4.2	-1.183	.1407	.0000*	.1997
4.4	-1.083	.1278	.5096+	.1675
4.6	-.965	.1113	.9968+	.1329
4.8	-.870	.0941	.0122	.1036
5.0	-.792	.0764	.0124	.0784
5.2	-.723	.0586	.0112	.0565
5.4	-.668	.0408	.8705+	.0372
5.6	-.619	.0233	.5411+	.0202
5.7	-.597	.0146	.3525+	.0124
5.8	-.577	.6088+	.1515+	.5050+
5.9	-.558	-.2319+	-.0594+	-.1884+
6.0	-.562	-.0962+	-.0245+	-.0784+
6.2	-.508	-.0268	-.7343+	-.0206
6.4	-.481	-.0424	-.0120	-.0315
6.6	-.454	-.0575	-.0168	-.0414
6.8	-.432	-.0719	-.0215	-.0504
7.0	-.412	-.0860	-.0262	-.0587
7.2	-.394	-.0993	-.0308	-.0662
7.4	-.378	-.1123	-.0353	-.0732
7.6	-.362	-.1247	-.0397	-.0796
7.8	-.349	-.1366	-.0439	-.0856
8.0	-.337	-.1480	-.0480	-.0911
8.2	-.325	-.1589	-.0519	-.0962
8.4	-.314	-.1695	-.0558	-.1010
8.6	-.305	-.1796	-.0594	-.1055
8.8	-.296	-.1896	-.0630	-.1098
9.0	-.287	-.1990	-.0664	-.1137
9.2	-.280	-.2082	-.0697	-.1175
9.4	-.273	-.2168	-.0728	-.1210
9.6	-.265	-.2253	-.0759	-.1243
9.8	-.259	-.2335	-.0788	-.1275
10.0	-.254	-.2414	-.0816	-.1305
10.2	-.248	-.2490	-.0843	-.1333
10.4	-.242	-.2564	-.0870	-.1360
10.6	-.237	-.2636	-.0895	-.1386
10.8	-.233	-.2704	-.0919	-.1410
11.0	-.228	-.2772	-.0943	-.1434
11.2	-.224	-.2837	-.0966	-.1456
11.4	-.220	-.2901	-.0988	-.1478

$\alpha_3 = 1.30$

α_4	LAM 1	LAM 2	LAM 3	LAM 4
4.6	-1.156	.1244	.0000*	.1679
4.8	-1.084	.1129	.3174+	.1435
5.0	-.975	.0968	.7225+	.1130
5.2	-.886	.0802	.9035+	.0870
5.4	-.812	.0634	.9148+	.0645
5.6	-.749	.0466	.7959+	.0447
5.8	-.695	.0300	.5783+	.0273
6.0	-.604	.0286+	.6619$*	.0239+
6.1	-.617	.0446+	.0100+	.0375+*
6.2	-.616	-.0526+	-.0118+	-.0442+
6.3	-.589	-.0104	-.2450+	-.8504+
6.4	-.572	-.0182	-.4399+	-.0146
6.6	-.539	-.0333	-.8469+	-.0258
6.8	-.510	-.0480	-.0127	-.0360
7.0	-.485	-.0622	-.0170	-.0453
7.2	-.463	-.0758	-.0213	-.0538
7.4	-.442	-.0890	-.0256	-.0616
7.6	-.424	-.1017	-.0298	-.0688
7.8	-.407	-.1140	-.0340	-.0754
8.0	-.392	-.1258	-.0380	-.0816
8.2	-.378	-.1372	-.0420	-.0873
8.4	-.365	-.1480	-.0458	-.0926
8.6	-.353	-.1584	-.0495	-.0975
8.8	-.342	-.1687	-.0531	-.1022
9.0	-.332	-.1784	-.0566	-.1065
9.2	-.322	-.1878	-.0600	-.1106
9.4	-.314	-.1969	-.0632	-.1145
9.6	-.305	-.2057	-.0664	-.1181
9.8	-.298	-.2141	-.0694	-.1215
10.0	-.291	-.2223	-.0723	-.1248
10.2	-.284	-.2304	-.0752	-.1279
10.4	-.277	-.2379	-.0779	-.1308
10.6	-.272	-.2453	-.0805	-.1336
10.8	-.266	-.2525	-.0831	-.1362
11.0	-.261	-.2595	-.0855	-.1388*
11.2	-.256	-.2662	-.0879	-.1412
11.4	-.251	-.2728	-.0902	-.1435
11.6	-.246	-.2792	-.0925	-.1457
11.8	-.242	-.2852	-.0946	-.1478

$\alpha_3 = 1.40$

α_4	LAM 1	LAM 2	LAM 3	LAM 4
5.0	-1.132	.1092	.0000*	.1411
5.2	-1.106	.1011	.0787+	.1268
5.4	-1.001	.0855	.4546+	.0991
5.6	-.916	.0697	.6296+	.0754
5.8	-.844	.0538	.6530+	.0547
6.0	-.782	.0379	.5603+	.0365
6.2	-.729	.0222	.3785+	.0204
6.3	-.706	.0145	.2611+	..0130
6.4	-.683	.6822+	.1292+	.5987+
6.5	-.660	-.1226+	-.0244+	-.1052+
6.6	-.643	-.8266+	-.1702+	-.6968+
6.8	-.607	-.0230	-.5060+	-.0187
7.0	-.575	-.0373	-.8670+	-.0293
7.2	-.547	-.0510	-.0124	-.0389
7.4	-.521	-.0645	-.0163	-.0478
7.6	-.498	-.0775	-.0202	-.0559
7.8	-.475	-.0900	-.0242	-.0633*
8.0	-.458	-.1020	-.0280	-.0702
8.2	-.440	-.1137	-.0319	-.0766
8.4	-.423	-.1250	-.0357	-.0825*
8.6	-.410	-.1358	-.0393	-.0881*
8.8	-.395	-.1463	-.0430	-.0932
9.0	-.383	-.1564	-.0465	-.0980
9.2	-.372	-.1662	-.0499	-.1026
9.4	-.361	-.1756	-.0532	-.1068
9.6	-.351	-.1846	-.0564	-.1108
9.8	-.342	-.1935	-.0595	-.1146
10.0	-.333	-.2018	-.0625	-.1181
10.2	-.325	-.2102	-.0655	-.1215
10.4	-.317	-.2181	-.0683	-.1247
10.6	-.310	-.2257	-.0710	-.1277
10.8	-.303	-.2332	-.0737	-.1306
11.0	-.297	-.2405	-.0762	-.1334*
11.2	-.291	-.2475	-.0787	-.1360
11.4	-.285	-.2542	-.0811	-.1385*
11.6	-.279	-.2609	-.0835	-.1409
11.8	-.274	-.2671	-.0857	-.1431
12.0	-.269	-.2734	-.0879	-.1453
12.2	-.265	-.2794	-.0900	-.1474

TABLE 4—*Continued*

α_3 = 1.50

α_4	LAM 1	LAM 2	LAM 3	LAM 4
5.4	-1.112	.0951	.0000*	.1182
5.6	-1.103	.0886	.0000*	.1083
5.8	-1.042	.0773	.1949+	.0899
6.0	-.957	.0622	.3907+	.0677
6.2	-.885	.0471	.4441+	.0483
6.4	-.824	.0321	.3885+	.0313
6.6	-.688	.0566+	.0104+*	.0494+*
6.7	-.747	.9962+	.1538+	.9059+
6.8	-.714	-.0290+	-.4897$	-.0256+
6.9	-.704	-.4446+	-.0768+	-.3882+
7.0	-.684	-.0115	-.2088+	-.9875+
7.2	-.647	-.0254	-.4989+	-.0210
7.4	-.615	-.0390	-.8156+	-.0312
7.6	-.585	-.0520	-.0115	-.0404
7.8	-.558	-.0648	-.0150	-.0489
8.0	-.536	-.0767	-.0184	-.0565
8.2	-.514	-.0891	-.0221	-.0640
8.4	-.494	-.1007	-.0257	-.0707
8.6	-.476	-.1118	-.0292	-.0769
8.8	-.459	-.1225	-.0327	-.0826
9.0	-.443	-.1330	-.0362	-.0880
9.2	-.429	-.1431	-.0396	-.0931
9.4	-.416	-.1528	-.0429	-.0978
9.6	-.404	-.1622	-.0461	-.1022
9.8	-.392	-.1713	-.0493	-.1064
10.0	-.382	-.1803	-.0524	-.1104
10.2	-.372	-.1887	-.0553	-.1141
10.4	-.363	-.1969	-.0582	-.1176
10.6	-.354	-.2049	-.0611	-.1209
10.8	-.346	-.2127	-.0638	-.1241
11.0	-.338	-.2202	-.0665	-.1271
11.2	-.331	-.2273	-.0690	-.1299
11.4	-.325	-.2339	-.0713	-.1325
11.6	-.317	-.2414	-.0740	-.1353
11.8	-.311	-.2478	-.0763	-.1377
12.0	-.305	-.2544	-.0786	-.1401
12.2	-.300	-.2607	-.0808	-.1424
12.4	-.295	-.2662	-.0827	-.1444
12.6	-.289	-.2726	-.0851	-.1466

α_3 = 1.60

α_4	LAM 1	LAM 2	LAM 3	LAM 4
6.0	-1.086	.0757	.0000*	.0896
6.2	-1.078	.0698	.0000	.0814
6.4	-1.011	.0573	.1699+	.0634
6.6	-.937	.0430	.2684+	.0449
6.8	-.875	.0287	.2597+	.0285
7.0	-.746	.0422+	.6356$*	.0378+*
7.1	-.796	.77738+	.0969+	.7177+
7.2	-.771	-.0341+	-.4634$	-.0309+
7.3	-.751	-.5924+	-.0858+	-.5279+
7.4	-.731	-.0127	-.1942+	-.0111
7.6	-.693	-.0258	-.4383+	-.0218
7.8	-.659	-.0386	-.7111+	-.0316
8.0	-.630	-.0511	-.0100	-.0406
8.2	-.602	-.0633	-.0131	-.0489
8.4	-.577	-.0752	-.0163	-.0566*
8.6	-.553	-.0866	-.0196	-.0636
8.8	-.534	-.0972	-.0227	-.0699
9.0	-.515	-.1084	-.0261	-.0763
9.2	-.496	-.1187	-.0294	-.0819
9.4	-.480	-.1288	-.0326	-.0872
9.6	-.465	-.1385	-.0358	-.0922
9.8	-.452	-.1480	-.0389	-.0969
10.0	-.438	-.1572	-.0420	-.1013
10.2	-.426	-.1659	-.0450	-.1054
10.4	-.415	-.1745	-.0479	-.1093
10.6	-.404	-.1828	-.0508	-.1130
10.8	-.394	-.1908	-.0536	-.1165
11.0	-.385	-.1986	-.0563	-.1198
11.2	-.377	-.2062	-.0589	-.1230
11.4	-.368	-.2135	-.0615	-.1260
11.6	-.360	-.2206	-.0640	-.1288
11.8	-.352	-.2275	-.0665	-.1315
12.0	-.346	-.2341	-.0688	-.1341
12.2	-.339	-.2407	-.0711	-.1366
12.4	-.333	-.2471	-.0734	-.1390
12.6	-.328	-.2527	-.0753	-.1411
12.8	-.321	-.2592	-.0777	-.1434
13.0	-.316	-.2650	-.0797	-.1455
13.2	-.311	-.2706	-.0817	-.1475

α_3 = 1.70

α_4	LAM 1	LAM 2	LAM 3	LAM 4
6.6	-1.064	.0580	.0000*	.0657
6.8	-1.057	.0525	.0000	.0588
7.0	-1.001	.0412	.1027+	.0441
7.2	-.935	.0275	.1513+	.0280
7.4	-.878	.0142	.1142+	.0138
7.5	-.852	.7546+	.0696+	.7179+
7.6	-.825	-.0250+	-.2601$	-.0232+*
7.7	-.806	-.5469+	-.0619+	-.5000+
7.8	-.784	-.0119	-.1463+	-.0107
8.0	-.745	-.0245	-.3423+	-.0212
8.2	-.709	-.0367	-.5705+	-.0308
8.4	-.678	-.0487	-.8225+	-.0397
8.6	-.650	-.0603	-.0109	-.0478
8.8	-.622	-.0717	-.0138	-.0553*
9.0	-.598	-.0827	-.0167	-.0623
9.2	-.578	-.0933	-.0196	-.0688
9.4	-.557	-.1036	-.0226	-.0748
9.6	-.538	-.1136	-.0256	-.0804
9.8	-.521	-.1233	-.0286	-.0857
10.0	-.505	-.1329	-.0316	-.0907
10.2	-.489	-.1420	-.0346	-.0953
10.4	-.476	-.1509	-.0375	-.0997
10.6	-.463	-.1594	-.0403	-.1038
10.8	-.451	-.1677	-.0431	-.1077
11.0	-.440	-.1758	-.0458	-.1114
11.2	-.429	-.1837	-.0485	-.1149
11.4	-.419	-.1913	-.0511	-.1182
11.6	-.410	-.1988	-.0537	-.1214
11.8	-.401	-.2059	-.0562	-.1244
12.0	-.392	-.2128	-.0586	-.1272
12.2	-.384	-.2195	-.0610	-.1299
12.4	-.377	-.2261	-.0633	-.1325
12.6	-.369	-.2326	-.0656	-.1350
12.8	-.362	-.2388	-.0678	-.1374
13.0	-.356	-.2450	-.0700	-.1397
13.2	-.350	-.2508	-.0720	-.1419
13.4	-.344	-.2566	-.0741	-.1440
13.6	-.338	-.2622	-.0761	-.1460
13.8	-.333	-.2675	-.0780	-.1479*

α_3 = 1.80

α_4	LAM 1	LAM 2	LAM 3	LAM 4
7.2	-1.045	.0417	.0000*	.0456
7.4	-1.039	.0367	.0000*	.0396
7.6	-1.007	.0284	.0378+	.0298
7.8	-.945	.0155	.0646+	.0155
7.9	-.918	.9177+	.0498+	.9006+
8.0	-.892	.2914+	.0193+	.2801+
8.1	-.868	-.3291+	-.0254+	-.3102+
8.2	-.846	-.9427+	-.0826+	-.8721+
8.4	-.804	-.0215	-.2289+	-.0192
8.6	-.767	-.0333	-.4103+	-.0288
8.8	-.733	-.0448	-.6190+	-.0376
9.0	-.702	-.0559	-.8489+	-.0456
9.2	-.675	-.0668	-.0109	-.0531
9.4	-.649	-.0774	-.0135	-.0601
9.6	-.625	-.0877	-.0162	-.0665
9.8	-.604	-.0978	-.0189	-.0726
10.0	-.583	-.1075	-.0217	-.0782
10.2	-.565	-.1169	-.0244	-.0835
10.4	-.548	-.1260	-.0272	-.0884
10.6	-.532	-.1349	-.0299	-.0931
10.8	-.517	-.1436	-.0327	-.0975
11.0	-.503	-.1520	-.0354	-.1016
11.2	-.490	-.1600	-.0380	-.1055
11.4	-.478	-.1679	-.0406	-.1092
11.6	-.467	-.1757	-.0432	-.1128
11.8	-.456	-.1831	-.0457	-.1161
12.0	-.445	-.1904	-.0482	-.1193
12.2	-.436	-.1974	-.0506	-.1223
12.4	-.427	-.2043	-.0530	-.1252
12.6	-.418	-.2109	-.0553	-.1279
12.8	-.410	-.2175	-.0576	-.1306
13.0	-.402	-.2238	-.0598	-.1331
13.2	-.395	-.2299	-.0619	-.1355
13.4	-.388	-.2359	-.0640	-.1378
13.6	-.381	-.2417	-.0661	-.1400
13.8	-.374	-.2473	-.0681	-.1421
14.0	-.368	-.2530	-.0701	-.1442
14.2	-.362	-.2583	-.0720	-.1461
14.4	-.357	-.2632	-.0737	-.1479

α_3 = 1.90

α_4	LAM 1	LAM 2	LAM 3	LAM 4
8.0	-1.023	.0220	.0000*	.0230
8.2	-1.018	.0175	.0000	.0181
8.4	-.968	.6447+	.0150+	.6431+
8.5	-.946	.1239+	.4120$	.1215+
8.6	-.917	-.5444+	-.0257+	-.5220+
8.7	-.893	-.0113	-.0657+	-.0106
8.8	-.871	-.0171	-.1167+	-.0158
9.0	-.831	-.0284	-.2475+	-.0254
9.2	-.794	-.0395	-.4100+	-.0343
9.4	-.761	-.0503	-.5975+	-.0424
9.6	-.731	-.0609	-.8046+	-.0500
9.8	-.703	-.0712	-.0103	-.0570
10.0	-.679	-.0811	-.0126	-.0635
10.2	-.656	-.0907	-.0150	-.0695
10.4	-.634	-.1002	-.0175	-.0752
10.6	-.614	-.1093	-.0200	-.0805
10.8	-.595	-.1183	-.0226	-.0855
11.0	-.578	-.1269	-.0251	-.0902
11.2	-.562	-.1355	-.0277	-.0947
11.4	-.547	-.1437	-.0302	-.0989
11.6	-.533	-.1515	-.0327	-.1028
11.8	-.520	-.1594	-.0352	-.1066
12.0	-.508	-.1665	-.0375	-.1100
12.2	-.495	-.1742	-.0401	-.1135
12.4	-.485	-.1811	-.0423	-.1166
12.6	-.474	-.1883	-.0448	-.1198
12.8	-.464	-.1950	-.0471	-.1227
13.0	-.455	-.2015	-.0493	-.1255
13.2	-.446	-.2080	-.0515	-.1282*
13.4	-.437	-.2142	-.0537	-.1307
13.6	-.429	-.2203	-.0558	-.1332
13.8	-.421	-.2262	-.0579	-.1355
14.0	-.414	-.2320	-.0599	-.1378
14.2	-.407	-.2376	-.0619	-.1399
14.4	-.400	-.2431	-.0638	-.1420
14.6	-.394	-.2485	-.0657	-.1440
14.8	-.388	-.2537	-.0676	-.1459
15.0	-.382	-.2589	-.0694	-.1478
15.2	-.376	-.2636	-.0711	-.1495

α_3 = 2.00

α_4	LAM 1	LAM 2	LAM 3	LAM 4
8.6	-1.009	.8397+	.0000*	.8541+
8.8	-1.004	.4147+	.0000*	.4182+
8.9	-1.002	.2061+	.0001$*	.2070+
9.0	-.993	-.1081+	-.0407$	-.1076+
9.1	-.974	-.5675+	-.7075$	-.5567+
9.2	-.950	-.0113	-.0272+	-.0109
9.4	-.905	-.0222	-.1012+	-.0207
9.6	-.865	-.0331	-.2125+	-.0298
9.8	-.828	-.0435	-.3537+	-.0381
10.0	-.796	-.0538	-.5187+	-.0458
10.2	-.766	-.0637	-.7027+	-.0529
10.4	-.738	-.0734	-.9016+	-.0595
10.6	-.713	-.0829	-.0111	-.0657
10.8	-.690	-.0920	-.0133	-.0714
11.0	-.670	-.1005	-.0154	-.0766
11.2	-.647	-.1097	-.0179	-.0819
11.4	-.629	-.1181	-.0202	-.0867
11.6	-.611	-.1264	-.0226	-.0912
11.8	-.595	-.1345	-.0249	-.0955
12.0	-.579	-.1423	-.0273	-.0995
12.2	-.565	-.1498	-.0296	-.1033
12.4	-.557	-.1555	-.0312	-.1062
12.6	-.539	-.1644	-.0342	-.1104
12.8	-.527	-.1715	-.0365	-.1137
13.0	-.515	-.1784	-.0388	-.1168
13.2	-.504	-.1851	-.0410	-.1198
13.4	-.495	-.1914	-.0431	-.1226
13.6	-.485	-.1979	-.0453	-.1254
13.8	-.475	-.2041	-.0474	-.1280
14.0	-.466	-.2101	-.0495	-.1305
14.2	-.458	-.2160	-.0515	-.1329
14.4	-.450	-.2216	-.0535	-.1351*
14.6	-.443	-.2271	-.0554	-.1373
14.8	-.436	-.2321	-.0571	-.1393
15.0	-.428	-.2380	-.0592	-.1415
15.2	-.422	-.2432	-.0610	-.1435*
15.4	-.415	-.2481	-.0628	-.1453*
15.6	-.409	-.2532	-.0646	-.1472
15.8	-.403	-.2580	-.0663	-.1489*

AMERICAN JOURNAL OF MATHEMATICAL AND MANAGEMENT SCIENCES

THE SERIES METHOD FOR RANDOM VARIATE GENERATION AND ITS APPLICATION TO THE KOLMOGOROV - SMIRNOV DISTRIBUTION

Luc Devroye
McGill University
School of Computer Science
805 Sherbrooke Street West
Montreal, Canada H3A 2K6

SYNOPTIC ABSTRACT

This paper presents a series method for the computer generation of a random variable X with density f when $f = \lim_{n\to\infty} f_n = \lim_{n\to\infty} g_n$ and f_n and g_n are given sequences of functions satisfying $f_n \geq f \geq g_n$; f is never evaluated. This method can be used when f is given as an infinite series. Three complete examples are given, and a computer program is included for the generation of random variates from the Kolmogorov-Smirnov distribution.

1. INTRODUCTION

Consider the problem of computer generation of a random variable X with density f, where f is a (complicated) function which can be approximated from above and below by simpler functions f_n and g_n . In particular, assume that:

(i) there exist sequences of functions f_n and g_n such that

$$f_n \geq f \geq g_n \text{ , for all } n \text{ ,} \tag{1}$$

where

(ii) $$f_n \to f,\ g_n \to f \text{ as } n \to \infty \text{ ;}$$

and

(iii) there exists an integrable nonnegative function h such that

Key Words and Phrases: random number generation; Kolmogorov-Smironov distribution; rejection method; alternating series.

1981, VOL. 1, NO. 4, 359-379
0196-6324/81/040359-21 $5.65

$$h \geq f \ . \tag{2}$$

Here h is proportional to a density which is easy to sample from, and f_n and g_n are sequences of functions which are easy to evaluate. (Note that g_n is not necessarily positive, and that f_n need not be integrable.) The following rejection-type algorithm can be used to generate X.

Algorithm S0 (The Series Method).

0.1 Generate X with density ch (c is a constant), generate an independent uniform (0,1) random variate U, set n = 0 and T = Uh(X).

0.2 n = n+1. If $T \leq f_n(X)$, exit with X.

0.3 If $T > g_n(X)$, go to 0.1.

0.4 Go to 0.2.

If f can be written as an alternating series

$$f(x) = h(x)\ (1-a_1(x)+a_2(x)-a_3(x)+\ldots) \tag{3}$$

(where a_n is a sequence of functions satisfying

$$a_n(x) \downarrow 0 \ , \text{ for all } x, \tag{4}$$

and h is a nonnegative integrable function) then X can be generated by:

Algorithm S1 (The Alternating Series Method).

1.1 Generate X with density ch (c is a constant), generate an independent uniform (0,1) random variate U, set n = 0 and T = 0.

1.2 n = n+1 , $T = T+a_n(X)$. If $U \geq T$, exit with X.

1.3 n = n+1 , $T = T-a_n(X)$. If $U < T$, go to 1.1.

1.4 Go to 1.2

It is clear that S1 is a special case of S0 because $f \leq h$ and

$$1 + \sum_{j=1}^{k} (-1)^j a_j(x) \leq \frac{f(x)}{h(x)} \leq 1 + \sum_{j=1}^{k+1} (-1)^j a_j(x) \ , \ k \text{ odd.}$$

S0 and S1 can be considered as generalizations of the acceptance/rejection method with squeezing, with the special feature that f or f/h need never be computed. S0 requires the evaluation of h, however. For recent detailed descriptions of the acceptance/rejection method, see Vaduva (1977), Tadikamalla (1978), or Tadikamalla and Johnson (1981). For the squeeze method, see Schmeiser and Lal (1980).

Different special cases are often encountered in practice. For example, if

$$f(x) = h(x)\ \exp(-a_1(x) + a_2(x)\ldots)$$

where $h \geq 0$ is integrable, and $a_n(x) \downarrow 0$ for all x , then X can be generated by

Algorithm S2 (The Exponential Series Method).

2.1 Generate X with density ch (c is a constant), generate an independent exponential random variate E, set $n = 0$ and $T = 0$.

2.2 $n = n+1$, $T = T + a_n(X)$. If $E \geq T$, exit with X.

2.3 $n = n+1$, $T = T - a_n(X)$. If $E < T$, go to 2.1.

2.4 Go to 2.2.

Some examples are given in Sections 3 and 4, and the average time taken by algorithm S1 is analyzed in Section 2. Experimental timings are included and show that for some distributions where f contains trigonometric and/or exponential/logarithmic functions some savings can be obtained via our methods unless random variates for the distribution in question can be obtained in a very simple fast way by other means (e.g., as for the exponential distribution). An example is included for which the series method seems the only feasible method of random variate generation (i.e., the Kolmogorov-Smirnov distribution).

2. ANALYSIS OF ALGORITHM S1.

Let X be a random variate generated by algorithm S1, and let N_i be the number of times step 1.i in the algorithm was executed, $1 \leq i \leq 4$. We will show that when $a_0 \equiv 1$, and a_n and h satisfy (4) and $a_1(x) \leq 1$,

$$\text{(i)} \quad P(N_2 < \infty) = 1 \quad \text{and}$$

$$\text{(ii)} \quad E(N_2) = \sum_{i=1}^{\infty} i \int h(x)\,(a_{2i-2}(x) - a_{2i}(x))dx.$$

Since $N_i \leq N_2$ for all i, it is clear that the properties of N_2 are essential for the study of the time taken by S1. Note that (i) is necessary in order for S1 to halt with probability one. It is possible however that $E(N_2) = \infty$.

Proof.

(i) is true in view of (4) and steps 1.2 and 1.3 of S1. For (ii), let (X,U) be a pair of random variates generated in step 1.1. Define event A by "(X,U) will be accepted; i.e., $Uh(X) \leq f(X)$". Now

$$P(N_2=i|A) = \int h(x)(a_{2i-2}(x) - a_{2i-1}(x))dx = p_i$$

and

$$P(N_2=i|\bar{A})=[\int h(x)\,(a_{2i-1}(x) - a_{2i}(x))dx]/[\int h(x)dx-1] = q_i ,$$

where $\bar{A}$ denotes the complement of A. Since on the average $\int h(x)dx$ pairs (X,U) are needed, we have

$$E(N_2) = \sum_i ip_i + [\ h(x)dx-1] \sum_i iq_i ,$$

from which (ii) follows.

3. THE EXPONENTIAL AND RAAB-GREEN DISTRIBUTIONS.

The exponential distribution. It is known that for all odd k,

$$\sum_{j=0}^{k-1} (-1)^j \frac{x^j}{j!} \ge e^{-x} \ge \sum_{j=0}^{k} (-1)^j \frac{x^j}{j!}, \quad x > 0.$$

For the generation of exponential random variates, we can use SO with a well-chosen function h. We choose h from the family of densities

$$\frac{na^n}{(x+a)^{n+1}}, \quad x > 0, \ n \ge 1 \text{ integer},$$

where $a > 0$ is a parameter. Note that this is the density of

$$a(U^{-\frac{1}{n}} - 1)$$

when U is a uniform (0,1) random variable. It is also the density of $a \max^{-1}(U_1,\ldots,U_n) - 1)$ when $U_1,\ldots,U_n$ are independent and identically distributed uniform (0,1) random variables. Since $e^{-x}(x+a)^{n+1}$ is maximal when $(n+1) = x+a$, i.e. $x = n+1-a$, we see that

$$e^{-x} \le \frac{\gamma\, na^n}{(x+a)^{n+1}}$$

where $\gamma = (\frac{n+1}{e})^{n+1} \frac{e^a a^{-n}}{n}$. Now, $e^a a^{-n}$ is minimal when $a = n$.

Thus, choosing $a = n$, $\gamma \frac{(n+1)^{n+1}}{n+1} \frac{e^n}{e^{n+1}} = (1+\frac{1}{n})^{n+1} \frac{1}{e} \to 1$ as $n \to \infty$. In particular, for $n = 1,2,3$ we obtain $\frac{4}{3} \approx 1.471518$, $\frac{(3/2)^3}{e} \approx 1.241592$, $\frac{(4/3)^4}{e} \approx 1.162679$. Since the rejection rate decreases with n, but the time needed to obtain a random variate with density $1/(1+x/n)^{n+1}$ (as $n(\max^{-1}(U_{1,\ldots,}U_n) - 1)$) increases with n, we expect that the best performance is obtained for some small n (the value $n = 2$ will be suggested below). The algorithm version for the exponential distribution with mean 1 is:

0.1.E. Generate $n \ge 2$ independent uniform (0,1) random variates $U_1,\ldots,U_n$. Let $U = \max(U_1,\ldots,U_n)$. If $U_1 = U$, interchange U_2 and U_1. Set $X = n(1/U - 1)$,

$$T = \gamma(\frac{U_1}{U})U^{n+1} - 1 = \gamma\, U_1 U^n - 1 \text{ where } \gamma = (\frac{n+1}{n})^{n+1} \cdot \frac{1}{e}.$$

[Note: U_1/U is uniform (0,1) and independent of U; $U^{n+1} = 1/(1+\frac{X}{n})^{n+1}$]. If $T > 0$, go to 0.1.E.

Set $j = 0$, $P = 1$.

0.2.E. $j = j+1$. $P = PX/j$. $T = T+P$. If $T \le 0$, exit with X .

0.3.E. $j = j+1$. $P = PX/j$. $T = T-P$. If $T > 0$, go to 0.1.E.

0.4.E. Go to 0.2.E.

Several remarks here will illustrate some dangers of the series method.

Remark 1. The computation of $E(N_2)$ is usually not possible by analytical means. However, much can still be said about this average. Clearly, $E(N_1) = \gamma$. Also, $E(N_4) = \gamma\, E(N_4^*)$ where N_4^* is the number of visits of step 4 before an exit (step 2) or a return to step 1 (see step 3). For the proof of this, let $N_4(r)$ ($N_4(a)$) be the number of visits of step 4 given ultimate rejection (ultimate acceptance). Then,

$$E(N_4) = (E(N_1) - 1)E(N_4(r)) + E(N_4(a))$$

$$= \frac{(1-1/\gamma)\ E(N_4(r))}{1/\gamma} + \frac{(1/\gamma)E(N_4(a))}{1/\gamma} = \gamma\ E(N_4^*) .$$

But

$$P(N_4^* \ge i) \le \int_0^\infty \frac{1}{(1+\frac{x}{n})^{n+1}} \cdot \min\left(1, \frac{x^{2i}/(2i)!}{1/(1+\frac{x}{n})^{n+1}}\right) dx$$

$$\le \int_0^c \frac{x^{2i}}{(2i)!}\, dx + \int_c^\infty \frac{1}{(1+x/n)^{n+1}}\, dx$$

$$= \frac{c^{2i+1}}{(2i+1)!} + \frac{1}{(1+\frac{c}{x})^n}$$

for any $c > 0$. If we take $c = \left(\frac{2i+1}{e}\right)^{\frac{2i+1}{2i+1+n}} \sim \frac{2i}{e}$, then the upper bound becomes asymptotic to

$$\frac{1}{(2i)^n\sqrt{2\pi 2i}} + \left(\frac{ne}{2i}\right)^n \sim \left(\frac{ne}{2i}\right)^n \text{ as } i \to \infty ,$$

by use of Stirling's formula. Thus, $E(N_4^*) = \sum_1^\infty P(N_4^* \ge i) < \infty$ for $n \ge 2$. For $n = 1$, with some work, one can show that $E(N_4^*) = \infty$ (step 1 has to be replaced by: generate U_1, U_2 independent uniform (0,1) random variates; set $X = 1/U_1 - 1$, $T = \gamma\, U_2\, U_1^2 - 1$ where $\gamma = 4/e$. If $T > 0$, go to 0.1.E. Set $j = 0$, $P = 1$.). This is due to the fact that the tail of the dominating function $1/(x+1)^2$ is too heavy, and that the series approximation of e^{-x} is too slow in the tail.

Remark 2. If one uses the given algorithm with $n = 2$, one will run into overflow problems in steps 2 and 3 since X is too large with too high a probability. To guard against this, step 1 could be extended as follows:

if $X > \alpha$ (a threshold): exit with X when $T+1 < \exp(-X)$, and go to 0.1.E. otherwise .

The average time (in microseconds) varies from 30.7 ($\alpha=0$) down to 26.2 ($\alpha=1$) and monotonically back up to 30.1 ($\alpha=5$). The case $\alpha=0$ corresponds to a complete bypass of steps 2, 3, 4 in the algorithm, and the case $\alpha \to \infty$ corresponds to no bypass. The average time (in microseconds) for the random variate generator $-\log(U)$ where U is uniform (0,1) is much lower: 17.8 . (All timings here and below were done on an AMDAHL V7 computer using 10000 observations and FORTRAN coding. The uniform random variate generator was taken from the Super-Duper random number package (see Dudewicz and Ralley (1981) for the code, and Marsaglia, Ananthanarayanan and Paul (1973) for additional explanation).

Remark 3. A similar experiment to that described in Remark 2 was carried out by the author for the normal distribution: the tails were taken care of by rejection from the Rayleigh density $x\,e^{-x^2/2}$ properly truncated, and the main body was treated using the series method, with rejection from the uniform density. The average computer time was mdiway between the average time for the polar method and the average times for the algorithms of Kinderman and Ramage (1976) and Marsaglia and Bray (1964). The space requirements were also about midway between those of the polar method and the Kinderman-Ramage, Marsaglia-Bray algorithms. This situates the method in an area of the time/space map (figure B of Kinderman and Ramage (1976)) practically by itself. No method known to us is both shorter and faster.

The Raab-Green distribution. Consider the density

$$f(x) = \frac{1+\cos x}{2\pi} \qquad -\pi < x < \pi .$$

$$= \frac{1}{\pi}\left(1 - \frac{x^2}{2.2!} + \frac{x^4}{2.4!} - \dots\right), \quad -\pi < x < \pi.$$

Thus, f can be put into form (3) with $h(x) = \frac{1}{\pi}$, $-\pi < x < \pi$, and $a_n(x) = \frac{x^{2n}}{2.(2n)!}$. It is easy to check that $a_n(x) \downarrow 0$ as $n \to \infty$, since

$$\frac{a_{n+1}(x)}{a_n(x)} = \frac{x^2}{(2n+2)\ (2n+1)} \le \frac{\pi^2}{12} < 1 .$$

Density f was suggested by Raab and Green (1961) as an approximation for the normal density. Algorithm S1 is, for the Raab-Green density:

1.1RG. Generate two independent uniform (0,1) random variates U and V . Set $X = \pi(2V-1)$, $n = 0$, $T = 0$, $P = \frac{1}{2}$.

1.2RG. $n = n+1$, $P = PX^2/[(2n)\ (2n-1)]$, $T = T+P$. If $U \ge T$, exit with X .

1.3RG. $n = n+1$, $P = PX^2/[(2n)(2n-1)]$, $T = T-P$. If $U < T$, go to 1.1RG.

1.4RG. Go to 1.2RG.

It is easy to see that $E(N_1) = 2$. However, we may apply the following alias principle (Walker (1977); see also Kronmal and Peterson (1979)) or band rejection method (Payne (1977)): generate (X,U) uniformly in $[-\frac{\pi}{2}, +\frac{\pi}{2}]\times[0,1]$. If $\frac{U}{\pi} \le f(X)$, exit with X , and otherwise, exit with $\pi \text{ sign } X - X$. X will have density f because (for $0 < x < \frac{\pi}{2}$) we have $\frac{1}{\pi} - f(x) = f(\pi-x)$. Thus, algorithm S1 can, for the Raab-Green density, be improved to:

1.1RGA. Generate two independent uniform (0,1) random variates U and V . Set $X = \frac{\pi}{2}(2V-1)$, $n = 0$, $T = 0$, $P = \frac{1}{2}$.

1.2RGA. $n = n+1$, $P = PX^2/[(2n)(2n-1)]$, $T = T+P$. If $U \ge T$, exit with X .

1.3RGA. $n = n+1$, $P = PX^2/[(2n)(2n=1)]$, $T = T-P$. If $U < T$, exit with $\pi \text{ sign } X - X$.

1.4RGA. Go to 1.2RGA.

Then we will have $N_1 = 1$ (no rejections), and furthermore

$$P(N_2>1) = 2 \int_0^{\pi/2} \frac{1}{\pi} \frac{x^4}{48} dx = \frac{\pi^4}{3840} \cong 0.0254 .$$

In other words, not only is no cosine evaluation necessary with this algorithm, but step 4 is reached only about 2.54% of the time.

The last algorithm RGA takes 18.6 μs per random variate on the average. If steps 2-4 are replaced by the direct method "If $2U \le 1 + \cos X$, exit with X . Otherwise, exit with $\pi \text{ sign } X-X$", then it takes 20.9 μs per random variate on the average. Thus in this example it is desirable to use the series method rather than the direct method.

4. THE KOLMOGOROV-SMIRNOV DISTRIBUTION.

The Kolmogorov-Smirnov distribution function

$$F(x) = \sum_{n=-\infty}^{\infty} (-1)^n e^{-2n^2x^2} , \quad x > 0 , \qquad (6)$$

appears as the limit distribution of the Kolmogorov-Smirnov test statistic (Kolmogorov (1933), Smirnov (1939), Feller (1948)). No simple procedure for inverting F is known, hence the inversion method is likely to be slow. The density f corresponding to F is

$$f(x) = 8 \sum_{n=1}^{\infty} (-1)^{n+1} n^2 x e^{-2n^2x^2} , \quad x > 0 , \qquad (7)$$

which is of form (3) when

$$h(x) = 8xe^{-2x^2} \quad , \; x > 0 \; ,$$

$$a_n(x) = (n+1)^2 \; e^{-2x^2[(n+1)^2-1]} \quad , \; x > 0 \tag{8}$$

It is known (for the equivalence of (7) and (9) see Whittaker and Watson (1963) or Byrd and Friedman (1964)) that F and f can also be written as

$$F(x) = \frac{\sqrt{2\pi}}{x} \sum_{n=1}^{\infty} e^{-(2n-1)^2\pi^2/8x^2} \quad , \; x > 0 \; , \tag{9}$$

and

$$f(x) = \frac{\sqrt{2\pi}}{x} \sum_{n=1}^{\infty} \left[\frac{(2n-1)^2\pi^2}{4x^3} - \frac{1}{x}\right] e^{-(2n-1)^2\pi^2/8x^2} \quad , \; x > 0 \; . \tag{10}$$

This also follows the format of (3), but now with

$$h(x) = \frac{\sqrt{2\pi}\;\pi^2}{4x^4} \; e^{-\pi^2/(8x^2)}$$

$$a_n(x) = \begin{cases} \dfrac{4x^2}{\pi^2} \; e^{-(n^2-1)\pi^2/8x^2} & , \; n \ge 1 \; , \; n \text{ odd} \; , \; x > 0 \; , \\ (n+1)^2 \; e^{-[(n+1)^2-1]\pi^2/8x^2} & , \; n \ge 1 \; , \; n \text{ even} \; , \; x > 0 \; . \end{cases} \tag{11}$$

<u>Lemma 1</u>. The terms $a_n(x)$ in (8) are monotone $\downarrow$ for $x > \sqrt{1/3}$. The terms $a_n(x)$ in (11) are monotone $\downarrow$ for $x < \pi/2$.

<u>Proof.</u> For (8), we have $\log(a_{n-1}(x)/a_n(x)) = -2\log(1+n^{-1}) + 2(2n+1)x^2 \ge -2n^{-1} + 2(2n+1)x^2 \ge -2 + 6x^2 > 0$. For (11), when n is even, we have $a_n(x)/a_{n+1}(x) = (n+1)^2\pi^2/4x^2 \ge \pi^2/4x^2 > 1$. Also, $\log(a_{n-1}(x)/a_n(x)) = -\log((n+1)^2\pi^2/4x^2) + n\pi^2/2x^2 = ny - 2\log(n+1) - \log(y/2)$ (where $y = \pi^2/2x^2$). The last expression is increasing in y for $y \ge 2$ and all $n \ge 2$. Thus it is not smaller than $2n-2\log(n+1) \ge 0$. This concludes the proof of Lemma 1.

The monotonicity condition (4) necessary to apply algorithm S1 is satisfied for (8) on $(\sqrt{1/3}, \infty)$ and for (11) on $(0, \pi/2)$. <u>The algorithm that we propose to generate a random variate X from The Kolmogorov-Smirnov f is a combination of S1 and the mixture method:</u>

<u>Algorithm S1M.</u>

1.0M. (Preparation.) Let c be a constant in $(\sqrt{1/3}, \pi/2)$ ($c = 0.75$ is suggested), and let $p = F(c)$.

1.1M. Generate a uniform (0,1) random variate U. If $U > p$, go to 1.2M.

Otherwise, exit with a random variate X from density

$$f_1(x) = f(x)/p \; , \; 0 < x < c \; .$$

To generate X, use S1 with (10) and (11).

1.2M. Exit with a random variate X from density

$$f_2(x) = f(x)/(1-p) \ , \ c < x \ ,$$

where X is generated by the alternating series method S1 applied to (7) and (8).

For details of the application of S1 to densities f_1 and f_2 we make use of the following lemmas.

Lemma 2. If G is a random variable with the truncated gamma($\frac{3}{2}$) density $c_1\sqrt{y}\ e^{-y}$, $y \geq c' = \pi^2/8c^2$, then $X = \pi/\sqrt{8G}$ has density

$$c_2\ x^{-4}\ e^{-\pi^2/8x^2} \ , \ x \leq c \ . \tag{12}$$

(Here c_1 and c_2 are normalization constants).

Proof. The Jacobian of the transformation $y = \pi^2/8x^2$ is $4\pi/(8y)^{3/2}$. If X has density (12), then $G = \pi^2/8X^2$ has density $c_1\sqrt{y}\ e^{-y}$, $y \geq c'$.

Lemma 3. If E is an exponential random variable, then $X = \sqrt{c^2+E/2}$ has density

$$c_3\ x\ e^{-2x^2} \ , \ x \geq c \ , \tag{13}$$

where c_3 is a normalization constant.

Proof. The distribution function of (13) is $1 - \exp(-2(x^2-c^2))$, $x \geq c$.
That of E is $1 - \exp(-x)$, $x \geq 0$.

We propose the following algorithms for the generation of random variates from f_1 and f_2 , respectively. The constant c is picked as in S1M and $c' = \pi^2/8c^2$.

(f_1) 1. 1A. Generate two independent exponential random variates E_0 and E_1 . Set $E_0 = E_0/(1-(2c')^{-1})$, $E_1 = 2E_1$, $G = c'+E_0$.

1B. If $E_0^2 > c'E_1(G+c')$, go to 1D.

1C. If $G/c' - 1 - \log(G/c') > E_1$, go to 1A.

1D. $X = \pi/\sqrt{8G}$, $T = 0$, $Z = (2G)^{-1}$, $P = e^{-G}$, $n = 1$, $Q = 1$.

Generate a uniform (0,1) random variate U .

2. $T = T+ZQ$. If $U \geq T$, exit with X .

3. $n = n+2$, $Q = P^{n^2-1}$, $T = T-n^2Q$. If $U < T$, go to 1.

4. Go to 2.

(f_2) 1. Generate an exponential random variate E and an independent uniform (0,1) random variate U . Set $X = \sqrt{c^2+E/2}$, $T \leftarrow 0$, $n \leftarrow 1$, $Z = \exp(-2X^2)$.

2. $n = n+1$, $T = T+n^2Z^{n^2-1}$. If $U \geq T$, exit with X .

3. $n = n+1$, $T = T=n^2Z^{n^2-1}$. If $U < T$, go to 1.

4. Go to 2.

The steps in both algorithms are numbered as in S1. In step 1 of the algorithm for f_1, a random variate X is generated that has density h (11) restricted to (0,c) . The algorithm uses rejection (step 1C) with squeezing (step 1B) (for details, see Dagpunar (1978) and Devroye (1980)). The computation of Z in step 1 of the f_2 algorithm requires exponentiation. This may be avoided some of the time by accepting X in step 1 "quickly" when $U \geq 4\exp(-6c^2)$. Similarly, we may accept X in step 1D of the f_1 algorithm when $U \geq 4c^2/\pi^2$, before Z or P are computed. In Table 1 values are given for these quick acceptance probabilities and for $p = F(c)$ as a function of c . Both probabilities are close to 0.80 when $c = 0.75$. A complete FORTRAN program with discussion is given at the end of this section.

TABLE 1. Values for Quick Acceptance Probabilities and $p \equiv F(c)$ as Functions of c.

c	$1-4e^{-6c^2}$	$1-4c^2/\pi^2$	p=F(c)	Average Time per Variate (μs*)
0.60	0.54	0.85	0.136	39
0.65	0.68	0.83	0.208	36
0.70	0.79	0.80	0.289	34
0.75	0.86	0.77	0.373	34
0.80	0.914	0.74	0.456	36
0.85	0.948	0.71	0.535	38
0.90	0.969	0.67	0.607	41
0.95	0.982	0.63	0.673	44
1.00	0.9901	0.59	0.730	48
1.05	0.9946	0.55	0.780	53

* 1 μs = 1 microsecond = 10^{-6} seconds.

Related limit distributions. The empiric distribution function $F_n(x)$ for a sample $X_1,\ldots,X_n$ of independent identically distributed random variables is defined by $F_n(x) = \sum_{i=1}^{n} n^{-1}[X_i \leq x]$ where I is the indicator function. If X_1 has distribution function $F(x)$, then the following statistics have been proposed for goodness-of-fit tests by various authors:

$$K_n^{+} = \sqrt{n} \sup_x F_n(x) - F(x) \ , \ K_n^{-} = \sqrt{n} \sup_x F(x) - F_n(x)$$

(the asymmetrical Kolmogorov-Smirnov statistics);

$K_n = \max(K_n^{+}, K_n^{-})$ (the Kolmogorov-Smirnov statistic) ;

$V_n = K_n^{+} + K_n^{-}$ (Kuiper's statistic) ;

$W_n^2 = n \int (F_n(x) - F(x))^2 dF(x)$ (von Mises' statistic) ;

$U_n^2 = n \int (F_n(x) - F(x) - \quad (F_n(y) - F(y))dF(y) \ )^2 dF(x)$

(Watson's statistic) ;

$A_n^{\ 2} = n \int \frac{(F_n(x)-F(x))^2}{F(x)\ (1-F(x))} \ dF(x)$ (the Anderson-Darling statistic).

For surveys of the properties and applications of these and other statistics, see Darling (1955), Barton and Mallows (1965), Sahler (1958), and Shapiro (1980). All the statistics mentioned here have limit distributions. The limit random variables will be denoted by K_∞^{+}, K_∞^{-}, K_∞, etc. In Table 2, the characteristic functions of these limit distributions are given. In all but one case they can be written as a countable product of characteristic functions of gamma random variables. Therefore the following conclusions about the generation of random variates from these distributions follow:

(1) $2K_\infty^{+2}$ and $2K_\infty^{-2}$ are exponentially distributed. For exponential random variate generation algorithms, see Maclaren, Marsaglia and Bray (1964), Sibuya (1961), Marsaglia (1961) and Ahrens and Dieter (1972).

(2) V_∞ is distributed as $K_\infty(1) + K_\infty(2)$, the sum of the two independent random variables distributed as K_∞ .

(3) U_∞ is distributed as K_∞/π^2.

(4) The sum of two independent W_∞^2 random variables $(W_\infty^2(1)+W_\infty^2(2))$ is distributed as K_∞/π^2 , but no simple function of K_∞ that would give a random variable distributed as W_∞^2 is known to us.

Note that the explicit form of the limit distribution of V_∞ is relatively simple (see Kuiper (1960), but A_∞^2 and W_∞^2 have complicated limit distributions (see Anderson and Darling ((1952)).

TABLE 2. Characteristic Functions of Some Limit Distributions for Tests of Fit.

Random Variable	Characteristic Function	Relevant References
K_∞	$\prod_{j=1}^{\infty} (1-it/2j^2)^{-1}$	Kolmogorov (1933), Smirnov (1939), Feller (1948)
K_∞^{+2}, K_∞^{-2}	$(1-it/2)^{-1}$	Smirnov (1939), Feller (1948)
V_∞	$\prod_{j=1}^{\infty} (1-it/2j^2)^{-2}$	Kuiper (1960)
W_∞^2	$\prod_{j=1}^{\infty} (1-2it/\pi^2 j^2)^{-1/2}$	Smirnov (1937), Anderson and Darling (1952)
U_∞^2	$\prod_{j=1}^{\infty} (1-it/2\pi^2 j^2)^{-1}$	Watson (1961, 1962)
A_∞^2	$\prod_{j=1}^{\infty}(1-it/j(j+1))^{-1/2}$	Anderson and Darling (1952)

```
      REAL FUNCTION SMIR(L)
C
C
C   THIS SUBPROGRAM PRODUCES VARIATES FROM THE KOLMOGOROV-SMIRNOV
C   LIMIT DISTRIBUTTON FUNCTION.
C
C   SOURCE:  LUC DEVROYE "THE SERIES METHOD FOR RANDOM VARIATE
C            GENERATION AND ITS APPLICATION  TO THE KOLMOGOROV-
C            SMIRNOV DISTRIBUTION," AMERICAN JOURNAL OF MATHEMATICAL
C            AND MANAGEMENT SCIENCES.
C   AT EACH CALL, THE ARGUMENT CAN BE GIVEN THE VALUE 0 AS IN
C   THE STATEMENT X = SMIR (0).
C   THE PROGRAM USES THE SUBPROGRAMS UNI AND REXP FOR THE GENERA-
C   TION OF UNIFORM (0,1) AND EXPONENTIAL RANDOM VARIATES.
C   THESE SUBPROGRAMS ARE PART OF THE SUPER-DUPER RANDOM NUMBER
C   GENERATOR PACKAGE OF MCGILL UNIVERSITY.
C
C   THE CONSTANTS IN THE PROGRAM DEPEND UPON C.  ALL THE VALUES IN
C   THE PROGRAM ARE FOR C=0.75.  FOR OTHER VALUES OF C IN THE
C   RANGE SQRT (1./3.)< C <3.14159265/2.  RECALCULATE THEM AS
    FOLLOWS :
C           P= VALUE OF KOLMOGOROV-SMIRNOV DISTRIBUTION FUNCTION
C              AT C
C           P1=1./P
C           P2=1./(1.-P)
C           CSQRE=C*C
C           P28C=PIE**2/(8*C*C), PIE=3.14159265
C           PINV=1./P28C
C           ALPHA=4.*EXP(-6.*C*C)
C           BETA=4*C*C/PIE**2
C           B=1./(1.-4*C*C/PIE**2)

      DATA C,P,P1,P2,CSQRE,P28C,PINV,ALPHA,BETA,B/0.75,0.3728330,
     + 2.682166,1.594471,0.5625,2.193245.0.4559454,0.1368725,
     + 0.2279727,1.295291/
1     DEC=UNI (0)
      IF(DEC.LT.P)GOTO59
```

```
C
C  GENERATE VARIATE FROM TAIL OF RAYLEIGH DENSITY
C
   V=(DEC-P)*P2
   GOTO21
2  V=UNI(0)
21 SMIR=CSQRE+0.5*REXP(0)
   SMIRSQ=SMIR+SMIR
C
C  CONSECUTIVE ACCEPTANCE/REJECTION STEPS
C
   IF (V.LT.ALPHA)GOTO4
3  SMIR=SQRT(SMIR)
   RETURN
4  IF (SMIRSQ.GT.174.) GOTO3
   T=EXP(-SMIRSQ)
   K=1
   NUM=0
   SUM=0
5  NUM=NUM+K+K+1
   K=K+1
   (IF(NUM*SMIRSQ.GT.174.)GOTO3
   SUM=SUM+(NUM+1)*T**NUM
   IF(V.GE.SUM)GOTO3
   NUM=NUM+K+K+1
   K=K+1
   IF(NUM*SMIRSQ.GT.174)GOTO3
   SUM=SUM=(NUM+1)*T**NUM
   IF(V.LT.SUM)GOTO2
   GOTO5
C
C  GENERATE VARIATE DISTRIBUTED AS INVERSE OF SQUARE ROOT OF TAIL
   OF CHI-SQUARE DENISTY WITH 3 DEGREES OF FREEDOM
C
59 V=DEC*P1
   GOTO7
6  V=UNI(0)
7  E=REXP(0)*B
   E1=E1+E1
   Y=P28C+E
   IF(E*E.LE.E1*P28C*(Y+P28C))GOTO8
   IF(PINV*Y-1.-ALOG(PINV*Y).GT.E1)GOTO6
   SMIR=1.1107206/SQRT(Y)
   Z=0.5/Y
   SUM=1.-Z
C
C  CONSECUTIVE ACCEPTANCE/REJECTION STEPS
C
   IF(V.LT.SUM)RETURN
   K=3
   KSQRE=8
   IF(Y.GT.21.7)RETURN
   T=EXP(-Y)
9  TU=T**KSQRE
   SUM=SUM+(KSQRE+1)*TU
   IF(V.GT.SUM)GOTO6
   SUM=SUM-Z*TU
   IF(V.LE.SUM)RETURN
   K=K+2
   KSQRE=K*K-1
   IF(Y*KSQRE.GT.174.)RETURN
   GOTO9
   END
```

Subprogram SMIR requires a uniform random variate generator (UNI), and an exponential random variate generator (REXP). Both UNI and REXP are part of the SUPER-DUPER random number generator package developed by Marsaglia, Anantharayanan and Paul (1973) at McGill University. The IBM Assembler code can be found in Dudewicz and Ralley (1981). It can also be obtained directly from McGill University. The average time required per variate changes with the parameter c (see Table 1), and on McGill University's AMDAHL V7 computer, c = 0.75 seems to be the best choice. The sequence of random variates produced by SMIR was also submitted to a Kolmogorov-Smirnov goodness-of-fit test for sample size 5000 (the p-values obtained were 0.12, 0.24 and 0.37). To compare the speed of SMIR (34 μs/variate on the average) with that of other algorithms, note that the statement X = -ALOG(UNI(0)) takes on the average 10.5 μs, and that the fastest FORTRAN coded gamma generators (see Tadikamalla and Johnson((1981)) take about 21 microseconds per variate for large values of the shape parameter.

REFERENCES

Ahrens, J.H. and Dieter, U (1972). Computer methods for sampling from the exponential and normal distributions. Communications of the ACM, 15, 873-882.

Anderson, T.W. and Darling D.A. (1974). Asymptotic theory of certain goodness of fit criteria based on stochastic processes. Annals of Mathematical Statistics, 23, 223-246.

Barton, D.E. and Mallows, C.L. (1965). Some aspects of the random sequence. Annals of Mathematical Statistics, 36, 236-260.

Byrd, P.F. and Friedman M.D. (1954). Handbook of Elliptic Integrals for Engineers and Physicists, Springer-Verlag, Berlin.

Dagpunar, J.S. (1978). Sampling of variates from a truncated gamma distribution. Journal of Statistical Computation and Simulation, 8, 59-64.

Darling, D.A. (1955). The Kolmogorov-Smirnov, Cramer-von Mises tests. Annals of Mathematical Statistics, 26, 1-20.

Devroye, L. (1980). Generating the maximum of independent identically distributed random variables. Computers and Mathematics with Applications, 6, 305-315.

Dudewicz, E.J. and Ralley, T.G. (1981). The Handbook of Random Number Generation and Testing with TESTRAND Computer Code. American Sciences Press, Inc., Columbus, Ohio.

Feller, W. (1948). On the Kolmogorov-Smirnov limit theorems for empirical distributions. Annals of Mathematical Statistics, 19, 177-189.

Kinderman, A.J. and Ramage, J.G. (1976). The computer generation of normal random variables. Journal of the American Statistical Association 71, 893-896.

Kolmogorov, A.N. (1933). Sulla determinazione empirica di una legge di distribuzione. Giorn. Inst. Ital. Actuari 4, 83-91.

Kronmal, R.A. and Peterson, A.V. (1979). On the alias method for generating random variables from a discrete distribution. The American Statistician, 33, 214-218.

Kuiper, N.H. (1960). Tests concerning random points on a circle. Proceedings Koninklijke Nederlandse Akademie van Wetenschappen, A 63, 38-47.

Maclaren, M.D., Marsaglia, G. and Bray, T.A. (1964). A fast procedure for generating exponential random variables. Communications of the ACM, 7, 298-300.

Marsaglia, G. (1961). Generating exponential random variables. Annals of Mathematical Statistics, 32, 899-902.

Marsaglia, G., Ananthanarayanan, K. and Paul, N. (1973). How to use the McGill random number package "SUPER-DUPER", four page typed description, School of Computer Science, McGill University, Montreal, Canada.

Marsaglia, G. and Bray, T.A. (1964). A convenient method for generating normal variables. SIAM Review, 6, 260-264.

Payne, W.H. (1977). Normal random numbers: using machine analysis to choose the best algorithm. ACM Transactions on Mathematical Software, 3, 346-358.

Raab, D.H. and Green, E.H. (1961). A cosine approximation to the normal distribution. Psychometrika, 26, 447-450.

Sahler, W. (1968). A survey of distribution-free statistics based on distances between distribution functions. Metrika, 13, 149-169.

Schmeiser, B.S. and Lal, R. (1980). Squeeze methods for generating gamma variates. Journal of the American Statistical Association, 75, 679-682.

Shapiro, S.S. (1980). How to Test Normality and Other Distributional Assumptions, Vol. 3 of the ASQC Basic References in Quality Control: Statistical Techniques (E.J. Dudewicz, ed.). American Society for Quality Control, Milwaukee, Wisconsin.

Sibuya, M. (1961). On exponential and other random variable generators. Annals of the Institute of Statistical Mathematics, 13, 231-237.

Smirnov, N.V. (1937). On the distribution of the ω^2 criterion of von Mises. Rec. Math. 2, 973-993.

Smirnov, N.V. (1939). On the estimation of the discrepancy between empirical curves of distribution for two independent samples. Bulletin Mathematique de l'Université de Moscou,2.

Tadikamalla, P.R. (1978). Computer generation of gamma random variables. Communications of the ACM, 21, 419-422.

Tadikamalla, P.R. and Johnson, M.E. (1981). Complete guide to gamma variate generation. American Journal of Mathematical and Management Sciences, 1.

Vaduva, I. (1977). On computer generation of gamma random variables by rejection and composition procedures. Mathematische Operations for schung-und Statistik, 8, 545-576.

von Mises, R. (1947). On the asymptotic distribution of differentiable statistical functions. Annals of Mathematical Statistics, 18, 309-348.

Walker, A.J. (1977). An efficient method for generating discrete random variables with general distributions. ACM Transactions on Mathematical Software, 3, 253-256.

Watson, G.S. (1961). Goodness-of-fit tests on a circle.I. Biometrika, 48, 109-114.

Watson, G. (1962). Goodness-of-fit tests on a circle.II. Biometrika, 49, 57-63.

Whittaker, E.T. and Watson, G.N. (1963). A Course of Modern Analysis. Cambridge University Press, Cambridge, England.

Received 5/16/80; Revised 2/17/82.

AMERICAN JOURNAL OF MATHEMATICAL AND MANAGEMENT SCIENCES

A COMPLETE GUIDE TO GAMMA VARIATE GENERATION

Pandu R. Tadikamalla
University of Pittsburgh
Pittsburgh, PA 15260, U.S.A.

Mark E. Johnson
Los Alamos National Laboratory
Los Alamos, NM 87545, U.S.A.

SYNOPTIC ABSTRACT

Considerable attention has recently been directed at developing simpler and faster algorithms for generating gamma random variates (with general, not necessarily integral, shape parameter α) on digital computers. This paper surveys the current state of the art, which includes fifteen gamma algorithms applicable for $\alpha \geq 1$ and six that are applicable for $\alpha < 1$. These algorithms are compared according to the criteria of speed and simplicity. General random variate generation techniques are explained with reference to these gamma algorithms. Computer simulation experiments on DEC and CDC computers are reported. Guidelines for some specific applications are given.

1. INTRODUCTION

The gamma distribution is a useful model for stochastic inputs in a wide variety of simulation applications: computer generated gamma variates have been used to model interarrival and service times in queueing problems, lead times and demand in inventory control, and failure times in reliability models. The gamma distribution's popularity can be traced to the properties it obtains by the appropriate selection of its shape parameter. The density of the three-parameter gamma distribution is

$$g(y) = (y-\xi)^{\alpha-1}\exp(-(y-\xi)/\gamma)/(\Gamma(\alpha)\cdot\gamma^{\alpha}), \quad y \geq \xi, \tag{1}$$

where α, γ, and ξ are called the shape, scale and location parameters, respectively. It is sufficient to generate the random variates from the standardized gamma distribution having density

$$f(x) = x^{\alpha-1}\exp(-x)/\Gamma(\alpha), \quad x \geq 0, \tag{2}$$

and make the transformation $y = \gamma x + \xi$, to obtain the general gamma distribution. Thus, the gamma distribution considered in this paper is (2).

Key Words and Phrases: gamma distribution; random numbers; simulation; random variate generation.

1981, VOL. 1, NO. 3, 213-236
0196-6324/81/030213-24 $6.10

Several parameter values of α are particularly important. If $\alpha = 1$, then f is the density of the exponential distribution. If $\alpha = k$, an integer, then a k-Erlang distribution is obtained. Some simple transformations of gamma variates lead to other well-known distributions. If X has the density f, then 2X has a chi-square distribution with 2α degrees of freedom. The ratio of independent chi-square variates is an F variate; $X_1/(X_1 + X_2)$ is a beta variate if X_1 and X_2 are independent gamma variates with the same scale. Finally, the limiting distribution of a gamma variate as $\alpha \to \infty$ is normal. Because of the gamma distribution's versatility and its appropriateness in simulation applications, considerable attention has rightfully been directed to improving methods for generating gamma variates on digital computers.

The purpose of this paper is to examine various gamma generation techniques in view of new developments since Atkinson and Pierce (1973) and Tadikamalla and Johnson (1978). The various gamma algorithms (applicable for $\alpha \geq 1$) in our study are Ahrens and Dieter's (1974) GO, Atkinson's (1977) ATK, Best's (1978) XG, Cheng's (1977) GB, Cheng and Feast's (1979) GKM3, Cheng and Feast's (1980) GT ($\alpha > 0.5$) and GBH ($\alpha > 0.25$), Fishman's (1975) GF, Greenwood's (1974) GG, Kinderman and Monahan's (1978) GRUB, Marsaglia's (1977) MSA ($\alpha > 1/3$), Schmeiser's (1978) G4PE, Tadikamalla's (1978a) TG1, Tadikamalla's (1978b) TG2 and Wallace's (1974) GW. In addition to GT, GBH and MSA which apply for some $\alpha < 1$, we consider Ahrens and Dieter's (1974) GS, Johnk's (1964) GJ and Vaduva's (1974) GV which are valid for $0 < \alpha < 1$. Whittekar's (1974) algorithm is identical to Johnk's algorithm.

2. GENERAL METHODS

The most common variate generation techniques can be roughly described as one of the following (see Tadikamalla (1975)): inverse probability integral transform, transformation, rejection, and mixture. We will briefly describe each of these techniques in this section.

The inverse function method is based on the following well-known result (see Halton, 1970): If X has distribution function F(x), then U = F(X) is distributed with uniform density on [0, 1]; conversely, if U is uniform in [0, 1], then $G(U) = \inf\{x: U \leq F(x)\}$ has distribution function F(x). Application of this method is limited to variates having $G = F^{-1}$ in simple closed form. For the gamma family, only the exponential distribution enjoys this property. This leads to the exponential generation formula $X = -\ln(1 - U)$, or alternatively (since both U and 1 - U have a uniform distribution) $X = -\ln(U)$. For arbitrary shape parameter $\alpha \neq 1$, this approach is inefficient since the evaluation of F^{-1} must be done numerically. However, for integer $\alpha = k$, we can apply the general transformation method.

In particular, $X = -\ln(\Pi_{i=1}^{k} U_i)$ has a gamma distribution with shape parameter k if the U_i's are independent uniform 0-1. The transformation method can also be used to obtain chi-square, F and beta variates as indicated in Section 1.

The rejection method (see Tadikamalla (1978a)) has recently been the most fruitful approach to developing new algorithms for generating gamma variates. The idea of the rejection method is to generate variates from a density $h(x; \underline{\theta})$ which somewhat resembles the desired density $f(x)$. Occasionally, variates generated from h are rejected in such a way that the accepted variates have a distribution corresponding to f. Formally, let $f(x)$, $x \in \Omega$, be the density from which samples are required. Let $h(x; \underline{\theta})$ be another density which is easy to generate, where $\underline{\theta}$ is a vector of unknown parameters. Also, let h have the same support (set of values with positive density) as f and assume it satisfies $f(x) < \delta h(x; \underline{\theta})$ for all $x \in \Omega$ and for some $\delta > 1$. The vector $\underline{\theta}$ of unknown parameters is to be determined as described below. The rejection method algorithm is:

1. Generate X having density $h(x; \underline{\theta})$.
2. Generate U which is uniform 0-1.
3. If $U > T(X) = f(X)/\delta h(X; \underline{\theta})$, go to 1. Otherwise, return X.

The variable δ is the expected number of "trials" until acceptance of X. The variable $1/\delta$ is generally referred to as the "efficiency" of the procedure. Several conflicting considerations sometimes enter into the selection of $h(x; \underline{\theta})$. They are summarized as follows:

1. A fast, simple algorithm for generating variates from $h(x; \underline{\theta})$ must be available (i.e., step 1 of the algorithm should be executed quickly).
2. The efficiency $1/\delta$ should be close to 1 (i.e., $h(x; \underline{\theta})$ should resemble f).
3. The acceptance-rejection test in step 3 should be simple (i.e., $T(X)$ should be easy to evaluate).
4. δ must be computable from $\delta = \min_{\underline{\theta}}[\max\{f(x)/h(x; \underline{\theta})\}]$.

We will return to these considerations in relation to specific gamma algorithms in Section 3.

The mixture method or composition technique (Butler (1956) and Butcher (1961)) is based upon representing the density f from which variates are to be generated as $f(x) = p_1 f_1 + p_2 f_2 + \ldots + p_n f_n$, where $p_1 + p_2 + \ldots + p_n = 1$ and each of the f_i's are densities. The rule of thumb in developing mixtures for f is to select the f_i's so that f_1 is the fastest f_i to generate, p_1 is close to 1, and the other f_i's are not unduly difficult to generate. The corresponding mixture method algorithm is simply to generate variates from each f_i with probability p_i. The mixture method has not received the same attention for generating gamma variates as it has for exponential and normal variates (Marsaglia (1961), Marsaglia, MacLaren, and Bray (1964), and Kinderman and Ramage (1976)). This is primarily due to the awkward problem that a different mixture must be used for each α value. Only in very large simulation studies could the effort in determining mixtures for the various values be justified.

3. ALGORITHMS.

In this section we briefly describe the gamma algorithms mentioned in Section 1. Most of these algorithms are based on the rejection method described in Section 2. Some of these algorithms are direct applications of the generalized rejection technique and differ only in their choices of $h(x; \underline{\theta})$. The simple choices for $h(x; \underline{\theta})$ include the exponential (GF), the k-Erlang (TG1), the double exponential (TG2), the student's t distribution with two degrees of freedom (XG) and the log-logistic (GB). Several other algorithms use a mixture of two densities as their choices for $h(x; \underline{\theta})$. These include a normal and an exponential (GO), a uniform and an exponential (ATK), and two k-Erlangs (GW). The algorithms MSA and GG generate the cube root of a gamma variate using a normal density for $h(x; \underline{\theta})$. The algorithm G4PE uses several uniforms and an exponential (for the tail) as an envelope. The more recent algorithms GRUB, GKM3, GT and GBH use the "ratio of uniforms" method as developed by Kinderman and Monahan (1977); this is essentially a rejection technique which can be applied in diverse settings.

Of the algorithms applicable for $\alpha < 1$, GV is a direct application of the generalized rejection technique where $h(x; \underline{\theta})$ is a Weibull density. Algorithm GS uses the mixture method where $f_1(x)$ is the power function density and $f_2(x)$ is the truncated exponential distribution. Algorithm GJ can be considered as a combination of the rejection and the transformation techniques.

Many of these algorithms have been streamlined considerably by their inventors to improve their relative performance. Preliminary fast acceptance tests to avoid evaluations of $T(X)$ are employed in the published versions of ATK, GB, GKM3, GRUB, GT, GBH, MSA and G4PE. For formal statements of the algorithms, the reader is referred to the cited papers; our overview is sufficient to elucidate the commonality of the methods. In the next section we compare the algorithms on the basis of speed and simplicity.

4. COMPARATIVE STUDY: SPEED AND SIMPLICITY

Three important criteria used in judging computing algorithms are speed, simplicity and accuracy. All the methods discussed so far in this paper are exact methods subject only to the accuracy of the machine and of the standard machine functions logarithm, exponential and square root. Thus, we do not attempt to compare the algorithms according to the quality of the output sequences. It is difficult to measure <u>simplicity</u>, but it can be considered as a combination of the effort required to implement the algorithm (which in turn can be measured in terms of the length of the program and the special functions or subprograms required, if any) and the core storage required for the algorithm. All of the algorithms in this study were coded in FORTRAN. If the published versions of the algorithms have a listing, these were used. Otherwise we have coded the algorithms ourselves and have followed the various authors' suggestions.

Table 1 gives the <u>average CPU time</u> for selected values of α on a DEC 10 computer, the setup times, the core storage in words, and the subroutine length (the number of FORTRAN statements including heading, return and end statements) for each of the fifteen gamma algorithms applicable for $\alpha \geq 1$. Figure 1 illustrates the execution time results. The enclosed dashed regions indicate the extent of certain algorithms. Table 2 gives analogous quantities for the six gamma algorithms applicable for $\alpha < 1$ on the DEC 10 computer. Table 3 and Figure 2 give the CPU times on a CDC 7600 computer for the gamma algorithms applicable for $\alpha \geq 1$. The timings reported on the DEC 10 were based on generating 30,000 variates in each case in a DO loop as a call to the subroutine. The uniform random number generator RAN which is

<u>TABLE 1: Comparison of Gamma Algorithms (for $\alpha > 1$) on a DEC 10 Computer.</u>

Algorithm	Length (FORTRAN statements)	Core storage (36 bit words)	CPU Times (based on 30,000 variates)										
			Setup	α = 1.25	1.75	2.5	3.5	4.5	5.5	7.5	10.5	25.5	50.5
GF	10	70	0	237	284	339	394	462	506	579	702	1090	1517
TG1	11	110	0	267	324	294	302	321	341	379	440	750	1282
GW	13	120	0	318	422	322	332	354	373	415	478	795	1324
XG	14	130	0	313	282	276	275	276	281	281	283	284	286
GB	17	136	51	295	275	260	255	250	250	249	246	244	240
GRUB	21	232	424	190	191	193	197	196	199	200	198	202	201
ATK	21	237	399	276	313	345	376	408	439	479	518	702	937
TG2	25	213	134	261	283	303	308	318	319	325	326	323	320
GKM3	26	200	33	181	185	206	208	206	209	209	211	211	214
G4PE	80	626	667	143	140	134	124	120	120	116	117	113	113
GG	15[1]	147[2]	103+64[3]	272	265	261	260	259	257	257	255	255	254
MSA	20[1]	163[2]	75	226	216	211	206	206	207	204	203	201	200
GO	32[1]	262[2]	80	n/a	n/a	n/a	253	252	249	244	235	203	189
GT	22	157	78	183	183	186	188	188	190	192	194	194	194
GBH	22	161	79	190	195	197	199	201	201	204	205	205	205

[1] The program length does not include 44 statements for the normal algorithm KR.

[2] The core storage does not include 251 words for the normal algorithm KR.

[3] The time for generating one normal variate using the algorithm KR.

n/a indicates that this algorithm is not applicable for $\alpha < 2.53$.

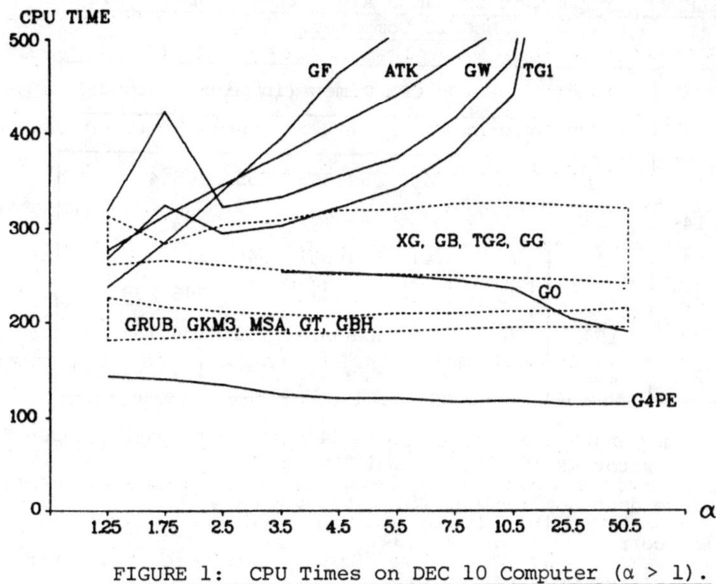

FIGURE 1: CPU Times on DEC 10 Computer ($\alpha > 1$).

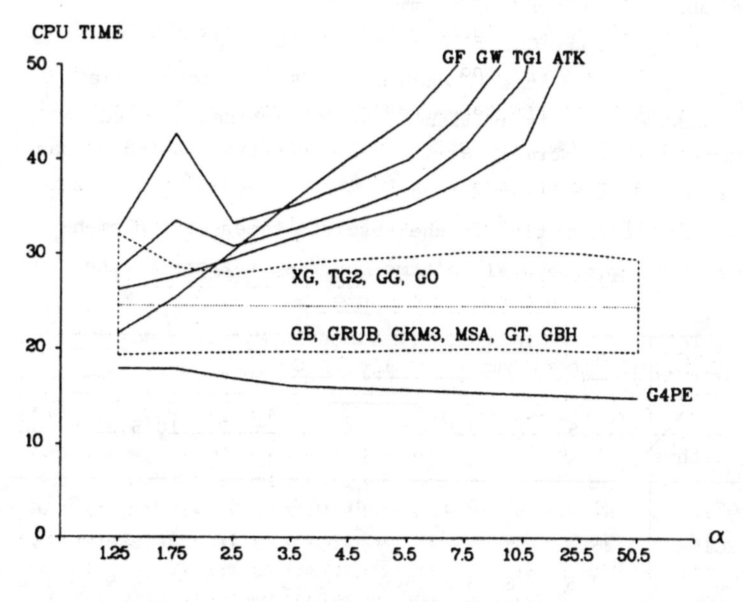

FIGURE 2: CPU Times on CDC 7600 Computer ($\alpha > 1$).

available on the DEC 10 computer was used in all of these computations. RAN is a Lehmer-type multiplicative congruential random number generator, $X_{i+1} = 630360016\ X_i \pmod{2^{31} - 1}$, given by Payne, Rabung and Bogyo (1969). It has been implemented in machine language and tested on the DEC 10 System by the University of Pittsburgh. The normal algorithm KR based on the composition rejection technique of Kinderman and Ramage (1976) is used to generate normal variates for the algorithms GO, GG and MSA. Exponential variates were generated using $-\ell n(U)$, where U is a uniform 0-1 variate. Subject to FORTRAN implementation, the $-\ell n(U)$ method of generating exponential variates (Ahrens and Dieter (1972)) and the Kinderman and Ramage normal algorithm com-

TABLE 2: Comparison of Gamma Algorithms (for $\alpha < 1$) on a DEC 10 Computer.

Algorithms	Length	Storage	CPU Times (in micro seconds)							
			Setup	α = .01	.1	.25	.5	.75	.9	.99
GJ	13	111	0	402	352	365	255	366	436	411
GS	14	126	0	226	243	269	298	307	306	304
GV	14	131	170	412	397	381	346	310	283	265
MSA	20[1]	163[2]	75	n/a	n/a	416[3]	281	246	239	233
GT	22	157	78	n/a	n/a	n/a	199[4]	184	184	184
GBH	22	161	79	n/a	n/a	175[5]	188	191	191	193

[1] The program length does not include 44 statements used by the normal generator KR (Kinderman and Ramage (1976)).

[2] The storage does not include 251 words used by KR.

[3] This time corresponds to $\alpha = 0.34$.

[4] This time corresponds to $\alpha = 0.51$.

[5] This time corresponds to $\alpha = 0.26$.

pare favorably with other algorithms in terms of speed.

The timings on the CDC 7600 computer (Table 3) were based on generating 100,000 variates in each case. The uniform variate generator RANNUM which is available on CDC machines is used in these computations. Normal variates were generated using Marsaglia's (1962) modified polar method.

Considerable variation in the absolute times of implementing these (or any other) algorithms can result from (i) the com-

TABLE 3: The Average CPU Times in Microseconds for Gamma Algorithms ($\alpha > 1$) on a CDC 7600 Computer.

Algorithms \ α	1.25	1.75	2.5	3.5	4.5	5.5	7.5	10.5	25.5	50.5
GF	21.6	25.4	30.3	35.5	40.0	44.2	51.1	60.1	92.3	129.5
TG1	28.3	33.5	30.8	32.6	34.5	36.9	41.4	48.8	85.5	147.9
GW	32.5	42.6	33.2	35.0	37.3	39.9	45.2	53.4	93.4	158.7
XG	32.1	28.6	27.8	27.7	27.8	27.9	28.0	28.1	28.3	28.2
GB	24.0	22.4	21.6	21.0	20.6	20.5	20.3	20.2	19.8	19.7
GRUB	20.4	20.7	21.2	21.3	21.4	21.6	21.7	21.8	22.1	22.2
ATK	26.2	27.6	29.5	31.6	33.4	35.0	37.9	41.6	55.1	70.5
TG2	24.5	26.3	27.8	28.8	29.4	29.7	30.2	30.3	30.1	29.5
GKM3	20.7	20.9	23.3	23.6	23.5	23.4	23.6	23.6	24.0	24.1
G4PE	17.9	17.9	16.9	16.1	15.9	15.7	15.5	15.3	15.1	14.9
GG	25.3	24.7	24.4	24.2	24.2	24.1	24.0	23.9	23.7	23.5
MSA	23.2	22.4	22.0	21.8	21.4	21.5	21.4	21.2	21.1	20.8
GO	n/a	n/a	n/a	28.5	28.3	28.0	27.2	26.3	23.4	21.6
GT	19.3	19.5	19.6	19.7	19.7	19.9	20.0	20.0	20.3	20.5
GBH	19.4	19.6	19.8	20.0	20.3	20.3	20.3	20.6	20.8	20.9

n/a indicates that this algorithm is not applicable for $\alpha < 2.53$.

puter and its operating system, (ii) the language of implementation, and (iii) the use of different uniform, exponential and normal variate generators. The use of different normal and exponential generators may also result in a different relative performace of these algorithms as shown in Tables 1 and 3.

A comparison of the gamma algorithms applicable for $0 < \alpha < 1$ is given in Table 2. The three algorithms GJ, GV and GS are of the same length and require no setup, while algorithm GV requires 170 microseconds for setup. Algorithm GJ is the fastest for α values close to 0.5. Algorithm GV is the fastest for $\alpha > 0.8$, and algorithm GS is the fastest for small values of α. <u>In general algorithm GS is recommended for $0 < \alpha < 1$.</u> However, if fast normal and exponential variate generators are available and are already implemented on the user's system, then algorithm MSA becomes a viable candidate for $\alpha > 1/3$.

The comparison of the gamma algorithms for $\alpha \geq 1$ is more complex. As can be seen from Tables 1 and 3 no single algorithm is uniformly superior in terms of both speed and simplicity. However, certain algorithms are dominated by others in terms of speed and simplicity and can be eliminated from further consideration. For example, algorithm MSA dominates algorithms GO and GG. Although the simple algorithms GF, TG1, GW and ATK may be competitive for small values of α $(1 \leq \alpha \leq 2)$, they are outperformed by other equally simple algorithms GB, GKM3, GRUB, GT and GBH for almost all values of α. Similarly, algorithms TG2, GB and XG are dominated by the algorithms GKM3, GRUB, GT and GBH. These four algorithms (GKM3, GRUB, GT and GBH) are based on the ratio of uniforms method and except for $1 \leq \alpha \leq 2$, GRUB is faster than GKM3. The algorithms GT and GBH dominate GRUB on CDC but are comparable on DEC computers. Algorithm GT is simpler than MSA and marginally faster. This leaves <u>three competing algorithms for $\alpha > 1$: GT, GBH, and G4PE.</u> Algorithms GT and GBH are simple; G4PE is fast. Some computing installations may have fast versions of normal and exponential generators available to the users, in which case the implementation of the algorithm MSA may be worthwhile. Two such random number packages (including very fast uniform, normal and exponential generators) are Marsaglia's (1973) SUPER-DUPER package and Learmouth and Lewis' (1973) random number package LLRANDOM, both of which are available in the TESTRAND package of Dudewicz and Ralley (1981). A pilot study by Devroye (1980) showed that the times for MSA and G4PE are comparable on an AMDAHL V7 computer using the SUPER-DUPER package.

For $\alpha \geq 1$, if fast versions of normal and exponential generators are readily available on the user's system, then MSA can be recommended. If speed is an important criterion, <u>algorithm G4PE is outstanding</u> with GT and GBH being viable contenders where simplicity and speed are of equal concern. If setup cost plus execution time is the primary consideration, the GKM3 should be used. Guidelines for some specific applications are discussed in Section 4.

4. GUIDELINES IN SELECTING A GAMMA GENERATOR

The execution times and storage requirements reported in the previous section are useful for selecting the appropriate gamma generation algorithm. In many applications, however, other criteria should be considered. In this section we discuss some applications which require gamma variates and suggest the appropriate generator(s).

One of the common uses of the gamma distribution is to model service times in queueing problems. If the queueing model is fairly complex, the contribution of the gamma generation would be a small fraction of the total execution time. Thus the savings from using the fastest generator might be quite small, and practically any of the algorithms would suffice. However, if there are many servers each with a different shape parameter governing their times, then an algorithm such as GKM3 would be most appropriate. On the other hand, if a very large number of gamma variates are required (with a fixed shape parameter), then the algorithms GT, GBH, and G4PE are the leading candidates. We can summarize the key considerations in selecting gamma algorithms for practical implementations as follows: (i) Will the shape parameter vary in the course of the simulation? (ii) Are extremely large number of gamma variates with the same shape parameter required? (iii) Will the simplest generator suffice? ("Simplest" could be defined here as "already available on the user's system.)

These considerations are also pertinent to diverse statistical studies. For example, in investigating the small sample behavior of the maximum likelihood estimators of the shape parameter of the gamma distribution, estimation will undoubtedly be much more time consuming then generating the random samples. The study of simpler estimation methods, on the other hand, could favor selection of the faster algorithms--GT, GBH or G4PE.

Gamma variates are also quite useful in constructing other distributions, for which careful selection of an algorithm can lead to savings in execution time as well as possible savings in implementation. Johnson, Tietjen and Beckman (1980) derived a new family of distributions constructed by $X = Z^T U$, where Z has a gamma distribution with shape parameter α and independent of U which is uniform 0-1. Algorithm G4PE can be recommended for generating variates from their distribution. Moreover, the structure of their distribution is immediately amenable to variance reducing simulation designs for fixed α.

The Wishart distribution of dimension n can be generated using Bartlett's decomposition (Kshirsagar (1959)). Here n gammas with different shape parameters are required. The algorithm GKM3 which requires minimal setup time but executes fairly quickly can be recommended. If the dimension of n is small, one could employ n gamma generators (one of the fast ones--GT, GBH or G4PE) at the expense of increased storage requirements.

As indicated by Gentle and Kennedy (1980), the generation of multivariate distributions has received relatively little attention. Schmeiser (1979) provides a useful survey of bivariate gamma generation methods as well as a broad family of algorithms. His work can be viewed as an extension of Cheriyan (1941), who studied the joint distribution of $Y_i = X_0 + X_i$, where X_i are gamma distributed with shape parameter α_i and the Y_i's are then marginally distributed as gamma with shape parameter $\alpha_i + \alpha_0$. Other multivariate gamma distributions can be obtained by judiciously compounding gammas with a negative binomial. The Beta-Stacy distribution (Mihram and Hultquist (1967)) results from generating a beta on the interval 0 to a, where a has a gamma distribution. In each of the above situations, the individual gamma distributions used in the constructions have shape parameters which are fixed (do not depend on the outcome of a random event), so that the considerations of the previous paragraph are appropriate. Namely, the simplest implementation will use GKM3. Faster execution times can be attained by using several gamma generators (with one setup per generator).

A summary of recommended generators by application is given in Table 4. The Appendix provides FORTRAN listings of the leading algorithms with a main driver to assist in verifying correct implementation.

TABLE 4: Summary of Recommended Generators.

Application	Recommended Generators
1. $\alpha < 1$.	GS
2. Many distinct α, few variates generated for each α.	GKM3
3. Large number of variates generated for each α.	GT, GBH, G4PE
4. Generation time is a small contributor to total execution time.	GT, GBH
5. Generating Johnson, Tietjen and Beckman's (1980) distribution.	G4PE
6. Generating Wishart distribution or multivariate gamma distributions.	GKM3

APPENDIX.

Given below are FORTRAN listings of the leading algorithms GS ($\alpha < 1$), GKM3, GT, GBH and G4PE. The arguments of these subroutines are ALP (the gamma shape parameter) and X (the returned variate). To facilitate implementation and verification on other computer systems, we use a table of four-digit uniform numbers that are supplied by the function program RANN. The main program initializes RANN and generates 10 gamma variates from each of the algorithms. These values are printed following the listing.

```
      PROGRAM MAIN
      COMMON INIT
      DIMENSION A(10)
C
C     THIS PROGRAM GENERATES 10 GAMMA VARIATES FROM
C       EACH OF THE LEADING ALGORITHMS:
C
C       GS     (ALPHA = 0.5)
C       GKM3   (ALPHA = 1.5)
C       GT     (ALPHA = 1.5)
C       GBH    (ALPHA = 1.5)
C       G4PE   (ALPHA = 1.5)
C
C     THE UNIFORM GENERATOR RANN CONSISTS OF A
C       TABLE OF 50 4-DIGIT UNIFORM 0-1 NUMBERS.
C       THIS SHOULD FACILITATE REPRODUCING THE SAME
C       STREAMS OF GAMMAS ON OTHER SYSTEMS.
C       THE GENERATED OUTPUT APPEARS IMMEDIATELY
C       FOLLOWING THIS LISTING.  A CAREFUL
C       LINE-BY-LINE CHECK OF THE SUBROUTINES IS
C       NEVERTHELESS STRONGLY RECOMMENDED.
C
      INIT = 0
      DO 10 I=1,10
 10   CALL GS(.5,A(I))
      PRINT *,'ALGORITHM GS'
      PRINT *,(A(I),I=1,4)
      PRINT *,(A(I),I=5,7)
      PRINT *,A(8),A(9),A(10)
      INIT = 0
      DO 20 I=1,10
 20   CALL GKM3(1.5,A(I))
      PRINT *,'ALGORITHM GKM3'
      PRINT *,(A(I),I=1,4)
      PRINT *,(A(I),I=5,7)
      PRINT *,A(8),A(9),A(10)
      INIT = 0
      DO 30 I=1,10
 30   CALL GT(1.5,A(I))
      PRINT *,'ALGORITHM GT'
      PRINT *,(A(I),I=1,4)
      PRINT *,(A(I),I=5,7)
      PRINT *,A(8),A(9),A(10)
      INIT = 0
      DO 40 I=1,10
 40   CALL GBH(1.5,A(I))
      PRINT *,'ALGORITHM GBH'
      PRINT *,(A(I),I=1,4)
      PRINT *,(A(I),I=5,7)
      PRINT *,A(8),A(9),A(10)
      DO 50 I=1,10
 50   CALL G4PE(1.5,A(I))
```

```
      PRINT *,'ALGORITHM G4PE'
      PRINT *,(A(I),I=1,4)
      PRINT *,(A(I),I=5,7)
      PRINT *,A(8),A(9),A(10)
      CALL EXIT
      END
      FUNCTION RANN(DUMM)
      COMMON INIT
      DIMENSION PHONYU(50)
       DATA PHONYU/.3696,.4063,.4288,.4741,.9532,.7739,
     + .5599,.3393,.3645,.2671,.4897,.0054,.2894,.1330,
     + .6579,.7370,.9692,.0050,.2025,.6676,.3381,.2154,
     + .2131,.2369,.7517,.6939,.9784,.1088,.9697,.6918,
     + .0166,.6353,.3403,.6003,.4080,.5856,.7464,.4686,
     + .0651,.2250,.1181,.4150,.8955,.9737,.6950,.1032,
     + .4913,.7699,.0840,.1233/
      IF (INIT .NE. 0) GO TO 10
      INIT = 1
      INUM = 0
   10 CONTINUE
      INUM = INUM + 1
      RANN = PHONYU(INUM)
      RETURN
      END
      SUBROUTINE GS (ALP,X)
C
C     AHRENS AND DIETER (COMPUTING, 1972)
C       ALPHA .LT. 1
C
    1 U1 = RANN(DUMM)
      B = (2.71828 + ALP) / 2.71828
      P = B * U1
      IF (P .GT. 1.0) GO TO 3
    2 X = EXP(ALOG(P)/ALP)
      U2 = RANN(DUMM)
      IF (U2 .GT. EXP(-X)) GO TO 1
      RETURN
    3 X = -ALOG( (B-P) / ALP )
      U3 = RANN(DUMM)
      IF (ALOG(U3) .GT. (ALP-1.)*ALOG(X) ) GO TO 1
      RETURN
      END
      SUBROUTINE GBH (ALP,X)
C
C     CHENG AND FEAST (COMM. OF THE ACM, 1980)
C       ALPHA .GT. 0.25
C
      DATA ASET/-1./
      IF (ASET .EQ. ALP) GO TO 1
      ASET = ALP
      A = ALP - 0.25
      B = ALP / A
      C = 2.0 / A
      D = C + 2.0
      T = 1. / SQRT(ALP)
      H1 = (0.4417 + 0.0245*T/ALP) * T
      H2 = (0.222 - 0.043*T) * T
    1 U1 = RANN(DUMM)
      U = RANN(DUMM)
      U2 = U1 + H1*U -H2
      IF (U2 .LE. 0.) GO TO 1
      IF (U2 .GE. 1.) GO TO 1
    2 W = B * (U1/U2)**4
      IF ( (C*U2-D+W+1./W) .LE. 0.) GO TO 4
    3 IF ( (C*ALOG(U2)-ALOG(W)+W-1.) .GE. 0.) GO TO 1
    4 X = A * W
      RETURN
      END
      SUBROUTINE GT (ALP,X)
```

```
C
C      CHENG AND FEAST (COMM. OF THE ACM, 1980)
C        ALPHA .GT. 0.5
C
       DATA ASET/-1./
       IF (ASET .EQ. ALP) GO TO 1
       ASET = ALP
       A = ALP - 0.5
       B = ALP / A
       C = 2.0 / A
       D = C + 2.0
       S = SQRT(ALP)
       H1 = (0.865 + 0.064/ALP) /S
       H2 = (0.4343 - 0.105/S) / S
  1    U1 = RANN(DUMM)
       U = RANN(DUMM)
       U2 = U1 + H1*U - H2
       IF (U2 .LE. 0.) GO TO 1
       IF (U2 .GE. 1) GO TO 1
  2    W = B * (U1/U2) * (U1/U2)
       IF ( (C*U2-D+W+1./W) .LE. 0.) GO TO 4
  3    IF ( (C*ALOG(U2)-ALOG(W)+W-1.) .GE. 0.) GO TO 1
  4    X = A * W
       RETURN
       END
       SUBROUTINE GKM3 (ALP,X)
C
C      CHENG AND FEAST (APPLIED STAT, 1979)
C        ALPHA .GE. 1
C
       DATA ASET/-1.0/
       DATA K/1/
       IF (ASET .EQ. ALP) GO TO 100
       ASET = ALP
       IF (ALP .GT. 2.5) K = 2
       A = ALP - 1
       B = (ALP-1./(6.*ALP)) / A
       C = 2. / A
       D = C + 2.
       GO TO (1,11),K
   1   U1 = RANN(DUMM)
       U2 = RANN(DUMM)
       GO TO 2
  11   F = SQRT(ALP)
  10   U1 = RANN(DUMM)
       U = RANN(DUMM)
       U2 = U1 + (1.0-1.86*U)/F
       IF ( (U2.LE.0.) .OR. (U2.GE.1.) ) GO TO 10
   2   W = B * U1 / U2
       IF ( (C*U2-D+W+1./W) .LE. 0.) GO TO 4
   3   IF ( (C*ALOG(U2)-ALOG(W)+W-1.) .LT. 0.) GO TO 4
 100   GO TO (1,10),K
   4   X = A * W
       RETURN
       END
       SUBROUTINE G4PE (ALPHA,X)
C
C      SCHMEISER AND LAL (JASA, 1980)
C        ALPHA .GT. 1
C        THIS CODE IS FURNISHED BY BRUCE
C        SCHMEISER, PURDUE UNIVERSITY
C
       DATA ASAVE/-1./,XLL/-1./
C
  ***  INITIALIZATION
       IF (ALPHA .EQ. ASAVE) GO TO 100
       ASAVE = ALPHA
       X1 = 0.
       X2 = 0.
       F1 = 0.
```

```
      F2 = 0.
      X3 = ALPHA - 1.
      D = SQRT(X3)
      IF (D .GE. X3) GO TO 10
      X2 = X3 - D
      X1 = X2 * (1.-1./D)
      XLL = 1. - X3/X1
      F1 = EXP(X3 * ALOG(X1/X3) + X3 - X1)
      F2 = EXP(X3 * ALOG(X2/X3) + X3 - X5)
   10 X4 = X3 + D
      X5 = X4 * (1.+1./D)
      XLR = 1.0 - X3 / X5
      F4 = EXP(X3 * ALOG(X4/X3) + X3 - X4)
      F5 = EXP(X3 * ALOG(X5/X3) + X3 - X5)
      P1 = F2 * (X3-X2)
      P2 = F4 * (X4-X3) + P1
      P3 = F1 * (X2-X1) + P2
      P4 = F5 * (X5-X4) + P3
      P5 = (1.-F2) * (X3-X2) + P4
      P6 = (1.-F4) * (X4-X3) + P5
      P7 = (F2-F1) * (X2-X1) * 0.5 + P6
      P8 = (F4-F5) * (X5-X4) * 0.5 + P7
      P9 = -F1 / XLL + P8
      P10 = F5 / XLR + P9
  100 U = RANN(DUMM) * P10
      IF (U .GT. P4) GO TO 500
      IF (U .GT. P1) GO TO 200
      X = X2 + U / F2
      GO TO 1400
  200 IF (U .GT. P2) GO TO 300
      X = X3 + (U-P1) / F4
      GO TO 1400
  300 IF (U .GT. P3) GO TO 400
      X = X1 + (U-P2) / F1
      GO TO 1400
  400 X = X4 + (U-P3) / F5
      GO TO 1400
C
C     THE TWO REGIONS USING RECTANGULAR
C       REJECTION
C
  500 W = RANN(DUMM)
      IF (U .GT. P5) GO TO 600
      X = X2 + (X3-X2) * W
      IF ( (U-P4)/(P5-P4) .LE. W) GO TO 1400
      V = F2 + (U-P4) / (X3-X2)
      GO TO 1300
  600 IF (U .GT. P6) GO TO 700
      X = X3 + (X4-X3) * W
      IF ( (P6-U)/(P6-P5) .GE. W) GO TO 1400
      V = F4 + (U-P5) / (X4-X3)
      GO TO 1300
C
C     THE TWO TRIANGULAR REGIONS
C
  700 IF (U .GT. P8) GO TO 900
      W2 = RANN(DUMM)
      IF (W2 .GT. W) W = W2
      IF (U .GT. P7) GO TO 800
      X = X1 + (X2-X1) * W
      V = F1 + 2.0 * W * (U-P6) / (X2-X1)
      IF (V .LE. F2*W) GO TO 1400
      GO TO 1300
  800 X = X5 - W * (X5-X4)
      V = F5 + 2.0 * W * (U-P7) / (X5-X4)
      GO TO 1300
C
C     THE TWO EXPONENTIAL REGIONS
C
  900 IF (U .GT. P9) GO TO 1000
```

```
      U = (P9-U) / (P9-P8)
      X = X1 - ALOG(U) / XLL
      IF (X .LE. 0.) GO TO 100
      IF (W .LT. (XLL*(X1-X)+1.)/U ) GO TO 1400
      V = W * F1 * U
      GO TO 1300
 1000 U = (P10-U) / (P10-P9)
      X = X5 - ALOG(U) / XLR
      IF (W .LT. (XLR*(X5-X)+1.)/U ) GO TO 1400
      V = W * F5 * U
C
C     PERFORM THE STANDARD REJECTION
C
 1300 IF (ALOG(V) .GT. X3*ALOG(X/X3)+X3-X ) GO TO 100
 1400 RETURN
      END
```

```
ALGORITHM GS
 0.1914799       0.2577323        0.4394203       0.1862320
 0.3361395       0.1173968         2.618247
 5.7479013E-02   0.1602321        6.3654073E-02
ALGORITHM GKM3
  1.263434        1.256181         2.291892        1.895357
  3.022139        1.239824        0.4212852
  2.180053        1.249355         1.504579
ALGORITHM GT
  1.374483        1.123100         1.694720        2.581551
 0.7679198       0.3612102         3.774790
  6.290128       0.8801019        0.6932100
ALGORITHM GBH
  1.526933        1.211195         1.821506        2.825200
 0.7626055       0.3200865         3.942728
  6.357284       0.8882249        0.7067778
ALGORITHM G4PE
 0.2928000        2.114264         1.177237       0.8554744
 0.4477500        4.674262        0.8106262
 0.3849500       0.7528353        0.8711261
```

REFERENCES

Ahrens, J.H. and Dieter, U. (1972). Computer methods for sampling from the exponential and normal distributions. Communications of the ACM, 15, 873-882.

Ahrens, J.H. and Dieter, U. (1974). Computer methods for sampling from gamma, beta, Poisson and binomial distributions. Computing, 12, 223-246.

Atkinson, A.C. (1977). An easily programmed algorithm for generating gamma random variables. Applied Statistics, 140, 232-234.

Atkinson, A.C. and Pierce, M.C. (1976). The computer generation of beta, gamma, and normal random variables. Journal of the Royal Statistical Society, 139A, 431-461.

Best, D.J. (1978). Letter to the Editor. Applied Statistics, 27, 181.

Butcher, J.C. (1961). Random sampling from the normal distribution. Computer Journal, 3, 251-253.

Butler, J.W. (1956). Machine sampling from given probability distributions. Symposium on Monte Carlo Methods, Ed. H.A. Meyer, John Wiley, New York, 249-264.

Cheng, R.C. (1977). The generation of gamma variables with non-integral shape parameter. Applied Statistics, 26, 71-75.

Cheng, R.C. and Feast, G.M. (1979). Some simple gamma variate generators. Applied Statistics, 28, 290-295.

Cheng, R.C. and Feast, G.M. (1980). Gamma variate generators with increased shape parameter range. Communications of the ACM, 23, 389-393.

Cheriyan, K.C. (1941). A bivariate correlated gamma-type distribution function. Journal of the Indian Mathematical Society, 64, 194-206.

Devroye, L. (1980). Personal communication.

Dudewicz, E.J. and Ralley, T.G. (1981). The Handbook of Random Number Generation and Testing with TESTRAND Computer Code. American Sciences Press, Inc., Box 21161, Columbus, Ohio 43221.

Fishman, G.S. (1975). Sampling from the gamma distribution on a computer, Communications of the ACM, 19, 407-409.

Greenwood, A.J. (1974). A fast generator for gamma distributed random variables. COMPSTAT (G. Bruckman, F. Ferschl, L. Schmetterer, eds.). Physica Verlag, Vienna, 19-27.

Halton, J.H. (1970). A retrospective and prospective survey of the Monte Carlo method. SIAM Review, 12, 1-63.

Johnk, M.D. (1964). Erzeugung von Betaverteilten und Gammaverteilten Zufallszahlen. Metrika, 8, 5-15.

Johnson, M.E., Tietjen, G.L. and Beckman, R.J. (1980). A new family of probability distributions with applications to Monte Carlo studies. Journal of the American Statistical Association, 75, 276-279.

Kennedy, W.J. and Gentle, J.E. (1980). Statistical Computing. Marcel Dekker, Inc., New York.

Kinderman, A.J. and Monahan, J.F. (1977). Computer generation of random variables using the ratio of uniform deviates. ACM Transactions on Mathematical Software, 3, 257-260.

Kinderman, A.J. and Monahan, J.F. (1978). Recent developments in the computer generation of student's t and gamma random variables. Proceedings of the ASA Statistical Computing Section, 90-94.

Kinderman, A.J. and Ramage, J.G. (1976). Computer generation of normal random variables. Journal of the American Statistical Association, 71, 893-896.

Kshirsagar, A.M. (1959). Bartlett decomposition and the Wishart distribution. Annals of Mathematical Statistics, 30, 239-241.

Learmonth, G.P. and Lewis, P.A.W. (1973). Naval Postgraduate School random number generator package LLRANDOM. Naval Postgraduate School technical report, Monterey, California.

Marsaglia, G. (1961). Generating exponential random variables. Annals of Mathematical Statistics, 32, 899-902.

Marsaglia, G. (1962). Random variables and computers. Transactions of the Third Prague Conference (J. Kozesnik, ed.). Czechoslovak Academy of Sciences, Prague, 499-510.

Marsaglia, G. (1977). A squeeze method for generating gamma random variables. Computers and Mathematics with Applications, 4, 321-326.

Marsaglia, G., Ananthanarayanan, K. and Paul, N. (1973). How to use the McGill random number package. McGill, Montreal, Canada.

Marsaglia, G., MacLaren, M.D., and Bray, T.A. (1964). A fast procedure for generating normal random variables. Communications of the ACM, 7, 4-10.

Mihram, G.A. and Hultquist, R.A. (1967). A bivariate warning-time/failure-time distribution. Journal of the American Statistical Association, 62, 589-599.

Payne, W.H., Rabung, J.R., and Bogyo, T.P. (1969). Coding the Lehmer pseudo random number generator. Communications of the ACM, 12, 85-86.

Schmeiser, B.W. (1978). Squeeze methods for generating gamma variates, Southern Methodist University technical report OREM 78009, Dallas, Texas.

Schmeiser, B.W. and Lal, R. (1979). Computer generation of bivariate gamma random vectors. Technical Report OREM 79009, Southern Methodist University, Dallas, Texas.

Schmeiser, B.W. and Lal, R. (1980). Squeeze methods for generating gamma variates. J. Amer. Stat. Assoc., 75, 679-682.

Tadikamalla, P.R. (1975). Modeling and Generating Stochastic Inputs for Simulation Studies. Unpublished Ph.D. dissertation, The University of Iowa, Iowa City, Iowa.

Tadikamalla, P.R. (1978a). Computer generation of gamma random variables. Communications of the ACM, 21, 419-422.

Tadikamalla, P.R. (1978b). Computer generation of gamma random variables--II. Communications of the ACM, 21, 925-929.

Tadikamalla, P.R. and Johnson, M.E. (1978). A survey of computer methods for sampling from the gamma distribution. Proceedings of the Winter Simulation Conference, 1, 131-134.

Vaduva, I. (1977). On computer generation of gamma random variables by rejection and composition techniques. Proceedings of the Fifth Conference on Probability Theory, Bucharest, 131-142.

Wallace, N.D. (1974). Computer generation of gamma random variables with non-integral shape parameters. Communications of the ACM, 17, 691-695.

Whittekar, J. (1974). Generating gamma and beta random variables with non-integral shape parameters. Applied Statistics, 23, 210-213.

Received 11/21/80; Revised 7/27/81.

AMERICAN JOURNAL OF MATHEMATICAL AND MANAGEMENT SCIENCES

CORRECTION NOTE

Tadikamalla, P.R. and Johnson, M.E. (1981). A complete guide to gamma variate generation. American Journal of Mathematical and Management Sciences, 1, 213-236.

On page 231 line 18 should read as follows:

```
F2 = EXP(X3 * ALOG(X2/X3) + X3 - X2)
```

The timings reported in the paper were based on a correct version of G4PE.

1982, VOL. 2, NO. 1, 93
0196-6324/82/010093-01 $2.65

AMERICAN JOURNAL OF MATHEMATICAL AND MANAGEMENT SCIENCES

GENERATION OF CONTINUOUS MULTIVARIATE DISTRIBUTIONS FOR STATISTICAL APPLICATIONS

Mark E. Johnson, Chiang Wang and John S. Ramberg
Systems and Industrial Engineering Department
The University of Arizona
Tucson, Arizona 85721

SYNOPTIC ABSTRACT

Two general and several specific schemes are described for generating variates from continuous multivariate distributions. Algorithms are provided for the multivariate normal, Johnson system, Cauchy, elliptically contoured (including Pearson Types II and VII), Morgenstern, Plackett, Ali, Gumbel, Burr (and related), Beta-Stacy and Khintchine distributions. Issues in designing multivariate Monte Carlo studies are discussed.

Key Words and Phrases: Cauchy, elliptical distributions, Morgenstern, Plackett, Ali, Gumbel, Burr, Beta-Stacy, Khintchine.

1984, VOL. 4, NOS. 3 & 4, 225-248
0196-6324/84/030225-24 $7.80

1. INTRODUCTION.

Many multivariate procedures suffer from a lack of empirical testing outside the realm in which they were originally designed -- frequently the multivariate normal distribution context. This deficiency is due not so much to negligence but rather to a lack of suitable and available distributions for Monte Carlo studies. Very little attention has been devoted to the generation of multivariate random vectors, so it seems that every investigator who decides to conduct empirical tests must effectively start from scratch. The purpose of this paper is to ease this burden by providing generation algorithms for a variety of distributions in multivariate Monte Carlo studies. This treatment is set initially in a general context so that algorithms for multivariate distributions not covered in this paper specifically may be developed along similar lines.

Aside from the issue of what multivariate distributions can be generated, a more pertinent question relates to what distributions should be generated. The answer to this question depends heavily on the particular multivariate procedure under investigation. Thus, at most only general guidelines can be given. A brief discussion of two research areas may serve to identify for the reader some possible uses of the distributions described in this paper.

Tests for multivariate normality provide a mechanism for assessing the appropriateness of this common distributional assumption. For associated references and specific tests, see Malkovich and Afifi (1973) and Hawkins (1981). The better tests have not been extensively scrutinized with respect to power as a function of (small) sample size, dimension, and (of course) the non-normality. The results given in this paper should provide a convenient starting point for pursuing such investigations.

Intuitively appealing classification rules are readily derived in discriminant analysis applications (Beckman and

Johnson (1981)). Detailed study of new classification rules invariably necessitates a Monte Carlo study, in which case the algorithms given herein can be exploited. In the classification rule context the multivariate normal distribution need not be the baseline case for comparison. Simulation studies with multivariate distributions in discriminant analysis have been conducted, although additional areas warrant attention (Johnson, Ramberg and Wang (1982)).

These two examples suggest the types of studies we have in mind in terms of the distributions to be covered below. For some investigations, our coverage of distributions may be inadequate. In such situations, the general methods described in Section 2 may be used to develop specific algorithms. In Section 3 algorithms are given for the multivariate normal, Johnson system, Cauchy, elliptically contoured (including Pearson Types II and VII), Morgenstern, Plackett, Ali, Gumbel, Burr (and related), Beta-Stacy and Khintchine distributions. A limited comparison of some of these distributions, which can have normal marginal distributions, is also provided by means of contour plots. More detailed comparisons of these distributions will be available in a forthcoming monograph (Johnson (1986)). The issue of distribution selection in Monte Carlo studies is addressed in Section 4.

2. GENERAL METHODS.

To establish a framework for the specific generation algorithms (given in Section 3 below), a discussion of two general methods, the transformation and conditional distribution techniques, will now be given. Strictly speaking, the conditional distribution method is a special case of transformation, since it involves a function of certain univariate random variables. However, because of its many successful applications (illustrated in Section 3) this special designation seems appropriate. On the other hand, "acceptance-rejection" (AR) methods, which are extensively used in univariate generation, are not examined here. Although in theory the AR method applies in multivariate situations, practical difficulties have stifled its use. In the parlance of AR methodology, these difficulties include a lack of suitable dominating functions, complications in optimizing the choice of parameters in the dominating function, and low efficiencies. The conditional distribution and transformation methods will now be described.

The Conditional Distribution Method reduces the multivariate generation problem to a sequence of univariate generation problems. Let $\underline{X} = (X_1,\ldots,X_p)'$ be a p-dimensional random vector distributed according to probability density function (pdf) $f(\underline{x})$. Rosenblatt (1952) and many subsequent authors suggest the following conditional distribution method for generating a random vector $\underline{X}$ having density f:

1. Generate $X_1 = x_1$ from the marginal distribution of X_1.
2. Generate $X_2 = x_2$ from the conditional distribution of X_2 given $X_1 = x_1$.
3. Generate $X_3 = x_3$ from the conditional distribution of X_3 given $X_1 = x_1$ and $X_2 = x_2$.

.
.
.

p. Generate $X_p = x_p$ from the conditional distribution of X_p given $X_1 = x_1$, $X_2 = x_2,\ldots, X_{p-1} = x_{p-1}$.

Therefore, two major tasks must be performed to generate random vectors using the conditional distribution method. First, the marginal distribution of X_1 and each of the above conditional univariate distributions must be derived. Second, a generation procedure for each of the univariate distributions must be determined. The efficiency of this method for a given multivariate distribution depends on how the univariate distributions needed can be generated. This method can be considered as a "transformation" method in the sense that the multivariate distribution is obtained by a transformation of specified univariate distributions. Eight examples of this approach are illustrated in Section 3.

The Transformation Method. The basic concept of the transformation method is as follows. Let $\underline{X} = (X_1, X_2, \ldots, X_p)'$ be a p-dimensional random vector to be generated having a certain multivariate distribution. Also, let $\underline{Y} = (Y_1, Y_2, \ldots, Y_q)'$ be a q-dimensional ($q > p$) random vector having a different distribution than $\underline{X}$. If there exists a function $g(\underline{Y}) = (g_1(\underline{Y}), \ldots, g_p(\underline{Y}))'$ such that $g(\underline{Y})$ has the same distribution as $\underline{X}$, then the random vector $\underline{X}$ can be generated directly by first generating the random vector $\underline{Y}$ and then evaluating $g(\underline{Y})$.

The transformation method is most attractive when the specific transformation is already "known." For a variety of distributions, this is the case (see Section 3). Given an arbitrary multivariate density $f(\underline{x})$, it is rarely obvious what transformations of easy-to-generate random vectors will lead to this distribution. The following guidelines may facilitate the effort:

1. Conduct a thorough literature search for an appropriate construction scheme in one of the references to the distribution.
2. Attempt (invertible) transformations to $\underline{X}$. Is it possible to obtain a recognizable result? Start with component transformations such as the probability integral transformation.
3. If 1 and 2 are unsuccessful, look again at the conditional distributions.

3. EXAMPLES.

This section describes specific generation algorithms for a variety of bivariate and multivariate distributions using the conditional distribution and transformation methods of Section 2. More detailed treatments of the distributions themselves can be found in the cited references. Johnson (1986) provides a unified treatment for Monte Carlo applications.

Multivariate Normal Distribution. The use of the conditional distribution and the transformation approaches for generating the multivariate normal distribution has been available for some time (Scheuer and Stoller (1962)). Let $\underline{X} \sim N_p(\underline{\mu},\Sigma)$ denote a p-dimensional multivariate normal random vector with mean vector $\underline{\mu}$ and covariance matrix Σ. Let L be the lower triangular (Choleski) decomposition of Σ, that is a matrix such that $LL' = \Sigma$. Given p-independent univariate standard normals, $\underline{Y}' = (Y_1,Y_2,\ldots,Y_p)$, transform them via $\underline{X} = L\,\underline{Y} + \underline{\mu}$ to achieve the desired $N_p(\underline{\mu},\Sigma)$ distribution. Simple formulas for determining the components of L are available in Kennedy and Gentle (1980, pp.294-296). The computer package LINPACK (Dongarra, Moler, Bunch and Stewart (1979)) also contains the subroutine SCHDC which can compute L.

In many situations, consideration of 2 or 3 dimensions will suffice. Here, the entries of L can be given explicity. Suppose the covariance matrix of $\underline{X}$ is given by

$$\Sigma = \begin{bmatrix} \sigma_1^2 & \rho_{12}\sigma_1\sigma_2 & \rho_{13}\sigma_1\sigma_3 \\ \rho_{12}\sigma_1\sigma_2 & \sigma_2^2 & \rho_{23}\sigma_2\sigma_3 \\ \rho_{13}\sigma_1\sigma_3 & \rho_{23}\sigma_2\sigma_3 & \sigma_3^2 \end{bmatrix}.$$

The Choleski decomposition of Σ is

$$L = \begin{bmatrix} \sigma_1 & 0 & 0 \\ \sigma_2\rho_{12} & \sigma_2\sqrt{(1-\rho_{12}^2)} & 0 \\ \sigma_3\rho_{13} & \dfrac{\sigma_3(\rho_{23}-\rho_{12}\rho_{13})}{\sqrt{(1-\rho_{12}^2)}} & \dfrac{\sigma_3\sqrt{[(1-\rho_{12}^2)(1-\rho_{13}^2)-(\rho_{23}-\rho_{12}\rho_{13})^2]}}{\sqrt{(1-\rho_{12}^2)}} \end{bmatrix}$$

Note that the submatrix obtained by deleting the third row and column of L is the Choleski decomposition of the covariance matrix for the first two components of $\underline{X}$. Univariate normal generators are widely available (two methods are given below when we consider elliptically contoured distributions).

Johnson Translation System. Johnson's (1949) system of distributions is easy to generate on a computer. Given a multivariate normal random vector $\underline{X}' = (X_1,\ldots,X_p)$, simply apply one of the following transformations to each component

lognormal: $Y_i = \exp(X_i)$

$\sinh^{-1}$-normal: $Y_i = [\exp(X_i) - \exp(-X_i)]/2$

logit-normal: $Y_i = [1 + \exp(X_i)]^{-1}$.

This scheme is so simple that frequently researchers have adapted it (especially in discriminant analysis Monte Carlo studies) without giving its implications much thought. In particular, the moment structure of $\underline{Y}$ will be considerably different from the moments of $\underline{X}$. Johnson, Ramberg and Wang (1982) derive some results for controlling such characteristics with the Johnson system and redo parts of previous simulation studies which had confounded the effects of non-normality and moment structure.

Multivariate Cauchy. Consider a multivariate Cauchy distribution having density (Johnson and Kotz (1972), p. 294)

$$f(\underline{x}) = \pi^{-(p+1)/2}\Gamma[(p+1)/2]\ [1+\underline{x}'\underline{x}]^{-(p+1)/2} . \qquad (1)$$

Variates having density (1) are readily generated using the conditional distribution approach since: the distribution of X_1 is Cauchy and can be obtained as $X_1 = \tan(\pi(U-0.5))$ where U is uniform 0-1 (see Dudewicz and Ralley, 1981); and the conditional distribution of $[m^{1/2}(1 + \sum_{j=1}^{m-1} X_j^2)^{-1/2}]\cdot X_m$ given $X_1 = x_1$, $X_2 = x_2,\ldots, X_{m-1} = x_{m-1}$ is univariate Student's t with m degrees of freedom. The $t_{(m)}$ distribution can be readily sampled from $Y/\sqrt{(Z/m)}$ where Y is standard normal and Z is chi-square with m degrees of freedom (see Tadikamalla and Johnson (1981)) or by the "ratio-of-uniforms" technique (Kinderman, Monahan and Ramage (1977)).

The distribution as given in (1) is a multivariate t with 1 degree of freedom and can be alternately generated, as follows: Generate $\underline{Y} \sim N_p(\underline{0}, I)$, where I is the pxp identity matrix, and take $\underline{X} = (Y_1/Z, Y_2/Z, \ldots Y_p/Z)'$ where Z is an independent $\chi^2_{(1)}$ variate. More generally, a multivariate t with k degrees of freedom is obtained if Z is $\chi^2_{(k)}$. These multivariate Cauchy and t distributions are special cases of elliptically contoured distribution for which some convenient methods of generation are given next.

Elliptically Contoured Distributions. An interesting and useful class of multivariate distributions, including the multivariate normal as a special case, is the set of elliptically contoured distributions. A valuable mathematical treatment of these distributions is given by Cambanis, Huang and Simons (1981). Before considering their approach, we set the stage by describing two simple methods for generating independent normal variates; the underlying principle of these two techniques is a precursor to the representation of the elliptically-contoured distributions given by Cambanis, Huang and Simons.

The well-known Box-Muller (1958) method for generating two independent standard normals X_1 and X_2 from two independent uniform 0-1's U_1 and U_2 sets

$$X_1 = (-2 \ln U_1)^{1/2} \cos(2\pi U_2)$$
$$X_2 = (-2 \ln U_1)^{1/2} \sin(2\pi U_2). \tag{2}$$

This method can be verified using standard change-of-variable arguments (e.g., Hogg and Craig (1978), p. 141). As an informal argument note that the point $(\cos 2\pi U_2, \sin 2\pi U_2)$ is uniformly distributed on the boundary of the unit circle. Also, if X_1 and X_2 are independent standard normals, $X_1^2 + X_2^2$ is distributed $\chi^2_{(2)}$. From (2)

$$X_1^2 + X_2^2 = (-2 \ln U_1)\cos^2(2\pi U_2)+(-2 \ln U_1)\sin^2(2\pi U_2)$$
$$= -2 \ln U_1,$$

which is distributed $\chi^2_{(2)}$ (an exponential with scale parameter 2).

A modification of (2) can be used to avoid trigonometric evaluations. This technique (Marsaglia and Bray (1964)) sets

$$X_1 = \left(\frac{-2 \ln W}{W}\right)^{1/2} U_1$$
$$X_2 = \left(\frac{-2 \ln W}{W}\right)^{1/2} U_2,$$

where U_1 and U_2 are independent uniform 0-1's constrained by $W \equiv U_1^2 + U_2^2 \leq 1$. This method is quite similar to the Box-Muller transformation. A simple change-of-variable argument shows that W is distributed uniform 0-1, and hence $-2 \ln W$ is $\chi^2_{(2)}$ as above. Also, $(U_1/W^{1/2}, U_2/W^{1/2})$ is the projection of the point (U_1,U_2) from the interior of the unit circle to the boundary of the unit circle. Thus, both methods (eventually) sample uniformly from the boundary of the unit circle and then adjust the generated point's distance from the origin according to the square root of a $\chi^2_{(2)}$ distribution.

More generally, Cambanis, Huang and Simons gave a useful representation of p-dimensional elliptically contoured distributions using some fundamental results of Schoenberg (1938). For notation, let L be a p x p lower triangular factorization of Σ, $\underline{U}^{(p)}$ denote a random vector with a uniform distribution on the surface of the p-dimensional unit hypersphere, and R be a positive random variable independent of $\underline{U}^{(p)}$. A p-dimensional random vector $\underline{X}$ is elliptically contoured with scaling matrix Σ and location vector $\underline{\mu}$ if

$$\underline{X} = R \cdot L \cdot \underline{U}^{(p)} + \underline{\mu}. \quad (3)$$

The random variable R^2 has the distribution of $(\underline{X}-\underline{\mu})'\Sigma^{-1}(\underline{X}-\underline{\mu})$. In some cases of practical interest, R^2 has a recognizable, easy-to-generate distribution. These cases include the Pearson Type II and Pearson Type VII multivariate distributions.

The generation of $\underline{U}^{(p)}$ is straightforward. First, generate $Z_1, Z_2, \ldots, Z_p$ independent standard normals and let $Z_0 = (Z_1^2 + Z_2^2 + \ldots + Z_p^2)^{1/2}$. Set the ith component of $\underline{U}^{(p)}$ to be Z_i/Z_0, $i = 1,2,\ldots,p$. (This method is generally credited to Muller (1959)). Alternate methods for generating $\underline{U}^{(p)}$ have been surveyed recently by Tashiro (1977).

The Pearson Type II distribution has density function

$$f(\underline{x}) = \frac{\Gamma(p/2 + m+1)}{\Gamma(m+1)\pi^{p/2}} |\Sigma|^{-1/2} [1-(\underline{x}-\underline{\mu})'\Sigma^{-1}(\underline{x}-\underline{\mu})]^m \tag{4}$$

where the support of the distribution is restricted to the region $(\underline{x}-\underline{\mu})'\Sigma^{-1}(\underline{x}-\underline{\mu}) \leq 1$; $\underline{\mu}$ is a location vector; Σ is the scaling matrix; the additional parameter $m > -1$; and, p is the dimension. Variate generation is quite easy since the distribution of the quadratic form $R^2 = (\underline{X} - \underline{\mu})'\Sigma^{-1}(\underline{X} - \underline{\mu})$ is beta with parameters $p/2$ and $m+1$. (Beta generation is described later in the discussion of the Beta-Stacy distribution.)

The Pearson Type VII distribution has density function:

$$f(\underline{x}) = \frac{\Gamma(m)}{\Gamma(m-p/2)\ \pi^{p/2}} |\Sigma|^{-1/2} [1+(\underline{x}-\underline{\mu})'\Sigma^{-1}(\underline{x}-\underline{\mu})]^{-m} , \tag{5}$$

where $\underline{\mu}$ is a location vector; Σ is the scaling matrix; and, m is an additional parameter restricted by $m > p/2$ with p the dimension. Variate generation is slightly less apparent here than for the Pearson Type II. The distribution of $Z=R^2$ corresponding to (5) has density function

$$f(z) = \frac{\Gamma(m)}{\Gamma(p/2)\ \Gamma(m-p/2)} z^{p/2-1} (1+z)^{-m}, \ z>0 . \tag{6}$$

This density is that of a Pearson Type VI distribution, and can be generated via $Z = Y/(1-Y)$, where Y is beta with parameters $p/2$ and $m-p/2$. If $m = (p+1)/2$, Σ is an identity matrix and $\underline{\mu}=\underline{0}$, then the density in (5) reduces to the multivariate Cauchy in (1).

In the bivariate case ($p=2$), explicit generation formulas for the Pearson II and VII distributions can be given (Johnson and Ramberg (1977)). Let U_1 and U_2 be independent uniform 0-1 variates. Then

Pearson II: $X_1 = (1-U_1^{1/(m+1)})^{1/2} \cos(2\pi U_2)$

$X_2 = (1-U_1^{1/(m+1)})^{1/2} \sin(2\pi U_2)$

Pearson VII: $X_1 = (U_1^{\frac{1}{1-m}}-1)^{1/2} \cos(2\pi U_2)$

$X_2 = (U_1^{\frac{1}{1-m}}-1)^{1/2} \sin(2\pi U_2)$

Cauchy: $X_1 = (U_1^{-2}-1)^{1/2} \cos(2\pi U_2)$

$X_2 = (U_1^{-2}-1)^{1/2} \sin(2\pi U_2)$

The resulting distributions have location vectors at the origin and identity matrix as scaling matrices.

These schemes for generating elliptically contoured distributions are easy to implement and have been used by Chmielewski (1981a) in a study of tests for normality. A guide to references on these distributions has also been published by Chmielewski (1981b).

Morgenstern's Bivariate Uniform Distribution. This distribution, which could be credited to Eyraud (1936), has the simple density function

$$f(x_1,x_2) = 1 + \alpha(2x_1-1)(2x_2-1), \quad 0\leq x_1,x_2\leq 1, \quad -1\leq\alpha\leq 1. \tag{7}$$

Each of the components of (X_1,X_2) has uniform 0-1 distribution so that the functional form of the density of $X_2|X_1 = x_1$ is also given by (7). Since this function is linear in x_2, the corresponding distribution function is quadratic in x_2. Setting the conditional distribution function equal to a generated uniform 0-1, say U_2, and solving for X_2, is the basis of the following algorithm.

1. Generate U_1, U_2 independent uniform 0-1 and set $X_1 = U_1$.
2. Compute:

$$A = \alpha(2U_1-1)-1$$
$$B = [1-2\alpha(2U_1-1) + \alpha^2(2U_1-1)^2 + 4\alpha U_2(2U_1-1)]^{1/2}$$

3. Set $X_2 = 2U_2/(B-A)$.

Multivariate extensions of (7) have been explored, although this distribution can model only weak dependencies. Johnson (1982) cites some relevant papers and describes the generation of multivariate forms of (7). The recent work of Cambanis (1977) may also be of interest.

Plackett's Bivariate Uniform Distribution. Plackett's (1965) distribution is given by the density function

$$f(x_1,x_2) = \frac{\alpha(\alpha+1)(x_1+x_2-2x_1x_2) + 1}{[\{1+(\alpha-1)(x_1+x_2)\}^2 - 4\alpha(\alpha-1)x_1x_2]^{3/2}} \tag{8}$$

$$\alpha > 0,\ 0 < x_1,\ x_2 < 1.$$

The components X_1 and X_2 are marginally uniform 0-1. For a detailed description of distribution (8), see Mardia (1970); for a recent application, see Wahrendorf (1980). Mardia (1967) developed the conditional distribution approach for sampling from (8) which is summarized, as follows:

1. Generate U_1, U_2 independent uniform 0-1 variates.
2. Set $X_1 = U_1$.
3. Set $X_2 = [B-(1-2U_2)C]/2A$, where

$$A = \alpha + (\alpha-1)^2\, U_2\, (1-U_2)$$

$$B = 2\, U_2(1-U_2)(\alpha^2 U_1-U_1+1) - 2\alpha U_2(1-U_2) + \alpha$$

$$C = [\alpha^2 + 4\alpha(1-\alpha)^2 U_1U_2(1-U_1)(1-U_2)]^{1/2}.$$

Then the pair (X_1,X_2) is distributed according to (8).

Ali's Bivariate Distribution. A recently developed distribution due to Ali, Mikhail and Haq (1978) can be easily generated via the conditional distribution approach. A simple representation of this bivariate distribution is given by its distribution function

$$F(u_1,u_2) = u_1u_2/[1-\alpha(1-u_1)(1-u_2)],\quad 0\leq u_1,u_2\leq 1,\ -1\leq\alpha\leq 1. \tag{9}$$

The pair (U_1,U_2) with this distribution function has uniform 0-1 marginal distributions. The equation $F(u_2|U_1=u_1) = p$, $0 < p < 1$, is quadratic in u_2 which leads to the following straightforward algorithm for sampling from (9):

1. Generate independent uniform 0-1 variates V_1, V_2, and set $U_1 = V_1$.
2. Compute
$$b = 1 - V_1$$
$$A = \alpha(2bV_2 + 1) + 2\alpha^2 b^2 V_2 + 1$$
$$B = \alpha^2(4b^2V_2 - 4bV_2 + 1) + \alpha^2(4V_2 - 4bV_2 - 2) + 1$$
3. Set $U_2 = 2V_2(\alpha b - 1)^2/(A + \sqrt{B})$.

Gumbel's Bivariate Exponential Distribution. The following bivariate density is due to Gumbel (1960):

$$f(x_1, x_2) = [(1+\theta x_1)(1+\theta x_2) - \theta] e^{-x_1 - x_2 - \theta x_1 x_2},$$
$$x_1, x_2 > 0, \ 0 \leqslant \theta \leqslant 1. \tag{10}$$

The pair (X_1, X_2) distributed as (10) has standard exponential components. The conditional distribution of X_2 given $X_1 = x_1$ has density

$$f(x_2 | X_1 = x_1) = e^{-x_2(1+\theta x_1)} [(1+\theta x_1)(1+\theta x_2) - \theta],$$

which can be rewritten as

$$f(x_2 | X_1 = x_1) = p\beta e^{-\beta x_2} + (1-p)\beta^2 x_2 e^{-\beta x_2}, \tag{11}$$

where $\beta = 1 + \theta x_1$ and $p = (\beta - \theta)/\beta$. The form in (11) admits an immediate probabilistic interpretation: a variate with density (11) arises with probability p as exponential with mean $1/\beta$, and with probability 1-p the variate is the sum of two independent exponentials each with mean $1/\beta$. Generation is therefore immediate since these exponentials can be obtained as $-\beta^{-1} \ell n(U)$, where U is uniform 0-1.

Multivariate Burr (and Related) Distributions. Takahasi (1965) describes the multivariate Burr distribution having the density function

$$f(\underline{x}) = \frac{\Gamma(k+p)}{\Gamma(k)} \left(1 + \sum_{i=1}^{p} \alpha_i x_i^{c_i}\right)^{-(k+p)} \prod_{i=1}^{p} (\alpha_i c_i x_i^{c_i - 1}), \tag{12}$$

$$x_i, \alpha_i, c_i > 0, \; i=1,2,\ldots,p.$$

The marginal distribution functions are

$$F(x_i) = 1 - [1+(\alpha_i x_i)^{c_i}]^{-k}, \; i = 1,2,\ldots,p, \tag{13}$$

which corresponds to Burr's (1942) distribution. This distribution can also be viewed as a power of an $F_{2,k}$ variate. The form in (13) is ideally suited for variate generation -- the equation $F(x_i) = U$ is readily solved for x_i, where U is uniform 0-1. Also, the conditional distributions in (12) are themselves distributed as scaled Burr variates. Thus, a scheme based on the conditional distribution approach for sampling from (12) could be developed (see Johnson and Kotz (1972), p. 288-290 for details).

Following Cook and Johnson (1981), a direct transformation approach will be described. The distribution in (12) can be constructed via $X_i = (\alpha_i^{-1} Y_i/Z)^{1/c_i}$, $i=1,2,\ldots,p$, where $Y_1, Y_2, \ldots, Y_p$ are independent standard exponentials and Z is an independent gamma variate with shape parameter k and unit scale parameter. Similarly, using the same Y_i's and Z, a <u>multivariate Pareto</u> (Johnson and Kotz (1972), p. 286) and a <u>multivariate logistic</u> (Johnson and Kotz (1972), p. 291) can be constructed:

Pareto: $X_i = \theta_i(1+Y_i/Z)$

logistic: $X_i = -\ell n(Y_i/Z)$

It should be evident that the multivariate Burr, Pareto and logistic distributions are intrinsically related. Cook and Johnson (1981) identified the standard form density

$$f(u_1,u_2,\ldots,u_n) = \frac{\Gamma(\alpha+n)}{\Gamma(\alpha)\alpha^n} \prod_{i=1}^{n} u_i^{-(1/\alpha)-1} \left[\sum_{i=1}^{n} u_i^{-1/\alpha} - n + 1\right]^{-(\alpha+n)},$$

$$0 < u_i \leq 1, \; i=1,2,\ldots,n,$$

which has uniform 0-1 marginals. The three special cases are obtained by applying the appropriate inverse distribution function transformation to each component.

<u>Beta-Stacy Bivariate Distribution.</u> Mihram and Hultquist (1967) developed a distribution whose actual construction follows

the conditional distribution approach. The first component X_1 has Stacy's (1962) generalized gamma distribution (a gamma $(\alpha,1)$ variate raised to the power $1/c$). The conditional distribution of X_2 given $X_1 = x_1$ is then specified as beta with parameters θ_1 and θ_2 on the interval $(0,x_1)$. Thus, both X_1 and $X_2|X_1 = x_1$ can be generated by use of a gamma generator, as follows:

1. Generate Y_1 having a $\Gamma(\alpha,1)$ distribution.
2. Set $X_1 = Y_1^{1/c}$.
3. Generate Z_i having a $\Gamma(\theta_i,1)$ distribution, $i = 1, 2$.
4. Set $X_2 = X_1[Z_1/(Z_1+Z_2)]$.

The resulting (unconditional) joint distribution of (X_1,X_2) has density

$$f(x_1,x_2) = \frac{c\Gamma(\theta_1+\theta_2)x_1^{\alpha c-\theta_1-\theta_2}x_2^{\theta_1-1}(x_1-x_2)^{\theta_2-1}}{\Gamma(\alpha)\Gamma(\theta_1)\Gamma(\theta_2)a^{\alpha c}}\exp[-(x_1/a)^c]$$

$$0 < x_2 < x_1 \quad .$$

<u>Khintchine-Normal Distribution</u>. Bryson and Johnson (1982) derived a variety of multivariate distributions by appealing to a theorem of Khintchine (1938) on the unimodality of continuous univariate distributions. One form of potential value to Monte Carlo work has density function

$$f(x_1,x_2) = \frac{\alpha x_1x_2}{2A}\phi(A) + \frac{1-\alpha x_1x_2}{2}\Phi(A),$$

where $A = \max\{|x_1|,|x_2|\}$, $\phi(x) = (2\pi)^{-1/2}\exp(-x^2/2)$, and $\Phi(x) = \int_{-\infty}^{x}\phi(t)dt$.

This distribution has standard normal marginals and is readily generated by the algorithm:

1. Generate (U_1, U_2) having Morgenstern's bivariate uniform distribution with dependence parameter α (Equation (7)).
2. Generate Y having a $\chi^2_{(3)}$ distribution.
3. Set

$$X_1 = Y^{1/2}(2U_1-1)$$
$$X_2 = Y^{1/2}(2U_2-1).$$

Other Multivariate Distributions can be generated by transformations of (easy-to-generate) variates. The Wishart distribution has received some attention (Gleser (1976), Johnson and Hegemann (1974), Odell and Feiveson (1966)), although its use is really limited to normal theory settings. Johnson and Tenenbein (1981) developed some distributions based on linear transformations of independent exponential, Laplace and uniform variates. Multivariate gamma distributions have been studied by number of authors. Schmeiser and Lal (1982) discuss much of the previous work and present a new approach. Schmeiser and Lal (1980) also give a survey of multivariate modeling in simulation with special reference to the field of quality control. Multivariate beta and Dirichlet distributions can be generated by appropriate transformations of independent gamma variates. There seems to be no limit to the types of distributions that can be constructed or generated. An important consideration and one which is frequently overlooked is the appropriateness of particular distributions in simulation studies. The topic of selecting multivariate distributions for inclusion in a simulation study is examined in Section 4.

Comparisons in the Normal Marginals Case. Eight of the bivariate forms given above are compared in the setting in which they have standard normal marginal distributions. In general, this is accomplished by considering a bivariate distribution having density $f(x_1,x_2)$ and component distribution functions F_1 and F_2, which are non-normal. Starting with (X_1,X_2) distributed according to f, the pair (Y_1,Y_2) defined by

$$Y_1 = \Phi^{-1}(F_1(X_1))$$
$$Y_2 = \Phi^{-1}(F_2(X_2))$$

has standard normal marginals, where

$$\Phi(t) = \int_{-\infty}^{t} \frac{1}{\sqrt{2\pi}} \exp(-x^2/2)\, dx.$$

The densities of several such (Y_1,Y_2) variates were derived where the starting distributions were taken from Section 3. These distributions and their abbreviations include the bivariate normal (BVN), Burr-Pareto-logistic (B-P-L), Ali-Mikhail-Haq (A-M-H), Khintchine (Хинчин), Morgenstern, Plackett, Gumbel and Cauchy. Figure 1 provides contours for these densities, some for more than one parameter value. The contours in Figure 1 are not labeled but correspond to .03 increments in f. Obviously, a diverse set of shapes were obtained and some interesting interrelationships were revealed. The cases BVN, Morgenstern, Plackett and Gumbel all look roughly the same, although among these, Gumbel's contours demonstrate some slight asymmetry. The B-P-L and A-M-H cases are similar in general and identical for $\alpha = 1$. Other than that parameter value, the distributions are however different. The Khintchine case is unique in that it reveals a sharp edge along the positive diagonal for $\alpha = 1$ or .5 and along both diagonals for $\alpha = 0$. In fact, for the $\alpha = 0$ case, the components are uncorrelated but not independent. Also, in the Cauchy example, the standard normal components are also uncorrelated but not independent. Figure 1 provides an unusual albeit limited first look at comparison plots. Other intriguing comparison plots, and a more detailed discussion of distributional properties, can be found in Johnson (1986).

4. DESIGN OF SIMULATION EXPERIMENTS.

The purpose of this paper has been to provide generation algorithms for a variety of multivariate distributions for possible use in statistical Monte Carlo studies. We must emphasize the word "possible" since the thoroughness of an investigation should not be measured by the number of distributions included, but rather by the extent to which the original motivating questions have been addressed. Thus, selection of distributions is critical in the initial stages of a study if the final analysis is to avoid being a hodge podge of tabled values. This section

Figure 1. Constant–Density Contours of Various Bivariate Distributions for which Each Marginal Distribution is Standard Normal.

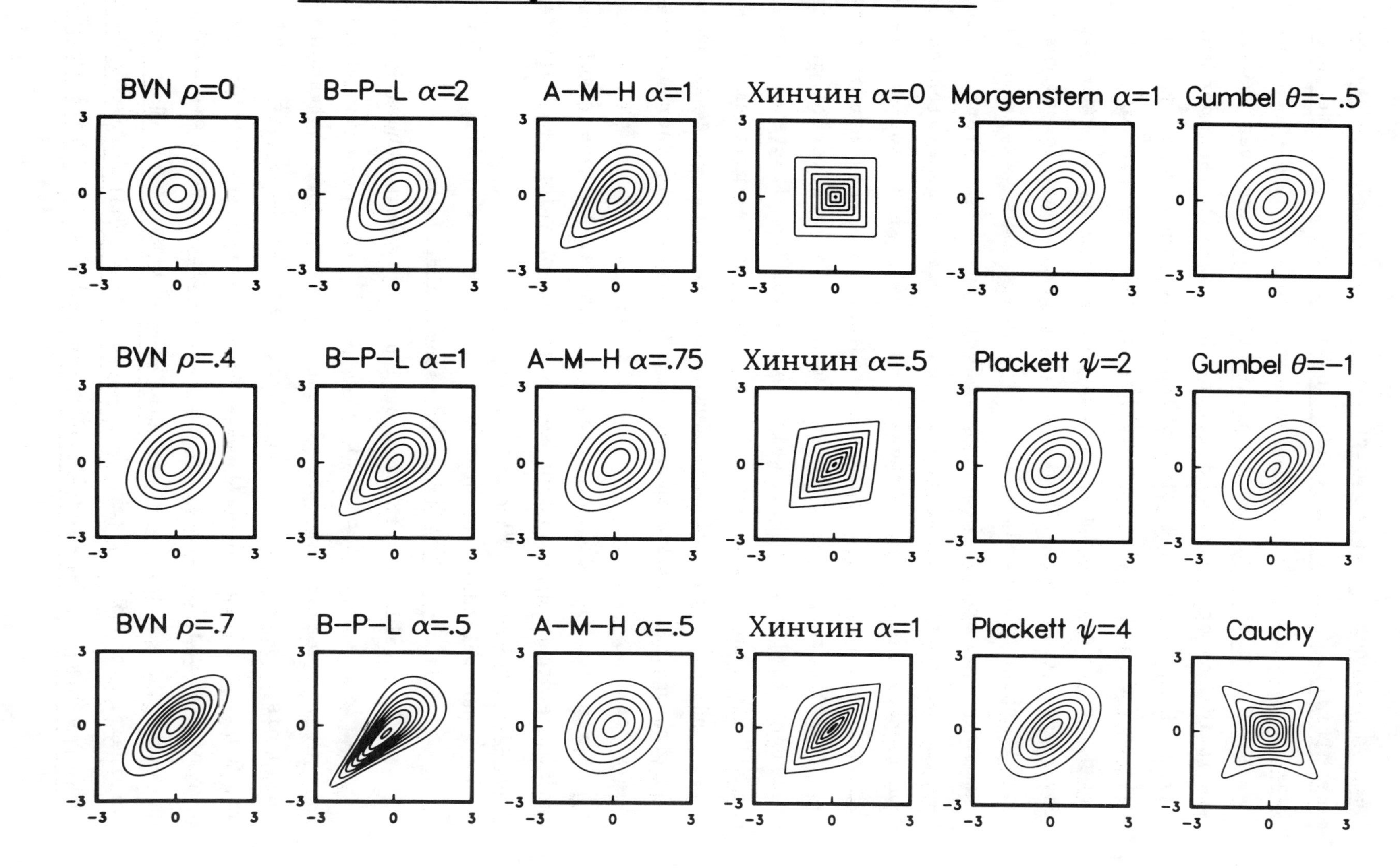

presents a number of issues which must be considered to avoid the many pitfalls inherent in conducting a multivariate Monte Carlo study.

The first step in selecting distributions involves the new (statistical or other) procedure itself. Are there important factors besides distribution which if overlooked or ignored, could become potential confounders? Typically these factors will include sample size, dimension, location (mean vectors) and scale (covariance structure). Sample size is certainly a relevant concern if the new procedure is valid only asymptotically. If multiple groups are involved then effects due to imbalance and individual group sample size may also need to be evaluated. Covariance structure across these groups has been a troublesome concern in discriminant analysis studies. Johnson, Ramberg and Wang (1982) noted several earlier simulation studies of discriminant analysis methods which confounded the effects of non-normality and unequal covariance matrices. Another problem with these studies is that the dimensions considered were large (4 and 10), although the results in two dimensions were not thoroughly understood.

Having identified other relevant factors, an investigator can now return to distribution selection. Two typical motivating questions are the following:

1. How well does the proposed procedure work locally -- for small departures from the assumed model?
2. How well does the proposed procedure work generally -- over a broad set of circumstances?

In many investigations, the assumed model is the multivariate normal distribution. Addressing these questions in this context is facilitated by the distributions in section 3. For question 1, some of the elliptically symmetric distributions can be employed, as well as the Burr-Pareto-logistic distribution with normal marginals. For more extreme departures, the Johnson translation system can be helpful, with covariance structures specified if necessary. A study related to question 2 was reported by Beckman and Johnson (1981), who examined a ranking procedure in discriminant analysis. They included bivariate normal, lognormal and Pearson VII (m = 5/2) distributions and examined the effects of sample size, mean location and covariance structure.

It is certainly difficult to give more than broad guidelines for distribution selection. Serious scrutiny of a proposed method is required to determine the appropriate cases. However, this effort can be rewarded in the long run. A carefully designed and implemented multivariate Monte Carlo study offers the following advantages:

1. Attention can be focused on the results of the study and not its methodology.
2. Confounding is avoided.
3. A framework for comparison with future methods is established.
4. The original questions that motivated the study are addressed.

ACKNOWLEDGMENTS.

We are grateful to the referee for comments leading specifically to fine tuning of the Morgenstern and Ali-Mikhail-Haq algorithms and generally to improvements in the presentation. This paper was written while Dr. Johnson was on sabbatical leave from the Statistics Group, Los Alamos National Laboratory, Los Alamos, New Mexico. Dr. Wang's present affiliation is Department of Management, California State University, Sacremento, California.

REFERENCES

Ali, M. M., Mikhail, N. N. and Haq, M. S. (1978). A class of bivariate distributions including the bivariate logistic. Journal of Multivariate Analysis, 8, 405-412.

Beckman, R. J. and Johnson, M. E. (1981). A ranking procedure for partial discriminant analysis. Journal of the American Statistical Association, 76, 671-675.

Box, G. E. P. and Muller, M. E. (1958). A note on the generation of random normal deviates. The Annals of Mathematical Statistics, 29, 610-611.

Bryson, M. C. and Johnson, M. E. (1982). Constructing and simulating multivariate distributions using Khintchine's theorem. Journal of Statistical Computation and Simulation, 16, 129-137.

Burr, I. W. (1942). Cumulative frequency functions. Annals of Mathematical Statistics, 13, 215-232.

Cambanis, S. (1977). Some properties and generalizations of multivariate Eyraud-Gumbel-Morgenstern distributions. Journal of Multivariate Analysis, 7, 551-559.

Cambanis, S., Huang, S. and Simons, G. (1981). On the theory of elliptically contoured distributions. Journal of Multivariate Analysis, 11, 368-385.

Chmielewski, M. A. (1981a). A re-appraisal of tests for normality. Communications in Statistics -- Theory and Methods tMultivariate Analysis, 10, 343-350.

Chmielewski, M. A. (1981b). A re-appraisal of tests for normality. Communications in Statistics -- Theory and Methods, 10, 2005-2014.

Chmielewski, M. A. (1981b). Elliptically symmetric distributions: a bibliography and review. International Statistical Review, 49, 67-74.

Cook, R. D. and Johnson, M. E. (1981). A family of distributions for modelling non-elliptically symmetric multivariate data. Journal of the Royal Statistical Society, Series B, 43, 210-218.

Dongarra, J. J., Moler, C. B., Bunch, J. R., Stewart, G. W. (1979). LINPACK User's Guide, SIAM, Philadelphia, Pennsylvania.

Dudewicz, E. J. and Ralley, T. G. (1981). The Handbook of Random Number Generation and Testing with TESTRAND Computer Code. American Sciences Press, Inc., Box 21161, Columbus, Ohio 43221.

Eyraud, H. (1936). Les principes de la mesure des correlations. Annales Universite Lyons, Series A, Sciences Mathematiques et Astronomie, 1, 30-47.

Gleser, L. J. (1976). A canonical representation for the non-central Wishart distribution useful for simulation. Journal of the American Statistical Association, 71, 690-695.

Gumbel, E. J. (1960). Bivariate exponential distributions. Journal of the American Statistical Association, 55, 698-707.

Hawkins, D. M. (1981). A new test for multivariate normality and homoscedasticity. Technometrics, 23, 105-110.

Hogg, R. V. and Craig, A. T. (1978). Introduction to Mathematical Statistics, Fourth Edition, MacMillian Publishing Co., Inc. New York.

Johnson, D. E. and Hegemann, V. (1974). Procedures to generate random matrices with noncentral distributions. Communications in Statistics, 3, 691-699.

Johnson, M. E. (1982). Computer generation of the generalized Eyraud distribution. Journal of Statistical Computation and Simulation, 15, 333-335.

Johnson, M. E. (1986). Multivariate Statistical Simulation. In preparation.

Johnson, M. E. and Ramberg, J. S. (1977). Elliptically symmetric distributions: characterizations and random variate generation. Statistical Computing Section Proceedings of the American Statistical Association, 262-265.

Johnson, M. E., Ramberg, J. S., and Wang, C. (1982). The Johnson translation system in Monte Carlo studies. Communications in Statistics -- Simulation and Computation, 11, 521-525.

Johnson, M. E. and Tenenbein, A. (1981). A bivariate distribution family with specified marginals. Journal of the American Statistical Association, 76, 198-201.

Johnson, N. L. (1949). Bivariate distributions based on simple translation systems. Biometrika, 36, 297-304.

Johnson, N. L. and Kotz, S. (1972). Distributions in Statistics: Continuous Multivariate Distributions. John Wiley & Sons, Inc., New York.

Kennedy, W. J. Jr., and Gentle, J. E. (1980). Statistical Computing. Marcel Dekker, Inc., New York.

Khintchine, A. Y. (1938). On unimodal distributions. Tomsk. Universitet. Nauchno-issledovatel´skii institut matematiki i mekhaniki IZVESTIIA, 2, 1-7.

Kinderman, A. J., Monahan, J. F. and J. G. Ramage (1977). Computer methods for sampling from Student´s t distribution. Mathematics of Computation, 31, 1009-1018.

Malkovich, J. F. and Afifi, A. A. (1973). On tests for multivariate normality. Journal of the American Statistical Association, 68, 176-179.

Mardia, K. V. (1967). Some contributions to contingency-type distributions. Biometrika, 54, 235-249. (Corrections: 1968, 55, 597.)

Mardia, K. V. (1970). Families of Bivariate Distributions. Hafner Publishing Company, Darien, Conneticut.

Marsaglia, G. and Bray, T. A. (1964). A convenient method for generating normal variables. SIAM Review, 6, 101-102.

Mihram, G. A. and Hultquist, R. A. (1967). A bivariate warning-time/failure-time distribution. Journal of the American Statistical Association, 62, 589-599.

Muller, M. E. (1959). A note on a method for generating points uniformly on n-dimensional spheres. Communications of the ACM, 2, 19-20.

Odell, P. L. and Feiveson, A. H. (1966). A numerical procedure to generate a sample covariance matrix. Journal of the American Statistical Association, 61, 199-203.

Plackett, R. L. (1965). A class of bivariate distributions. Journal of the American Statistical Association, 60, 516-522.

Rosenblatt, M. (1952). Remarks on a multivariate transformation. Annals of Mathematical Statistics, 23, 470-472.

Scheuer, E. M. and Stoller, D. S. (1962). On the generation of normal random vectors. Technometrics, 4, 278-281.

Schmeiser, B. and Lal, R. (1980). Multivariate modeling in simulation: a survey. ASQC Technical Conference Transactions, 252-261.

Schmeiser, B. and Lal, R. (1982). Bivariate gamma random vectors. Operations Research, 30, 358-374.

Schoenberg, I. J. (1938). Metric spaces and completely monotone functions. Annals of Mathematics, 39, 811-841.

Stacy, E. W. (1962). A generalization of the gamma distribution. The Annals of Mathematical Statistics, 33, 1187-1192.

Tadikamalla, P. R. and Johnson, M. E. (1981). A complete guide to gamma variate generation. American Journal of Mathematical and Management Sciences, 1, 213-236.

Takahasi. K. (1965). Note on the multivariate Burr's distribution. Annals of the Institute of Statistical Mathematics, 17, 257-260.

Tashiro, Y. (1977). On methods for generating uniform points on the surface of a sphere. Annals of the Institute of Statistical Mathematics, 29, 295-300.

Wahrendorf, J. (1980). Inference in contingency tables with ordered categories using Plackett's coefficient of association for bivariate distributions. Biometrika, 67, 15-21.

Received 2/14/83; Revised 7/25/84.

Part III: Efficient Design Techniques for the Choice and Reduction of Simulation Run Time

Choice of simulation run length should be **dictated by the goals** the simulation experimenter wishes to achieve. One common goal is the **selection of that one of several proposed systems (or system configurations) which is "best"** with respect to some specified criterion. For this goal, the statistical area of "ranking and selection procedures" tells us how to proceed; an overview of the area is given in the first two articles (Dudewicz (1976) and Dudewicz (1980)). Consideration of **how to determine if one's systems (or system configurations) have unequal variabilities** is given in the third article (Dudewicz (1983)), while the fourth by Dudewicz and Zaino (1977) discusses **how to proceed if observations are correlated within each system**—as they will be if, for example, a single long run is broken up into a number of parts ("observations"). While often parameters which statisticians assume to be known are in practice unknown, **the probability of correctly selecting the best system with the simulation can still be estimated, by Monte Carlo techniques** as shown for one important case in Dudewicz (1982). Final considerations along this line are directed to **experiments run in a factorial structure** (Dudewicz and Taneja (1982)) and to experiments run to select the best random number generators (Dudewicz and van der Meulen (1983)).

If one has **a setting where several factors are to be varied** in a simulation, the temptation is to run once at each of the possible factor settings. If one has three factors at 10 levels each, this means 1000 simulation runs, which would in most cases be prohibitively expensive and/or time-consuming. Faced with such prohibition, experimenters often resort to **the (invalid) one-at-a-time-method**. These problems are outlined in the next section. The next article then gives an application of **modern design in the context of simulation** (Dudewicz and Mishra (1983)), where it is shown how, in a setting with 10 factors, 149 simulation runs allowed the fitting of a full quadratic model to the output as a function of the factors; a full factorial experiment with just two levels for each factor would have required 2**10 = 1024 experiments (not feasible in our case), and it would not have yielded curvature in the factors in the resulting model.

This part concludes with an extensive treatment of **the specification of designs (specifications of which simulation runs to make)**.

STATISTICS IN SIMULATION: HOW TO DESIGN FOR SELECTING THE BEST ALTERNATIVE

Edward J. Dudewicz
Department of Statistics
The Ohio State University

ABSTRACT

In many simulation studies the experimenter (the person running the simulation) has under consideration several (two or more) proposed procedures (e.g., for running a real-world system), and is simulating in order to determine which is the best procedure (with regard to certain specified criteria of "goodness"). Such an experimenter does not wish basically to test hypotheses, or construct confidence intervals, or perform regression analyses (though these may be appropriate minor parts of his analysis); he does wish basically to select the best of several procedures, and the major part of his analysis should therefore be directed towards this goal.

It is precisely for this problem that ranking-and-selection procedures were developed. These procedures set sample size (in simulation this means run-length) explicitly so as to guarantee that the probability that "the procedure actually selected by the experimenter is the best procedure" is suitably large.

In this paper we first review the background ideas of ranking-and-selection and contrast them to other approaches to multi-population problems (which, while sometimes appropriate in such areas as social science experimentation, are almost wholly inappropriate for use in statistical design and analysis of simulation experiments). Recommended procedures for several common situations are then outlined in detail. References where further theoretical details may be obtained are provided, along with information on current developments in the area. It is intended that the motivation and technical detail given be sufficient for intelligent application in many common situations (though other situations will still require supplementary consultation).

I. BACKGROUND OF MULTI-POPULATION PROBLEMS

Statistics for many years concerned itself to a large extent with problems in which the basic observations came from one source or population (one-population problems). Two-population problems were well-known (if unsolved, for example the Behrens-Fisher problem), but for the most part it was a one-population world until some time in the 1950's when R. E. Bechhofer, by pioneering work (see reference (1) for the first published account of this work, a major event in statistical thought) in ranking-and-selection, brought the subject to full light of day with a context other than the type described by saying (as in classical ANOVA) "We have k populations, but would like to test the hypothesis that we really only have one." The relevance of the pioneering ranking-and-selection work to statistical design and analysis of simulation experiments was soon recognized by workers in the field. For example, on p. 53 of (3), Conway stated in 1963 that "...the analysis of variance seems a completely inappropriate approach to these problems. It is centered upon the test of the hypothesis that all of the alternatives are equivalent. Yet the alternatives are actually different and it is reasonable to expect some difference in performance, however slight. Thus, the failure to reject the null hypothesis only indicates that the test was not sufficiently powerful to detect the difference - e.g., a longer run would have to be employed. Moreover, even when the investigator rejects the hypothesis, it is highly likely that he is more interested in identifying the best alternative than in simply concluding that the alternatives are not equivalent. Recently proposed ranking[-and-selection] procedures...seem more appropriate to the problem than the conventional analysis of variance techniques...." This recognition has continued to the present day, as is exemplified by the fact that in Kleijnen's (9) treatise on statistical aspects of simulation 77 pages (out of 390 which are non-introductory) are devoted to ranking-and-selection procedures (which are also often called "multiple ranking procedures" by Kleijnen and others). Nevertheless it was true, as pointed out by Conway (3) in 1963, that "...the investigator is still going to have difficulty satisfying the assumptions (normality, common variance, independence) that the statistician will require." However in recent years this difficulty has also largely been removed. While in Section II below we will introduce the ranking-and-selection area with an example using the traditional assumptions (normality, common variance, and independence of observations) for simplicity, in Section III work of recent years allows us to recommend procedures given recently for much more general situations.

II. RANKING-AND-SELECTION

In order to introduce the area of ranking-and-selection, let us talk in terms of a simple explicit problem, that of choosing (i.e., selecting) the job shop precedence rule which yields the highest output on the average. (This particular example is chosen only for ease of reference, and we could as easily talk of selecting the queue discipline which yields the highest output on the average, or the investment strategy which yields the highest return on the average. The reader is encouraged is think of an example pertinent to his field and rephrase the considerations given below in terms of that example.)

To be specific, suppose that it is desired to select that one of 10 job shop precedence rules which has the highest average output per period. If we run the shop with rule one for one period, we will observe some output, say X_{11}. However, in a second period, still using rule one, we will observe a different output, say X_{12}. Similarly we obtain output X_{13} in a third period,...,output X_{1N} in an Nth period. Thus, in each period the output using rule one differs.

However, it is reasonable to assume that it varies about some value, say μ_1, in the sense that if we average the output per period using rule one over many periods the number so obtained will be close to μ_1. To take into account the variability in output, assume that X_{11} obeys a normal probability distribution, has mean value μ_1, and has variance σ^2. Similarly for $X_{12},\ldots,X_{1N}$. Then, if one period's output doesn't affect another's,

$$\bar{X}_1 = \frac{X_{11}+X_{12}+\ldots+X_{1N}}{N}$$

will obey a normal probability distribution with mean value μ_1 and variance σ^2/N, i.e. its variability from μ_1 is decreased and (if N is large) we expect $\bar{X}_1 \approx \mu_1$.

Now, considerations like the above hold for each of the 10 rules. Thus, we may observe the outputs over N periods of each of the 10 rules and obtain average outputs $\bar{X}_1, \bar{X}_2, \ldots, \bar{X}_{10}$. Since we expect these to be close to the mean values $\mu_1,\mu_2,\ldots,\mu_{10}$ of the 10 rules, we select the rule yielding the largest average as having the highest output (see Table 1).

TABLE 1

Rule 1		Rule 2		...	Rule 10	
Period 1	X_{11}	Period 1	X_{21}		Period 1	$X_{10,1}$
Period 2	X_{12}	Period 2	X_{22}		Period 2	$X_{10,2}$
.		.			.	
.		.			.	
.		.			.	
Period N	X_{1N}	Period N	X_{2N}		Period N	$X_{10,N}$
	$\bar{X}_1$		$\bar{X}_2$			$\bar{X}_{10}$

X_{ij} = output in period j using rule i. (1)

Select rule yielding $\max(\bar{X}_1,\bar{X}_2,\ldots,\bar{X}_{10})$.

However, it will be hard to distinguish the best rule (i.e., the one with the highest mean output) when the mean outputs of the other rules are very close to the largest one, since although $\bar{X}_1, \bar{X}_2, \ldots, \bar{X}_{10}$ will be close to $\mu_1, \mu_2, \ldots, \mu_{10}$ the values $\mu_1, \mu_2, \ldots, \mu_{10}$ are also close to each other. (E.g., although we may be 95% sure that $|\bar{X}_1-\mu_1| \le 0.5$, $|\bar{X}_2-\mu_2| \le 0.5$, ..., $|\bar{X}_{10}-\mu_{10}| \le 0.5$, if in reality $\mu_1=\mu_2=\ldots=\mu_9=233.40$ and $\mu_{10}=233.41$ we may fail to pick the best rule, that with mean output μ_{10}, since $\bar{X}_1, \bar{X}_2, \ldots, \bar{X}_9$ as well as $\bar{X}_{10}$ will be close to μ_{10}. This can be remedied by raising the sample size N so that we are, e.g., 95% sure that $|\bar{X}_1-\mu_1| \le 0.01$, $|\bar{X}_2-\mu_2| \le 0.01$, ..., $|\bar{X}_{10}-\mu_{10}| \le 0.01$.) Thus, if N isn't large enough, the probability of selecting the best rule may be unacceptably small; here "large enough" depends on the closeness of $\mu_1, \mu_2, \ldots, \mu_{10}$ (see Illustration 1, where $\mu_{[1]} \le \cdots \le \mu_{[10]}$ denote the mean outputs in numerical order from smallest to largest).

ILLUSTRATION 1

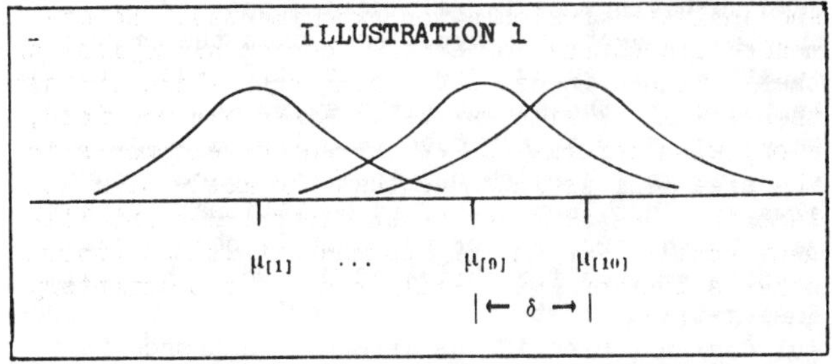

If it is the case, however, that $\mu_{[9]}$ is very close to $\mu_{[10]}$ then we may not care whether we select the rule with mean output $\mu_{[10]}$ or the rule with mean output $\mu_{[9]}$ (which is almost as good). In some cases we may only care about our chances of selecting the best rule when $\mu_{[10]} - \mu_{[9]} \ge 0.1$, in which case we wish to be 90% sure that we make a "Correct Selection" (abbreviated "CS"); i.e. we desire

$$\text{Prob}\{CS\} \geq 0.90 \text{ if } \mu_{[10]} - \mu_{[9]} \geq 0.1.$$

In general, this desired statement is of the form

$$\text{Prob}\{CS\} \geq P^* \text{ if } \delta \equiv \mu_{[10]} - \mu_{[9]} \geq \delta^*, \quad (2)$$

where δ^* and P^* are pre-set by the experimenter (e.g., $\delta^* = 0.1$ and $P^* = 0.90$). Note that δ^* must be positive (a requirement with $\delta^* \leq 0$ is meaningless since $\mu_{[10]} - \mu_{[9]} \geq 0$ always, by the definition of $\mu_{[10]}$ as the largest of $\mu_1, \mu_2, \ldots, \mu_{10}$) and that $0.10 < P^* < 1$ ($P^* < 1$ since we can never be absolutely certain that the best rule yielded the largest of $\bar{X}_1, \bar{X}_2, \ldots, \bar{X}_{10}$ -- there is always a chance, however small, that another rule did; $P^* > 0.10$, since we can be assured of a 10% chance of correct rule selection by picking one of the rules at random). It is clear that the Prob{CS} is minimized when rules other than the best have mean outputs as large as possible. When we "care" (i.e., when $\mu_{[10]} - \mu_{[9]} \geq \delta^*$) this means the Prob{CS} is a minimum when $\mu_{[1]} = \ldots = \mu_{[9]} = \mu_{[10]} - \delta^*$, which is therefore called the least-favorable configuration (LFC). Since available tables (1) allow us to choose N so that Prob{CS} is at least P^* when the LFC $\mu_{[1]} = \ldots = \mu_{[9]} = \mu_{[10]} - \delta^*$ is the case, we can choose N so as to achieve (2).

This example suggests certain conclusions about selection procedures. First, they are precise; that is, the selection approach can give us a rational basis for choosing N (the number of periods to be observed) and tell us (e.g.) how the Prob{CS} varies as we change N, and how large the Prob{CS} is if in fact $\mu_{[1]} = \ldots = \mu_{[9]} = \mu_{[10]} - \delta$ for some value δ other than δ^*. Contrast this to a typical old-style approach: testing the hypothesis that the 10 mean outputs are equal (perhaps by running an Analysis of Variance on an elaborately-designed experiment) and then selecting the rule yielding the largest sample mean as having the largest mean output if the test accepts the hypothesis that they're unequal (while saying "there's no difference" if the test accepts the hypothesis they're equal). Not only does such an approach make little sense because we know they're not equal and should thus always select, but it offers no rational (with regard to the problem for which it is being used), precise grounds for choice of N. Note that this does not mean that one should neglect to use careful design choice if the selection approach is appropriate to his problem (see p. 25 of (1) for further details).

Second, selection procedures are practical in two ways. First, they are applicable to problems often arising in practice, and second, they are feasible because quantities such as necessary sample sizes N have been tabled or can be computed. This may be contrasted to the situation in other branches of statistics where some quantities are almost impossible to compute.

The essential problem formulation of 1954 is thus that we have:

> populations (sources of observations) $\Pi_1, \ldots, \Pi_k$ ($k \geq 2$) with respective unknown means $\mu_1, \ldots, \mu_k$ for their observations, and whose observations obey a normal probability distribution with a common known variance σ^2 about their respective means; a goal of selecting the population associated with $\mu_{[k]} = \max(\mu_1, \ldots, \mu_k)$; a probability requirement that $\text{Prob}\{CS\} \geq P^*$ ($1/k < P^* < 1$) if $\mu_{[k]} - \mu_{[k-1]} \geq \delta^*$ ($\delta^* > 0$); and a procedure of selecting the population yielding $\bar{X}_{MAX} = \max(\bar{X}_1, \bar{X}_2, \ldots, \bar{X}_k)$.

While the above example assumed normality of observations, common variance of output per period for each rule, and independence of observations across periods (all somewhat difficult to justify), the procedures given in Section III weaken these assumptions to an extent sufficient to make this approach feasible in a significant number of simulation studies.

As a numerical example, if we have $k = 10$ rules with $\sigma = 5$ units per period and wish to have probability of correct selection at least $P^* = 0.95$ whenever the average output of the best rule is at least $\delta^* = 2$ units larger than the other outputs, then we will need a sample size of at least

$$N = \frac{h_k^2(P^*)\sigma^2}{(\delta^*)^2} = \frac{(3.4182)^2(5)^2}{(2)^2} = 73.03 \quad (3)$$

periods, and so would take $N = 74$ periods in the simulation. The factor $h_k(P^*)$ needed is obtained from tables originally given by Bechhofer (1) which are reprinted on p. 347 of (5). Note that it is known (4) that a good approximation to the required N is given by

$$N_1 = \frac{-4\sigma^2 \ln(1-P^*)}{(\delta^*)^2}, \quad (4)$$

which in our example is $N_1 = 74.89$. Also note that as long as we have a common variance σ^2, if we have correlations within periods (i.e., $X_{11}, X_{21}, \ldots, X_{10,1}$ are correlated) but not across periods (i.e. X_{i1} and X_{i2} are uncorrelated), and if all correlations are positive or zero, then the N of equation (3) is conservative. This was shown in (4) and partially justifies the traditional wisdom that "positive correlation within a block is helpful in simulation".

III. RECOMMENDED SELECTION PROCEDURES

If we are faced with $k \geq 2$ normal populations with unknown means $\mu_1, \ldots, \mu_k$ and a common known variance σ^2, then the sample size N calculated from equation (3) will suffice when the observations are independent (and, in fact, is sufficient and conservative even if one has positive correlations within periods).

More often in simulation the rules to be evaluated will have unequal variances $\sigma_1^2, \ldots, \sigma_k^2$ which are also unknown. This problem was solved recently by Dudewicz and Dalal (6), and the solution has been applied in simulations for selecting water resource system alternatives by Vicéns and Schaake (11) and in accounting system simulations by Lin (10). This solution is very appropriate if no correlations are present. Since correlations within a population are often present (e.g., due to lack of frequent regeneration points), a heuristic procedure recently developed and studied by Dudewicz and Zaino (8) will be given for this more general problem.

The recommended Procedure $A(\hat{\rho}_i, s_i^2)$ is as follows. Take an initial sample of $N_0 = 30$ observations using each rule. Calculate

$$M_i = \max\left(N_0, \left[\frac{s_i^2 h^2}{(\delta^*)^2}\right]\right) \tag{5}$$

(which is the number of observations which would be needed if we had all correlations $\rho_i = 0$ (zero correlation), where h depends on k and P* and is given in Table 2 below, extracted from (7)). Calculate

$$\hat{\rho}_i = \frac{\sum_{n=2}^{N_0}(X_{in}-\bar{X}_i)(X_{i,n-1}-\bar{X}_i)}{\sum_{n=1}^{N_0}(X_{in}-\bar{X}_i)^2} \tag{6}$$

and form the 100(1-α)% confidence interval for ρ_i from

$$(\rho_i-\hat{\rho}_i)^2 \leq \frac{N_0-1}{N_0(N_0-3)}(1-\hat{\rho}_i^2)t^2_{N_0-3}(1-\alpha/2) \tag{7}$$

with $\alpha = .05$. (Here $t_r(q)$ is the 100q percent point of Student's-t distribution with r degrees of freedom.) If this 95% confidence interval contains $\rho_i = 0$, judge the sample size N_i as being adequate for population i. Otherwise calculate

$$N_{2i} = \left[M_i\left(\frac{1+\hat{\rho}_i}{1-\hat{\rho}_i}\right)\right] \tag{8}$$

and continue the run until we have N_{2i} observations from π_i. Finally calculate $\bar{X}_1, \ldots, \bar{X}_k$ based on all available observations and select (as being best) that population which produced the largest of $\bar{X}_1, \ldots, \bar{X}_k$.

While Procedure $A(\hat{\rho}_i, s_i^2)$ is heuristic (unlike the procedure of Dudewicz and Dalal for $\rho_i = 0$, which is entirely rigorously derived), studies show it should be sufficient to preclude gross errors due to significant correlations. (Work in progress presently studies properties of Procedure $A(\hat{\rho}_i, s_i^2)$ in further detail by Monte Carlo, compares $A(\hat{\rho}_i, s_i^2)$ with procedures based on the regenerative methods of Iglehart, and develops a corresponding fully rigorous mathematical procedure by utilizing the Heteroscedastic Method recently developed by Dudewicz and Bishop (see (2)).)

TABLE 2

Quantity h Needed in Equation (5)

	P* = .95	P* = .99
k = 2	2.41	3.45
k = 3	2.81	3.81
k = 4	3.03	4.01
k = 5	3.18	4.14
k = 6	3.30	4.25
k = 7	3.39	4.33
k = 8	3.46	4.40
k = 9	3.53	4.46
k = 10	3.58	4.51
k = 15	3.79	4.71
k = 20	3.92	4.84
k = 25	4.03	4.94

BIBLIOGRAPHY

1. Bechhofer, R. E. "A Single-Sample Multiple Decision Procedure for Ranking Means of Normal Populations with Known Variances," in Annals of Mathematical Statistics, Vol. 25 (1954), pp. 16-39.

2. Bishop, Thomas A. Heteroscedastic ANOVA, MANOVA, and Multiple-Comparisons. Unpublished Ph. D. Dissertation, Department of Statistics, The Ohio State University, Columbus, Ohio, 1976.

3. Conway, R. W. "Some Tactical Problems in Digital Simulation," in Management Science, Vol. 10, No. 1 (October 1963), pp. 47-61.

4. Dudewicz, Edward J. "An Approximation to the Sample Size in Selection Problems,' in Annals of Mathematical Statistics, Vol. 40, No. 2 (1969), pp. 492-497.

5. Dudewicz, Edward J. Introduction to Statistics and Probability. Holt, Rinehart and Winston, New York, 1976.

6. Dudewicz, Edward J. and Dalal, Siddhartha R. "Allocation of Observations in Ranking and Selection with Unequal Variances," in Sankhyā, Series B, Vol. 37, Part 1 (1975), pp. 28-78.

7. Dudewicz, E. J., Ramberg, J. S., and Chen, H. J. "New Tables for Multiple Comparisons With a Control (Unknown Variances)," in Biometrische

Zeitschrift, Vol. 17, Part 1 (1975), pp. 13-26.

8. Dudewicz, Edward J. and Zaino, Nicholas A., Jr. "Allowance for Correlation in Setting Simulation Run-length via Ranking-and-Selection Procedures," Technical Report, Department of Statistics, Stanford University, Stanford, California, 1976. To appear in Management Science.

9. Kleijnen, Jack P. C. Statistical Techniques in Simulation, Part II. Marcel Dekker, Inc., New York, 1975.

10. Lin, W. T. "Multiple Objective Budgeting Models: A Simulation," USC Working Paper #12-01-75, Department of Accounting, Graduate School of Business Administration, University of Southern California, Los Angeles, California.

11. Vicéns, Guillermo J. and Schaake, John C., Jr. "Simulation Criteria for Selecting Water Resource System Alternatives," Report No. 154, Ralph M. Parsons Laboratory for Water Resources and Hydrodynamics, Department of Civil Engineering, Massachusetts Institute of Technology, Cambridge, Massachusetts, September 1972.

Ranking (Ordering) and Selection: An Overview of How to Select the Best

Edward J. Dudewicz

Department of Statistics
The Ohio State University
Columbus, OH 43210

In many studies the experimenter has under consideration several (two or more) alternatives, and is studying them in order to determine which is the best (with regard to certain specified criteria of "goodness"). Such an experimenter does not wish basically to test hypotheses, or construct confidence intervals, or perform regression analyses (though these may be appropriate parts of his analysis); he does wish to select the best of several alternatives, and the major part of his analysis should therefore be directed towards this goal. It is precisely for this problem that ranking and selection procedures were developed. This paper presents an overview of some recent work in this field, with emphasis on aspects important to experimenters confronted with selection problems.

KEY WORDS

Selection
Ordering
Ranking
Complete ranking
Indifference zone
Nonparametric selection
Subset selection
Estimation
Factorial design

1. INTRODUCTION

In many studies the experimenter's basic goal is to determine which of several alternatives is the best. The experimenter in such a situation will desire to use procedures which explicitly guarantee that the probability that "the alternative selected is the best alternative" is suitably large.

Procedures for this situation began to be developed in the 1950's by R. E. Bechhofer (1954) (assuming normality and equal known variance). In the ensuing 25 years (1954–1979) such procedures have been developed for more complex (and more realistic) settings. As a result of the recentness of this work, it has been mainly in the theoretical literature (see Dudewicz, 1976, chapter 11, for an introduction) and largely unavailable to experimenters in easily-usable form until very recently (see Gibbons, Olkin and Sobel, 1977).

This article presents an overview of seven papers which were presented at the Advanced Practitioner Sessions on "Selection of Alternatives through Simulation" at the 1977 Winter Simulation Conference (held at the National Bureau of Standards, Gaithersburg, Maryland) and at the 1978 Winter Simulation Conference (held in Miami Beach, Florida) organized and chaired by the current author. The papers discuss various aspects of the same real (traffic fatality) data set. Taken together, they comprise a fairly comprehensive study of this data set (including aspects of selection, ordering, estimation, design, nonparametric analysis, and choice of measure) which should be almost invaluable as a benchmark for selection studies. In conjunction with the methods book by Gibbons, Olkin, and Sobel (1977) (reviewed in Dudewicz, 1979) the motivation and technical detail is intended to be sufficient for intelligent application in many situations.

2. BACKGROUND OF MULTIPOPULATION PROBLEMS

Theoretical statistics concerned itself too little with problems in which the basic observations come from several sources or populations until the 1950's and R. E. Bechhofer's pioneering work in ranking and selection (see Bechhofer, 1954, for the first published account of this work, a major event in statistical thought). Bechhofer brought the subject to full light

Received May 1979; revised August 1979

of day with a context other than the type described by saying (as in classical ANOVA) "We have k populations, but would like to test the hypothesis that we really only have one." The relevance of the pioneering ranking and selection work to statistical design and analysis of experiments was soon recognized by workers in the field. For example, R. W. Conway (1963, p. 53) stated that

> ... the analysis of variance seems a completely inappropriate approach to these problems. It is centered upon the test of the hypothesis that all of the alternatives are equivalent. Yet the alternatives are actually different and it is reasonable to expect some difference in performance, however slight. Thus, the failure to reject the null hypothesis only indicates that the test was not sufficiently powerful to detect the difference—e.g., a longer run would have to be employed. Moreover, even when the investigator rejects the hypothesis, it is highly likely that he is more interested in identifying the best alternative than in simply concluding that the alternatives are not equivalent. Recently proposed ranking [and selection] procedures ... seem more appropriate to the problem than the conventional analysis of variance techniques....

This recognition has continued to the present day, as is exemplified by the fact that in Kleijnen's (1975) treatise on statistical aspects of simulation, 77 pages (out of 390 which are nonintroductory) are devoted to ranking and selection procedures (which are also often called "multiple ranking procedures" by Kleijnen and others). Nevertheless it was true, as pointed out by Conway (1963), that "... the investigator is still going to have difficulty satisfying the assumptions (normality, common variance, independence) that the statistician will require." However, in recent years this difficulty has also largely been removed. While for simplicity in Section 3 below we will introduce the ranking and selection area with an example using the traditional assumptions (normality, common variance, and independence of observations), in Section 4 the work of recent years allows us to recommend procedures for much more general situations under normality, and Gibbons, Olkin and Sobel (1977) contains details for nonnormal situations. The case study papers of the specific traffic fatality data set are then discussed in Section 5, in light of the material of Sections 3 and 4.

3. RANKING AND SELECTION

In order to introduce the area of ranking and selection, let us talk in terms of a simple explicit problem, that of choosing (i.e., selecting) the job shop precedence rule which yields the highest output on the average. To be specific, suppose that it is desired to select that one of 10 job shop precedence rules which has the highest average output per period. If we run the shop with rule 1 for one period, we will observe some output, say X_{11}. In a second period, still using rule 1, we will observe a different output, say X_{12}. Similarly we obtain output X_{13} in a third period, $\cdots$, output X_{1N} in an Nth period. Thus, in each period the output using rule 1 differs.

However, it is reasonable to assume that it varies about some value, say μ_1, in that if we average the output per period using rule 1 over many periods the number so obtained will be close to μ_1. To take account of variability in output, assume that X_{11} obeys a normal probability distribution, has mean value μ_1, and has variance σ^2, and so on for $X_{12}, \cdots, X_{1N}$. Then, if one period's output does not affect another's, $\overline{X}_1 = (X_{11} + X_{12} + \cdots + X_{1N})/N$ will obey a normal probability distribution with mean value μ_1 and variance σ^2/N, and (if N is large) we expect $\overline{X}_1 \approx \mu_1$.

Now, considerations like the above hold for each of the 10 rules. Thus, we may observe the outputs over N periods of each of the 10 rules and obtain average outputs $\overline{X}_1, \overline{X}_2, \cdots, \overline{X}_{10}$. Since we expect these to be close to the mean values $\mu_1, \mu_2, \cdots, \mu_{10}$ of the 10 rules, we select the rule yielding the largest average as having the highest output.

However, it will be hard to distinguish the best rule (i.e., the one with the highest mean output) when the mean outputs of the other rules are very close to the largest one, since although $\overline{X}_1, \overline{X}_2, \cdots, \overline{X}_{10}$ will be close to $\mu_1, \mu_2, \cdots, \mu_{10}$, the values $\mu_1, \mu_2, \cdots, \mu_{10}$ are also close to each other. (E.g., although we may be 95% sure that $|\overline{X}_1 - \mu_1| \leq 0.5$, $|\overline{X}_2 - \mu_2| \leq 0.5, \cdots, |\overline{X}_{10} - \mu_{10}| \leq 0.5$, if in reality $\mu_1 = \mu_2 = \cdots = \mu_9 = 233.40$ and $\mu_{10} = 233.41$, we may fail to pick the best rule, that with mean output μ_{10}, since $\overline{X}_1, \overline{X}_2, \cdots, \overline{X}_9$ as well as $\overline{X}_{10}$ will be close to μ_{10}. This can be remedied by raising the sample size N so that we are, e.g., 95% sure that $|\overline{X}_1 - \mu_1| \leq 0.01$, $|\overline{X}_2 - \mu_2| \leq 0.01, \cdots, |\overline{X}_{10} - \mu_{10}| \leq 0.01$.) Thus, if N is not large enough, the probability of selecting the best rule may be unacceptably small; here "large enough" depends on the closeness of $\mu_1, \mu_2, \cdots, \mu_{10}$.

If it is the case, however, that $\mu_{[9]}$ is very close to $\mu_{[10]}$ (where $\mu_{[1]} \leq \cdots \leq \mu_{[10]}$ denote the mean outputs in numerical order from smallest to largest) then we may not care whether we select the rule with mean output $\mu_{[10]}$ or the rule with mean output $\mu_{[9]}$ (which is almost as good). In some cases we may only care about our chances of selecting the best rule when $\mu_{[10]} - \mu_{[9]} \geq 0.1$, in which case we might wish to be 90% sure that we make a "Correct Selection" (abbreviated "CS"); i.e. we desire

$$\text{Prob}\{\text{CS}\} \geq 0.90 \quad \text{if} \quad \mu_{[10]} - \mu_{[9]} \geq 0.1. \tag{1}$$

In general, this desired statement is of the form

$$\text{Prob}\{\text{CS}\} \geq P^* \quad \text{if} \quad \delta \equiv \mu_{[10]} - \mu_{[9]} \geq \delta^*, \tag{2}$$

where δ^* and P^* are preset by the experimenter (e.g., $\delta^* = 0.1$ and $P^* = 0.90$). Note that δ^* must be positive (since $\mu_{[10]} - \mu_{[9]} \geq 0$ by the definition of $\mu_{[10]}$) and that $0.10 < P^* < 1$ ($P^* < 1$ since we can never be absolutely certain that the best rule yielded the largest of $\overline{X}_1, \overline{X}_2, \cdots, \overline{X}_{10}$—there is always a chance, however small, that another rule did; $P^* > 0.10$, since we can be assured of a 10% chance of correct rule selection by randomly picking one of the rules). It is clear that the Prob{CS} is minimized when rules other than the best have mean outputs as large as possible. When we "care" (i.e., when $\mu_{[10]} - \mu_{[9]} \geq \delta^*$) this means the Prob{CS} is a minimum when $\mu_{[1]} = \cdots = \mu_{[9]} = \mu_{[10]} - \delta^*$, which is therefore called the *least favorable configuration* (LFC). Available tables (Bechhofer, 1954) allow us to choose N so that Prob{CS} is P^* in the LFC $\mu_{[1]} = \cdots = \mu_{[9]} = \mu_{[10]} - \delta^*$.

This example suggests certain conclusions about selection procedures. *First*, they are precise; that is, the selection approach can give us a rational basis for choosing N(the number of periods to be observed) and tell us (e.g.) how the Prob{CS} varies as we change N, and how large the Prob{CS} is if in fact $\mu_{[1]} = \cdots = \mu_{[9]} = \mu_{[10]} - \delta$ for some value δ other than δ^*. Contrast this to a typical old-style approach: testing the hypothesis that the 10 mean outputs are equal (perhaps by running an Analysis of Variance on an elaborately designed experiment) and then selecting the rule yielding the largest sample mean as having the largest mean output *if* the test accepts the hypothesis that they are unequal (while saying "there is no difference" if the test accepts the hypothesis they are equal). Not only does such an approach make little sense because we *know* the outputs are not equal and we should thus always select, but it offers no rational (with regard to the problem for which it is being used), precise grounds for choice of N. Note that this does not mean that one should neglect to use careful design choice if the selection approach is appropriate to his problem (see Bechhofer, 1954, p. 25 for further details).

Second, selection procedures are practical in two ways. First, they are applicable to problems often arising in practice, and second, they are feasible because quantities such as necessary samples sizes N have been tabled or can be computed.

The essential problem formulation of Bechhofer in 1954 is:

> There exist *populations* (sources of observations) $\pi_1, \cdots, \pi_k$ ($k \geq 2$) with respective unknown means $\mu_1, \cdots, \mu_k$ for their observations, and whose observations obey a normal probability distribution with a common known variance σ^2; a *goal* of selecting the population associated with $\mu_{[k]} \equiv \max(\mu_1, \cdots, \mu_k)$; a *probability requirement* that Prob{CS} $\geq P^*$ ($1/k < P^* < 1$) if $\mu_{[k]} - \mu_{[k-1]} \geq \delta^*$ ($\delta^* > 0$); and a *procedure* of selecting the population yielding $\overline{X}_{MAX} = \max(\overline{X}_1, \overline{X}_2, \cdots, \overline{X}_k)$.

While the above formulation assumed normality of observations, common variance of output per period for each rule, and independence of observations across periods (all somewhat difficult to justify), the procedures given in Section 4 below weaken these assumptions to an extent sufficient to make a ranking and selection approach feasible in a significant number of studies.

As a numerical example, if we have $k = 10$ rules with $\sigma = 5$ units per period and wish to have probability of correct selection at least $P^* = 0.95$ whenever the average output of the best rule is at least $\delta^* = 2$ units larger than the other outputs, then we will need a sample size of at least

$$N = \frac{h_k^2(P^*)\sigma^2}{(\delta^*)^2} = \frac{(3.4182)^2(5)^2}{(2)^2} = 73.03 \qquad (3)$$

periods, and so would take $N = 74$ periods in the study. The factor $h_k(P^*)$ needed is obtained from tables given by Bechhofer (1954) (reprinted in Dudewicz, 1976, p. 347). Note that it is known (see Dudewicz, 1969) that a good approximation to the required N is given by

$$N_1 = \frac{-4\sigma^2 \ln(1 - P^*)}{(\delta^*)^2}, \qquad (4)$$

which in our example is $N_1 = 74.89$. Also note that as long as we have a common variance σ^2, if we have correlations within periods (i.e., $X_{11}, X_{21}, \cdots, X_{10,1}$ are correlated) but not across periods (i.e., X_{i1} and X_{i2} are uncorrelated), and if all correlations are positive or zero, then the N of equation (3) is conservative. This was shown in Dudewicz (1969) and partially justifies the traditional wisdom that "positive correlation within a block is helpful."

The approach to the ranking and selection problem which has just been described is often called (Bechhofer's) *indifference zone approach* since it does not explicitly seek to control the Prob{CS} at parameter points where $\mu_{[k]} - \mu_{[k-1]} < \delta^*$ (which hence has obtained the name of a "zone" where one is "indifferent" to which population is selected). In fact, one will often not be "indifferent" to which population is selected when $\mu_{[k]} - \mu_{[k-1]} < \delta^*$ (one still prefers the best to the others), but rather specifies such a zone in recognition of the fact (see equation (3)) that as $\delta^* \to 0$ the required sample size $N \to \infty$, requiring large samples for assurance of trivial mean gains. (The same situation is well known in classical statistical testing of hypotheses, where the power is controlled only at and beyond a certain distance from the null hypothesis.)

This indifference zone approach is feasible (in terms of the N required) only if two conditions are met: first, k (the number of populations) is not extremely large (e.g., $k < 50$); and, second, one has some *design control* (via choice of N, the sample size per population). If either k is very large (resulting in an N which is impractical), or one has no control over N (as when a data set is brought to a consultant for analysis), an alternate *subset selection approach* pioneered by S. S. Gupta (1956) is often used. This uses the available N observations to cut the original set of k populations $\{\pi_1, \pi_2, \cdots, \pi_k\}$ to a smaller subset which, with probability at least P^*, contains the best population; this smaller subset can (if additional observations are available for this (often much smaller) number of populations in the subset) be studied further in a search for the one best population. This approach is largely needed in the data set studied in Section 5 below, since k is large ($k = 49$) and N is fairly small ($N = 17$) and the experimenter cannot obtain additional observations. Additional discussion of subset selection is given (at an introductory level) in Section 2 of Gupta and Hsu (1977) and (as applied to the traffic fatality data set) in McDonald (1979), Gupta and Hsu (1977), Sobel (1977), and Lee (1977).

The indifference zone problem was first solved, by telling how many observations N to take from each population, by Bechhofer (1954), who assumed that $\sigma_1^2 = \cdots = \sigma_k^2 = \sigma^2$ with σ^2 known and that all observations were independent. This case is discussed and extended now since the papers of the 1977 and 1978 Winter Simulation Conferences (see Section 5 below) deal largely with other cases, and both situations are

TABLE 1—*Quantity h used in setting the sample size N_{1i}, (5), of the Dudewicz-Zaino procedure, with $N_0 = 30$ initial observations, as a function of number of populations (k) and desired probability of correct selection (P^*).*

	P* = .95	P* = .99
k = 2	2.41	3.45
k = 3	2.81	3.81
k = 4	3.03	4.01
k = 5	3.18	4.14
k = 6	3.30	4.25
k = 7	3.39	4.33
k = 8	3.46	4.40
k = 9	3.53	4.46
k = 10	3.58	4.51
k = 15	3.79	4.71
k = 20	3.92	4.84
k = 25	4.03	4.94

important for practitioners who wish to use selection techniques. The case of $\sigma_1^2, \cdots, \sigma_k^2$ unknown and unequal was solved more recently by Dudewicz and Dalal (1975), and requires a two-stage procedure which has been applied in simulations for selecting water resource system alternatives by Vicéns and Schaake (1972) and in accounting system simulations by Lin (1975). This solution is appropriate if no correlations are present. For the case where the observations are correlated within each population (but not across populations), a heuristic procedure was recently given. The recommended *Dudewicz-Zaino* (1977) *procedure* $A(\hat{\rho}_i, s_i^2)$ is as follows. Take an initial sample of $N_0 = 30$ observations from each population. Calculate

$$N_{1i} = max\left(N_0, \left[\frac{s_i^2h^2}{(\delta^*)^2}\right]\right) \quad (5)$$

(which is the number of observations which would be needed if we had all correlations $\rho_i = 0$ (zero correlation), where h depends on k and P^* and is given in Table 1, extracted from Dudewicz, Ramberg and Chen, 1975). Calculate

$$\hat{\rho}_i = \frac{\sum_{n=2}^{N_0} (X_{i,n} - \overline{X}_i)(X_{i,n-1} - \overline{X}_i)}{\sum_{n=2}^{N_0} (X_{i,n} - \overline{X}_i)^2} \quad (6)$$

and form the $100(1 - \alpha)\%$ confidence interval for ρ_i from

$$(\rho_i - \hat{\rho}_i)^2 \leq \frac{N_0 - 1}{N_0(N_0 - 3)} (1 - \hat{\rho}_i^2) t_{n_0-3}^2(1 - \alpha/2) \quad (7)$$

with $\alpha = .05$. (Here $t_r(q)$ is the $100q$ percent point of Student's-t distribution with r degrees of freedom.) If this 95% confidence interval contains $\rho_i = 0$, judge the sample size N_{1i} as being adequate for population i. Otherwise calculate

$$N_{2i} = \left[N_{1i} \frac{1 + \hat{\rho}_i}{1 - \hat{\rho}_i}\right] \quad (8)$$

and continue until we have N_{2i} observations from π_i. Finally calculate $\overline{X}_1, \cdots, \overline{X}_k$ based on all available observations and select (as being best) that population which produced the largest of $\overline{X}_1, \cdots \overline{X}_k$.

While procedure $A(\hat{\rho}_i, s_i^2)$ is heuristic (unlike the procedure of Dudewicz and Dalal for $\rho_i = 0$, which is rigorously derived), studies in Dudewicz and Zaino (1977) show it should be sufficient to preclude gross errors due to significant correlations.

5. SELECTION PROCEDURES FOR THE TRAFFIC FATALITY DATA SET

In Section 4 of McDonald (1979), the author specifies a data set X_{ij} ($i = 1, \cdots, 49; j = 1, \cdots, 17$) of motor vehicle traffic fatality rates (MFRs) per 10^8 vehicle miles in the U.S.A. (48 states plus the District of Columbia for the years 1960 to 1976 (17 years), and poses the questions: Which state has the worst fatality rate? Which state has the best fatality rate? (Reasons for interest in the answers are given in his Section 4.)

Considering years as blocks, he asks if there is an interaction between blocks and states. An ANOVA and Tukey's test for nonadditivity indicate so, but once transformed via $Y_{ij} = \ln(X_{ij} - 1)$ an additive model is indicated as appropriate (which suggests a multiplicative model for the X_{ij}'s such as $EX_{ij} = 1 + MA_iB_j$, but this is not explored by McDonald since additivity in monotone transformation of the data is sufficient for his purposes; such models were studied directly for variances (but not for means) by Bechhofer in two *Technometrics* papers (1968a, b), but subset selection in such a model is still an open problem).

Since a large k ($k = 49$), small N ($N = 17$) and high P^* ($P^* = .90$) are present, McDonald chooses to use the subset selection approach. The basic distribution of the data being unknown, he applies *nonparametric* (block-effect) *selection procedures.* For the selection of the best state, the two rules considered are called R_1' and R_2'. Only R_2' is known to satisfy the desired probability requirement Prob $\{CS\} \geq .90$, and it yields a subset containing 31 states. (Rule R_1' reduces this to 12 states, but its use is not fully justified yet.)

In Gupta and Hsu (1977), the authors apply *Gupta's subset selection procedure*, even though analysis of the Y_{ij} residuals reveals heteroscedasticity and a (two-sided Kolmogorov-Smirnov) test for normality leads to rejection of normality at the 1.2% or higher level. Their interest in results assuming normality is still a proper one, however: in applied statistics one often gains insight from comparing the results of several methods applied to a data set (even if some assumptions are clearly violated). The results indicate that if one could take account of magnitudes of differences, which the nonparametric procedures based on ranks cannot, the subset size could be reduced substantially. This information could lead one to lean towards McDonald's rule R_2' (which yields a subset of size 12) for analysis of this data set. (Note that if we had design control, the subset selection procedures in Dudewicz and Dalal (1975) could be used validly even in the presence of heteroscedasticity.)

Sobel (1977) gives new results on *selecting the population which is* most consistent (*least variable*), with consistency being measured by smallness of the inter (α, β)-range. This is done for both indifference zone and subset selection approaches. Due to a lack of tables, these are not applied to the MFR data (and, if they were, would approach a different question than McDonald raised originally). However, Sobel has developed procedures for *selection based on α-quantiles*, and he applies these to the MFR data. Taking $\alpha = 0.5$ (selection based on medians; means also, if one assumes normality), this procedure yields a subset of 3 states as including the best one with Prob{CS} at least .90. Since Sobel's procedure assumes that the $N = 17$ observations are a random sample within each state, however, his use of X_{ij} data is inappropriate: instead he must in fact use the transformed Y_{ij} adjusted to remove time effects, i.e. use $Z_{ij} = Y_{ij} - \hat{\beta}_j$. It would be good to see the subset yielded by this analysis.

Lee, in a revised version of Lee (1977), has a clear discussion of the approaches of McDonald, and of Gupta and Hsu, for indifference zone as well as subset selection. He also develops indifference zone and subset *selection* for *procedures based on a (Bechhofer-Sobel) maximum statistic.* Lee's analysis of the data with regard to its normality, or if not normality then some other location parameter family, makes the use of these procedures seem attractive. (His complete rankings, however, would seem to have a Prob{CS} quite close to zero, and should be approached carefully indeed; as Bishop and Gibbons indicate in a revised version of Bishop (1978), complete ranking of all $k = 49$ with $N = 17$ observations is hardly reliable here.)

Chen (1977) discusses the related *estimation* problem, where one asks "How good is the best alternative?", and gives new procedures for two-factor experiments with no interaction. He precedes this with a comprehensive survey of known results for the one-way classification, and applies them to the MFR data. He also details how one applies indifference zone procedures and *estimates* the *true* Prob{CS}; this is feasible (and one is not restricted to subset selection only) because he is dealing with only $k = 10$ southeastern states.

Bishop and Gibbons (revision of Bishop, 1978) show how to apply *complete ranking* theory (to the six New England states), and indicate how the results would be of interest in the insurance industry. Ranking in terms of variability is also covered. While Bishop and Gibbons indicate their achieved Prob{CS} of .607 is low, note that it is in fact much larger than the $1/6!$ a random ranking would yield. (While in selecting the one best an analyst compares with a random choice Prob{CS} of $1/k$, in complete ranking one compares with $1/k!$. Hence a Prob{CS} of .607 seems quite reasonable in light of the difficulty of the problem.)

Bechhofer's results (1977) deal with the case of *factorial design* where one has design control (which one does not in the MFR data set). This is worth some discussion, because in the physical, chemical, and engineering sciences one often has such control.

When one's experiment is being run to select the optimum levels of two or more factors, one is dealing with a *factorial experiment*. (For example, one could be considering the best combination of the two factors "number of vehicles" and "routing algorithm" for a transportation system where loads are generated at random.) In the setting where there is no interaction (i.e., where the effects due to each factor simply add to produce the effect of the combination) the problem was discussed by Bechhofer (1954, 1977), and by Bawa (1972). When the factors do interact, Lun (1977) showed that one might want to select based on cell means of factor combinations (rather than based on marginal means for each factor), and that to do otherwise could lead to arbitrarily low Prob{CS}.

To be specific, suppose that we have a two-factor experiment with k_1 levels of factor 1 (e.g., k_1 possibilities for "number of vehicles" in the example of the above paragraph) and k_2 levels of factor 2 (k_2 would then be the number of routing algorithms under consideration). Assume our observations are normally distributed with known common variance σ^2 and that if Y_{ij} is an observation taken at level i of factor 1 and level j of factor 2 then

$$E(Y_{ij}) = \mu + \alpha_i + \beta_j + \gamma_{ij} \qquad (9)$$

($i = 1, \cdots, k_1; j = 1, \cdots, k_2$). The "best" population is that one of the $K = k_1k_2$ combinations of factor levels which maximizes the system's mean yield $E(Y_{ij})$. If $\gamma_{ij} = 0$ for all i and j, we have no "interaction" (in which case the same routing algorithm yields the best results regardless of the number of vehicles available to the system).

One can formulate many procedures for dealing with the above problem, especially if all $\gamma_{ij} = 0$.

Procedure SP1 (Bechhofer, 1954) takes N independent observations from each of the K populations, and selects the level associated with the largest marginal sample mean of each of the two factors; the combination of these two levels is asserted to be the best factor combination.

Procedure SP2 (traditional "one-at-a-time" method) takes N_1 independent observations at each level of factor 1, with factor 2 held fixed at one of its levels. The level of factor 1 yielding the largest sample mean is selected. That selected level is then used in experimentation taking N_2 observations at each level of factor 2, after which the level of factor 2 yielding the largest sample mean is selected. The combination of the levels of factor 1 and factor 2 so selected is asserted to be best.

Procedure SP3 takes M independent observations from each of the $K = k_1k_2$ populations and selects the factor combination (population) yielding the largest sample mean as best. (See Gibbons, Olkin and Sobel, 1977 and Lun, 1977.)

Procedures SP1, SP2, and SP3 are respectively the "Factorial", "One-at-a-time", and "Interaction" methods. Bawa (1972) compared SP1 and SP2 when there is no interaction. Lun (1977) showed that if there is interaction, then the inf (over $\mu_{[K]} - \mu_{[K-1]} \geq \delta^*$) of the P(CS) for SP1 is $\leq 1/k_1$ and $\leq 1/k_2$, hence $\leq 1/\max(k_1,k_2)$. We believe it can also be shown that this inf of the P(CS) is, for SP1, $\leq 1/(k_1k_2)$, which shows SP1 to be unreasonable for situations where interaction may be present in unknown magnitude (since one can achieve P(CS) of $1/(k_1k_2)$ by a totally random selection). If one estimates the interactions after the experiment, a reasonable SP4 should be able to be developed which acts as SP1 does if the interactions are "small", and otherwise acts as does SP3. The above considerations generalize to any fixed number of factors.

6. CONCLUSIONS

Selection procedures are now available which allow one to treat a selection problem *as* a selection problem, with appropriate evaluation of the Prob{CS}, and not as some other sort of problem. This paper has given an overview of this area, and concludes with its application to a traffic fatality data set.

7. ACKNOWLEDGMENTS

This research was supported in part by the NATO Research Grants Programme, NATO Research Grant No. 1674.

Comments by S. S. Shapiro and the Editor are acknowledged with thanks.

The article was completed while the author was a Visiting Professor with the Department of Mathematics, Katholieke Universiteit Leuven, Leuven, Belgium.

REFERENCES

BAWA, V. S. (1972). Asymptotic efficiency of one R-factor experiment relative to R one-factor experiments for selecting the best normal population. *J. Amer. Statist. Assoc.*, *67*, 660–661.

BECHHOFER, R. E. (1954). A single-sample multiple decision procedure for ranking means of normal populations with known variances. *Ann. Math. Statist.*, *25*, 16–39.

BECHHOFER, R. E. (1968a). Single-stage procedures for ranking multiply-classified variances of normal populations. *Technometrics*, *10*, 693–714.

BECHHOFER, R. E. (1968b). Multiple comparisons with a control for multiply-classified variances of normal populations. *Technometrics*, *10*, 715–718.

BECHHOFER, R. E. (1977). Selection in factorial experiments. *Proceedings of the 1977 Winter Simulation Conference*, 67–70.*

BISHOP, T. A. (1978). Designing simulation experiments to completely rank alternatives. *Proceedings of the 1978 Winter Simulation Conference*, 203–205.*

CHEN, H. J. (1977). Estimation of the best alternative. *Proceedings of the 1977 Winter Simulation Conference*, 73–79.*

CONWAY, R. W. (1963). Some tactical problems in digital simulation. *Management Science*, *10*, 47–61.

DUDEWICZ, E. J. (1969). An approximation to the sample size in selection problems. *Ann. Math. Statist., 40*, 492–497.

DUDEWICZ, E. J. (1976). *Introduction to Statistics and Probability*. New York: Holt, Rinehart and Winston.

DUDEWICZ, E. J. (1979). Review of *Selecting and Ordering Populations: A New Statistical Methodology*, by J. Gibbons, I. Olkin, and M. Sobel. *Technometrics, 21*, 582–583.

DUDEWICZ, E. J. and DALAL, S. R. (1975). Allocation of observations in ranking and selection with unequal variances. *Sankhyā, 37B*, 28-78.

DUDEWICZ, E. J., RAMBERG, J. S., and CHEN, H. J. (1975). New tables for multiple comparisons with a control (unknown variances). *Biometrische Zeitschrift, 17*, 13–26.

DUDEWICZ, E. J. and ZAINO, N. A., Jr. (1977). Allowance for correlation in setting simulation run-length via ranking-and-selection procedures. *TIMS Studies in the Management Sciences, 7*, 51–61.

GIBBONS, J., OLKIN, I., and SOBEL, M. (1977). *Selecting and Ordering Populations: A New Statistical Methodology*. New York: John Wiley & Sons, Inc.

GUPTA, S. S. (1956). On a decision rule for a problem in ranking means. Ph.D. thesis, University of North Carolina, Chapel Hill, North Carolina.

GUPTA, S. S. and HSU, J. C. (1977). Subset selection procedures with special reference to the analysis of two-way layout: Application to motor-vehicle fatality data. *Proceedings of the 1977 Winter Simulation Conference*, 81–85.*

KLEIJNEN, J. P. C. (1975). *Statistical Techniques in Simulation, Part II*. New York: Marcel Dekker, Inc.

LEE, Y. J. (1977). Winner selection. *Proceedings of the 1977 Winter Simulation Conference*, 87–91.*

LIN, W. T. (1975). Multiple objective budgeting models: A simulation. USC Working Paper #12-01-75, Department of Accounting, Graduate School of Business Administration, University of Southern California, Los Angeles, California.

LUN, F. W. (1977). Selection in factorial experiments. Unpublished Project Report, Statistics 828 (Ranking, Selection, and Multiple-Decision), offered Spring Quarter 1977, Department of Statistics, The Ohio State University, Columbus, Ohio. Available upon request.

McDONALD, G. C. (1979). Nonparametric selection procedures applied to state traffic fatality rates. *Technometrics, 21*, 515–523. (Presented at the 1977 Winter Simulation Conference.)

SOBEL, M. (1977). Selecting the population with the smallest dispersion in a nonparametric setting. *Proceedings of the 1977 Winter Simulation Conference*, 103–114.*

VICÉNS, G. J. and SCHAAKE, J. C., Jr. (1972). Simulation criteria for selecting water resource system alternatives. Report No. 154, Ralph M. Parsons Laboratory for Water Resources and Hydrodynamics, Department of Civil Engineering, Massachusetts Institute of Technology, Cambridge, Massachusetts.

*Available from Institute of Electrical and Electronic Engineers (IEEE), New Jersey Service Center, 445 Hoes Lane, Piscataway, NJ 08854.

HETEROSCEDASTICITY

EDWARD J. DUDEWICZ

If one has observations from several sources, say $X_{i1}, X_{i2}, \ldots$ (which are independent and identically distributed random variables) from source i ($i = 1, 2, \ldots, k$), interest is often in the means $\mu_i = E(X_{i1})$ and variances $\sigma_i^2 = \text{var}(X_{i1})$. One talks of *homoscedasticity* if $\sigma_1^2 = \cdots = \sigma_k^2$, and of *heteroscedasticity* otherwise.

Until recently, the procedures available for these problems assumed normality and $\sigma_1^2 = \cdots = \sigma_k^2 = \sigma^2$ and provided performance characteristics (e.g., power* for a test, confidence coefficient or length for a confidence interval*, probability of correct selection for a ranking-and-selection* procedure) which depended on the unknown σ^2. The solutions given below for these problems do not assume equal variances, yet do allow full control of such performance characteristics as power*, confidence interval* length, and probability of correct selection. These solutions, and those of a large number of other problems involving heteroscedasticity, have as a cornerstone the solution of a simpler problem given below.

When the X_{ij} have nonnormal distributions,

$$\bar{X}_i = \sum_{j=1}^{n} X_{ij}/n \quad \text{and}$$

$$s_i^2 = \sum_{j=1}^{n} \left(X_{ij} - \bar{X}_i\right)^2/(n-1)$$

are still unbiased estimators of μ_i and σ_i^2, respectively. However, $\bar{X}_i$ and s_i^2 are no longer independent random variables and confidence intervals for μ_i (even with random length) are not generally available. Moreover, $\bar{X}_i$ and s_i^2 (while each asymptotically normal by the central limit theorem) are no longer jointly minimal sufficient statistics for (μ_i, σ_i^2) in this setting (*see* ASYMPTOTIC NORMALITY; SUFFICIENCY). Here transformations* are often used; i.e., one lets $Y_{ij} = \xi(X_{ij})$, $1 \leqslant j \leqslant n$, for some function $\xi(\cdot)$ such that $\xi(X_{ij})$ is normally distributed. Then $E(Y_{ij})$ and $\text{var}(Y_{ij})$ can be estimated as before, and confidence intervals provided. Methods are available for relating these estimates and intervals back to μ_i and σ_i^2, the quantities of primary interest.

The procedures given require the experimenter to have design control, but generalize to any statistical problem via the heteroscedastic method (to be discussed in the section "Heteroscedastic Method"). Problems of nonnormality, comparison with the usual variance-stabilizing-transformation approach, and other comparisons and questions that arise in practical implementation are discussed throughout.

BASIC SAMPLING RULE $\mathscr{S}_B(n_0, w)$

If we are able to observe independent and identically distributed normal random variables $X_1, X_2, \ldots$ with mean μ and variance σ^2 (both unknown), and wish to make inferences about μ, recall that $(\bar{X} - \mu)/(s/\sqrt{n})$ has Student's t-distribution* with $n - 1$ degrees of freedom, which gives not only a point estimate, but also a $100(1-\alpha)\%$ confidence interval for μ:

$$\bar{X} - t_{n-1, 1-\alpha/2}\frac{s}{\sqrt{n}} \leqslant \mu \leqslant \bar{X} + t_{n-1, 1-\alpha/2}\frac{s}{\sqrt{n}}.$$

But the interval length, $2t_{n-1,1-\alpha/2}s/\sqrt{n}$, is a random variable and cannot be controlled to be $\leqslant 2L$, for example, by choice of the sample size n. Instead, therefore, we use the following:

Sampling Rule $\mathscr{S}_B(n_0, w)$. Take an initial sample $X_1, \ldots, X_{n_0}$ of size $n_0 (\geqslant 2)$, and calculate

$$\bar{X}(n_0) = \sum_{j=1}^{n_0} X_j/n_0,$$

$$s^2 = \sum_{j=1}^{n_0} \left[X_j - \bar{X}(n_0)\right]^2/(n_0 - 1),$$

$$N = \max\{n_0 + 1, [(ws)^2]\},$$

where $w > 0$ (depends on the problem under

Table 1 $w = t_{n_0-1,1-\alpha/2}/L$ **for** $L = 1$, $1-\alpha = 0.95$

n_0	2	3	4	5	6	7	8	9	10	11	12	∞
w	12.706	4.303	3.182	2.776	2.571	2.447	2.365	2.306	2.262	2.228	2.201	1.960

consideration) and $[y]$ denotes the smallest integer $\geqslant y$ (e.g., $[5.1] = 6$ . . . introduced because sample sizes must be integers). Take $N - n_0$ additional observations $X_{n_0+1}, \ldots, X_N$ and calculate

$$\overline{X}(N - n_0) = \sum_{j=n_0+1}^{N} X_j/(N - n_0),$$

$$\tilde{\overline{X}} = b\overline{X}(n_0) + (1 - b)\overline{X}(N - n_0),$$

$$\overline{X} = \sum_{j=1}^{N} X_j/N,$$

where

$$b = \frac{n_0}{N}\left[1 - \sqrt{1 - \frac{N}{n_0}\left[1 - \frac{N - n_0}{(ws)^2}\right]}\,\right].$$

Stein showed in 1945 [22] that, for his sampling rule $\mathscr{S}$, $(\tilde{\overline{X}} - \mu)/(1/w) \sim t_{n_0-1}$ (*see* FIXED-WIDTH CONFIDENCE INTERVALS). Therefore,

$$\tilde{\overline{X}} - t_{n_0-1,1-\alpha/2}\frac{1}{w} \leqslant \mu \leqslant \tilde{\overline{X}} + t_{n_0-1,1-\alpha/2}\frac{1}{w}$$

is an exact $100(1-\alpha)\%$ confidence interval for μ, and its half-length can be fully controlled to a preset number $L > 0$ by choosing w in $\mathscr{S}_B(n_0, w)$ such that

$$t_{n_0-1,1-\alpha/2}\frac{1}{w} = L, \quad \text{i.e., } w = \frac{t_{n_0-1,1-\alpha/2}}{L}.$$

Note that, since N is an increasing function of w, the total sample size N required is larger for small values of half-length L as well as for high confidence coefficients $1 - \alpha$. (While use of $\overline{X}$ in this setting would yield a slight uniform improvement over use of $\tilde{\overline{X}}$, it is noted in the section "Heteroscedasticity (Several Sources): Tests" that this is not usually the case when dealing with heteroscedasticity of several sources. Attempts to show the opposite retarded development of procedures for heteroscedasticity for many years.)

The procedure described above is valid for any preliminary sample size $n_0 \geqslant 2$. Since $t_{n_0-1,1-\alpha/2}$ decreases as n_0 increases, it is reasonable to make n_0 large if possible. The decrease is negligible after $n_0 \geqslant 12$ or so; hence it is reasonable to take $n_0 = 12$ (or larger, for example, if one is sure to take $n_0 \geqslant B$ for some positive integer $B \geqslant 12$).

The validity of the procedure for any n_0 (with no "optimal" choice of n_0 being obvious) bothered early workers in the field and led to disuse of these early procedures. The realization that $n_0 \geqslant 12$ is all that is required for good results in practice as far as n_0 is concerned is a factor leading to great current interest in these procedures and their extensions as will be discussed later. One may think of the situation as follows: one's total sample size N is approximately w^2s^2, and taking n_0 very small will force a large total sample size simply because of a poor initial estimate s^2 (see Table 1).

BASIC NONNORMALITY AND TRANSFORMATIONS

If X_1 in the preceding section is nonnormal, one often uses such transformations as

$$\xi_1(X_1) = \sqrt{X_1 - a}$$

$$\xi_2(X_1) = X_1^{1/3}$$

$$\xi_3(X_1) = \log_{10}(X_1)$$

$$\xi_4(X_1) = \arcsin\sqrt{X_1}$$

$$\xi_5(X_1) = \sinh^{-1}\sqrt{X_1}\,.$$

If one of these, say $\xi(X_1)$, is normally distributed, then the mean and variance of $Y_i = \xi(X_i)$ may be estimated by

$$\overline{Y} = \sum_{j=1}^{n} Y_j/n, \quad s_Y^2 = \sum_{j=1}^{n}\left(Y_j - \overline{Y}\right)^2/(n-1).$$

However, interest in many cases is not in $E\xi(X_1)$ and $\operatorname{var}\xi(X_1)$, but in the original problem units, EX_1 and $\operatorname{var}(X_1)$. Simply using the inverse transformation, for example to estimate EX_1 by $\overline{Y}^2 + a$ in the case of ξ_1, results in a biased estimate. However, Neyman* and Scott showed in 1960 that the unique minimum variance unbiased estimators* (MVUEs) of $E(X_1)$ are as shown in

Table 2 Transformations and MVUEs[a] of $E(X_1)$

$\xi(X_1)$	MVUE of EX_1
$\sqrt{X_1 - a}$	$\bar{Y}^2 + a + (1 - \frac{1}{n})s_Y^2$
$\log_{10}(X_1)$	$10^{\bar{Y}}S[(\ln 10)^2(1 - \frac{1}{n})(n-1)s_Y^2, n-1]$
$\arcsin\sqrt{X_1}$	$(\sin^2\bar{Y} - 0.5)S[4(\frac{1}{n} - 1)(n-1)s_Y^2, n-1] + 0.5$
$\sinh^{-1}\sqrt{X_1}$	$(\sinh^2\bar{Y} + 0.5)S[4(1 - \frac{1}{n})(n-1)s_Y^2, n-1] - 0.5$

[a] Here $S(a,b) = \sum_{i=0}^{\infty}[(1/i!)(\Gamma(b/2)/\Gamma(i+(b/2)))(a/4))^i]$; this series converges faster than the series for the exponential function.

Table 3 MVUEs[a] of Variances of Table 2 Estimators of EX_1

$\xi(X_1)$	MVUE of Variance of MVUE of EX_1
$\sqrt{X_1 - a}$	$\frac{4}{n}s_Y^2\bar{Y}^2 + s_Y^4\{(1 - \frac{1}{n})^2 - \frac{n-1}{n+1}[1 - 2(1 - \frac{1}{n})^2 + 3(1 - \frac{1}{n})^4]\}$
$\log_{10}(X_1)$	$10^{2\bar{Y}}\{S^2[(\ln 10)^2(1 - \frac{1}{n}), n-1] - S[2(\ln 10)^2(1 - \frac{2}{n}), n-1]\}$
$\arcsin\sqrt{X_1}$	$(\widehat{EX_1})^2 - \frac{1}{4} - \frac{1}{8}S(-8, n-1) + \frac{1}{2}\cos(2\bar{Y})S[-4(1 - \frac{1}{n}), n-1] - \frac{1}{8}\cos(4\bar{Y})S[-8(1 - \frac{2}{n}), n-1]$
$\sinh^{-1}\sqrt{X_1}$	$(\widehat{EX_1})^2 - \frac{1}{4} - \frac{1}{8}S(8, n-1) + \frac{1}{2}\cosh(2\bar{Y})S[4(1 - \frac{1}{n}), n-1] - \frac{1}{8}\cosh(4\bar{Y})S[8(1 - \frac{2}{n}), n-1]$

[a] For $S(a,b)$, see the footnote to Table 2.

Table 2, assuming that $\xi(X_1)$ is normally distributed. General results for second-order entire functions were also given by Neyman and Scott.

In 1968, Hoyle provided the MVUEs of $\mathrm{var}(X_1)$ and, more important, of the variances of the estimators of EX_1 given in Table 2 (see Table 3). [Here it is also assumed that $\xi(X_1)$ is normally distributed.] The latter can be used to obtain approximate 95% confidence intervals for EX_1; for example, when using $\sqrt{X_1 - a}$,

$$\mu \in \bar{Y}^2 + a + \left(1 - \frac{1}{n}\right)s_Y^2 \pm 2\sqrt{\lambda},$$

$$\lambda = \frac{4}{n}s_Y^2\bar{Y}^2 + s_Y^4\left\{\left(1 - \frac{1}{n}\right)^2 - \frac{n-1}{n+1}\left[1 - 2\left(1 - \frac{1}{n}\right)^2 + 3\left(1 - \frac{1}{n}\right)^4\right]\right\}.$$

More recent work and comparisons with other methods are given by Land [18].

HETEROSCEDASTICITY (SEVERAL SOURCES): TESTS

Let $X_{i1}, X_{i2}, \ldots$ be independent and identically distributed normal random variables with mean μ_i and variance σ_i^2 $(i = 1, 2, \ldots, k)$. Experimenters have often been cautioned that "the assumption of equal variability should be investigated" (e.g., by Cochran and Cox in 1957 [6], by Juran et al. in 1974 [17]). For some tests for homoscedasticity, see, e.g., Harrison and McCabe [13] or Bickel [2]; typically such tests have low power, and may not detect even substantial heteroscedasticity. However, no exact statistical procedures have been available for dealing with cases where one finds that variabilities are unequal. [A variance-stabilizing transformation is commonly employed (e.g., arcsin for binomial data; *see* EQUALIZATION OF VARIANCES); however, if X_{ij} *is* normal, then $\xi(X_{ij})$ will be nonnormal. The transformation method has not been developed to handle this problem except in special cases, and even there one deals not with the parameters $\mu_1, \ldots, \mu_k$ of basic interest if one uses such a transformation, but rather with some transform whose meaning (i.e., interpretability) will not often be clear. We do not therefore regard transformations as of general use for $k \geqslant 2$ when $\mu_1, \ldots, \mu_k$ are parameters of natural interest (not arbitrary parametrizations).]

It was first developed in the 1970s by E. J. Dudewicz that, applying sampling procedure $\mathscr{S}_B(n_0, w)$ from the section "Basic Sampling Rule $\mathscr{S}_B(n_0, w)$" separately to each source of observations, one would obtain the ability to control fully the performance characteristics of statistical procedures even in the presence of heteroscedasticity. Let $\tilde{\tilde{X}}_i$ result from applying the sampling procedure to X_{i1}, $X_{i2}, \ldots$ $(i = 1, 2, \ldots, k)$. When $k = 1$, one can develop procedures (as in the preceding section) using $\tilde{\tilde{X}}$, but if one replaces this by $\bar{X}$ at the end, the procedure is still valid; it has slightly better performance characteristics (higher power), and is even simpler ($\bar{X}$ being simpler than $\tilde{\tilde{X}}$, which is a random-weighted combination of the sample means of the first and second stages of sampling). However, this improvement is not large: approximately the amount that increasing sample size from N to $N+1$ will buy. This improvement of $\bar{X}$ over $\tilde{\tilde{X}}$ has been shown *not* to hold generally when $k \geqslant 2$: in most such cases, if $\bar{X}_1, \ldots, \bar{X}_k$ are used to replace $\tilde{\tilde{X}}_1, \ldots, \tilde{\tilde{X}}_k$, the procedure no longer has the desired performance characteristics.

We describe the new analysis-of-variance* procedures for the one-way* layout; similar procedures are available for r-way layouts, $r > 1$. In the one-way layout, we might want to test the null hypothesis

$$H_0 : \mu_1 = \mu_2 = \cdots = \mu_\kappa .$$

Define

$$\tilde{F} = \sum_{i=1}^{k} w^2 \left(\tilde{\tilde{X}}_i - \tilde{\tilde{X}}.\right)^2,$$

$$\tilde{\tilde{X}}. = \frac{1}{k} \sum_{i=1}^{k} \tilde{\tilde{X}}_i ;$$

reject H_0 if and only if

$$\tilde{F} > \tilde{F}^{\alpha}_{k,n_0},$$

where $\tilde{F}^{\alpha}_{k,n_0}$ is the upper αth percent point of the null distribution of $\tilde{F}$. This distribution is also that of $Q = \sum_{i=1}^{k}(t_i - \bar{t}.)^2$, where the $\{t_i\}$ are independent identically distributed Student's-t variates with $n_0 - 1$ degrees of freedom and $\bar{t}. = (1/k)\sum_{i=1}^{k} t_i$.

Values of $\tilde{F}^{\alpha}_{k,n_0}$ obtained by a Monte Carlo* sampling experiment, together with the power attained at various alternatives measured by $\delta = \sum_{i=1}^{k}(\mu_i - \bar{\mu}.)^2$, for various given $1/w^2$ values, appear in Bishop and Dudewicz [3]. There is a need for approximations to the percentage points of the $\tilde{F}$ statistics under the null and alternative distributions. Such approximations are available in the general setting (see Dudewicz and Bishop [9]), and have been studied in special cases (see Bishop et al. [4]). Consider first the limiting distribution of $\tilde{F}$ as $n_0 \rightarrow \infty$. This is noncentral chi-square* with $k - 1$ degrees of freedom and noncentrality parameter $\Delta = \sum_{i=1}^{k} w^2(\mu_i - \bar{\mu}.)^2$, denoted by $\chi^2_{k-1}(\Delta)$. However, numerical results indicate that for small n_0 the tails of this distribution are too light to give a good approximation. One therefore approximates by a random variable with a $[(n_0 - 1)/(n_0 - 3)]\chi^2_{k-1}(\Delta)$ distribution (in which case $\tilde{F}$ and its approximating distribution have the same expected value under H_0).

Example. Suppose that we wish to test the hypothesis that four different chemicals are equivalent in their effects. Suppoe that we decide to take initial samples of size 10 with each treatment, that we want only a 5% chance of rejecting H_0 if in fact H_0 is true, and an 85% chance of rejecting H_0 if the spread among μ_1, μ_2, μ_3, and μ_4 is at least 4.0 units. We then proceed, step by step, as follows.

Step 1: Problem Specification. With $k = 4$ sources of observations, we desire an $\alpha = 0.05$ level test of H_0: $\mu_1 = \mu_2 = \mu_3 = \mu_4$, and if the spread among μ_1, μ_2, μ_3, and μ_4 is $\delta = 4.0$ units or more, we desire power (probability of then rejecting the false hypothesis H_0) of at least $P^* = 0.85$.

Step 2: Choice of Procedure. Assuming we do not know that $\sigma_1^2 = \sigma_2^2 = \sigma_3^2 = \sigma_4^2$, only procedure $\mathscr{S}_B(n_0, w)$ can guarantee the specifications. It requires that we sample n_0 observations in our first stage, and recommends that n_0 be at least 12 (although any $n_0 \geqslant 2$ will work). Suppose that the experimenter only wants to invest 40 units in first-stage experimentation and sets $n_0 = 10$.

Step 3: First Stage. Draw $n_0 = 10$ independent observations from each source, with results as shown in Table 4.

Step 4: Analysis of First-Stage Data. Calculate the first-stage sample variances s_1^2, s_2^2,

s_3^2, s_4^2, the total sample sizes N_1, N_2, N_3, N_4, needed from the four sources, and the factors b_1, b_2, b_3, b_4 to be used in the second-stage analysis. These quantities appear in Table 5. The value of w is found as follows.

Table 4 First-Stage Samples

Chemical 1	Chemical 2	Chemical 3	Chemical 4
77.199	80.522	79.417	78.001
74.466	79.306	78.017	78.358
82.746	81.914	81.596	77.544
76.208	80.346	80.802	77.364
82.876	78.385	80.626	77.554
76.224	81.838	79.011	75.911
78.061	82.785	80.549	78.043
76.391	80.900	78.479	78.947
76.155	79.185	81.798	77.146
78.045	80.620	80.923	77.386

We desire power $P^* = 0.85$ (step 1 above) when

$$\Delta = \frac{w^2\delta^2}{4} = \frac{w^2(4.0)^2}{4} = 4.0w^2.$$

To set w for this requirement, we first need to know when we reject. We will later reject H_0 if $\tilde{F} > \tilde{F}_{4,10}^{0.05}$, where, approximately,

$$\tilde{F}_{4,10}^{0.05} = \frac{n_0 - 1}{n_0 - 3}(7.81) = 10.04,$$

and where a chi-square variable with $k - 1 = 3$ degrees of freedom exceeds 7.81 with probability $\alpha = 0.05$ (see standard tables, e.g., Pearson and Hartley [20, p. 137], Dudewicz [8, p. 459]). The power will be, approximately,

$$P\left[\chi_3^2(\Delta) > 7.81\right] = 0.85$$

if $\Delta = 12.301$ (see the tables in Haynam et al. [14, p. 53]), so $w^2 = 12.301/4.0 = 3.075$.

Step 5: Second Stage. Draw $N_i - n_0$ observations from source i ($i = 1,2,3,4$), yielding Table 6.

Table 6 Second-Stage Samples

Chemical 1	Chemical 2	Chemical 3	Chemical 4
82.549	79.990	80.315	78.037
78.970			
78.496			
78.494			
80.971			
80.313			
76.556			
80.115			
78.659			
77.697			
80.590			
79.647			
82.733			
80.552			
79.098			

Step 6: Final Calculations. Calculate the $\tilde{\tilde{X}}_i$ and $\tilde{F}$, and find

$$\tilde{\tilde{X}}_1 = 79.079, \quad \tilde{\tilde{X}}_2 = 80.688,$$
$$\tilde{\tilde{X}}_3 = 80.197, \quad \tilde{\tilde{X}}_4 = 77.597;$$
$$\tilde{\tilde{X}}_. = 79.390, \quad \tilde{F} = 17.38.$$

Step 7: Final Decision. Since $\tilde{F} = 17.38$ exceeds $\tilde{F}_{4,10}^{0.05} = 10.04$, we reject the null hypothesis and decide that the chemicals differ.

HETEROSCEDASTICITY (SEVERAL SOURCES): CONFIDENCE INTERVALS

The case of a confidence interval for the mean when $k = 1$ was considered in the section "Basic Sampling Rule $\mathscr{S}_B(n_0, w)$". When $k = 2$, a two-sided confidence interval for the difference $\mu_1 - \mu_2$, of half-length $L > 0$ and with confidence coefficient $1 - \alpha$, is given by

$$(\tilde{\tilde{X}}_1 - \tilde{\tilde{X}}_2) - L \leqslant \mu_1 - \mu_2 \leqslant (\tilde{\tilde{X}}_1 - \tilde{\tilde{X}}_2) + L$$

Table 5 Analysis of First Stage

	Chemical 1	Chemical 2	Chemical 3	Chemical 4
n_0	10	10	10	10
Sample mean	77.837	80.580	80.122	77.625
s_i^2	7.9605	1.8811	1.7174	0.6762
w	1.754	1.754	1.754	1.754
N_i	25	11	11	11
b_i	0.330	0.936	0.939	0.969

if we choose [in $\mathscr{S}_B(n_0, w)$]

$$w = \frac{c_{1-\alpha/2}(n_0)}{L},$$

where c is as given in Table 7.

Note that the corresponding test solves the Behrens–Fisher problem* exactly in two stages, with controlled level and power. This was first noted by Chapman [5].

For $k > 2$, multiple-comparison* procedures are also available for many of the usual multiple-comparison confidence interval goals see, e.g., Dudewicz et al. [12]).

HETEROSCEDASTICITY (SEVERAL SOURCES): RANKING AND SELECTION*

Here $k \geqslant 2$ and, in the indifference-zone formulation of the problem, we wish to select that source having mean value $\max(\mu_1, \dots, \mu_k)$. Let $\mu_{[1]} \leqslant \cdots \leqslant \mu_{[k]}$ denote the ordered values of $\mu_1, \dots, \mu_k$; thus $\mu_{[k]}$ denotes $\max(\mu_1, \dots, \mu_k)$, etc.

The performance characteristic of interest is the probability that we will make a correct selection (CS), i.e., that the population selected is the one that has mean $\mu_{[k]}$. Following Bechhofer [1], we require Pr(CS) to have at least a specified value P^* $(1/k < P^* < 1)$ whenever the largest mean is at least δ^* more than the next-to-largest mean; i.e. we require

$$\Pr(\mathrm{CS}) \geqslant P^*$$

whenever

$$\mu_{[k]} - \mu_{[k-1]} \geqslant \delta^* > 0.$$

The procedure (see Dudewicz and Dalal [10]) is to select that source which yields the largest of $\tilde{\tilde{X}}_1, \dots, \tilde{\tilde{X}}_k$; i.e.,

$$\text{select } \pi_i \text{ iff } \tilde{\tilde{X}}_i = \max\left(\tilde{\tilde{X}}_1, \dots, \tilde{\tilde{X}}_k\right).$$

In the sampling rule $\mathscr{S}_B(n_0, w)$ one chooses $w = c_{P^*}(n_0)/\delta^*$, where $c_{P^*}(n_0)$ for specified values of P^* and n_0 is given in Table 7 for $k = 2$, and in Dudewicz et al. [12] for $k > 2$. Approximations for $k > 25$ are given by Dudewicz and Dalal [10], as are procedures for the subset-selection formulation of the problem.

THE HETEROSCEDASTIC METHOD

The special-case solutions described above have been placed into a general theory with the heteroscedastic method of Dudewicz and Bishop [9]. In a general decision-theoretic setting, they show how to develop procedures like the one above in any problem. It is also shown that no single-stage procedure can solve most such problems.

Some questions one might ask about the procedures thus produced are as follows. First, how do they perform under violation of normality? Iglehart [15] has shown, in some computational settings, that replacing s^2 by a jackknife* estimator is sufficient to preserve the main properties of the procedures. Other recent work [11] shows asymptotic validity under asymptotic normality*. Second, are they preferable to comparable sequential procedures? In most cases there are no "comparable" sequential procedures: those of Chow–Robbins type (*see* FIXED-WIDTH CONFIDENCE INTERVALS) which are usually mentioned only have asymptotic validity even under exact normality, while the $\mathscr{S}_B(n_0, w)$-based two-stage procedures have exact known properties. It is sometimes claimed that the sequential procedures are more efficient, but this is only as, for example, $\sigma_i^2 \to 0$. The so-called inefficiency of $\mathscr{S}_B(n_0, w)$ in this situation is because it then

Table 7 $c_{1-\gamma}(n_0)$

$1-\gamma$ \ n_0	10	11	12	13	14	15	20	25	30
0.75	1.03	1.02	1.02	1.01	1.01	1.00	0.99	0.98	0.98
0.80	1.29	1.28	1.27	1.26	1.26	1.25	1.24	1.23	1.22
0.85	1.60	1.59	1.57	1.56	1.56	1.55	1.53	1.51	1.51
0.90	2.00	1.98	1.96	1.95	1.94	1.93	1.90	1.88	1.87
0.95	2.61	2.58	2.56	2.53	2.52	2.50	2.45	2.42	2.41
0.975	3.18	3.13	3.09	3.06	3.04	3.02	2.95	2.91	2.88
0.99	3.89	3.82	3.76	3.71	3.67	3.64	3.54	3.48	3.45
0.995	4.41	4.31	4.24	4.18	4.13	4.09	3.96	3.89	3.85
0.999	5.61	5.45	5.32	5.22	5.14	5.07	4.86	4.74	4.67

requires $N = n_0 + 1$ (since $N \geqslant n_0 + 1$ always) and in fact (as $\sigma_i^2 \to 0$) and $N \to 1$ will suffice. This appears to have little practical relevance, as one usually knows that trivial sample sizes will be insufficient for one's problems; it is rather a curiosity of mathematical interest only.

As a final note, we mention that while variance-stabilizing transformations and other approximate methods have existed for many years, most experimental situations are such that the problem is far from solved by these approximate methods. For example, such methods misallocate sample size by taking the same sample size from a treatment with relatively small variability, as from a treatment with relatively large variability, even though the need for observations on the latter is substantially greater and they have a greater beneficial effect on performance characteristics of the overall analysis. Also, procedures based on $\mathscr{S}_B(n_0, w)$ behave acceptably even if variances are equal; hence the equality-of-variances tests, which are known to be weak in power, can be skipped and these new procedures can be applied directly without regard to equality or inequality of variances.

References

[1] Bechhofer, R. E. (1954). *Ann. Math. Statist.*, **25**, 16–39. (The original paper on ranking and selection methods.)

[2] Bickel, P. J. (1978). *Ann. Statist.*, **6**, 266–291. (A theoretical study of asymptotic power functions of tests for heteroscedasticity, especially in linear models under nonnormality.)

[3] Bishop, T. A. and Dudewicz, E. J. (1978). *Technometrics*, **20**, 419–430. (The original ANOVA procedures for heteroscedastic situations, with tables and approximations needed for implementation.)

[4] Bishop, T. A., Dudewicz, E. J., Juritz, J., and Stephens, M. A. (1978). *Biometrika*, **65**, 435–439. (Considers approximating the $\tilde{F}$ distribution.)

[5] Chapman, D. G. (1950). *Ann. Math. Statist.*, **21**, 601–606. [Considered the $k = 2$ test of H_0: $\mu_1 = \mu_2$ vs. H_1: $|\mu_1 - \mu_2| = d$, and also H_0: $\mu_1 = r\mu_2$. Tabled c for $n_0 = 2(2)12$ and $1 - \alpha = 0.975, 0.995$, correct to 0.1 (except for a gross error when $n_0 = 4$).]

[6] Cochran, W. G. and Cox, G. M. (1957). *Experimental Designs* 2nd ed. Wiley, New York. (On p. 91, notes the need to test for heteroscedasticity.)

[7] Dantzig, G. B. (1940). *Ann. Math. Statist.*, **11**, 186–191. (First to show that one-stage procedures could not solve many practical problems.)

[8] Dudewicz, E. J. (1976). *Introduction to Statistics and Probability*. American Sciences Press, Columbus, Ohio.

[9] Dudewicz, E. J. and Bishop, T. A. (1979). In *Optimizing Methods in Statistics*, J. S. Rustagi, ed. Academic Press, New York, pp. 183–203. (Develops the heteroscedastic method as a unifying procedure in a general setting, and shows how the procedures referred to here fit in as special cases.)

[10] Dudewicz, E. J. and Dalal, S. R. (1975). *Sankhyā B*, **37**, 28–78. (Solves the heteroscedastic ranking and selection problem in indifference-zone and subset-selection settings. Gives extensive tables and suggestions on solutions of other problems with similar methods.)

[11] Dudewicz, E. J. and van der Meulen, E. C. (1980). *Entropy-Based Statistical Inference, II: Selection-of-the-Best/Complete Ranking for Continuous Distributions on* (0, 1), *with Applications to Random Number Generators*, Communication No. 123. Mathematical Institute, Katholieke Universiteit, Leuven, Belgium. [New results on validity of $\mathscr{S}_B(n_0, w)$ under asymptotic (rather than exact) normality.]

[12] Dudewicz, E. J., Ramberg, J. S., and Chen, H. J. (1975). *Biom. Zeit.*, **17**, 13–26. [Gives procedures and theory for one-sided multiple comparisons with a control, plus extensive tables of $c_{1-\gamma}(n_0)$ useful in many problems.]

[13] Harrison, M. J. and McCabe, B. P. M. (1979). *J. Amer. Statist. Ass.*, **74**, 494–499. (Introduces and compares tests for heteroscedasticity, especially in linear regression models.)

[14] Haynam, G. E., Govindarajulu, Z., and Leone, F. C. (1970). *Selected Tables in Mathematical Statistics*, Vol. 1, H. L. Harter and D. B. Owen, eds. Markham, Chicago, pp. 1–78. (Tables of the cumulative noncentral chi-square distribution.)

[15] Iglehart, D. L. (1977). *TIMS Stud. Manag. Sci.*, **7**, 37–49. [Suggested using a jackknife variance estimator with $\mathscr{S}_B(n_0, w)$, and indicated it solves nonnormality problems in his context.]

[16] Johnson, N. L. and Kotz, S. (1970). *Distributions in Statistics: Continuous Univariate Distributions*, Vol. 2. Wiley, New York.

[17] Juran, J. M., Gryna, F. M., Jr., and Bingham, R. S., Jr., eds. (1974). *Quality Control Handbook*, 3rd ed. McGraw-Hill, New York. (Recommend testing for heteroscedasticity on p. 46 of Section 27.)

[18] Land, C. E. (1974). *J. Amer. Statist. Ass.*, **69**, 795–802. (Considers and compares several methods for confidence interval estimation for original means after data transformations to normality, including the method considered in the section "Basic Nonnormality and Transformations.")

[19] Miller, R. G. (1974). *Biometrika*, **61**, 1–15. (Recent survey of jackknife methods, with an extensive bibliography.)

[20] Pearson, E. S. and Hartley, H. O., eds. (1970). *Biometrika Tables for Statisticians*, Vol. 1, 3rd ed. Cambridge University Press, Cambridge (reprinted with additions).

[21] Ruben, H. (1962). *Sankhyā A*, **24**, 157–180. (Looks at testing $H_0: \mu_1 = \mu_2$ when $k = 2$, concentrating attention on $\bar{X}_1$ and $\bar{X}_2 \ldots$ hence missing the generalizations found in the 1970s.)

[22] Stein, C. (1945). *Ann. Math. Statist.*, **16**, 243–258. [The original reference to $\mathscr{S}_B(n_0, w)$, but did not consider heteroscedasticity, perhaps due to emphasis on $\bar{X}$ as a replacement for $\tilde{\bar{X}}$.]

Acknowledgment

This research was supported by Office of Naval Research Contract N00014-78-C-0543.

(EQUALIZATION OF VARIANCE
FIXED-WIDTH CONFIDENCE INTERVALS
MULTIPLE COMPARISONS
RANKING AND SELECTION)

TIMS Studies in the Management Sciences 7 (1977) 51–61

ALLOWANCE FOR CORRELATION IN SETTING SIMULATION RUN–LENGTH VIA RANKING-AND-SELECTION PROCEDURES *

Edward J. DUDEWICZ
Departments of Statistics, The Ohio State University and Stanford University

and

Nicholas A. ZAINO, Jr.
Department of Statistics, The University of Rochester

Dudewicz and Zaino (1973) investigated for the first time the effect of correlation, commonly present in simulation experiments, on ranking and selection procedures. These procedures are often the most appropriate statistical approach to the design and analysis of simulation experiments, and numerous authors (e.g. [3,8,12,13,16,17]) have suggested their use to set the length of computer simulation runs. Since ranking and selection procedures have generally assumed the observations are independent and since in simulation the data encountered are often correlated (e.g., within a population through lack of frequent regeneration points, and across populations through use of common random number sequences for various alternatives being simulated) those procedures may not be fully appropriate. In this paper we review the effect of correlated data on the sample size (simulation run-length) requirements of a selection procedure, detail the use of ranking and selection procedures in simulation experiments with reference to specific examples, and give methods for using ranking and selection procedures to set simulation run-length in the presence of correlation.

1. Ranking and selection in simulation

Today simulation studies play an important and ever more significant role in virtually every field of human endeavor. For example, such studies arise in business (including such areas as jobshop scheduling and marketing), the humanities (psychoanalysis), science (air pollution, chemistry, genetics), and social science (social conflict, housing policies); for detailed references to each of these applications see [6, pp. 25–26]. In many cases the experimenter has under consideration several (two or more) proposed procedures for running a real-world system, and is simulating in order to determine which is the best procedure (with regard to certain specified criteria of "goodness"). In other words, often the experimenter does not wish basically to test hypotheses, or construct confidence intervals, or perform regression analyses (though these may be appropriate minor parts of his analysis): he wishes basically to select the best of several procedures (and the major part of his analysis should therefore be directed towards this goal). This is precisely the prob-

* Received March 30, 1976, revised August 18, 1976.

lem which ranking and selection procedures were developed for, and these procedures explicitly set sample size (in simulation this means run-length) so as to guarantee that the probability the experimenter actually selects the best procedure (called the "probability of correct selection" or P(CS)) is suitably large; for further details see [7].

2. The model studied

Assume given k populations $\pi_1, ..., \pi_k$ (sources of observations which, in the context of simulation, arise from k rules or procedures which are to be evaluated via a simulation experiment), the observations from π_i denoted by $\{X_{in}\}$ where

$$X_{in} = \rho X_{i,n-1} + Z_{in} = \sum_{j=0}^{\infty} \rho^j Z_{i,n-j} \tag{1}$$

for some ρ ($|\rho| < 1$), and $\{Z_{in}\}$ are sequences of uncorrelated random variables with means $(1 - \rho)\mu_i$ and variances σ^2. (In our notation X_{in} is to be thought of as the nth observation from population π_i.) The problem is to select a population associated with the largest mean yield $\mu_{[k]} = \max(\mu_1, ..., \mu_k)$. Effects due to having $\rho \neq 0$ (instead of the usually-assumed $\rho = 0$) were first discussed by Dudewicz and Zaino [9] and will be summarized in section 3, and their implications incorporated into the recommended ranking and selection procedures of section 4.

Before discussing the assumptions behind model (1) further, we will give an example from simulation to illustrate the X_{ij}'s and the need for ranking and selection concepts. Lin [14] investigated the effects of four modes of budgeting for a firm on the firm's actual profit and sales per period. The basic goal was to select that mode which yielded the highest mean value of actual profit per period (though there were subsidiary goals which required use of confidence intervals and analysis of variance procedures, the sample size was set using ranking and selection procedures, so that the basic goal of selecting the best mode would be guaranteed with high probability). In that context X_{in} is the actual profit per period in period n using mode i (i = 1, 2, 3, 4 since there are k = 4 modes of budgeting available), and, since past decisions affect present profit, $X_{i1}, X_{i2}, ...$ are *not* independent random variables. The problem is [1] to select the mode (population) with the largest mean profit per period.

For model (1), it can easily be shown (see, e.g. [1] or [4]) that

$$E(X_{in}) = \mu_i, \tag{2}$$

[1] While the data generated by Lin were as specified, he was only interested in the 24-period (two year) performance of the new firm and therefore used 126 replicate simulation runs of 24 periods each, rather than one long run of (e.g.) 126 × 24 = 3024 periods.

$$\mathrm{Var}(X_{in}) = \frac{\sigma^2}{1-\rho^2} \equiv \sigma_X^2 \quad \text{(say)}, \tag{3}$$

$$R_s = \mathrm{Cov}(X_{in}, X_{i,n+s}) = \frac{\sigma^2}{1-\rho^2}\rho^{|s|} = \sigma_X^2 \rho^{|s|}. \tag{4}$$

(Note that the alternative assumptions that the observations $\{X_{in}\}$ have means μ_i, common variances σ_X^2, and spacing-dependent correlations $\rho^{|s|}$... which yield higher absolute associations between observations of closer indices ... are actually equivalent to assuming the first-order autoregressive process model (1).) We assume that the k populations are independent (hence there are effects on run-length of correlation within a population, but there is no correlation across populations), and that Z_{in} is normally distributed. Since the sample means $\overline{X}_i$ are unbiased estimators for μ_i and are asymptotically normal under general assumptions about Z_{in} (see, e.g. [5]) the results of section 3 should be indicative for large samples even if Z_{in} is not normal.

3. The "sample means" ranking and selection procedure's performance under correlation

One possible procedure in the context of section 2 (*Procedure* A) is to take some number N_3 of observations from each population, and choose as "best" the population which yields the largest sample mean (where "best" refers to any population with mean yield $\mu_{[k]} = \max(\mu_1, ..., \mu_k)$).

Adopting the usual indifference-zone type formulation of ranking and selection originated by Bechhofer [2], let (λ^*, P^*) $(0 < \lambda^* < \infty, 1/k < P^* < 1)$ be two specified constants, denote the ranked means $\mu_1, ..., \mu_k$ by $\mu_{[1]} \leqslant ... \leqslant \mu_{[k]}$, and try to set N_3 as the smallest sample size for which the probability of correct selection of a population associated with $\mu_{[k]}$ (say P(CS)) is $\geqslant P^*$ whenever $\mu_{[k]} - \mu_{[k-1]} \geqslant \delta^* \equiv \lambda^* \sigma_X$. Then N_3 is the smallest integer satisfying

$$\frac{1}{N_3}\left\{\frac{1+\rho}{1-\rho} - \frac{2\rho(1-\rho^{N_3})}{N_3(1-\rho)^2}\right\} \leqslant \frac{1}{N} \tag{5}$$

where N is the sample size required by the procedure of Bechhofer [2] for the case of *independent* observations ($\rho = 0$).

The fact that N_3 in Procedure A should be set as in (5) follows from the fact that our problem reduces to the framework of [2] since

$$\begin{aligned}\mathrm{Var}(\overline{X}_i) &= \frac{1}{N_3}\sum_{s=1-N_3}^{N_3-1}\left(1-\frac{|s|}{N_3}\right)R_s \\ &= \frac{\sigma_X^2}{N_3}\left\{\frac{1+\rho}{1-\rho} - \frac{2\rho(1-\rho^{N_3})}{N_3(1-\rho)^2}\right\} = Q(N_3) \quad \text{(say)},\end{aligned} \tag{6}$$

Table 1
Evaluation of approximation N_2 to the N_3 of procedure A.

	P^*	$\rho = -0.95$		$\rho = -0.50$		$\rho = -0.10$		$\rho = 0.10$		$\rho = 0.50$		$\rho = 0.95$	
		N_2	N_3	N_2	N_3	N_2	N_3	N_2	N_3	N_2	N_3	N_2	N_3
$k = 2$	0.7500	3	6	31	32	75	75	112	112	273	272	3549	3549
$\lambda^* = 0.10$	0.9000	9	16	110	111	269	269	402	402	986	985	12 811	12 791
	0.9900	28	40	361	363	886	886	1323	1323	3248	3246	42 213	42 194
$k = 25$	0.7662	20	30	254	255	623	623	930	930	2282	2281	26 696	26 676
$\lambda^* = 0.10$	0.9108	31	44	401	402	983	983	1468	1468	3602	3601	46 820	46 800
	0.9926	58	74	749	750	1837	1837	2744	2744	6734	6733	87 536	87 516
$k = 2$	0.7500	1	1	1	1	1	1	2	1	3	1	36	1
$\lambda^* = 1.00$	0.9000	1	2	2	2	3	3	5	4	10	9	129	105
	0.9900	1	2	4	5	9	10	14	14	33	32	423	402
$k = 25$	0.7662	1	2	3	4	7	7	10	10	23	22	297	276
$\lambda^* = 1.00$	0.9108	1	2	5	6	10	11	15	15	37	35	469	448
	0.9926	1	2	8	9	19	19	28	28	68	66	876	856

so that X_{in} being $N(\mu_i, \sigma_X^2)$ implies $\overline{X}_i$ is $N(\mu_i, Q(N_3))$ and this is essentially the problem of Bechhofer [2]. With $\rho = 0$ the necessary sample size is called N instead of N_3, and the quantity N is found from the relation

$$P^* = h_k\left(\frac{\delta^*\sqrt{N}}{\sigma_X}\right) \tag{7}$$

where $h_k(\cdot)$ is a function tabulated by (e.g.) Bechhofer [2] and Milton [15]. With $\rho \neq 0$, one needs the smallest N_3 such that $Q(N_3) \leqslant \sigma_X^2/N$. (Note that although σ_X^2 is a function of ρ we delete $\sigma_X^2(\rho)/\sigma_X^2(\rho = 0)$ via $\delta^* = \lambda^*\sigma_X$ in order to standardize our comparisons.) For *positive correlation* ($\rho > 0$), one can show that N_3 is an increasing function of N, hence we can find N_3 by: solving (7) for N, solving for N_3 in (5) with the inequality replaced by an equality (e.g., use an iterative procedure such as the Newton-Raphson method, or the method indicated below), and rounding N_3 up to the next largest integer. Alternatively one could approximate (instead of iterating) by e.g.

$$N_2 = N\frac{1+\rho}{1-\rho}, \tag{8}$$

which can be shown to be conservative and (for moderate or large N) "close" to the exact solution N_3.

For *negative correlation* ($\rho < 0$) equation (5) cannot be used for nonintegral N_3, hence a search procedure is required to find N_3 (see below). Approximation (8) would underestimate the necessary sample size N_3 (see below).

An extensive comparison [2] of the exact and approximate solutions to (5) was made and is summarized in Table 1. For $\rho > 0$, N_3 was determined by solving equa-

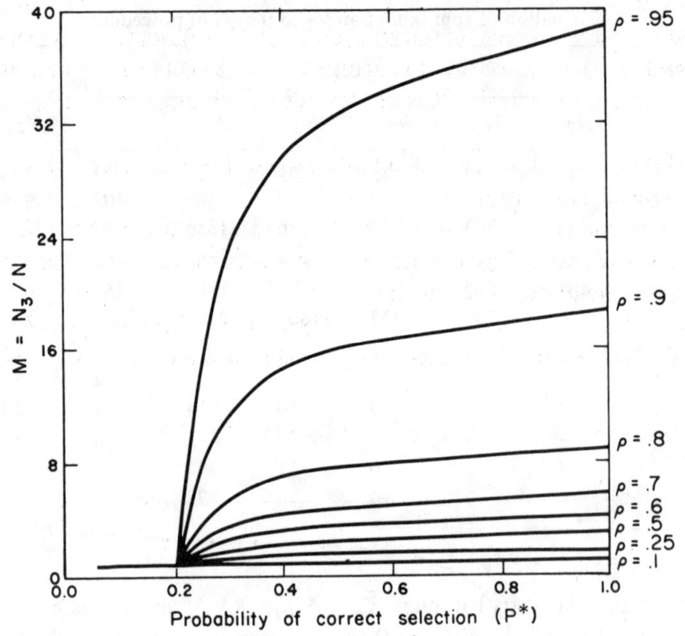

Figure 1. $M = N_3/N$ as a function of P^* for $\lambda^* = 1.0$, $k = 25$.

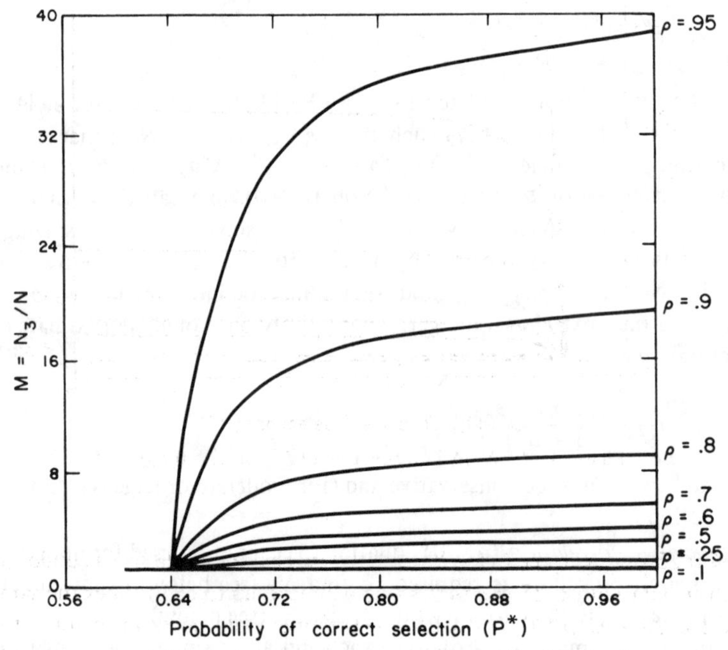

Figure 2. $M = N_3/N$ as a function of P^* for $\lambda^* = 0.5$, $k = 2$.

tion (5) using IBM's Scientific Subroutine Package subroutine RTWI. For $\rho < 0, N_3$ was found by a search procedure starting either at $N_3 = 0$ or at $N_3 = N_1$ (here N_1, to be discussed in [10], is known to satisfy $N_1 \leqslant N_3$) and incrementing by 1 until equation (5) was satisfied. (In all cases, N was determined using the tables of Milton [15].)

From Table 1 it is clear that, for a large range of the parameters involved, N_2 is a very adequate approximation to N_3.

In order to compare the sample size N needed without correlation ($\rho = 0$) with the sample size N_3 needed by the means procedure in the case of correlation, graphs of $M = N_3/N$ were plotted (as a function of P^*) for: $\rho = 0.1, 0.25, 0.5, 0.6, 0.7, 0.8, 0.9, 0.95$; $\lambda^* = 0.5, 1.0$; $k = 2(1)\ 5, 10, 25$. Three of these graphs ($\lambda^* = 1.0$ with $k = 25$ and $\lambda^* = 0.5$ with $k = 2, 25$) are given below and are indicative: for data with

[2] For: 40 values of P^*; $\rho = \pm 0.1, \pm 0.25, \pm 0.5, \pm 0.75, \pm 0.9, \pm 0.95$; $\lambda^* = 0.1, 0.25, 0.5, 0.75, 1.0, 2.0$; $k = 2(1)5, 10, 25$.

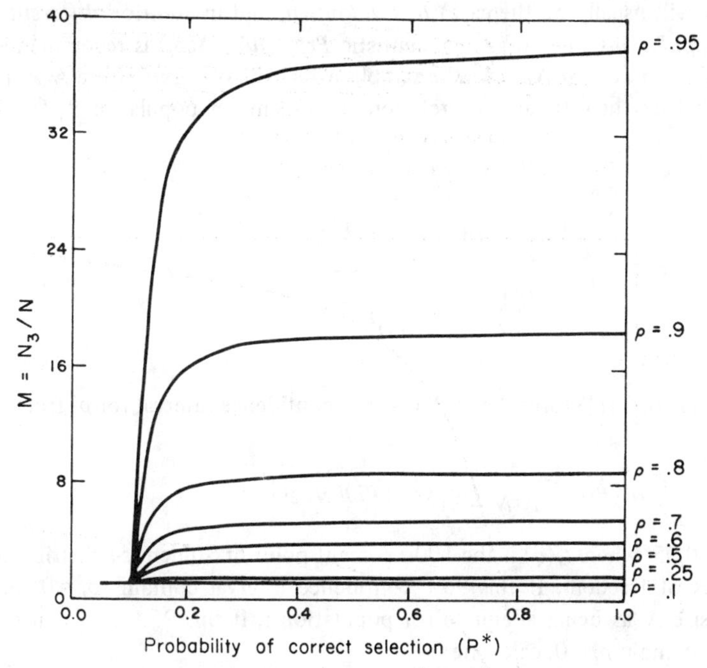

Figure 3. $M = N_3/N$ as a function of P^* for $\lambda^* = 0.5$, $k = 25$.

$0 < \rho \leqslant 0.25$ one will need (for "high" P^*) twice the sample size as under independence, for data with $0.25 < \rho \leqslant 0.8$ one will need up to eight times the sample size needed for the case $\rho = 0$, and with $\rho > 0.8$ one could need up to forty times the sample size required by $\rho = 0$. This sort of behavior is reasonable: as $\rho > 0 \uparrow$, a fixed number of samples will yield "less" information about μ; in fact in the extreme case $\rho = 1$, a sample of size one has the same information as any sample of size >1.

4. Recommended ranking and selection procedures

Based on the above analysis over a large range of the parameters involved in any ranking and selection problem using the indifference-zone formulation of Bechhofer [2], we believe that in cases in which simulation run-length is to be set using such procedures the possibility of nonzero correlation ρ within the samples should merit serious consideration by the experimenter. Of course *if ρ is known* one may use Procedure A of section 3 directly, with sample size equal to the smallest integer N_3 satisfying equation (5) (or, more simply, with sample size approximately equal to the N_2 of equation (8)).

If (as will usually be the case) *ρ is unknown,* and in addition *different for different populations,* the following heuristic *Procedure* A($\hat{\rho}_i$) is recommended. Use an initial sample of size $N_0 = N$ (the sample size needed for Procedure A with $\rho = 0$). Then calculate the estimated correlation coefficient for population π_i ($i = 1, 2, ..., k$) by

$$\hat{\rho}_i = \frac{\sum_{n=2}^{N_0} (X_{in} - \bar{X}_i)(X_{i,n-1} - \bar{X}_i)}{\sum_{n=1}^{N_0} (X_{in} - \bar{X}_i)^2} \tag{9}$$

and see [11, p. 191]) form the $100(1-\alpha)\%$ confidence interval for ρ_i from

$$(\rho_i - \hat{\rho}_i)^2 \leq \frac{N-1}{N(N-3)} (1 - \hat{\rho}_i^2) t_{N-3}^2 (1 - \alpha/2) \tag{10}$$

with $\alpha = 0.05$, where $t_r(q)$ is the $100q$ percent point of Student's-t distribution with r degrees of freedom. If this 95% confidence interval contains $\rho_i = 0$, judge the sample size N as being adequate for population i. If this 95% confidence interval does not contain $\rho_i = 0$, calculate

$$N_{2i} = \left[N \left(\frac{1 + \hat{\rho}_i}{1 - \hat{\rho}_i} \right) \right] \tag{11}$$

and continue the run until we have N_{2i} observations from π_i. Finally calculate $\bar{X}_1, ..., \bar{X}_k$ based on all available observations and select (as being best) that population which produced the largest of $\bar{X}_1, ..., \bar{X}_k$.

If (as will often be the case) the variances of $\pi_1, ..., \pi_k$ are *unknown and unequal,* the following heuristic *Procedure* $A(\hat{\rho}_i, s_i^2)$ is recommended. Take an initial sample of $N_0 = 30$ observations from each process. Calculate the number of observations which would be needed if we had $\rho_i = 0$ (zero correlation), namely (from results of Dudewicz and Dalal [8] and their tables of h)

$$M_i = \max\left(N_0, \left[\frac{s_i^2 h^2}{\delta^{*2}}\right]\right). \tag{12}$$

If, as in Procedure $A(\hat{\rho}_i)$, ρ_i is determined to be significantly different from zero, set the required sample size as

$$N_{2i} = \left[M_i\left(\frac{1+\hat{\rho}_i}{1-\hat{\rho}_i}\right)\right] \tag{13}$$

and continue the run. If ρ_i is determined not to be significantly different from zero, set the required sample size as M_i.

Table 2
Simulation results on procedure $A(\hat{\rho}_i)$, k autoregressive processes, correlation ρ, 1000 replications ($\mu_1 = 0, \mu_2 = 0.1$, and if $k = 4, \mu_3 = 0.1, \mu_4 = 0.1$).

α	k	P^*	λ^*	ρ	N	$\hat{P}_N$	$\hat{E}(N_2)$	$\hat{V}(N_2)$	$\hat{P}(CS)$
0.05	2	0.75	0.1	0.1	91	0.745 (0.718, 0.772)	99.0	435.6	0.748 (0.722, 0.776)
0.10	2	0.75	0.1	0.5	91	0.680 (0.651, 0.709)	259.7	3975.4	0.757 (0.730, 0.784)
0.05	2	0.9	0.2	0.5	83	0.749 (0.722, 0.776)	237.0	3502.7	0.864 (0.843, 0.885)
0.10	2	0.9	0.2	0.5	83	0.756 (0.730, 0.783)	238.1	3501.1	0.875 (0.855, 0.895)
0.10	2	0.9	0.1	0.1	329	0.892 (0.874, 0.912)	384.9	3308.4	0.911 (0.893, 0.928)
0.10	2	0.95	0.2	0.5	136	0.820 (0.796, 0.843)	397.0	6216.2	0.947 (0.934, 0.961)
0.10	2	0.95	0.5	0.5	22	0.812 (0.787, 0.836)	48.5	862.7	0.871 (0.850, 0.891)
0.10	4	0.95	0.5	0.5	35	0.748 (0.721, 0.775)	92.0	1609.9	0.952 (0.939, 0.965)

N = Number of observations required by procedure of Bechhofer (ignoring correlation).
$\hat{P}_N$ = Estimate of probability of correct selection based on N observations (point estimate and 95% confidence interval).
$\hat{E}(N_2)$ = Estimate (based on 1000 simulated observations) of expected number of observations.
$\hat{V}(N_2)$ = Estimate of variance of number of observations.
$\hat{P}(CS)$ = Estimate of probability of correct selection using N_2 observations.

These heuristic procedures are now being investigated in an attempt to put them on firm theoretical ground along the lines of Dudewicz and Dalal [8]. We believe (see section 5 for details) they should be sufficient to preclude gross errors due to significant correlations.

5. Simulation results on procedures $A(\hat{\rho}_i)$, $A(\hat{\rho}_i, s_i^2)$

In Table 2 we give simulation results (based on 1000 replicate runs) on the performance of Procedure $A(\hat{\rho}_i)$. From this table we see the following facts. When the autoregressive model is valid, Procedure $A(\hat{\rho}_i)$ satisfactorily meets the probability requirement, since the attained $P(CS)$ $\hat{P}(CS)$ is close to the desired level P^*. This holds whether 95% ($\alpha = 0.05$) or 90% ($\alpha = 0.10$) intervals are used in the basic procedure $A(\hat{\rho}_i)$. However if the initial sample size N required (by the $\rho_i = 0$ case) is small, $\hat{P}(CS)$ may be further from P^* than one would like; this does not seem serious, but can be avoided by modifying α.

In Table 3 simulation results on the performance of Procedure $A(\hat{\rho}_i, s_i^2)$ are given, with $\alpha = 0.10$. Again the attained $P(CS)$ is close to the desired level P^*.

Further study of these procedures, from both simulation and theoretical

Table 3

Simulation results on procedure $A(\hat{\rho}_i, s_i^2)$, k autoregressive processes, correlation ρ, 1000 replications, initial sample size $N_0 = 30$.

k	P^*	δ^*	Variances	ρ	h	$\hat{P}_N$	$\hat{P}(CS)$
2	0.892	0.5	$\sigma_{X_1}^2 = 1.$ $\sigma_{X_2}^2 = 2.$	0.5	1.8	0.812 (0.788, 0.836)	0.895 (0.876, 0.914)
2	0.9495	0.2	$\sigma_{X_1}^2 = 1.$ $\sigma_{X_2}^2 = 2.$	0.5	2.5	0.828 (0.805, 0.851)	0.890 (0.871, 0.909)
2	0.892	0.5	$\sigma_{X_1}^2 = 2.$ $\sigma_{X_2}^2 = 1.$	0.75	1.8	0.709 (0.682, 0.738)	0.866 (0.845, 0.887)
4	0.9474	0.5	$\sigma_{X_1}^2 = \sigma_{X_2}^2 = 1.$ $\sigma_{X_3}^2 = \sigma_{X_4}^2 = 2.$	0.5	3.0	0.732 (0.705, 0.759)	0.960 (0.948, 0.972)
4	0.8952	0.5	$\sigma_{X_1}^2 = \sigma_{X_2}^2 = 2.$ $\sigma_{X_3}^2 = \sigma_{X_4}^2 = 1.$	0.5	2.5	0.687 (0.658, 0.716)	0.895 (0.876, 0.914)
2	0.892	0.2	$\sigma_{X_1}^2 = 3.$ $\sigma_{X_2}^2 = 1.$	0.2	1.8	0.843 (0.820, 0.865)	0.858 (0.836, 0.880)

$\hat{P}_N$ is estimate of probability of correct selection using procedure of Dudewicz and Dalal (ignoring correlation).

approaches, is in progress, but at this point it seems clear they can be used in practice and will alleviate gross errors due to correlation within sequences of observations. Their behavior for nonautoregressive processes, and the behavior of modifications (such as: eliminate the test of $\rho_i = 0$ and *always* use a sample size based on either $\hat{\rho}_i$ or on the upper end point $\hat{\rho}_i^+$ of some appropriate confidence interval on ρ_i), are also under study.

References

[1] Anderson, T.W., The Statistical Analysis of Time Series, John Wiley and Sons, New York (1971).

[2] Bechhofer, R.E., A single-sample multiple-decision procedure for ranking means of normal populations with known variances, Annals of Mathematical Statistics, 25 (1954) 16–39.

[3] Burdick, D.S. and Naylor, T.H., Design of computer simulation experiments for industrial systems, Communications of the ACM, 9 (1966) 329–339.
[4] Cox, D.R. and Miller, H.D., The Theory of Stochastic Processes, John Wiley and Sons, New York (1965).
[5] Diananda, P.H., Some probability limit theorems with statistical applications, Proc. Cambridge Phil. Soc., 49 (1953), 239–246.
[6] Dudewicz, E.J., Random numbers: The need, the history, the generators, Statistical Distributions in Scientific Work, Vol. 2: Model Building and Model Selection, edited by G.P. Patil, S. Kotz, and J.K. Ord, D. Reidel Publishing Company, Dordrecht, Holland, (1975) 25–36.
[7] Dudewicz, E.J., Statistics in simulation: How to design for selecting the best alternative, Proceedings of the 1976 Bicentennial Winter Simulation Conference, (1976) 67–71.
[8] Dudewicz, E.J. and Dalal, S.R., Allocation of observations in ranking and selection with unequal variances, Sankhyā, Series B, 37 (1975) 28–78.
[9] Dudewicz, E.J. and Zaino, N.A., Jr., Ranking and selection procedures with correlated observations and simulation, Proceedings of the Computer Science and Statistics Seventh Annual Symposium on the Interface, edited by W.J. Kennedy, Statistical Laboratory, Iowa State University, Ames, Iowa, (1973) 150–155.
[10] Dudewicz, E.J. and Zaino, N.A., Jr., Ranking and selection procedures with correlated observations and simulation, II, in preparation.
[11] Jenkins, G.M. and Watts, D.G., Spectral Analysis and Its Applications, Holden-Day, Inc., San Francisco (1968).
[12] Kleijnen, J.P.C., Statistical Techniques in Simulation, Part II, Marcel Dekker, Inc., New York (1975).
[13] Kleijnen, J.P.C. and Naylor, T.H., The use of multiple ranking procedures to analyze simulations of business and economic systems, 1969 Business and Economic Statistics Section Proceedings of the American Statistical Association, (1969) 605–615.
[14] Lin, W.T., Multiple objective budgeting models: A simulation, USC Working Paper #12-01-75, Department of Accounting, Graduate School of Business Administration, University of Southern California, Los Angeles, California (1975).
[15] Milton, R.C., Tables of the equally correlated multivariate normal probability integral, Technical Report No. 27, Department of Statistics, University of Minnesota (1963).
[16] Naylor, T.H., Wertz, K. and Wonnacott, T.H., Methods for analyzing data from computer simulation experiments, Communications of the ACM, 11 (1967) 703–710.
[17] Naylor, T.H., Wertz, K. and Wonnacott, T.H., Some methods of evaluating the effects of economic policies using simulation experiments, Review of the International Statistical Institute, 36 (1968) 184–200.

Biom. J. vol. 24 · 1982 · no. 3, pp. 297–307

Department of Statistics, The Ohio State University

Estimation of the P(CS) with a Computer Program

Edward J. Dudewicz[1]

Abstract

In the area of ranking and selection, recent papers have presented procedures which attempt to allow experimenters to take at least partial advantage of more-favorable configurations of the population parameters (compared to the least-favorable configuration). The current paper presents a computer program which allows experimenters to dispense with all of inequalities, an indifference-zone, and known equal variabilities in their assessment of the probability of correct selection. It has additional uses in implementation of recently-proposed sequential adaptive selection procedures.

Key words: Ranking and selection, estimation of P(CS), probability of correct selection, inequalities, Monte Carlo, sequential, adaptive, heteroscedastic.

1. Introduction

When an experimenter uses the indifference-zone ranking-and-selection procedures pioneered by Bechhofer (1954) and recently explicated by Gibbons, Olkin, and Sobel (1977), he sets his sample size(s) to guarantee a P(CS) (probability of correct selection) at least P^* whenever a minimal separation (of the best population from the other populations) of at least δ^* exists; a separation of δ^* is then least-favorable. The actual separation is likely to be more favorable than least-favorable (and the P(CS) thus $> P^*$), and experimenters would like to take advantage of this favorableness after the experiment has been run. Previous approaches to this problem by Olkin, Sobel, and Tong (1976) and Anderson, Bishop, and Dudewicz (1977) all involved inequalities. In this paper we give a computer program which allows assessment of the P(CS) without use of inequalities, is simple and easily implemented on even small computers, and does not use

[1] This research was planned and initiated while Edward J. Dudewicz was a Visiting Scholar, Department of Mathematics and Statistics, University of Nebraska, Lincoln. Edward J. Dudewicz is Professor, Department of Statistics, The Ohio State University, Columbus, Ohio 43210, U.S.A.

This research was supported by Office of Naval Research Contract No. N00014-78-C-0543.

an indifference-zone. It is also directly useful when (as is often the case) the samples are drawn before the statistician is consulted, and he must evaluate their adequacy. It has additional uses in implementation of recently proposed sequential adaptive selection procedures of TONG (1978).

While our approach is oriented towards the practitioner and computer software provider in this paper, it should be noted that the area is also one of recent intensive theoretical investigation by OLKIN, SOBEL, and TONG (1979), BOFINGER (1980), and FALTIN (1980a, b).

Note that OLKIN, SOBEL, and TONG (1979) consider only the case of σ^2 known. Work of BOFINGER (1980) contradicts their result on asymptotic normality of their P(CS) estimator. FALTIN (1980a) shows this estimator is biased on the high side when $k=2$, while FALTIN (1980b) gives an alternate quantile unbiased estimator (also for the case $k=2$). By contrast, our estimator allows for unknown and heteroscedastic variances and does not require either bounds or tables: the experimenter makes one inexpensive run of a simple computer program (presented in full detail with numerical examples below).

2. The P(CS) Function

If the experimenter has obtained a random sample of size n_i from population π_i which yields normally distributed observations with unknown mean μ_i and variance σ_i^2, $1 \leqq i \leqq k$, and wishes to evaluate the adequacy of these sample sizes when the means procedure ("select the population yielding the largest sample mean as having the largest population mean") is used (and all $n_1 + \ldots + n_k$ observations are independent), he will need to assess the function

$$\mathrm{P(CS)} = \int_{-\infty}^{\infty} \left\{ \prod_{i=1}^{k-1} \Phi\left(\frac{\sigma_{(k)}}{\sigma_{(i)}} \sqrt{\frac{n_{(i)}}{n_{(k)}}}\, z + \frac{\mu_{[k]} - \mu_{[i]}}{\sigma_{(i)}/\sqrt{n_{(i)}}} \right) \right\} \varphi(z)\, \mathrm{d}z$$

where: $\mu_{[1]} \leqq \ldots \leqq \mu_{[k]}$ denote $\mu_1, \ldots, \mu_k$ in numerical order; $\sigma_{(i)}$, $n_{(i)}$ are the standard deviation and sample size of the population with true mean $\mu_{[i]}$; Φ, φ are the standard normal distribution function and density function. Note that, with error $\varepsilon (0 < \varepsilon < .005$ since $\Phi\ (2.81) = .9975)$,

$$\mathrm{P(CS)} = \int_{-2.81}^{2.81} \left\{ \prod_{i=1}^{k-1} \Phi\left(\frac{\sigma_{(k)}}{\sigma_{(i)}} \sqrt{\frac{n_{(i)}}{n_{(k)}}}\, z + \frac{\mu_{[k]} - \mu_{[i]}}{\sigma_{(i)}/\sqrt{n_{(i)}}} \right) \right\} \varphi(z)\, \mathrm{d}z\, .$$

Once the $\mu_{[i]}$, $\sigma_{(i)}$ are estimated from the samples, this can be estimated simply by Monte Carlo methods on even small computers, with ease compared to quadrature methods, and with accuracy compared to methods involving inequalities. A computer program is given in Section 3, with an example in Section 4.

3. Computer Program

The following computer program reads K (the number of populations) and XM(I), S(I), N(I) (the sample mean, sample standard deviation, and sample size of population I), $1 \leq I \leq K$, from data cards and then uses 40,000 Monte Carlo trials to estimate P(CS) with a standard error not exceeding .005/2, so that (with 95% confidence, also taking account of the truncation error ε) we estimate the P(CS) integral within .01. A numerical example is given in section 4.

```
C
C     THE FOLLOWING COMPUTER PROGRAM:
C       1. READS K (THE NUMBER OF POPULATIONS) FROM A DATA
C       CARD;
C       2. READS XM(I), S(I), N(I) (THE SAMPLE MEAN, STANDARD
C          DEVIATION AND SIZE OF POPULATION I), 1<=I<=K, FROM
C          K DATA CARDS;
C       3. USES 40,000 MONTE CARLO TRIALS TO ESTIMATE THE P(CS)
C          WITH A STANDARD ERROR NOT EXCEEDING .005/2, SO THAT
C          (WITH 95 PERCENT CONFIDENCE, ALSO TAKING ACCOUNT
C          OF THE TRUNCATION ERROR) WE ESTIMATE THE P(CS)
C          INTEGRAL WITHIN .01.
      INTEGER R, SEED
      DIMENSION XM (50), S(50), N(50), R(50)
C
C       PERSONS WITH K>50 SHOULD INCREASE ALL FOUR DIMEN-
C       SIONS ABOVE TO THAT NUMBER BEFORE RUNNING THIS
C       PROGRAM. THEY SHOULD ALSO INCREASE THE DIMENSIONS
C       OF YM(.) AND S(.) IN SUBROUTINE VSORT.
C
      READ (5, 20) K
   20 FORMAT (I10)
      DO 30 I=1, K
      READ (5, 21) XM(I), S(I), N(I)
   21 FORMAT (F10.0, F10, 0, I10)
   30 CONTINUE
C
C       WE NOW FIND INTEGERS R(1), . . . , R(K) SUCH THAT XM(R(1))
C       <=XM (R(2))<=. . .<=XM(R(K)) BY A VIRTUAL SHELL SORT.
C
      CALL VSORT(XM, R, K)
C
      SEED=987021
```

```
      I=0
      JCNT=0
C
C     WE NOW ENTER THE MONTE CARLO LOOP.
C
   31 CONTINUE
      IF(I.GE.40000)GO TO 34
      CALL URN20(SEED, RANX, 1)
      CALL URN20 (SEED, RANY, 1)
      RANX=(RANX-0.5)*2.0*2.81
      RANY=RANY*(1.0/SQRT(2.0*3.14159))
C
      PROD=1.0
      KK=K-1
      DO 32 L=1, KK
      TERM=XM(R(K))-XM(R(L))
      TERM=TERM/S(R(L))
      AN=FLOAT(N(R(L)))
      TERM=TERM*SQRT(AN)
      FACT=S(R(K))/S(R(L))
      CN=FLOAT(N(R(L)))
      DN=FLOAT(N(R(K)))
      BN=CN/DN
      FACT=FACT*SQRT(BN)
      FACT=FACT*RANX+TERM
      TERM=DCDFN(FACT)
      PROD=PROD*TERM
   32 CONTINUE
      FN=PROD/SQRT(2.0*3.14159)
      FN=FN*EXP(-0.5*RANX*RANX)
      IF(FN.LE.RANY)GO TO 33
      JCNT=JCNT+1
   33 CONTINUE
      I=I+1
      GO TO 31
C
C     WE HAVE DONE 40000 MONTE CARLO TRIALS, JCNT OF WHICH
C     YIELDED A VALUE OF THE INTEGRAND ABOVE RANY. WE NOW
C     CALCULATE AND REPORT THE P(CS) ESTIMATE AND ITS STAN-
C     DARD DEVIATION.
C
   34 CONTINUE
      PCSEST=(JCNT/40000.)*2.81*2.0*(1.0/SQRT(2.0*3.14159))
```

```
      STDER = (JCNT/40000.)*(1.0 - (JCNT/40000.))
      STDER = SQRT(STDER)/200.
      STDER = STDER*2.81*2.*(1.0/SQRT(2.0*3.14159))
C
      WRITE(6,41)
   41 FORMAT(1H1)
      WRITE(6,42)
   42 FORMAT (1H0, 6X, 54HTHE PROBABILITY OF CORRECT SELECTION
     1 IN A PROBLEM WITH)
      WRITE (6,43) K
   43 FORMAT (1H0, 6X, 5H K= , I 2, 12H POPULATIONS)
      DO 35 I = 1, K
      WRITE (6,44) I, XM(I), S(I), N(I)
   44 FORMAT (1H0, 6X, 13H POPULATION, 12, 10H HAS MEAN, F10.4
     1 12H, STD. DEV., F10.4, 19H, AND SAMPLE SIZE, I 4)
   35 CONTINUE
      WRITE (6,45) PCSEST
   45 FORMAT (1H0, 6X, 11HIS P(CS)= , F6.4, 34H BASED ON 40000 MONTE
     1 CARLO TRIALS)
      WRITE(6,46) STDER
   46 FORMAT (1H0, 6X, 25HAND HAS STANDARD ERROR= ,F6.4)
      WRITE (6,47)
   47 FORMAT (1H1)
      END
```

```
0001         SUBROUTINE VSORT(XM, R, K)
0002         INTEGER R, S, SAV 2
0003         DIMENSION XM(1), R(1), YM(50), S(50)
0004         DO 10 I = 1, K
0005         YM(I) = XM(I)
0006         S(I) = I
0007      10 CONTINUE
      C
0008         M = K
0009       1 M = M/2
0010         IF(M.EQ.0) GO TO 5
0011         J = 1
0012       2 I = J
0013       4 IF(YM(I).LE.YM (I+M)) GO TO 3
0014         SAV1 = YM(I)
```

```
0015          SAV2=S(I)
0016          YM(I)=YM (I+M)
0017          S(I)=S(I+M)
0018          YM(I+M)=SAV1
0019          S(I+M)=SAV2
0020          I=I-M
0021          IF(I.GE.1) GO TO 4
0022        3 J=J+1
0023          IF(J.GT.K-M) GO TO 1
0024          GO TO 2
0025        5 CONTINUE
0026          DO 11 I=1, K
0027          R(I)=S(I)
0028       11 CONTINUE
0029          RETURN
0030          END
```

```
C                                                          00011040
C    FUNCTION DCDFN(X)                                     00011050
C                                                          00011060
C  ******************************************************* 00011070
C                                                          00011080
C  A SUBROUTINE TO CALCULATE PROBABILITIES FOR             00011090
C    STANDARD NORMAL RANDOM VARIABLES.                     00011100
C                                                          00011110
C  PURPOSE                                                 00011120
C    GENERATES THE PROBABILITY THAT Z IS LESS THAN         00011130
C    OR EQUAL TO X, WHERE Z IS A STANDARD NORMAL           00011140
C    RANDOM VARIABLE                                       00011150
C  USAGE                                                   00011160
C    "LET P=DCDFN(X)"                                      00011170
C                                                          00011180
C  DESCRIPTION OF PARAMETERS                               00011190
C    X . . . IS THE CUTOFF POINT FOR WHICH THE PROBA-      00011200
C    BILITY THAT Z IS LESS THAN OR EQUAL TO X IS           00011210
C    DESIRED                                               00011220
C  REMARKS                                                 00011230
C    THIS IS A DOUBLE PRECISION FUNCTION AND CAN           00011240
C    ACCOMMODATE A DOUBLE PRECISION ARGUMENT               00011250
C                                                          00011260
```

```
C  SUBROUTINES AND FUNCTION SUBPROGRAMS REQUI-   00011270
C     RED NONE                                    00011280
C                                                 00011290
C  METHOD                                         00011300
C     FOR THE ALGORITHM USED AND ITS ACCURACY, SEE 00011310
C     REFERENCE 2. FOR SPEED AND COMPARISONS WITH 00011320
C     OTHER ALGORITHMS (WITH REGARD TO SPEED AND  00011330
C     ACCURACY), SEE REFERENCE 1.                 00011340
C     1. DUDEWICZ, E. J. AND RALLEY, T.: "THE HANDBOOK 00011350
C        OF RANDOM NUMBER GENERATION AND TESTING  00011360
C        WITH TESTRAND COMPUTER CODE", AMERICAN   00011365
C        SCIENCES PRESS, INC., COLUMBUS, OHIO 43221-0161 00011370
C        U.S.A., 1981.                            00011380
C     2. MILTON, R. C. AND HOTCHKISS, R.: "COMPUTER 00011390
C        EVALUATION OF THE NORMAL AND INVERSE NOR- 00011400
C        MAL DISTRIBUTION FUNCTIONS," TECHNOMETRICS, 00011410
C        VOL. 11 (1969), PP- 817 822.             00011420
C  ********************************************************** 00011430
      DOUBLE PRECISION FUNCTION DCDFN(X)          00011440
      DIMENSION A(6), D(6), C(6, 10)              00011450
      REAL*8 A, C, D, X, Y, T, SGNY               00011460
      DATA A/.625D0, 1.25D0, 2.0D0, 2.45D0, 3.5D0, 4.62D0/ 00011470
      DATA D/.3D0,. 925D0, 1.625D0, 2.225D0, 2.95D0,4. 15D0/ 00011480
      DATA C/6.7982403291D-04, -1.2709753598D-03, 6.7964525797 00011490
     * D-04,                                      00011500
     *-8.8500314297D-05, 1.4791554152D-07, 5.4879072878D-07, 00011510
     *5.1028640888D-03,-1.4822649744D-03,-3.4589883732D-04, 00011520
     *5.6826070991D-04,-6.1965013125D-05,-8.6656125337D-07, 00011530
     *-6.0160862379D-03,8.7718965604D-03,-3.0883401043D-03, 00011540
     *-4.5015531032D-04,2.2266710841D-04,6.6877769441D-07, 00011550
     *-2.2725611373D-02,-2.0200415601D-03,7.2141082947D-03, 00011560
     *-1.3872389725D-03,-4.4390018986D-04,-6.4084653798D-07, 00011570
     *3.3555772127D-02,-3.4681674488D-02,-1.0677039440D-03, 00011580
     *5.5422185916D-03, 6.2978750848D-04, 6.2471782478D-07, 00011590
     *7.2703311825D-02,4.7642403207D-02, -2.4859750804D-02, 00011600
     *-1.0221129286D-02,-6.6382630882D-04,-4.2105004094D-07, 00011610
     *-1.4093895854D-01, 5.6850293056D-02, 5.7420315819D-02, 00011620
     *1.1850195831D-02, 5.1265678121D-04, 2.0742133302D-07, 00011630
     *-1.5468912970D-01,-2.2180615138D-01,-6.5383705170D-02, 00011640
     *-8.8864030978D-03,-2.7657162532D-04,-7.6856040218D-08, 00011650
     *5.1563045460D-01, 2.3979043073D-01, 4.0236129479D-02, 00011660
     *3.9938885701D-03, 9.3751413487D-05, 1.8722020822D-08, 00011670
     *1.6431337971D-01, 4.0458836515D-01, 4.8922186659D-01, 00011680
```

```
     *4.9917416726D-01, 4.9998489849D-01, 4.9999999779D-01/  00011690
      Y=X*0.70710678119D0                                    00011700
      SGNY=1.D0                                              00011710
      IF(Y)2, 1, 3                                           00011720
      1 DCDFN=.5D0                                           00011730
      RETURN                                                 00011740
    2 SGNY=-1.D0                                             00011750
      Y=-Y                                                   00011760
    3 D0 4 I=1,6                                             00011770
      IF(Y-A(I)) 5, 5, 4                                     00011780
    4 CONTINUE                                               00011790
      Z=.5D0                                                 00011800
      GO TO 7                                                00011810
    5 Y=Y-D(I)                                               00011820
      Z=C(I, 1)                                              00011830
      DO 6 J=2, 10                                           00011840
    6 Z=Z*Y+C(I, J)                                          00011850
    7 DCDFN=.5D0+SGNY*Z                                      00011860
      RETURN                                                 00011870
      END                                                    00011880
```

```
URN20    CSECT                                                   00000010
         USING URN20, 15                                         00000020
         STM   14, 12, 12(13)                                    00000030
         ST    13, SAVE+4                                        00000040
         LR    11, 13          COPY ADDR OF CALLING
*                              SAVE INTO REG 11                  00000050
         LA    13, SAVE        ADDR OF SAVE IN REG 13            00000060
         ST    13, 8(11)       STORE ADDR OF SAVE
*                              IN CALLING PROG SAVE              00000070
*                                                                00000080
         LM    2, 4, 0(1)      REG(2)=ADDR(IX) REG(3)
*                              =ADDR(X) REG(4)=ADDR
*                              (NBATCH)
         L     11,0(4)         VALUE OF NBATCH IN
*                              REG 11
         L     10,=F'0'        INIT INDEX                        00000110
         L     9,=X'40000000'  EXPONENT FOR FLOATING
```

```
*                             POINT CONVERSION              00000120
LOOP    L      7, 0(2)        VALUE OF IX IN REG7           00000130
        M      6, MLT         REG6=QUOT, REG7=REM
*                             MOD 2**32                     00000140
        SLDL   6,1            PUTS QUOT MOD 2**31 IN
*                             REG6                          00000150
        SRL    7,1            PUTS REM MOD 2**31 IN
*                             REG7                          00000160
        AR     6,7            REG6=REM MOD 2**31-1
*                             =REM MOD 2**31+QUOT
*                             MOD 2**31                     00000171
        C      6,=F'0'        CHECK FOR OVERFLOW            00000172
        BNL    UP1            BRANCH TO UP 1 IF SUM
*                             IS NOT NEG                    00000174
        SLL    6, 1           DIVIDE REG6 BY 2**31          00000176
        SRL    6, 1           REM MOD 2**31 IN REG6         00000177
        A      6,=F'1'        ADD QUOT FROM DIVISION
*                             TO REM TO                     00000178
*                             OBTAIN REM MOD 2**31-1
*                             IN REG6                       00000179
UP1     ST     6, 0(2)        STORE VALUE OF IX FOR
*                             NEXT RND NOS                  00000190
        LPR    6, 6           ABSOL VALUE OF IX IN
*                             REG6                          00000200
        L      5,=X'4F000000' EXPONENT IN REG5 (16**
*                             14)                           00000210
        STM    5, 6, DOUBLE   STORE FLT PT NOS IN
*                             REGS5 & 6                     00000220
        LD     2, DOUBLE      LOAD FLT PT REG2 WITH
*                             FLT PT NOS                    00000230
        AD     2,=D'0.0'      NORMALIZE BY ADDING 0         00000240
        DD     2, MODUL       DIVIDE BY 2**31-1             00000250
        STD    2, 0(10, 3)    STORE NORMALIZED FLT
*                             PT NOS IN X(INDEX)            00000260
        LA     10, 4(10)      INCREMENT INDEX               00000270
        BCT    11, LOOP       BOTTOM OF LOOP                00000280
        L      13, SAVE+4     RESTORE REGISTERS             00000290
        LM     14, 12 12(13)                                00000300
        BCR    15, 14         RETURN TO CALLING PRO-
*                             GRAM                          00000310
MLT     DC     F'2027812808'  MULTIPLIER FOR URN20
DOUBLE  DS     D                                            00000340
MODUL   DC     X'487FFFFFFF000000'                          00000350
```

```
SAVE     DS     18F                                  00000360
         END                                         00000370
                =D'0.0'
                =F'0'
                =X'40000000'
                =F'1'
                =X'4E000000'
```

4. Example and Comparison

As an example, when we have $k=8$ populations with means all 0.0 except for one which is 2.34, with standard deviations all 9.0, and sample sizes all 81, we know (from the tables of BECHHOFER, 1954) that the true P(CS) = .80. Our computer program yields the output

THE PROBABILITY OF CORRECT SELECTION IN A PROBLEM WITH K = 8 POPULATIONS
POPULATION 1 HAS MEAN 2.3400, STD. DEV. 9.0000, AND SAMPLE SIZE 81
POPULATION 2 HAS MEAN 0.0 , STD. DEV. 9.0000, AND SAMPLE SIZE 81
POPULATION 3 HAS MEAN 0.0 , STD. DEV. 9.0000, AND SAMPLE SIZE 81
POPULATION 4 HAS MEAN 0.0 , STD. DEV. 9.0000, AND SAMPLE SIZE 81
POPULATION 5 HAS MEAN 0.0 , STD. DEV. 9.0000, AND SAMPLE SIZE 81
POPULATION 6 HAS MEAN 0.0 , STD. DEV. 9.0000, AND SAMPLE SIZE 81
POPULATION 7 HAS MEAN 0.0 , STD. DEV. 9.0000, AND SAMPLE SIZE 81
POPULATION 8 HAS MEAN 0.0 , STD. DEV. 9.0000, AND SAMPLE SIZE 81
IS P(CS) = 0.8030 BASED ON 40000 MONTE CARLO TRIALS AND HAS STANDARD ERROR = 0.0054

which agrees with the theoretical value, thus furnishing a check on the computer program.

References

ANDERSON, P. O., BISHOP, T. A., and DUDEWICZ, E. J., 1977: Indifference-zone ranking and selection: confidence intervals for true achieved P(CD). Comm. Statist. – Theory and Methods **A6**, 1121–1132.

BECHHOFER, R. E., 1954: A single-sample multiple decision procedure for ranking means of normal populations with known variances. Ann. Math. Statist. **25**, 16–39.

BOFINGER, E., 1980: On the non-existence of consistent estimators for P{C.S.}. Unpublished manuscript.

FALTIN, F. W., 1980a: Performance of the Sobel-Tong estimator of the probability of correct selection achieved by Bechhofer's single-stage procedure for the normal means problem. Abstract, IMS Bull. **9**, 180.

FALTIN, F. W., 1980b: A quantile unbiased estimator of the probability of correct selection achieved by Bechhofer's single-stage procedure for the two population normal means problem. Abstract, IMS Bull. **9**, 180–181.

GIBBONS, J. D., OLKIN, I., and SOBEL, M., 1977: Selecting and Ordering Populations: A New Statistical Methodology. New York: John Wiley and Sons, Inc.

OLKIN, I., SOBEL, M., and TONG, Y. L., 1976: Estimating the true probability of correct selection for location and scale parameter families. Technical Report No. 110, Department of Statistics, Stanford University, Stanford, California.

OLKIN, I., SOBEL, M., and TONG, Y. L., 1979: Bounds for a k-fold integral for location and scale parameter models with applications to statistical ranking and selection problems. Technical Report No. 141, Department of Statistics, Stanford University, Stanford, California.

TONG, Y. L., 1978: An adaptive solution to ranking and selection problems. Ann. Statist. **6**, 658–672.

RANKING AND SELECTION IN DESIGNED EXPERIMENTS: COMPLETE FACTORIAL EXPERIMENTS*

Edward J. Dudewicz and Baldeo K. Taneja**

As early as the first paper of Bechhofer on ranking and selection in 1954 it was recognized that, with a ranking and selection goal as well as with other goals such as estimation or hypothesis testing, it might be desirable to carry out one's experiment in some design other than the completely randomized design. Nevertheless, over the years almost all of the papers in the area have developed their methods and theory explicitly only for the completely randomized design. In this paper we review what is known about ranking and selection in design settings other than the completely randomized design, and then proceed to new results on the complete factorial experiment setting.

1. Introduction

In the first paper on ranking and selection, Bechhofer [2] recognized the importance of allowing for experiments carried out in some design other than the completely randomized design. For full references see Dudewicz and Koo [9]. In this paper we: briefly survey the state of knowledge of ranking and selection in designed experiments (Section 2); discuss factorial experiments without interaction in some detail (Section 3); and give new results for factorial experiments where interaction may be present (Section 4).

2. Ranking and selection in designed experiments

In the basic formulation of the ranking and selection problem, there are K populations (sources of observations) $\pi_1,\cdots,\pi_K$ with respective unknown means $\mu_1,\cdots,\mu_K$ for their observations, normally distributed with common known variance $\sigma^2>0$, and the goal is to select a population whose mean is $\mu_{[K]}$, where $\mu_{[1]}\leqq\mu_{[2]}\leqq\cdots\leqq\mu_{[K]}$ are the ordered $\mu_1,\cdots,\mu_K$. Achieving this goal is called making a Correct Selection (abbreviated as CS). Bechhofer [2] gave a single-stage procedure which guaranteed that $P(\mathrm{CS})\geqq P^*$ whenever $\mu_{[K]}-\mu_{[K-1]}\geqq\delta^*$ $(1/K<P^*<1,\ 0<\delta^*)$. There have been many papers written in the area, with extensions in many directions; see Dudewicz [8].

Bechhofer [2] dealt mainly with the completely randomized design, but also considered the 2-factor factorial experiment without interaction, giving the Factorial Procedure SP1 of Section 3 below. He briefly noted how to similarly solve the r-factor factorial experiment (again without interaction) with SP1. Also see pp. 77–82 of Gibbons, Olkin, and Sobel [10] for examples, restricted to the no-interaction case.

Bawa [1] compared Bechhofer's SP1 to the traditional One-at-a-time Procedure

Received January 12, 1981. Revised November 24, 1981.
* This research was supported by Office of Naval Research Contract No. N00014-78-C-0543.
** Department of Statistics, The Ohio State University, Columbus, Ohio 43210, U.S.A.

SP2 in a no-interaction setting, and found SP1 is superior in asymptotic efficiency.

A procedure for the factorial experiment with interaction (Interaction Procedure SP3 in Section 3 below) was given in Dudewicz [7], and independently simultaneously by Bechhofer [3].

For the problem of selecting the best regression, Dudewicz ([6], Section 14.4) noted the solution (also see Chapter 9 of Gibbons, Olkin, and Sobel [10], with an erroneous claim of originality on p. 241). The problem of selecting the largest interaction was addressed by Bechhofer, Santner, and Turnbull [4]. Rinott and Santner [11] applied inequalities to design an experiment to select the best treatment in an analysis of covariance model.

3. Ranking and selection in factorial experiments: no interaction

Suppose an r-factor ($r \geqq 2$) factorial experiment with k_j ($k_j \geqq 2$) levels of factor j ($j=1, 2, \cdots, r$). Assume observations are normally distributed with known common variance $\sigma^2>0$ and that for an observation $X_{i_1 i_2 \cdots i_r}$ taken at {level i_1 of factor 1, level i_2 of factor 2, $\cdots$, level i_r of factor r}

$$E(X_{i_1 i_2 \cdots i_r})=\mu+\sum_{j=1}^{r} \alpha_{i_j}^{(j)}$$

($i_j=1, 2, \cdots, k_j$; $j=1, 2, \cdots, r$) where μ and $\alpha_{i_j}^{(j)}$ are unknown parameters with

$$\sum_{i_j=1}^{k_j} \alpha_{i_j}^{(j)}=0 \qquad (j=1, 2, \cdots, r) .$$

For each fixed j ($j=1, 2, \cdots, r$), let

$$\alpha_{[1]}^{(j)} \leqq \alpha_{[2]}^{(j)} \leqq \cdots \leqq \alpha_{[k_j]}^{(j)}$$

denote the $\alpha_1^{(j)}, \cdots, \alpha_{k_j}^{(j)}$ in numerical order. Each factor-level combination ($i_1, i_2, \cdots, i_r$) is called a *population*, and a "best" population (among all $K=k_1 k_2 \cdots k_r$ populations) is any one with maximum $E(X_{i_1 i_2 \cdots i_r})$, i.e. any one with mean

$$\mu_{[K]}=\mu+\alpha_{[k_1]}^{(1)}+\alpha_{[k_2]}^{(2)}+\cdots+\alpha_{[k_r]}^{(r)} .$$

Define (for $j=1, 2, \cdots, r$)

$$\delta_j=\alpha_{[k_j]}^{(j)}-\alpha_{[k_j-1]}^{(j)} ,$$

and let $\boldsymbol{\delta}=(\delta_1, \delta_2, \cdots, \delta_r)$. Then set $\boldsymbol{\delta}^*=(\delta_1^*, \delta_2^*, \cdots, \delta_r^*)$ and P^* ($0<\delta_1^*, \delta_2^*, \cdots, \delta_r^*<\infty$; $1/K<P^*<1$) and seek procedures $\mathcal{P}$ which guarantee the *probability requirement*

$$P(\mathrm{CS} \mid \mathcal{P}) \geqq P^* \qquad \text{whenever } \boldsymbol{\delta} \geqq \boldsymbol{\delta}^* \text{ (componentwise)} .$$

Three procedures proposed in the literature are the following.

Procedure SP1 (*Factorial Procedure*). Take N independent observations from each of the K populations. Let $\bar{X}_{i_j}^{(j)}$ denote the (marginal) sample mean of all observations having the jth factor at level i_j ($j=1, 2, \cdots, r$). For each j, let $\bar{X}_{[1]}^{(j)} \leqq \cdots \leqq \bar{X}_{[k_j]}^{(j)}$ denote $\bar{X}_1^{(j)}, \cdots, \bar{X}_{k_j}^{(j)}$ in numerical order. Select the level associated with the largest (marginal) sample mean of each factor, i.e. with $\bar{X}_{[k_1]}^{(1)}, \bar{X}_{[k_2]}^{(2)}, \cdots, \bar{X}_{[k_r]}^{(r)}$, and assert that that population (factor combination) is best. (NK observations are used.)

Procedure SP2 (*One-at-a-time Procedure*). Fix the level of each factor except factor j, take N_j observations at each level of factor j, and compute the sample mean at each such level. Then select the level (of factor j) yielding the largest such sample mean. Proceed similarly for $j=1,\cdots,r$ (keeping "selected" levels fixed for the corresponding factors), taking a set of independent observations at each stage of experimentation. Finally state that the population corresponding to the selected factor levels is best. $\left(\sum_{j=1}^{r} k_j N_j \text{ observations are used.}\right)$

Procedure SP3 (*Interaction Procedure*). Take M independent observations from each of the K populations. Select the population yielding the largest sample mean as best. (MK observations are used.)

In order to compare procedures SP1, SP2, SP3, let n_1, n_2, n_3 denote the respective smallest number of observations each procedure needs in order to guarantee the probability requirement. From Bawa [1] we know

$$n_1=\min\left\{NK\colon N \text{ such that } \prod_{j=1}^{r}(1-\exp(-a_j N_j'))\geqq P^*\right\}$$

and

$$n_2=\min\left\{\sum_{j=1}^{r} k_j N_j\colon (N_1,\cdots,N_r) \text{ such that } \prod_{j=1}^{r}(1-\exp(-a_j N_j))\geqq P^*\right\},$$

where (for $j=1,2,\cdots,r$)

$$a_j=\frac{(\delta_j^*)^2}{4\sigma^2}, \qquad N_j'=\frac{NK}{k_j}.$$

Using asymptotic calculus as in Bawa [1] and Dudewicz [5] it follows that

$$n_1\sim-4\sigma^2\log(1-P^*)\max_{1\leq j\leq r}\left\{\frac{k_j}{(\delta_j^*)^2}\right\}+o(1-P^*),$$

$$n_2\sim-4\sigma^2\log(1-P^*)\sum_{j=1}^{r}\frac{k_j}{(\delta_j^*)^2}+B$$

where $a\sim b$ means the limit of a/b is 1 as $P^*\to1$ and B does not depend on P^*. Also (Dudewicz [5])

$$n_3\sim-4\sigma^2\log(1-P^*)\frac{K}{(\delta^*)^2},$$

where $P(\text{CS}|\text{SP3})\geqq P^*$ is guaranteed for $\delta\equiv\mu_{[K]}-\mu_{[K-1]}\geqq\delta^*$. Therefore

$$\text{ARE}(\text{SP1},\text{SP2})=\lim_{P^*\to1}\frac{n_1}{n_2}=\frac{\max_{1\leq j\leq r}\left\{\frac{k_j}{(\delta_j^*)^2}\right\}}{\sum_{j=1}^{r}\frac{k_j}{(\delta_j^*)^2}},$$

$$\text{ARE}(\text{SP1},\text{SP3})=\lim_{P^*\to1}\frac{n_1}{n_3}=\frac{(\delta^*)^2}{K}\max_{1\leq j\leq r}\left\{\frac{k_j}{(\delta_j^*)^2}\right\},$$

$$\text{ARE}(\text{SP2},\text{SP3})=\lim_{P^*\to1}\frac{n_2}{n_3}=\frac{(\delta^*)^2}{K}\sum_{j=1}^{r}\frac{k_j}{(\delta_j^*)^2}.$$

Note that

$$\delta=\mu_{[K]}-\mu_{[K-1]}=\left\{\mu+\sum_{j=1}^{r}\alpha_{[k_j]}^{(j)}\right\}-\max_{1\leq i\leq r}\left\{\mu+\sum_{j=1}^{r}\alpha_{[k_j]}^{(j)}-\delta_i\right\}$$
$$=\min\{\delta_1,\cdots,\delta_r\},$$

hence $\boldsymbol{\delta}\geqq\boldsymbol{\delta}^*$ (componentwise) and $\delta^*=\min(\delta_1^*,\cdots,\delta_r^*)$ furnish comparable probability requirements for SP1, SP2, and SP3. Letting c be that j $(1\leqq j\leqq r)$ where $k_j/(\delta_j^*)^2$ is maximized,

$$\text{ARE (SP1, SP3)}=\frac{(\delta^*)^2}{(\delta_c^*)^2}\frac{k_c}{K}\leqq 1$$

(since $\delta^*\leqq\delta_c^*$ and $k_c\leqq K$), and

$$\text{ARE (SP2, SP3)}\leqq\frac{(\delta^*)^2}{(\delta_c^*)^2}\frac{rk_c}{K}\leqq 1\,.$$

Hence SP1 and SP2 are each more efficient than SP3. In the "symmetric" case $k_1=k_2=\cdots=k_r=k$ and $\delta_1^*=\delta_2^*=\cdots=\delta_r^*=\delta^*$

$$\text{ARE (SP1, SP2)}=\frac{1}{r}\,,\qquad \text{ARE (SP1, SP3)}=\frac{1}{k^{r-1}}\,,$$

$$\text{ARE (SP2, SP3)}=\frac{r}{k^{r-1}}\,.$$

Thus, when there is no interaction, SP1 is best and SP3 is worst. However, we see (Section 4) that SP3 is fully resistant to interaction, whereas SP1 and SP2 are not.

4. Ranking and selection in factorial experiments: interaction

We will now study factorial experiments with interactions. So, we suppose an r-factor $(r\geqq 2)$ factorial experiment with k_j levels of factor j $(j=1,\cdots,r)$ and assume our observations are normally distributed with known common variance $\sigma^2>0$ and that if $X_{i_1 i_2\cdots i_r}$ is an observation taken at {level i_1 of factor 1, level i_2 of factor 2,$\cdots$, level i_r of factor r} then

$$E(X_{i_1 i_2\cdots i_r})=\mu+\sum_{j_1=1}^{r}\alpha_{i_{j_1}}^{(j_1)}+\sum_{j_2=1}^{r}\sum_{j_1=1}^{r}\alpha_{i_{j_1}i_{j_2}}^{(j_1 j_2)}+\cdots+\sum_{j_{r-1}=1}^{r}\sum_{j_{r-2}=1}^{r}\cdots\sum_{j_1=1}^{r}\alpha_{i_{j_1}i_{j_2}\cdots i_{j_{r-1}}}^{(j_1 j_2\cdots j_{r-1})}+\alpha_{i_1 i_2\cdots i_r}^{(1\,2\cdots r)}$$

$(i_j=1,2,\cdots,k_j$ for $j=1,2,\cdots,r)$ where μ and the α's are unknown parameters with

$$\sum_{i_{j_1}=1}^{k_{j_1}}\alpha_{i_{j_1}}^{(j_1)}=0\qquad\text{for } j_1=1,2,\cdots,r\,;$$

$$\sum_{i_{j_1}=1}^{k_{j_1}}\alpha_{i_{j_1}i_{j_2}}^{(j_1 j_2)}=\sum_{i_{j_2}=1}^{k_{j_2}}\alpha_{i_{j_1}i_{j_2}}^{(j_1 j_2)}=0$$

(for $j_1\,(=1,2,\cdots,r)<j_2\,(=1,2,\cdots,r)$);

$$\sum_{i_{j_1}=1}^{k_{j_1}}\alpha_{i_{j_1}i_{j_2}i_{j_3}}^{(j_1 j_2 j_3)}=\sum_{i_{j_2}=1}^{k_{j_2}}\alpha_{i_{j_1}i_{j_2}i_{j_3}}^{(j_1 j_2 j_3)}=\sum_{i_{j_3}=1}^{k_{j_3}}\alpha_{i_{j_1}i_{j_2}i_{j_3}}^{(j_1 j_2 j_3)}=0$$

(for $j_1\,(=1,2,\cdots,r)<j_2\,(=1,2,\cdots,r)<j_3\,(=1,2,\cdots,r)$); and so on; and lastly

$$\sum_{i_1=1}^{k_1}\alpha_{i_1 i_2\cdots i_r}^{(1\,2\cdots r)}=\sum_{i_2=1}^{k_2}\alpha_{i_1 i_2\cdots i_r}^{(1\,2\cdots r)}=\cdots=\sum_{i_r=1}^{k_r}\alpha_{i_1 i_2\cdots i_r}^{(1\,2\cdots r)}=0$$

($i_j=1, 2,\cdots, k_j$ for $j=1, 2,\cdots, r$). For each fixed j ($j=1, 2,\cdots, r$), let

$$\alpha_{[1]}^{(j)} \leqq \alpha_{[2]}^{(j)} \leqq \cdots \leqq \alpha_{[k_j]}^{(j)}$$

denote the $\alpha_1^{(j)}, \alpha_2^{(j)},\cdots, \alpha_{k_j}^{(j)}$ in numerical order. This r-factor experiment may be conceived of as an r-dimensional cuboid. Each factor-level combination $(i_1, i_2,\cdots, i_r)$ is called a *population*, and a "best" population (among all $K=k_1k_2\cdots k_r$ populations) is any one with $E(X_{i_1i_2\ldots i_r})$ maximized over all r-tuples $(i_1, i_2,\cdots, i_r)$ (called *cells*).

Once observations have been taken, let $\bar{X}_{i_1i_2\ldots i_r}$ denote the sample mean of all observations in cell $(i_1, i_2,\cdots, i_r)$, and define the (marginal) sample means

$$\bar{X}_{i_1\cdots\cdot}=\frac{\sum_{i_2=1}^{k_2}\sum_{i_3=1}^{k_3}\cdots\sum_{i_r=1}^{k_r}\bar{X}_{i_1i_2\ldots i_r}}{k_2k_3\cdots k_r} \qquad (i_1=1, 2,\cdots, k_1),$$

$$\bar{X}_{\cdot i_2\cdots\cdot}=\frac{\sum_{i_1=1}^{k_1}\sum_{i_3=1}^{k_3}\cdots\sum_{i_r=1}^{k_r}\bar{X}_{i_1i_2\ldots i_r}}{k_1k_3\cdots k_r} \qquad (i_2=1, 2,\cdots, k_2),$$

$$\bar{X}_{\cdots\cdot i_r}=\frac{\sum_{i_1=1}^{k_1}\sum_{i_2=1}^{k_2}\cdots\sum_{i_{r-1}=1}^{k_{r-1}}\bar{X}_{i_1i_2\ldots i_r}}{k_1k_2\cdots k_{r-1}} \qquad (i_r=1, 2,\cdots, k_r).$$

We now wish to prove (as indicated briefly in Section 3) that SP1 and SP2 are vitiated by interaction, while SP3's validity remains unaltered.

THEOREM 1. *Let n_1 be the smallest sample size for which procedure* SP1 *guarantees $P(\text{CS}\,|\,\text{SP1})\geqq P^*$ whenever $\boldsymbol{\delta}\geqq\boldsymbol{\delta}^*$ (componentwise) in a model with zero interactions. Then in a general (non-zero interaction) model,*

$$\inf_{\Omega(\delta^*)} P(\text{CS}\,|\,\text{SP1})\leqq\frac{1}{K}$$

where $\Omega(\delta^*)=\{(\mu_1, \mu_2,\cdots, \mu_K)\colon\ \mu_{[K]}-\mu_{[K-1]}\geqq\delta^*,\ \delta^*=\min\{\delta_1^*,\cdots, \delta_r^*\}\}$.

Before proceeding to prove Theorem 1, we will show that (in the presence of interaction) not only may the cell selected by SP1 not be the best cell, but it may be the case (with high probability) that, for each of the r factors, no column with the respective largest (marginal) sample mean contains the best cell. The proof will utilize an addition of interaction to achieve such a situation.

To verify the claims of the above paragraph, consider a model with no interactions. By labeling and relabeling the cells, we may arrange our experiment so that

$$E(X_{i_1i_2\ldots i_r})=\mu+\alpha_{[i_1]}^{(1)}+\alpha_{[i_2]}^{(2)}+\cdots+\alpha_{[i_r]}^{(r)}$$

($i_1=1, 2,\cdots, k_1$; $i_2=1, 2,\cdots, k_2$; $\cdots$; $i_r=1, 2,\cdots, k_r$), still with $\sum_{i_{j_1}=1}^{k_{j_1}}\alpha_{i_{j_1}}^{(j_1)}=0$ for $j_1=1, 2,\cdots, r$. (Thus, cell $(1, 1,\cdots, 1)$ is now the worst cell, while cell $(k_1, k_2,\cdots, k_r)$ is the best.)

Now, to achieve the desired situation, we assume without loss of generality that $k_j \geqq k_1 \geqq 2$ ($j=2, 3, \cdots, r$) and that all interactions are zero except for highest-order interactions. For the case $k_1=2$, see Remark 2 following the proof of Theorem 1. In the case $k_1>2$, we add highest-order interactions to each $E(X_{i_1 i_2 \cdots i_r})$ as multiples of a number "a" (yet to be chosen) according to a scheme such that cell $(1, 1, \cdots, 1)$ becomes the best cell for large positive "a". To motivate this scheme for r-factor experiments, we first illustrate it for two-factor experiments, next show how to extend it to three-factor experiments, and finally explain the general case ($r>3$).

For a *two-factor experiment* ($r=2$), add multiples of a (yet to be chosen) to each $E(X_{i_1 i_2})$ as specified in Figure 1 to form model $E(X_{i_1 i_2}(a))$. Note that there is a simple pattern over most of the table of Figure 1 (except for row 1 and column 1) and that (except for cell (1, 1)) row 1 and column 1 entries are chosen so the appropriate row or column sum is zero. Finally, the cell (1, 1) entry is chosen so that the row 1 and column 1 sums are each zero also. The added multiples of "a" are the interactions $\alpha^{(12)}_{i_1 i_2}(a)$, and satisfy $\sum_{i_1=1}^{k_1} \alpha^{(12)}_{i_1 i_2}(a) = \sum_{i_2=1}^{k_2} \alpha^{(12)}_{i_1 i_2}(a) = 0$ for all a. (Note that cell (1, 1) is best for large positive a.)

	1	2	$\cdots$	k_2-k_1+1	k_2-k_1+2	k_2-k_1+3	$\cdots$	k_2-2	k_2-1	k_2
					k_1-1 columns					
1	$\frac{(k_1-1)k_1(k_1+1)}{6}$	0	$\cdots$	0	-1	-3	$\cdots$	$-\frac{(k_1-3)(k_1-2)}{2}$	$-\frac{(k_1-2)(k_1-1)}{2}$	$-\frac{(k_1-1)k_1}{2}$
2	$-\frac{(k_1-1)k_1}{2}$	0	$\cdots$	0	1	2	$\cdots$	k_1-3	k_1-2	k_1-1
3	$-\frac{(k_1-2)(k_1-1)}{2}$	0	$\cdots$	0	0	1	$\cdots$	k_1-4	k_1-3	k_1-2
$\vdots$	$\vdots$	$\vdots$		$\vdots$	$\vdots$	$\vdots$	$\vdots$	$\vdots$	$\vdots$	$\vdots$
k_1-2	-6	0	$\cdots$	0	0	0	$\cdots$	1	2	3
k_1-1	-3	0	$\cdots$	0	0	0	$\cdots$	0	1	2
k_1	-1	0	$\cdots$	0	0	0	$\cdots$	0	0	1

Fig. 1. Multiples of "a" to be added to each $E(X_{i_1 i_2})$ to form model $E(X_{i_1 i_2}(a))$.

Now "a" can be chosen so that, letting $f_{i_1 i_2}(k_1, k_2)$ be the Figure 1 entry in row i_1 and column i_2, we simultaneously satisfy

$$\mu+\alpha^{(1)}_{[1]}+\alpha^{(2)}_{[1]}+\alpha^{(12)}_{11}(a) \geqq \mu+\alpha^{(1)}_{[i_1]}+\alpha^{(2)}_{[i_2]}+\alpha^{(12)}_{i_1 i_2}(a)+\delta^* \qquad (\forall\, (i_1, i_2) \neq (1, 1)),$$

i.e.

$$\mu+\alpha^{(1)}_{[1]}+\alpha^{(2)}_{[1]}+f_{11}(k_1, k_2)a \geqq \mu+\alpha^{(1)}_{[i_1]}+\alpha^{(2)}_{[i_2]}+f_{i_1 i_2}(k_1, k_2)a+\delta^* \qquad (\forall\, (i_1, i_2) \neq (1, 1)),$$

i.e.

$$a \geqq \frac{\{\alpha^{(1)}_{[i_1]}+\alpha^{(2)}_{[i_2]}\}-\{\alpha^{(1)}_{[1]}+\alpha^{(2)}_{[1]}\}+\delta^*}{f_{11}(k_1, k_2)-f_{i_1 i_2}(k_1, k_2)} \qquad (\forall\, (i_1, i_2) \neq (1, 1))$$

(where it is evident from Figure 1 that the denominator is >0). For this purpose it suffices to take

$$a=6\frac{\{\alpha^{(1)}_{[k_1]}+\alpha^{(2)}_{[k_2]}\}-\{\alpha^{(1)}_{[1]}+\alpha^{(2)}_{[1]}\}+\delta^*}{(k_1-1)(k_1-2)(k_1+3)}\,.$$

For a *three-factor experiment* ($r=3$),

$$E(X_{i_1i_2i_3})=\mu+\alpha^{(1)}_{i_1}+\alpha^{(2)}_{i_2}+\alpha^{(3)}_{i_3}+\alpha^{(12)}_{i_1i_2}+\alpha^{(13)}_{i_1i_3}+\alpha^{(23)}_{i_2i_3}+\alpha^{(123)}_{i_1i_2i_3}$$

($i_1=1, 2,\cdots, k_1$; $i_2=1, 2,\cdots, k_2$; $i_3=1, 2,\cdots, k_3$), where μ and α's are unknown parameters with

$$\sum_{i_1=1}^{k_1}\alpha^{(1)}_{i_1}=\sum_{i_2=1}^{k_2}\alpha^{(2)}_{i_2}=\sum_{i_3=1}^{k_3}\alpha^{(3)}_{i_3}=0\,,$$

$$\sum_{i_1=1}^{k_1}\alpha^{(12)}_{i_1i_2}=\sum_{i_2=1}^{k_2}\alpha^{(12)}_{i_1i_2}=\sum_{i_1=1}^{k_1}\alpha^{(13)}_{i_1i_3}=\sum_{i_3=1}^{k_3}\alpha^{(13)}_{i_1i_3}=\sum_{i_2=1}^{k_2}\alpha^{(23)}_{i_2i_3}=\sum_{i_3=1}^{k_3}\alpha^{(23)}_{i_2i_3}=0\,,$$

$$\sum_{i_1=1}^{k_1}\alpha^{(123)}_{i_1i_2i_3}=\sum_{i_2=1}^{k_2}\alpha^{(123)}_{i_1i_2i_3}=\sum_{i_3=1}^{k_3}\alpha^{(123)}_{i_1i_2i_3}=0\,.$$

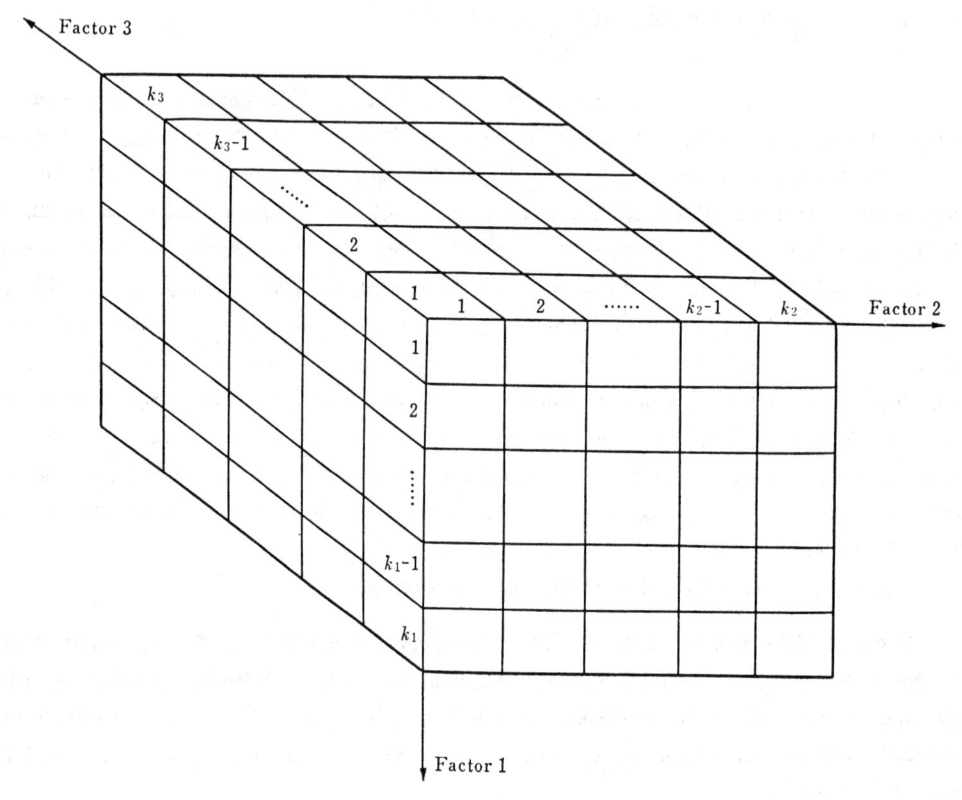

Fig. 2. Three-factor experiment in the form of a cuboid. (Factor j has k_j levels ($j=1, 2, 3$).)

Figure 2 shows the three-factor experiment as a cuboid. We now add second-order interactions to the cells as multiples of a (yet to be chosen), as follows. In the last k_1-1 of the k_3 levels of factor 3 (i.e. levels $k_3-k_1+2,\cdots, k_3$ of factor 3) add $-f_{i_1i_2}(k_1, k_2)a$ in cell (i_1, i_2, i_3), where $f_{i_1i_2}(k_1, k_2)$ is as specified in Figure 1. The other levels of factor 3 have zero second-order interactions in each cell, except for the first level, in which we add $f_{i_1 i_2}(k_1, k_2)a(k_1-1)$ in cells $(i_1, i_2, 1)$, where again $f_{i_1 i_2}(k_1, k_2)$ is as specified in Figure 1. This makes the column sums (corresponding to factor 3) zero, and the interactions all satisfy the model restrictions. (Note that cell (1, 1, 1) is best for large positive a.)

Let $f_{i_1 i_2 i_3}(k_1, k_2, k_3)$ denote the number of multiples of a added to cell (i_1, i_2, i_3) $(i_1=1, \cdots, k_1;\ i_2=1, \cdots, k_2;\ i_3=1, \cdots, k_3)$. Then a can be chosen so that we simultaneously satisfy

$$\mu+\alpha_{[1]}^{(1)}+\alpha_{[1]}^{(2)}+\alpha_{[1]}^{(3)}+f_{111}(k_1, k_2, k_3)a \geqq \mu+\alpha_{[i_1]}^{(1)}+\alpha_{[i_2]}^{(2)}+\alpha_{[i_3]}^{(3)}+f_{i_1 i_2 i_3}(k_1, k_2, k_3)a+\delta^* \qquad (\forall\,(i_1, i_2, i_3)\neq(1,1,1)),$$

i.e.

$$a \geqq \frac{\{\alpha_{[i_1]}^{(1)}+\alpha_{[i_2]}^{(2)}+\alpha_{[i_3]}^{(3)}\}-\{\alpha_{[1]}^{(1)}+\alpha_{[1]}^{(2)}+\alpha_{[1]}^{(3)}\}+\delta^*}{f_{111}(k_1, k_2, k_3)-f_{i_1 i_2 i_3}(k_1, k_2, k_3)} \qquad (\forall\,(i_1, i_2, i_3)\neq(1,1,1))$$

(where the denominator is >0). For this purpose it suffices to take

$$a=6\frac{\{\alpha_{[k_1]}^{(1)}+\alpha_{[k_2]}^{(2)}+\alpha_{[k_3]}^{(3)}\}-\{\alpha_{[1]}^{(1)}+\alpha_{[1]}^{(2)}+\alpha_{[1]}^{(3)}\}+\delta^*}{(k_1-2)(k_1-1)k_1(k_1+2)}.$$

The general case of an *r-factor experiment* $(r\geqq 2)$ is handled by induction on r, using the method shown to proceed from $r=2$ to $r=3$ above. Namely, suppose we have an interaction scheme for an $(r-1)$-factor experiment where all the interactions are zero except the highest-order ones. In order to extend it to an r-factor experiment, suppose that the rth factor has k_r levels. In the last (k_1-1) levels of factor r (that is, from k_r-k_1+2 through k_r, both inclusive), we add interactions of highest-order as in the $(r-1)$-factor experiment scheme, but with sign of each entry reversed. The rest of the levels (except for the first one) receive zero highest-order interaction in each cell. Then to make appropriate interaction sums zero, we add highest-order interactions in the first level in such a way that each entry is (k_1-1) times the corresponding entry of the $(r-1)$-factor experiment scheme. As before we may find a large positive number a such that the cell (1, 1, $\cdots$, 1) is best.

We are now in a position to prove Theorem 1.

Proof of Theorem 1. Let n_1 for SP1 satisfy $P(\mathrm{CS}\,|\,\mathrm{SP1})\geqq P^*$ whenever $\boldsymbol{\delta}\geqq\boldsymbol{\delta}^*$ (componentwise) in a model with zero interactions (which, of course, implies a sample size N per cell, with $n_1=NK$). Now $\bar{X}_{1\cdot\cdots\cdot}, \bar{X}_{2\cdot\cdots\cdot}, \cdots, \bar{X}_{k_1\cdot\cdots\cdot}$ are independent normal random variables with (possibly different) means and the same variance $k_1\sigma^2/(NK)$, hence

$$\bar{X}^*_{1\cdot\cdots\cdot}, \bar{X}^*_{2\cdot\cdots\cdot}, \cdots, \bar{X}^*_{k_1\cdot\cdots\cdot}$$

are independent $N(0, 1)$ random variables when we define

$$\bar{X}^*_{i_1\cdot\cdots\cdot}=\frac{\bar{X}_{i_1\cdot\cdots\cdot}-E(\bar{X}_{i_1\cdot\cdots\cdot})}{\sqrt{\sigma^2 k_1/(NK)}}, \qquad (i_1=1, 2, \cdots, k_1).$$

Now consider model $E(X_{i_1 i_2 \cdots i_r}(a))$ with "a" chosen so large positive that cell (1,

1, ⋯, 1) is best, according to the interaction scheme previously given. Then, in this model with interaction,

$$\begin{aligned} P(\mathrm{CS}|\mathrm{SP1}) &= P[\bar{X}_{1\cdots\cdots} = \max(\bar{X}_{1\cdots\cdots}, \bar{X}_{2\cdots\cdots}, \cdots, \bar{X}_{k_1\cdots\cdots}); \\ &\qquad \bar{X}_{\cdot 1\cdots\cdots} = \max(\bar{X}_{\cdot 1\cdots\cdots}, \bar{X}_{\cdot 2\cdots\cdots}, \cdots, \bar{X}_{\cdot k_2\cdots\cdots}); \\ &\qquad \cdots \\ &\qquad \bar{X}_{\cdots\cdots 1} = \max(\bar{X}_{\cdots\cdots 1}, \bar{X}_{\cdots\cdots 2}, \cdots, \bar{X}_{\cdots\cdots k_r})] \\ &= P[Y_{11}>0,\ Y_{12}>0, \cdots, Y_{1,k_1-1}>0; \\ &\qquad Y_{21}>0,\ Y_{22}>0, \cdots, Y_{2,k_2-1}>0; \\ &\qquad \cdots \\ &\qquad Y_{r1}>0,\ Y_{r2}>0, \cdots, Y_{r,k_r-1}>0] \end{aligned}$$

where

$$\begin{aligned} Y_{1i_1} &= \bar{X}_{1\cdots\cdots} - \bar{X}_{i_1+1,\cdots\cdots} && (i_1=1, 2, \cdots, k_1-1), \\ Y_{2i_2} &= \bar{X}_{\cdot 1\cdots\cdots} - \bar{X}_{\cdot, i_2+1,\cdots\cdots} && (i_2=1, 2, \cdots, k_2-1), \\ &\cdots \\ Y_{ri_r} &= \bar{X}_{\cdots\cdots 1} - \bar{X}_{\cdots\cdots, i_r+1} && (i_r=1, 2, \cdots, k_r-1). \end{aligned}$$

Now $(Y_{11}, \cdots, Y_{1,k_1-1};\ Y_{21}, \cdots, Y_{2,k_2-1}; \cdots;\ Y_{r1}, \cdots, Y_{r,k_r-1})$ has a $(k_1+k_2+\cdots+k_r-r)$-variate normal distribution with

$$\begin{aligned} E(Y_{1i_1}) &= \alpha^{(1)}_{[1]} - \alpha^{(1)}_{[i_1+1]}, \quad \mathrm{Var}(Y_{1i_1}) = \frac{2\sigma^2 k_1}{NK} && (i_1=1, 2, \cdots, k_1-1), \\ E(Y_{2i_2}) &= \alpha^{(2)}_{[1]} - \alpha^{(2)}_{[i_2+1]}, \quad \mathrm{Var}(Y_{2i_2}) = \frac{2\sigma^2 k_2}{NK} && (i_2=1, 2, \cdots, k_2-1), \\ &\cdots \\ E(Y_{ri_r}) &= \alpha^{(r)}_{[1]} - \alpha^{(r)}_{[i_r+1]}, \quad \mathrm{Var}(Y_{ri_r}) = \frac{2\sigma^2 k_r}{NK} && (i_r=1, 2, \cdots, k_r-1), \\ &\mathrm{Cov}(Y_{1i_1}, Y_{1i_1'}) = \frac{\sigma^2 k_1}{NK} && (i_1 \neq i_1'), \\ &\mathrm{Cov}(Y_{2i_2}, Y_{2i_2'}) = \frac{\sigma^2 k_2}{NK} && (i_2 \neq i_2'), \\ &\cdots \\ &\mathrm{Cov}(Y_{ri_r}, Y_{ri_r'}) = \frac{\sigma^2 k_r}{NK} && (i_r \neq i_r'), \end{aligned}$$

$\mathrm{Cov}(Y_{ji_r}, Y_{li_l})=0$ $(j \neq l;\ j=1, \cdots, r;\ l=1, \cdots, r)$ $(i_1=1, 2, \cdots, k_1-1;\ i_2=1, 2, \cdots, k_2-1;\ \cdots;\ i_r=1, 2, \cdots, k_r-1)$. Thus $(Y_{11}, \cdots, Y_{1,k_1-1}), (Y_{21}, \cdots, Y_{2,k_2-1}), \cdots, (Y_{r1}, \cdots, Y_{r,k_r-1})$ are independent, so

$$P(\mathrm{CS}|\mathrm{SP1}) = \prod_{j=1}^{r} P[Y_{j1}>0, Y_{j2}>0, \cdots, Y_{j,k_j-1}>0].$$

Considering the first factor,

$$\begin{aligned} &P[Y_{11}>0, Y_{12}>0, \cdots, Y_{1,k_1-1}>0] \\ &\quad = P[\bar{X}_{1\cdots\cdots} = \max(\bar{X}_{1\cdots\cdots}, \bar{X}_{2\cdots\cdots}, \cdots, \bar{X}_{k_1\cdots\cdots})] \\ &\quad \leq P[\bar{X}^*_{1\cdots\cdots} = \max(\bar{X}^*_{1\cdots\cdots}, \bar{X}^*_{2\cdots\cdots}, \cdots, \bar{X}^*_{k_1\cdots\cdots})] \\ &\quad = \frac{1}{k_1}. \end{aligned}$$

Proceeding similarly, we find that

$$P[Y_{j1}>0, Y_{j2}>0, \cdots, Y_{j,k_j-1}>0] \leqq \frac{1}{k_j} \qquad (j=2, \cdots, r),$$

hence

$$P(\mathrm{CS} \mid \mathrm{SP1}) \leqq 1/K$$

as was to be proven.

Remark 1. Note that for all schemes of adding interactions and for all $k_j \geqq 2$ $(j=1, 2, \cdots, r)$, procedure SP1 still achieves probability at least P^* for selecting the cell associated with $\mu + \sum_{j=1}^{r} \alpha_{[k_j]}^{(j)}$ (which may not be the "best" cell after adding interactions). This is true because the sum of interactions in each row and in each column is zero and so the (marginal) sample means do not change.

Remark 2. In the case $k_1=2$, we add interactions in such a way that cell $(k_1, k_2, \cdots, k_r)$ does not remain best. From Remark 1, it then follows that

$$P(\mathrm{CS} \mid \mathrm{SP1}) \leqq 1-P^* .$$

THEOREM 2. *Let n_2 be the smallest sample size for which procedure* SP2 *guarantees $P(\mathrm{CS} \mid \mathrm{SP2}) \geqq P^*$ whenever $\boldsymbol{\delta} \geqq \boldsymbol{\delta}^*$ (componentwise) in a model with zero interactions. Then in a general (non-zero interaction) model,*

$$\inf_{\Omega(\delta^*)} P(\mathrm{CS} \mid \mathrm{SP2}) \leqq \frac{1}{\min\limits_{1 \leq j \leq r} \{k_j\}}$$

where $\Omega(\delta^*) = \{(\mu_1, \mu_2, \cdots, \mu_K) \colon \mu_{[K]} - \mu_{[K-1]} \geqq \delta^*,\ \delta^* = \min\{\delta_1^*, \cdots, \delta_r^*\}\}$.

PROOF OF THEOREM 2. Let n_2 for SP2 satisfy $P(\mathrm{CS} \mid \mathrm{SP2}) \geqq P^*$ whenever $\boldsymbol{\delta} \geqq \boldsymbol{\delta}^*$ (componentwise) in a model with zero interactions. Let factors be ordered as in the discussion following the statement of Theorem 1, and consider model $E(X_{i_1 i_2 \dots i_r}(a))$ with a set so that cell $(1, 1, \cdots, 1)$ is best. Suppose, without loss of generality, that the experimenter decides to first fix levels of each factor except factor j (and hence experiments across the levels of factor j). Define events

$$F_{ji_j} = \{\text{Experimenter starts with level } i_j \text{ of factor } j\},$$

$$E_{ji_j} = \{\text{Experimenter selects level } i_j \text{ of factor } j\},$$

$(i_j=1, \cdots, k_j;\ j=1, \cdots, r)$ so that event $E_{11}E_{21} \cdots E_{r1}$ corresponds to a correct selection. Now, for any $\varepsilon>0$, a may be taken sufficiently large that

$$P(E_{11}E_{21} \cdots E_{r1} \mid F_{ji_j}) \leqq \varepsilon \qquad (i_j \neq 1).$$

Now, letting $F_j = F_{j1} \cup \cdots \cup F_{jk_j}$ (the event that the experimenter starts with factor j) and taking $P(F_{j1} \mid F_j) = \cdots = P(F_{jk_j} \mid F_j) = 1/k_j$ as a reasonable model in light of no prior knowledge (on the part of the experimenter) as to which levels are better than others, we find

$$\begin{aligned} P(\mathrm{CS} \mid \mathrm{SP2}, F_j) &= \sum_{i_j=1}^{k_j} P(E_{11}E_{21} \cdots E_{r1} F_{ji_j} \mid F_j) \\ &= \sum_{i_j=1}^{k_j} P(F_{ji_j} \mid F_j) P(E_{11}E_{21} \cdots E_{r1} \mid F_{ji_j}) \end{aligned}$$

$$=P(F_{j1}\,|\,F_j)P(E_{11}E_{21}\cdots E_{r1}\,|\,F_{j1})$$

$$+\sum_{i_j=2}^{k_j} P(F_{ji_j}\,|\,F_j)P(E_{11}E_{21}\cdots E_{r1}\,|\,F_{ji_j})$$

$$\leqq \frac{1}{k_j}+\sum_{i_j=2}^{k_j}\frac{\varepsilon}{k_j}\leqq\frac{1}{k_j}+\varepsilon\ ,$$

and the theorem follows.

THEOREM 3. *Let n_3 be the smallest sample size for which procedure* SP3 *guarantees* $P(\text{CS}\,|\,\text{SP3})\geqq P^*$ *whenever* $\boldsymbol{\delta}\geqq\boldsymbol{\delta}^*$ *(componentwise) in a model with zero interactions. Then in a general (non-zero interaction) model,* $P(\text{CS}\,|\,\text{SP3})\geqq P^*$ *whenever* $\mu_{[K]}-\mu_{[K-1]}\geqq\delta^*=\min\{\delta_1^*,\cdots,\delta_r^*\}$.

PROOF OF THEOREM 3. In a model with zero interactions, suppose that cell $(m_1, m_2,\cdots, m_r)$ is best. Then, if M observations per cell have been taken, whenever $\boldsymbol{\delta}\geqq\boldsymbol{\delta}^*$ (componentwise) we have

$$P(\text{CS}\,|\,\text{SP3})=P[\bar{X}_{m_1m_2\cdots m_r}=\max_{i_1,i_2,\cdots,i_r}\{\bar{X}_{i_1i_2\cdots i_r}\},$$

$$i_j=1,2,\cdots,k_j\ (j=1,2,\cdots,r)]$$

$$=P\left[\bar{X}^*_{i_1i_2\cdots i_r}\leqq\bar{X}^*_{m_1m_2\cdots m_r}+\frac{E(\bar{X}_{m_1m_2\cdots m_r})-E(\bar{X}_{i_1i_2\cdots i_r})}{\sqrt{\sigma^2/M}},\right.$$

$$\left.i_j=1,2,\cdots,k_j\ (j=1,2,\cdots,r)\right]$$

$$\geqq P\left[\bar{X}^*_{i_1i_2\cdots i_r}\leqq\bar{X}^*_{m_1m_2\cdots m_r}+\frac{\min\{\delta_1^*,\cdots,\delta_r^*\}}{\sqrt{\sigma^2/M}},\right.$$

$$\left.i_j=1,2,\cdots,k_j\ (j=1,2,\cdots,r)\right]$$

and

$$\inf_{\boldsymbol{\delta}\geqq\boldsymbol{\delta}^*}P(\text{CS}\,|\,\text{SP3})=\int_{-\infty}^{\infty}\Phi^{K-1}\left(x+\frac{\sqrt{M}\,\min\{\delta_1^*,\cdots,\delta_r^*\}}{\sigma}\right)\phi(x)dx\ ,$$

where $\Phi(\cdot)$ and $\phi(\cdot)$ denote the standard normal distribution and density functions, respectively. For this inf to be $\geqq P^*$ we need

$$M\geqq\frac{h_K^2(P^*)\sigma^2}{(\min\{\delta_1^*,\cdots,\delta_r^*\})^2}$$

where $h_K(P^*)$ is the solution h of the equation

$$\int_{-\infty}^{\infty}\Phi^{K-1}(x+h)\phi(x)dx=P^*\ .$$

Now in a general (non-zero interaction) model suppose that cell $(c_1, c_2,\cdots, c_r)$ is best, M as above is used, and $\mu_{[K]}-\mu_{[K-1]}\geqq\delta^*=\min\{\delta_1^*,\cdots,\delta_r^*\}$. Then

$$P(\text{CS}\,|\,\text{SP3})=P\left[\bar{X}^*_{i_1i_2\cdots i_r}\leqq\bar{X}^*_{c_1c_2\cdots c_r}+\frac{E(\bar{X}_{c_1c_2\cdots c_r})-E(\bar{X}_{i_1i_2\cdots i_r})}{\sqrt{\sigma^2/M}},\right.$$

$$\left.i_j=1,2,\cdots,k_j\ (j=1,2,\cdots,r)\right]$$

$$\geqq P\left[\bar{X}^*_{i_1i_2\cdots i_r}\leqq\bar{X}^*_{c_1c_2\cdots c_r}+\frac{\delta^*\sqrt{M}}{\sigma},\right.$$

$$i_j=1,2,\cdots,k_j\ (j=1,2,\cdots,r)\Big]$$

$$=\int_{-\infty}^{\infty}\Phi^{K-1}\left(x+\frac{\delta^*\sqrt{M}}{\sigma}\right)\phi(x)dx\geqq P^*$$

and the theorem follows.

Thus, while procedure SP3 is fully robust to the presence of interaction, procedures SP1 and SP2 may be fully vitiated by interaction. One may therefore wish to estimate interaction size and choose accordingly between SP1 and SP3 in practice. An SP4 which incorporates this idea (acting as does SP1 when interactions are "negligible" and as does SP3 when interactions are "large") is now being developed.

References

[1] Bawa, V. S. (1972). Asymptotic efficiency of one R-factor experiment relative to R one-factor experiments for selecting the best normal population, *J. Amer. Statist. Assoc.*, **67**, 660-661.

[2] Bechhofer, R. E. (1954). A single-sample multiple decision procedure for ranking means of normal populations with known variances, *Ann. Math. Statist.*, **25**, 16-39.

[3] Bechhofer, R. E. (1977). Selection in factorial experiments, *Proceedings of the 1977 Winter Simulation Conference*, 65-70.

[4] Bechhofer, R. E., Santner, T. J. and Turnbull, B. W. (1977). Selecting the largest interaction in a two-factor experiment, *Statistical Decision Theory and Related Topics*—II (Eds. S. S. Gupta and D. S. Moore), Academic Press, New York, 1-18.

[5] Dudewicz, E. J. (1969). An approximation to the sample size in selection problems, *Ann. Math. Statist.*, **40**, 492-497.

[6] Dudewicz, E. J. (1976). *Introduction to Statistics and Probability*, New York: Holt, Rinehart and Winston.

[7] Dudewicz, E. J. (1977). New procedures for selection among (simulated) alternatives, *Proceedings of the 1977 Winter Simulation Conference*, 59-62.

[8] Dudewicz, E. J. (1980). Ranking (ordering) and selection: an overview of how to select the best, *Technometrics*, **22**, 113-119.

[9] Dudewicz, E. J. and Koo, J. O. (1982). *The Complete Categorized Guide to Statistical Selection and Ranking Procedures*, Columbus, Ohio: American Sciences Press, Inc., P.O. Box 21161.

[10] Gibbons, J. D., Olkin, I. and Sobel, M. (1977). *Selecting and Ordering Populations: A New Statistical Methodology*, New York: John Wiley & Sons, Inc.

[11] Rinott, Y. and Santner, T. J. (1977). An inequality for multivariate normal probabilities with application to a design problem, *Ann. Statist.*, **5**, 1228-1234.

ENTROPY-BASED STATISTICAL INFERENCE, II: SELECTION-OF-THE-BEST/COMPLETE RANKING FOR CONTINUOUS DISTRIBUTIONS ON (0,1), WITH APPLICATIONS TO RANDOM NUMBER GENERATORS*

Edward J. Dudewicz and Edward C. van der Meulen

Received : Revised version: June 16, 1982

ABSTRACT

In I of this series we discussed entropy and its estimation, defined an "entropy test" and notions of "e-distinguishable" and "e-unique", discussed the testing of composite (e.g., normal) and simple (e.g., uniform on (0,1)) hypotheses, and in the latter case showed by extensive Monte Carlo calculations that the entropy test possesses good power properties; asymptotic normality of the test statistic was discussed, and closeness in entropy was related to closeness in density. We now develop methods allowing use of the Dudewicz-Dalal selection procedure to select the best-entropy population, and develop an extension for complete ranking based on entropy, for continuous distributions on (0,1). (Other ranking and selection goals could be achieved by similar methods.) Both indifference-zone and preferred-population formulations are considered. Previous asymptotic normality results are used extensively, and new results on validity of Stein-type sampling under asymptotic (rather than exact) normality are given. Applications to random number generators are considered.

1. INTRODUCTION

Let f be a density function with variance $\sigma^2(f) < \infty$ and distribution function F. The entropy H(f) of f is defined as

$$H(f) = - \int_{-\infty}^{\infty} f(x) \log f(x) dx.$$

Let $X_1, \ldots, X_n$ ($n \geq 3$) be independent random variables, each with density f

---

*AMS Subject Classifications: G2F07, G2F05, G2F35, G5C10.

Key Words and Phrases: Entropy, selection, complete ranking, random number generators, asymptotic normality, Monte Carlo.

Research supported by the NATO Research Grants Programme, NATO Research Grant N° 1674.

as above, let $Y_1 \leq Y_2 \leq \ldots \leq Y_n$ denote the ordered values of $X_1, \ldots, X_n$, and define $Y_j = Y_1$ if $j < 1$, $Y_j = Y_n$ if $j > n$. Then (Vasicek [13]) the estimator

$$H_{m,n} = \frac{\sum_{i=1}^{n} \log\{\frac{n}{2m}(Y_{i+m} - Y_{i-m})\}}{n} \xrightarrow{p} H(f)$$

(either finite or $-\infty$) as $n \to \infty$, $m \to \infty$, $\frac{m}{n} \to 0$. It is also known (Dudewicz and van der Meulen [6], Theorem 6) that if f is a bounded (above and below) positive step function on $0 \leq x \leq 1$ with at most a finite number of discontinuities, then

$$n^{\frac{1}{2}}\left(\frac{n}{n+2-2m}\right)^{\frac{1}{2}}[H_{m,n} + \log(2m) + \gamma - R(1,2m-1)-H(f)]$$

is asymptotically normal with mean zero and variance $\tau(2m) + \mathrm{Var}_f \log f(X)$, i.e. $\sim N(0,\tau(2m) + \mathrm{Var}_f \log f(X))$, where

$$\tau(m) = \begin{cases} (2m^2-2m+1)\{\frac{1}{6}\pi^2-(1+\frac{1}{2^2}+\frac{1}{3^2}+\ldots+\frac{1}{(m-1)^2})\}-2m+1 & \text{if } m \geq 2 \\ \frac{1}{6}\pi^2-1 & \text{if } m = 1, \end{cases}$$

$\gamma \doteq .5772$, and $R(1,j) = j^{-1} + (j-1)^{-1} + \ldots + 2^{-1} + 1^{-1}$ $(j \geq 1)$. These facts are basic to our development below of ranking and selection procedures for continuous distributions on (0,1).

2. STEIN-TYPE SAMPLING UNDER ASYMPTOTIC NORMALITY

In this section we wish to establish results on Stein-type sampling under asymptotic (rather than exact) normality. These results are then used to develop statistical ranking and selection procedures under asymptotic normality (such as we have for $H_{m,n}$ as $n \to \infty$), rather than exact normality, in succeeding sections. The results also justify use of Stein-type sampling under general non-normal distributions. Some early studies under Edgeworth series distributions were given by Bhattacharjee [1], who indicated some lack of robustness; his assertion is refuted by Ramkaran [10], who corrects an error in [1] and also includes the gamma distribution. More recent work includes that of Blumenthal and Govindarajulu [3], who investigated robustness under mixtures of normal populations differing in means and derived explicit level and power expressions as well as approximate expansions and numerical studies, and found the results indicated robustness against mixtures of normal populations differing in location. A number of other recent contributions also point in the same direction [9], [11].

Stein-type sampling for one source of observations (due to Stein [12] and recently utilized in a new way to solve previously unsolved k-population problems under heteroscedasticity following work of Dudewicz and Dalal [4]) is in a setting where $X = (X_1, X_2, X_3, \ldots)$ is a sequence of independent $N(\mu,\sigma^2)$ random variables (normal with mean μ and variance $\sigma^2 > 0$). Let n_0 (≥ 2) and $h > 0$ be fixed. Using $X_1, \ldots, X_{n_0}$, calculate

$$\bar{X}(n_0) = \sum_{i=1}^{n_0} X_i/n_0, \; s^2 = \sum_{i=1}^{n_0} (X_i - \bar{X}(n_0))^2/(n_0-1),$$

$$n_1 = \max\{n_0 + 1, [s^2h^2]\}$$

where $[y]$ denotes the smallest integer $\geq y$. Let $a_1, \ldots, a_{n_1}$ be random variables (through n_1 and s^2 only) such that

$$a_1 + \ldots + a_{n_1} = 1, \; a_1 = \ldots = a_{n_0}, \; s^2 \sum_{i=1}^{n_1} a_i^2 = (1/h)^2.$$

Define

$$\tilde{X} = \sum_{i=1}^{n_1} a_i X_i .$$

Then

$$\frac{\tilde{X} - \mu}{1/h} = \frac{\tilde{X} - \mu}{s\sqrt{\sum_{i=1}^{n_1} a_i^2}}$$

has Student's-t distribution with n_0-1 degrees of freedom, i.e. is t_{n_0-1}.

In our more general setting, let $X(n^*) = (X_1(n^*), X_2(n^*), \ldots)$ be a sequence of independent random variables such that $X_i(n^*) \xrightarrow{D} X_i$ as $n^* \to \infty$ (i.e., $X_i(n^*)$ converges in distribution to a $N(\mu,\sigma^2)$ random variable as $n^* \to \infty$). Then all finite-dimensional distributions of $X(n^*)$ converge to the corresponding finite-dimensional distributions of X as $n^* \to \infty$, hence (see pp. 19, 30 of Billingsley [2]) $X(n^*) \xrightarrow{D} X$ as $n^* \to \infty$. Now define function $g(\cdot)$ on an infinite sequence so that

$$g(X) = \frac{\tilde{X} - \mu}{1/h} .$$

Then under X the set D_g of discontinuities of $g(\cdot)$ has probability zero, hence by Corollary 1 of p. 31 of Billingsley [2], $g(X(n^*)) \xrightarrow{D} g(X)$ as $n^* \to \infty$; i.e., as $n^* \to \infty$ we find

$$\frac{\tilde{\bar{X}}(n^*)-\mu}{1/h}$$

converges in distribution to t_{n_0-1}. We state this as

<u>THEOREM 1</u>. Let $X = (X_1, X_2, \ldots)$ be a sequence of independent $N(\mu,\sigma^2)$ random variables, and let $X(n^*) = (X_1(n^*), X_2(n^*),\ldots)$ be a sequence of independent random variables such that $X_i(n^*) \xrightarrow{D} X_i$ as $n^* \to \infty$ ($i = 1, 2, \ldots$). Then,

as $n^* \to \infty$,

$$\frac{\tilde{X}(n^*) - \mu}{1/h}$$

converges to a t_{n_0-1} random variable.

In our applications of Theorem 1 below, we will be dealing with random variables of the form

$$T = H_{m,n^*} + \log(2m) + \gamma - R(1, 2m-1)$$

which are asymptotically $N(\mu, \sigma^2/n^*)$ with $\mu = H(f)$ and

$$\sigma^2 = \tau(2m) + \mathrm{Var}_f \log f(X) .$$

In that setting $X_i(n^*) = \sqrt{n^*}(T_i - H(f))$ is asymptotically $N(0,\sigma^2)$, and by Theorem 1

$$\frac{\tilde{X} - 0}{1/h}$$

is asymptotically t_{n_0-1}. However, it is easy to check that this is exactly the statistic which arises if one applies Stein-type sampling to the sequence T_i with h replaced by $h\sqrt{n^*}$. Hence, in that latter setting,

$$\frac{\tilde{T} - h(f)}{1/(h\sqrt{n^*})} \xrightarrow{D} t_{n_0-1} .$$

3. SELECTION OF THE BEST ENTROPY ON (0,1): IZ FORMULATION

Let $\pi_1, \pi_2, \ldots, \pi_k$ represent $k(\geq 2)$ independent sources of random variables. Assume observations from π_i are normally distributed $N(\mu_i, \sigma_i^2)$ $(1 \leq i \leq k)$ with $\mu_1, \ldots, \mu_k, \sigma_1^2, \ldots, \sigma_k^2$ all unknown. Assume (*Indifference zone* formulation) that the goal is to select that population which has the largest mean $\mu_{[k]}$ (where $\mu_{[1]} \leq \cdots \leq \mu_{[k]}$ denote the ordered means) in such a way that the probability of correct selection P(CS) satisfies

$$P(CS) \geq P^* \text{ whenever } \mu_{[k]} - \mu_{[k-1]} \geq \delta^*$$

where P^* and δ^* $(1/k < P^* < 1, 0 < \delta^*)$ are specified in advance by the experimenter. Then Dudewicz and Dalal [4] proved that the following procedure $\mathcal{T}_{DD}(IZ)$ achieves the goal.

PROCEDURE $\mathcal{T}_{DD}(IZ)$. Take an initial sample $X_{i1}, \ldots, X_{in_0}$ of size $n_0(\geq 2)$ from π_i and define

$$\bar{X}_i(n_0) = \sum_{j=1}^{n_0} X_{ij}/n_0, \; s_i = \sum_{j=1}^{n_0} (X_{ij} - \bar{X}_i(n_0))^2/(n_0-1),$$

$$n_i = \max\{n_0 + 1, [(s_i h/\delta^*)^2]\}$$

where $h = h_{n_0}(k,P^*)$ solves

$$\int_{-\infty}^{\infty} (F_{n_0}(z+h))^{k-1} f_{n_0}(z)dz = P^*$$

(here $F_{n_0}(\cdot)$ and $f_{n_0}(\cdot)$ are respectively the distribution function and density function of a Student's-t random variable with $n_0-1 \geq 1$ degrees of freedom) and $[y]$ denotes the smallest integer $\geq y$ $(i = 1, \ldots, k)$. Take n_i-n_0 additional observations $X_{i,n_0+1}, \ldots, X_{in_i}$ from π_i, and define

$$\tilde{X}_i = \sum_{j=1}^{n_i} a_{ij}X_{ij}$$

$(i = 1, \ldots, k)$. (Here the a_{ij}'s, $j = 1, \ldots, n_i$, $1 \leq i \leq k$, are chosen so that

$$\sum_{j=1}^{n_i} a_{ij} = 1, \; a_{i1} = \ldots = a_{in_0}, \text{ and } s_i^2 \sum_{j=1}^{n_i} a_{ij}^2 = (\delta^*/h)^2.$$

To be specific, take the solution with $a_{i,n_0+1} = \ldots = a_{ij_i}$.) Finally, select that population which yielded $\tilde{X}_{[k]}$ (where $\tilde{X}_{[1]} \leq \ldots \leq \tilde{X}_{[k]}$ denote the $\tilde{X}_1, \ldots, \tilde{X}_k$ in numerical order).

In our setting, we are given $k(\geq 2)$ basic independent populations $\pi_1^*, \ldots, \pi_k^*$. Let f_i denote the density function (with finite variance) of observations from π_i^*, and assume that f_i is a bounded (above and below) positive finite-step function on $0 \leq x \leq 1$ $(1 \leq i \leq k)$. Our basic (IZ) goal is to select that population which has the largest entropy $H(f_{[k]})$ (where $H(f_{[i]})$ denotes the $k - i+1^{st}$ largest of the entropies $H(f_1), \ldots, H(f_k)$ $(1 \leq i \leq k)$, so that $H(f_{[1]}) \leq H(f_{[2]}) \leq \ldots \leq H(f_{[k]})$) in such a way that the probability of correct selection satisfies

$$P(CS) \geq P^* \text{ whenever } H(f_{[k]}) - H(f_{[k-1]}) \geq \delta^*$$

where P^* and δ^* $(k/k < P^* < 1, 0 < \delta^*)$ are specified in advance by the experimenter. Now fix an m. If we take a random sample of size n^* from π_i^*, then (for $n^* \to \infty$) the resulting H_{m,n^*} will be asymptotically normal once standardized, or

$$T_i \equiv H_{m,n*} + \log(2m) + \gamma - R(1, 2m-1)$$

will be asymptotically $N(\mu_i, \sigma_i^2)$ where $\mu_i = H(f_i)$ and $\sigma_i^2 = (\tau(2m) + \mathrm{Var}_{f_i} \log f_i(X))\ (n* + 2 - 2m)/(n*)^2$. Suppose this normality were exact (not just asymptotic). Then our IZ selection problem would be solved by

PROCEDURE $\mathcal{T}_{DV}(IZ)$. From any given basic population π_i^*, sample in batches of $n*$ observations at a time, and consider the resulting statistic

$$T_i = H_{m,n*} + \log(2m) + \gamma - R(1, 2m-1)$$

to be one observation from π_i $(1 \leq i \leq k)$. Then apply procedure $\mathcal{T}_{DD}(IZ)$ with h replaced by $h\sqrt{n*}$ to $\pi_1, \ldots, \pi_k$ to obtain a solution to the selection problem involving $\pi_1, \ldots, \pi_k$ (which is also a solution to the problem involving $\pi_1^*, \ldots, \pi_k^*$ since $\mu_i = H(f_i)$ for all i).

That procedure $\mathcal{T}_{DV}(IZ)$ in fact (for large $n*$) approximately solves the problem even when T_i is only asymptotically normal (as $n* \to \infty$) was shown by the results of Section 2 (which guaranteed that each $\tilde{X}_i$ converges in distribution to a Student's-t random variable with n_0-1 degrees of freedom as $n* \to \infty$).

4. SELECTION OF THE BEST ENTROPY ON (0,1): PPF FORMULATION

Let $\pi_1, \pi_2, \ldots, \pi_k$ represent $k(\geq 2)$ independent sources of random variables. Assume observations from π_i are normally distributed $N(\mu_i, \sigma_i^2)$ $(1 \leq i \leq k)$ with $\mu_1, \ldots, \mu_k, \sigma_1^2, \ldots, \sigma_k^2$ all unknown. The *Preferred Population Formulation* (PPF) due to Hooper and Santner (1979) has as its goal selection of a *preferred population* (call any such selection a "correct selection") in such a way that the probability of correct selection P(CS) satisfies

$$P(CS) \geq P*$$

for all possible parameter configurations, where $P*$ $(1/k < P* < 1)$ is specified in advance by the experimenter. The experimenter must also specify his *preference function* $p(\cdot)$, which is required to satisfy (in our setting): $p(\cdot)$ is continuous, strictly increasing, and $p(\mu) < \mu$ for $-\infty < \mu < +\infty$. Let $\mu_{[1]} \leq \cdots \leq \mu_{[k]}$ denote the ordered means. Then population π_i is *preferred* iff

$$\mu_i \geq p(\mu_{[k]})$$

$(1 \leq i \leq k)$. Of course the population with mean $\mu_{[k]}$ is preferred, since $p(\mu_{[k]}) < \mu_{[k]}$. For each fixed $\mu_{[k]}$ define $\delta_p(\mu_{[k]})$ via

$$p(\mu_{[k]}) = \mu_{[k]} - \delta_p(\mu_{[k]}).$$

Then, at level $\mu_{[k]}$ of the best mean, we prefer population π_i iff

$$\mu_i \geq \mu_{[k]} - \delta_p(\mu_{[k]}).$$

Now it can be shown easily that the Dudewicz-Dalal [4] procedure $\mathcal{T}_{DD}(IZ)$ given in Section 3, as well as satisfying

$$P(CS) \geq P^* \text{ whenever } \mu_{[k]} - \mu_{[k-1]} \geq \delta^*.$$

also satisfies

P[The selected population's mean μ_s satisfies $\mu_{[k]} - \mu_s \leq \delta^*] \geq P^*$.

Hence it follows that if we use the Dudewicz-Dalal procedure with the choice

$$\delta^* = \delta_p^* \equiv \inf_\mu \delta_p(\mu)$$

then it satisfies $P(CS) \geq P^*$ for all possible parameter configurations, as long of course as $\delta_p^* > 0$. If $\delta_p^* = 0$, then there is a sequence of $\mu_{[k]}$ such that in the limit we perfer only $\mu_{[k]}$. However, since with arbitrarily close μ's no procedure (except perhaps a sequential one) probably exists which can make the $P(CS) \geq P^*$ for *all* parameter configurations, it is reasonable to also require the experimenter's $p(\cdot)$ satisfy $\delta_p^* > 0$. (Certainly in single-stage procedures, as in Hooper and Santner [8], the restriction $\delta_p^* > 0$ is essential for dealing with location cases.) Note that, although Hooper and Santner argue to the special functions $p(\mu_{[k]}) = \mu_{[k]} - d$ and $p(\mu_{[k]}) = c\mu_{[k]}$, our solution is valid in much greater generality.

To summarize the above development, let $\pi_1, \pi_2, \ldots, \pi_k$ be $k(\geq 2)$ independ sources of random variables. Assume observations from π_i are normally distributed $N(\mu_i, \sigma_i^2)$ $(1 \leq i \leq k)$ with $\mu_1, \ldots, \mu_k, \sigma_1^2, \ldots, \sigma_k^2$ all unknown. Assume (PPF formulation) the goal is to select a preferred population (call such a selection a correct selection) with $P(CS) \geq P^*$ for all possible parameter configurations, where P^* $(1/k < P^* < 1)$ and preference function $p(\cdot)$ (which is continuous, strictly increasing, with $p(\mu) < \mu$ for all μ, and has $\delta_p^* \equiv \inf_\mu \delta_p(\mu) > 0$ where $\delta_p(\mu) = \mu - p(\mu)$) are specified in advance by the experimenter. Then the following procedures $\mathcal{T}_{DD}(PPF)$ achieves the goal.

PROCEDURE $\mathcal{T}_{DD}(PPF)$. Define $\delta^* = \delta_p^*$ and proceed according to procedure $\mathcal{T}_{DD}(IZ)$.

In our setting, we are given $k(\geq 2)$ basic independent populations $\pi_1^*, \ldots, \pi_k^*$. Let f_i denote the density function (with finite variance) of observations from π_i^*, and assume that f_i is a bounded (above and below) positive finite-step function on $0 \leq x \leq 1$ $(1 \leq i \leq k)$. Let $H(f_{[i]})$ denote the $k-i+1^{st}$ largest of the entropies $H(f_1), \ldots, H(f_k)$ $(1 \leq i \leq k)$, so that $H(f_{[1]}) \leq H(f_{[2]}) \leq \ldots \leq H(f_{[k]})$. Our PPF goal is to have probability $\geq P^*$ of selecting a preferred population, where P^* $(1/k < P^* < 1)$ and preference function $p(\cdot)$ (which is continuous, strictly increasing, with $p(H) < H$ for all H, and has $\delta_p^* \equiv \inf_H \delta_p(H) > 0$ where $\delta_p(H) = H - p(H)$) are specified in advance by the experimenter. Then, proceeding as in Section 3 (using again the results of Section 2) we find our PPF selection problem is solved (approximately) by

PROCEDURE T_{DV}(PPF). From any given basic population π_i^*, sample in batches of n^* observations at a time, and consider the resulting statistic

$$T_i = H_{m,n^*} + \log(2m) + \gamma - R(1,2m-1)$$

to be one observation from π_i $(1 \le i \le k)$. Then apply procedure T_{DD}(PPF) with h replaced by $h\sqrt{n^*}$ to $\pi_1, \ldots, \pi_k$ to obtain a solution to the selection problem involving $\pi_1, \ldots, \pi_k$ (which is also a solution to the problem involving $\pi_1^*, \ldots, \pi_k^*$ since $\mu_i = H(f_i)$ for all i).

5. COMPLETE RANKING OF ENTROPIES ON (0,1): IZ FORMULATION

Let $\pi_1, \pi_2, \ldots, \pi_k$ represent $k(\ge 2)$ independent sources of random variables. Assume observations from π_i are normally distributed $N(\mu_i, \sigma_i^2)$ $(1 \le i \le k)$ with $\mu_1, \ldots, \mu_k, \sigma_1^2, \ldots, \sigma_k^2$ all unknown. Assume (*Indifference Zone* formulation) that the goal is to completely rank the poupulations from that with the largest mean $\mu_{[k]}$ down through that with the smallest mean $\mu_{[1]}$ (where $\mu_{[1]} \le \cdots \le \mu_{[k]}$ denote the ordered means) in such a way that the probability of correct ranking P(CR) satisfies

$$P(CR) \ge P^* \text{ whenever } \mu_{[j]} - \mu_{[j-1]} \ge \delta_j^* \ (j = 2, \ldots, k)$$

where P^* and $\delta_2^*, \ldots, \delta_k^*$ $(1/k! < P^* < 1; 0 < \delta_2^*, \ldots, \delta_k^*)$ are specified in advance by the experimenter. Consider the following procedure

PROCEDURE T_{DD}(CR-IZ). Take an initial sample $X_{i1}, \ldots, X_{in_0}$ of size $n_0(\ge 2)$ from π_i and define

$$\bar{X}_i(n_0) = \sum_{j=1}^{n_0} X_{ij}/n_0, \ s_i^2 = \sum_{j=1}^{n_0} (X_{ij} - \bar{X}_i(n_0))^2/(n_0-1),$$

$$n_i = \max\{n_0 + 1, [(s_i h/\delta^*)^2]\}$$

where $\delta^* > 0$ is arbitrary (it turns out we may, without loss of generality, take $\delta^* = 1$) and $h > 0$ is yet to be specified, while [y] denotes the smallest integer $\ge$ y (i = 1, ..., k). Take $n_i - n_0$ additional observations $X_{i,n_0+1}, \ldots, X_{in_i}$ from π_i and define

$$\tilde{X}_i = \sum_{j=1}^{n_i} a_{ij} X_{ij}$$

(i = 1, ..., k). (Here the a_{ij}'s, j = 1, ..., n_i, $1 \le i \le k$, are chosen so that

$$\sum_{j=1}^{n_i} a_{ij} = 1, \ a_{i1} = \cdots = a_{in_0}, \text{ and } s_i^2 \sum_{j=1}^{n_i} a_{ij}^2 = (\delta^*/h)^2.$$

To be specific, take the solution with $a_{i,n_0+1} = \cdots = a_{in_i}$.) Finally, assert (for i = 1, ..., k) that the population which yielded $\tilde{X}_{[i]}$ has mean $\mu_{[i]}$ (where $\tilde{X}_{[1]} \le \cdots \le \tilde{X}_{[k]}$ denote $\tilde{X}_1, \ldots, \tilde{X}_k$ in numerical order).

Although this problem (CR-IZ) was not explicitly considered by Dudewicz and Dalal [4], it is easy to see from that paper that (letting $\tilde{X}_{(i)}$ denote the $\tilde{X}$-statistic produced by the population with mean $\mu_{[i]}$, $1 \le i \le k$)

$$P(CR|\mathcal{T}_{DD}(CR - IZ)) = P[\tilde{X}_{(1)} < \tilde{X}_{(2)} < \ldots < \tilde{X}_{(k-1)} < \tilde{X}_{(k)}]$$

$$= P\left[\frac{\tilde{X}_{(\ell)}-\mu_{[\ell]}}{\delta^*/h} < \frac{\tilde{X}_{(\ell+1)}-\mu_{[\ell+1]}}{\delta^*/h} + \frac{\mu_{[\ell+1]}-\mu_{[\ell]}}{\delta^*/h}, \ \ell = 1, \ldots, k-1\right]$$

$$= P\left[Y_\ell < Y_{\ell+1} + \frac{\mu_{[\ell+1]}-\mu_{[\ell]}}{\delta^*/h}, \ \ell = 1, \ldots, k-1\right]$$

where $Y_1, \ldots, Y_k$ are independent Student's-t random variables each with n_0-1 degrees of freedom. Over the preference zone $\mu_{[j]}-\mu_{[j-1]} \geq \delta_j^*$ ($j = 2, \ldots, k$), it is easy to see this P(CR) is minimized when $\mu_{[j]}-\mu_{[j-1]} = \delta_j^*$ ($j = 2, \ldots, k$), in which case it equals

$$P_{LFC}(CR|\mathcal{T}_{DD}(CR - IZ)) = P[Y_\ell < Y_{\ell+1} + \delta^*_{\ell+1}h/\delta^*, \ \ell = 1, \ldots, k-1]$$

$$= P[Y_1<Y_2+\delta_2^*h/\delta^*, \ Y_2<Y_3+\delta_3^*h/\delta^*, \ Y_3<Y_4+\delta_4^*h/\delta^*, \ldots, Y_{k-1}<Y_k+\delta_k^*h/\delta^*]$$

$$= \int_{-\infty}^{\infty} \int_{-\infty}^{y_k+\delta_k^*h/\delta^*} \cdots \int_{-\infty}^{y_4+\delta_4^*h/\delta^*} \int_{-\infty}^{y_3+\delta_3^*h/\delta^*}$$

$$F_{n_0}(y_2+\delta_2^*h/\delta^*)f_{n_0}(y_2)f_{n_0}(y_3) \ldots f_{n_0}(y_{k-1})f_{n_0}(y_k)dy_2dy_3\ldots dy_{k-1}dy_k$$

(where $F_{n_0}(\cdot)$ and $f_{n_0}(\cdot)$ are respectively the distribution and density function of a Student's-t random variable with n_0-1 degrees of freedom), which $\uparrow$ from $1/k!$ to 1 as h from 0 to ∞, hence there is a unique $h = h_{n_0,k}(P^*,\delta_2^*/\delta^*,\ldots,\delta_k^*/\delta^*)$ such that

$$P_{LFC}(CR|\mathcal{T}_{DD}(CR\text{-}IZ)) = P^* .$$

Since this result depends only on the ratios $\delta_2^*/\delta^*, \ldots, \delta_k^*/\delta^*$, <u>we may</u> without loss of generality <u>take</u> $\delta^* = 1$ in both theory and practice. In practice one would usually choose $\delta_2^* = \ldots = \delta_k^*$, though this is to be decided in consultation with the experimenter on his/her aims and requirements/resources. *If* one's indifference zones are fully symmetric in that $\delta_2^* = \ldots = \delta_k^* = \delta^*$ (say), then the $h = h_{n_0,k}(P^*)$ to be used is the unique solution of

$$\int_{-\infty}^{\infty} \int_{-\infty}^{y_k+h} \cdots \int_{-\infty}^{y_4+h} \int_{-\infty}^{y_3+h} F_{n_0}(y_2+h)f_{n_0}(y_2)f_{n_0}(y_3)$$

$$\ldots f_{n_0}(y_{k-1})f_{n_0}(y_k)dy_2dy_3 \ldots dy_{k-1}dy_k = P^*.$$

In what follows, we suppose $\mathcal{T}_{DD}$(CR-IZ) is always used with

$$h = h_{n_0,k}(P^*, \ \delta_2^*/\delta^*, \ \ldots, \ \delta_k^*/\delta^*),$$

which is currently being tabulated using both numerical integration (for small k) and Monte Carlo techniques (for moderate and large k).

In our setting, we are given $k(\geq 2)$ basic independent populations $\pi_1^*, \ldots, \pi_k^*$. Let f_i denote the density function (with finite variance) of observations from π_i^*, and assume that f_i is a bounded (above and below) positive finite-step function on $0 \leq x \leq 1$ $(1 \leq i \leq k)$. Our basic (IZ) goal is to completely rank the population from that which has the largest entropy $H(f_{[k]})$ down through that with the smallest entropy $H(f_{[1]})$ (where $H(f_{[i]})$ denotes the $k-i+1^{st}$ largest of the entropies $H(f_1), \ldots, H(f_k)$ $(1 \leq i \leq k)$, so that $H(f_{[1]}) \leq H(f_{[2]}) \leq \ldots \leq H(f_{[k]})$) in such a way that the probability of correct ranking satisfies

$$P(CR) \geq P^* \text{ whenever } H(f_{[j]})-H(f_{[j-1]}) \geq \delta_j^* \quad (j = 2, \ldots, k)$$

where P^* and $\delta_2^*, \ldots, \delta_k^*$ $(1/k! < P^* < 1; 0 < \delta_2^*, \ldots, \delta_k^*)$ are specified in advance by the experimenter. Now fix an m. If we take a random sample of size n^* from π_i^*, then (for $n^* \to \infty$) the resulting H_{m,n^*} will be asymptotically normal once standardized, or

$$T_i \equiv H_{m,n^*} + \log(2m) + \gamma - R(1,2m-1)$$

will be asymptotically $N(\mu_i, \sigma_i^2)$ where $\mu_i = H(f_i)$ and

$$\sigma_i^2 = (\tau(2m) + \mathrm{Var}_{f_i} \log f_i(X))(n^* + 2 - 2m)/(n^*)^2 .$$

Suppose this normality were exact (not just asymptotic). Then our IZ ranking problem would be solved by

<u>PROCEDURE</u> $\mathcal{T}_{DV}$(CR-IZ). From any given basic population π_i^*, sample in batches of n^* observations at a time, and consider the resulting statistic

$$T_i = H_{m,n^*} + \log(2m) + \gamma - R(1, 2m-1)$$

to be one observation from π_i $(1 \leq i \leq k)$. Then apply procedure $\mathcal{T}_{DD}$(CR-IZ) with h replaced by $h\sqrt{n^*}$ to $\pi_1, \ldots, \pi_k$ to obtain a solution to the ranking problem involving $\pi_1, \ldots, \pi_k$ (which is also a solution to the problem involving $\pi_1^*, \ldots, \pi_k^*$ since $\mu_i = H(f_i)$ for all i).

That procedure $\mathcal{T}_{DV}$(CR-IZ) in fact (for large n^*) approximately solves the problem even when T_i is only asymptotically normal (as $n^* \to \infty$) was shown by the results of Section 2 (which guaranteed that each $\tilde{X}_i$ converges in distribution to a Student's-t random variable with n_0-1 degrees of freedom as $n^* \to \infty$).

6. COMPLETE RANKING OF ENTROPIES ON (0,1): PPF FORMULATION

Let $\pi_1, \pi_2, \ldots, \pi_k$ represent $k(\geq 2)$ independent sources of random variables. Assume observations from π_i are normally distributed $N(\mu_i, \sigma_i^2)$ $(1 \leq i \leq k)$ with $\mu_1, \ldots, \mu_k, \sigma_1^2, \ldots, \sigma_k^2$ all unknown. Our *Preferred Population Formulation* (PPF) of the complete ranking problem has as its goal statement of a complete ranking of the populations in such a way that

$$P(CR) \geq P^*$$

for all possible parameter configurations, where P^* $(1/k! < P^* < 1)$ is specified in advance by the experimenter and event "CR" is considered to occur if the order specified is correct or can be made correct by one or more interchanges of assertions involving populations whose true means differ by at most δ^* ($\delta^* > 0$).

Now it can easily be shown that procedure $\mathcal{T}_{DD}$(CR-IZ) with $\delta_2^* = \ldots = \delta_k^* = \delta^*$, given in Section 5, satisfies $P(CR) \geq P^*$ with this new definition of event CR. Thus the following procedure achieves the goal.

PROCEDURE $\mathcal{T}_{DD}$(CR-PPF). Choose $\delta_2^* = \ldots = \delta_k^* = \delta^*$ and proceed according to procedure $\mathcal{T}_{DD}$(CR-IZ).

In our setting, we are given $k(\geq 2)$ basic independent populations $\pi_1^*, \ldots, \pi_k^*$. Let f_i denote the density function (with finite variance) of observations from π_i^*, and assume that f_i is a bounded (above and below) positive finite-step function on $0 \leq x \leq 1$ $(1 \leq i \leq k)$. Let $H(f_{[i]})$ denote the $k+i+1^{st}$ largest of the entropies $H(f_1), \ldots, H(f_k)$ $(1 \leq i \leq k)$, so that $H(f_{[1]}) \leq H(f_{[2]}) \leq \ldots \leq H(f_{[k]})$. Our PPF goal is to have probability $\geq P^*$ of a stated complete ranking being correct modulo interchanges of statements about populations whose true means differ by δ^* or less, where P^* and δ^* $(1/k! < P^* < 1, 0 < \delta^*)$ are specified in advance by the experimenter. Then, proceeding as in Section 5 (using again the results of Section 2) we find our PPF ranking problem is solved (approximately) by

PROCEDURE $\mathcal{T}_{DV}$(CR-PPF). From any given basic population π_i^*, sample in batches of n^* observations at a time, and consider the resulting statistic

$$T_i = H_{m,n^*} + \log(2m) + \gamma - r(1,2m-1)$$

to be one observation from π_i $(1 \leq i \leq k)$. Then apply procedure $\mathcal{T}_{DD}$(CR-PPF) with h replaced by $h\sqrt{n^*}$ to $\pi_1, \ldots, \pi_k$ to obtain a solution to the ranking problem involving $\pi_1, \ldots, \pi_k$ (which is also a solution to the problem involving $\pi_1^*, \ldots, \pi_k^*$ since $\mu_i = H(f_i)$ for all i).

7. APPLICATIONS TO RANDOM NUMBER GENERATORS

In this section we wish to briefly indicate the applicability of the previous results to random number generators. The typical output of a random number generator (e.g., see Dudewicz and Ralley [5]) is a real number between 0 and 1, represented up to the limits of machine accuracy. The distribution of the output of such a random number generator may therefore be represented by a density function f (of course with finite variance). In many cases it will be reasonable to assume f is bounded (above and below) and is positive finite-step function on $0 \leq x \leq 1$. A computer center or scientist with $k(\geq 2)$ such generators available could desire to (1) select the best (closest to uniform on (0,1)) for use in applications requiring high-qualtiy random numbers, or (2) rank the generators from best to worst and let this information be made available to users. In this paper we have shown how to achieve these goals, with an entropy measure of deviation from true uniformity on (0,1) (see Section 8 of Dudewicz and van der Meulen [6], or Dudewicz and van der Meulen [7], for the relation between closeness in entropy and closeness in density), with probability $\geq P^*$ (experimenter-specified) of success (under both IZ and PPF formulations of the problem). Implementation for a set of random number generators is contemplated.

REFERENCES

[1] Bhattacharjee, G. P.: Effect of non-normality on Stein's two sample test. Ann. Math. Statist. 36, 651-663 (1965).

[2] Billingsley, P.: Convergence of Probability Measures, New York, Wiley (1968).

[3] Blumenthal, S. & Govindarajulu, Z.: Robustness of Stein's two-stage procedure for mixtures of normal populations. J. Amer. Statist. Assn. 72, 192-196 (1977).

[4] Dudewicz, E. J. & Dalal, S. R.: Allocation of observations in ranking and selection with unequal variances. Sankhyā B 37, 28-78 (1975).

[5] Dudewicz, E. J. & Ralley, T. G.: The Handbook of Random Number Generation and Testing with TESTRAND Computer Code, Columbus, Ohio, American Sciences Press, Inc. (1981).

[6] Dudewicz, E. J. & van der Meulen, E. C.: Entropy-Based Statistical Inference, I: Testing Hypotheses on Continuous Probability Densities, with Special Reference to Uniformity, Communication N° 120, Mathematical Institute, Katholieke Universiteit Leuven, Leuven, Belgium (1979).

[7] Dudewicz, E. J. & van der Meulen, E. C.: Entropy-based tests of uniformity. J. Amer. Statist. Assn. 76, 967-974 (1981).

[8] Hooper, J. H. & Santner, T. J.: Design of experiments for selection from ordered families of distributions. Ann. Statist. 7, 615-643 (1979).

[9] Mukhopadhyay, N.: Stein's two-stage procedure and exact consistency. Scand. Actuarial J. 1982, to appear (1982).

[10] Ramkaran, D.: Robustness of Stein's two-stage procedure. Abstract, IMS Bulletin 10, 155 (1981).

[11] Sen, P. K.: Sequential Nonparametrics, New York, Wiley (1981).

[12] Stein, C. M.: A two-sample test for a linear hypothesis whose power is independent of the variance. Ann. Math. Statist. 16, 243-258 (1945).

[13] Vasicek, O.: A test for normality based on sample entropy. J. R. Statist. Soc. B 38, 54-59 (1976).

E. J. Dudewicz

Department of Mathematics
Syracuse University
Syracuse, New York 13210, U.S.A.

and

E. C. van der Meulen

Department of Mathematics
Katholieke Universiteit Leuven
Celestijnenlaan 200B
B-3030 Leuven, Belgium

MODERN DESIGN OF SIMULATION EXPERIMENTS*

Introduction

A "simulation" is an experiment run inside the computer (instead of in what is generally, in the field, termed the "real world") in order to obtain information about a system (existing or proposed), about modifications to a system, or about several competing systems. Since it is well-known that **much more information can be extracted from experiments that have been carefully designed statistically** than can be extracted from experiments that were not statistically designed, statistics has a large role to play in simulation. This article deals with that role. This is an especially important article for those not accustomed to thinking of their simulations in statistical terms (e.g., those who "run until the money runs out" or who believe that "one run will show me how the system behaves"), as it will enable them to obtain valid results where invalid ones were previously obtained, to obtain those results at a lower cost than would have been possible had the experiment not been carefully designed, and to obtain valid results for systems where (without using statistical design) the study is computationally infeasible.

As we have said, since simulation data are essentially obtained from an experiment run "inside the computer," we need to know how to design the experiment whereby these data are generated, and we need to know how to analyze the resulting data. Now this comprises a large part of the field commonly termed **statistics**, to which whole university departments are devoted. Hence, only the most important aspects of design and analysis will be covered here, with a caveat that, if a simulation study is to be properly designed and executed, in many cases it will be desirable to have associated with the study from the outset a statistician competent in simulation applications.

It is a fact that statistics grew up in a largely agricultural setting where such assumptions as homoscedasticity (equality of variances of observations taken from diverse sources, such as weights of tomatoes from different types of tomato plants) commonly held true, while in simulation these are often violated (e.g., serious heteroscedasticity—greatly unequal variances—is more often the rule in simulation studies, as when a number of diverse alternative job-shop scheduling rules are being simulated). The theory required to handle these new situations has only recently been developed (e.g., for heteroscedasticity [1] and for autocorrelated observations [2]). Thus, many parts of the topics covered here will, for some time to come, be unavailable in almost any texts or books and manuals on statistical methods.

*This research was supported by Office of Naval Research Contract No. N00014-78-C-0543.

1. Design

1.a. Factorial, fractional-factorial, and one-at-a-time experiments: why and why/not, efficiency.

The pitfalls of lack of statistical design in a simulation experiment include invalid inferences; valid inferences at substantially increased costs; and inability to complete the study.

For example, suppose a simulation model has been built of the output $Y(x_1, x_2)$ of a system which depends on two input variables x_1 (temperature in °F) and x_2 (reaction time in minutes [each reader will know similar examples in his/her own field, and may find it helpful to re-cast this example in the terms applicable to one such example]. Suppose that it is standard to operate at $x_1 = 300$, $x_2 = 10$, and that the simulation is to evaluate possible gains of increasing x_1 (by 25°F) or x_2 (by 2 minutes). Commonly one would run the model at (x_1, x_2) equal to (300, 10), (325, 10), and (300, 12) to evaluate the effect of the proposed changes (**one-at-a-time method**). Suppose one does so, and the results (each based on one run) are as shown in Table 1. Let us suppose that no variability is present (so, untypically, results of one simulation run are typical), that past experience with the system has typically yielded 100.0 output units at $(x_1, x_2) =$ (300, 10), that expensive plant changes are required to implement changes of this magnitude (so we must simulate before making changes in the plant), and that to be economically feasible we must obtain at least a 5% process yield increase from any recommended changes. We estimate the gain from incrementing x_1 to be

$$102.5 - 100.0 = 2.5$$

and the gain from incrementing x_2 to be

$$101.0 - 100.0 = 1.0$$

for a total estimated gain of 2.5 + 1.0 = 3.5 units, which is less than 5%, so the proposed improvement in process is discarded.

Table 1. Results of one-at-a-time experiment.

(x_1, x_2)	$Y(x_1, x_2)$ from simulation run
(300, 10)	100.0
(325, 10)	102.5
(300, 12)	101.0

Table 2. Results of 2^2 experiment.

(x_1, x_2)	$Y(x_1, x_2)$ from simulation run
(300, 10)	100.0
(325, 10)	102.5
(300, 12)	101.0
(325, 12)	110.0

Now the above argument assumes an output linear in each of x_1 and x_2 (perhaps reasonable in a small range about the usual process operating conditions), but with no **synergistic effects** (no **interaction**)—which is often false in the real-world (at least, we should not believe such an assertion without validation, and the design above allows no such verification). For example, an experiment run at (325, 12) might complete our data set to that shown in Table 2. This set of data has a far different interpretation than does that of Table 1; we now see it will be very profitable (double the needed 5% gain in output) to make the plant modifications needed to run at the higher temperature (x_1 = 325°F) and longer time (x_2 = 12 min.), thus **avoiding an invalid inference**. The results of Table 2 would be better displayed as presented in Table 3, which shows the 2^2 (two factors, each at two levels) design more clearly. Such results as these can be obtained from as simple an underlying true function as

$$Y(x_1,x_2) = 100.0 + 0.1(x_1 - 300.0) + 0.5(x_2 - 10.0) + 0.13(x_1 - 300.0)(x_2 - 10.0).$$

As the number n of variables increases, the (invalid) one-at-a-time method will require a number of simulation runs equal to the number of variables; for example, if in

Table 3. Results of 2^2 experiment.

	$Y(x_1,x_2)$ from simulation run	
x_1 \ x_2	$x_2 = 10$	$x_2 = 12$
$x_1 = 300$	100.0	101.0
$x_1 = 325$	102.5	110.0

addition to x_1 (temperature) and x_2 (reaction time) we also have variables x_3 (acid concentration) and x_4 (pH) present, then $n = 4$ simulation runs would be needed by this (invalid) method. In order to assess effects with each variable at the traditional or a higher level, the analog of the experiment in Table 2 would require $2^4 = 16$ simulation runs. Because these runs can be very expensive (e.g., a nuclear simulation run can consume several hours of computer time), we now wish to investigate **more sophisticated designs which yield comparably correct inferences at a substantially reduced cost.**

To perform this task, we must first consider models for yield Y as a function of the variables $x_1, x_2, x_3, x_4, \ldots$. It is reasonable to assume that the true mean yield $E(Y)$ may be accurately represented by a polynomial equation of sufficiently high order:

$$EY(x_1, x_2, x_3, x_4, \ldots)$$

$$= \beta_0 + \beta_1 x_1 + \beta_2 x_2 + \beta_3 x_3 + \beta_4 x_4 + \beta_{12}x_1x_2 + \beta_{13}x_1x_3 + \beta_{14}x_1x_4 + \beta_{23}x_2x_3 + \beta_{24}x_2x_4 + \beta_{34}x_3x_4$$

$$+ \beta_{11}x_1^2 + \beta_{22}x_2^2 + \beta_{33}x_3^2 + \beta_{44}x_4^2$$

$$+ \beta_{123}x_1x_2x_3 + \beta_{124}x_1x_2x_4 + \beta_{134}x_1x_3x_4 + \beta_{234}x_2x_3x_4$$

$$+ \beta_{1234}x_1x_2x_3x_4 + \ldots.$$

Different choices of the points $(x_1, x_2, x_3, x_4, \ldots)$ at which simulation runs should be made (each such choice is called an **experimental design**) allow us to estimate various of the β's (singly or in combination), while assuming others are negligible (which can often be tested statistically). Typically one assumes that terms of higher than second order are negligible ($0 = \beta_{123} = \beta_{124} = \beta_{134} = \beta_{234} = \beta_{1234} = \ldots$). The coefficients $\beta_1, \beta_2, \beta_3, \beta_4$ are called **main effects** of variables x_1, x_2, x_3, x_4, while $\beta_{12}, \beta_{13}, \beta_{14}, \beta_{23}, \beta_{24}, \beta_{34}$ are called **2-factor interactions** (2 **f.i.**'s) of the variables involved in the subscripts. An experimental design is called a **Resolution III design** if no main effect is **confounded** with (not able to be estimated separately from) any other main effect, but main effects are confounded with 2 f.i. and 2 f.i. with each other. In a **Resolution IV design** no main effect is confounded with any other main effect or 2 f.i., but 2 f.i.'s are confounded with each other. Thus, a Resolution III design will allow us to fit a model $\beta_0 + \Sigma\beta_i x_i$ in factors x_i, but interaction will bias the fit. A Resolution IV design will allow a fit unbiased by 2 f.i., i.e. $\hat{\beta}_i$ will estimate β_i (not $\beta_i \pm$ some of the β_{ij}'s of the $x_i x_j$ interactions). The one-at-a-time method yields a resolution IV design, and its disadvantages have been noted above.

1.b. Screening experiments.

In **screening experiments** one attempts to use designs which allow one to find out (at relatively modest cost, i.e. no. of simulation runs) which of the variables x_1, x_2, x_3, x_4, . . . are most influential as to one's system's output in very few experiments by using a Resolution III design (of course one then gives up almost all model-fitting ability, the intention being to run a more extensive experiment later on the few highly influential variables identified). For example, one design commonly used is the Resolution III Plackett-Burman design which can study 7 variables—x_1, x_2, x_3, x_4, x_5, x_6, x_7—in 8 simulation runs, as shown in Table 4. In this design, each experimental variable has 2 levels, low (−) and high (+) (e.g., 300(−) and 325(+) for x_1). In experiment 1, variables 1, 2, 3, and 5 are set to their high levels, while variables 4, 6, and 7 are set to their low levels. If desired later, this design can be augmented to a Resolution IV design by adding 8 more runs (for a total of 16 runs). One can also study 6 factors in a 12-experiment (Webb) experimental design, at Resolution IV. As another example of a screening design, if we have 5 independent variables then a Webb design of Resolution IV is available (see Table 5) which involves 10 experiments.

Table 4. Resolution III Plackett-Burman design.

		Variable						
		x_1	x_2	x_3	x_4	x_5	x_6	x_7
	1	+	+	+	−	+	−	−
	2	−	+	+	+	−	+	−
Experiment	3	−	−	+	+	+	−	+
	4	+	−	−	+	+	+	−
	5	−	+	−	−	+	+	+
	6	+	−	+	−	−	+	+
	7	+	+	−	+	−	−	+
	8	−	−	−	−	−	−	−

Note that the first 5 experiments form a Resolution III design, but the 10-run experiment is much more **efficient** than the 5-run experiment (it yields estimates with variance $\sigma^2/9$ for main effects, vs. σ^2 for the 5-run one-at-a-time experiment—a 9-fold reduction at a price of 5 more runs, and with 2 f.i. elimination; here σ^2 is the variability inherent in each simulation run's outcome). As a final example of a screening design, with the 4 variables x_1, x_2, x_3, x_4, a Webb Resolution IV design with 8 experiments is available (Table

Table 5. 5-variable screening design.

		Variable				
		x_1	x_2	x_3	x_4	x_5
	1	+	−	−	−	−
	2	−	+	−	−	−
	3	−	−	+	−	−
	4	−	−	−	+	−
Experiment	5	−	−	−	−	+
	6	−	+	+	+	+
	7	+	−	+	+	+
	8	+	+	−	+	+
	9	+	+	+	−	+
	10	+	+	+	+	−

6) (which is a "fractional factorial" design, i.e., a fraction of the 2^4 = 16-experiment factorial design with confounding scheme $I = ABCD$) which allows one to fit β_0, β_1, β_2, β_3, and β_4 unbiased by two-factor interactions (though those interactions themselves are not estimable).

Table 6. 4-variable screening design.

		Variable			
		x_1	x_2	x_3	x_4
	1	−	−	−	−
	2	+	−	−	+
	3	−	+	−	+
	4	+	+	−	−
Experiment	5	−	−	+	+
	6	+	−	+	−
	7	−	+	+	−
	8	+	+	+	+

1.c. Central composite designs and full quadratic models.

Assuming that screening has been completed, we will wish to use a **design which allows for assessment of all main effects, all 2 f.i.'s, and (perhaps) all quadratic effects.** We will illustrate with a 4-variable example.

Here one wishes **not simply to find the most important factors, but rather to model system output as a function of design settings.** Of course the appropriate design to use depends on one's goals and scope of study, as well as on one's budget. One possibility is an **11-experiment Webb design** (Table 7). This design allows one to fit β_0, β_1, β_2, β_3, β_4, β_{12}, β_{13}, β_{14}, β_{23}, β_4, β_{24}, and β_{34}, assuming no curvature.

Since curvature cannot often be ruled out *a priori*, one will usually desire another design (which, however, will require more experiments) unless one's budget is severely constrained. The **full factorial 2^4 design (16 experiments)** allows one to estimate $\beta_0 + \beta_{11} + \beta_{22} + \beta_{33} + \beta_{44}$, β_1, β_2,

Table 7. 4-variable Webb design.*

		Variable			
		x_1	x_2	x_3	x_4
	1	0	0	0	0
	2	0	0	1	1
	3	0	1	0	1
	4	0	1	1	0
	5	1	0	0	1
Experiment	6	1	0	1	0
	7	1	1	0	0
	8	0	1	1	1
	9	1	0	1	1
	10	1	1	0	1
	11	1	1	1	0

* "0" denotes low (−), "1" denotes high (+).

β_3, β_4, β_{12}, β_{13}, β_{14}, β_{23}, β_{24}, β_{34}, β_{123}, β_{124}, β_{134}, β_{234} and (assuming that β_{1234} is negligible) have an estimate of experimental error σ^2. [No suitable "fractions" of this design exist, as they all confound 1- and/or 2-factor effects, so no such easy reduction of the number of experiments needed is possible in the 4-factor case.] By adding one experiment, one obtains the **$2^4 + 1$ design (17 points)**, with which one can now separately estimate β_0, $\beta_{11} + \beta_{22} + \beta_{33} + \beta_{44}$, hence (barring cancelling magnitudes) assess the total quadratic effect independently of the response at the center (β_0).

While the above designs are in common use, more recently the **Central Composite Design (CCD)** has been used to good advantage in such situations. This design requires that 8 **star (or axial) points** be added to the $2^4 + 1$ design, for a total of 25 experiments, and it allows a full quadratic model to be fitted. **Suitable fractions** of the 2^4 **may be used (i.e., a fraction with at most one 2-factor interaction in any alias set)**, and (via the confounding scheme $I = ABC$) one can obtain the full quadratic model estimation with

$$\frac{1}{2}2^4 + 1 + 8 = 17 \text{ points.}$$

As an example, suppose we have six variables to consider, say x_1, x_2, x_3, x_4, x_5, x_6. A full-factorial approach would require $2^6 = 64$ experiments, which in many cases would not be feasible. However, a CCD can be implemented with only $2^{6-2} + (2)(6) + 1 = 29$ experiments, as shown in Table 8. There the levels are "coded" so that "1" means the "high" level of the variable, "−1" means the "low" level of the variable, and "0" means the average of the high and low levels. "$-\alpha$" and "α" represent multiples of the "low" and "high" levels; e.g., **if one takes $\alpha = 1$ (face-centered star points)** these are equal to the respective "low" and "high" levels of the variable in question. Some typical possibilities are given in Table 9.

If one has bounds L_i and U_i on variable x_i and wishes to explore the full space, $\alpha = 1$ is often recommended. In

Table 8. CCD studying 6 variables, in 29 experiments.

		Variable					
Name	Expt. No.	x_1	x_2	x_3	x_4	x_5	x_6
	1	1	−1	−1	1	−1	−1
	2	−1	1	−1	1	−1	−1
	3	−1	−1	1	1	−1	−1
	4	1	1	1	1	−1	−1
	5	1	−1	−1	−1	1	−1
	6	−1	1	−1	−1	1	−1
	7	−1	−1	1	−1	1	−1
2^{6-2} Points	8	1	1	1	−1	1	−1
	9	1	−1	−1	−1	−1	1
	10	−1	1	−1	−1	−1	1
	11	−1	−1	1	−1	−1	1
	12	1	1	1	−1	−1	1
	13	1	−1	−1	1	1	1
	14	−1	1	−1	1	1	1
	15	−1	−1	1	1	1	1
	16	1	1	1	1	1	1
	17	$-\alpha$	0	0	0	0	0
	18	α	0	0	0	0	0
	19	0	$-\alpha$	0	0	0	0
	20	0	α	0	0	0	0
	21	0	0	$-\alpha$	0	0	0
Star	22	0	0	$-\alpha$	0	0	0
	23	0	0	0	$-\alpha$	0	0
	24	0	0	0	α	0	0
	25	0	0	0	0	$-\alpha$	0
	26	0	0	0	0	α	0
	27	0	0	0	0	0	$-\alpha$
	28	0	0	0	0	0	α
Center	29	0	0	0	0	0	0

Table 9. Coding of variable x_1 levels, CCD.

	Level				
	$-\alpha$	1	0	1	α
$\alpha = 1$	300	300	312.5	325	325
$\alpha = 2$	287.5	300	312.5	325	337.5

other settings $1 < \alpha \leq 2$ is often used. The specifics vary from setting to setting, with $\alpha = 1.5$ being a reasonable compromise for experimenters who do not have access to a statistical design expert.

1.d. Efficient implementation.

It is important to note that the ordering of experiments in the above tables is *not* the recommended order in which the simulation runs should be made. Since often the "next" run starts with the end of the "last" run's random number stream, **runs should be in random order** to prevent systematic effects (from possible deviations from randomness of one's random number generator) from systematically biasing the results. **A carefully tested and chosen random number generator is essential**, and an existing generator should not be used without extensive validation (available in [3]).

Note that (unless one has prior knowledge of how the system will use random numbers and correlation, which is rare) use of variance reduction techniques (see [6]) will not often be appropriate.

1.e. Two-stage and sequential designs.

The designs explored so far are reasonable ones to use in homoscedastic systems. In heteroscedastic systems (where σ^2 is a function of the levels of $x_1, x_2, \ldots$), other designs are called for. Most commonly one finds that **if $x_1, x_2, \ldots$ represent levels of a continuous variable (such as temperature), then the designs given thus far are appropriate** even if σ^2 varies slightly as $x_1, x_2, \ldots$ are changed.

However **if $x_1, x_2, \ldots$ represent the presence (1) or absence (0) of an attribute, then new designs for selecting the best combination of attributes re: system performance are called for.** See [4] for some details.

2. Analysis.

The designs presented in "Design" are appropriate for situations where one wants to know **"which of k factors are most important** in determining my system's output?" or **"what model for system output, in k input factors, should be subjected to optimization"** ([5])? Design and analysis of the simulation experiment when one's goal is not to answer the above questions, but is rather to answer others such as a) **What is the long-run mean of my system?** (which may be answered using a transformation-based analysis, and involves questions of normality and run-in time, and leads us to 2-stage and regenerative approaches); b) **Which system parameters have significant effects?** (which leads us into analysis of variance (ANOVA), and where transformations should generally not be used); c) **How different are the various systems' performances?** (which requires simultaneous interval estimates); and d) **Which is the best system (or set of system parameters)?** (which requires the new methodology of ranking-and-selection procedures)—is a subject now undergoing rapid development. While such problems have traditionally been approached in the past with transformations, 1-stage procedures, or ANOVA, the pitfalls of some of these traditions and of equal-sample-sizes, and the new methods recently developed which should be used in the future, are explored in [1].

Towards the future, ongoing developments in multivariate analogs of the design and analysis procedures presented should lead to future procedures which are able to simultaneously consider several output characteristics.

REFERENCES

[1] Dudewicz, Edward J., "Heteroscedasticity," *Encyclopedia of Statistical Sciences, Volume 3* (edited by N. L. Johnson, S. Kotz, and C. B. Read), John Wiley & Sons, Inc., New York, 1982.

[2] Dudewicz, Edward J. and Zaino, N. A., Jr., "Allowance for Correlation in Setting Simulation Run-Length via Ranking-and-Selection Procedures," *TIMS Studies in the Management Sciences*, Volume 7, 1977, pp. 51–61.

[3] Dudewicz, Edward J. and Ralley, Thomas G., *The Handbook of Random Number Generation and Testing With TESTRAND Computer Code*, American Sciences Press, Inc., Columbus, Ohio, 1981.

[4] Dudewicz, Edward J., "Ranking (Ordering) and Selection: An Overview of How to Select the Best," *Technometrics*, Volume 22, 1980, pp. 113–119.

[5] Golden, Bruce, Assad, Arjang, and Zanakis, Stelios H., *Statistics and Optimization: The Interface*, American Sciences Press, Inc., Columbus, Ohio, 1984.

[6] Tadikamalla, Pandu R. (Editor), *Modern Digital Simulation Methodology: Input, Modeling, and Output*, American Sciences Press, Inc., Columbus, Ohio, 1984.

MODERN DESIGN OF EXPERIMENTS, WITH AN APPLICATION TO SIMULATION EXPERIMENTS: ALLOWANCE FOR CORRELATION IN SETTING SIMULATION RUN-LENGTH VIA RANKING-AND-SELECTION PROCEDURES II

by

Dr. Edward J. Dudewicz, Professor
Department of Mathematics
Syracuse University
Syracuse, New York 13210, U.S.A.

and

Dr. Satya Narayan Mishra, Assistant Professor
Department of Mathematics and Statistics
University of South Alabama
Mobile, Alabama 36688, U.S.A.

ABSTRACT

Modern statistical design of experiments allows study of many factors (settings which influence the outcome of the experiment) with practical numbers of experiments. For example, while a traditional two-level factorial experiment with 9 factors would require $2^9 = 512$ experiments, and even a quarter-replicate or fraction of it would require 128 experiments, a modern experimental design can provide more information with substantially fewer experiments. An outline of the material of modern design of experiments is given. Traditionally, this material has been accessible only to near-Ph.D.'s in mathematics and statistics. The first author, however, offers (both at universities through Continuing Education, and In-Plant) a course on this material which emphasizes applications and is accessible to most experimenters. An application to a (simulation) experiment is detailed, with an indication of savings over the "run all possible experiments" and "run until the money runs out" approaches.

1. INTRODUCTION; IMPORTANCE OF MODEL BUILDING AND DESIGN

Statistical model building, design, and analysis is the most important part of the statistical method, and understanding of current methodology is important to virtually everyone involved in experimental investigations. In the past much of the material needed for modern statistical model building, design, and analysis was accessible only at the advanced graduate level. However, now a course is available which presents "state of the art" information in a ready-for-application form, emphasizing how-to aspects for the most important cases typically enountered in conducting experiments and management analysis. (For further information contact the first author; both regular university Continuing-Education, and In-Plant, offerings are available.) By utilizing this information and these strategies, investigators are able to obtain valid results where invalid ones were obtained previously, and to secure those results at lower cost than it takes to secure valid results without a designed experiment.

An outline of this modern material, essential if we are to avoid the Edisonian approach (recall that Edison ran over 30,000 experiments in developing the lightbulb filament, which is working "harder": statistical design is working "smarter") consists of:

I. Basic Ideas of Model Building
Overview: One-at-a-time experiments; Factorial experiments, Fractional factorial experiments; Composite experiments; Mixture experiments; Pitfalls.

II. Statistical Principals on Which Model Building is Based
Frequency distributions; Distribution characteristics (mean, variance); Normal distribution; Estimates from samples (mean, standard deviation); Distributions of sample mean; Significance tests; Confidence intervals; Prediction intervals.

III. Linear Models
Linear regression; Best fit; Least-squares; Optimality; Standard error; Parameter estimation; Ridge regression.

IV. Model Interpretation
Testing; Confidence bands; Prediction bands; Extrapolation; Curvilinear models; Tolerance intervals.

V. Experimentation Strategies
Randomized blocks; Latin squares; One-at-a-time method; Interaction; Complete factorial experiments; Quantitative vs. Qualitative factors; Analysis and comparison of completely randomized and randomized block designs in qualitative factors.

VI. Analysis of Experiments
Error rates per-comparison, per-experiment, experiment-wise; Multiple comparisons procedures; Selection procedures.

VII. Quantitative Variable Strategies
Models of resolution III, IV, V; Screening experiments (including catalog of designs); Fractional factorial designs (including computer program); Central composite designs; Efficient implementation; Heteroscedasticity; Computer programs for analysis; Interpretation; Optimization (including minimal cost goal achievement).

VIII. Mixture Experimentation Strategies
Mixture experiments; Constraints; Mixture and process variables (design catalog).

Upon completion of such a course participants should:

- understand the validity and cost/feasibility of statistical model building;
- be able to evaluate one-at-a-time experimental methods;
- be able to design factorial and fractional-factorial experiments in from one to eight variables;
- be able to design screening experiments to find out (at relatively modest cost) which of a large number of variables are most influential on results;
- be able to design central composite experiments, in up to ten variables, allowing for fitting of an optimizable model;
- be able to detect and correct many typical "bad" data points;
- be able to build models with up to four variables which are subject to constraints;
- know how to use computer programs to perform related analyses;
- understand the role of randomization of order or experiments, how it may be violated for experimental ease and the resulting interpretation;
- know how to optimize a resulting model, possibly subject to constraints;
- understand multiple comparisons, curve fitting, and ridge regression methods.

2. APPLICATION TO A (SIMULATION) EXPERIMENT

A new statistical procedure for "ranking and selection" (see Section 3 for a brief description of this area) was given by Dudewicz and Zaino (1977). The probability that procedure produces correct results depends on 10 factors (called "parameters"). If each factor were studied at 2 levels, $2^{10} = 1024$ simulation runs would be needed to study the influence of all 10 factors on the output of interest. Since that would be exorbitantly costly (in terms of both computer time and experimenters' effort), modern statistical design and analysis was used. This allowed a valid study with 85 percent fewer experiments.

3. RANKING AND SELECTION

"Ranking and selection" is an alternative approach to classical analysis of variance. Many analysis of variance related problems have been solved using this new methodology. (For an extensive and exhaustive bibliography see Dudewicz and Koo (1982).) The techniques of ranking and selection answer questions raised in many investigations which are not answered by a traditional analysis of variance approach, for example:

(1) Which brand of cigarette, among all available brands, has maximum nicotine?

(2) Which university in the United States has the "best" I.Q. undergraduate students?

(3) Among available brands of cereals, which has the highest sugar content?

In these problems we are faced with the problem of choosing the "best" from among several alternatives. Ranking and selection procedures have been developed for many such problems. On the other hand, analysis of variance and multiple comparison procedures do not answer this question of "bestness." Analysis of variance and ranking and selection both often assume equality of variances, normality of errors, and independence of observations. Since in simulation, time series studies and other problems, the data encountered are often dependent, such procedures are not fully adequate. Below we deal with dependent observations and unequal variances simultaneously, studying a heuristic procedure due to Dudewicz and Zaino (1977).

4. STATISTICAL PROCEDURE TO BE STUDIED

Suppose we have a set, $\Pi = \{\pi_1, \pi_2, \ldots, \pi_k\}$, of k different sources (populations). Let the observations from population π_i be denoted by $\{X_{im}\}$, which is not necessarily independent (the problem in case of independence has been fully solved by Dudewicz and Dalal (1975)), that is observations are dependent and their normal distribution depends on $\theta_i = (\mu_i, \sigma_i^2, \rho_i)$. Except for the values of θ_i, the distributions are assumed not to differ from population to population. For correlation we assume

$$X_{i,n} = \rho_i X_{i,n-1} + Z_{i,n} = \sum_{j=0}^{\infty} \rho_i^j Z_{i,n-j}, \tag{1}$$

i.e. observations in the ith population are obtained according to a first order autoregressive process AR(1), where $\{Z_{i,n}\}$ is a sequence of uncorrelated random variables with mean $(1-\rho_i)\mu_i$ and variance σ_i^2. The problem is to select a population which has the largest mean $\mu_{[k]} = \max(\mu_1, \ldots, \mu_k)$, where $\mu_{[1]} \leq \mu_{[2]} \leq \cdots \leq \mu_{[k]}$ is the true (unknown) numerical ordering of the means. Dudewicz and Zaino (1977) have discussed the effect of $\rho \neq 0$ and its implications, and given a heuristic procedure for this problem. We will study that procedure under the parameter configurations

$$\left.\begin{aligned} &\mu_{[1]} = \mu_{[2]} = \cdots = \mu_{[k-1]} = \mu_{[k]}-\delta^* \\ &\sigma^2_{[1]} = \sigma^2_{[2]} = \cdots = \sigma^2_{[k-1]} = \frac{1}{a}\sigma^2_{[k]} \\ &\rho_{[1]} = \rho_{[2]} = \cdots = \rho_{[k-1]} = \frac{1}{b}\rho_{[k]} \end{aligned}\right\} \qquad (2)$$

where $\sigma^2_{[1]} \leq \cdots \leq \sigma^2_{[k]}$ is the numerical ordering of variances, $\rho_{[1]} \leq \cdots \leq \rho_{[k]}$ is the numerical ordering of correlations, and a, b, δ^* are prefixed constants chosen by the experimenter. Without loss of generality, we can and will assume

$$\mu_{[1]} = \mu_{[2]} = \cdots = \mu_{[k-1]} = 0, \quad \mu_{[k]} = \delta^*.$$

These configurations are believed to be nearly "least-favorable" (LFC) for studying the "probability of correct selection" (P(CS)) of the Dudewicz-Zaino procedure.

5. EXPERIMENTAL FACTORS

As mentioned earlier, we consider 10 factors as influencing P(CS) in this experiment:

(1) Number of populations (k): It is intuitively clear that k is an important factor in any ranking and selection problem since P(CS) is a non-increasing function of k. In our experiment we consider k from 2 to 10.

(2) The guaranteed probability P*: Any researcher would like to make a correct selection with high probability, and therefore P* is an important factor; we consider P* from 0.85 to 0.95.

(3) Configuration of means and δ^*: The LFC of means has been (see (2)) noted earlier. δ^* is an important factor because P(CS) is a non-decreasing function of δ^*, i.e. if we separate the best population from the next best population by δ^* units and let δ^* tend to infinity, then the truly best population has no competition and any sampling procedure will choose this pupulation to be the "best" with probability 1. Similarly by making $\delta^* \downarrow 0$, populations are not well separated, and any procedure will be only as good as picking a population at random. δ^* is chosen to be from 0.5 to 1.5.

(4) and (5) Heterogeneity of variance ($\sigma^2_{[k]}$, b): The equality of variances, in real life situations, does not occur. Therefore, heterogeneity is a potentially important factor. As noted in configuration (2), heterogeneity can be broken into two parts; $\sigma^2_{[k]}$, which is chosen from 0.5 to 1.5, and a constant "a" which multiplies $\sigma^2_{[k]}$ to give variance of indifferent populations; "a" varies from 0.25 to 4.

(6) and (7) Heterogeneity of correlation ($\rho_{[k]}$, b): Dudewicz and Zaino (1977) made simulation runs on homogeneous correlations. To extend their idea, we take our correlations for "best" and "indifferent" populations unequal. The configuration (2) of correlation involves two factors, $\rho_{[k]}$ and b. $\rho_{[k]}$ is considered from 0.13 to 0.87 (almost no correlation to almost perfect correlation when multiplied by b), and "b," which multiplies $\rho_{[k]}$ to give correlations of indifferent populations, from 0.9 to 1.1.

(8) Initial sample size (N_0): As we will see, the procedure takes an initial sample of size N_0 to compute estimates of correlation and variance in order to set the final run length. Since the weak law of large numbers guarantees that estimates of variances and correlation will tend to true values of the parameters as sample size increases, it seems N_0 is potentially important factor. We consider N_0 from 30 to 60.

(9) Alpha (α): In the procedure, after taking a sample of size N_0 we compute estimates of correlations and variances and proceed to compute the run length in the second stage (that run length depends greatly

on whether correlation tests as significantly different from zero). This significance level will be an important factor since it can change the final sample size considerably. We consider α from 0.05 to 0.15.

(10) Number of warm-ups needed before sampling can start (MWARM): As Anderson (1979) states, for the first order autoregressive process AR(1),

$$X_i = \rho X_{i-1} + Z_i, \quad |\rho| \leq 1, \tag{3}$$

which can be written as an mth order moving average process

$$X_i = \sum_{j=0}^{m} \rho^j Z_{i-j}, \tag{4}$$

one should "warm-up" any simulation with m discarded terms, where m is chosen such that the variance of induction error $\sum_{j=m+2}^{\infty} \rho^{2j}\sigma^2$ for the first recorded time series value (an observation on X_{m+1}) is sufficiently small.

Using his formula we obtain

$$H_i(m,n,\rho) = \rho^{m+1}\{\rho^{i-1} - \frac{(1-\rho^n)}{n(1-\rho)}\}, \tag{5}$$

where

ρ is the correlation

m is the number of observations needed to start first sample

n is the length of simulation.

Here m is to be chosen so that $H_1(m,n,\rho)$ is negligible, where

$$H_1(m,n,\rho) = \rho^{m+1}\{1 - \frac{(1-\rho^n)}{n(1-\rho)}\},$$

i.e.

$$\{[\log H_1 - \log(1 - \frac{(1-\rho)^n}{n(1-\rho)})]/\log \rho\} - 1 = m .$$

For different run lengths (at two levels of N_0, namely N_0=30 and N_0=60) we found the number of discarded units before a first sample is taken should be almost 48. Thus we consider MWARM from 0 (no warm-up) to 50 (warming-up done according to Anderson (1979)); in many cases more warming-up is done than is required.

6. STATISTICAL PROCEDURE $A(\hat{\rho}_i, s_i^2)$

After necessary warm-up, take an initial sample of size N_0 from each of the k processes. For $i=1,\ldots,k$, calculate estimates of σ_i^2 and correlation ρ_i of the ith process by

$$s_i^2 = \sum_{n=1}^{N_0} (X_{i,n} - \bar{X}_i)^2/(N_0-1)$$

and

$$\hat{\rho}_i = \frac{\sum_{n=2}^{N_0} (X_{i,n}-\bar{X}_i)(X_{i,n-1}-\bar{X}_i)}{\sum_{n=1}^{N_0} (X_{i,n}-\bar{X}_i)^2} .$$

Calculate

$$M_i = \max(N_0 + 1, [\frac{s_1^2 h^2}{\delta^{*2}}]),$$

where $[y]$ is the smallest integer $\leq y$ and h comes from (interpolation in) Bechhofer's (1954) table. Now construct a $100(1-\alpha)\%$ confidence interval for ρ_i

$$(\rho_i-\hat{\rho}_i)^2 \leq \frac{N-1}{N(N-3)} (1-\hat{\rho}_i^2) t_{N-3}(1-\alpha/2).$$

If ρ_i is determined to be significantly different from zero at level of significance α, then set the required sample size as

$$N_{2i} = [M_i(\frac{1+\hat{\rho}_i}{1-\hat{\rho}_i})],$$

and continue the run; however if ρ_i is not significantly different from zero at this level, choose sample size M_i for π_i. Now, once again consider $\bar{X}_i$ $(i=1,2,\ldots,k)$ based on sample of size M_i or N_{2i} (according to the procedure) and select the population that yields largest of the sample means $\bar{X}_1, \bar{X}_2, \ldots, \bar{X}_k$.

7. A DESIGN OF RESOLUTION V

For ten factors at two levels each, a full factorial experiment has $2^{10} = 1024$ runs; running this many combinations will not be economical. To avoid astronomical computer time, one may seek a fraction of the 2^{10} factorial design. Since we wish to develop a second degree response surface, we will confound in such a way that main effects and first order interactions can be estimated (second degree terms will be considered in Section 8). Such a design is said to be of resolution V. Box and Hunter (1961) have given a scheme which confounds interactions such that the resulting fractional factorial is of resolution V; there does not exist a higher fraction than $\frac{1}{2^3}$ which is of resolution V for ten factors, hence we will confound three independent interactions such that the resulting design is of resolution V. To use the confounding pattern given in Box and Hunter (1961), we generated a 2^{10} design via computer and used (see Cochran and Cox (1957)) the "odd-even letters rule" to separate the blocks by confounding the three independent interactions

$$I = ABCGH = BCDEJ = ACDFK.$$

This process was carried using a computer program which works as follows: generate the 2^n design $(n \geq 2)$; check how many interactions are to be confounded (interactions may be dependent or independent); pick the first interaction; divide the 2^n combinations in two parts each of size 2^{n-1} according to the even-odd rule; retain only the principal block (the "odd" case); pick the second interaction and proceed similarly on the retained principal block; similarly proceed through all interactions to be confounded. Therefore the principal block here changes its size from 2^{n-1} to 2^{n-2} due to the second interaction (if that interaction was independent of the first, otherwise the block size will remain 2^{n-1}) and so on. (The entire computer program listing is given in the course noted in Section 1.) Since our three interactions are independent, the final outcome consists of 128 combinations (out of all 1024 possible combinations).

8. SECOND ORDER RESPONSE SURFACE AND CENTRAL COMPOSITE DESIGN

For the problem of finding the effects of different levels of input quantitative variables $x_1, x_2, \ldots, x_n$ on a response variable y, such classical methods such as regression analysis, correlation analysis, and analysis of variance are often used. A major breakthrough for such studies appeared in the pioneering paper of Box and Wilson in 1951 (see Davies (1956)).

We have 10 factors, all quantitative, the importance of which has been discussed in Section 5. The output estimate of P(CS) depends on these 10 factors, and we desire to estimate P(CS) as a function of these 10 variables

$$P(CS) = f(k, P^*, \delta^*, \sigma^2_{[k]}, a, \rho_{[k]}, b, \alpha, N_0, MWARM) \tag{6}$$

where $f(\cdot)$ is a polynomial of degree 2. According to the design of Section 7, main effects and interactions can be estimated. For square terms to be estimable, one could use factors at three levels each, however then instead of $2^{10} = 1024$ runs we need $3^{10} = 59049$ runs and even a $1/3^5$ fraction needs $3^5 = 243$ runs, much larger than can be afforded. Economical statistical design still allowing estimates of square terms' coefficients leads us to a central composite design.

The composite designs are first order factorial designs augmented by additional points to allow estimation of coefficients of a second order surface (mainly quadratic terms). The central composite design is a 2^k factorial or fractional factorial design (the two levels coded as -1 or 1 and center level coded as 0) augumented by the 2n+1 points

$$\begin{array}{cccc} x_1 & x_2 & \cdots & x_n \end{array}$$
$$\begin{bmatrix} 0 & 0 & \cdots & 0 \\ \alpha & 0 & \cdots & 0 \\ -\alpha & 0 & \cdots & 0 \\ 0 & \alpha & \cdots & 0 \\ 0 & -\alpha & \cdots & 0 \\ \cdots & \cdots & \cdots & \cdots \\ \cdots & \cdots & \cdots & \cdots \\ \cdots & \cdots & \cdots & \cdots \\ 0 & 0 & \cdots & \alpha \\ 0 & 0 & \cdots & -\alpha \end{bmatrix} \qquad (7)$$

In our case, we take $\alpha=1$. A set of n_a = 2x (number of factors) star or axial points and a center point is attached to a fractional factorial or factorial design. Thus in order to estimate coefficients of linear, quadratic and interaction terms (assuming all higher order interactions are negligible), we need to run a total of

$$\underset{\text{fractional factorial runs}}{2^{10-3}} + \underset{\text{star points}}{2\text{x}10} + \underset{\text{center point}}{1} = 149$$

possible treatment combinations. To make this idea geometrically clear, for a three factor case we need to run 2^3 + 2x3 + 1 = 15 treatment combinations, represented on a cube in Figure 1. The axial points provide three levels for each factor (not all possible 3 level combinations, but enough for estimation).

In our experiment the factors at their different quantitative levels have been considered as in Table 1.

FACTOR	LOW LEVEL	CENTER LEVEL	HIGH LEVEL
(1) k	2	6	10
(2) P*	0.85	0.90	0.95
(3) δ^*	0.5	1.0	1.5
(4) $\sigma^2_{[k]}$	0.75	1.0	1.25
(5) a	0.25	1.25	2.25
(6) $\rho_{[k]}$	0.13	0.5	0.87
(7) b	0.9	1.0	1.1
(8) N_0	30	45	60
(9) α	0.05	0.10	0.15
(10) MWARM	0	25	50

TABLE 1.

9. COMPUTER RUNS

Keeping in mind that we want to find a second order response surface and are using a central composite design, we ran 149 different possible combinations of 10 factors.

The computer program was written in FORTRAN WATFOR. A random deviate generator was used to generate uncorrelated normal deviates. Each run originated its respective k autoregressive processes of order 1 at its respective means and, after necessary warm-up (if needed), a sample of size N_0 from each of the processes was taken and necessary calculations were performed.

The program yields a printout which contains numerical values of the factors used, the estimated value of P(CS) and its 95 percent lower and

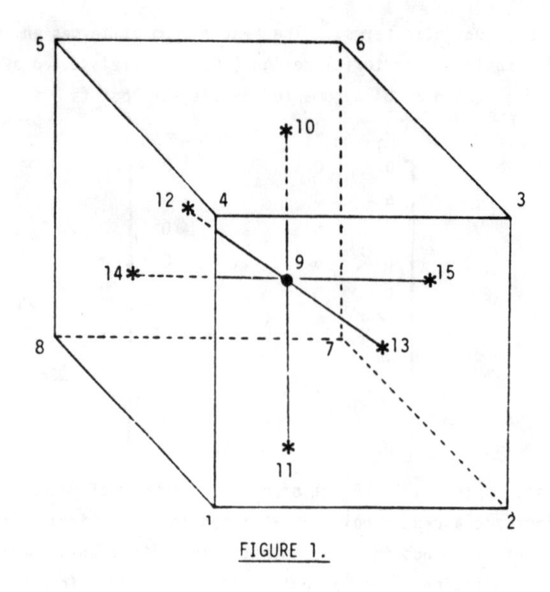

FIGURE 1.

upper confidence limits (LCL and UCL, respectively). A typical printout appears in Table 2.

Test runs were made to reproduce Table III of Dudewicz and Zaino (1977). Certain intermediate runs were made to see the effect of correlation alone and one possible analysis appears in Section 10.

10. INTERMEDIATE RUNS

Two runs were made on identical parameters except for the value of the correlation. In one run, correlation was 0.5 for all the populations, while in the other run correlation was 0.9534. The change in correlation from 0.5 to 0.9534 changed the estimated P(CS) from 0.886 to 0.435. From Table III of Dudewicz and Zaino (1977) and our runs, it appears that P(CS) is a nonincreasing function of ρ . One possible reason is as follows.

When ρ is estimated, then in the process of estimation, sometimes we overestimate ρ and sometimes underestimate it. When the ρ value (theoretical) is so high (as 0.50 or 0.9534) then $100(1-\alpha)\%$ of the time we are going to find that ρ is significantly different from zero and therefore, in a such case, the final sample size changes from

$$N_2 = \max(N_0+1,\ [\frac{s^2h^2}{\delta^2}])$$

to

$$N_3 = N_2[\frac{1+\hat{\rho}}{1-\hat{\rho}}] \ . \qquad (8)$$

When ρ is overestimated and ρ in fact is "very high," then P(CS) reaches 1 (not much increase in P(CS), because P* = .95 was chosen); while, when it is underestimated, the multiplying factor of N_3 is considerably smaller, we sample fewer units, and the P(CS) dips below (possibly much below) the required level P*. Thus, while overestimating ρ, there is not much gain in P(CS), but underestimation of ρ results in large loss of P(CS) and yields a value lower (at times much lower) than P*.

11. STATISTICAL DATA ANALYSIS

We will now consider final aspects of the problem. 149 data points were obtained from FORTRAN runs and SAS (Statistical Analysis System) program GLM (General Linear Model) on 65 independent variables (10 principal variables, 10 square terms, and $\binom{10}{2}$ = 45 interactions), with P(CS) as dependent variable, was run for preliminary screening purposes.

NUMBER	K	PSTAR	DELTA	SIGMASQ	ACONST	RHO	BCONST	NSTART	ALPHA	MWARM	P(CS)	LCL	UCL
129	6	0.90	1.0	1.00	1.25	0.50	1.0	45	0.10	25	0.998	0.9952	1.0008
130	10	0.90	1.0	1.00	1.25	0.50	1.0	45	0.10	25	0.994	0.9892	0.9988
131	2	0.90	1.0	1.00	1.25	0.50	1.0	45	0.10	25	0.999	0.9970	1.0010
132	6	0.95	1.0	1.00	1.25	0.50	1.0	45	0.10	25	0.998	0.9952	1.0008
133	6	0.85	1.0	1.00	1.25	0.50	1.0	45	0.10	25	0.996	0.9921	0.9999
134	6	0.90	1.5	1.00	1.25	0.50	1.0	45	0.10	25	1.000	1.0000	1.0000
135	6	0.90	0.5	1.00	1.25	0.50	1.0	45	0.10	25	0.844	0.8215	0.8665
136	6	0.90	1.0	1.25	1.25	0.50	1.0	45	0.10	25	0.998	0.9952	1.0008
137	6	0.90	1.0	0.75	1.25	0.50	1.0	45	0.10	25	1.000	1.0000	1.0000
138	6	0.90	1.0	1.00	2.25	0.50	1.0	45	0.10	25	0.984	0.9762	0.9918
139	6	0.90	1.0	1.00	0.25	0.50	1.0	45	0.10	25	1.000	1.0000	1.0000
140	6	0.90	1.0	1.00	1.25	0.87	1.0	45	0.10	25	0.800	0.7752	0.8248
141	6	0.90	1.0	1.00	1.25	0.13	1.0	45	0.10	25	0.999	0.9970	1.0010
142	6	0.90	1.0	1.00	1.25	0.50	1.1	45	0.10	25	0.996	0.9921	0.9999
143	6	0.90	1.0	1.00	1.25	0.50	0.9	45	0.10	25	0.995	0.9906	0.9994
144	6	0.90	1.0	1.00	1.25	0.50	1.0	60	0.10	25	1.000	1.0000	1.0000
145	6	0.90	1.0	1.00	1.25	0.50	1.0	30	0.10	25	0.981	0.9725	0.9895
146	6	0.90	1.0	1.00	1.25	0.50	1.0	45	0.15	25	0.999	0.9970	1.0010
147	6	0.90	1.0	1.00	1.25	0.50	1.0	45	0.05	25	0.995	0.9906	0.9994
148	6	0.90	1.0	1.00	1.25	0.50	1.0	45	0.10	50	1.000	1.0000	1.0000
149	6	0.90	1.0	1.00	1.25	0.50	1.0	45	0.10	0	0.997	0.9936	1.0004

TABLE 2. TYPICAL OUTPUT PAGE.

Only 21 variables were found to be significant at a 10% level of significance. At this point we employed Mallows' C_p-criterion to look at 10 "best" subsets yielding the minimum C_p-value via the BMDP package, which looks at all 21 independent variables at a time, computes corresponding C_p-values and regression coefficients, selects one variable at a time, and does so for 10 best subsets of p-variables (p=1,2,3,...,21). Since in the preliminary stage we did not find P* to be significant, therefore, we made another BMDP run on P(CS)-P* as dependent variable with 21 originally screened independent variables, using the C_p-criterion. The residual analysis, C_p-criterion, normal probability plots and multiple correlations were used in the second stage for detecting important factors.

Applying these techniques, we found 15 outliers, based on residual analysis and normal probability plots. These 15 runs have three factors in common, k=10, ρ=.87, b=1.1, and have estimated P(CS) $\leq$ .56. The presence of outliers made normal plots consisting of 2 straight line segments. (As we know, for a good normal fit we should have one straight line.) The same outliers make plots of predicted versus residuals not random but with some trend.

We made another set of analysis runs, one consisting of 134 points and one consisting of 15 "outliers," to fit the response surface in piecewise parts. Again, residual analysis, C_p-criterion, normal probability plots and multiple correlation on 21 retained variables (retained after the first stage) were considered on the run which had 134 points. The response surface for the outliers will be discussed below; first we discuss the response surface on the major part of the data.

12. FITTED RESPONSE SURFACE, 134 DATA POINTS

Two sets of equations were obtained, one by considering P(CS) alone as dependent variable, the second with P(CS)-P* as dependent variable. (This further analysis improved our residual analysis, normal probability plots, and multiple correlations considerably.) Response surfaces were as follows:

$$\begin{aligned}
\widehat{P(CS)} = {} & 0.98521 - 0.00727535k - 0.0455273\,\sigma^2_{[k]} \\
& + .761686\,\rho_{[k]} + 0.00555625\,k\,\delta^* \\
& - 0.00225318\,k\,a - 0.642227\,\rho^2_{[k]} \\
& - 0.0100561\,k\,\rho_{[k]} + 0.172978\,\delta^*\,\rho_{[k]} \\
& - 0.0392481\,a\,\rho_{[k]} + 0.0000626093\,k\,N_0 \\
& - 0.515162\,\rho_{[k]}\,b + 0.00282329\,\rho_{[k]}\,N_0
\end{aligned} \tag{9}$$

with R = 0.94185. (Here each variable entered the regression equation at the 10% level of significance.)

$$\begin{aligned}
\widehat{P(CS)-P^*} = {} & 0.481069 - 0.00449128k - 0.0451177\,\sigma^2_{[k]} \\
& + 0.726928\,\rho_{[k]} + 0.00550579\,k\,\delta^* - 0.00222795\,k\,a \\
& - 0.630337\,\delta^*\,\rho_{[k]} - 0.0389418\,a\,\rho_{[k]} \\
& - 0.485064\,P^{*2} - 0.511685\,\rho_{[k]}\,b \\
& + 0.00324164\,\rho_{[k]}\,N_0
\end{aligned} \tag{10}$$

with multiple correlation R = .94937.

13. INTERPRETATION OF SURFACES

The coefficient of k is negative in both (9) and (10), indicating P(CS) is a nonincreasing function of the number of populations. The coefficients of $\rho_{[k]}$ in both the equations are positive and coefficients of $b\,\rho_{[k]}$ in those equations are negative, which means by simultaneously decreasing correlation of indifferent populations and increasing correlation of best population, we can increase estimate of P(CS). However, $\rho_{[k]}$ in both the equations interacts with quite a few other factors. Incorporation of all the interactions of $\rho_{[k]}$ and interpretation requires contour plots. (In our earlier analysis, the values of ρ in all the populations were the same and variances were not in the same configuration as in this experiment, therefore we cannot say that our experiment contradicts earlier claims that by decreasing ρ we can substantially increase our estimate of P(CS).) In both equations (9) and (10) interaction $\delta^*\rho_{[k]}$ explains at least 15.67% of variability; $\rho^2_{[k]}$ is also an important factor, as it explains at least 8.6% of variability. (Figures 2-7 give some details.) Overall (10) seems a suitable response surface; its variables are entered at the 1.6% level of significance.

14. ANALYSIS OF OUTLIERS

Earlier we discussed and described two estimates of response surface to predict P(CS) for future use, based on 134 points. Based on normal probability plots and residual analysis, we saw indicated that the observations were coming from a contaminated normal population. We now try to identify the remaining 15 points from another normal population. To this end, since k, $\rho_{[k]}$ and b were constant in all the cases (and since the number of observations was only 15), we ran BMDP programs with C_p-criterion on the ten independent variables $\sigma^2_{[k]}$, a, P^*, δ^{*2}, $\delta^*\sigma^2_{[k]}$, δ^*a, $\sigma^2_{[k]}a$, $P^*\delta^*$, N_0, $P^*\sigma^2_{[k]}$ with P(CS) as dependent variable. The C_p-criterion selects seven variables out of these ten variables at .7% level of significance with multiple correlation R = .99055, giving estimated response surface for outliers:

$$\begin{aligned}
\widehat{P(CS)} = {} & -4.21918 + 8.92882\,P^* + 4.62197\,\delta^{*2} - 0.306096\,\delta^*\,\sigma^2_{[k]} \\
& - 0.261900\,\delta^*\,a + 0.0795723\,\sigma^2_{[k]}\,a \\
& - 9.18508\,P^*\,\delta^* - 0.00807093\,N_0\,.
\end{aligned} \tag{11}$$

For a complete response surface (11) should be used in conjunction with (9) and (10). Figures 8 and 9 explain the residual analysis for outliers.

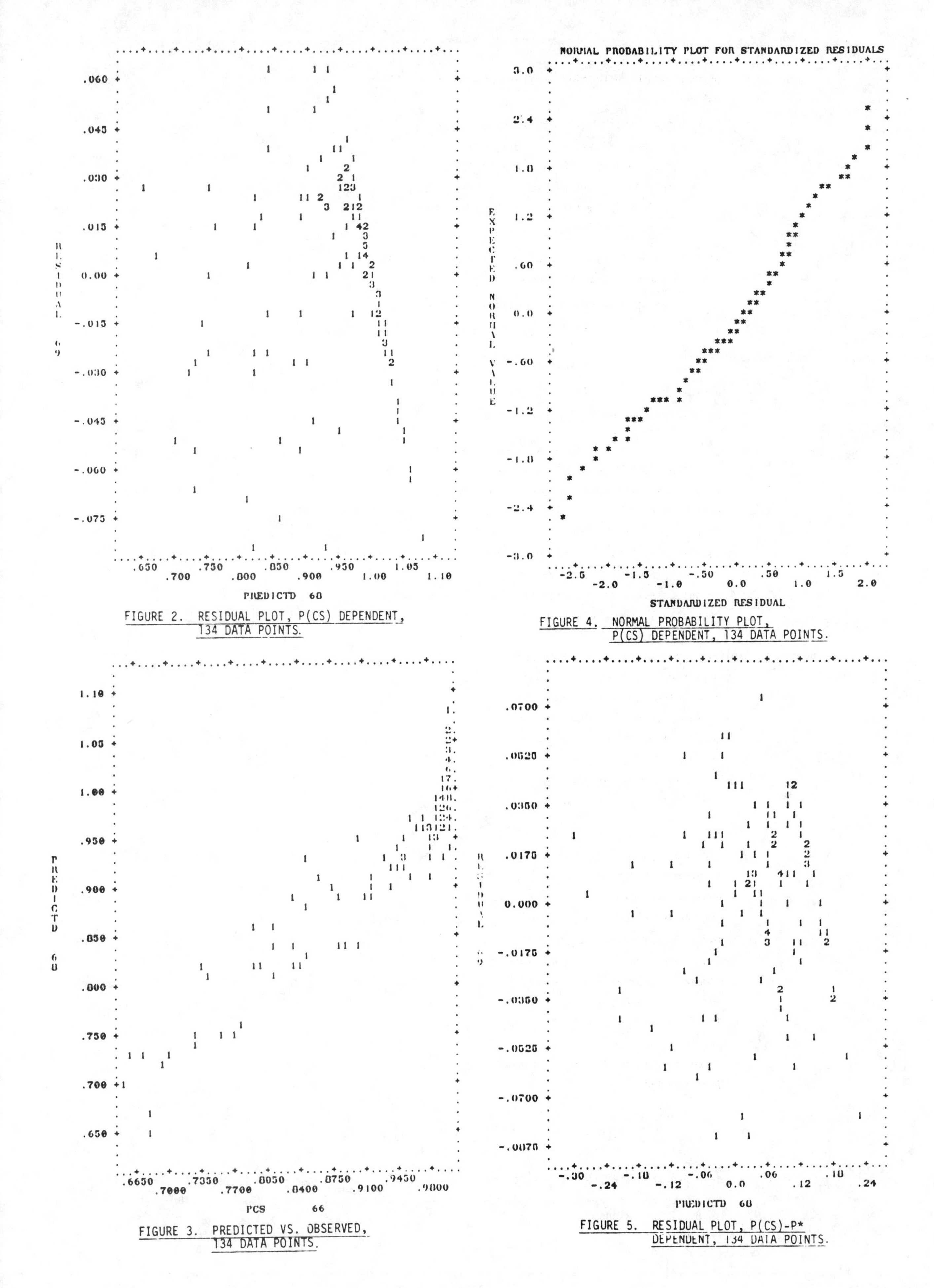

FIGURE 2. RESIDUAL PLOT, P(CS) DEPENDENT, 134 DATA POINTS.

FIGURE 3. PREDICTED VS. OBSERVED, 134 DATA POINTS.

FIGURE 4. NORMAL PROBABILITY PLOT, P(CS) DEPENDENT, 134 DATA POINTS.

FIGURE 5. RESIDUAL PLOT, P(CS)-P* DEPENDENT, 134 DATA POINTS.

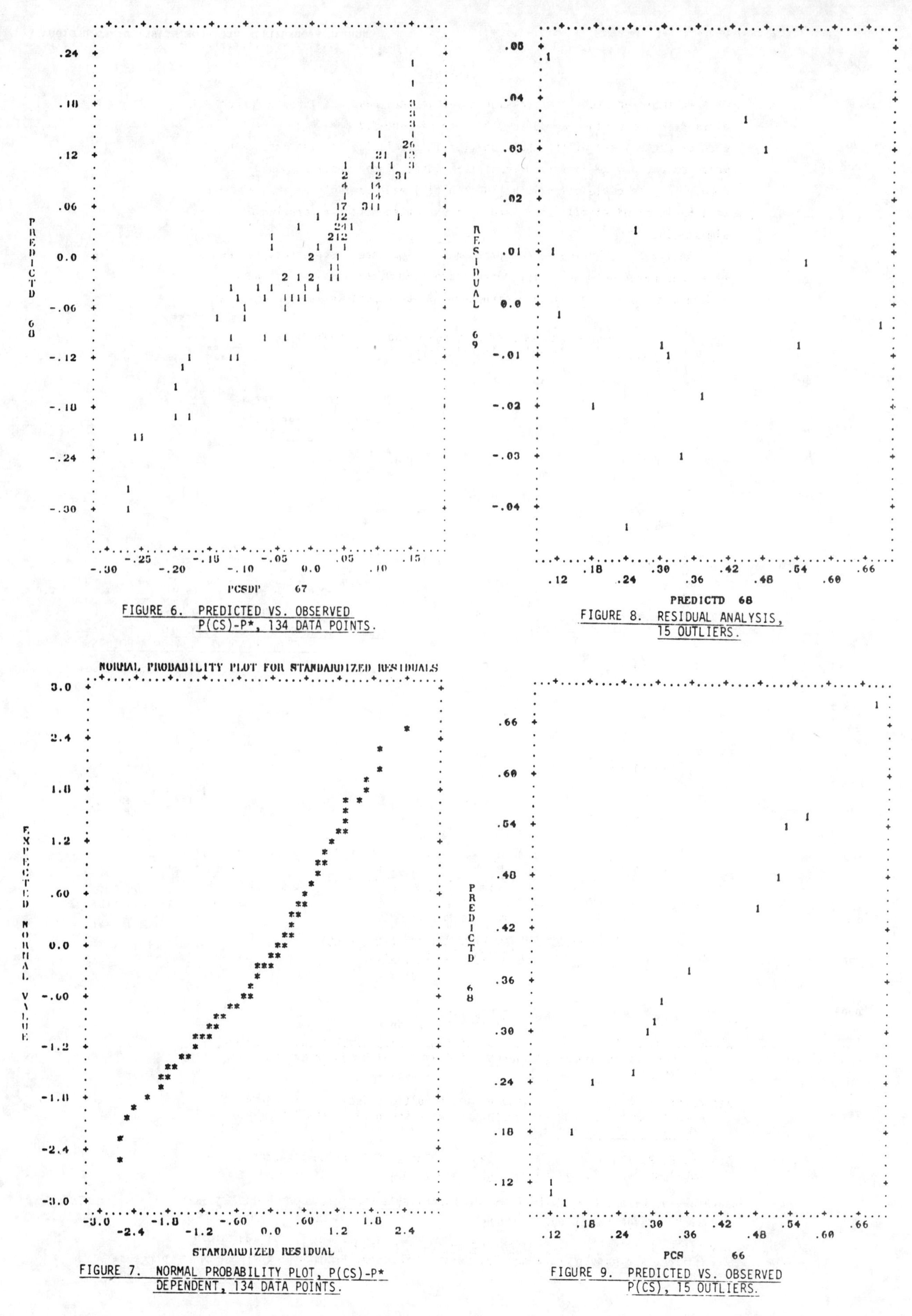

FIGURE 6. PREDICTED VS. OBSERVED P(CS)-P*, 134 DATA POINTS.

FIGURE 8. RESIDUAL ANALYSIS, 15 OUTLIERS.

FIGURE 7. NORMAL PROBABILITY PLOT, P(CS)-P* DEPENDENT, 134 DATA POINTS.

FIGURE 9. PREDICTED VS. OBSERVED P(CS), 15 OUTLIERS.

15. COMMENTS

We employed statistical design to estimate a response surface for prediction in use of the heuristic procedure of Dudewicz and Zaino (1977), when parameters are within the scope (of interpolation) of the experiment, which covers a wide variety of configurations. Experimenters should choose (11) in conjunction with (10); in (10) each variable enters, at most at 1.6% level of significance, and with estimated multiple correlation almost .95.

Analytical solution to the problem is still under consideration. We have considered only AR(1) for correlation structure; behavior of the procedure under other correlation structures is a subject for further studies.

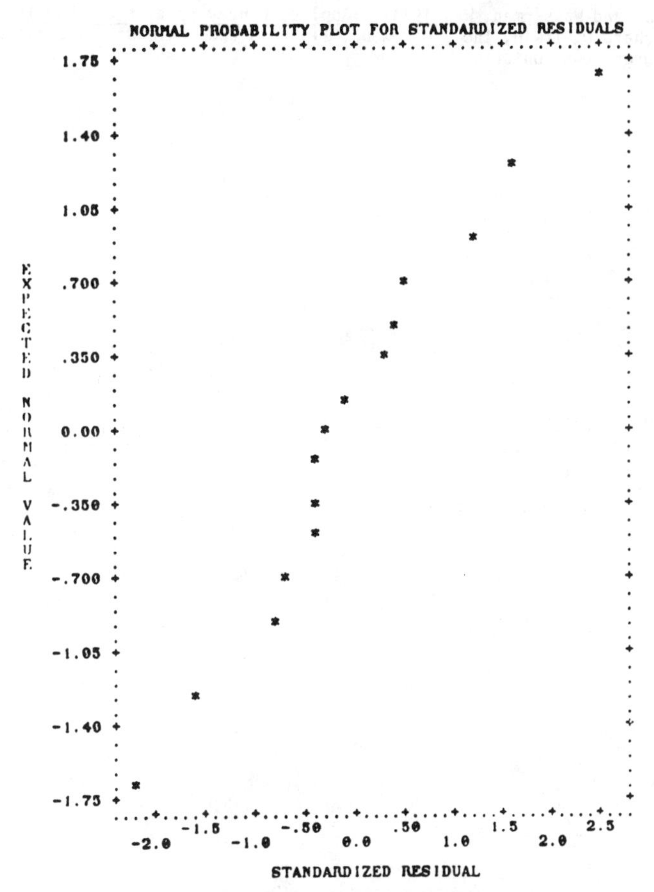

FIGURE 10. NORMAL PROBABILITY PLOT, P(CS) DEPENDENT, 15 OUTLIERS.

REFERENCES

Anderson, O. D. (1979): "On warming-up time series simulations generated by Box-Jenkins models," Journal of Operational Research Society, 30, 587-589.

Bechhofer, R. E. (1954): "A single-sample multiple-decision procedure for ranking means of normal populations with known variances," Annals of Mathematical Statistics, 25, 16-39.

Box, G. E. P. and Hunter, J. S. (1961): "The 2^{k-p} fractional factorial designs Part II," Technometrics, 3, 449-458.

Cochran, W. G. and Cox, G. M. (1957): Experimental Designs, John Wiley & Sons, New York.

Davies, O. L. (1956): The Design and Analysis of Industrial Experiments, Oliver and Boyd, London and Edinburgh, and Hafner Publishing Company, New York.

Dudewicz, E. J. and Dalal, S. R. (1975): "Allocation of observations in ranking and selection with unequal variances," Sankhya, 37B, 28-78.

Dudewicz, E. J. and Zaino, N. A. (1977): "Allowance for correlation in setting simulation run-length via ranking-and-selection procedures," The Institute of Management Sciences Studies in the Management Sciences, 7, 51-61.

Dudewicz, E. J. and Koo, J. O. (1982): The Complete Categorized Guide to Statistical Selection and Ranking Procedures, American Sciences Press, Columbus, Ohio.

John, P. W. M. (1971): Statistical Design and Analysis of Experiments, The Macmillan Company, New York.

Kleijnen, J. P. C. (1975): Statistical Techniques in Simulation, Part II, Marcel Dekker, Inc., New York.

Neter, J. and Wasserman, W. (1974): Applied Linear Statistical Models, Richard D. Irwin, Inc., Homewood, Illinois and Irwin-Dorsey Limited, George Town, Ontario.

Specification of Designs for Simulation

Models of Resolution III, IV, V. How would a statistician recommend that experimentation be carried out? How can a statistically valid experiment be run to study the effects of many factors on a desired output (or on several characteristics of output, all of which are of interest)? How should pilot experiments be run if we do not wish to hold many of $x_1, x_2, \ldots$ constant and vary a few, but wish to vary many and to explore their effects? (Here, the x's may be such variables as chemical concentration, reaction time, temperature, pH, etc.)

We will phrase the answer to this question in fairly general terms (which will apply to each of the three settings mentioned, as well as to others), and we will illustrate it with a specific design. While this particular design will not be the one which should be used in all cases, it should be used in some; similar designs are available to meet situations with variations in the numbers of variables to be studied, probable interactions present, size of experiment desired, and so on.

If we wish to study the effects of k factors at two levels each, a full factorial experiment will require 2^k observations; this number increases rapidly with increasing k:

k	1	2	3	4	5	6	7	8
2^k	2	4	8	16	32	64	128	256

However, if many interactions of high order may be assumed to be zero, then a smaller fraction of the complete factorial may be used. (If the assumption that these interactions are zero is not satisfied, then the values calculated for some of the effects will actually measure the sum of those effects and some of the ones assumed to be zero; these effects are said to be **confounded** or to be **aliases** of each other.) A design is said to be of:

> **Resolution III** if no main effect is confounded with any other main effect (but main effects are confounded with 2-factor interactions and 2-factor interactions are confounded with each other);
>
> **Resolution IV** if no main effect is confounded with any other main effect or 2-factor interaction (but 2-factor interactions are confounded with each other);
>
> **Resolution V** if no main effect or 2-factor interaction is confounded with any other main effect or 2-factor interaction (but 2-factor interactions are confounded with 3-factor interactions).

For example, with $k = 8$ factors, a resolution IV design can be obtained with $2^{8-4} = 16$ experiments, as follows (where "+" denotes one level of a factor, the so-called "high" level, and "−" denotes its other, or "low", level):

	Factor							
Run	1	2	3	8	4	5	6	7
1	−	−	−	−	−	−	−	−
2	+	−	−	−	+	+	−	+
3	−	+	−	−	+	−	+	+
4	+	+	−	−	−	+	+	−
5	−	−	+	−	−	+	+	+
6	+	−	+	−	+	−	+	−
7	−	+	+	−	+	+	−	−
8	+	+	+	−	−	−	−	+
9	−	−	−	+	+	+	+	−
10	+	−	−	+	−	−	+	+
11	−	+	−	+	−	+	−	+
12	+	+	−	+	+	−	−	−
13	−	−	+	+	+	−	−	+
14	+	−	+	+	−	+	−	−
15	−	+	+	+	−	−	+	−
16	+	+	+	+	+	+	+	+

With this design the main effects and 2-factor interactions have aliases as follows: 1, 2, 3, 4, 5, 6, 7, 8 are estimated without (main or 2-factor) alias; among 2-factor interactions we estimate

$$12 + 37 + 48 + 56$$

$$13 + 27 + 58 + 46$$

$$14 + 28 + 36 + 57$$

$$15 + 38 + 26 + 47$$

$$16 + 78 + 34 + 25$$

$$17 + 23 + 68 + 45$$

$$18 + 24 + 35 + 67.$$

(Of course, if 3-factor or higher interactions are also present, then their effects will be added to the main and 2-factor effects estimated above.)

A design such as the one above would often be appropriate for screening out the few most important main effects and interactions, which could then be studied further in an experiment which would **augment** the first one. At later stages (when one has a substantial amount of information available about which effects and 2-factor or higher order interactions are, or are not, zero), one would tailor-make a design so as not to confound non-zero effects and interactions. At that time, one might also have determined fixed levels for certain factors (not to be subjected to further variation) and those would be dropped from the experiment (though they should be kept if the fixing is only tentative).

Models. Above we have referred to "main effects," "2-factor interactions," "3-factor interactions," etc.; we now wish to make this notion (justified only intuitively thus far) precise. In the following, without further discussion, assume that: **Replications** may have been made at each design point; and that all **regions** of the space of the x's are capable of experimentation (i.e., experiments can be run there; if this is not the case, one will be dealing with constrained experimentation, and the details will depend on the nature of the constraints).

Generally, mean yield is a smooth function of the experimental conditions $x_1, x_2, x_3, x_4, \ldots$ and true mean yield may be accurately represented by a polynomial equation of sufficiently high order:

$$\begin{aligned}\eta = \beta_0 &+ \beta_1x_1 + \beta_2x_2 + \beta_3x_3 + \beta_4x_4 \\ &+ \beta_{12}x_1x_2 + \beta_{13}x_1x_3 + \beta_{14}x_1x_4 + \beta_{23}x_2x_3 \\ &\qquad + \beta_{24}x_2x_4 + \beta_{34}x_3x_4 \\ &+ \beta_{11}x_1^2 + \beta_{22}x_2^2 + \beta_{33}x_3^2 + \beta_{44}x_4^2 \\ &+ \beta_{123}x_1x_2x_3 + \beta_{124}x_1x_2x_4 + \beta_{134}x_1x_3x_4 \\ &\qquad + \beta_{234}x_2x_3x_4 \\ &+ \beta_{1234}x_1x_2x_3x_4 + \ldots .\end{aligned}$$

Different experimental designs allow us to estimate various of the β's (singly or in combination), while assuming others are negligible (which can often be tested). Typically one assumes terms of higher than second order are negligible ($0 = \beta_{123} = \beta_{124} = \beta_{134} = \beta_{234} = \beta_{1234} = \ldots$). By the following:

"main effects" we denote $\beta_1, \beta_2, \beta_3\ \beta_4$;

"2-factor interactions" we denote $\beta_{12}, \beta_{13}, \beta_{14}, \beta_{23}, \beta_{24}, \beta_{34}$;

"quadratic effects" we denote $\beta_{11}, \beta_{22}, \beta_{33}, \beta_{44}$;

etc.

If the number of factors (i.e., x's) is relatively small (1, 2, 3, 4, 5), then there are a number of **designs which are widely used**, depending on what assumptions (if any) about the model one can justifiedly make in advance of experimentation. For example, **for 4 factors** x_1, x_2, x_3, x_4:

2^4 design (16 points). With this design one can estimate $\beta_0 + \beta_{11} + \beta_{22} + \beta_{33} + \beta_{44}$, $\beta_1, \beta_2, \beta_3, \beta_4, \beta_{12}, \beta_{13}, \beta_{14}, \beta_{23}, \beta_{24}, \beta_{34}, \beta_{123}, \beta_{124}, \beta_{134}, \beta_{234}$ and (assuming β_{1234} negligible) have an estimate of experimental error. (No suitable fractions of this design exist, as they all confound one- and/or two-factor effects.)

$2^4 + 1$ design (17 points). With this design one can now separately estimate β_0, $\beta_{11} + \beta_{22} + \beta_{33} + \beta_{44}$ and hence (barring canceling magnitudes) assess the total quadratic effect independently of the response at the center (β_0).

Central Composite Design (CCD) (17 + 8 = 25 points). Through adding 8 star (or axial) points, the central composite design allows a full quadratic model to be fitted. Suitable fractions of the 2^4 may be used (i.e., a fraction with at most one 2-factor interaction in any alias set), and (via confounding scheme $I = ABC$) one can obtain the full quadratic model estimation with

$$\frac{1}{2}2^4 + 1 + 8 = 17 \text{ points}.$$

Webb (1968) resolution IV design, 8 points. For 4 factors, this is a fractional factorial design, and one fits β_0, $\beta_1, \beta_2, \beta_3, \beta_4$ unbiased by 2-factor interactions (though those interactions themselves are not estimable). (Confounding scheme $I = ABCD$ will yield such a design.)

Webb (1971) design, 11 points. For 4 factors, this design allows one to fit $\beta_0, \beta_1, \beta_2, \beta_3, \beta_4, \beta_{12}, \beta_{13}, \beta_{14}, \beta_{23}, \beta_{24}, \beta_{34}$, assuming no curvature.

Webb (1968) 8-point design

Run*	Factor A	B	C	D	Name
1	−	−	−	−	$(I)\surd$
2	+	−	−	+	$ad\surd$
3	−	+	−	+	$bd\surd$
4	+	+	−	−	$ab\surd$
5	−	−	+	+	$cd\surd$
6	+	−	+	−	$ac\surd$
7	−	+	+	−	$bc\surd$
8	+	+	+	+	$abcd$

"−" denotes low level; "+" denotes high level.

*: Experimental (field) order should be randomized (or randomized within blocks).

$\surd$: In 11-point Webb (1971) design.

For **2-factor** and **3-factor** cases, similar **designs are as follows:**

a. 3-Factors. Here, the designs

$2^3 \ldots 8$ points
$2^3 + 1 \ldots 9$ points
$CCD \ldots 15$ points (9 + 6)

Webb (1971) 11-point design

Run*	Factor A	B	C	D	Name
1	0	0	0	0	(*I*)√
2	0	0	1	1	*cd*√
3	0	1	0	1	*bd*√
4	0	1	1	0	*bc*√
5	1	0	0	1	*ad*√
6	1	0	1	0	*ac*√
7	1	1	0	0	*ab*√
8	0	1	1	1	*bcd*
9	1	0	1	1	*acd*
10	1	1	0	1	*abd*
11	1	1	1	0	*abc*

"0" denotes low level; "1" denotes high level.

*: Experimental (field) order should be randomized (or randomized within blocks).

: In 8-point Webb (1968) design.

are available with properties similar to those in the 4-factor case. In addition, via the confounding scheme $I = ABC$ we can obtain a half-replicate of the 2^3 which has at most one 2 f.i. in any alias set, hence can fit a full quadratic model with a

$$CCD \ldots 11 \text{ points } (5 + 6).$$

This half-replicate contains treatments *a, b, c,* and *abc*.

A Webb (1968) resolution IV design, with 6 points and properties as in the 4-factor case except that β_0 is not estimable unless a 2 f.i. is suppressed, is given below.

Webb (1968) 6-point design

Run*	Factor A	B	C	Name
1	+	–	–	*a*
2	–	+	–	*b*
3	–	–	+	*c*
4	–	+	+	*bc*
5	+	–	+	*ac*
6	+	+	–	*ab*

"–" denotes low levels; "+" denotes high level.

*: Experimental (field) order should be randomized.

Webb (1971) does not allow for all linear and interaction terms with anything less than the 8-point 2^3 full factorial experiment.

For 3 factors, the 11-point CCD is strongly recommended, since it allows a full quadratic model at a number of treatment formulations which will often be feasible.

b. 2-Factors. Here the only Resolution IV design is the 2^2 4-point full factorial. A CCD with 9 points (5 + 4) will allow a full quadratic model to be fitted, and is recommended. (The 2-point fraction (*I*), *ab* could also be utilized, allowing for a 7-point (3 + 4) CCD and still permitting a full quadratic model.)

Note that many of these designs conduct experiments with each of the **factors** $x_1, x_2, x_3, \ldots$ (also often called the factors $A, B, C, \ldots$) at **two levels**, called **high and low** (**"+" and "–"**; or, **"1" and "0"**). Each experiment which is run in such a setting has a special name, depending on which of the factors are at the high level in that experiment. For example, in the Webb (1968) 6-point design, factor *A* might be temperature (either 300 or 350 degrees), factor *B* might be reaction time (5 or 10 minutes), and factor *C* might be stirring rate (30 or 60 revolutions per minute). Then Run 4 has name *bc* because *A* is at its low level (300 degrees), while B and C are each at their high levels (10 minutes and 60 revolutions, respectively).

Screening Experiments. When the number of factors k is small (1, 2, 3, 4, and perhaps 5), one may be able to run 2^k experiments of the complete factorial experiment (32 if $k=5$); however, when k is larger than 5, this rapidly becomes infeasible; e.g., if $k=8$, 256 experiments would be required. One possibility is to use another type of design, a fractional factorial (yet to be studied) or a Central Composite Design (also yet to be studied). *However*, it is not advisable to run experiments with large numbers of factors (6 or more) unless it is absolutely necessary, as there are many pitfalls which can befall such an experiment (such as lack of easy interpretation of the results, problems in its conduct, etc.). Hence, **it is strongly recommended that a screening experiment be run first, to reduce the number of factors** to those with the strongest influence on the output(s) of interest, when k is 6 or more. Such screening is almost mandatory when k is 10 or more. We now wish to discuss such screening designs, with emphasis on the *Plackett-Burman* screening designs, of which a catalog (first given in *Biometrika*, volume *33* (1946): 305–325 by R. L. Plackett and J. P. Burman) will be given and explained.

Recall that in a **Resolution III design**, no main effect is confounded with any other main effect, but main effects are confounded with 2 f.i. and 2 f.i.'s are confounded with each other. In a **Resolution IV design**, no main effect is confounded with any other main effect or 2 f.i., but 2 f.i.'s are confounded with each other. Thus, a Resolution III will allow us to fit a model $\beta_0 + \sum_i \beta_i Z_i$ in factors Z_i, but interaction will bias the fit. A Resolution IV design will allow a fit unbiased by 2 f.i., i.e. β_i will estimate β_i (not $\beta_i \pm$ some of the β_{ij}'s of the Z_iZ_j interactions).

It is possible to study $k=7$ variables in $N=8$ runs with a

Resolution III (Plackett-Burman) design. If desired later, this design can be augmented to a Resolution IV design by adding 8 more runs (for a total of 16 runs). One can also study $k=6$ factors in an $N=12$ run (Webb) experimental design, at Resolution IV.

The 8-run Resolution III design in 7 variables is given in Table 1.

Table 1. A Screening Design.

Run	Variable 1	2	3	4	5	6	7
1	+	+	+	−	+	−	−
2	−	+	+	+	−	+	−
3	−	−	+	+	+	−	+
4	+	−	−	+	+	+	−
5	−	+	−	−	+	+	+
6	+	−	+	−	−	+	+
7	+	+	−	+	−	−	+
8	−	−	−	−	−	−	−

Here each experimental variable has 2 levels: low (−) and high (+). In Run 1, variables 1, 2, 3, and 5 are set to their high levels, while variables 4, 6, and 7 are set to their low levels.

Note that the runs would not be made in the order given, but they would be randomized in order, so as to obviate the possibility of (e.g.) systematic changes over time biasing the analysis.

A **table of such designs, for 4, 8, 12, . . ., or 100 experiments** is as follows:

TABLE OF DESIGNS

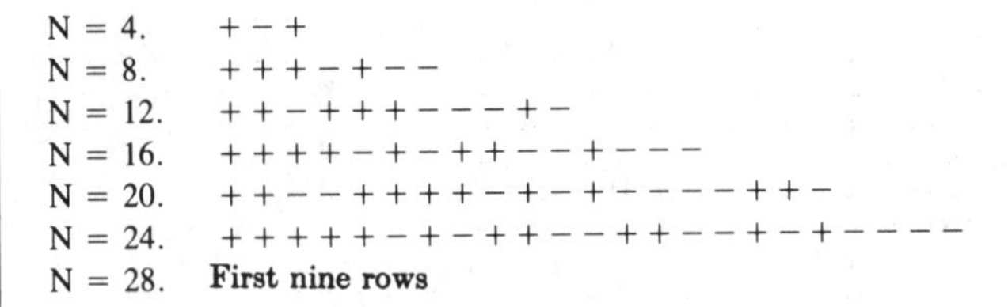

```
N = 4.   + − +
N = 8.   + + + − + − −
N = 12.  + + − + + + − − − + −
N = 16.  + + + + − + − + + − − + − − −
N = 20.  + + − − + + + + − + − + − − − − + + −
N = 24.  + + + + + − + − + + − − + + − − + − + − − − −
N = 28.  First nine rows
```

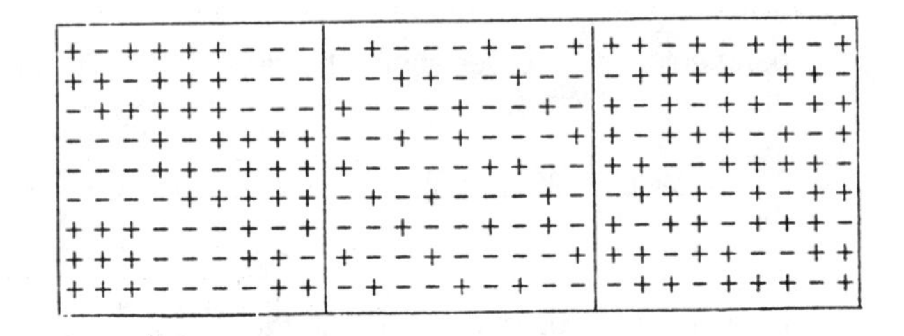

```
+ − + + + + − − − | − + − − − + − − + | + + − + − + + − +
+ + − + + + − − − | − − + + − − + − − | − + + + + − + + −
− + + + + + − − − | + − − − + − − + − | + − + − + + − + +
− − − + − + + + + | − − + − + − − − + | + − + + + − + − +
− − − + + − + + + | + − − − − + + − − | + + − − + + + + −
− − − − + + + + + | − + − + − − − + − | − + + + − + − + +
+ + + − − − + − + | − − + − − + − + − | + − + + − + + + −
+ + + − − − + + − | + − − + − − − − + | + + − + + − − + +
+ + + − − − − + + | − + − − + − + − − | − + + − + + + − +
```

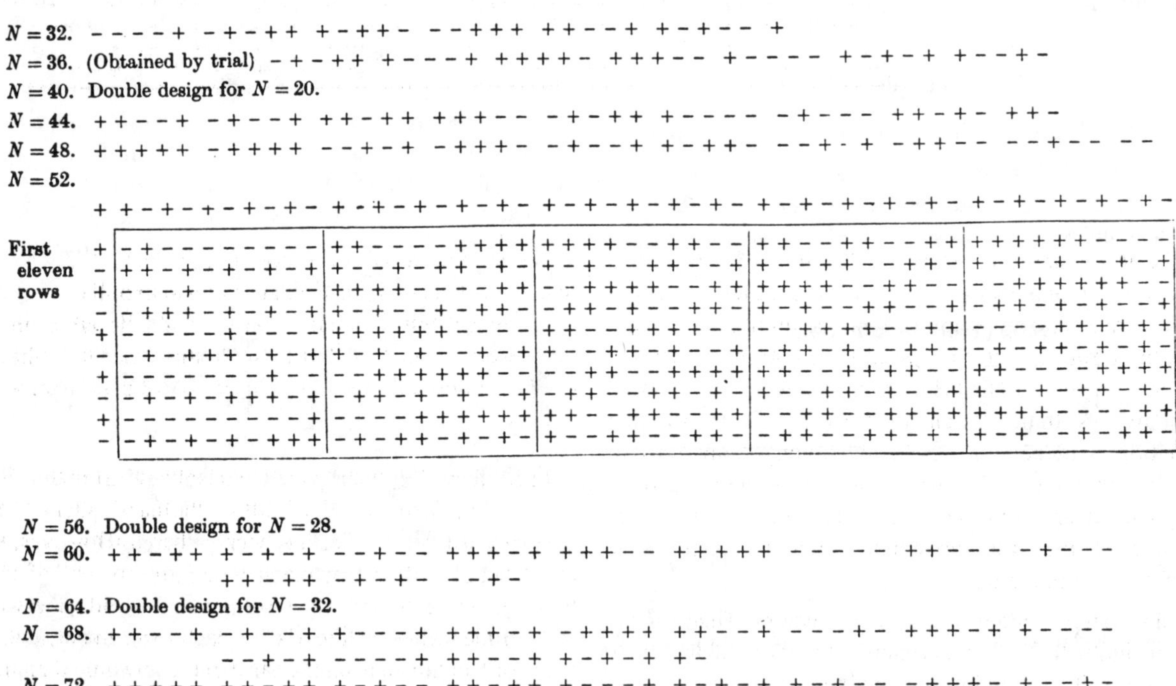

```
N = 32.  − − − − +  − + − + +  + − + + −  − − + + +  + + − − +  + − + − −  +
N = 36.  (Obtained by trial)  − + − + +  + − − − +  + + + + −  + + + − −  + − − − −  + − + − +  + − − + −
N = 40.  Double design for N = 20.
N = 44.  + + − − +  − + − − +  + + − + +  + + + − −  − + − + +  + − − − −  − + − − −  + + − + −  + + −
N = 48.  + + + + +  − + + + +  − − + − +  − + + + −  − + − − +  + − + + −  − − + − +  − + + − −  − − + − −  − −
N = 52.

+  + − + − + − + − + −  + − + − + − + − + −  + − + − + − + − + −  + − + − + − + − + −  + − + − + − + − + −

First     +  | − + − − − − − − − − | + + − − − − + + + + | + + + + − − + + − − | + + − − + + − − + + | + + + + + + − − − − |
eleven    −  | + + − + − + − + − + | + − − + − + + − + − | + − + − − + + − − + | + − − + + − − + + − | + − + − + − − + − + |
rows      +  | − − − + − − − − − − | + + + + − − − − + + | − − + + + + − − + + | + + + + − − + + − − | − − + + + + + + − − |
          −  | − + + + − + − + − + | + − + − − + − + + − | − + + − + − − + + − | + − + − − + + − − + | − + + − + − + − − + |
          +  | − − − − − + − − − − | + + + + + + − − − − | + + − − + + + + − − | − − + + + + − − + + | − − − − + + + + + + |
          −  | − + − + + + − + − + | + − + − + − − + − + | + − − + + − + − − + | − + + − + − − + + − | − + − + + − + − + − |
          +  | − − − − − − − + − − | − − + + + + + + − − | − − + + − − + + + + | + + − − + + + + − − | + + − − − − + + + + |
          −  | − + − + − + + + − + | − + + − + − + − − + | − + + − − + + − + − | + − − + + − + − − + | + − − + − + + − + − |
          +  | − − − − − − − − − + | − − − − + + + + + + | + + − − + + − − + + | − − + + − − + + + + | + + + + − − − − + + |
          −  | − + − + − + − + + + | − + − + + − + − + − | + − − + + − − + + − | − + + − − + + − + − | + − + − − + − + + − |

N = 56.  Double design for N = 28.
N = 60.  + + − + +  + − + − +  − − + − −  + + + − +  + + + − −  + + + + +  − − − − −  + + − − −  − + − − −
                    + + − + +  − + − + −  − − + −
N = 64.  Double design for N = 32.
N = 68.  + + − − +  − + − − +  + − − − +  + + + − +  − + + + +  + + − − +  − − − + −  + + + − +  + − − − −
                    − − + − +  − − − − +  + + − − +  + − + − +  + −
```

```
N = 72.  + + + + +  + + − + +  + − + − −  + + − + +  + − − − +  + − + − +  + − + − −  − + + + −  + − − + −
                    + − − + +  + − − − +  − − + + −  + − − − +  − − − − −  −
N = 76.  + + − + − + − + − + − + − + − + − + − + − + − + − + − + − + − + − + − + − + − + − + −

+ | − + | − − | + + | − − | − − | + + | + + | − − | + + | − − | − − | − − | − − | + + | + + | + + | − − | + + | + + | + + | + + | − − |
− | + + | − + | + − | − + | − + | + − | + − | − + | + − | − + | − + | − + | − + | + − | + − | + − | − + | + − | + − | + − | + − | − + |

+ − + − + − + − + − + − + − + − + − + − + − + − + − + − + − + −

| + + | + + | + + | − − | − − | − − | − − | + + | − − | + + | + + | − − | − − | + + | − − |
| + − | + − | + − | − + | − + | − + | − + | + − | − + | + − | + − | − + | − + | + − | − + |
```

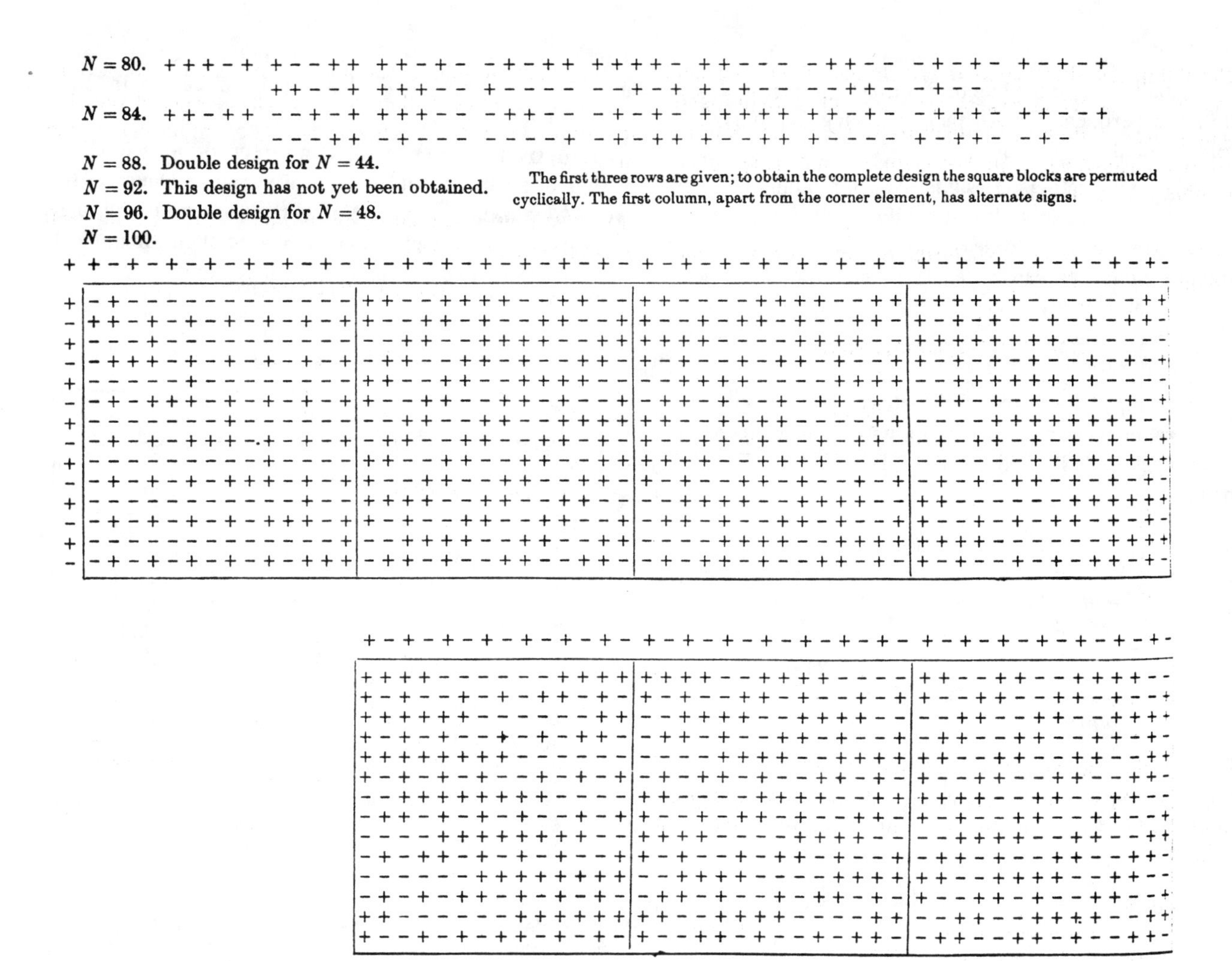

$N = 80.$ + + + − + + − − + + + + − + − − + − + + + + + + − + + − − − − + + − − − + − + − + − + − +
+ + − − + + + + − − + − − − − − − + − + + − + − − − − + + − − + − −

$N = 84.$ + + − + + − − + − + + + + − − − + + − − − + − + − + + + + + + + − + − − + + + − + + − − +
− − − + + − + − − − − − − − + − + − + + + − − + + + − − − − + − + + − − + −

$N = 88.$ Double design for $N = 44$.
$N = 92.$ This design has not yet been obtained.
$N = 96.$ Double design for $N = 48$.
$N = 100.$

The first three rows are given; to obtain the complete design the square blocks are permuted cyclically. The first column, apart from the corner element, has alternate signs.

+	+-+-+-+-+-+-+-	+-+-+-+-+-+-+-	+-+-+-+-+-+-+-	+-+-+-+-+-+-+-
+	-+------------	++--++++--++--	++----++++--++	++++++------++
-	++-+-+-+-+-+-+	+--++-+--++--+	+--+-++-+--++-	+-+-+--+-+-++-
+	---+----------	--++--++++--++	++++----++++--	++++++++------
-	-+++-+-+-+-+-+	-++--++-+--++-	+-+--+-++-+--+	+-+-+-+--+-+-+
+	-----+--------	++--++--++++--	--++++----++++	--++++++++----
-	-+-+++-+-+-+-+	+--++--++-+--+	-++-+--+-++-+-	-++-+-+-+--+-+
+	-------+------	--++--++--++++	++--++++----++	----++++++++--
-	-+-+-+++-+-+-+	-++--++--++-+-	+--++-+--+-++-	-+-++-+-+-+--+
+	---------+----	++--++--++--++	++++--++++----	------++++++++
-	-+-+-+-+++-+-+	+--++--++--++-	+-+--++-+--+-+	-+-+-++-+-+-+-
+	-----------+--	++++--++--++--	--++++--++++--	++------++++++
-	-+-+-+-+-+++-+	+-+--++--++--+	-++-+--++-+--+	+--+-+-++-+-+-
+	-------------+	--++++--++--++	----++++--++++	++++------++++
-	-+-+-+-+-+-+++	-++-+--++--++-	-+-++-+--++-+-	+-+--+-+-++-+-

+-+-+-+-+-+-+-	+-+-+-+-+-+-+-	+-+-+-+-+-+-+-
++++------++++	++++--++++----	++--++--++++--
+-+--+-+-++-+-	+-+--++-+--+-+	+--++--++-+--+
++++++------++	--++++--++++--	--++--++--++++
+-+-+--+-+-++-	-++-+--++-+--+	-++--++--++-+-
++++++++------	----++++--++++	++--++--++--++
+-+-+-+--+-+-+	-+-++-+--++-+-	+--++--++--++-
--++++++++----	++----++++--++	++++--++--++--
-++-+-+-+--+-+	+--+-++-+--++-	+-+--++--++--+
----++++++++--	++++----++++--	--++++--++--++
-+-++-+-+-+--+	+-+--+-++-+--+	-++-+--++--++-
------++++++++	--++++----++++	++--++++--++--
-+-+-++-+-+-+-	-++-+--+-++-+-	+--++-+--++--+
++------++++++	++--++++----++	--++--++++--++
+--+-+-++-+-+-	+--++-+--+-++-	-++--++-+--++-

In the above tables, if a design is "cyclic" (i.e., cycles via shifting its row right by one symbol to create each new row), then only the first row is given. N is the number of experiments to be run, and it must be at least one more than the number of factors k. The last row of any design is always a row of all −'s. In the designs for $N = 28, 52, 76, 100$, the square blocks are permuted cyclically among themselves; in the three latter cases, the extra column has alternate signs throughout, apart from the corner element. The larger designs are grouped for convenience. These designs have certain optimal properties.

As an example, suppose we have 7 factors. Then we can use a design with $N = 8$ experiments, given in Table 1; note how it is obtained from the design table via cycling on

$$+ + + - + - -$$

and then adding a row of

$$- - - - - - -$$

If we are able to perform more experiments, e.g., $N = 12$, then we will have more accurate estimation of the effects and of the model. One similarly constructs the design as though there were 11 factors, but the columns other than the first seven are ignored, resulting in a set of 12 experiments in 7 factors.

In Section 10 of their paper, Plackett and Burman (1946) gave simple formulas which allow the hand calculation of the analysis after the data has been gathered. However, with the statistical software available today, this method of calculation is not necessary for most users. Instead, one would feed in the values of the x's for each experiment, and the value of Y obtained in that experiment, and would then fit the complete model

$$\beta_0 + \beta_1 x_1 + \beta_2 x_2 + \ldots + \beta_k x_k.$$

Significance of each of the main effects can then be assessed via the customary tests for significance of coefficients in a regression model. This will, in many cases, allow one to substantially reduce the number k of factors to be studied in further experimentation (the others being carefully controlled, not allowed to run free).

Fractional Factorial Designs. When running all 2^k experiments of a factorial design would involve more experiments than can be afforded, or when it is believed that suitable inferences can be made with fewer experiments (e.g., because many interactions are known not to exist), often what is called a "fraction" of a factorial design is used, denoted 2^{k-p} for some integer p (at least 1).

For example, for $k=5$ a full factorial experiment would have $2^5=32$ experiments. The following 2^{5-1} design has 16 experiments and is called a "half-replicate" of the 2^5 experiment:

Treatment combination	Observations Y
(1)	24.62
a(e)	49.08
b(e)	10.51
ab	28.89
c(e)	11.71
ac	44.97
bc	13.05
abc(e)	25.14
d(e)	19.19
ad	60.18
bd	15.69
abd(e)	29.72
cd	26.43
acd(e)	43.26
bcd(e)	14.37
abcd	34.73

This design has been constructed using the so-called "defining contrast" $I=ABDCE$; this will be discussed in more detail later.

Analysis from this point **can proceed in two ways. Either one can use an analysis of variance program (or work by hand, e.g., with Yates' method)**, resulting in an analysis or variance table and its interpretation, **or one can use regression programs**. With the first alternative, we find Tables 2 and 3.

Table 3. Analysis of Variance

Effect*	Degrees of Freedom	Mean Squares
B	1	*720.12*
C	1	36.66
A	1	*2034.01*
E	1	129.85
D	1	79.21
BC	1	53.22
AB	1	*160.53*
BE	1	25.86
BD	1	.19
AC	1	14.67
CE	1	.91
CD	1	9.36
AE	1	.37
AD	1	1.01
DE	1	14.82
Total	15	3280.79

*The aliases of the main effects and two-factor interactions are, respectively, four-factor and three-factor interactions which are regarded as negligible.

Table 2. Analysis of *Y* by Yates' Method

Tr. comb.	Obs.	(1)	(2)	(3)	(4)	Mean effect (4)/8	SS $(4)^2/16$	Effects measured: Aliases
(1)	24.62	73.70	113.10	207.97	451.54	= Total	———	———
a(e)	49.08	39.40	94.87	243.57	180.40	22.55	2034.01	*A, −BCDE*
b(e)	10.51	56.68	124.78	88.19	−107.34	−13.42	720.12	*B, −ACDE*
ab	28.89	38.19	118.79	92.21	−50.68	−6.34	160.53	*AB, −CDE*
c(e)	11.71	79.37	42.84	−52.79	−24.22	−3.03	36.66	*C, −ABDE*
ac	44.97	45.41	45.35	−54.55	−15.32	−1.92	14.67	*AC, −BDE*
bc	13.05	69.69	55.02	−27.25	29.18	3.65	53.22	*BC, −ADE*
abc(e)	25.14	49.10	37.19	−23.43	15.40	1.93	14.82	*ABC, −DE*
d(e)	19.19	24.46	−34.30	−18.23	35.60	4.45	79.21	*D, −ABCE*
ad	60.18	18.38	−18.49	−5.99	4.02	.50	1.01	*AD, −BCE*
bd	15.69	33.26	−33.96	2.51	−1.76	−.22	.19	*BD, −ACE*
abd(e)	29.72	12.09	−20.59	−17.83	3.82	.48	.91	*ABD, −CE*
cd	26.43	40.99	−6.08	15.81	12.24	1.53	9.36	*CD, −ABE*
acd(e)	43.26	14.03	−21.17	13.37	−20.34	−2.54	25.86	*ACD, −BE*
bcd(e)	14.37	16.83	−26.96	−15.09	−2.44	−.31	.37	*BCD −AE*
abcd	34.73	20.36	3.53	30.49	45.58	5.70	129.85	*ABCD, −E*
		Total sum of squares					3280.79	

$A=x_1$, $B=x_2$, $C=x_3$, $D=x_4$, $E=x_5$

From this analysis we see that, in the experimental range used (which we have not given), Y is strongly related (in order) to $A(x_1)$, to $B(x_2)$, and (to a lesser extent) to their interaction. The effect of x_1 is positive: as x_1 increases, Y increases. The effect of x_2 is negative: as x_2 decreases, Y increases. (See the "mean effect" column in Table 2 for these last inferences on sign.)

Since only two variables are involved here, we may ignore the others and compute Table 4 of means:

Table 4. Interaction of x_1 with x_2

A / B	$x_1 = 10.5$	$x_1 = 20.0$
$x_2 = 68$	13.41	29.62
$x_2 = 40$	20.49	49.37

Since the difference in Y gain due to x_1 at high x_2 (29.62 − 13.41 = 16.21), and the difference in Y gain due to x_1 at low x_2 (49.37 − 20.49 = 28.88) differ, we say **interaction** is present; but it is mild in amount here. (Similar analyses could be performed on any other variables which might be measured in addition to y.)

Construction of Fractional Factorial Designs. There are rules for the construction of fractional factorial designs. However, while those rules are easy enough to apply when the number of variables is not more than 5 or 6, they become cumbersome in the cases of real interest (7 to 10 or more variables), where the basic 2^k design one must deal with in the design construction has many basic experiments in its list (e.g., 1024 if $k = 10$). For those who desire to use such hand construction, we now give **a list of the 2^k for k up to 10, and a computer program for generating this list.** We then give **a program which will, for any p defining contrasts, develop a 2^{k-p} fraction** (assuming the defining contrasts are independent).

In modern experimental design, the real interest of these designs for large k is in many ways due to the fact that they are the basis for Central Composite Designs of usable size (i.e., relatively small numbers of experiments).

FORTRAN Program for Printing of 2^k Designs (Full Factorial)

```
0001             DIMENSION NX(1024,10),NY(1024,10)
0002             IN=2
0003        11   I=1                                   }STARTS WITH 2
0004             NX(1,1)=0                              FACTORS (IN=2)
0005             NX(2,1)=1
0006        20   IF( I .EQ. IN) GO TO 100
0007             I=I+1
0008             I1=I-1
0009             I2=2**I1
0010             DO 30 J=1,I2
0011             J1=2*J-1
0012             J2=2*J
0013             DO 40 K=1,I1
0014             NY(J1,K)=NX(J,K)
0015             NY(J2,K)=NX(J,K)
0016        40   CONTINUE
0017             NY(J1,I)=0
0018             NY(J2,I)=1
0019        30   CONTINUE
0020             I3=2**I
0021             DO 60 L=1,I3
0022             DO 70  M=1,I
0023             NX(L,M)= NY(L,M)
0024        70   CONTINUE
0025        60   CONTINUE
0026             GO TO 20
0027       100   WRITE(6,300) IN
0028       300   FORMAT('1',64X,'IN=',I2,//)
0029             DO 80 L1=1,I3
0030             WRITE(6,200) (NX(L1,M1),M1=1,IN)
0031       200   FORMAT(' ',T44,10(I1,2X))
0032        80   CONTINUE
0033             IN=IN+1
0034             IF(IN .LE. 10) GO TO 11               }ENDS WITH 10
0035             RETURN                                 FACTORS (IN=10)
0036             END
```

The design for 2 factors, say x_1 and x_2 with values (low and high) of 80 and 90, and 3.2 and 4.6, respectively, would look like

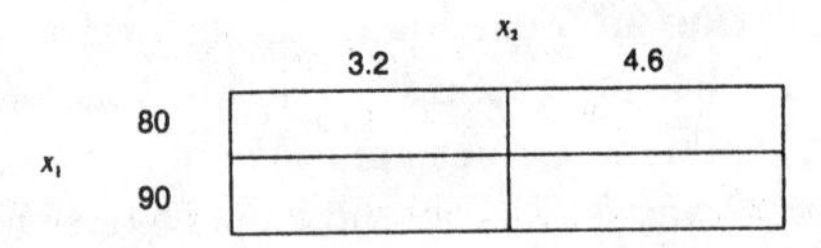

Running an experiment at "0 0" means at $x_1 = 80$ with $x_2 = 3.2$ (low level each), perhaps (as in Table 4) yielding a value of 20.49. After the experiment, we have Table 5.

Table 5. Data

x_1 \ x_2	3.2	4.6
80	20.49	13.41
90	49.37	29.62

and can fit a model such as

$$E(Y_{ij}) = \mu + \alpha_i + \beta_j + \gamma_{ij} \quad (i,j = 0,1).$$

Since (in the usual model)

$$\alpha_0 + \alpha_1 = 0,\ \beta_0 + \beta_1 = 0,\ \gamma_{00} + \gamma_{01} = \gamma_{10} + \gamma_{11} = \gamma_{00} + \gamma_{10} = \gamma_{01} + \gamma_{11} = 0,$$

this model has cell means

x_1 \ x_2	$j = 0$	$j = 1$
$i = 0$	$\mu - \alpha_1 - \beta_1 + \gamma_{11}$	$\mu - \alpha_1 + \beta_1 - \gamma_{11}$
$i = 1$	$\mu + \alpha_1 - \beta_1 - \gamma_{11}$	$\mu + \alpha_1 + \beta_1 + \gamma_{11}$

and we find the estimates

$$\mu = 28.2225,\ \alpha_1 = 11.2725$$
$$\beta_1 = -6.7075$$
$$\gamma_{11} = -3.1675$$

Sum of all 4 cell means $= 4\mu$, so estimate μ by

$$\frac{Y_{00} + Y_{01} + Y_{10} + Y_{11}}{4}$$

	$j=0$	$j=1$
$i=0$	Y_{00}	Y_{01}
$i=1$	Y_{10}	Y_{11}

$$EY_{10} - EY_{00} = 2\alpha_1 - 2\gamma_{11}$$
$$EY_{11} - EY_{01} = 2\alpha_1 + 2\gamma_{11}$$

so

$$E(Y_{10} - Y_{00}) + E(Y_{11} - Y_{01}) = 4\alpha_1$$

and estimate α_1 by

$$\frac{Y_{10} - Y_{00} + Y_{11} - Y_{01}}{4} = 11.2725.$$

Similarly estimate β_1 by

$$\frac{(Y_{01} - Y_{00}) + (Y_{11} - Y_{10})}{4}$$

and γ_{11} by $\dfrac{-(Y_{01} - Y_{00}) + (Y_{11} - Y_{10})}{4}$.

It can easily be verified that the cells then have observations as in Table 5 (no "degrees of freedom" remain so σ^2 cannot be estimated).

Now, if we delete one (or more) of the observations, we will not be able to estimate all of the parameters. (Since with all of the observations we were just able to estimate all of the parameters above, it is clear that with one or more observations missing we will not be able to estimate all of the parameters.) If, in advance of experimentation, we delete observations in such a way that the parameters which cannot be estimated are ones which we have reason to believe are 0 anyway, then little will be lost, and the experimentation will be less costly.

Recall that each experiment (treatment combination) has a name corresponding to the factors in it which are at their "high" levels. Thus, the above experiment has the four experiments (denoting x_1 as factor A and x_2 as factor B)

$x_1 = 80$, $x_2 = 3.2$, or (1)
$x_1 = 90$, $x_2 = 3.2$, or a
$x_1 = 80$, $x_2 = 4.6$, or b
$x_1 = 90$, $x_2 = 4.6$, or ab.

To estimate the main effect of factor A (which is 2 times the α_1 defined earlier; i.e., the main effect is usually taken as the change in going from the low level to the high level), a sum of the observations (now denoted (1), a, b, ab) is used, with a sign on each, i.e., a + if a appears in the expression for the treatment combination, and a − otherwise. **For a general effect in a general factorial experiment, we estimate the effect by the "Even versus Odd Rule,"** i.e. put a + sign on those treatment combinations which have an even number of letters in common with the effect to be estimated, and put a − sign on those which have an odd number of letters in common with the effect to be estimated. (However, for the main effects, add all treatment combinations with the corresponding letter and subtract the rest.) Thus, we estimate

A by $-(1) + a + b + ab$
B by $-(1) - a + b + ab$, and
AB by $(1) - a - b + ab$.

(*Note* that these are the same as before, except for the divisor, which is 2 here rather than $4 = 2^2$. In a general 2^k factorial experiment, the divisor 2^{k-1} is used.)

This can be put in a table as follows:

Treatment Combination	Factorial Effects A	B	AB
(1)	−	−	+
a	+	−	−
b	−	+	−
ab	+	+	+

Now suppose we do not wish to estimate the interaction effect AB (e.g., because we know it is zero, and that the effects of the factors add with no interaction). Then we make AB the **defining contrast** and seek to eliminate experiments so that the ones remaining allow the estimation of the other parameters. We will end up with a **½ fraction** if we use one defining contrast. To do this, construct the table just presented. Then, split the experiments into those where the defining contrast has a + sign, and those where it has a − sign. Either of these halves may be used as the experiment, and we will have as our 2^{2-1} experiments

either (*1*) and *ab*
or *a* and *b*.

However, since there were before 2^2 effects estimated using 2^2 observations, now we can estimate only 2 with 2 observations, and **those remaining are CONFOUNDED WITH OTHERS**. If we write our defining contrast as

$$I = AB,$$

then by multiplying using the rule that a squared letter equals I (and $I^2 = I$) we find the **alias structure** is $A = B$. Thus, the fraction of a 2^2 is useless, since the effects of factors A and B are confounded with each other (and cannot be separated from each other). **Therefore a fraction of a 2^2 experiment is never used in practice.**

With a 2^3 experiment, the situation is different, and a suitable fraction (one with an aliasing structure which is usable) **does exist**. Here we construct the table

Treatment Combination	Factorial Effects A	B	C	AB	AC	BC	ABC
(1)	−	−	−	+	+	+	−
a	+	−	−	−	−	+	+
b	−	+	−	−	+	−	+
c	−	−	+	+	−	−	+
ab	+	+	−	+	−	−	−
ac	+	−	+	−	+	−	−
bc	−	+	+	−	−	+	−
abc	+	+	+	+	+	+	+

Since the highest-way interaction is that most likely to be zero, we may make it our defining contrast. We then find that we will run the $2^{3-1} = 4$ experiments

either (*1*), *ab*, *ac*, *bc*
or *a*, *b*, *c*, *abc*.

The confounding (aliasing) which will exist is determined as before, and is

$$I = ABC$$
$$A = BC$$
$$B = AC$$
$$C = AB.$$

Thus, no main effects are confounded with each other, and the design may be used to estimate main effects in the absence of (large) interactions. On the other hand, if we have reason to believe that items are confounded which both do exist, then this will not be a suitable choice of a fraction.

Higher fractions. To find a higher fraction, e.g. a 2^{k-p} design, **one specifies p defining contrasts** (which must be independent, else they will not cut the design appropriately) and uses each in turn to cut in half the design points which are left; thus they are cut in half p times for 2^{k-p} points at the end of the process. This is most conveniently done with a computer program, such as that given below (for which non-independence will yield a number of design points greater than that desired, as a cue to reexamine the defining contrast set).

```
IN= 2

0 0
0 1
1 0
1 1

IN= 3

0 0 0
0 0 1
0 1 0
0 1 1
1 0 0
1 0 1
1 1 0
1 1 1

IN= 4

0 0 0 0
0 0 0 1
0 0 1 0
0 0 1 1
0 1 0 0
0 1 0 1
0 1 1 0
0 1 1 1
1 0 0 0
1 0 0 1
1 0 1 0
1 0 1 1
1 1 0 0
1 1 0 1
1 1 1 0
1 1 1 1
```

IN = 5

0	0	0	0	0
0	0	0	0	1
0	0	0	1	0
0	0	0	1	1
0	0	1	0	0
0	0	1	0	1
0	0	1	1	0
0	0	1	1	1
0	1	0	0	0
0	1	0	0	1
0	1	0	1	0
0	1	0	1	1
0	1	1	0	0
0	1	1	0	1
0	1	1	1	0
0	1	1	1	1
1	0	0	0	0
1	0	0	0	1
1	0	0	1	0
1	0	0	1	1
1	0	1	0	0
1	0	1	0	1
1	0	1	1	0
1	0	1	1	1
1	1	0	0	0
1	1	0	0	1
1	1	0	1	0
1	1	0	1	1
1	1	1	0	0
1	1	1	0	1
1	1	1	1	0
1	1	1	1	1

IN = 6

0	0	0	0	0	0
0	0	0	0	0	1
0	0	0	0	1	0
0	0	0	0	1	1
0	0	0	1	0	0
0	0	0	1	0	1
0	0	0	1	1	0
0	0	0	1	1	1
0	0	1	0	0	0
0	0	1	0	0	1
0	0	1	0	1	0
0	0	1	0	1	1
0	0	1	1	0	0
0	0	1	1	0	1
0	0	1	1	1	0
0	0	1	1	1	1
0	1	0	0	0	0
0	1	0	0	0	1
0	1	0	0	1	0
0	1	0	0	1	1
0	1	0	1	0	0
0	1	0	1	0	1
0	1	0	1	1	0
0	1	0	1	1	1
0	1	1	0	0	0
0	1	1	0	0	1

0	1	1	0	1	0
0	1	1	0	1	1
0	1	1	1	0	0
0	1	1	1	0	1
0	1	1	1	1	0
0	1	1	1	1	1
1	0	0	0	0	0
1	0	0	0	0	1
1	0	0	0	1	0
1	0	0	0	1	1
1	0	0	1	0	0
1	0	0	1	0	1
1	0	0	1	1	0
1	0	0	1	1	1
1	0	1	0	0	0
1	0	1	0	0	1
1	0	1	0	1	0
1	0	1	0	1	1
1	0	1	1	0	0
1	0	1	1	0	1
1	0	1	1	1	0
1	0	1	1	1	1
1	1	0	0	0	0
1	1	0	0	0	1
1	1	0	0	1	0
1	1	0	0	1	1
1	1	0	1	0	0
1	1	0	1	0	1
1	1	0	1	1	0
1	1	0	1	1	1
1	1	1	0	0	0
1	1	1	0	0	1
1	1	1	0	1	0
1	1	1	0	1	1
1	1	1	1	0	0
1	1	1	1	0	1
1	1	1	1	1	0
1	1	1	1	1	1

IN = 7

0	0	0	0	0	0	0
0	0	0	0	0	0	1
0	0	0	0	0	1	0
0	0	0	0	0	1	1
0	0	0	0	1	0	0
0	0	0	0	1	0	1
0	0	0	0	1	1	0
0	0	0	0	1	1	1
0	0	0	1	0	0	0
0	0	0	1	0	0	1
0	0	0	1	0	1	0
0	0	0	1	0	1	1
0	0	0	1	1	0	0
0	0	0	1	1	0	1
0	0	0	1	1	1	0
0	0	0	1	1	1	1
0	0	1	0	0	0	0
0	0	1	0	0	0	1
0	0	1	0	0	1	0
0	0	1	0	0	1	1
0	0	1	0	1	0	0
0	0	1	0	1	0	1
0	0	1	0	1	1	0
0	0	1	0	1	1	1
0	0	1	1	0	0	0
0	0	1	1	0	0	1
0	0	1	1	0	1	0
0	0	1	1	0	1	1
0	0	1	1	1	0	0
0	0	1	1	1	0	1
0	0	1	1	1	1	0
0	0	1	1	1	1	1
0	1	0	0	0	0	0
0	1	0	0	0	0	1
0	1	0	0	0	1	0
0	1	0	0	0	1	1

0	1	0	0	1	0	0
0	1	0	0	1	0	1
0	1	0	0	1	1	0
0	1	0	0	1	1	1
0	1	0	1	0	0	0
0	1	0	1	0	0	1
0	1	0	1	0	1	0
0	1	0	1	0	1	1
0	1	0	1	1	0	0
0	1	0	1	1	0	1
0	1	0	1	1	1	0
0	1	0	1	1	1	1
0	1	1	0	0	0	0
0	1	1	0	0	0	1
0	1	1	0	0	1	0
0	1	1	0	0	1	1
0	1	1	0	1	0	0
0	1	1	0	1	0	1
0	1	1	0	1	1	0
0	1	1	0	1	1	1
0	1	1	1	0	0	0
0	1	1	1	0	0	1
0	1	1	1	0	1	0
0	1	1	1	0	1	1
0	1	1	1	1	0	0
0	1	1	1	1	0	1
0	1	1	1	1	1	0
0	1	1	1	1	1	1
1	0	0	0	0	0	0
1	0	0	0	0	0	1
1	0	0	0	0	1	0
1	0	0	0	0	1	1
1	0	0	0	1	0	0
1	0	0	0	1	0	1
1	0	0	0	1	1	0
1	0	0	0	1	1	1
1	0	0	1	0	0	0
1	0	0	1	0	0	1
1	0	0	1	0	1	0
1	0	0	1	0	1	1
1	0	0	1	1	0	0
1	0	0	1	1	0	1
1	0	0	1	1	1	0
1	0	0	1	1	1	1
1	0	1	0	0	0	0
1	0	1	0	0	0	1
1	0	1	0	0	1	0
1	0	1	0	0	1	1
1	0	1	0	1	0	0
1	0	1	0	1	0	1
1	0	1	0	1	1	0
1	0	1	0	1	1	1
1	0	1	1	0	0	0
1	0	1	1	0	0	1
1	0	1	1	0	1	0
1	0	1	1	0	1	1
1	0	1	1	1	0	0
1	0	1	1	1	0	1
1	0	1	1	1	1	0
1	0	1	1	1	1	1
1	1	0	0	0	0	0
1	1	0	0	0	0	1
1	1	0	0	0	1	0
1	1	0	0	0	1	1
1	1	0	0	1	0	0
1	1	0	0	1	0	1
1	1	0	0	1	1	0
1	1	0	0	1	1	1
1	1	0	1	0	0	0
1	1	0	1	0	0	1
1	1	0	1	0	1	0
1	1	0	1	0	1	1
1	1	0	1	1	0	0
1	1	0	1	1	0	1
1	1	0	1	1	1	0
1	1	0	1	1	1	1
1	1	1	0	0	0	0
1	1	1	0	0	0	1
1	1	1	0	0	1	0
1	1	1	0	0	1	1

1	1	1	0	1	0	0
1	1	1	0	1	0	1
1	1	1	0	1	1	0
1	1	1	0	1	1	1
1	1	1	1	0	0	0
1	1	1	1	0	0	1
1	1	1	1	0	1	0
1	1	1	1	0	1	1
1	1	1	1	1	0	0
1	1	1	1	1	0	1
1	1	1	1	1	1	0
1	1	1	1	1	1	1

IN = 8

0	0	0	0	0	0	0	0
0	0	0	0	0	0	0	1
0	0	0	0	0	0	1	0
0	0	0	0	0	0	1	1
0	0	0	0	0	1	0	0
0	0	0	0	0	1	0	1
0	0	0	0	0	1	1	0
0	0	0	0	0	1	1	1
0	0	0	0	1	0	0	0
0	0	0	0	1	0	0	1
0	0	0	0	1	0	1	0
0	0	0	0	1	0	1	1
0	0	0	0	1	1	0	0
0	0	0	0	1	1	0	1
0	0	0	0	1	1	1	0
0	0	0	0	1	1	1	1
0	0	0	1	0	0	0	0
0	0	0	1	0	0	0	1
0	0	0	1	0	0	1	0
0	0	0	1	0	0	1	1
0	0	0	1	0	1	0	0
0	0	0	1	0	1	0	1
0	0	0	1	0	1	1	0
0	0	0	1	0	1	1	1
0	0	0	1	1	0	0	0
0	0	0	1	1	0	0	1
0	0	0	1	1	0	1	0
0	0	0	1	1	0	1	1
0	0	0	1	1	1	0	0
0	0	0	1	1	1	0	1
0	0	0	1	1	1	1	0
0	0	0	1	1	1	1	1
0	0	1	0	0	0	0	0
0	0	1	0	0	0	0	1
0	0	1	0	0	0	1	0
0	0	1	0	0	0	1	1
0	0	1	0	0	1	0	0
0	0	1	0	0	1	0	1
0	0	1	0	0	1	1	0
0	0	1	0	0	1	1	1
0	0	1	0	1	0	0	0
0	0	1	0	1	0	0	1
0	0	1	0	1	0	1	0
0	0	1	0	1	0	1	1
0	0	1	0	1	1	0	0
0	0	1	0	1	1	0	1
0	0	1	0	1	1	1	0
0	0	1	0	1	1	1	1
0	0	1	1	0	0	0	0
0	0	1	1	0	0	0	1
0	0	1	1	0	0	1	0
0	0	1	1	0	0	1	1
0	0	1	1	0	1	0	0
0	0	1	1	0	1	0	1
0	0	1	1	0	1	1	0
0	0	1	1	0	1	1	1
0	0	1	1	1	0	0	0
0	0	1	1	1	0	0	1
0	0	1	1	1	0	1	0
0	0	1	1	1	0	1	1
0	0	1	1	1	1	0	0
0	0	1	1	1	1	0	1
0	0	1	1	1	1	1	0

0	0	1	1	1	1	1	1
0	1	0	0	0	0	0	0
0	1	0	0	0	0	0	1
0	1	0	0	0	0	1	0
0	1	0	0	0	0	1	1
0	1	0	0	0	1	0	0
0	1	0	0	0	1	0	1
0	1	0	0	0	1	1	0
0	1	0	0	0	1	1	1
0	1	0	0	1	0	0	0
0	1	0	0	1	0	0	1
0	1	0	0	1	0	1	0
0	1	0	0	1	0	1	1
0	1	0	0	1	1	0	0
0	1	0	0	1	1	0	1
0	1	0	0	1	1	1	0
0	1	0	0	1	1	1	1
0	1	0	1	0	0	0	0
0	1	0	1	0	0	0	1
0	1	0	1	0	0	1	0
0	1	0	1	0	0	1	1
0	1	0	1	0	1	0	0
0	1	0	1	0	1	0	1
0	1	0	1	0	1	1	0
0	1	0	1	0	1	1	1
0	1	0	1	1	0	0	0
0	1	0	1	1	0	0	1
0	1	0	1	1	0	1	0
0	1	0	1	1	0	1	1
0	1	0	1	1	1	0	0
0	1	0	1	1	1	0	1
0	1	0	1	1	1	1	0
0	1	0	1	1	1	1	1
0	1	1	0	0	0	0	0
0	1	1	0	0	0	0	1
0	1	1	0	0	0	1	0
0	1	1	0	0	0	1	1
0	1	1	0	0	1	0	0
0	1	1	0	0	1	0	1
0	1	1	0	0	1	1	0
0	1	1	0	0	1	1	1
0	1	1	0	1	0	0	0
0	1	1	0	1	0	0	1
0	1	1	0	1	0	1	0
0	1	1	0	1	0	1	1
0	1	1	0	1	1	0	0
0	1	1	0	1	1	0	1
0	1	1	0	1	1	1	0
0	1	1	0	1	1	1	1
0	1	1	1	0	0	0	0
0	1	1	1	0	0	0	1
0	1	1	1	0	0	1	0
0	1	1	1	0	0	1	1
0	1	1	1	0	1	0	0
0	1	1	1	0	1	0	1
0	1	1	1	0	1	1	0
0	1	1	1	0	1	1	1
0	1	1	1	1	0	0	0
0	1	1	1	1	0	0	1
0	1	1	1	1	0	1	0
0	1	1	1	1	0	1	1
0	1	1	1	1	1	0	0
0	1	1	1	1	1	0	1
0	1	1	1	1	1	1	0
0	1	1	1	1	1	1	1
1	0	0	0	0	0	0	0
1	0	0	0	0	0	0	1
1	0	0	0	0	0	1	0
1	0	0	0	0	0	1	1
1	0	0	0	0	1	0	0
1	0	0	0	0	1	0	1
1	0	0	0	0	1	1	0
1	0	0	0	0	1	1	1
1	0	0	0	1	0	0	0
1	0	0	0	1	0	0	1
1	0	0	0	1	0	1	0
1	0	0	0	1	0	1	1
1	0	0	0	1	1	0	0
1	0	0	0	1	1	0	1
1	0	0	0	1	1	1	0

1	0	0	0	1	1	1	1
1	0	0	1	0	0	0	0
1	0	0	1	0	0	0	1
1	0	0	1	0	0	1	0
1	0	0	1	0	0	1	1
1	0	0	1	0	1	0	0
1	0	0	1	0	1	0	1
1	0	0	1	0	1	1	0
1	0	0	1	0	1	1	1
1	0	0	1	1	0	0	0
1	0	0	1	1	0	0	1
1	0	0	1	1	0	1	0
1	0	0	1	1	0	1	1
1	0	0	1	1	1	0	0
1	0	0	1	1	1	0	1
1	0	0	1	1	1	1	0
1	0	0	1	1	1	1	1
1	0	1	0	0	0	0	0
1	0	1	0	0	0	0	1
1	0	1	0	0	0	1	0
1	0	1	0	0	0	1	1
1	0	1	0	0	1	0	0
1	0	1	0	0	1	0	1
1	0	1	0	0	1	1	0
1	0	1	0	0	1	1	1
1	0	1	0	1	0	0	0
1	0	1	0	1	0	0	1
1	0	1	0	1	0	1	0
1	0	1	0	1	0	1	1
1	0	1	0	1	1	0	0
1	0	1	0	1	1	0	1
1	0	1	0	1	1	1	0
1	0	1	0	1	1	1	1
1	0	1	1	0	0	0	0
1	0	1	1	0	0	0	1
1	0	1	1	0	0	1	0
1	0	1	1	0	0	1	1
1	0	1	1	0	1	0	0
1	0	1	1	0	1	0	1
1	0	1	1	0	1	1	0
1	0	1	1	0	1	1	1
1	0	1	1	1	0	0	0
1	0	1	1	1	0	0	1
1	0	1	1	1	0	1	0
1	0	1	1	1	0	1	1
1	0	1	1	1	1	0	0
1	0	1	1	1	1	0	1
1	0	1	1	1	1	1	0
1	0	1	1	1	1	1	1
1	1	0	0	0	0	0	0
1	1	0	0	0	0	0	1
1	1	0	0	0	0	1	0
1	1	0	0	0	0	1	1
1	1	0	0	0	1	0	0
1	1	0	0	0	1	0	1
1	1	0	0	0	1	1	0
1	1	0	0	0	1	1	1
1	1	0	0	1	0	0	0
1	1	0	0	1	0	0	1
1	1	0	0	1	0	1	0
1	1	0	0	1	0	1	1
1	1	0	0	1	1	0	0
1	1	0	0	1	1	0	1
1	1	0	0	1	1	1	0
1	1	0	0	1	1	1	1
1	1	0	1	0	0	0	0
1	1	0	1	0	0	0	1
1	1	0	1	0	0	1	0
1	1	0	1	0	0	1	1
1	1	0	1	0	1	0	0
1	1	0	1	0	1	0	1
1	1	0	1	0	1	1	0
1	1	0	1	0	1	1	1
1	1	0	1	1	0	0	0
1	1	0	1	1	0	0	1
1	1	0	1	1	0	1	0
1	1	0	1	1	0	1	1
1	1	0	1	1	1	0	0
1	1	0	1	1	1	0	1
1	1	0	1	1	1	1	0

1 1 0 1 1 1 1 1
1 1 1 0 0 0 0 0
1 1 1 0 0 0 0 1
1 1 1 0 0 0 1 0
1 1 1 0 0 0 1 1
1 1 1 0 0 1 0 0
1 1 1 0 0 1 0 1
1 1 1 0 0 1 1 0
1 1 1 0 0 1 1 1
1 1 1 0 1 0 0 0
1 1 1 0 1 0 0 1
1 1 1 0 1 0 1 0
1 1 1 0 1 0 1 1
1 1 1 0 1 1 0 0
1 1 1 0 1 1 0 1
1 1 1 0 1 1 1 0
1 1 1 0 1 1 1 1
1 1 1 1 0 0 0 0
1 1 1 1 0 0 0 1
1 1 1 1 0 0 1 0
1 1 1 1 0 0 1 1
1 1 1 1 0 1 0 0
1 1 1 1 0 1 0 1
1 1 1 1 0 1 1 0
1 1 1 1 0 1 1 1
1 1 1 1 1 0 0 0
1 1 1 1 1 0 0 1
1 1 1 1 1 0 1 0
1 1 1 1 1 0 1 1
1 1 1 1 1 1 0 0
1 1 1 1 1 1 0 1
1 1 1 1 1 1 1 0
1 1 1 1 1 1 1 1

IN = 9

0 0 0 0 0 0 0 0 0
0 0 0 0 0 0 0 0 1
0 0 0 0 0 0 0 1 0
0 0 0 0 0 0 0 1 1
0 0 0 0 0 0 1 0 0
0 0 0 0 0 0 1 0 1
0 0 0 0 0 0 1 1 0
0 0 0 0 0 0 1 1 1
0 0 0 0 0 1 0 0 0
0 0 0 0 0 1 0 0 1
0 0 0 0 0 1 0 1 0
0 0 0 0 0 1 0 1 1
0 0 0 0 0 1 1 0 0
0 0 0 0 0 1 1 0 1
0 0 0 0 0 1 1 1 0
0 0 0 0 0 1 1 1 1
0 0 0 0 1 0 0 0 0
0 0 0 0 1 0 0 0 1
0 0 0 0 1 0 0 1 0
0 0 0 0 1 0 0 1 1
0 0 0 0 1 0 1 0 0
0 0 0 0 1 0 1 0 1
0 0 0 0 1 0 1 1 0
0 0 0 0 1 0 1 1 1
0 0 0 0 1 1 0 0 0
0 0 0 0 1 1 0 0 1
0 0 0 0 1 1 0 1 0
0 0 0 0 1 1 0 1 1
0 0 0 0 1 1 1 0 0
0 0 0 0 1 1 1 0 1
0 0 0 0 1 1 1 1 0
0 0 0 0 1 1 1 1 1
0 0 0 1 0 0 0 0 0
0 0 0 1 0 0 0 0 1
0 0 0 1 0 0 0 1 0
0 0 0 1 0 0 0 1 1
0 0 0 1 0 0 1 0 0
0 0 0 1 0 0 1 0 1
0 0 0 1 0 0 1 1 0
0 0 0 1 0 0 1 1 1
0 0 0 1 0 1 0 0 0
0 0 0 1 0 1 0 0 1
0 0 0 1 0 1 0 1 0

0 0 0 1 0 1 0 1 1
0 0 0 1 0 1 1 0 0
0 0 0 1 0 1 1 0 1
0 0 0 1 0 1 1 1 0
0 0 0 1 0 1 1 1 1
0 0 0 1 1 0 0 0 0
0 0 0 1 1 0 0 0 1
0 0 0 1 1 0 0 1 0
0 0 0 1 1 0 0 1 1
0 0 0 1 1 0 1 0 0
0 0 0 1 1 0 1 0 1
0 0 0 1 1 0 1 1 0
0 0 0 1 1 0 1 1 1
0 0 0 1 1 1 0 0 0
0 0 0 1 1 1 0 0 1
0 0 0 1 1 1 0 1 0
0 0 0 1 1 1 0 1 1
0 0 0 1 1 1 1 0 0
0 0 0 1 1 1 1 0 1
0 0 0 1 1 1 1 1 0
0 0 0 1 1 1 1 1 1
0 0 1 0 0 0 0 0 0
0 0 1 0 0 0 0 0 1
0 0 1 0 0 0 0 1 0
0 0 1 0 0 0 0 1 1
0 0 1 0 0 0 1 0 0
0 0 1 0 0 0 1 0 1
0 0 1 0 0 0 1 1 0
0 0 1 0 0 0 1 1 1
0 0 1 0 0 1 0 0 0
0 0 1 0 0 1 0 0 1
0 0 1 0 0 1 0 1 0
0 0 1 0 0 1 0 1 1
0 0 1 0 0 1 1 0 0
0 0 1 0 0 1 1 0 1
0 0 1 0 0 1 1 1 0
0 0 1 0 0 1 1 1 1
0 0 1 0 1 0 0 0 0
0 0 1 0 1 0 0 0 1
0 0 1 0 1 0 0 1 0
0 0 1 0 1 0 0 1 1
0 0 1 0 1 0 1 0 0
0 0 1 0 1 0 1 0 1
0 0 1 0 1 0 1 1 0
0 0 1 0 1 0 1 1 1
0 0 1 0 1 1 0 0 0
0 0 1 0 1 1 0 0 1
0 0 1 0 1 1 0 1 0
0 0 1 0 1 1 0 1 1
0 0 1 0 1 1 1 0 0
0 0 1 0 1 1 1 0 1
0 0 1 0 1 1 1 1 0
0 0 1 0 1 1 1 1 1
0 0 1 1 0 0 0 0 0
0 0 1 1 0 0 0 0 1
0 0 1 1 0 0 0 1 0
0 0 1 1 0 0 0 1 1
0 0 1 1 0 0 1 0 0
0 0 1 1 0 0 1 0 1
0 0 1 1 0 0 1 1 0
0 0 1 1 0 0 1 1 1
0 0 1 1 0 1 0 0 0
0 0 1 1 0 1 0 0 1
0 0 1 1 0 1 0 1 0
0 0 1 1 0 1 0 1 1
0 0 1 1 0 1 1 0 0
0 0 1 1 0 1 1 0 1
0 0 1 1 0 1 1 1 0
0 0 1 1 0 1 1 1 1
0 0 1 1 1 0 0 0 0
0 0 1 1 1 0 0 0 1
0 0 1 1 1 0 0 1 0
0 0 1 1 1 0 0 1 1
0 0 1 1 1 0 1 0 0
0 0 1 1 1 0 1 0 1
0 0 1 1 1 0 1 1 0
0 0 1 1 1 0 1 1 1
0 0 1 1 1 1 0 0 0
0 0 1 1 1 1 0 0 1
0 0 1 1 1 1 0 1 0

0	0	1	1	1	1	0	1	1
0	0	1	1	1	1	1	0	0
0	0	1	1	1	1	1	0	1
0	0	1	1	1	1	1	1	0
0	0	1	1	1	1	1	1	1
0	1	0	0	0	0	0	0	0
0	1	0	0	0	0	0	0	1
0	1	0	0	0	0	0	1	0
0	1	0	0	0	0	0	1	1
0	1	0	0	0	0	1	0	0
0	1	0	0	0	0	1	0	1
0	1	0	0	0	0	1	1	0
0	1	0	0	0	0	1	1	1
0	1	0	0	0	1	0	0	0
0	1	0	0	0	1	0	0	1
0	1	0	0	0	1	0	1	0
0	1	0	0	0	1	0	1	1
0	1	0	0	0	1	1	0	0
0	1	0	0	0	1	1	0	1
0	1	0	0	0	1	1	1	0
0	1	0	0	0	1	1	1	1
0	1	0	0	1	0	0	0	0
0	1	0	0	1	0	0	0	1
0	1	0	0	1	0	0	1	0
0	1	0	0	1	0	0	1	1
0	1	0	0	1	0	1	0	0
0	1	0	0	1	0	1	0	1
0	1	0	0	1	0	1	1	0
0	1	0	0	1	0	1	1	1
0	1	0	0	1	1	0	0	0
0	1	0	0	1	1	0	0	1
0	1	0	0	1	1	0	1	0
0	1	0	0	1	1	0	1	1
0	1	0	0	1	1	1	0	0
0	1	0	0	1	1	1	0	1
0	1	0	0	1	1	1	1	0
0	1	0	0	1	1	1	1	1
0	1	0	1	0	0	0	0	0
0	1	0	1	0	0	0	0	1
0	1	0	1	0	0	0	1	0
0	1	0	1	0	0	0	1	1
0	1	0	1	0	0	1	0	0
0	1	0	1	0	0	1	0	1
0	1	0	1	0	0	1	1	0
0	1	0	1	0	0	1	1	1
0	1	0	1	0	1	0	0	0
0	1	0	1	0	1	0	0	1
0	1	0	1	0	1	0	1	0
0	1	0	1	0	1	0	1	1
0	1	0	1	0	1	1	0	0
0	1	0	1	0	1	1	0	1
0	1	0	1	0	1	1	1	0
0	1	0	1	0	1	1	1	1
0	1	0	1	1	0	0	0	0
0	1	0	1	1	0	0	0	1
0	1	0	1	1	0	0	1	0
0	1	0	1	1	0	0	1	1
0	1	0	1	1	0	1	0	0
0	1	0	1	1	0	1	0	1
0	1	0	1	1	0	1	1	0
0	1	0	1	1	0	1	1	1
0	1	0	1	1	1	0	0	0
0	1	0	1	1	1	0	0	1
0	1	0	1	1	1	0	1	0
0	1	0	1	1	1	0	1	1
0	1	0	1	1	1	1	0	0
0	1	0	1	1	1	1	0	1
0	1	0	1	1	1	1	1	0
0	1	0	1	1	1	1	1	1
0	1	1	0	0	0	0	0	0
0	1	1	0	0	0	0	0	1
0	1	1	0	0	0	0	1	0
0	1	1	0	0	0	0	1	1
0	1	1	0	0	0	1	0	0
0	1	1	0	0	0	1	0	1
0	1	1	0	0	0	1	1	0
0	1	1	0	0	0	1	1	1
0	1	1	0	0	1	0	0	0
0	1	1	0	0	1	0	0	1
0	1	1	0	0	1	0	1	0

0	1	1	0	0	1	0	1	1
0	1	1	0	0	1	1	0	0
0	1	1	0	0	1	1	0	1
0	1	1	0	0	1	1	1	0
0	1	1	0	0	1	1	1	1
0	1	1	0	1	0	0	0	0
0	1	1	0	1	0	0	0	1
0	1	1	0	1	0	0	1	0
0	1	1	0	1	0	0	1	1
0	1	1	0	1	0	1	0	0
0	1	1	0	1	0	1	0	1
0	1	1	0	1	0	1	1	0
0	1	1	0	1	0	1	1	1
0	1	1	0	1	1	0	0	0
0	1	1	0	1	1	0	0	1
0	1	1	0	1	1	0	1	0
0	1	1	0	1	1	0	1	1
0	1	1	0	1	1	1	0	0
0	1	1	0	1	1	1	0	1
0	1	1	0	1	1	1	1	0
0	1	1	0	1	1	1	1	1
0	1	1	1	0	0	0	0	0
0	1	1	1	0	0	0	0	1
0	1	1	1	0	0	0	1	0
0	1	1	1	0	0	0	1	1
0	1	1	1	0	0	1	0	0
0	1	1	1	0	0	1	0	1
0	1	1	1	0	0	1	1	0
0	1	1	1	0	0	1	1	1
0	1	1	1	0	1	0	0	0
0	1	1	1	0	1	0	0	1
0	1	1	1	0	1	0	1	0
0	1	1	1	0	1	0	1	1
0	1	1	1	0	1	1	0	0
0	1	1	1	0	1	1	0	1
0	1	1	1	0	1	1	1	0
0	1	1	1	0	1	1	1	1
0	1	1	1	1	0	0	0	0
0	1	1	1	1	0	0	0	1
0	1	1	1	1	0	0	1	0
0	1	1	1	1	0	0	1	1
0	1	1	1	1	0	1	0	0
0	1	1	1	1	0	1	0	1
0	1	1	1	1	0	1	1	0
0	1	1	1	1	0	1	1	1
0	1	1	1	1	1	0	0	0
0	1	1	1	1	1	0	0	1
0	1	1	1	1	1	0	1	0
0	1	1	1	1	1	0	1	1
0	1	1	1	1	1	1	0	0
0	1	1	1	1	1	1	0	1
0	1	1	1	1	1	1	1	0
0	1	1	1	1	1	1	1	1
1	0	0	0	0	0	0	0	0
1	0	0	0	0	0	0	0	1
1	0	0	0	0	0	0	1	0
1	0	0	0	0	0	0	1	1
1	0	0	0	0	0	1	0	0
1	0	0	0	0	0	1	0	1
1	0	0	0	0	0	1	1	0
1	0	0	0	0	0	1	1	1
1	0	0	0	0	1	0	0	0
1	0	0	0	0	1	0	0	1
1	0	0	0	0	1	0	1	0
1	0	0	0	0	1	0	1	1
1	0	0	0	0	1	1	0	0
1	0	0	0	0	1	1	0	1
1	0	0	0	0	1	1	1	0
1	0	0	0	0	1	1	1	1
1	0	0	0	1	0	0	0	0
1	0	0	0	1	0	0	0	1
1	0	0	0	1	0	0	1	0
1	0	0	0	1	0	0	1	1
1	0	0	0	1	0	1	0	0
1	0	0	0	1	0	1	0	1
1	0	0	0	1	0	1	1	0
1	0	0	0	1	0	1	1	1
1	0	0	0	1	1	0	0	0
1	0	0	0	1	1	0	0	1
1	0	0	0	1	1	0	1	0

1	0	0	0	1	1	0	1	1
1	0	0	0	1	1	1	0	0
1	0	0	0	1	1	1	0	1
1	0	0	0	1	1	1	1	0
1	0	0	0	1	1	1	1	1
1	0	0	1	0	0	0	0	0
1	0	0	1	0	0	0	0	1
1	0	0	1	0	0	0	1	0
1	0	0	1	0	0	0	1	1
1	0	0	1	0	0	1	0	0
1	0	0	1	0	0	1	0	1
1	0	0	1	0	0	1	1	0
1	0	0	1	0	0	1	1	1
1	0	0	1	0	1	0	0	0
1	0	0	1	0	1	0	0	1
1	0	0	1	0	1	0	1	0
1	0	0	1	0	1	0	1	1
1	0	0	1	0	1	1	0	0
1	0	0	1	0	1	1	0	1
1	0	0	1	0	1	1	1	0
1	0	0	1	0	1	1	1	1
1	0	0	1	1	0	0	0	0
1	0	0	1	1	0	0	0	1
1	0	0	1	1	0	0	1	0
1	0	0	1	1	0	0	1	1
1	0	0	1	1	0	1	0	0
1	0	0	1	1	0	1	0	1
1	0	0	1	1	0	1	1	0
1	0	0	1	1	0	1	1	1
1	0	0	1	1	1	0	0	0
1	0	0	1	1	1	0	0	1
1	0	0	1	1	1	0	1	0
1	0	0	1	1	1	0	1	1
1	0	0	1	1	1	1	0	0
1	0	0	1	1	1	1	0	1
1	0	0	1	1	1	1	1	0
1	0	0	1	1	1	1	1	1
1	0	1	0	0	0	0	0	0
1	0	1	0	0	0	0	0	1
1	0	1	0	0	0	0	1	0
1	0	1	0	0	0	0	1	1
1	0	1	0	0	0	1	0	0
1	0	1	0	0	0	1	0	1
1	0	1	0	0	0	1	1	0
1	0	1	0	0	0	1	1	1
1	0	1	0	0	1	0	0	0
1	0	1	0	0	1	0	0	1
1	0	1	0	0	1	0	1	0
1	0	1	0	0	1	0	1	1
1	0	1	0	0	1	1	0	0
1	0	1	0	0	1	1	0	1
1	0	1	0	0	1	1	1	0
1	0	1	0	0	1	1	1	1
1	0	1	0	1	0	0	0	0
1	0	1	0	1	0	0	0	1
1	0	1	0	1	0	0	1	0
1	0	1	0	1	0	0	1	1
1	0	1	0	1	0	1	0	0
1	0	1	0	1	0	1	0	1
1	0	1	0	1	0	1	1	0
1	0	1	0	1	0	1	1	1
1	0	1	0	1	1	0	0	0
1	0	1	0	1	1	0	0	1
1	0	1	0	1	1	0	1	0
1	0	1	0	1	1	0	1	1
1	0	1	0	1	1	1	0	0
1	0	1	0	1	1	1	0	1
1	0	1	0	1	1	1	1	0
1	0	1	0	1	1	1	1	1
1	0	1	1	0	0	0	0	0
1	0	1	1	0	0	0	0	1
1	0	1	1	0	0	0	1	0
1	0	1	1	0	0	0	1	1
1	0	1	1	0	0	1	0	0
1	0	1	1	0	0	1	0	1
1	0	1	1	0	0	1	1	0
1	0	1	1	0	0	1	1	1
1	0	1	1	0	1	0	0	0
1	0	1	1	0	1	0	0	1
1	0	1	1	0	1	0	1	0

1	0	1	1	0	1	0	1	1
1	0	1	1	0	1	1	0	0
1	0	1	1	0	1	1	0	1
1	0	1	1	0	1	1	1	0
1	0	1	1	0	1	1	1	1
1	0	1	1	1	0	0	0	0
1	0	1	1	1	0	0	0	1
1	0	1	1	1	0	0	1	0
1	0	1	1	1	0	0	1	1
1	0	1	1	1	0	1	0	0
1	0	1	1	1	0	1	0	1
1	0	1	1	1	0	1	1	0
1	0	1	1	1	0	1	1	1
1	0	1	1	1	1	0	0	0
1	0	1	1	1	1	0	0	1
1	0	1	1	1	1	0	1	0
1	0	1	1	1	1	0	1	1
1	0	1	1	1	1	1	0	0
1	0	1	1	1	1	1	0	1
1	0	1	1	1	1	1	1	0
1	0	1	1	1	1	1	1	1
1	1	0	0	0	0	0	0	0
1	1	0	0	0	0	0	0	1
1	1	0	0	0	0	0	1	0
1	1	0	0	0	0	0	1	1
1	1	0	0	0	0	1	0	0
1	1	0	0	0	0	1	0	1
1	1	0	0	0	0	1	1	0
1	1	0	0	0	0	1	1	1
1	1	0	0	0	1	0	0	0
1	1	0	0	0	1	0	0	1
1	1	0	0	0	1	0	1	0
1	1	0	0	0	1	0	1	1
1	1	0	0	0	1	1	0	0
1	1	0	0	0	1	1	0	1
1	1	0	0	0	1	1	1	0
1	1	0	0	0	1	1	1	1
1	1	0	0	1	0	0	0	0
1	1	0	0	1	0	0	0	1
1	1	0	0	1	0	0	1	0
1	1	0	0	1	0	0	1	1
1	1	0	0	1	0	1	0	0
1	1	0	0	1	0	1	0	1
1	1	0	0	1	0	1	1	0
1	1	0	0	1	0	1	1	1
1	1	0	0	1	1	0	0	0
1	1	0	0	1	1	0	0	1
1	1	0	0	1	1	0	1	0
1	1	0	0	1	1	0	1	1
1	1	0	0	1	1	1	0	0
1	1	0	0	1	1	1	0	1
1	1	0	0	1	1	1	1	0
1	1	0	0	1	1	1	1	1
1	1	0	1	0	0	0	0	0
1	1	0	1	0	0	0	0	1
1	1	0	1	0	0	0	1	0
1	1	0	1	0	0	0	1	1
1	1	0	1	0	0	1	0	0
1	1	0	1	0	0	1	0	1
1	1	0	1	0	0	1	1	0
1	1	0	1	0	0	1	1	1
1	1	0	1	0	1	0	0	0
1	1	0	1	0	1	0	0	1
1	1	0	1	0	1	0	1	0
1	1	0	1	0	1	0	1	1
1	1	0	1	0	1	1	0	0
1	1	0	1	0	1	1	0	1
1	1	0	1	0	1	1	1	0
1	1	0	1	0	1	1	1	1
1	1	0	1	1	0	0	0	0
1	1	0	1	1	0	0	0	1
1	1	0	1	1	0	0	1	0
1	1	0	1	1	0	0	1	1
1	1	0	1	1	0	1	0	0
1	1	0	1	1	0	1	0	1
1	1	0	1	1	0	1	1	0
1	1	0	1	1	0	1	1	1
1	1	0	1	1	1	0	0	0
1	1	0	1	1	1	0	0	1
1	1	0	1	1	1	0	1	0

```
1 1 0 1 1 1 0 1 1
1 1 0 1 1 1 1 0 0
1 1 0 1 1 1 1 0 1
1 1 0 1 1 1 1 1 0
1 1 0 1 1 1 1 1 1
1 1 1 0 0 0 0 0 0
1 1 1 0 0 0 0 0 1
1 1 1 0 0 0 0 1 0
1 1 1 0 0 0 0 1 1
1 1 1 0 0 0 1 0 0
1 1 1 0 0 0 1 0 1
1 1 1 0 0 0 1 1 0
1 1 1 0 0 0 1 1 1
1 1 1 0 0 1 0 0 0
1 1 1 0 0 1 0 0 1
1 1 1 0 0 1 0 1 0
1 1 1 0 0 1 0 1 1
1 1 1 0 0 1 1 0 0
1 1 1 0 0 1 1 0 1
1 1 1 0 0 1 1 1 0
1 1 1 0 0 1 1 1 1
1 1 1 0 1 0 0 0 0
1 1 1 0 1 0 0 0 1
1 1 1 0 1 0 0 1 0
1 1 1 0 1 0 0 1 1
1 1 1 0 1 0 1 0 0
1 1 1 0 1 0 1 0 1
1 1 1 0 1 0 1 1 0
1 1 1 0 1 0 1 1 1
1 1 1 0 1 1 0 0 0
1 1 1 0 1 1 0 0 1
1 1 1 0 1 1 0 1 0
1 1 1 0 1 1 0 1 1
1 1 1 0 1 1 1 0 0
1 1 1 0 1 1 1 0 1
1 1 1 0 1 1 1 1 0
1 1 1 0 1 1 1 1 1
1 1 1 1 0 0 0 0 0
1 1 1 1 0 0 0 0 1
1 1 1 1 0 0 0 1 0
1 1 1 1 0 0 0 1 1
1 1 1 1 0 0 1 0 0
1 1 1 1 0 0 1 0 1
1 1 1 1 0 0 1 1 0
1 1 1 1 0 0 1 1 1
1 1 1 1 0 1 0 0 0
1 1 1 1 0 1 0 0 1
1 1 1 1 0 1 0 1 0
1 1 1 1 0 1 0 1 1
1 1 1 1 0 1 1 0 0
1 1 1 1 0 1 1 0 1
1 1 1 1 0 1 1 1 0
1 1 1 1 0 1 1 1 1
1 1 1 1 1 0 0 0 0
1 1 1 1 1 0 0 0 1
1 1 1 1 1 0 0 1 0
1 1 1 1 1 0 0 1 1
1 1 1 1 1 0 1 0 0
1 1 1 1 1 0 1 0 1
1 1 1 1 1 0 1 1 0
1 1 1 1 1 0 1 1 1
1 1 1 1 1 1 0 0 0
1 1 1 1 1 1 0 0 1
1 1 1 1 1 1 0 1 0
1 1 1 1 1 1 0 1 1
1 1 1 1 1 1 1 0 0
1 1 1 1 1 1 1 0 1
1 1 1 1 1 1 1 1 0
1 1 1 1 1 1 1 1 1
```

IN = 10

```
0 0 0 0 0 0 0 0 0 0
0 0 0 0 0 0 0 0 0 1
0 0 0 0 0 0 0 0 1 0
0 0 0 0 0 0 0 0 1 1
0 0 0 0 0 0 0 1 0 0
0 0 0 0 0 0 0 1 0 1
0 0 0 0 0 0 0 1 1 0
```

```
0 0 0 0 0 0 0 1 1 1
0 0 0 0 0 0 1 0 0 0
0 0 0 0 0 0 1 0 0 1
0 0 0 0 0 0 1 0 1 0
0 0 0 0 0 0 1 0 1 1
0 0 0 0 0 0 1 1 0 0
0 0 0 0 0 0 1 1 0 1
0 0 0 0 0 0 1 1 1 0
0 0 0 0 0 0 1 1 1 1
0 0 0 0 0 1 0 0 0 0
0 0 0 0 0 1 0 0 0 1
0 0 0 0 0 1 0 0 1 0
0 0 0 0 0 1 0 0 1 1
0 0 0 0 0 1 0 1 0 0
0 0 0 0 0 1 0 1 0 1
0 0 0 0 0 1 0 1 1 0
0 0 0 0 0 1 0 1 1 1
0 0 0 0 0 1 1 0 0 0
0 0 0 0 0 1 1 0 0 1
0 0 0 0 0 1 1 0 1 0
0 0 0 0 0 1 1 0 1 1
0 0 0 0 0 1 1 1 0 0
0 0 0 0 0 1 1 1 0 1
0 0 0 0 0 1 1 1 1 0
0 0 0 0 0 1 1 1 1 1
0 0 0 0 1 0 0 0 0 0
0 0 0 0 1 0 0 0 0 1
0 0 0 0 1 0 0 0 1 0
0 0 0 0 1 0 0 0 1 1
0 0 0 0 1 0 0 1 0 0
0 0 0 0 1 0 0 1 0 1
0 0 0 0 1 0 0 1 1 0
0 0 0 0 1 0 0 1 1 1
0 0 0 0 1 0 1 0 0 0
0 0 0 0 1 0 1 0 0 1
0 0 0 0 1 0 1 0 1 0
0 0 0 0 1 0 1 0 1 1
0 0 0 0 1 0 1 1 0 0
0 0 0 0 1 0 1 1 0 1
0 0 0 0 1 0 1 1 1 0
0 0 0 0 1 0 1 1 1 1
0 0 0 0 1 1 0 0 0 0
0 0 0 0 1 1 0 0 0 1
0 0 0 0 1 1 0 0 1 0
0 0 0 0 1 1 0 0 1 1
0 0 0 0 1 1 0 1 0 0
0 0 0 0 1 1 0 1 0 1
0 0 0 0 1 1 0 1 1 0
0 0 0 0 1 1 0 1 1 1
0 0 0 0 1 1 1 0 0 0
0 0 0 0 1 1 1 0 0 1
0 0 0 0 1 1 1 0 1 0
0 0 0 0 1 1 1 0 1 1
0 0 0 0 1 1 1 1 0 0
0 0 0 0 1 1 1 1 0 1
0 0 0 0 1 1 1 1 1 0
0 0 0 0 1 1 1 1 1 1
0 0 0 1 0 0 0 0 0 0
0 0 0 1 0 0 0 0 0 1
0 0 0 1 0 0 0 0 1 0
0 0 0 1 0 0 0 0 1 1
0 0 0 1 0 0 0 1 0 0
0 0 0 1 0 0 0 1 0 1
0 0 0 1 0 0 0 1 1 0
0 0 0 1 0 0 0 1 1 1
0 0 0 1 0 0 1 0 0 0
0 0 0 1 0 0 1 0 0 1
0 0 0 1 0 0 1 0 1 0
0 0 0 1 0 0 1 0 1 1
0 0 0 1 0 0 1 1 0 0
0 0 0 1 0 0 1 1 0 1
0 0 0 1 0 0 1 1 1 0
0 0 0 1 0 0 1 1 1 1
0 0 0 1 0 1 0 0 0 0
0 0 0 1 0 1 0 0 0 1
0 0 0 1 0 1 0 0 1 0
0 0 0 1 0 1 0 0 1 1
0 0 0 1 0 1 0 1 0 0
0 0 0 1 0 1 0 1 0 1
```

0 0 0 1 0 1 0 1 1 0
0 0 0 1 0 1 0 1 1 1
0 0 0 1 0 1 1 0 0 0
0 0 0 1 0 1 1 0 0 1
0 0 0 1 0 1 1 0 1 0
0 0 0 1 0 1 1 0 1 1
0 0 0 1 0 1 1 1 0 0
0 0 0 1 0 1 1 1 0 1
0 0 0 1 0 1 1 1 1 0
0 0 0 1 0 1 1 1 1 1
0 0 0 1 1 0 0 0 0 0
0 0 0 1 1 0 0 0 0 1
0 0 0 1 1 0 0 0 1 0
0 0 0 1 1 0 0 0 1 1
0 0 0 1 1 0 0 1 0 0
0 0 0 1 1 0 0 1 0 1
0 0 0 1 1 0 0 1 1 0
0 0 0 1 1 0 0 1 1 1
0 0 0 1 1 0 1 0 0 0
0 0 0 1 1 0 1 0 0 1
0 0 0 1 1 0 1 0 1 0
0 0 0 1 1 0 1 0 1 1
0 0 0 1 1 0 1 1 0 0
0 0 0 1 1 0 1 1 0 1
0 0 0 1 1 0 1 1 1 0
0 0 0 1 1 0 1 1 1 1
0 0 0 1 1 1 0 0 0 0
0 0 0 1 1 1 0 0 0 1
0 0 0 1 1 1 0 0 1 0
0 0 0 1 1 1 0 0 1 1
0 0 0 1 1 1 0 1 0 0
0 0 0 1 1 1 0 1 0 1
0 0 0 1 1 1 0 1 1 0
0 0 0 1 1 1 0 1 1 1
0 0 0 1 1 1 1 0 0 0
0 0 0 1 1 1 1 0 0 1
0 0 0 1 1 1 1 0 1 0
0 0 0 1 1 1 1 0 1 1
0 0 0 1 1 1 1 1 0 0
0 0 0 1 1 1 1 1 0 1
0 0 0 1 1 1 1 1 1 0
0 0 0 1 1 1 1 1 1 1
0 0 1 0 0 0 0 0 0 0
0 0 1 0 0 0 0 0 0 1
0 0 1 0 0 0 0 0 1 0
0 0 1 0 0 0 0 0 1 1
0 0 1 0 0 0 0 1 0 0
0 0 1 0 0 0 0 1 0 1
0 0 1 0 0 0 0 1 1 0
0 0 1 0 0 0 0 1 1 1
0 0 1 0 0 0 1 0 0 0
0 0 1 0 0 0 1 0 0 1
0 0 1 0 0 0 1 0 1 0
0 0 1 0 0 0 1 0 1 1
0 0 1 0 0 0 1 1 0 0
0 0 1 0 0 0 1 1 0 1
0 0 1 0 0 0 1 1 1 0
0 0 1 0 0 0 1 1 1 1
0 0 1 0 0 1 0 0 0 0
0 0 1 0 0 1 0 0 0 1
0 0 1 0 0 1 0 0 1 0
0 0 1 0 0 1 0 0 1 1
0 0 1 0 0 1 0 1 0 0
0 0 1 0 0 1 0 1 0 1
0 0 1 0 0 1 0 1 1 0
0 0 1 0 0 1 0 1 1 1
0 0 1 0 0 1 1 0 0 0
0 0 1 0 0 1 1 0 0 1
0 0 1 0 0 1 1 0 1 0
0 0 1 0 0 1 1 0 1 1
0 0 1 0 0 1 1 1 0 0
0 0 1 0 0 1 1 1 0 1
0 0 1 0 0 1 1 1 1 0
0 0 1 0 0 1 1 1 1 1
0 0 1 0 1 0 0 0 0 0
0 0 1 0 1 0 0 0 0 1
0 0 1 0 1 0 0 0 1 0
0 0 1 0 1 0 0 0 1 1
0 0 1 0 1 0 0 1 0 0
0 0 1 0 1 0 0 1 0 1

0 0 1 0 1 0 0 1 1 0
0 0 1 0 1 0 0 1 1 1
0 0 1 0 1 0 1 0 0 0
0 0 1 0 1 0 1 0 0 1
0 0 1 0 1 0 1 0 1 0
0 0 1 0 1 0 1 0 1 1
0 0 1 0 1 0 1 1 0 0
0 0 1 0 1 0 1 1 0 1
0 0 1 0 1 0 1 1 1 0
0 0 1 0 1 0 1 1 1 1
0 0 1 0 1 1 0 0 0 0
0 0 1 0 1 1 0 0 0 1
0 0 1 0 1 1 0 0 1 0
0 0 1 0 1 1 0 0 1 1
0 0 1 0 1 1 0 1 0 0
0 0 1 0 1 1 0 1 0 1
0 0 1 0 1 1 0 1 1 0
0 0 1 0 1 1 0 1 1 1
0 0 1 0 1 1 1 0 0 0
0 0 1 0 1 1 1 0 0 1
0 0 1 0 1 1 1 0 1 0
0 0 1 0 1 1 1 0 1 1
0 0 1 0 1 1 1 1 0 0
0 0 1 0 1 1 1 1 0 1
0 0 1 0 1 1 1 1 1 0
0 0 1 0 1 1 1 1 1 1
0 0 1 1 0 0 0 0 0 0
0 0 1 1 0 0 0 0 0 1
0 0 1 1 0 0 0 0 1 0
0 0 1 1 0 0 0 0 1 1
0 0 1 1 0 0 0 1 0 0
0 0 1 1 0 0 0 1 0 1
0 0 1 1 0 0 0 1 1 0
0 0 1 1 0 0 0 1 1 1
0 0 1 1 0 0 1 0 0 0
0 0 1 1 0 0 1 0 0 1
0 0 1 1 0 0 1 0 1 0
0 0 1 1 0 0 1 0 1 1
0 0 1 1 0 0 1 1 0 0
0 0 1 1 0 0 1 1 0 1
0 0 1 1 0 0 1 1 1 0
0 0 1 1 0 0 1 1 1 1
0 0 1 1 0 1 0 0 0 0
0 0 1 1 0 1 0 0 0 1
0 0 1 1 0 1 0 0 1 0
0 0 1 1 0 1 0 0 1 1
0 0 1 1 0 1 0 1 0 0
0 0 1 1 0 1 0 1 0 1
0 0 1 1 0 1 0 1 1 0
0 0 1 1 0 1 0 1 1 1
0 0 1 1 0 1 1 0 0 0
0 0 1 1 0 1 1 0 0 1
0 0 1 1 0 1 1 0 1 0
0 0 1 1 0 1 1 0 1 1
0 0 1 1 0 1 1 1 0 0
0 0 1 1 0 1 1 1 0 1
0 0 1 1 0 1 1 1 1 0
0 0 1 1 0 1 1 1 1 1
0 0 1 1 1 0 0 0 0 0
0 0 1 1 1 0 0 0 0 1
0 0 1 1 1 0 0 0 1 0
0 0 1 1 1 0 0 0 1 1
0 0 1 1 1 0 0 1 0 0
0 0 1 1 1 0 0 1 0 1
0 0 1 1 1 0 0 1 1 0
0 0 1 1 1 0 0 1 1 1
0 0 1 1 1 0 1 0 0 0
0 0 1 1 1 0 1 0 0 1
0 0 1 1 1 0 1 0 1 0
0 0 1 1 1 0 1 0 1 1
0 0 1 1 1 0 1 1 0 0
0 0 1 1 1 0 1 1 0 1
0 0 1 1 1 0 1 1 1 0
0 0 1 1 1 0 1 1 1 1
0 0 1 1 1 1 0 0 0 0
0 0 1 1 1 1 0 0 0 1
0 0 1 1 1 1 0 0 1 0
0 0 1 1 1 1 0 0 1 1
0 0 1 1 1 1 0 1 0 0
0 0 1 1 1 1 0 1 0 1

0 0 1 1 1 1 0 1 1 0
0 0 1 1 1 1 0 1 1 1
0 0 1 1 1 1 1 0 0 0
0 0 1 1 1 1 1 0 0 1
0 0 1 1 1 1 1 0 1 0
0 0 1 1 1 1 1 0 1 1
0 0 1 1 1 1 1 1 0 0
0 0 1 1 1 1 1 1 0 1
0 0 1 1 1 1 1 1 1 0
0 0 1 1 1 1 1 1 1 1
0 1 0 0 0 0 0 0 0 0
0 1 0 0 0 0 0 0 0 1
0 1 0 0 0 0 0 0 1 0
0 1 0 0 0 0 0 0 1 1
0 1 0 0 0 0 0 1 0 0
0 1 0 0 0 0 0 1 0 1
0 1 0 0 0 0 0 1 1 0
0 1 0 0 0 0 0 1 1 1
0 1 0 0 0 0 1 0 0 0
0 1 0 0 0 0 1 0 0 1
0 1 0 0 0 0 1 0 1 0
0 1 0 0 0 0 1 0 1 1
0 1 0 0 0 0 1 1 0 0
0 1 0 0 0 0 1 1 0 1
0 1 0 0 0 0 1 1 1 0
0 1 0 0 0 0 1 1 1 1
0 1 0 0 0 1 0 0 0 0
0 1 0 0 0 1 0 0 0 1
0 1 0 0 0 1 0 0 1 0
0 1 0 0 0 1 0 0 1 1
0 1 0 0 0 1 0 1 0 0
0 1 0 0 0 1 0 1 0 1
0 1 0 0 0 1 0 1 1 0
0 1 0 0 0 1 0 1 1 1
0 1 0 0 0 1 1 0 0 0
0 1 0 0 0 1 1 0 0 1
0 1 0 0 0 1 1 0 1 0
0 1 0 0 0 1 1 0 1 1
0 1 0 0 0 1 1 1 0 0
0 1 0 0 0 1 1 1 0 1
0 1 0 0 0 1 1 1 1 0
0 1 0 0 0 1 1 1 1 1
0 1 0 0 1 0 0 0 0 0
0 1 0 0 1 0 0 0 0 1
0 1 0 0 1 0 0 0 1 0
0 1 0 0 1 0 0 0 1 1
0 1 0 0 1 0 0 1 0 0
0 1 0 0 1 0 0 1 0 1
0 1 0 0 1 0 0 1 1 0
0 1 0 0 1 0 0 1 1 1
0 1 0 0 1 0 1 0 0 0
0 1 0 0 1 0 1 0 0 1
0 1 0 0 1 0 1 0 1 0
0 1 0 0 1 0 1 0 1 1
0 1 0 0 1 0 1 1 0 0
0 1 0 0 1 0 1 1 0 1
0 1 0 0 1 0 1 1 1 0
0 1 0 0 1 0 1 1 1 1
0 1 0 0 1 1 0 0 0 0
0 1 0 0 1 1 0 0 0 1
0 1 0 0 1 1 0 0 1 0
0 1 0 0 1 1 0 0 1 1
0 1 0 0 1 1 0 1 0 0
0 1 0 0 1 1 0 1 0 1
0 1 0 0 1 1 0 1 1 0
0 1 0 0 1 1 0 1 1 1
0 1 0 0 1 1 1 0 0 0
0 1 0 0 1 1 1 0 0 1
0 1 0 0 1 1 1 0 1 0
0 1 0 0 1 1 1 0 1 1
0 1 0 0 1 1 1 1 0 0
0 1 0 0 1 1 1 1 0 1
0 1 0 0 1 1 1 1 1 0
0 1 0 0 1 1 1 1 1 1
0 1 0 1 0 0 0 0 0 0
0 1 0 1 0 0 0 0 0 1
0 1 0 1 0 0 0 0 1 0
0 1 0 1 0 0 0 0 1 1
0 1 0 1 0 0 0 1 0 0
0 1 0 1 0 0 0 1 0 1

0 1 0 1 0 0 0 1 1 0
0 1 0 1 0 0 0 1 1 1
0 1 0 1 0 0 1 0 0 0
0 1 0 1 0 0 1 0 0 1
0 1 0 1 0 0 1 0 1 0
0 1 0 1 0 0 1 0 1 1
0 1 0 1 0 0 1 1 0 0
0 1 0 1 0 0 1 1 0 1
0 1 0 1 0 0 1 1 1 0
0 1 0 1 0 0 1 1 1 1
0 1 0 1 0 1 0 0 0 0
0 1 0 1 0 1 0 0 0 1
0 1 0 1 0 1 0 0 1 0
0 1 0 1 0 1 0 0 1 1
0 1 0 1 0 1 0 1 0 0
0 1 0 1 0 1 0 1 0 1
0 1 0 1 0 1 0 1 1 0
0 1 0 1 0 1 0 1 1 1
0 1 0 1 0 1 1 0 0 0
0 1 0 1 0 1 1 0 0 1
0 1 0 1 0 1 1 0 1 0
0 1 0 1 0 1 1 0 1 1
0 1 0 1 0 1 1 1 0 0
0 1 0 1 0 1 1 1 0 1
0 1 0 1 0 1 1 1 1 0
0 1 0 1 0 1 1 1 1 1
0 1 0 1 1 0 0 0 0 0
0 1 0 1 1 0 0 0 0 1
0 1 0 1 1 0 0 0 1 0
0 1 0 1 1 0 0 0 1 1
0 1 0 1 1 0 0 1 0 0
0 1 0 1 1 0 0 1 0 1
0 1 0 1 1 0 0 1 1 0
0 1 0 1 1 0 0 1 1 1
0 1 0 1 1 0 1 0 0 0
0 1 0 1 1 0 1 0 0 1
0 1 0 1 1 0 1 0 1 0
0 1 0 1 1 0 1 0 1 1
0 1 0 1 1 0 1 1 0 0
0 1 0 1 1 0 1 1 0 1
0 1 0 1 1 0 1 1 1 0
0 1 0 1 1 0 1 1 1 1
0 1 0 1 1 1 0 0 0 0
0 1 0 1 1 1 0 0 0 1
0 1 0 1 1 1 0 0 1 0
0 1 0 1 1 1 0 0 1 1
0 1 0 1 1 1 0 1 0 0
0 1 0 1 1 1 0 1 0 1
0 1 0 1 1 1 0 1 1 0
0 1 0 1 1 1 0 1 1 1
0 1 0 1 1 1 1 0 0 0
0 1 0 1 1 1 1 0 0 1
0 1 0 1 1 1 1 0 1 0
0 1 0 1 1 1 1 0 1 1
0 1 0 1 1 1 1 1 0 0
0 1 0 1 1 1 1 1 0 1
0 1 0 1 1 1 1 1 1 0
0 1 0 1 1 1 1 1 1 1
0 1 1 0 0 0 0 0 0 0
0 1 1 0 0 0 0 0 0 1
0 1 1 0 0 0 0 0 1 0
0 1 1 0 0 0 0 0 1 1
0 1 1 0 0 0 0 1 0 0
0 1 1 0 0 0 0 1 0 1
0 1 1 0 0 0 0 1 1 0
0 1 1 0 0 0 0 1 1 1
0 1 1 0 0 0 1 0 0 0
0 1 1 0 0 0 1 0 0 1
0 1 1 0 0 0 1 0 1 0
0 1 1 0 0 0 1 0 1 1
0 1 1 0 0 0 1 1 0 0
0 1 1 0 0 0 1 1 0 1
0 1 1 0 0 0 1 1 1 0
0 1 1 0 0 0 1 1 1 1
0 1 1 0 0 1 0 0 0 0
0 1 1 0 0 1 0 0 0 1
0 1 1 0 0 1 0 0 1 0
0 1 1 0 0 1 0 0 1 1
0 1 1 0 0 1 0 1 0 0
0 1 1 0 0 1 0 1 0 1

0 1 1 0 0 1 0 1 1 0
0 1 1 0 0 1 0 1 1 1
0 1 1 0 0 1 1 0 0 0
0 1 1 0 0 1 1 0 0 1
0 1 1 0 0 1 1 0 1 0
0 1 1 0 0 1 1 0 1 1
0 1 1 0 0 1 1 1 0 0
0 1 1 0 0 1 1 1 0 1
0 1 1 0 0 1 1 1 1 0
0 1 1 0 0 1 1 1 1 1
0 1 1 0 1 0 0 0 0 0
0 1 1 0 1 0 0 0 0 1
0 1 1 0 1 0 0 0 1 0
0 1 1 0 1 0 0 0 1 1
0 1 1 0 1 0 0 1 0 0
0 1 1 0 1 0 0 1 0 1
0 1 1 0 1 0 0 1 1 0
0 1 1 0 1 0 0 1 1 1
0 1 1 0 1 0 1 0 0 0
0 1 1 0 1 0 1 0 0 1
0 1 1 0 1 0 1 0 1 0
0 1 1 0 1 0 1 0 1 1
0 1 1 0 1 0 1 1 0 0
0 1 1 0 1 0 1 1 0 1
0 1 1 0 1 0 1 1 1 0
0 1 1 0 1 0 1 1 1 1
0 1 1 0 1 1 0 0 0 0
0 1 1 0 1 1 0 0 0 1
0 1 1 0 1 1 0 0 1 0
0 1 1 0 1 1 0 0 1 1
0 1 1 0 1 1 0 1 0 0
0 1 1 0 1 1 0 1 0 1
0 1 1 0 1 1 0 1 1 0
0 1 1 0 1 1 0 1 1 1
0 1 1 0 1 1 1 0 0 0
0 1 1 0 1 1 1 0 0 1
0 1 1 0 1 1 1 0 1 0
0 1 1 0 1 1 1 0 1 1
0 1 1 0 1 1 1 1 0 0
0 1 1 0 1 1 1 1 0 1
0 1 1 0 1 1 1 1 1 0
0 1 1 0 1 1 1 1 1 1
0 1 1 1 0 0 0 0 0 0
0 1 1 1 0 0 0 0 0 1
0 1 1 1 0 0 0 0 1 0
0 1 1 1 0 0 0 0 1 1
0 1 1 1 0 0 0 1 0 0
0 1 1 1 0 0 0 1 0 1
0 1 1 1 0 0 0 1 1 0
0 1 1 1 0 0 0 1 1 1
0 1 1 1 0 0 1 0 0 0
0 1 1 1 0 0 1 0 0 1
0 1 1 1 0 0 1 0 1 0
0 1 1 1 0 0 1 0 1 1
0 1 1 1 0 0 1 1 0 0
0 1 1 1 0 0 1 1 0 1
0 1 1 1 0 0 1 1 1 0
0 1 1 1 0 0 1 1 1 1
0 1 1 1 0 1 0 0 0 0
0 1 1 1 0 1 0 0 0 1
0 1 1 1 0 1 0 0 1 0
0 1 1 1 0 1 0 0 1 1
0 1 1 1 0 1 0 1 0 0
0 1 1 1 0 1 0 1 0 1
0 1 1 1 0 1 0 1 1 0
0 1 1 1 0 1 0 1 1 1
0 1 1 1 0 1 1 0 0 0
0 1 1 1 0 1 1 0 0 1
0 1 1 1 0 1 1 0 1 0
0 1 1 1 0 1 1 0 1 1
0 1 1 1 0 1 1 1 0 0
0 1 1 1 0 1 1 1 0 1
0 1 1 1 0 1 1 1 1 0
0 1 1 1 0 1 1 1 1 1
0 1 1 1 1 0 0 0 0 0
0 1 1 1 1 0 0 0 0 1
0 1 1 1 1 0 0 0 1 0
0 1 1 1 1 0 0 0 1 1
0 1 1 1 1 0 0 1 0 0
0 1 1 1 1 0 0 1 0 1

0 1 1 1 1 0 0 1 1 0
0 1 1 1 1 0 0 1 1 1
0 1 1 1 1 0 1 0 0 0
0 1 1 1 1 0 1 0 0 1
0 1 1 1 1 0 1 0 1 0
0 1 1 1 1 0 1 0 1 1
0 1 1 1 1 0 1 1 0 0
0 1 1 1 1 0 1 1 0 1
0 1 1 1 1 0 1 1 1 0
0 1 1 1 1 0 1 1 1 1
0 1 1 1 1 1 0 0 0 0
0 1 1 1 1 1 0 0 0 1
0 1 1 1 1 1 0 0 1 0
0 1 1 1 1 1 0 0 1 1
0 1 1 1 1 1 0 1 0 0
0 1 1 1 1 1 0 1 0 1
0 1 1 1 1 1 0 1 1 0
0 1 1 1 1 1 0 1 1 1
0 1 1 1 1 1 1 0 0 0
0 1 1 1 1 1 1 0 0 1
0 1 1 1 1 1 1 0 1 0
0 1 1 1 1 1 1 0 1 1
0 1 1 1 1 1 1 1 0 0
0 1 1 1 1 1 1 1 0 1
0 1 1 1 1 1 1 1 1 0
0 1 1 1 1 1 1 1 1 1
1 0 0 0 0 0 0 0 0 0
1 0 0 0 0 0 0 0 0 1
1 0 0 0 0 0 0 0 1 0
1 0 0 0 0 0 0 0 1 1
1 0 0 0 0 0 0 1 0 0
1 0 0 0 0 0 0 1 0 1
1 0 0 0 0 0 0 1 1 0
1 0 0 0 0 0 0 1 1 1
1 0 0 0 0 0 1 0 0 0
1 0 0 0 0 0 1 0 0 1
1 0 0 0 0 0 1 0 1 0
1 0 0 0 0 0 1 0 1 1
1 0 0 0 0 0 1 1 0 0
1 0 0 0 0 0 1 1 0 1
1 0 0 0 0 0 1 1 1 0
1 0 0 0 0 0 1 1 1 1
1 0 0 0 0 1 0 0 0 0
1 0 0 0 0 1 0 0 0 1
1 0 0 0 0 1 0 0 1 0
1 0 0 0 0 1 0 0 1 1
1 0 0 0 0 1 0 1 0 0
1 0 0 0 0 1 0 1 0 1
1 0 0 0 0 1 0 1 1 0
1 0 0 0 0 1 0 1 1 1
1 0 0 0 0 1 1 0 0 0
1 0 0 0 0 1 1 0 0 1
1 0 0 0 0 1 1 0 1 0
1 0 0 0 0 1 1 0 1 1
1 0 0 0 0 1 1 1 0 0
1 0 0 0 0 1 1 1 0 1
1 0 0 0 0 1 1 1 1 0
1 0 0 0 0 1 1 1 1 1
1 0 0 0 1 0 0 0 0 0
1 0 0 0 1 0 0 0 0 1
1 0 0 0 1 0 0 0 1 0
1 0 0 0 1 0 0 0 1 1
1 0 0 0 1 0 0 1 0 0
1 0 0 0 1 0 0 1 0 1
1 0 0 0 1 0 0 1 1 0
1 0 0 0 1 0 0 1 1 1
1 0 0 0 1 0 1 0 0 0
1 0 0 0 1 0 1 0 0 1
1 0 0 0 1 0 1 0 1 0
1 0 0 0 1 0 1 0 1 1
1 0 0 0 1 0 1 1 0 0
1 0 0 0 1 0 1 1 0 1
1 0 0 0 1 0 1 1 1 0
1 0 0 0 1 0 1 1 1 1
1 0 0 0 1 1 0 0 0 0
1 0 0 0 1 1 0 0 0 1
1 0 0 0 1 1 0 0 1 0
1 0 0 0 1 1 0 0 1 1
1 0 0 0 1 1 0 1 0 0
1 0 0 0 1 1 0 1 0 1

1	0	0	0	1	1	0	1	1	0
1	0	0	0	1	1	0	1	1	1
1	0	0	0	1	1	1	0	0	0
1	0	0	0	1	1	1	0	0	1
1	0	0	0	1	1	1	0	1	0
1	0	0	0	1	1	1	0	1	1
1	0	0	0	1	1	1	1	0	0
1	0	0	0	1	1	1	1	0	1
1	0	0	0	1	1	1	1	1	0
1	0	0	0	1	1	1	1	1	1
1	0	0	1	0	0	0	0	0	0
1	0	0	1	0	0	0	0	0	1
1	0	0	1	0	0	0	0	1	0
1	0	0	1	0	0	0	0	1	1
1	0	0	1	0	0	0	1	0	0
1	0	0	1	0	0	0	1	0	1
1	0	0	1	0	0	0	1	1	0
1	0	0	1	0	0	0	1	1	1
1	0	0	1	0	0	1	0	0	0
1	0	0	1	0	0	1	0	0	1
1	0	0	1	0	0	1	0	1	0
1	0	0	1	0	0	1	0	1	1
1	0	0	1	0	0	1	1	0	0
1	0	0	1	0	0	1	1	0	1
1	0	0	1	0	0	1	1	1	0
1	0	0	1	0	0	1	1	1	1
1	0	0	1	0	1	0	0	0	0
1	0	0	1	0	1	0	0	0	1
1	0	0	1	0	1	0	0	1	0
1	0	0	1	0	1	0	0	1	1
1	0	0	1	0	1	0	1	0	0
1	0	0	1	0	1	0	1	0	1
1	0	0	1	0	1	0	1	1	0
1	0	0	1	0	1	0	1	1	1
1	0	0	1	0	1	1	0	0	0
1	0	0	1	0	1	1	0	0	1
1	0	0	1	0	1	1	0	1	0
1	0	0	1	0	1	1	0	1	1
1	0	0	1	0	1	1	1	0	0
1	0	0	1	0	1	1	1	0	1
1	0	0	1	0	1	1	1	1	0
1	0	0	1	0	1	1	1	1	1
1	0	0	1	1	0	0	0	0	0
1	0	0	1	1	0	0	0	0	1
1	0	0	1	1	0	0	0	1	0
1	0	0	1	1	0	0	0	1	1
1	0	0	1	1	0	0	1	0	0
1	0	0	1	1	0	0	1	0	1
1	0	0	1	1	0	0	1	1	0
1	0	0	1	1	0	0	1	1	1
1	0	0	1	1	0	1	0	0	0
1	0	0	1	1	0	1	0	0	1
1	0	0	1	1	0	1	0	1	0
1	0	0	1	1	0	1	0	1	1
1	0	0	1	1	0	1	1	0	0
1	0	0	1	1	0	1	1	0	1
1	0	0	1	1	0	1	1	1	0
1	0	0	1	1	0	1	1	1	1
1	0	0	1	1	1	0	0	0	0
1	0	0	1	1	1	0	0	0	1
1	0	0	1	1	1	0	0	1	0
1	0	0	1	1	1	0	0	1	1
1	0	0	1	1	1	0	1	0	0
1	0	0	1	1	1	0	1	0	1
1	0	0	1	1	1	0	1	1	0
1	0	0	1	1	1	0	1	1	1
1	0	0	1	1	1	1	0	0	0
1	0	0	1	1	1	1	0	0	1
1	0	0	1	1	1	1	0	1	0
1	0	0	1	1	1	1	0	1	1
1	0	0	1	1	1	1	1	0	0
1	0	0	1	1	1	1	1	0	1
1	0	0	1	1	1	1	1	1	0
1	0	0	1	1	1	1	1	1	1
1	0	1	0	0	0	0	0	0	0
1	0	1	0	0	0	0	0	0	1
1	0	1	0	0	0	0	0	1	0
1	0	1	0	0	0	0	0	1	1
1	0	1	0	0	0	0	1	0	0
1	0	1	0	0	0	0	1	0	1

1	0	1	0	0	0	0	1	1	0
1	0	1	0	0	0	0	1	1	1
1	0	1	0	0	0	1	0	0	0
1	0	1	0	0	0	1	0	0	1
1	0	1	0	0	0	1	0	1	0
1	0	1	0	0	0	1	0	1	1
1	0	1	0	0	0	1	1	0	0
1	0	1	0	0	0	1	1	0	1
1	0	1	0	0	0	1	1	1	0
1	0	1	0	0	0	1	1	1	1
1	0	1	0	0	1	0	0	0	0
1	0	1	0	0	1	0	0	0	1
1	0	1	0	0	1	0	0	1	0
1	0	1	0	0	1	0	0	1	1
1	0	1	0	0	1	0	1	0	0
1	0	1	0	0	1	0	1	0	1
1	0	1	0	0	1	0	1	1	0
1	0	1	0	0	1	0	1	1	1
1	0	1	0	0	1	1	0	0	0
1	0	1	0	0	1	1	0	0	1
1	0	1	0	0	1	1	0	1	0
1	0	1	0	0	1	1	0	1	1
1	0	1	0	0	1	1	1	0	0
1	0	1	0	0	1	1	1	0	1
1	0	1	0	0	1	1	1	1	0
1	0	1	0	0	1	1	1	1	1
1	0	1	0	1	0	0	0	0	0
1	0	1	0	1	0	0	0	0	1
1	0	1	0	1	0	0	0	1	0
1	0	1	0	1	0	0	0	1	1
1	0	1	0	1	0	0	1	0	0
1	0	1	0	1	0	0	1	0	1
1	0	1	0	1	0	0	1	1	0
1	0	1	0	1	0	0	1	1	1
1	0	1	0	1	0	1	0	0	0
1	0	1	0	1	0	1	0	0	1
1	0	1	0	1	0	1	0	1	0
1	0	1	0	1	0	1	0	1	1
1	0	1	0	1	0	1	1	0	0
1	0	1	0	1	0	1	1	0	1
1	0	1	0	1	0	1	1	1	0
1	0	1	0	1	0	1	1	1	1
1	0	1	0	1	1	0	0	0	0
1	0	1	0	1	1	0	0	0	1
1	0	1	0	1	1	0	0	1	0
1	0	1	0	1	1	0	0	1	1
1	0	1	0	1	1	0	1	0	0
1	0	1	0	1	1	0	1	0	1
1	0	1	0	1	1	0	1	1	0
1	0	1	0	1	1	0	1	1	1
1	0	1	0	1	1	1	0	0	0
1	0	1	0	1	1	1	0	0	1
1	0	1	0	1	1	1	0	1	0
1	0	1	0	1	1	1	0	1	1
1	0	1	0	1	1	1	1	0	0
1	0	1	0	1	1	1	1	0	1
1	0	1	0	1	1	1	1	1	0
1	0	1	0	1	1	1	1	1	1
1	0	1	1	0	0	0	0	0	0
1	0	1	1	0	0	0	0	0	1
1	0	1	1	0	0	0	0	1	0
1	0	1	1	0	0	0	0	1	1
1	0	1	1	0	0	0	1	0	0
1	0	1	1	0	0	0	1	0	1
1	0	1	1	0	0	0	1	1	0
1	0	1	1	0	0	0	1	1	1
1	0	1	1	0	0	1	0	0	0
1	0	1	1	0	0	1	0	0	1
1	0	1	1	0	0	1	0	1	0
1	0	1	1	0	0	1	0	1	1
1	0	1	1	0	0	1	1	0	0
1	0	1	1	0	0	1	1	0	1
1	0	1	1	0	0	1	1	1	0
1	0	1	1	0	0	1	1	1	1
1	0	1	1	0	1	0	0	0	0
1	0	1	1	0	1	0	0	0	1
1	0	1	1	0	1	0	0	1	0
1	0	1	1	0	1	0	0	1	1
1	0	1	1	0	1	0	1	0	0
1	0	1	1	0	1	0	1	0	1

1	0	1	1	0	1	0	1	1	0
1	0	1	1	0	1	0	1	1	1
1	0	1	1	0	1	1	0	0	0
1	0	1	1	0	1	1	0	0	1
1	0	1	1	0	1	1	0	1	0
1	0	1	1	0	1	1	0	1	1
1	0	1	1	0	1	1	1	0	0
1	0	1	1	0	1	1	1	0	1
1	0	1	1	0	1	1	1	1	0
1	0	1	1	0	1	1	1	1	1
1	0	1	1	1	0	0	0	0	0
1	0	1	1	1	0	0	0	0	1
1	0	1	1	1	0	0	0	1	0
1	0	1	1	1	0	0	0	1	1
1	0	1	1	1	0	0	1	0	0
1	0	1	1	1	0	0	1	0	1
1	0	1	1	1	0	0	1	1	0
1	0	1	1	1	0	0	1	1	1
1	0	1	1	1	0	1	0	0	0
1	0	1	1	1	0	1	0	0	1
1	0	1	1	1	0	1	0	1	0
1	0	1	1	1	0	1	0	1	1
1	0	1	1	1	0	1	1	0	0
1	0	1	1	1	0	1	1	0	1
1	0	1	1	1	0	1	1	1	0
1	0	1	1	1	0	1	1	1	1
1	0	1	1	1	1	0	0	0	0
1	0	1	1	1	1	0	0	0	1
1	0	1	1	1	1	0	0	1	0
1	0	1	1	1	1	0	0	1	1
1	0	1	1	1	1	0	1	0	0
1	0	1	1	1	1	0	1	0	1
1	0	1	1	1	1	0	1	1	0
1	0	1	1	1	1	0	1	1	1
1	0	1	1	1	1	1	0	0	0
1	0	1	1	1	1	1	0	0	1
1	0	1	1	1	1	1	0	1	0
1	0	1	1	1	1	1	0	1	1
1	0	1	1	1	1	1	1	0	0
1	0	1	1	1	1	1	1	0	1
1	0	1	1	1	1	1	1	1	0
1	0	1	1	1	1	1	1	1	1
1	1	0	0	0	0	0	0	0	0
1	1	0	0	0	0	0	0	0	1
1	1	0	0	0	0	0	0	1	0
1	1	0	0	0	0	0	0	1	1
1	1	0	0	0	0	0	1	0	0
1	1	0	0	0	0	0	1	0	1
1	1	0	0	0	0	0	1	1	0
1	1	0	0	0	0	0	1	1	1
1	1	0	0	0	0	1	0	0	0
1	1	0	0	0	0	1	0	0	1
1	1	0	0	0	0	1	0	1	0
1	1	0	0	0	0	1	0	1	1
1	1	0	0	0	0	1	1	0	0
1	1	0	0	0	0	1	1	0	1
1	1	0	0	0	0	1	1	1	0
1	1	0	0	0	0	1	1	1	1
1	1	0	0	0	1	0	0	0	0
1	1	0	0	0	1	0	0	0	1
1	1	0	0	0	1	0	0	1	0
1	1	0	0	0	1	0	0	1	1
1	1	0	0	0	1	0	1	0	0
1	1	0	0	0	1	0	1	0	1
1	1	0	0	0	1	0	1	1	0
1	1	0	0	0	1	0	1	1	1
1	1	0	0	0	1	1	0	0	0
1	1	0	0	0	1	1	0	0	1
1	1	0	0	0	1	1	0	1	0
1	1	0	0	0	1	1	0	1	1
1	1	0	0	0	1	1	1	0	0
1	1	0	0	0	1	1	1	0	1
1	1	0	0	0	1	1	1	1	0
1	1	0	0	0	1	1	1	1	1
1	1	0	0	1	0	0	0	0	0
1	1	0	0	1	0	0	0	0	1
1	1	0	0	1	0	0	0	1	0
1	1	0	0	1	0	0	0	1	1
1	1	0	0	1	0	0	1	0	0
1	1	0	0	1	0	0	1	0	1

1	1	0	0	1	0	0	1	1	0
1	1	0	0	1	0	0	1	1	1
1	1	0	0	1	0	1	0	0	0
1	1	0	0	1	0	1	0	0	1
1	1	0	0	1	0	1	0	1	0
1	1	0	0	1	0	1	0	1	1
1	1	0	0	1	0	1	1	0	0
1	1	0	0	1	0	1	1	0	1
1	1	0	0	1	0	1	1	1	0
1	1	0	0	1	0	1	1	1	1
1	1	0	0	1	1	0	0	0	0
1	1	0	0	1	1	0	0	0	1
1	1	0	0	1	1	0	0	1	0
1	1	0	0	1	1	0	0	1	1
1	1	0	0	1	1	0	1	0	0
1	1	0	0	1	1	0	1	0	1
1	1	0	0	1	1	0	1	1	0
1	1	0	0	1	1	0	1	1	1
1	1	0	0	1	1	1	0	0	0
1	1	0	0	1	1	1	0	0	1
1	1	0	0	1	1	1	0	1	0
1	1	0	0	1	1	1	0	1	1
1	1	0	0	1	1	1	1	0	0
1	1	0	0	1	1	1	1	0	1
1	1	0	0	1	1	1	1	1	0
1	1	0	0	1	1	1	1	1	1
1	1	0	1	0	0	0	0	0	0
1	1	0	1	0	0	0	0	0	1
1	1	0	1	0	0	0	0	1	0
1	1	0	1	0	0	0	0	1	1
1	1	0	1	0	0	0	1	0	0
1	1	0	1	0	0	0	1	0	1
1	1	0	1	0	0	0	1	1	0
1	1	0	1	0	0	0	1	1	1
1	1	0	1	0	0	1	0	0	0
1	1	0	1	0	0	1	0	0	1
1	1	0	1	0	0	1	0	1	0
1	1	0	1	0	0	1	0	1	1
1	1	0	1	0	0	1	1	0	0
1	1	0	1	0	0	1	1	0	1
1	1	0	1	0	0	1	1	1	0
1	1	0	1	0	0	1	1	1	1
1	1	0	1	0	1	0	0	0	0
1	1	0	1	0	1	0	0	0	1
1	1	0	1	0	1	0	0	1	0
1	1	0	1	0	1	0	0	1	1
1	1	0	1	0	1	0	1	0	0
1	1	0	1	0	1	0	1	0	1
1	1	0	1	0	1	0	1	1	0
1	1	0	1	0	1	0	1	1	1
1	1	0	1	0	1	1	0	0	0
1	1	0	1	0	1	1	0	0	1
1	1	0	1	0	1	1	0	1	0
1	1	0	1	0	1	1	0	1	1
1	1	0	1	0	1	1	1	0	0
1	1	0	1	0	1	1	1	0	1
1	1	0	1	0	1	1	1	1	0
1	1	0	1	0	1	1	1	1	1
1	1	0	1	1	0	0	0	0	0
1	1	0	1	1	0	0	0	0	1
1	1	0	1	1	0	0	0	1	0
1	1	0	1	1	0	0	0	1	1
1	1	0	1	1	0	0	1	0	0
1	1	0	1	1	0	0	1	0	1
1	1	0	1	1	0	0	1	1	0
1	1	0	1	1	0	0	1	1	1
1	1	0	1	1	0	1	0	0	0
1	1	0	1	1	0	1	0	0	1
1	1	0	1	1	0	1	0	1	0
1	1	0	1	1	0	1	0	1	1
1	1	0	1	1	0	1	1	0	0
1	1	0	1	1	0	1	1	0	1
1	1	0	1	1	0	1	1	1	0
1	1	0	1	1	0	1	1	1	1
1	1	0	1	1	1	0	0	0	0
1	1	0	1	1	1	0	0	0	1
1	1	0	1	1	1	0	0	1	0
1	1	0	1	1	1	0	0	1	1
1	1	0	1	1	1	0	1	0	0
1	1	0	1	1	1	0	1	0	1

1	1	0	1	1	1	0	1	1	0
1	1	0	1	1	1	0	1	1	1
1	1	0	1	1	1	1	0	0	0
1	1	0	1	1	1	1	0	0	1
1	1	0	1	1	1	1	0	1	0
1	1	0	1	1	1	1	0	1	1
1	1	0	1	1	1	1	1	0	0
1	1	0	1	1	1	1	1	0	1
1	1	0	1	1	1	1	1	1	0
1	1	0	1	1	1	1	1	1	1
1	1	1	0	0	0	0	0	0	0
1	1	1	0	0	0	0	0	0	1
1	1	1	0	0	0	0	0	1	0
1	1	1	0	0	0	0	0	1	1
1	1	1	0	0	0	0	1	0	0
1	1	1	0	0	0	0	1	0	1
1	1	1	0	0	0	0	1	1	0
1	1	1	0	0	0	0	1	1	1
1	1	1	0	0	0	1	0	0	0
1	1	1	0	0	0	1	0	0	1
1	1	1	0	0	0	1	0	1	0
1	1	1	0	0	0	1	0	1	1
1	1	1	0	0	0	1	1	0	0
1	1	1	0	0	0	1	1	0	1
1	1	1	0	0	0	1	1	1	0
1	1	1	0	0	0	1	1	1	1
1	1	1	0	0	1	0	0	0	0
1	1	1	0	0	1	0	0	0	1
1	1	1	0	0	1	0	0	1	0
1	1	1	0	0	1	0	0	1	1
1	1	1	0	0	1	0	1	0	0
1	1	1	0	0	1	0	1	0	1
1	1	1	0	0	1	0	1	1	0
1	1	1	0	0	1	0	1	1	1
1	1	1	0	0	1	1	0	0	0
1	1	1	0	0	1	1	0	0	1
1	1	1	0	0	1	1	0	1	0
1	1	1	0	0	1	1	0	1	1
1	1	1	0	0	1	1	1	0	0
1	1	1	0	0	1	1	1	0	1
1	1	1	0	0	1	1	1	1	0
1	1	1	0	0	1	1	1	1	1
1	1	1	0	1	0	0	0	0	0
1	1	1	0	1	0	0	0	0	1
1	1	1	0	1	0	0	0	1	0
1	1	1	0	1	0	0	0	1	1
1	1	1	0	1	0	0	1	0	0
1	1	1	0	1	0	0	1	0	1
1	1	1	0	1	0	0	1	1	0
1	1	1	0	1	0	0	1	1	1
1	1	1	0	1	0	1	0	0	0
1	1	1	0	1	0	1	0	0	1
1	1	1	0	1	0	1	0	1	0
1	1	1	0	1	0	1	0	1	1
1	1	1	0	1	0	1	1	0	0
1	1	1	0	1	0	1	1	0	1
1	1	1	0	1	0	1	1	1	0
1	1	1	0	1	0	1	1	1	1
1	1	1	0	1	1	0	0	0	0
1	1	1	0	1	1	0	0	0	1
1	1	1	0	1	1	0	0	1	0
1	1	1	0	1	1	0	0	1	1
1	1	1	0	1	1	0	1	0	0
1	1	1	0	1	1	0	1	0	1
1	1	1	0	1	1	0	1	1	0
1	1	1	0	1	1	0	1	1	1
1	1	1	0	1	1	1	0	0	0
1	1	1	0	1	1	1	0	0	1
1	1	1	0	1	1	1	0	1	0
1	1	1	0	1	1	1	0	1	1

1	1	1	0	1	1	1	1	0	0
1	1	1	0	1	1	1	1	0	1
1	1	1	0	1	1	1	1	1	0
1	1	1	0	1	1	1	1	1	1
1	1	1	1	0	0	0	0	0	0
1	1	1	1	0	0	0	0	0	1
1	1	1	1	0	0	0	0	1	0
1	1	1	1	0	0	0	0	1	1
1	1	1	1	0	0	0	1	0	0
1	1	1	1	0	0	0	1	0	1
1	1	1	1	0	0	0	1	1	0
1	1	1	1	0	0	0	1	1	1
1	1	1	1	0	0	1	0	0	0
1	1	1	1	0	0	1	0	0	1
1	1	1	1	0	0	1	0	1	0
1	1	1	1	0	0	1	0	1	1
1	1	1	1	0	0	1	1	0	0
1	1	1	1	0	0	1	1	0	1
1	1	1	1	0	0	1	1	1	0
1	1	1	1	0	0	1	1	1	1
1	1	1	1	0	1	0	0	0	0
1	1	1	1	0	1	0	0	0	1
1	1	1	1	0	1	0	0	1	0
1	1	1	1	0	1	0	0	1	1
1	1	1	1	0	1	0	1	0	0
1	1	1	1	0	1	0	1	0	1
1	1	1	1	0	1	0	1	1	0
1	1	1	1	0	1	0	1	1	1
1	1	1	1	0	1	1	0	0	0
1	1	1	1	0	1	1	0	0	1
1	1	1	1	0	1	1	0	1	0
1	1	1	1	0	1	1	0	1	1
1	1	1	1	0	1	1	1	0	0
1	1	1	1	0	1	1	1	0	1
1	1	1	1	0	1	1	1	1	0
1	1	1	1	0	1	1	1	1	1
1	1	1	1	1	0	0	0	0	0
1	1	1	1	1	0	0	0	0	1
1	1	1	1	1	0	0	0	1	0
1	1	1	1	1	0	0	0	1	1
1	1	1	1	1	0	0	1	0	0
1	1	1	1	1	0	0	1	0	1
1	1	1	1	1	0	0	1	1	0
1	1	1	1	1	0	0	1	1	1
1	1	1	1	1	0	1	0	0	0
1	1	1	1	1	0	1	0	0	1
1	1	1	1	1	0	1	0	1	0
1	1	1	1	1	0	1	0	1	1
1	1	1	1	1	0	1	1	0	0
1	1	1	1	1	0	1	1	0	1
1	1	1	1	1	0	1	1	1	0
1	1	1	1	1	0	1	1	1	1
1	1	1	1	1	1	0	0	0	0
1	1	1	1	1	1	0	0	0	1
1	1	1	1	1	1	0	0	1	0
1	1	1	1	1	1	0	0	1	1
1	1	1	1	1	1	0	1	0	0
1	1	1	1	1	1	0	1	0	1
1	1	1	1	1	1	0	1	1	0
1	1	1	1	1	1	0	1	1	1
1	1	1	1	1	1	1	0	0	0
1	1	1	1	1	1	1	0	0	1
1	1	1	1	1	1	1	0	1	0
1	1	1	1	1	1	1	0	1	1
1	1	1	1	1	1	1	1	0	0
1	1	1	1	1	1	1	1	0	1
1	1	1	1	1	1	1	1	1	0
1	1	1	1	1	1	1	1	1	1

Listing of Program Deck (with sample data cards) for Fractional Factorials.

```
// TIME=(,25)
/*JOBPARM LINES=8000
/*ROUTE   PRINT PLOT11
/*JOBPARM DISKIO=1300
/*JOBPARM CARDS=200
// EXEC FORTRUN
//CMP.SYSIN DD *
      DIMENSION NX(1024,10),NY(1024,10)
      DIMENSION NR(1024), IQ(10)
      IN=10
      WRITE(6,555) IN
  555 FORMAT('1',64X,'IN=',I2,////)
      INTN=3
      I=1
      NX(1,1)=0
      NX(2,1)=1
   20 IF( I .EQ. IN) GO TO 100
      I=I+1
      I1=I-1
      I2=2**I1
      DO 30 J=1,I2
      J1=2*J-1
      J2=2*J
      DO 40 K=1,I1
      NY(J1,K)=NX(J,K)
      NY(J2,K)=NX(J,K)
   40 CONTINUE
      NY(J1,I)=0
      NY(J2,I)=1
   30 CONTINUE
      I3=2**I
      DO 60 L=1,I3
      DO 70  M=1,I
      NX(L,M)= NY(L,M)
   70 CONTINUE
   60 CONTINUE
      GO TO 20
  100 DO 80 L1=1,I3
      WRITE(6,200) (NX(L1,M1),M1=1,IN)
  200 FORMAT(' ',T44,10(I1,2X))
   80 CONTINUE
      WRITE(6,666) INTN
  666 FORMAT('1',50X, 'NUMBER OF INTERACTIONS USED=',I2,///)
      DO 300 IJ=1,I3
  300 NR(IJ)=0
      DO 310 IJK=1,INTN
      READ (5,96) (IQ(JK),JK=1,IN)
   96 FORMAT(10(I1))
      WRITE (6,66)  (IQ(JK),JK=1,IN)
   66 FORMAT('-',T44,10(I1,2X))
      DO 320 JR=1,I3
      ISUM=0
      IF( NR(JR) .NE. 0 ) GO TO 351
      DO 330 LR=1,IN
      IF( IQ(LR) .EQ. 1 .AND. NX(JR,LR) .EQ. 1) ISUM=ISUM+1
  330 CONTINUE
      IF ( MOD(ISUM,2) .EQ. 1) NR(JR)=1
  351 CONTINUE
  320 CONTINUE
  310 CONTINUE
      WRITE(6,777)
  777 FORMAT('1',32X,'USING ABOVE INTERACTIONS THE PRINCIPAL BLOCK IS
     C FOLLOWING:',////)
      IXC=0
      DO 500 IG=1,I3
      IF(NR(IG) .EQ. 1) GO TO 600
      IXC=IXC+1
      WRITE(6,444) IXC, (NX(IG,MG), MG=1,IN)
      WRITE(7,445) IXC,(NX(IG,MG),MG=1,IN)
  445 FORMAT(I3,10(1X,I1))
  444 FORMAT('-',16X,I3,20X,10(2X,I1,2X))
  600 CONTINUE
  500 CONTINUE
      RETURN
      END
/*
//GO.SYSIN DD *
1110001100
0111100010
1011010001
/*
//
```

Annotations:

- (JCL cards) COMPUTER CENTER JOB CONTROL LANGUAGE (JCL) CARDS.
- (DIMENSION statements) SET UP FOR UP TO 10 FACTORS.
- (IN=10) IN=NO. FACTORS DESIRED (10 IN THIS EXAMPLE).
- (INTN=3) INTN=NO. OF DEFINING EFFECTS (3 HERE FOR A 2**(10-3) DESIGN, WITH 2**7=128 EXPERIMENTS INSTEAD OF 2**10=1024).
- (statements 100–80) 2^{10} DESIGN IS PRINTED (OMITTED BELOW SINCE ALREADY GIVEN ABOVE IN THIS BOOK).
- (WRITE(6,444) and WRITE(7,445)) RESULTING DESIGN IS PRINTED AND PUNCHED ON CARDS.
- (data cards) 10 FACTORS A,B,C,D,E,F,G,H,J,K; DEFINING CONTRASTS ARE ABCGH, BCDEJ, and ACDFK.

Above, the defining contrasts were chosen for a Resolution V design (all main effects and 2-factor interactions may be estimated). P.W.M. John (*Statistical Design and Analysis of Experiments*, The Macmillan Company, New York, 1971, p. 157) gives the following **Resolution V designs:**

2^{9-2}: $I = ABCDEF = ABCGHJ$ ($= DEFGHJ$)

2^{10-3}: $I = ABCDEF = ABCGHJ = ABDEGHK$

2^{11-4}: $I = ABCDEF = ABFJK = AEFGKL = ACEHL$.

In general, it is a time-consuming and error-prone procedure to write out by hand the aliasing patterns of these designs, since there are, in a k-factor situation, 2^k effects to account for. It is therefore recommended that a suitable set of defining contrasts be used as above or from the literature. (The multiplication rules given on page 213 are used to determine the aliasing.)

IN = 10

```
0 0 0 0 0 0 0 0 0 0
0 0 0 0 0 0 0 0 0 1
0 0 0 0 0 0 0 0 1 0
0 0 0 0 0 0 0 0 1 1
0 0 0 0 0 0 0 1 0 0
0 0 0 0 0 0 0 1 0 1
0 0 0 0 0 0 0 1 1 0
0 0 0 0 0 0 0 1 1 1
0 0 0 0 0 0 1 0 0 0
0 0 0 0 0 0 1 0 0 1
0 0 0 0 0 0 1 0 1 0
0 0 0 0 0 0 1 0 1 1
0 0 0 0 0 0 1 1 0 0
0 0 0 0 0 0 1 1 0 1
0 0 0 0 0 0 1 1 1 0
0 0 0 0 0 0 1 1 1 1
0 0 0 0 0 1 0 0 0 0
0 0 0 0 0 1 0 0 0 1
0 0 0 0 0 1 0 0 1 0
0 0 0 0 0 1 0 0 1 1
0 0 0 0 0 1 0 1 0 0
0 0 0 0 0 1 0 1 0 1
0 0 0 0 0 1 0 1 1 0
0 0 0 0 0 1 0 1 1 1
0 0 0 0 0 1 1 0 0 0
0 0 0 0 0 1 1 0 0 1
0 0 0 0 0 1 1 0 1 0
0 0 0 0 0 1 1 0 1 1
0 0 0 0 0 1 1 1 0 0
0 0 0 0 0 1 1 1 0 1
0 0 0 0 0 1 1 1 1 0
0 0 0 0 0 1 1 1 1 1
0 0 0 0 1 0 0 0 0 0
0 0 0 0 1 0 0 0 0 1
0 0 0 0 1 0 0 0 1 0
0 0 0 0 1 0 0 0 1 1
0 0 0 0 1 0 0 1 0 0
```

```
0 0 0 0 1 0 0 1 0 1
0 0 0 0 1 0 0 1 1 0
0 0 0 0 1 0 0 1 1 1
0 0 0 0 1 0 1 0 0 0
0 0 0 0 1 0 1 0 0 1
0 0 0 0 1 0 1 0 1 0
0 0 0 0 1 0 1 0 1 1
0 0 0 0 1 0 1 1 0 0
0 0 0 0 1 0 1 1 0 1
0 0 0 0 1 0 1 1 1 0
0 0 0 0 1 0 1 1 1 1
0 0 0 0 1 1 0 0 0 0
0 0 0 0 1 1 0 0 0 1
0 0 0 0 1 1 0 0 1 0
0 0 0 0 1 1 0 0 1 1
0 0 0 0 1 1 0 1 0 0
0 0 0 0 1 1 0 1 0 1
0 0 0 0 1 1 0 1 1 0
0 0 0 0 1 1 0 1 1 1
0 0 0 0 1 1 1 0 0 0
0 0 0 0 1 1 1 0 0 1
0 0 0 0 1 1 1 0 1 0
0 0 0 0 1 1 1 0 1 1
0 0 0 0 1 1 1 1 0 0
0 0 0 0 1 1 1 1 0 1
0 0 0 0 1 1 1 1 1 0
```

etc. (2^{10} points)

NUMBER OF INTERACTIONS USED = 3

```
1 1 1 0 0 0 1 1 0 0

0 1 1 1 1 0 0 0 1 0

1 0 1 1 0 1 0 0 0 1
```

USING ABOVE INTERACTIONS THE PRINCIPAL BLOCK IS FOLLOWING:

```
1     0 0 0 0 0 0 0 0 0 0

2     0 0 0 0 0 0 1 1 0 0

3     0 0 0 0 0 1 0 0 0 1

4     0 0 0 0 0 1 1 1 0 1

5     0 0 0 0 1 0 0 0 1 0

6     0 0 0 0 1 0 1 1 1 0

7     0 0 0 0 1 1 0 0 1 1
```

8	0	0	0	0	1	1	1	1	1	1
9	0	0	0	1	0	0	0	0	1	1
10	0	0	0	1	0	0	1	1	1	1
11	0	0	0	1	0	1	0	0	1	0
12	0	0	0	1	0	1	1	1	1	0
13	0	0	0	1	1	0	0	0	0	1
14	0	0	0	1	1	0	1	1	0	1
15	0	0	0	1	1	1	0	0	0	0
16	0	0	0	1	1	1	1	1	0	0
17	0	0	1	0	0	0	0	1	1	1
18	0	0	1	0	0	0	1	0	1	1
19	0	0	1	0	0	1	0	1	1	0
20	0	0	1	0	0	1	1	0	1	0
21	0	0	1	0	1	0	0	1	0	1
22	0	0	1	0	1	0	1	0	0	1
23	0	0	1	0	1	1	0	1	0	0
24	0	0	1	0	1	1	1	0	0	0
25	0	0	1	1	0	0	0	1	0	0
26	0	0	1	1	0	0	1	0	0	0
27	0	0	1	1	0	1	0	1	0	1
28	0	0	1	1	0	1	1	0	0	1
29	0	0	1	1	1	0	0	1	1	0
30	0	0	1	1	1	0	1	0	1	0
31	0	0	1	1	1	1	0	1	1	1
32	0	0	1	1	1	1	1	0	1	1
33	0	1	0	0	0	0	0	1	1	0
34	0	1	0	0	0	0	1	0	1	0

35	0	1	0	0	0	1	0	1	1	1
36	0	1	0	0	0	1	1	0	1	1
37	0	1	0	0	1	0	0	1	0	0
38	0	1	0	0	1	0	1	0	0	0
39	0	1	0	0	1	1	0	1	0	1
40	0	1	0	0	1	1	1	0	0	1
41	0	1	0	1	0	0	0	1	0	1
42	0	1	0	1	0	0	1	0	0	1
43	0	1	0	1	0	1	0	1	0	0
44	0	1	0	1	0	1	1	0	0	0
45	0	1	0	1	1	0	0	1	1	1
46	0	1	0	1	1	0	1	0	1	1
47	0	1	0	1	1	1	0	1	1	0
48	0	1	0	1	1	1	1	0	1	0
49	0	1	1	0	0	0	0	0	0	1
50	0	1	1	0	0	0	1	1	0	1
51	0	1	1	0	0	1	0	0	0	0
52	0	1	1	0	0	1	1	1	0	0
53	0	1	1	0	1	0	0	0	1	1
54	0	1	1	0	1	0	1	1	1	1
55	0	1	1	0	1	1	0	0	1	0
56	0	1	1	0	1	1	1	1	1	0
57	0	1	1	1	0	0	0	0	1	0
58	0	1	1	1	0	0	1	1	1	0
59	0	1	1	1	0	1	0	0	1	1
60	0	1	1	1	0	1	1	1	1	1
61	0	1	1	1	1	0	0	0	0	0
62	0	1	1	1	1	0	1	1	0	0

63	0	1	1	1	1	1	0	0	0	1
64	0	1	1	1	1	1	1	1	0	1
65	1	0	0	0	0	0	0	1	0	1
66	1	0	0	0	0	0	1	0	0	1
67	1	0	0	0	0	1	0	1	0	0
68	1	0	0	0	0	1	1	0	0	0
69	1	0	0	0	1	0	0	1	1	1
70	1	0	0	0	1	0	1	0	1	1
71	1	0	0	0	1	1	0	1	1	0
72	1	0	0	0	1	1	1	0	1	0
73	1	0	0	1	0	0	0	1	1	0
74	1	0	0	1	0	0	1	0	1	0
75	1	0	0	1	0	1	0	1	1	1
76	1	0	0	1	0	1	1	0	1	1
77	1	0	0	1	1	0	0	1	0	0
78	1	0	0	1	1	0	1	0	0	0
79	1	0	0	1	1	1	0	1	0	1
80	1	0	0	1	1	1	1	0	0	1
81	1	0	1	0	0	0	0	0	1	0
82	1	0	1	0	0	0	1	1	1	0
83	1	0	1	0	0	1	0	0	1	1
84	1	0	1	0	0	1	1	1	1	1
85	1	0	1	0	1	0	0	0	0	0
86	1	0	1	0	1	0	1	1	0	0
87	1	0	1	0	1	1	0	0	0	1
88	1	0	1	0	1	1	1	1	0	1
89	1	0	1	1	0	0	0	0	0	1
90	1	0	1	1	0	0	1	1	0	1

91	1	0	1	1	0	1	0	0	0	0
92	1	0	1	1	0	1	1	1	0	0
93	1	0	1	1	1	0	0	0	1	1
94	1	0	1	1	1	0	1	1	1	1
95	1	0	1	1	1	1	0	0	1	0
96	1	0	1	1	1	1	1	1	1	0
97	1	1	0	0	0	0	0	0	1	1
98	1	1	0	0	0	0	1	1	1	1
99	1	1	0	0	0	1	0	0	1	0
100	1	1	0	0	0	1	1	1	1	0
101	1	1	0	0	1	0	0	0	0	1
102	1	1	0	0	1	0	1	1	0	1
103	1	1	0	0	1	1	0	0	0	0
104	1	1	0	0	1	1	1	1	0	0
105	1	1	0	1	0	0	0	0	0	0
106	1	1	0	1	0	0	1	1	0	0
107	1	1	0	1	0	1	0	0	0	1
108	1	1	0	1	0	1	1	1	0	1
109	1	1	0	1	1	0	0	0	1	0
110	1	1	0	1	1	0	1	1	1	0
111	1	1	0	1	1	1	0	0	1	1
112	1	1	0	1	1	1	1	1	1	1
113	1	1	1	0	0	0	0	1	0	0
114	1	1	1	0	0	0	1	0	0	0
115	1	1	1	0	0	1	0	1	0	1
116	1	1	1	0	0	1	1	0	0	1
117	1	1	1	0	1	0	0	1	1	0
118	1	1	1	0	1	0	1	0	1	0

119	1	1	1	0	1	1	0	1	1	1
120	1	1	1	0	1	1	1	0	1	1
121	1	1	1	1	0	0	0	1	1	1
122	1	1	1	1	0	0	1	0	1	1
123	1	1	1	1	0	1	0	1	1	0
124	1	1	1	1	0	1	1	0	1	0
125	1	1	1	1	1	0	0	1	0	1
126	1	1	1	1	1	0	1	0	0	1
127	1	1	1	1	1	1	0	1	0	0
128	1	1	1	1	1	1	1	0	0	0

= $2^7 = 2^{10-3}$, **verifying independence of defining contrasts specified as input.**

Central Composite Designs.

When there are large numbers of variables, or when there may be quadratic effects, it is not desirable to use factorial (or even fractional factorial) designs. **Suppose** we are presented with **a situation in which there are 7 variables (factors) and 50 or so experiments are possible;** quadratic effects, as well as interactions, cannot be ruled out. Then **four main options are the following:**

A. **Reducing to 6 variables** (around 6 or 7 variables is the largest number where one can, in a reasonable number of runs, fit a full quadratic model); this option will be discarded when the possible effect(s) of all variables are desired to be investigated at one time (with greater than 7 variables we might well go to a screening design as a first stage).

B. **Using a saturated composite design** (James M. Lucas, *The Optimum Design of Industrial Experiments*, Ph.D. Dissertation, Institute of Statistics, Texas A&M University, College Station, Texas, 1972); this would require $\binom{7}{2} + 2\times 7 + 1 = 36$ runs, and is useful if: experimentation is expensive, experimental error is small (or an independent estimate is available), and a quadratic model is adequate. Some of these constraints are often violated.

C. **Using a CCD based on a 2^{7-1} fraction**; this requires 79 runs, and so is discarded.

D. **Using a CCD based on a 2^{7-2} fraction**; this requires at least 47 runs and (in light of desires and constraints) seems a reasonable selection for a design.

Design D. has the points shown on p. 234

The coding used is usually given in a table, as in Table 6.

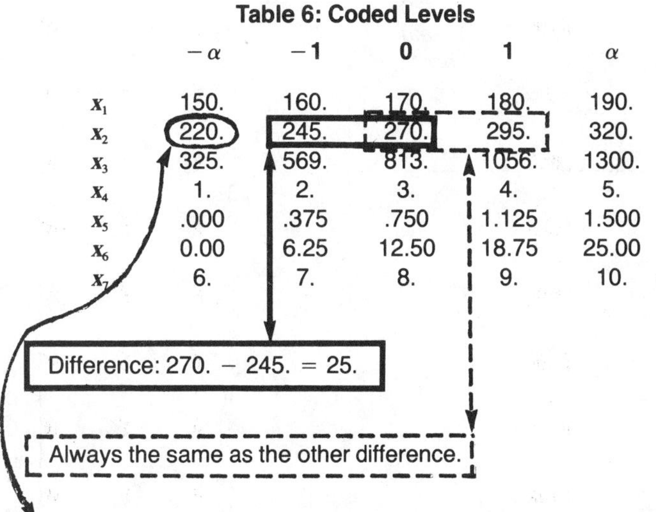

Table 6: Coded Levels

	$-\alpha$	-1	0	1	α
x_1	150.	160.	170.	180.	190.
x_2	220.	245.	270.	295.	320.
x_3	325.	569.	813.	1056.	1300.
x_4	1.	2.	3.	4.	5.
x_5	.000	.375	.750	1.125	1.500
x_6	0.00	6.25	12.50	18.75	25.00
x_7	6.	7.	8.	9.	10.

Difference: 270. − 245. = 25.
Always the same as the other difference.
Choice here is α = 2.00, so we go twice as far from the center as we did for the −1, +1 levels on each variable, e.g., 270. − 25. = the "−1" level; and 270. − α * (25.) = 270. − 2.00 * (25.) = 220. = the "−2" level.

There are rules by which the parameter alpha may be chosen, but there is no general agreement on which should be used. Most of these rules lead to an alpha between 1.00 and

Experimental Design (Coded Levels)

Name		Treatment No.	A x_1	B x_2	C x_3	D x_4	E x_5	F x_6	G x_7
	(1)	1	−1	−1	−1	−1	−1	−1	−1
	bc	2	−1	1	1	−1	−1	−1	−1
	bd	3	−1	1	−1	1	−1	−1	−1
	be	4	−1	1	−1	−1	1	−1	−1
	cd	5	−1	−1	1	1	−1	−1	−1
	ce	6	−1	−1	1	−1	1	−1	−1
	de	7	−1	−1	−1	1	1	−1	−1
	bcde	8	−1	1	1	1	1	−1	−1
	af	9	1	−1	−1	−1	−1	1	−1
	abcf	10	1	1	1	−1	−1	1	−1
	abdf	11	1	1	−1	1	−1	1	−1
	abef	12	1	1	−1	−1	1	1	−1
	acdf	13	1	−1	1	1	−1	1	−1
	acef	14	1	−1	1	−1	1	1	−1
	adef	15	1	−1	−1	1	1	1	−1
2^{7-2} Points	*abcdef*	16	1	1	1	1	1	1	−1
	abg	17	1	1	−1	−1	−1	−1	1
	acg	18	1	−1	1	−1	−1	−1	1
	adg	19	1	−1	−1	1	−1	−1	1
	aeg	20	1	−1	−1	−1	1	−1	1
	abcdg	21	1	1	1	1	−1	−1	1
	abceg	22	1	1	1	−1	1	−1	1
	abdeg	23	1	1	−1	1	1	−1	1
	acdeg	24	1	−1	1	1	1	−1	1
	bfg	25	−1	1	−1	−1	−1	1	1
	cfg	26	−1	−1	1	−1	−1	1	1
	dfg	27	−1	−1	−1	1	−1	1	1
	efg	28	−1	−1	−1	−1	1	1	1
	bcdfg	29	−1	1	1	1	−1	1	1
	bcefg	30	−1	1	1	−1	1	1	1
	bdefg	31	−1	1	−1	1	1	1	1
	cdefg	32	−1	−1	1	1	1	1	1
		33	$-\alpha$	0	0	0	0	0	0
		34	α	0	0	0	0	0	0
		35	0	$-\alpha$	0	0	0	0	0
		36	0	α	0	0	0	0	0
		37	0	0	$-\alpha$	0	0	0	0
		38	0	0	α	0	0	0	0
Star		39	0	0	0	$-\alpha$	0	0	0
		40	0	0	0	α	0	0	0
		41	0	0	0	0	$-\alpha$	0	0
		42	0	0	0	0	α	0	0
		43	0	0	0	0	0	$-\alpha$	0
		44	0	0	0	0	0	α	0
		45	0	0	0	0	0	0	$-\alpha$
		46	0	0	0	0	0	0	α
		47	0	0	0	0	0	0	0
Center		48	0	0	0	0	0	0	0
		49	0	0	0	0	0	0	0
		50	0	0	0	0	0	0	0

Here:

−1 denotes the low level of a factor

1 denotes the high level of a factor

0 denotes the average of the high and low levels of a factor

$-\alpha$ denotes a level below the −1 level

$+\alpha$ denotes a level above the 1 level.

Table 7: Experimental Order

First = 49	Eleventh = 22	21*st* = 12	31*st* = 50	41*st* = 25
38	13	29	19	35
2	1	31	41	43
39	16	3	46	45
48	40	42	34	9
6	36	11	27	26
37	5	23	7	10
17	44	24	33	4
20	8	18	21	47
14	32	28	30	Last = 15

2.00, and in this range it may fairly safely be chosen by the experimenter to give levels which seem to be desirable ones at which to experiment. The choice $\alpha = 1.00$ is called a **face-centered design** for reasons which are geometrically clear in the low-dimensional cases which will be drawn shortly.

Of course **experimental order should be randomized,** leading to an order of experimentation as given in Table 7. Thus, the first experiment performed would have all factors at the center level, i.e., $x_1 = 170, x_2 = 270, x_3 = 813$, etc., while the second experiment performed would have $x_1 = 170, x_2 = 270, x_3 = 1300$, etc.

Note that the one-quarter replicate of $2^{7-2} = 32$ runs in the fractional factorial part of the design utilizes the *defining contrasts* (also called *generators*) *ABCDEF* and *AFG* (hence their generalized interaction *ABCDEFAFG* = *BCDEG* also acts as a defining contrast).

The 32 treatment combinations (¼ of $2^7 = 128$) are derived from using *ABCDEF* to divide the 2^7 combinations into halves (the "+" half being those with an even number of letter matches), and from using the *AFG* interaction to again divide 2^7 into half (W. G. Cochran and G. M. Cox, *Experimental Designs* (Second Edition), John Wiley & Sons, Inc., New York, 1957, pp. 245, 253). This is efficiently accomplished for the *(1)*, *f*, *g*, and *fg* portions, the *(1)* starting portion of 32 being

(1)	*ad*	*abc*	*bde*
a	*ae*	*abd*	*cde*
b	*bc*	*abe*	*abcd*
c	*bd*	*acd*	*abce*
d	*be*	*ace*	*abde*
e	*cd*	*ade*	*acde*
ab	*ce*	*bcd*	*bcde*
ac	*de*	*bce*	*abcde*.

The general rule for constructing such a design is:

—You may use the full 2^k factorial as a base for the design

—To this are added the $2k$ "star" points

—To the above are added m "center" points (m must be at least 1, and in most cases it is recommended it be at least 3 so as to allow an estimate of error at that point, which often will represent the current operating conditions of a process).

Using this rule will require many observations; i.e., $2^k + 2k + m$ is large when k is large. However, **it is allowable to use as the base any 2^{k-p} fraction which has the property that no alias set contains more than one 2-factor interaction.** (How the main effects, 3-factor interactions, etc., are aliased is irrelevant. The point is that we cannot have any 2-factor interactions confounded with each other in the fraction chosen.) This often allows us to study many factors, fitting full quadratic models, with numbers of variables that would be prohibitive for other designs (e.g., if one had $k = 9$ factors and tried to run a 3-level full factorial, then $3^9 = 19{,}683$ experiments would be required).

In order to make the construction of this important design clear, let us now look at a number of examples.

5 variables. Setting the defining contrast for our design to $I = -ABCDE$, it follows (using $A^2 = B^2 = C^2 = D^2 = E^2 = 1$) that the following confounding exists:

A, BCDE	*AB, CDE*
B, ACDE	*AC, BDE*
C, ABDE	*AD, BCE*
D, ABCE	*AE, BCD*
E, ABCD	*BC, ADE*
	BD, ACE
	BE, ACD
	CD, ABE
	CE, ABD
	DE, ABC.

If, from theoretical considerations or prior experimentation, only the underscored effects are expected, then this fraction could be used as a design to study the effects of 5 variables with 16 experiments. Otherwise, a CCD could be used with $16+10+3=29$ experiments, since none of the 2-factor interactions are confounded with each other.

6 variables. With 6 factors, say A, B, C, D, E, and F, may we use the defining contrasts $I = ABDE = ACDF$ to find the 2^{6-2} experiments

(1), *aef*, *be*, *abf*, *cf*, *ace*, *bcef*, *abc*, *def*, *ad*, *bdf*, *abde*, *cde*, *acdf*, *bcd*, *abcdef*

to base a CCD on? Here, the confounding implied is

$$I = ABDE = ACDF = BCEF$$
$$A = BDE = CDF = ABCEF$$
$$B = ADE = ABCDF = CEF$$
$$C = ABCDE = ADF = BEF$$
$$D = ABE = ACF = BCDEF$$
$$E = ABD = ACDEF = BCF$$
$$F = ABDEF = ACD = BCE$$
$$AB = DE = BCDF = ACEF$$
$$AC = BCDE = DF = ABEF$$
$$AD = BE = CF = ABCDEF$$
$$AE = BD = CDEF = ABCF$$
$$AF = BDEF = CD = ABCE$$
$$BC = ACDE = ABDF = CF$$
$$BF = ADEF = ABCD = CE$$
$$ABC = CDE = BDF = AEF$$
$$ABF = DEF = BCD = ACE$$

which is a Resolution IV design (no main effect is confounded with any other main effect or 2-factor interaction, but 2-factor interactions *are* confounded with each other . . . so this cannot be used as a base for a CCD).

Analysis.

Once the data specified by the design has been gathered, analysis may proceed easily using the regression programs available in such packages as *BMDP* (in particular, *BMDP9R*) and *SAS*. Here one will (by whatever method is chosen, e.g., "all possible subsets" regression, "best subsets" regression, and "stepwise" regression are some of the most commonly used) **regress the data points on the $k(k+1)$ variables**

$x_1, \ldots, x_k$

$x_1x_2, x_1x_3, \ldots, x_1x_k, x_2x_3, \ldots, x_2x_k, \ldots, x_{k-1}x_k$

$x_1^2, x_2^2, \ldots, x_k^2$.

If several variables are measured as output, a separate regression is run for each. In this case, insight as to which variables are important to the system as a whole may be obtained from the regressions for each of the Y's of interest on all $k(k+1)$ variables. This is easy to obtain in most packages.

Often, many of the $k(k+1)$ variables of the full quadratic model are not significant and (if the model is to be used for prediction) will cause predictions with more variability than if the model had fewer variables. This is so because the mean squared error is the sum of variance and squared bias. **Use of the "best subsets" regression program from *BMDP9R* with the C_p criterion (rather than the widely used R^2 criterion) is appropriate** in this instance.

However, *BMDP9R* does not work (in a reasonable time) at many computer centers when the number $k(k+1)$ is larger than about 25—i.e., when the experiment has more than 5 factors. In such cases, the significances from the full quadratic model, and runs of other (e.g., stepwise) regression programs, may be made first and used to reduce the set of variables to be used in the "best subsets" analysis.

When the first regression runs are made, certain **checks should be made to detect incorrectly entered data, experiments that went awry, and other data problems.** One of the best methods is to **examine the residuals in several ways.** Here by a *residual* we mean the value observed in the experiment, subtracted from the value predicted by the regression model. One should

- Plot the observed *vs.* the predicted on a set of axes;
- Plot the residuals *vs.* the predicted on a set of axes;
- Plot the residuals *vs.* each of the $k(k+1)$ variables, whether they are included in the model that the regression runs are suggesting or not;
- Table the residuals against the observed values, and identify any which are a large number of estimated standard deviations away from the value 0;
- Make normal probability plots of the residuals; and
- Plot the residuals against the time order in which the experiments were run.

These plots and tables, carefully interpreted, can reveal possible data problems in ways in which no formulas yet known (nor any software yet available) can. Possible **data points which seem "odd" should be subjected to a careful examination and consultation** with the personnel who ran the experiments to determine if there is reason to suspect they may be in error. If there is such reason, the data points should be omitted and the analysis should be rerun (or, reanalyzed, with the new values used in the regression development, or else, the corresponding experiment could be rerun with the new values used to execute modeling). The data points **should not be omitted simply because they look odd**; this oddness can also be due to a change in the nature of the relationship by which Y is related to the x's in the neighborhood of the "odd" data point, which can be very valuable information if the Y thus obtained is of a desirable sort and perhaps even unattainable elsewhere in the space of the x's.

A typical plot of residuals against the time order of the experiments is shown below. In this case, we see three outlying observations for further investigation, and we may also think we see a time trend of decreasing residuals. The latter observation will become a real possibility to be investigated if the three "outliers" seem to have been correct values.

Such **time trends are not uncommon**, and they may have one or more causes, which include: increasing experience with the experiment yields increasingly accurate settings of variables and measurements, refined laboratory techniques yield more accurate measurements, knowledge that the experiment is being analyzed statistically increases worker care, and results in the range of later experimentation are less variable than they were in the ranges used in earlier experimentation.

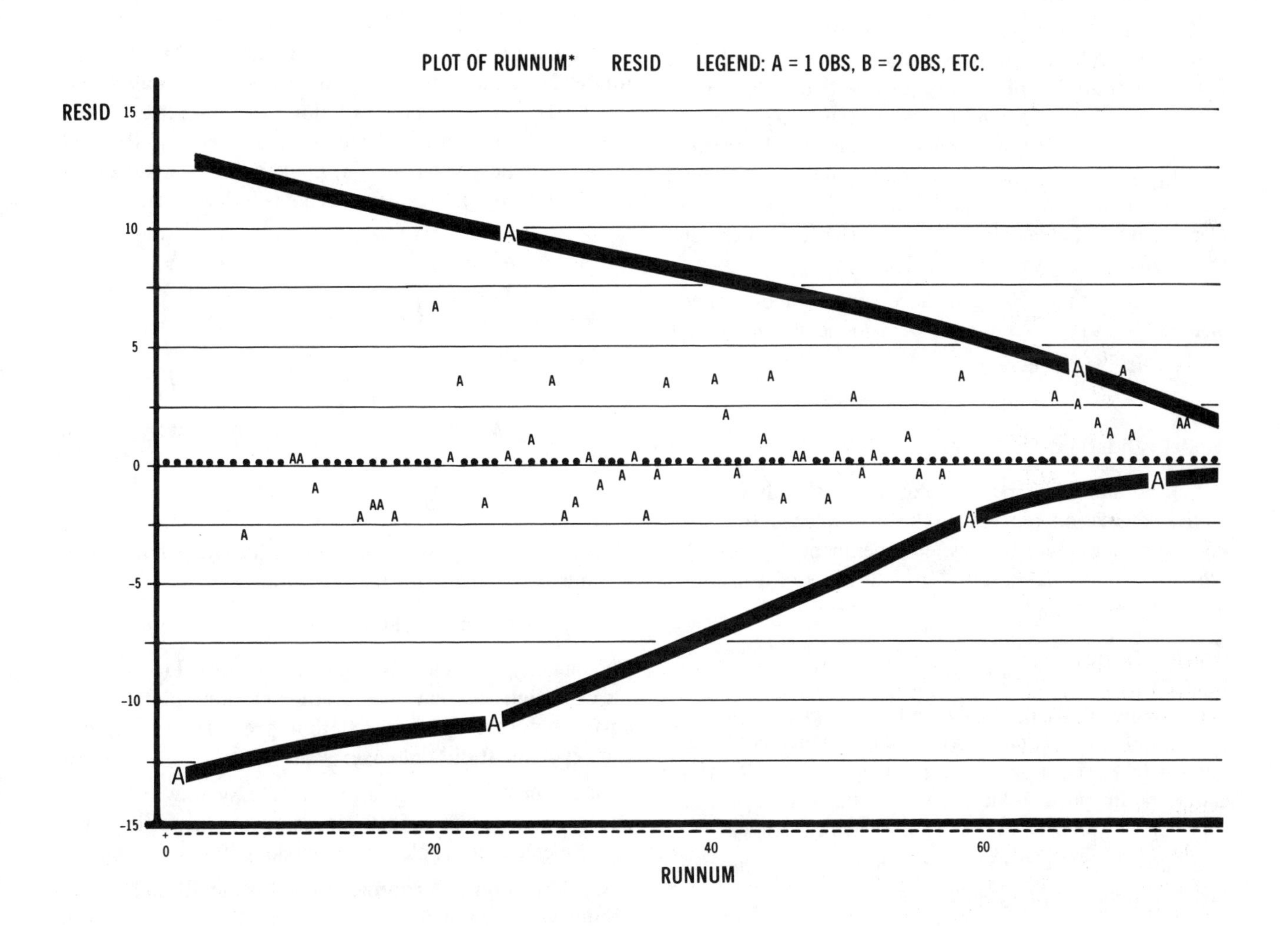

Plot of Run Number* vs. Residual
Legend: A = 1 obs, B = 2 obs, etc.

Outliers in terms of the residual as a multiple of the estimated standard deviation.

Broadly speaking, a certain observation (or group of observations) is an "outlier" if it deviates markedly from the other available data. In the regression case, we consider our "data" to be the residuals $R_1, \ldots, R_n$ (say). Numerous tests are available for detecting outliers in this set (*Journal of Quality Technology*, Vol. 9, pp. 38–41), e.g., those of Dixon, Nair, T, studentized range, Ferguson, coefficient of Kurtosis, Grubbs, and Tietjen-Moore. However, as a simple rule one generally would call a residual an outlier if it lies 2.5, 3, or 4 standard deviations, or further, from the mean of the residuals (N. R. Draper and H. Smith, *Applied Regression Analysis,* John Wiley & Sons, Inc., New York, 1966, p. 94).

It is generally agreed that **outliers should be deleted from the analysis only if they can be traced to specific causes** (such as recording errors or experimental errors, e.g., in apparatus set-up). Otherwise, one should carefully investigate to find whether the outlier may be providing information the other data points are not (e.g., it may arise from an unusual combination of circumstances, which could potentially be very valuable).

Note that, at the 5% significance level, Grubbs (*Technometrics*, 1969) would use a standard deviation multiple which depends on the sample size: 2.55 for 15 observations, 2.91 for 30, and 3.13 for 50.

Model validation.

Validation of the model developed **is desirable when** the model (regression model) has been developed from data derived from a carefully designed experiment. **It is essential when** the model has resulted from analysis of a historical data set.

One method of validation is to *compare model predictions and coefficients with theory*.

A second method of validation is to *collect new data, check model predictions*. Any set of x's which is of major interest (e.g., because of the Y's it is predicted to yield) should be included. (One can in fact allow in advance for such carefully chosen points in a designed experiment.)

A third method of validation is to *compare results with theoretical models and simulated data*.

A fourth method of validation (called *"cross-validation"*) is to develop the model on only a portion of the data, reserving some data to obtain an independent measure of model prediction accuracy. (A computer program for a logical split has been given by Snee, *Technometrics,* 19 (1977): 415–428.)

Model Use and Optimization

Once a model has been developed by the process described above (including analysis of residuals, outliers, etc.), the use to which it is put will depend on the setting in which the experimenter has been working.

One goal is to gain insight about the basic process from the model, and this may be difficult when there are more than three x factors, since 3 factors are the most easily amenable to **plotting of the Y as a function of the x's.** However, even in the case of 3 or more factors, one can gain insight by (e.g., when there are 6 factors in the resulting model) **plotting contours of equal predicted Y on (x_1,x_2) axes at each of a number of values of the factors (x_3, x_4, x_5, x_6)**. On such plots, one may also exclude points where another output Z, also modeled, has undesirable values; extensions to several other measurements on the output are clear. If one uses 2 values for each of the $x_3, \ldots, x_6$, then there will be just 16 plots.

One must **be careful not to put great faith in use of the model far outside the range of x's where it was developed.** This can be assured as follows. Suppose that one has N experiments, and that 4 variables have entered the final model; call those variables r, s, t, and u (e.g., r may be x_2, s may be x_1x_3, etc.). Then let M denote the $N \times 5$ matrix:

$$M = \begin{pmatrix} 1 & r_1 & s_1 & t_1 & u_1 \\ 1 & r_2 & s_2 & t_2 & u_2 \\ 1 & r_3 & s_3 & t_3 & u_3 \\ \vdots & & & & \\ 1 & r_N & s_N & t_N & u_N \end{pmatrix}$$

and $V = (M'M)^{-1}$, a 5×5 matrix. At any point (r, s, t, u), denote a 5-element vector by

$$p = (1 \;\; r \;\; s \;\; t \;\; u).$$

Then ± 2 std. dev. **confidence limits for the prediction of the mean of Y from our model at this (r, s, t, u) are**

$$\pm 2SD = \pm 2 * (\text{std. dev. in model}) * \sqrt{pVp'}\,.$$

The quantity 2 SD **can also be contour plotted (or used as an exclusion relation on the other contour plots) to assure one does not put great faith in predictions in regions the experiment did not cover well.**

Such contour plots are a relatively easy way to find the approximate region of the optimal Y in many problems. Other methods for exploring the model will now be noted.

Optimization of a complex model is facilitated by algorithms which search for the maximum of a function (if you desire the minimum, simply take the negative of the function), and in which undesirable regions can be avoided by simply assigning the function a large value in such a region. One such algorithm is the **Nelder-Mead Function Minimization Algorithm** for which a computer program is available in D. M. Olsson, "A Sequential Simplex Program for Solving Minimization Problems," *Journal of Quality Technology*, *6* (1974): p. 53–57. With this program one can take instances where complex models of several output variables have been developed, and optimize one while assuring that the others stay within desired bounds. One can also search for the most efficient (lowest cost) production process. While the basic program provides the search and takes the function(s) as subroutines, it can be modified to give substantially more output, which is desirable in many instances.

More recently **another method** has been proposed by A.I. Khuri and M. Conlon, "Simultaneous Optimization of Multiple Responses Represented by Polynomial Regression Functions," *Technometrics, 23* (1981): 363–375. This method is also said to be useful for mixture experiments, and (on page 367, section 4.3) the authors offer to send their program to interested parties upon request.

Mixture Experimentation Designs.

Returning to mixture designs, a number of **options for their construction** are now available (e.g., see "Designs for Mixture Experiments Involving Process Variables," *Technometrics, 21* (1979)). These are as follows.

A. A "maximal" 2-dimensional rectangle can be inscribed into the 3-dimensional mixture space, obtaining 2 independent mixture-related variables. These are then combined with the process variables (e.g., 3 others) to yield a space of 5 independent variables in which an appropriate design (e.g., central composite) can be constructed. With 30 points, one could in fact study 6 variables and fit a full quadratic model.

B. One could study the 5 with a lesser number of experimental points. One must, however, be careful not to lose a large part of the mixture space with the inscribing process, and interpretation is rendered more difficult by the process of reducing 3 dependent mixture variables to 2 independent mixture-related variables.

These options are especially valuable when one has both mixture and process variables, which occurs commonly. (In cases where there are only mixture variables, the *XVERT* algorithm of Snee and Marquardt (1974) is useful; it utilizes the extreme vertices, sometimes the constraint plane centroids, and the overall centroid. Centroids of the long edges are also sometimes added. The process is set up only for linear constraints, and custom-designed methods must be used for cases where there are constraints which are not linear, or where the constraints are developed as the experimentation proceeds.)

Recently, I.N. Vuchkov, D.L. Damgaliev, and Ch.A. Yontchev ("Sequentially Generated Second Order Quasi D-Optimal Designs for Experiments with Mixture and Process Variables," *Technometrics, 23* (1981): 233–238) have given **designs for *q* mixture variables and *r* process variables, with $q+r$ up to 7**, which require a number of experiments *N* which is very reasonable in comparison with past methods.

In their design tables, "*T*" is an abbreviation for 0.5 ("to save space" is their explanation). Once one reaches the number of experiments where the quantity D_{eff} starts to be given, the experiments may be terminated if the model is adequate as described in their paper.

Table 1. Discrete Quasi *D*-Optimal Design for $q=3$, $r=1$

Nr	x_1	x_2	x_3	x_4	Nr	x_1	x_2	x_3	x_4	D_{eff}
1	0	1	0	1	11	0	T	T	0	0.888
2	0	0	1	1	12	T	0	T	0	0.916
3	T	0	T	1	13	T	0	T	-1	0.942
4	0	T	T	1	14	0	T	T	-1	0.945
5	1	0	0	0	15	T	T	0	1	0.953
6	1	0	0	-1	16	0	1	0	0	0.958
7	0	1	0	-1	17	0	0	1	0	0.960
8	0	0	1	-1	18	T	0	T	0	0.961
9	T	T	0	-1	19	0	1	0	1	0.961
10	1	0	0	1	20	0	1	0	-1	0.963

Table 2. Discrete Quasi *D*-Optimal Design for $q=4$, $r=1$

Nr	x_1	x_2	x_3	x_4	x_5	Nr	x_1	x_2	x_3	x_4	x_5	D_{eff}
1	1	0	0	0	1	16	0	T	0	T	0	0.880
2	0	1	0	0	1	17	T	0	0	T	0	0.899
3	0	0	1	0	1	18	T	0	T	0	0	0.916
4	0	0	0	1	1	19	T	T	0	0	-1	0.932
5	T	T	0	0	1	20	0	0	T	T	-1	0.950
6	T	0	T	0	1	21	T	0	T	0	-1	0.949
7	0	T	0	T	1	22	0	T	T	0	1	0.949
8	0	0	T	T	1	23	T	0	0	T	1	0.952
9	0	T	T	0	0	24	0	T	0	T	-1	0.957
10	1	0	0	0	-1	25	0	1	0	0	0	0.958
11	0	1	0	0	-1	26	1	0	0	0	-1	0.958
12	0	0	1	0	-1	27	0	0	0	1	0	0.959
13	0	0	0	1	-1	28	0	0	1	0	1	0.962
14	T	0	0	T	-1	29	1	0	0	0	1	0.965
15	0	T	T	0	-1	30	0	0	1	0	-1	0.969

Table 3. Discrete Quasi *D*-Optimal Design for $q=5$, $r=1$

Nr	x_1	x_2	x_3	x_4	x_5	x_6	Nr	x_1	x_2	x_3	x_4	x_5	x_6	D_{eff}
1	1	0	0	0	0	1	22	T	0	0	T	0	0	0.866
2	0	1	0	0	0	1	23	0	T	T	0	0	0	0.879
3	0	0	1	0	0	1	24	0	0	T	T	0	-1	0.889
4	0	0	0	1	0	1	25	0	T	0	0	T	1	0.898
5	0	0	0	0	1	1	26	T	0	T	0	0	-1	0.909
6	T	0	T	0	0	1	27	0	0	0	T	T	-1	0.920
7	T	0	0	T	0	1	28	0	T	0	T	0	-1	0.931
8	0	T	T	0	0	1	29	T	0	0	0	T	1	0.944
9	0	T	0	T	0	1	30	0	0	T	0	T	1	0.956
10	0	0	0	T	T	1	31	T	0	T	0	0	0	0.956
11	T	T	0	0	0	0	32	0	0	T	T	0	1	0.956
12	0	0	T	T	0	0	33	T	T	0	0	0	1	0.957
13	1	0	0	0	0	-1	34	T	0	0	T	0	-1	0.959
14	0	1	0	0	0	-1	35	0	T	T	0	0	-1	0.962
15	0	0	1	0	0	-1	36	0	0	0	1	0	0	0.961
16	0	0	0	1	0	-1	37	0	0	1	0	0	1	0.961
17	0	0	0	0	1	-1	38	1	0	0	0	0	1	0.961
18	T	T	0	0	0	-1	39	0	1	0	0	0	-1	0.962
19	T	0	0	0	T	-1	40	0	0	0	0	1	1	0.964
20	0	T	0	0	T	-1	41	0	T	0	T	0	0	0.966
21	0	0	T	0	T	-1	42	0	0	1	0	0	-1	0.968

Table 4. Discrete Quasi *D*-Optimal Design for $q=6$, $r=1$

Nr	x_1	x_2	x_3	x_4	x_5	x_6	x_7	Nr	x_1	x_2	x_3	x_4	x_5	x_6	x_7	D_{eff}
1	1	0	0	0	0	0	1	25	0	T	0	0	T	0	-1	
2	0	1	0	0	0	0	1	26	0	0	T	T	0	0	-1	
3	0	0	1	0	0	0	1	27	0	0	T	0	T	0	-1	
4	0	0	0	1	0	0	1	28	0	0	0	T	0	T	-1	
5	0	0	0	0	1	0	1	29	0	T	T	0	0	0	0	0.860
6	0	0	0	0	0	1	1	30	T	0	T	0	0	0	0	0.869
7	T	T	0	0	0	0	1	31	T	0	0	0	T	0	-1	0.875
8	T	0	T	0	0	0	1	32	0	0	0	0	T	T	-1	0.881
9	T	0	0	T	0	0	1	33	0	0	T	T	0	0	1	0.888
10	0	T	0	T	0	0	1	34	T	T	0	0	0	0	-1	0.895
11	0	0	T	0	0	T	1	35	0	0	0	T	0	T	1	0.902
12	0	0	0	T	T	0	1	36	0	T	0	0	T	0	1	0.910
13	0	0	0	0	T	T	1	37	0	0	T	0	0	T	-1	0.918
14	T	0	0	0	T	0	0	38	T	0	0	T	0	0	-1	0.926
15	0	T	0	T	0	0	0	39	0	T	0	0	0	T	-1	0.935
16	0	T	0	0	0	T	0	40	0	0	T	0	T	0	1	0.944
17	1	0	0	0	0	0	-1	41	T	0	0	0	0	T	1	0.953
18	0	1	0	0	0	0	-1	42	0	0	0	T	T	0	-1	0.963
19	0	0	1	0	0	0	-1	43	0	T	0	0	0	T	1	0.964
20	0	0	0	1	0	0	-1	44	0	T	0	T	0	0	-1	0.964
21	0	0	0	0	1	0	-1	45	T	0	0	0	T	0	1	0.964
22	0	0	0	0	0	1	-1	46	0	T	T	0	0	0	1	0.965
23	T	0	0	0	0	T	-1	47	T	0	T	0	0	0	-1	0.967
24	0	T	T	0	0	0	-1	48	T	0	0	T	0	0	0	0.967

Table 5. Discrete Quasi *D*-Optimal Design for $q=3$, $r=2$

Nr	x_1	x_2	x_3	x_4	x_5	Nr	x_1	x_2	x_3	x_4	x_5	D_{eff}
1	1	0	0	0	1	16	T	0	T	1	1	0.842
2	0	1	0	1	1	17	0	0	1	-1	-1	0.870
3	0	0	1	0	1	18	T	T	0	-1	0	0.895
4	0	1	0	-1	1	19	1	0	0	1	1	0.910
5	T	0	T	-1	1	20	0	T	T	-1	1	0.924
6	0	T	T	1	0	21	0	0	1	1	1	0.938
7	T	0	T	0	0	22	1	0	0	-1	1	0.952
8	0	0	1	-1	0	23	T	T	0	0	1	0.962
9	1	0	0	1	-1	24	0	1	0	0	0	0.964
10	0	1	0	1	-1	25	T	0	T	-1	-1	0.966
11	0	0	1	1	-1	26	0	0	1	1	-1	0.966
12	T	T	0	1	-1	27	0	0	1	-1	1	0.967
13	0	T	T	0	-1	28	1	0	0	1	0	0.968
14	1	0	0	-1	-1	29	0	1	0	-1	-1	0.970
15	0	1	0	-1	-1	30	0	1	0	1	1	0.971

Table 6. Discrete Quasi *D*-Optimal Design for $q=4$, $r=2$

Nr	x_1	x_2	x_3	x_4	x_5	x_6	Nr	x_1	x_2	x_3	x_4	x_5	x_6	D_{eff}
1	1	0	0	0	1	1	22	0	0	T	T	0	0	0.816
2	1	0	0	0	-1	-1	23	0	1	0	0	1	-1	0.834
3	0	1	0	0	1	1	24	0	0	1	0	1	-1	0.850
4	0	1	0	0	-1	-1	25	1	0	0	0	1	-1	0.868
5	0	0	1	0	1	1	26	0	1	0	0	-1	0	0.880
6	0	0	1	0	-1	-1	27	0	0	1	0	-1	1	0.892
7	0	0	0	1	1	1	28	1	0	0	0	-1	1	0.906
8	0	0	0	1	-1	1	29	0	1	0	0	0	1	0.918
9	0	0	0	1	1	-1	30	T	T	0	0	0	0	0.926
10	0	0	0	1	-1	-1	31	0	0	T	T	-1	-1	0.932
11	T	T	0	0	-1	1	32	T	0	0	T	1	-1	0.938
12	T	T	0	0	1	-1	33	0	T	0	T	1	0	0.943
13	T	0	T	0	-1	1	34	0	T	0	T	0	-1	0.949
14	T	0	T	0	1	-1	35	0	T	T	0	0	0	0.952
15	T	0	0	T	0	1	36	T	0	T	0	-1	-1	0.955
16	T	0	0	T	-1	0	37	T	0	T	0	1	1	0.957
17	0	T	T	0	-1	1	38	1	0	0	0	0	-1	0.956
18	0	T	T	0	1	-1	39	1	0	0	0	1	0	0.956
19	0	T	0	T	-1	1	40	0	T	T	0	-1	-1	0.956
20	0	T	0	T	0	0	41	T	T	0	0	1	1	0.957
21	0	0	T	T	1	1	42	0	0	0	1	-1	-1	0.957

Table 7. Discrete Quasi *D*-Optimal Design for $q=5$, $r=2$

Nr	x_1	x_2	x_3	x_4	x_5	x_6	x_7	Nr	x_1	x_2	x_3	x_4	x_5	x_6	x_7	D_{eff}
1	1	0	0	0	0	1	0	25	T	T	0	0	0	0	0	
2	0	1	0	0	0	-1	1	26	0	0	T	T	0	0	0	
3	0	1	0	0	0	0	-1	27	0	0	T	0	T	1	0	
4	1	0	0	0	0	-1	-1	28	0	T	0	T	0	1	-1	
5	0	0	1	0	0	-1	1	29	0	T	0	0	T	0	1	0.786
6	0	0	0	1	0	1	0	30	0	0	0	T	T	1	1	0.804
7	0	0	1	0	0	1	-1	31	0	0	1	0	0	-1	-1	0.819
8	0	0	0	1	0	-1	-1	32	T	T	0	0	0	1	-1	0.831
9	0	0	0	0	1	-1	1	33	T	0	0	T	0	1	1	0.843
10	0	0	0	0	1	1	-1	34	0	0	0	0	1	-1	-1	0.852
11	0	1	0	0	0	1	1	35	0	0	T	T	0	1	-1	0.860
12	0	0	1	0	0	1	1	36	T	0	T	0	0	1	1	0.868
13	0	0	0	0	1	1	1	37	0	T	T	0	0	-1	-1	0.875
14	1	0	0	0	0	-1	1	38	T	0	0	0	T	1	-1	0.883
15	0	0	0	1	0	-1	1	39	0	1	0	0	0	1	-1	0.890
16	0	T	0	0	T	-1	-1	40	0	0	0	1	0	1	-1	0.895
17	T	0	0	T	0	0	-1	41	1	0	0	0	0	1	-1	0.901
18	T	0	0	T	0	-1	0	42	0	1	0	0	0	-1	-1	0.906
19	0	T	T	0	0	0	0	43	0	0	0	1	0	1	1	0.911
20	0	0	T	0	T	-1	-1	44	1	0	0	0	0	1	1	0.917
21	T	0	0	0	T	0	0	45	0	0	T	0	T	0	1	0.922
22	0	0	0	T	T	0	0	46	0	T	0	0	T	1	0	0.926
23	0	T	0	T	0	0	0	47	0	0	T	T	0	-1	1	0.931
24	T	0	T	0	0	0	0	48	T	0	0	0	T	-1	1	0.935

Table 8. Discrete Quasi *D*-Optimal Design for $q=3$, $r=3$

Nr	x_1	x_2	x_3	x_4	x_5	x_6	Nr	x_1	x_2	x_3	x_4	x_5	x_6	D_{eff}
1	0	0	1	1	1	1	22	T	T	0	-1	1	-1	0.826
2	1	0	0	-1	1	1	23	T	0	T	0	1	-1	0.844
3	0	1	0	-1	1	1	24	0	T	T	0	-1	1	0.859
4	1	0	0	1	-1	1	25	T	T	0	1	1	1	0.876
5	0	1	0	1	-1	1	26	1	0	0	-1	-1	1	0.891
6	T	0	T	1	-1	1	27	T	0	T	-1	0	1	0.903
7	0	0	1	-1	-1	1	28	0	1	0	0	1	-1	0.911
8	1	0	0	1	1	-1	29	1	0	0	1	-1	-1	0.916
9	0	1	0	1	1	0	30	1	0	0	-1	0	-1	0.921
10	0	0	1	1	1	-1	31	1	0	0	1	1	1	0.927
11	0	T	T	1	0	-1	32	0	T	T	-1	-1	0	0.930
12	0	T	T	-1	1	-1	33	0	0	1	0	-1	-1	0.933
13	T	T	0	1	-1	-1	34	0	0	1	-1	1	1	0.936
14	1	0	0	-1	-1	-1	35	0	1	0	-1	-1	1	0.940
15	0	0	1	-1	-1	-1	36	0	1	0	1	-1	-1	0.944
16	0	1	0	0	-1	-1	37	0	1	0	1	0	1	0.947
17	0	1	0	-1	0	-1	38	T	0	T	-1	-1	-1	0.951
18	0	0	1	-1	1	0	39	0	0	1	1	-1	1	0.954
19	0	0	1	1	-1	0	40	0	T	T	0	1	1	0.957
20	0	1	0	-1	-1	0	41	1	0	0	-1	1	0	0.960
21	T	T	0	0	0	0	42	0	1	0	1	1	-1	0.963

Table 9. Discrete Quasi *D*-Optimal Design for $q=4$, $r=3$

Nr	x_1	x_2	x_3	x_4	x_5	x_6	x_7	Nr	x_1	x_2	x_3	x_4	x_5	x_6	x_7	D_{eff}
1	1	0	0	0	0	1	1	25	T	0	0	T	1	1	0	
2	0	1	0	0	1	1	1	26	1	0	0	0	-1	-1	0	
3	0	0	0	1	0	1	1	27	0	0	0	1	-1	-1	0	
4	0	1	0	0	-1	1	1	28	0	T	0	T	0	0	0	
5	0	0	1	0	-1	1	1	29	0	0	0	1	1	-1	1	0.803
6	0	0	1	0	1	-1	1	30	T	T	0	0	0	0	0	0.822
7	1	0	0	0	1	-1	1	31	0	0	T	T	1	-1	-1	0.839
8	T	0	T	0	-1	-1	1	32	T	0	T	0	1	1	1	0.854
9	0	T	0	T	-1	-1	1	33	0	T	T	0	1	-1	0	0.865
10	0	0	1	0	1	1	-1	34	1	0	0	0	-1	1	1	0.877
11	T	T	0	0	1	1	-1	35	0	0	T	T	1	0	1	0.886
12	0	T	T	0	1	1	-1	36	T	T	0	0	-1	-1	-1	0.894
13	0	0	0	1	-1	1	-1	37	0	T	T	0	-1	-1	1	0.901
14	T	0	T	0	-1	1	-1	38	0	T	0	T	1	1	1	0.907
15	0	0	T	T	-1	1	-1	39	T	0	T	0	1	-1	-1	0.913
16	0	1	0	0	1	-1	-1	40	0	T	T	0	-1	0	-1	0.916
17	0	0	0	1	1	0	-1	41	T	T	0	0	-1	1	1	0.920
18	0	0	1	0	-1	-1	-1	42	0	0	0	1	-1	-1	-1	0.924
19	T	0	0	T	0	-1	-1	43	0	0	0	1	1	1	-1	0.927
20	0	1	0	0	0	-1	1	44	0	0	0	1	-1	1	1	0.931
21	0	1	0	0	0	1	-1	45	0	0	T	T	-1	-1	1	0.933
22	T	0	0	T	-1	0	1	46	0	1	0	0	-1	1	0	0.936
23	1	0	0	0	1	0	-1	47	0	0	1	0	1	1	1	0.938
24	0	1	0	0	-1	-1	-1	48	0	1	0	0	1	-1	1	0.940

Table 10. Discrete Quasi *D*-Optimal Design for $q=3$, $r=4$

Nr	x_1	x_2	x_3	x_4	x_5	x_6	x_7	Nr	x_1	x_2	x_3	x_4	x_5	x_6	x_7	D_{eff}
1	1	0	0	1	1	1	1	25	1	0	0	1	-1	1	0	
2	0	1	0	-1	1	1	1	26	1	0	0	-1	1	-1	0	
3	1	0	0	-1	-1	1	1	27	T	0	T	0	0	0	0	
4	0	T	T	1	1	-1	1	28	0	T	T	-1	-1	-1	0	
5	0	0	1	-1	0	-1	1	29	0	1	0	1	-1	1	1	0.830
6	T	T	0	-1	1	0	1	30	0	0	1	1	1	-1	0	0.846
7	T	0	T	1	-1	-1	1	31	T	0	T	-1	1	-1	1	0.860
8	0	1	0	1	1	1	-1	32	T	T	0	0	1	1	-1	0.872
9	0	0	1	1	1	1	-1	33	0	1	0	1	1	-1	1	0.881
10	1	0	0	-1	1	1	-1	34	1	0	0	1	-1	1	-1	0.890
11	0	T	T	1	-1	1	-1	35	0	0	1	-1	-1	1	1	0.897
12	0	1	0	-1	-1	1	-1	36	T	T	0	1	-1	-1	0	0.905
13	0	0	1	-1	-1	1	-1	37	T	0	T	1	0	-1	-1	0.911
14	1	0	0	1	1	-1	-1	38	0	T	T	0	1	-1	-1	0.917
15	0	1	0	-1	1	-1	-1	39	0	1	0	-1	-1	0	1	0.922
16	0	1	0	1	-1	-1	-1	40	0	T	T	-1	1	1	0	0.926
17	1	0	0	-1	-1	-1	-1	41	1	0	0	0	-1	-1	1	0.930
18	0	0	1	0	1	1	1	42	0	0	1	1	0	1	-1	0.933
19	0	1	0	0	-1	-1	1	43	T	0	T	1	1	1	1	0.935
20	0	0	1	0	-1	-1	-1	44	0	1	0	1	1	0	-1	0.938
21	T	T	0	1	0	1	1	45	1	0	0	-1	0	1	1	0.939
22	1	0	0	1	0	-1	1	46	T	T	0	-1	-1	0	-1	0.941
23	0	0	1	1	-1	0	1	47	0	0	1	-1	1	-1	-1	0.943
24	0	0	1	-1	1	0	-1	48	0	1	0	-1	0	-1	-1	0.945

Part IV: Fitting Distributions to Simulation Input/Output Data

The primary purpose of this Part is the investigation of some of the **methods for fitting probability density functions to input and output data** associated with computer simulation experiments. Some of the data that will be used for illustration was obtained from computer system performance and evaluation studies. The determination of a probability density function allows the analyst to verify that certain underlying distributions exist; thus, statistical methods (such as those discussed by Mamrak and DeRuyter [2]), which assume the existence of such distributions, can be used.

Considerable advantages can be gained if it could be shown that a known probability density function fits the data. In addition to settling such issues as "approximate normality," the known density function can be used as a model to generate data for further study. The use of a density function to generate data will, almost always, be less expensive and demand less computing resources than will the processes that yield the actual data.

A priori knowledge of the distributions of random variables related to computer simulation experiments is unusual. **It therefore is necessary to accumulate sufficient data** (through simulation or other means) **to permit proposing probability density functions for the random variables** in question. After a specific distribution is chosen, classical tests such as the chi-square goodness of fit test or the Kolmogorov-Smirnov test could be used to judge the quality of the approximation.

The identification of a suitable density function for a given data set is therefore a central problem in the modeling of systems as well as in the fitting of distributions to data. **Three dissimilar methods for fitting density functions to data will be considered** in subsequent articles. The *first*, **the kernel estimate,** approaches the problem in the most intuitive way by using a variation of the histogram of the data to produce a density function. The *second* method, due to **Ramberg and Schmeiser** [4], uses the generalized lambda distribution to fit density functions which have some of the same statistical properties (the first four population moments are set equal to the first four sample moments) as the original data. The *third* method, due to **Schmidt and Taylor** [6], uses a combination of polynomial and logarithmic functions to approximate the inverse distribution function. The following articles discuss these methods, their use, and their suitability in specific situations.

Kernel Estimates

The use of the sample histogram as an estimator of the population density function has intuitive appeal. Tarter and Kronmal [7], however, point out a number of serious shortcomings in the use of histograms. One problem is the dependence of the histogram on the number of intervals chosen for its construction. Degenerate cases result when too few or too many intervals are used. There are no universal guidelines for an appropriate number of intervals other than the general principle that it should depend on sample size.

Another problem, closely related to the decision of the number of intervals, is the choice of interval length. This is often decided on the basis of data range and the number of intervals. A third problem is that of the location of intervals. The usual recommendation of locating intervals at midpoints which are "round" numbers is too arbitrary to be of any use.

The kernel method for estimating probability density functions attempts to minimize the errors inherent in the use of histograms. First, the assumption that every data point within an interval has equal weight (i.e., is located at the midpoint) is dropped. Instead, data points are weighted at their actual locations. In terms of the histogram, this stipulates the location of rectangular blocks centered at the actual data points rather than at the class midpoint.

Another improvement results from the consideration of shapes other than fixed rectangles to be located at the data points. This is done through the use of a kernel function.

$$K((x - x_j)/h(n)) \qquad (1)$$

where $x_1, x_2, \ldots, x_n$ are independent and identically distributed random variables, and $h(n)$ corresponds to the width of the blocks. The estimate of the density function is given by

$$\hat{f}(x) = C(n) \sum_j K((x - x_j)/h(n)) \qquad (2)$$

where $C(n) = 1/(nh(n))$. Tarter and Kronmal [7] suggest experimenting with various values of $h(n)$ until the graph of $\hat{f}(x)$ is "smooth." As a starting point, they recommend choosing $h(n)$ proportional to $1/\sqrt[5]{n}$. This last recommendation has little merit since almost any number can be construed to be proportional to $1/\sqrt[5]{n}$. The "smoothness" condition is somewhat reminiscent of the problems related to the choice of the number and length of intervals of histograms.

The specification of the kernel function, K, is also critical to the quality of estimation. The function

$$K((x - x_j)/h(n)) = \begin{cases} 0.5 \text{ if } |x - x_j| < h(n) \\ 0 \text{ otherwise} \end{cases} \qquad (3)$$

is a possibility suggested by Tarter and Kronmal [7]. In this case, if each x_j were to be replaced by its corresponding class midpoint, the resulting density estimator, $\hat{f}$, would be the usual histogram.

The following seven examples are used to illustrate the use of the kernel function given in (3) for the construction of a probability density estimator, $\hat{f}$.

Example 1. The algorithm $x = \sum_{i=1}^{12} r_i - 6$; $x = 20x + 200$, where $r_1, r_2, \ldots, r_{12}$ are numbers from the interval [0,1] generated by a congruential random number generator, was used to obtain Sample 1 of **100 values from $N(200,400)$**, the normal population with a mean of 200 and a variance of 400.

Sample 1: DATA – $N(200,400)$

222	199	220	185	160	194	188	217	212	218
219	206	225	204	172	146	171	183	165	165
225	234	189	185	206	154	215	213	204	210
196	196	208	179	199	179	205	176	231	206
186	190	192	187	213	187	217	189	200	223
218	209	199	223	196	192	181	168	213	199
173	206	175	233	203	193	216	201	225	187
218	170	160	185	160	205	227	184	200	194
208	167	201	172	212	171	207	178	161	225
210	194	168	206	214	217	228	208	180	201

Figures 1a, 1b, 1c, and 1d illustrate $\hat{f}$ for choices of $h(n)$ = 5, 15, 20, and 50, respectively. It can be seen that as $h(n)$ increases from 5 to 20, $\hat{f}$ becomes smoother and more bell shaped. Values of $h(n)$ slightly larger than 20 do not improve the quality of $\hat{f}$, and values that are substantially larger than 20 (such as 50 in Figure 1d) produce "flattened out" curves.

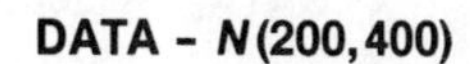

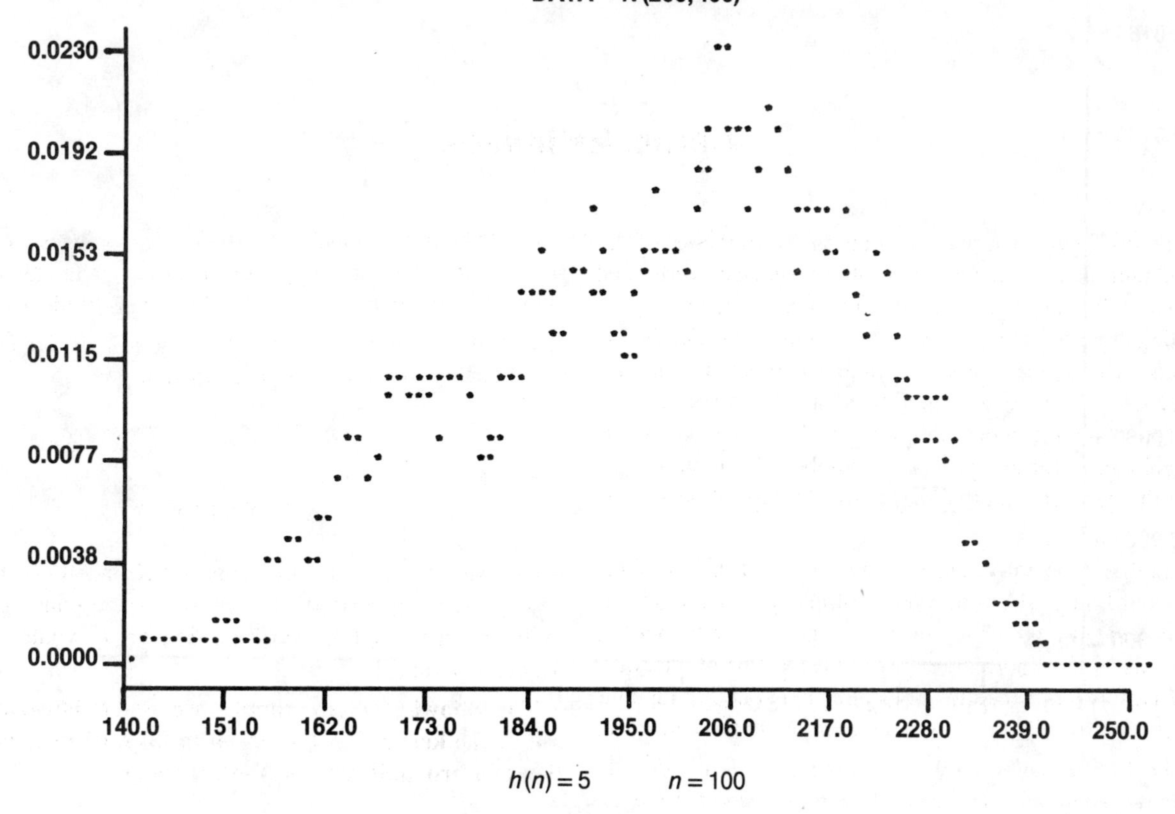

Fig. 1a

DATA - $N(200, 400)$

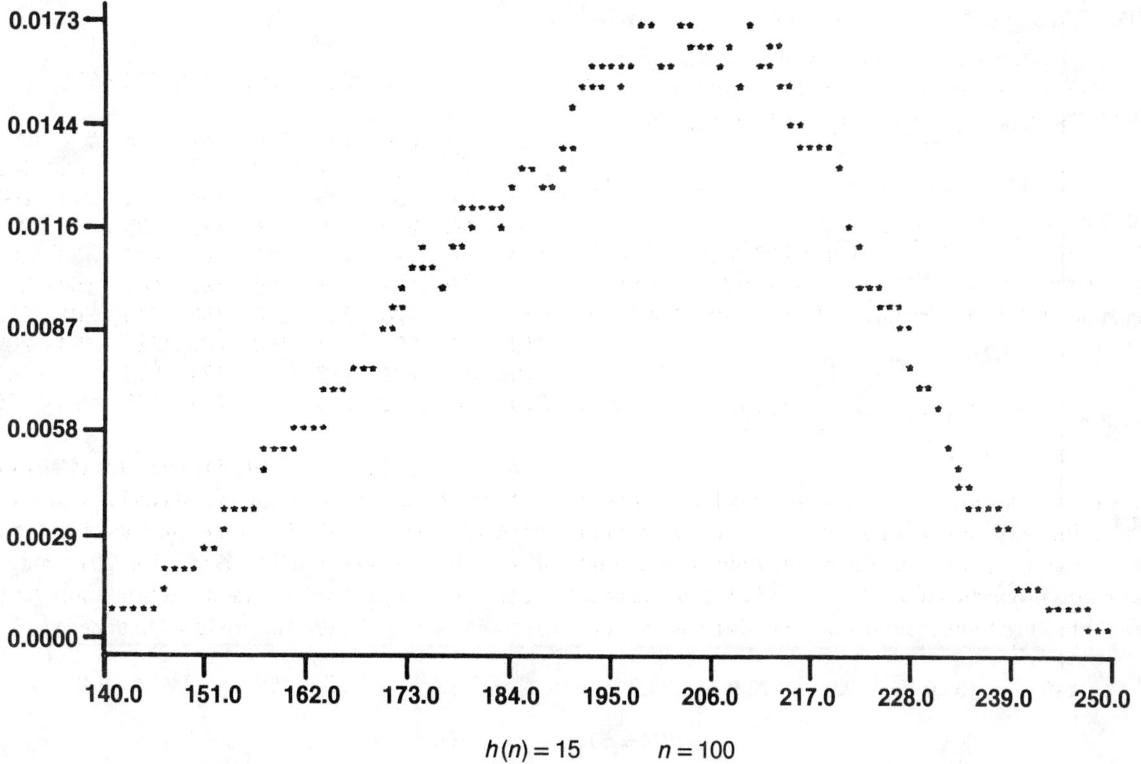

$h(n) = 15 \qquad n = 100$

Fig. 1b

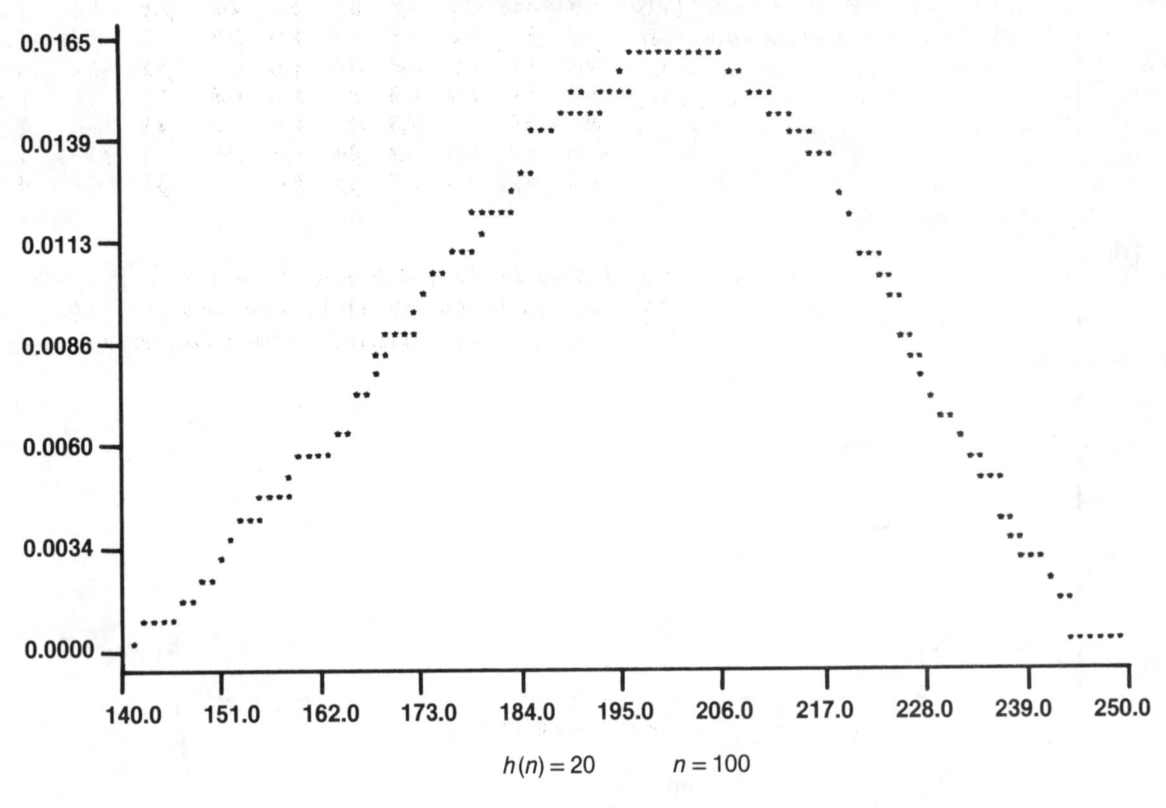
DATA - N(200,400)
0.0165
0.0139
0.0113
0.0086
0.0060
0.0034
0.0000
140.0
151.0
162.0
173.0
184.0
195.0
206.0
217.0
228.0
239.0
250.0
h(n) = 20
n = 100

Fig. 1c

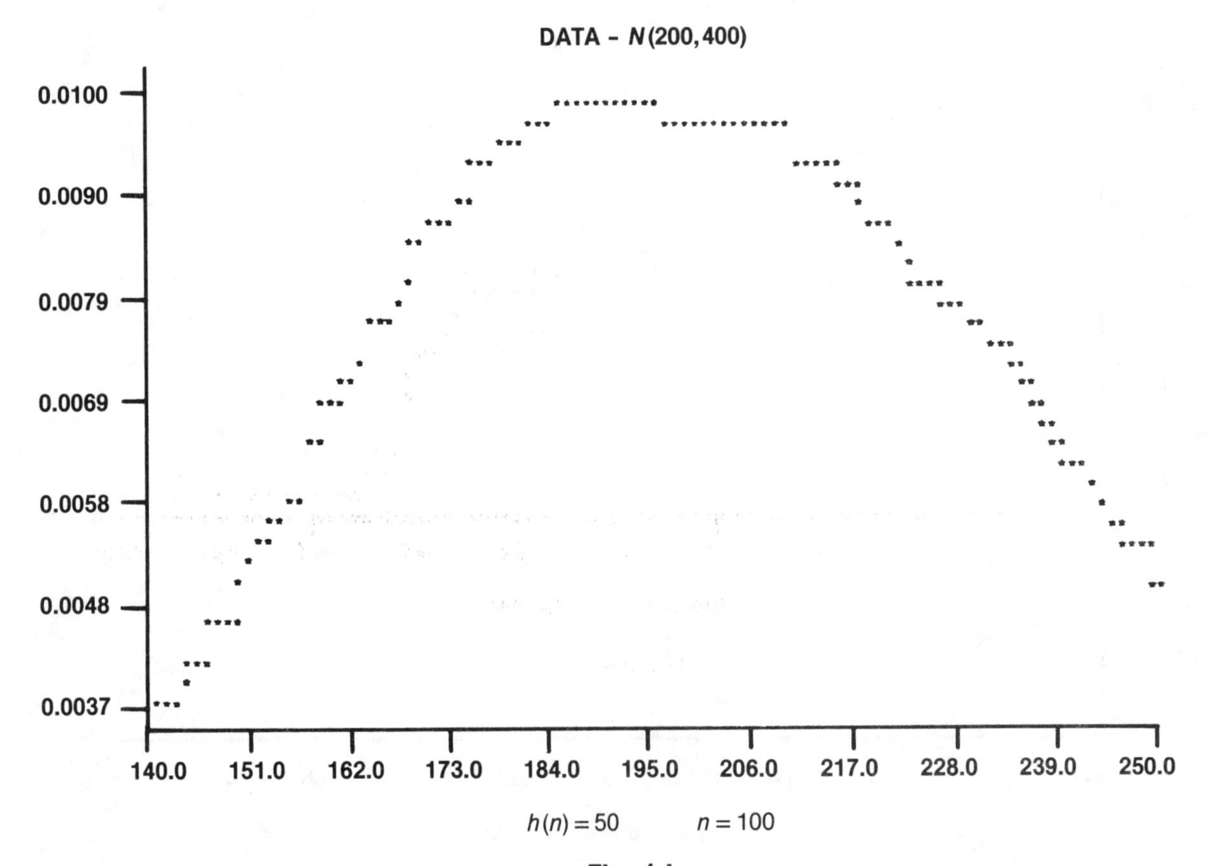
DATA - N(200,400)
0.0100
0.0090
0.0079
0.0069
0.0058
0.0048
0.0037
140.0
151.0
162.0
173.0
184.0
195.0
206.0
217.0
228.0
239.0
250.0
h(n) = 50
n = 100

Fig. 1d

Example 2. To try the method on **an asymmetric data set**, **Sample 2 of 100** was generated **from a chi-square population** with five degrees of freedom. The sum of the squares of five standard normal variates, generated by the method discussed in Example 1, are added to obtain the chi-square observations.

Sample 2: DATA-x^2, df = 5

5.2	8.1	2.8	9.5	4.5	1.9	7.3	3.7	6.6	4.9
1.1	4.6	4.6	4.3	5.8	2.1	5.1	2.1	1.6	5.9
6.0	2.5	2.7	6.8	4.2	4.8	2.5	3.0	1.1	3.9
5.7	6.9	3.9	8.2	3.5	5.4	2.9	2.8	9.0	2.3
3.0	2.7	8.3	5.6	6.1	2.0	2.3	6.3	6.8	6.0
13.9	1.0	6.7	8.8	9.9	4.0	8.6	13.2	2.5	10.6
6.2	5.8	3.0	6.8	3.5	8.9	0.8	1.8	9.0	6.3
8.1	6.0	6.1	2.8	9.1	2.6	1.2	3.9	4.6	5.6
4.8	4.4	6.6	3.2	3.8	4.6	2.9	4.8	12.1	2.0
4.1	5.2	7.5	2.7	3.2	3.8	6.3	5.7	3.0	8.7

Figures 2a, 2b, and 2c show $\hat{f}$ with $h(n) = 1, 3$, and 5, respectively. The best choice here seems to be $h(n) = 3$ since for $h(n) = 5$ (Figure 2c), $\hat{f}$ has already become too flat.

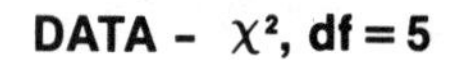

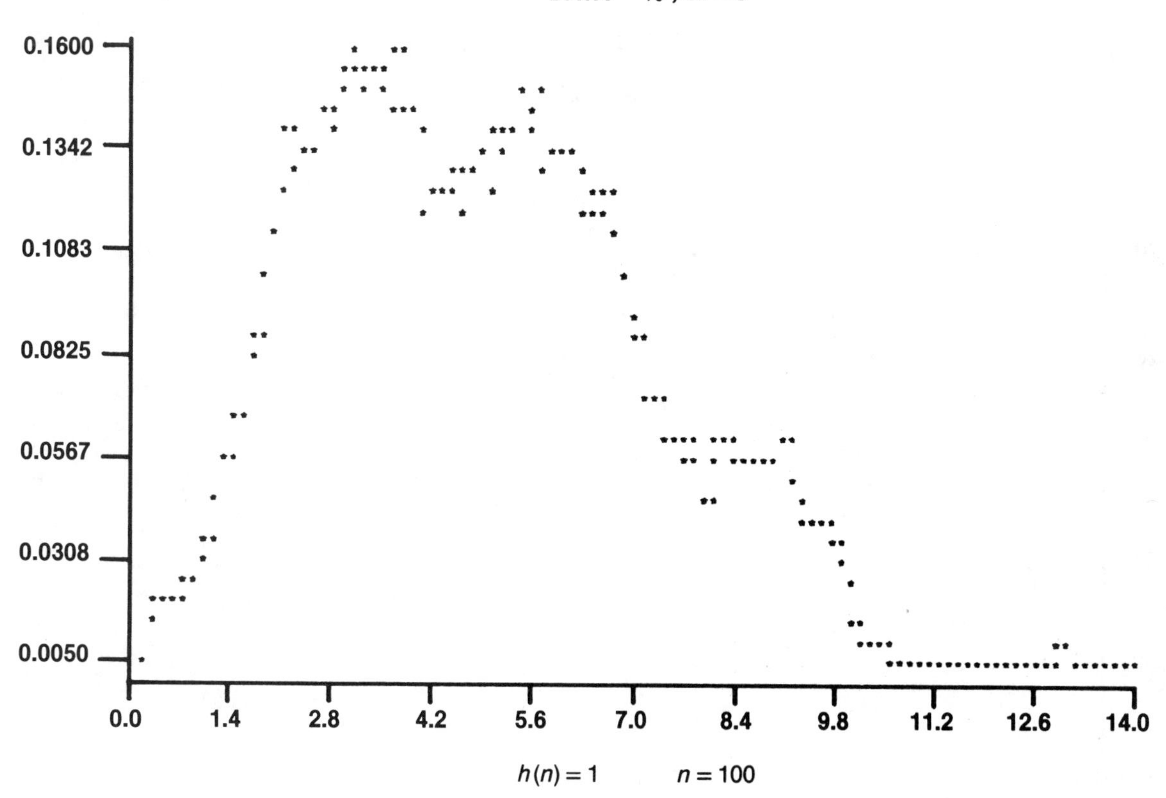

Fig. 2a

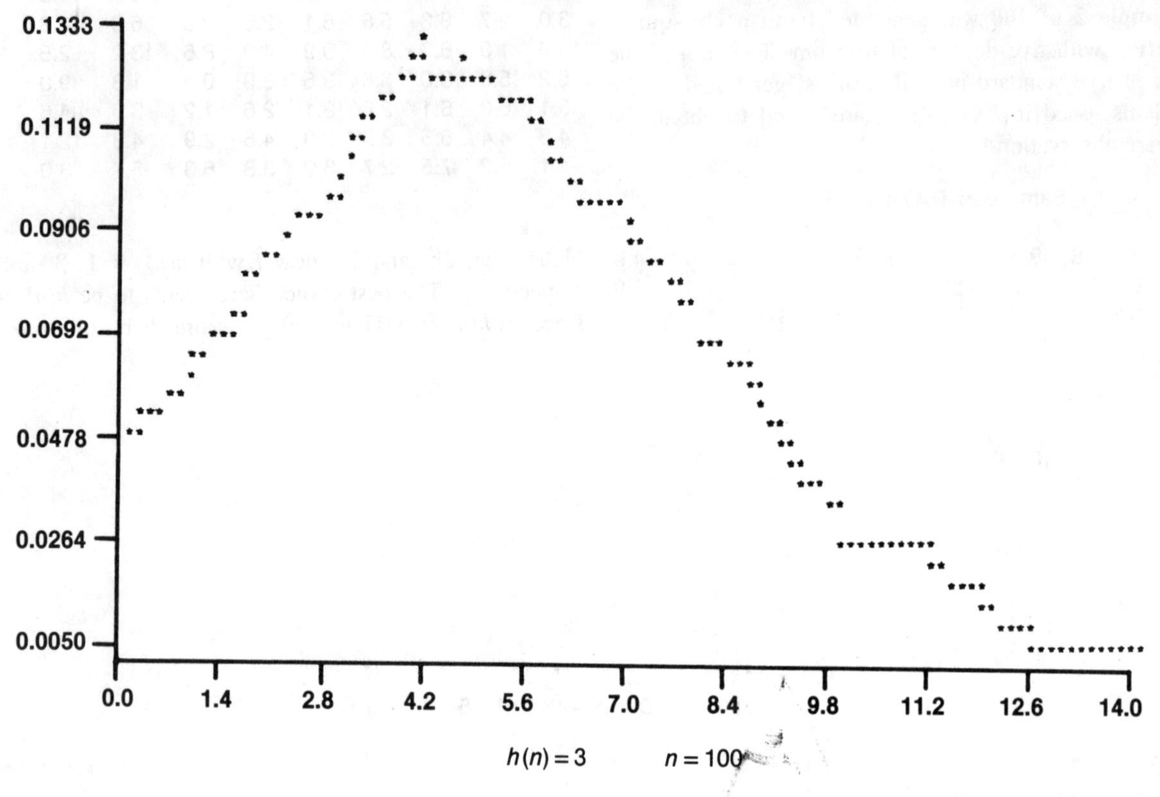
DATA - χ^2, df = 5
0.1333
0.1119
0.0906
0.0692
0.0478
0.0264
0.0050
0.0
1.4
2.8
4.2
5.6
7.0
8.4
9.8
11.2
12.6
14.0
$h(n) = 3$
$n = 100$

Fig. 2b

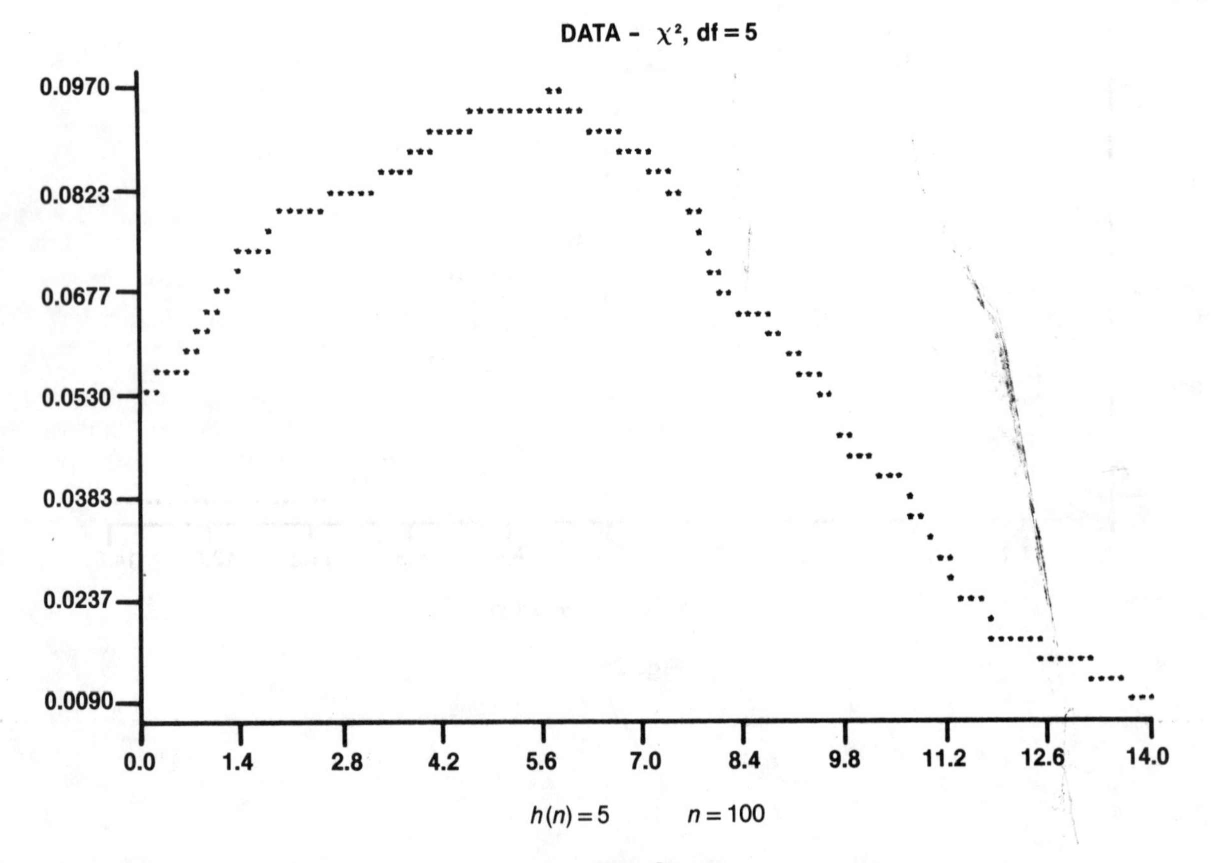
DATA - χ^2, df = 5
0.0970
0.0823
0.0677
0.0530
0.0383
0.0237
0.0090
0.0
1.4
2.8
4.2
5.6
7.0
8.4
9.8
11.2
12.6
14.0
$h(n) = 5$
$n = 100$

Fig. 2c

Example 3. In Sample 3, **100 random numbers** are generated on the interval [100,200] by computing $100r_i + 100$, where a congruential random number generator is used to produce r_i on the interval [0,1].

Sample 3: DATA – U(100,200)

165	170	113	131	154	132	168	165	183	197
144	145	170	152	151	100	137	184	182	148
164	114	116	170	167	146	139	147	110	131
122	151	158	170	114	110	191	162	105	110
195	156	176	135	114	186	115	193	109	113
138	100	116	155	125	157	110	115	188	130
192	110	112	179	174	101	140	134	159	198
126	145	176	193	125	150	138	134	181	187
100	151	191	162	116	130	105	165	172	171
113	162	160	165	121	181	138	165	197	148

Figures 3a, 3b, 3c, and 3d illustrate $\hat{f}$ for $h(n) = 5, 15, 25$, and 35, respectively. The value $h(n) = 5$ is too small to even indicate the presence of any type of curve. **As $h(n)$ gets large,** $\hat{f}$ does become smoother; however, it **never attains a shape that is suitable for $U(100, 200)$**, the uniform distribution on [100, 200].

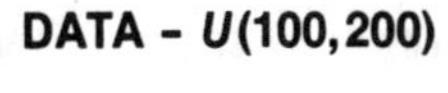

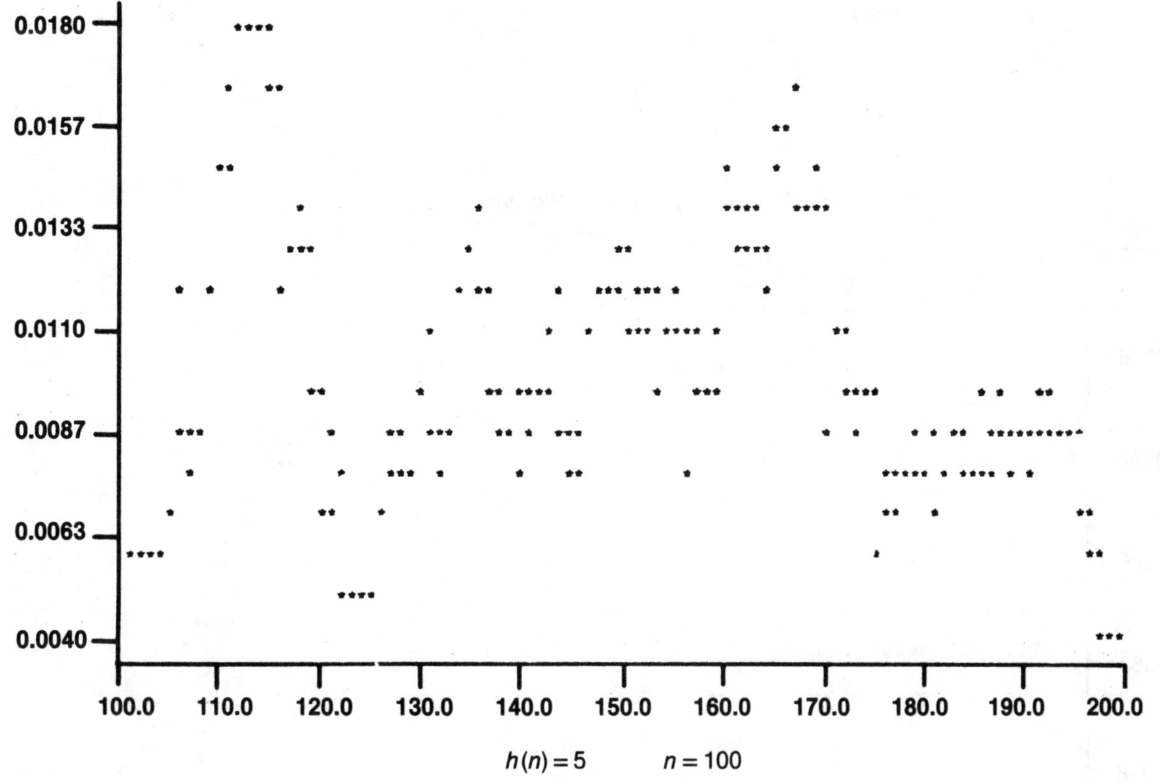

$h(n) = 5$ $n = 100$

Fig. 3a

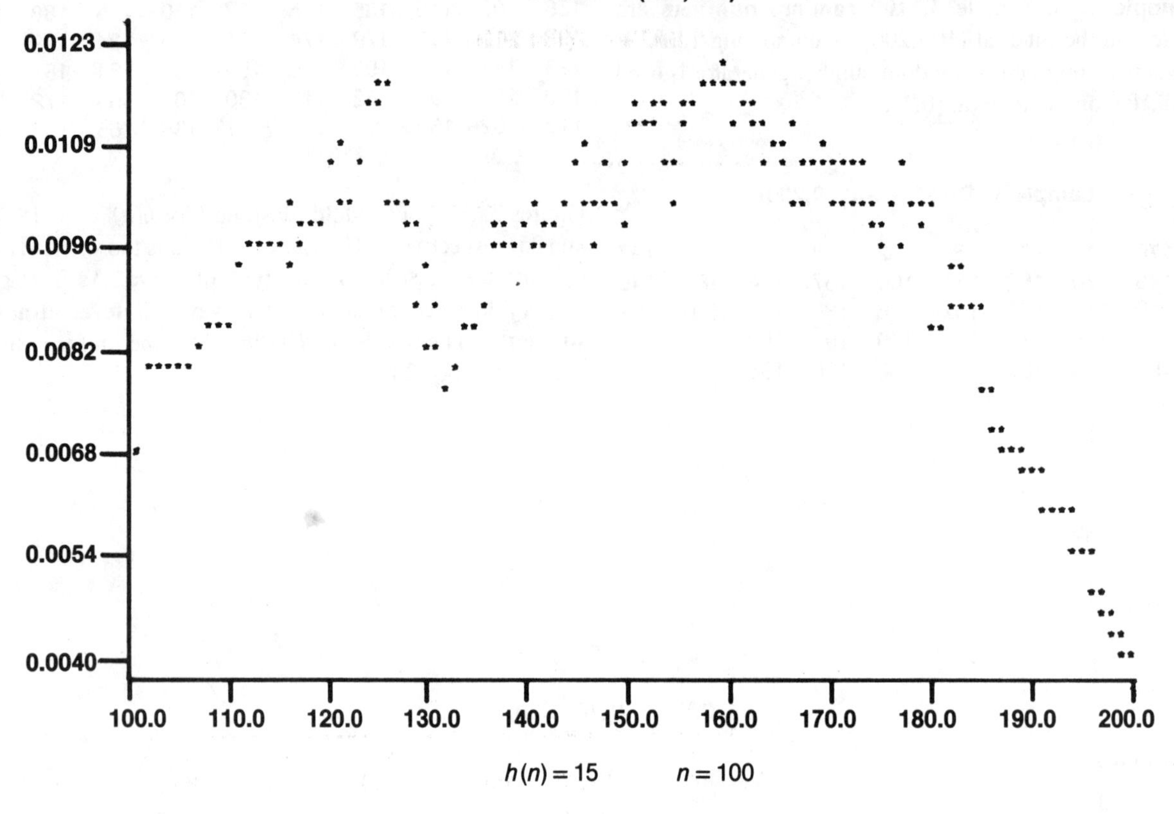
DATA - U(100,200)
0.0123
0.0109
0.0096
0.0082
0.0068
0.0054
0.0040
100.0
110.0
120.0
130.0
140.0
150.0
160.0
170.0
180.0
190.0
200.0
h(n) = 15
n = 100

Fig. 3b

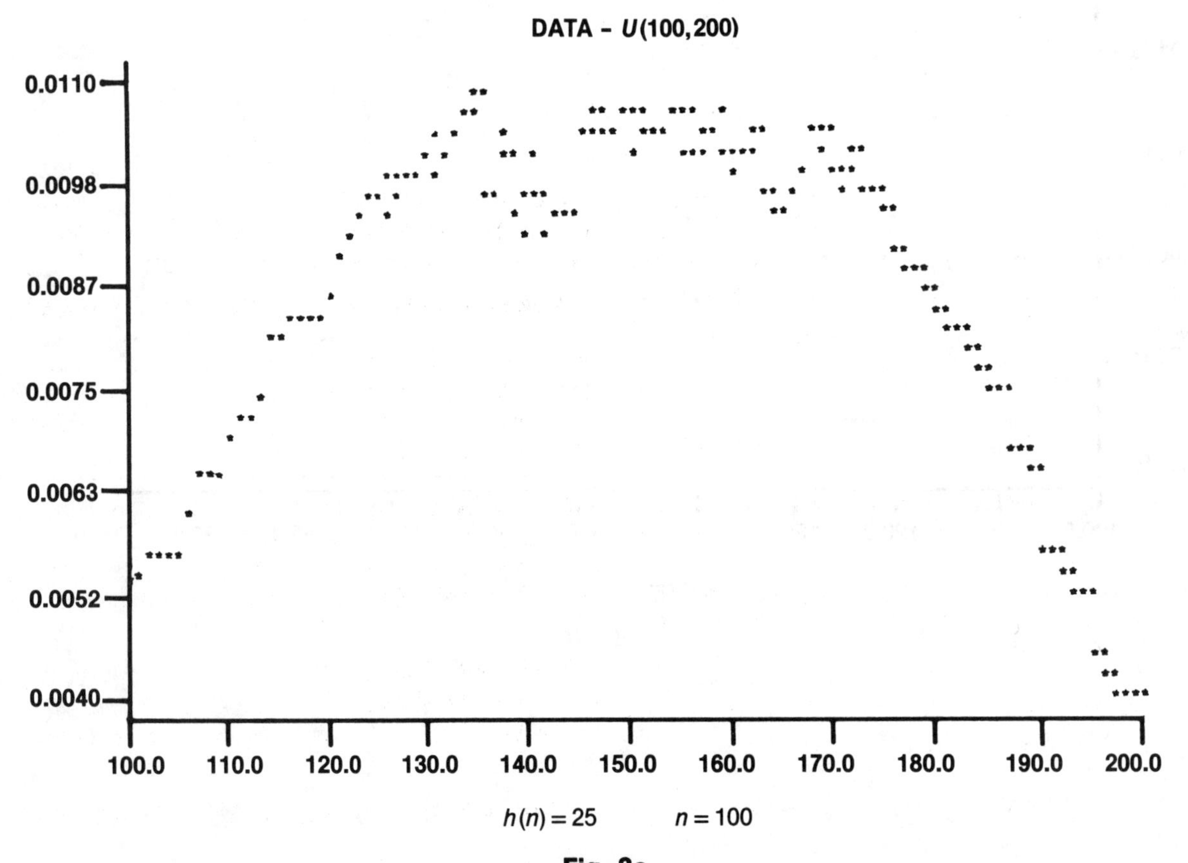
DATA - U(100,200)
0.0110
0.0098
0.0087
0.0075
0.0063
0.0052
0.0040
100.0
110.0
120.0
130.0
140.0
150.0
160.0
170.0
180.0
190.0
200.0
h(n) = 25
n = 100

Fig. 3c

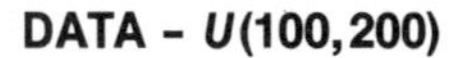

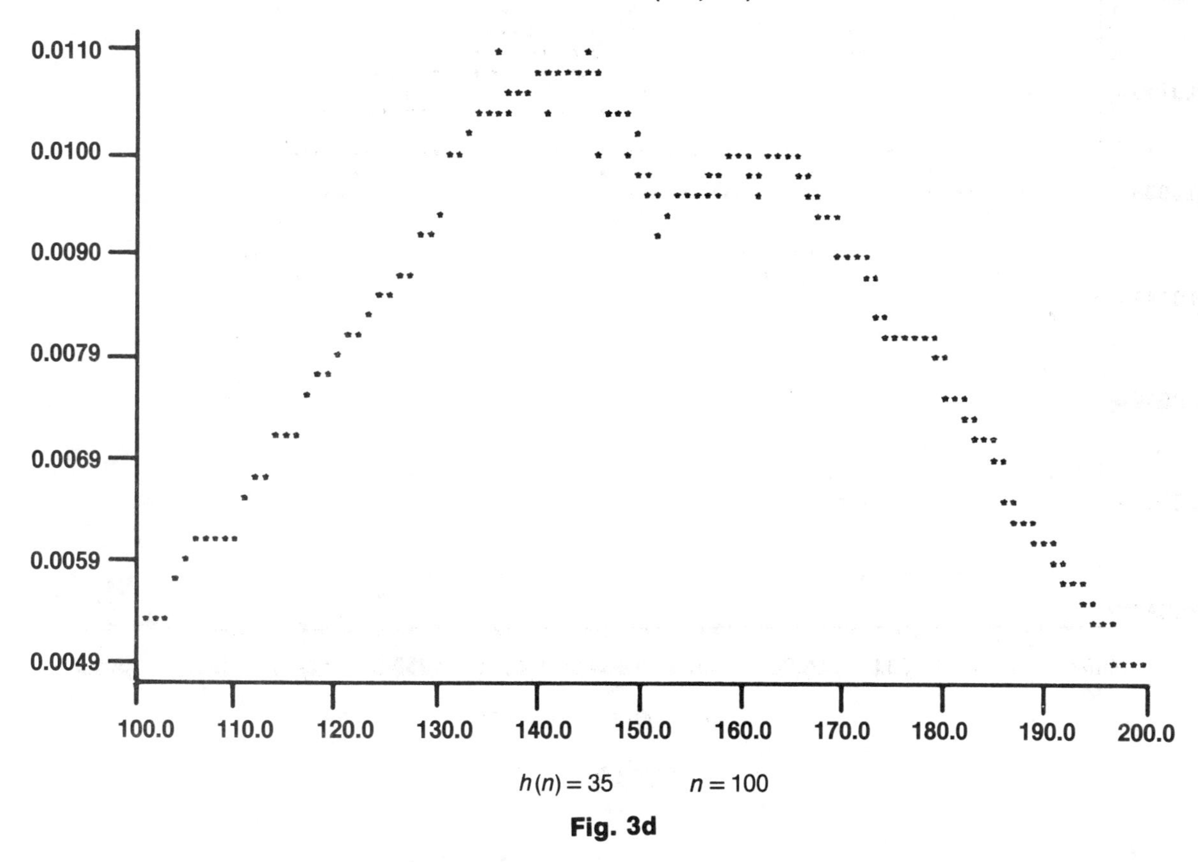

$h(n) = 35$ $\quad n = 100$

Fig. 3d

Example 4. The data used in this and the following three examples are **connect times (in seconds) and elapsed times (in seconds) for the execution of an interactive script** on an IBM 370 under the TSO option and a DEC System 10 running under a TOPS 10 monitor. The data in Sample 4 consists of connect times on the IBM 370.

Sample 4: DATA — connect times for IBM 370 (TSO)

1295	427	412	695	315	733	476	345	913	666
801	597	302	376	577	455	762	447	318	543
526	571	527	364	396	437	913	605	485	523
372	396	316	337	348	319	379	431	460	497
519	386	547	450	483	497	427	583	540	433
635	705	431	482	565	364	523	429	430	498
814	1136	440	875	813	1210	1089	609	604	894
607	742	748	357	429	336	371	320	369	881
835	1039	373	1827	1027	581	618	469	468	581

For $h(n) = 50$ and 100, $\hat{f}$ is shown in Figures 4a and 4b. Even though Figure 4b ($h(n) = 100$) seems rather rough, **larger values of $h(n)$ do not improve $\hat{f}$.**

Example 5. Sample 5 consists of connect times (in seconds) on the DEC System 10. As Figure 5 shows, $h(n) = 75$ seems to work reasonably well.

Sample 5: Data—connect times for DEC system 10

287	478	639	422	771	260	255	272	289	330
327	357	311	395	273	284	246	270	271	283
325	266	276	272	279	302	350	325	325	347
296	280	256	334	264	290	283	302	273	278
274	275	276	282	271	280	280	278	288	264
332	326	339	506	428	347	304	299	305	304
296	243	313	260	309	283	261	305	288	366
279	265	240	239	260	289	310	333	278	299
295	281	282	294	328	340	288	305	352	342

Example 6. The data in Sample 6 represents elapsed times (in seconds) on an IBM 370 (TSO). Although not very satisfactory, $h(n) = 75$ is as good a choice as can be made in this case.

Sample 6: Data—elapsed times for IBM 370 (TSO)

735	415	402	200	200	303	501	140	131	153

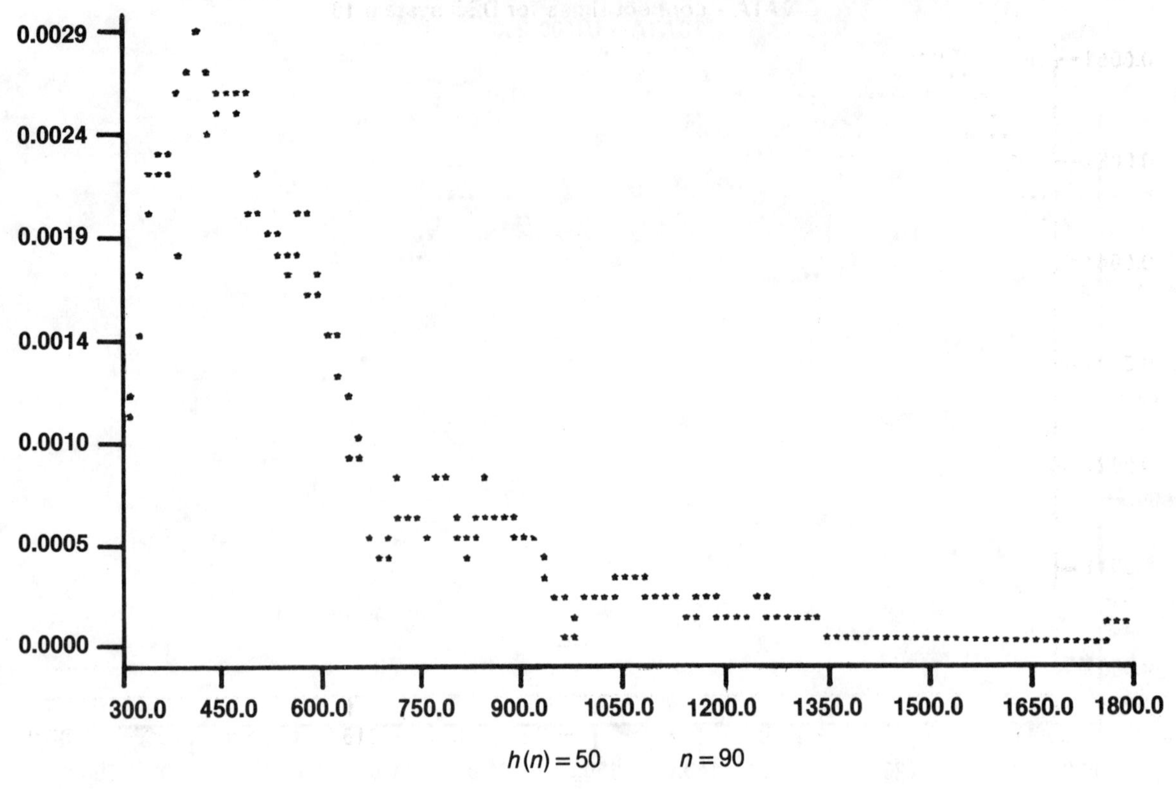
DATA - connect times for IBM 370 (TSO)
0.0029
0.0024
0.0019
0.0014
0.0010
0.0005
0.0000
300.0
450.0
600.0
750.0
900.0
1050.0
1200.0
1350.0
1500.0
1650.0
1800.0
$h(n) = 50$ $n = 90$

Fig. 4a

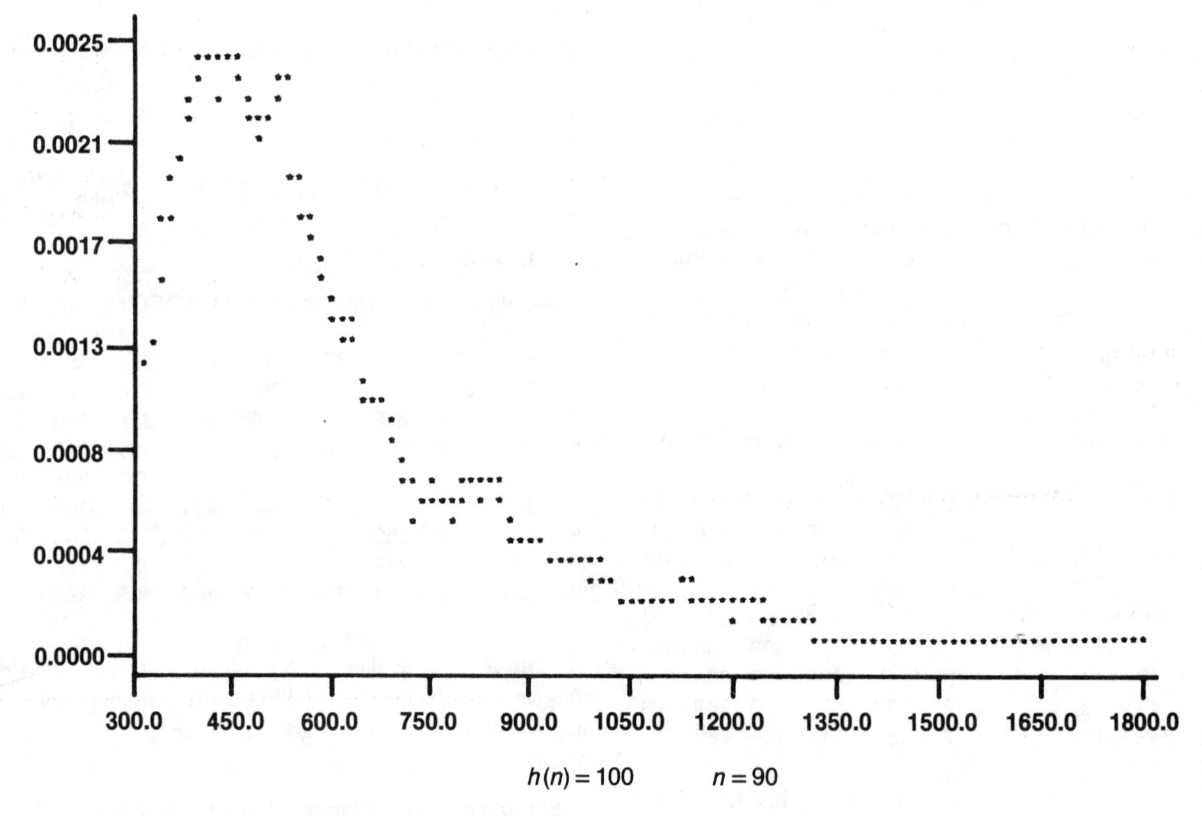
DATA - connect times for IBM 370 (TSO)
0.0025
0.0021
0.0017
0.0013
0.0008
0.0004
0.0000
300.0
450.0
600.0
750.0
900.0
1050.0
1200.0
1350.0
1500.0
1650.0
1800.0
$h(n) = 100$ $n = 90$

Fig. 4b

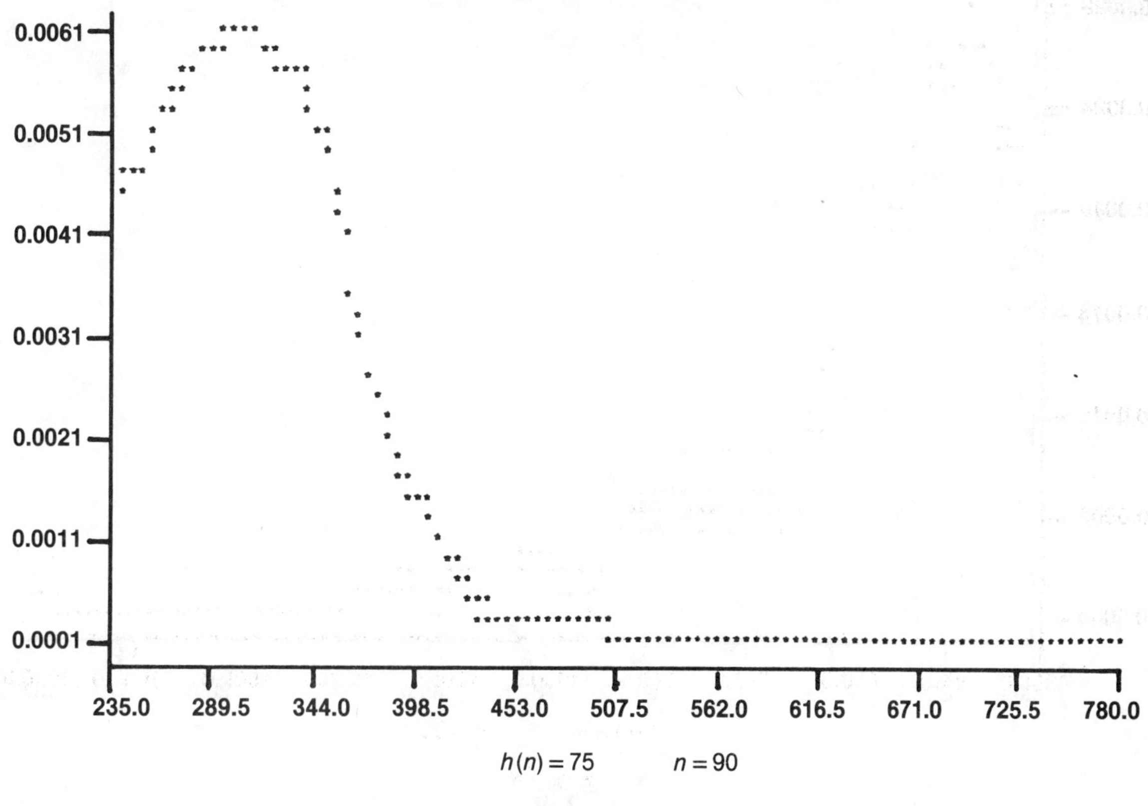

Fig. 5

137	167	132	128	432	375	572	180	574	397
106	425	287	231	188	205	567	294	256	287
138	278	292	401	288	243	164	638	250	234
311	516	150	1366	570	239	193	263	306	225
302	345	197	202	205	134	141	126	131	134
143	131	136	163	215	547	759	298	293	453
322	388	387	147	199	147	245	233	202	241
129	152	160	143	224	270	296	194	231	326

Example 7. These (Sample 7) are elapsed times (in seconds) for the DEC system 10. A good choice is $h(n) = 25$, which does moderately well.

Sample 7: Data—elapsed times for DEC system 10

114	231	193	175	407	110	114	110	129	150
127	160	118	144	94	105	97	104	112	105
116	116	119	120	117	143	139	131	129	135
133	126	114	176	112	102	112	123	103	109
106	98	97	102	98	120	133	118	126	121
129	132	132	209	180	111	110	118	119	117
120	95	143	97	115	113	104	112	117	178
111	113	116	112	113	110	111	121	104	104
125	111	105	134	111	152	120	138	138	146

It is clear from examples 1-7 that, for a variety of reasons, **even the best choice of $h(n)$ may not produce a satisfactory $\hat{f}$.** Figures 3d, 4c, and 6 show that for optimal $h(n)$, $\hat{f}$ is still "rough." The main reason for this is the use of (3) as the kernel function. With this choice of K, a data point x_j contributes 0.5 to $\hat{f}(x)$ as long as $|x - x_j| \leq h(n)$. A choice of K where the contribution of x_j to $\hat{f}$ would take a continuum of values depending on the proximity of x to x_j may resolve this difficulty. For a fixed x_j, the general shape of K should be something like

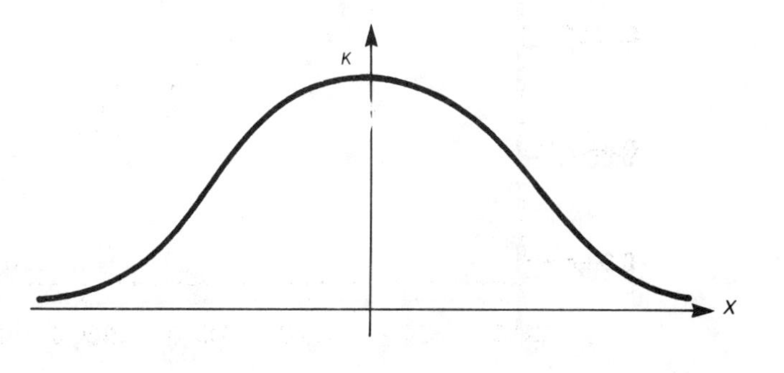

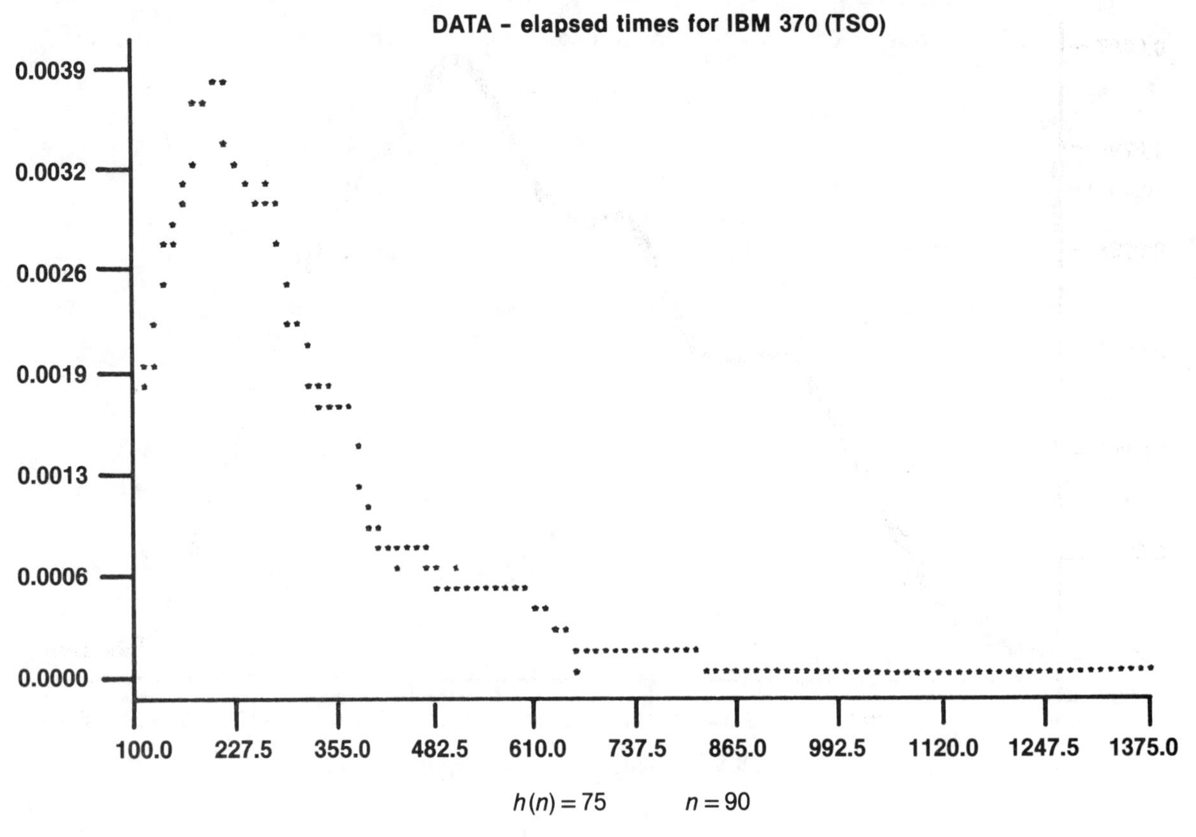
DATA - elapsed times for IBM 370 (TSO)
0.0039
0.0032
0.0026
0.0019
0.0013
0.0006
0.0000
100.0
227.5
355.0
482.5
610.0
737.5
865.0
992.5
1120.0
1247.5
1375.0
$h(n) = 75$
$n = 90$

Fig. 6

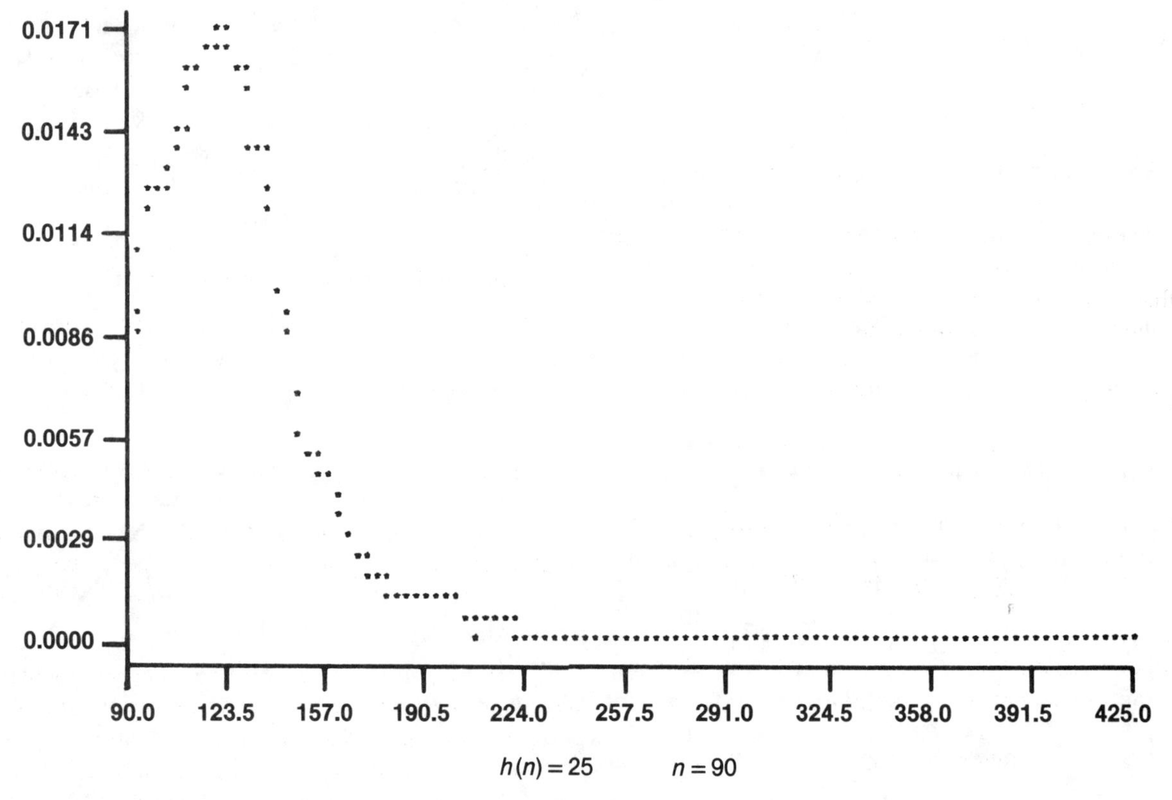
DATA - elapsed times for DEC system 10
0.0171
0.0143
0.0114
0.0086
0.0057
0.0029
0.0000
90.0
123.5
157.0
190.5
224.0
257.5
291.0
324.5
358.0
391.5
425.0
$h(n) = 25$
$n = 90$

Fig. 7

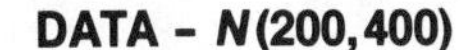

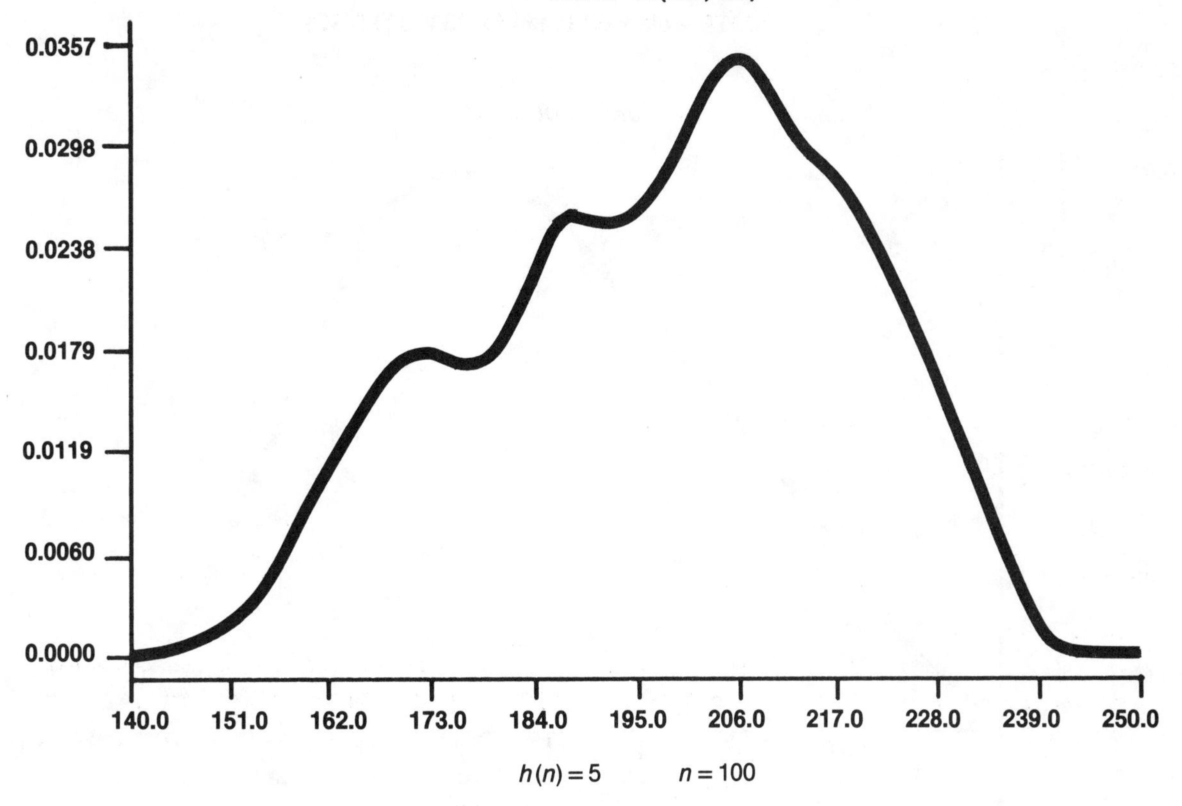

Fig. 1a′

Functions such as

$$K((x - x_j)/h(n)) = K(z) = \exp(-t|z|); \quad t \geqslant 0 \tag{4}$$

or

$$K((x - x_j)/h(n)) = K(z) = \exp(-tz^2); \quad t \geqslant 0 \tag{5}$$

have the desired property, but there is no assurance that the use of (4) or (5) will produce an $\hat{f}$ which is a density function (i.e., that $\hat{f}$ will be non-negative and that the area under $\hat{f}$ will be 1). The following result solves this problem.

THEOREM: *If* $z = (x - x_j)/h(n)$ *and* $C = 1/(nh(n)\sqrt{\pi})$, *then*

$$f = C \sum_j \exp(-z^2) \text{ is a density function.}$$

PROOF: Clearly $f \geqslant 0$ and we need to show that $A = 1$ where

$$A = \int_{\mathbb{R}} f(x)dx = \int_{\mathbb{R}} C \sum_j \exp(-(x - x_j)^2/h^2(n))dx. \tag{6}$$

Using the substitution $u = \sqrt{2}(x - x_j)/h(n)$ and changing the order of summation and integration,

$$A = C \sum_j \int_{\mathbb{R}} (h(n)/\sqrt{2}) \exp(-u^2/.2)du$$
$$= (Ch(n)/\sqrt{2}) \sum_j \int_{\mathbb{R}} \exp.(-u^2/2)du. \tag{7}$$

Since $g(x) = \exp(-x^2/2)/\sqrt{2\pi}$ is the density function for the standard normal distribution, then

$$\int_{\mathbb{R}} \exp(-x^2/2)dx = \sqrt{2\pi}.$$

Using this result in (7), we obtain

$$A = Cnh(n)\sqrt{\pi}.$$

The choice of $C = 1/(nh(n)\sqrt{\pi})$ makes $A = 1$.

The following figures illustrate the use of the kernel function

$$K((x - x_j/h(n)) = \exp(-(x - x_j)^2/h^2(n)) \tag{8}$$

on the data sets described in Examples 1-7. For purposes of comparison, Figure 1a′ corresponds to Figure 1a, etc. Clearly **this modification produces much smoother estimators of the density function.** This is particularly apparent when Figures 1a, 1c, 2a, 2b, 3a, 3b, 3d, 4a, 4b, and 6 are compared with their counterparts.

To get an understanding of how good an estimator $\hat{f}$ actually is, the best results obtained for each data are shown superimposed on the data histograms in figures 1″ through 7″. The kernel described in (8) was used since this turns out to be better in all cases. Figure 1″ refers to the data of Example 1, and so on.

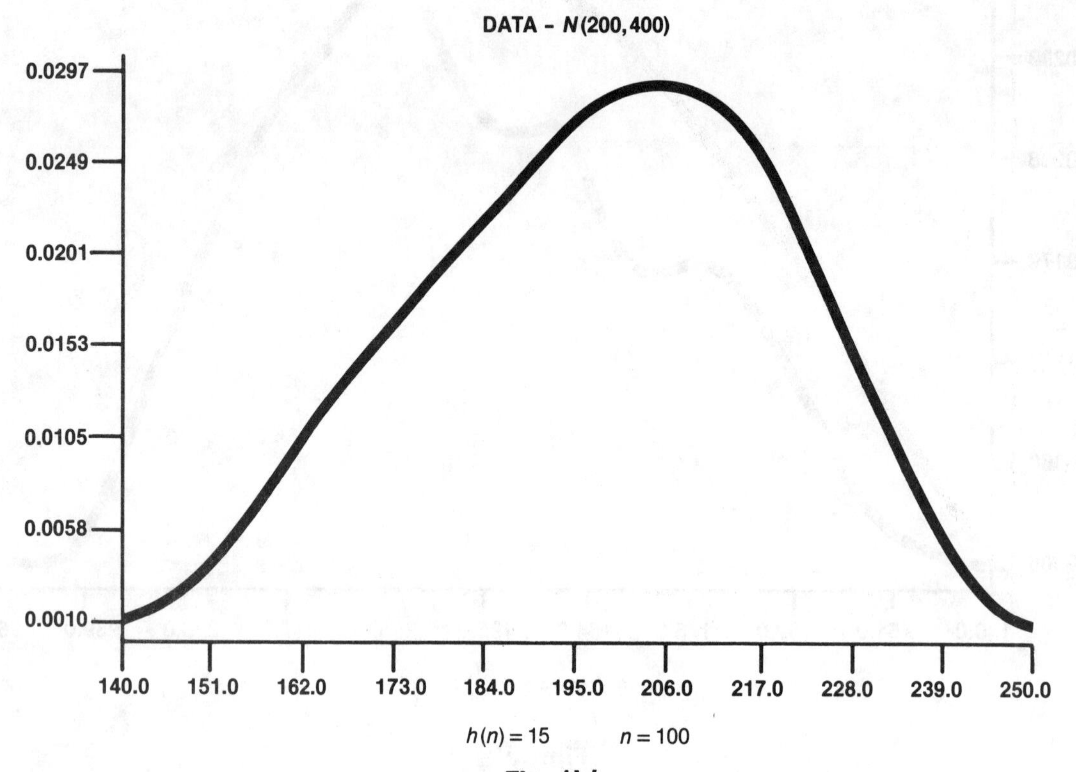
DATA - N(200,400)
0.0297
0.0249
0.0201
0.0153
0.0105
0.0058
0.0010
140.0
151.0
162.0
173.0
184.0
195.0
206.0
217.0
228.0
239.0
250.0
h(n) = 15
n = 100

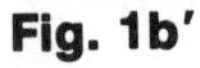

Fig. 1b′

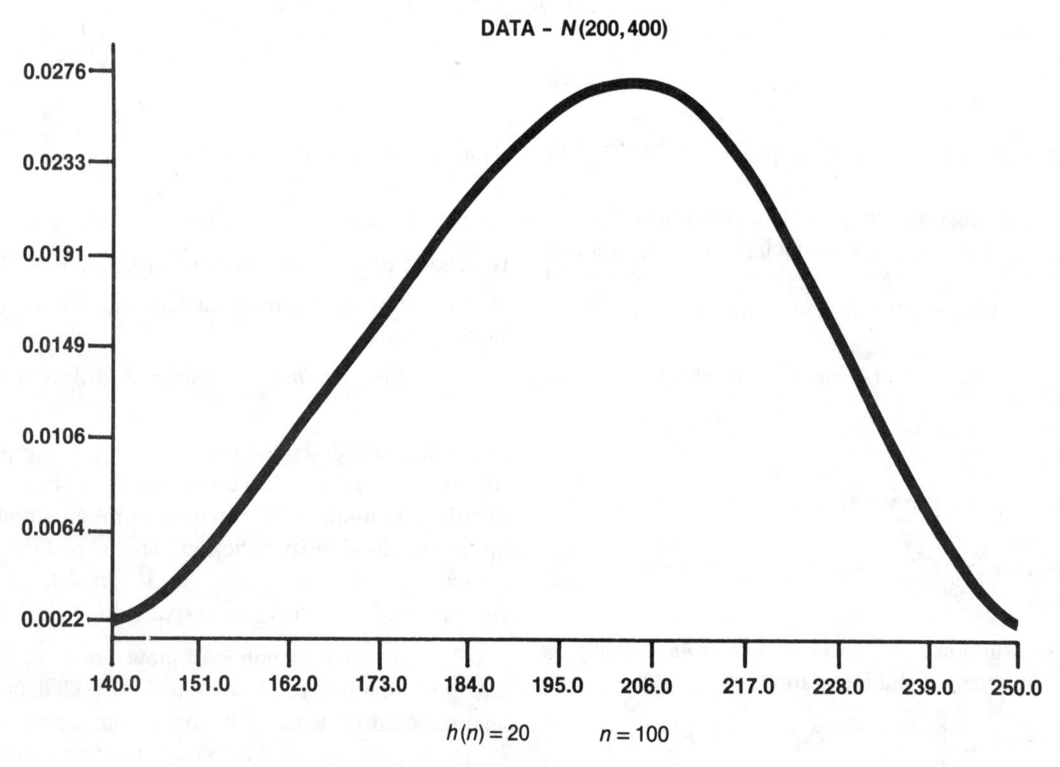
DATA - N(200,400)
0.0276
0.0233
0.0191
0.0149
0.0106
0.0064
0.0022
140.0
151.0
162.0
173.0
184.0
195.0
206.0
217.0
228.0
239.0
250.0
h(n) = 20
n = 100

Fig. 1c′

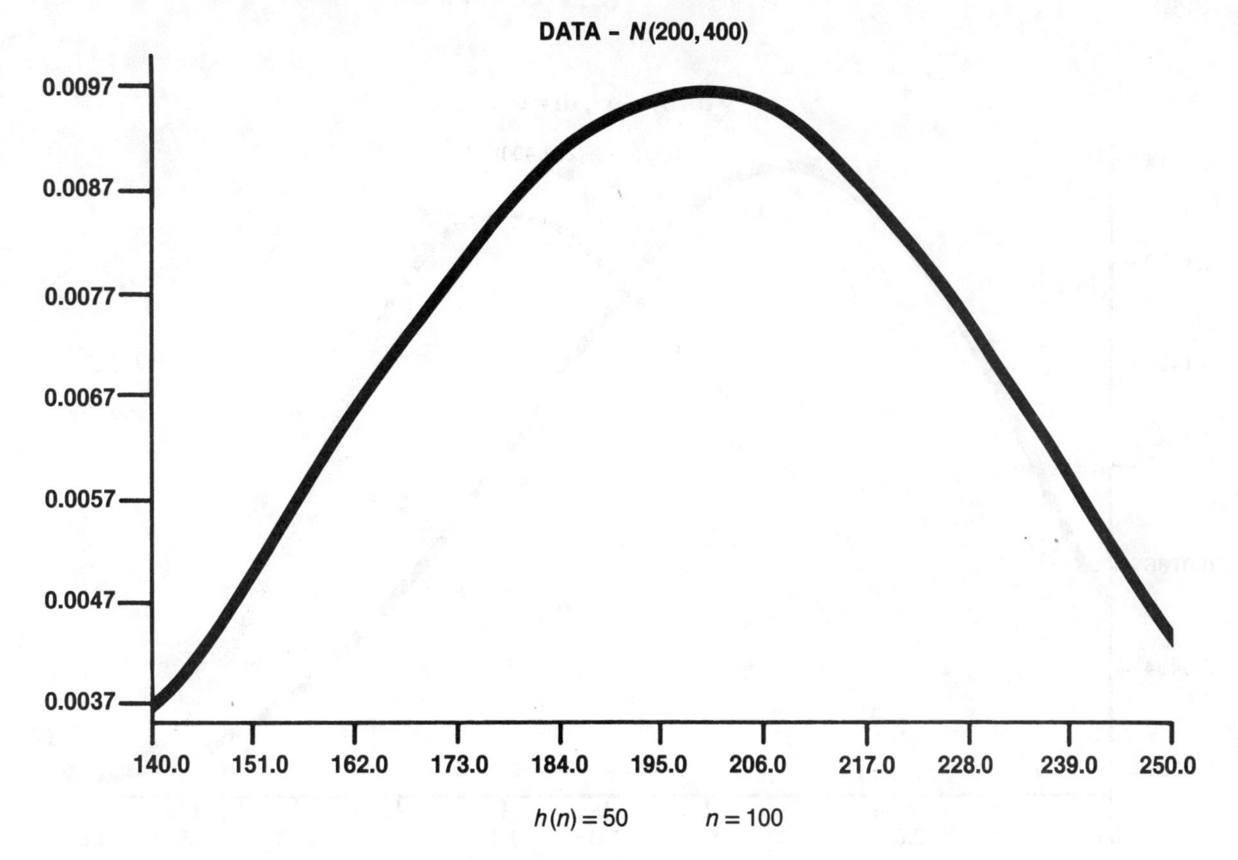

$h(n) = 50$ $n = 100$

Fig. 1d′

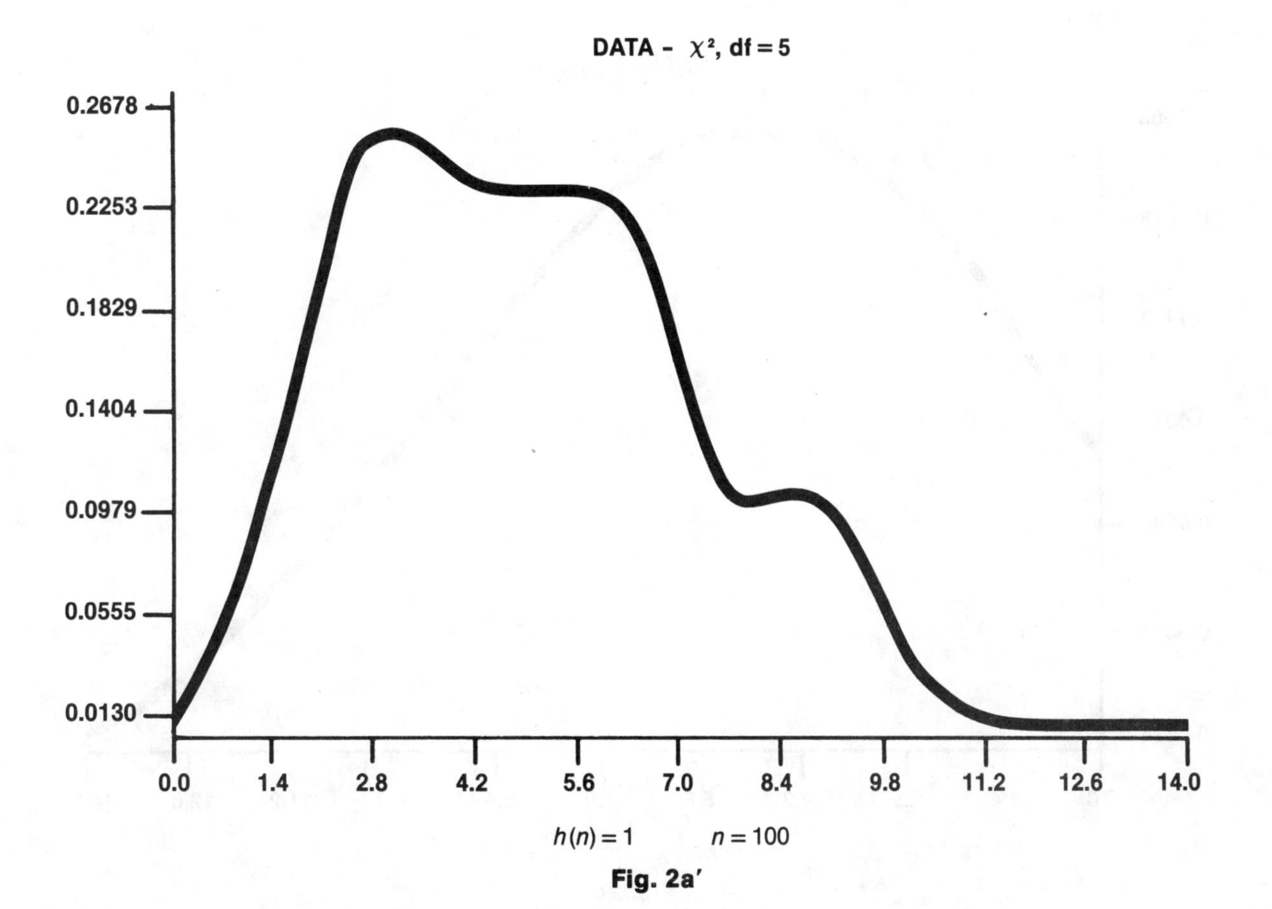

$h(n) = 1$ $n = 100$

Fig. 2a′

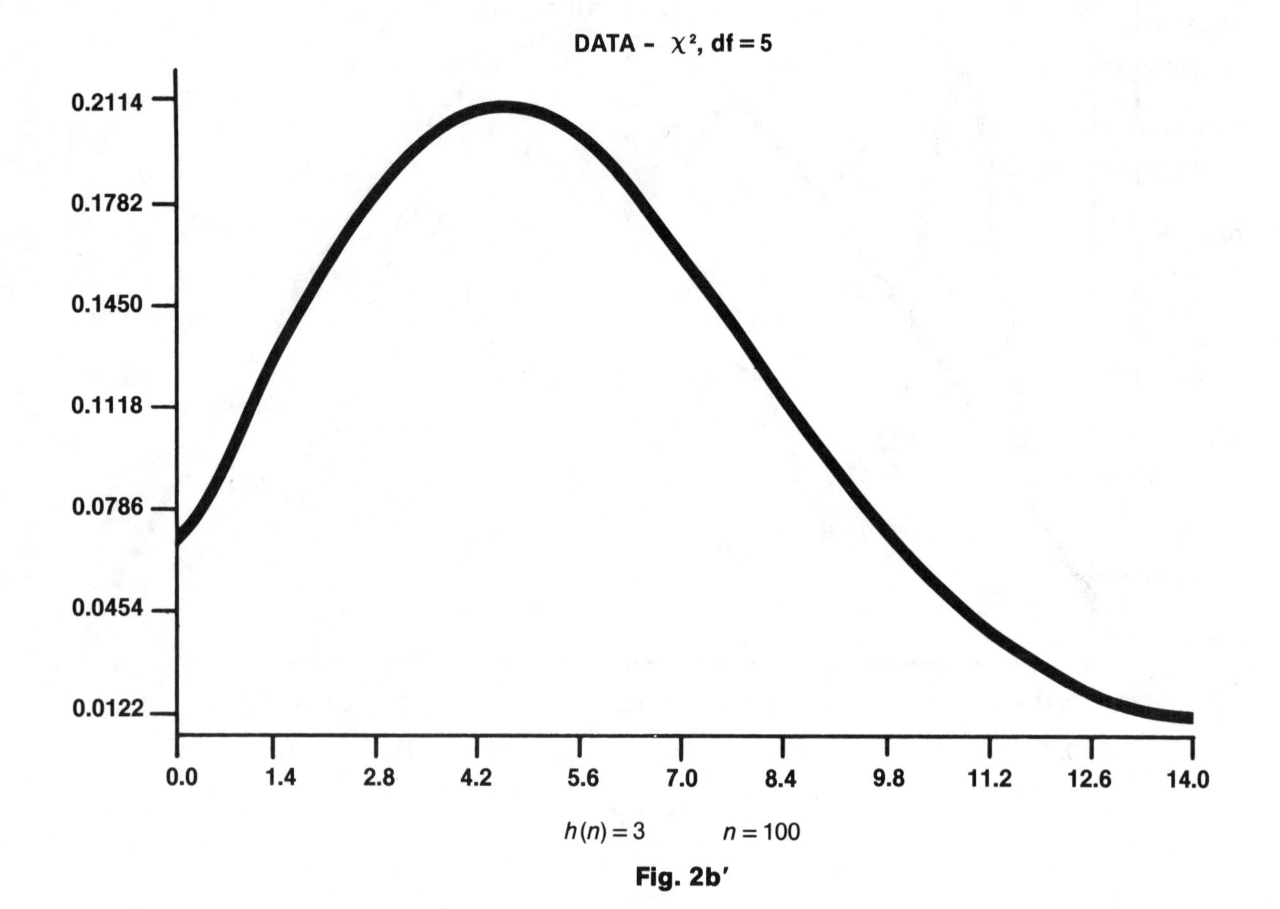
DATA - χ², df = 5
0.2114
0.1782
0.1450
0.1118
0.0786
0.0454
0.0122
0.0
1.4
2.8
4.2
5.6
7.0
8.4
9.8
11.2
12.6
14.0
h(n) = 3
n = 100

Fig. 2b′

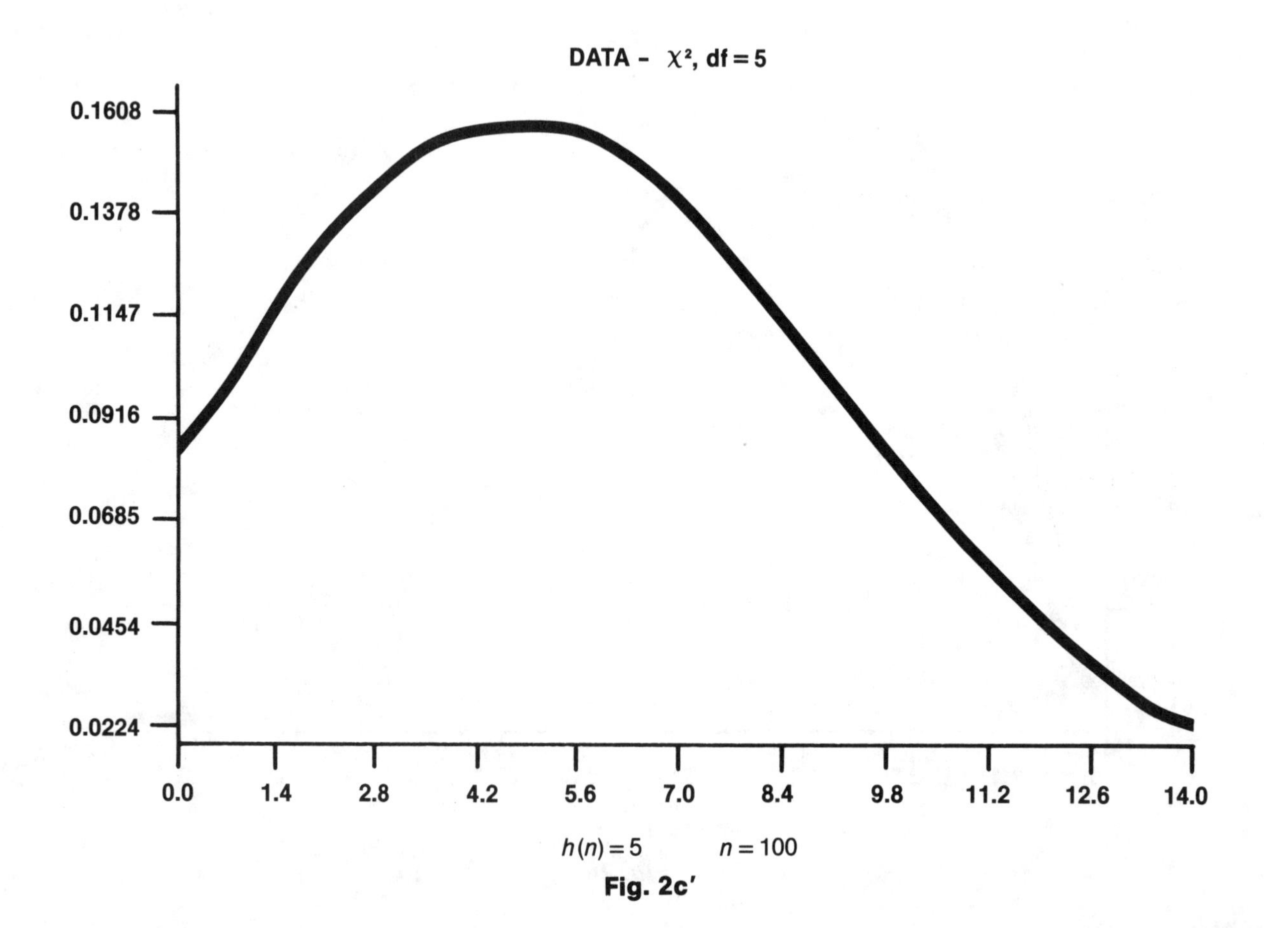
DATA - χ², df = 5
0.1608
0.1378
0.1147
0.0916
0.0685
0.0454
0.0224
0.0
1.4
2.8
4.2
5.6
7.0
8.4
9.8
11.2
12.6
14.0
h(n) = 5
n = 100

Fig. 2c′

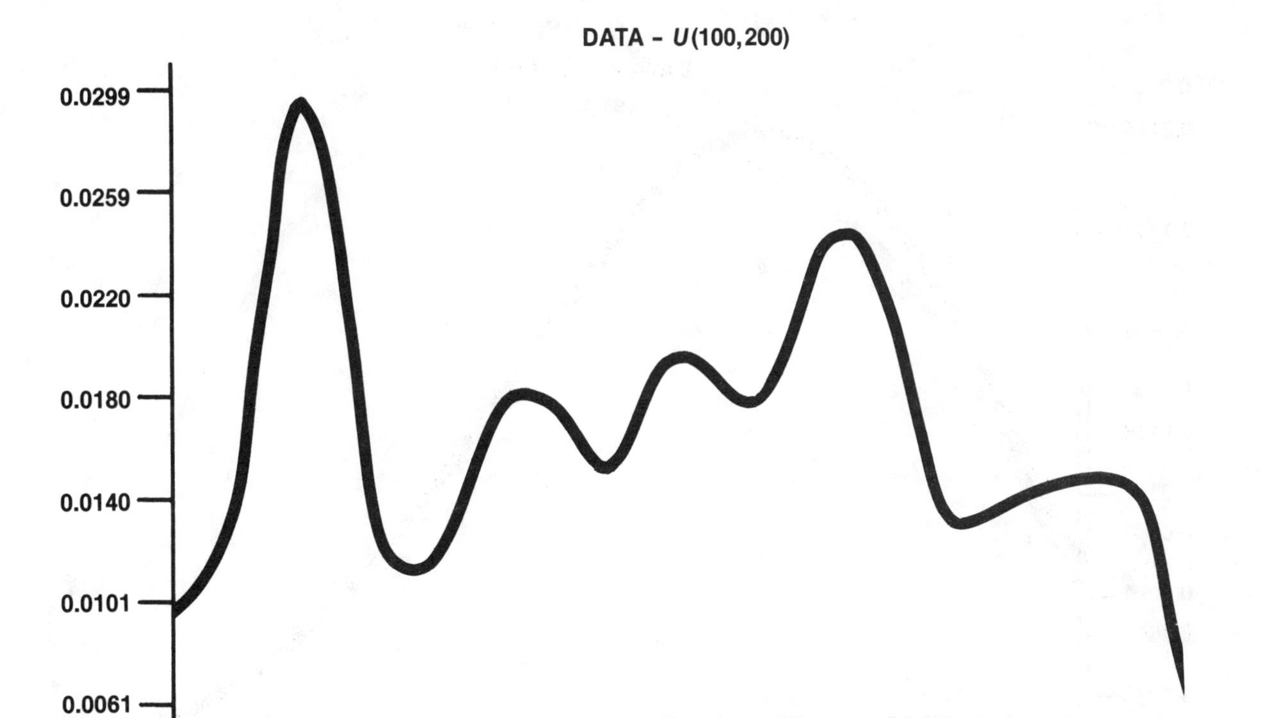

$h(n) = 5$ $n = 100$

Fig. 3a'

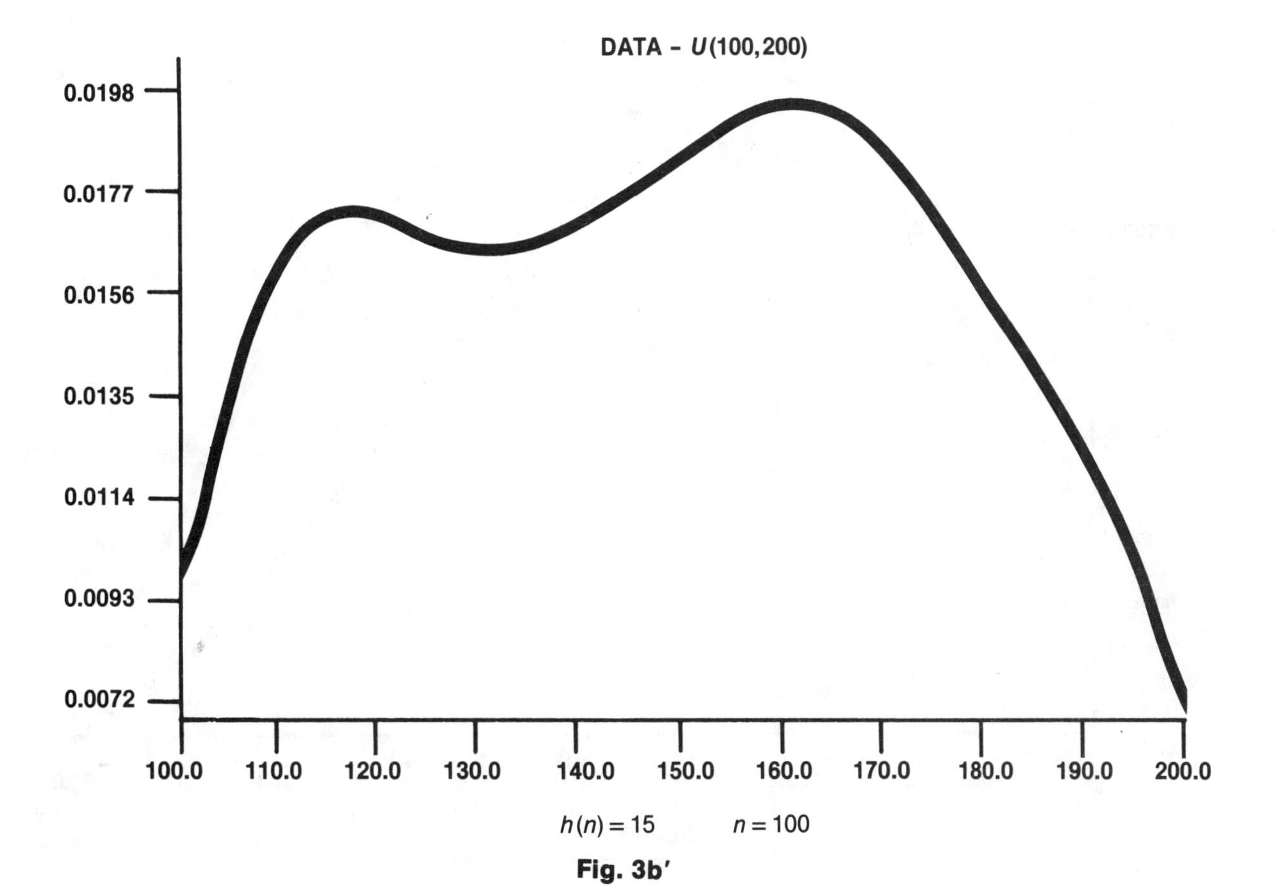

$h(n) = 15$ $n = 100$

Fig. 3b'

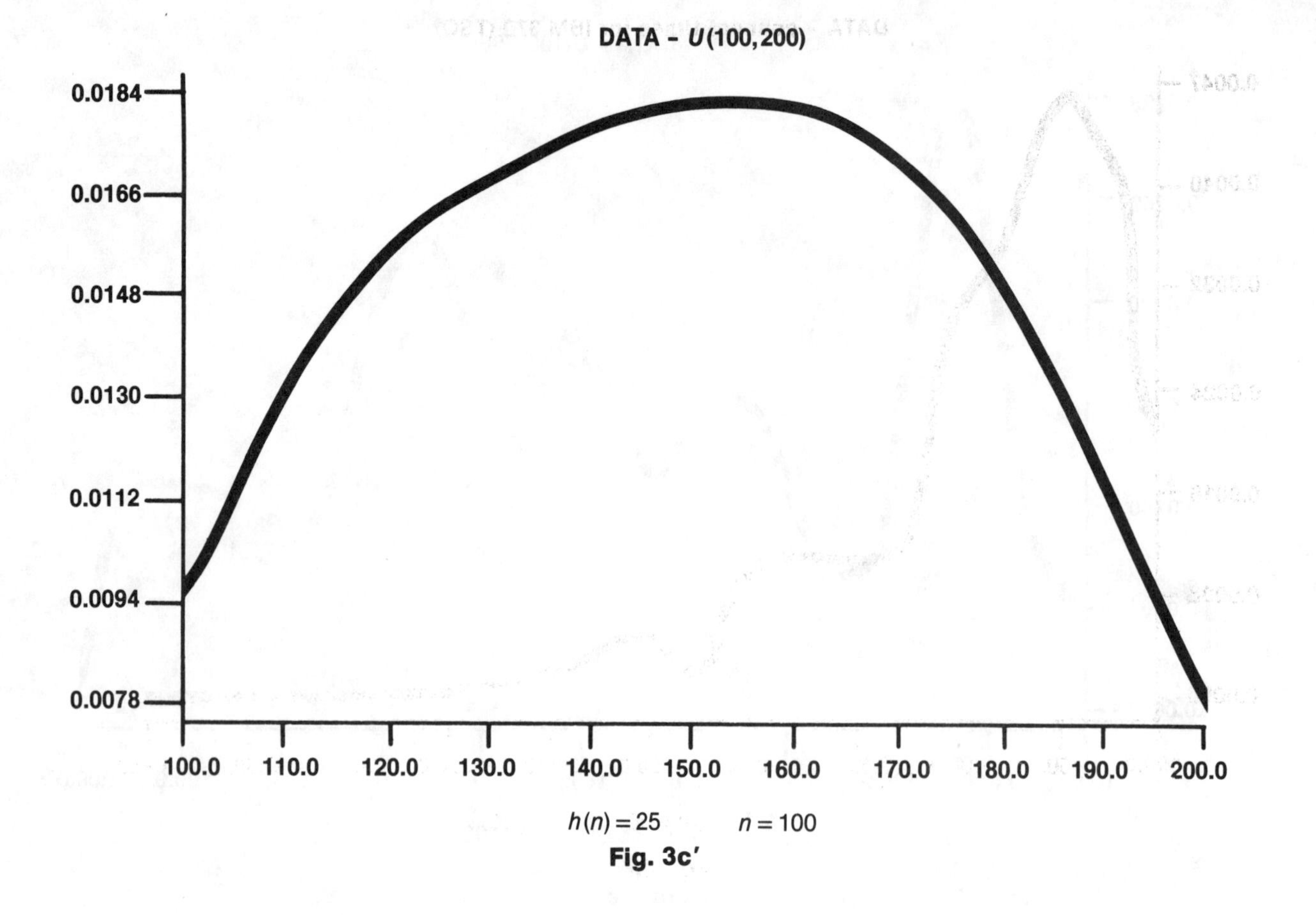

$h(n) = 25 \qquad n = 100$

Fig. 3c′

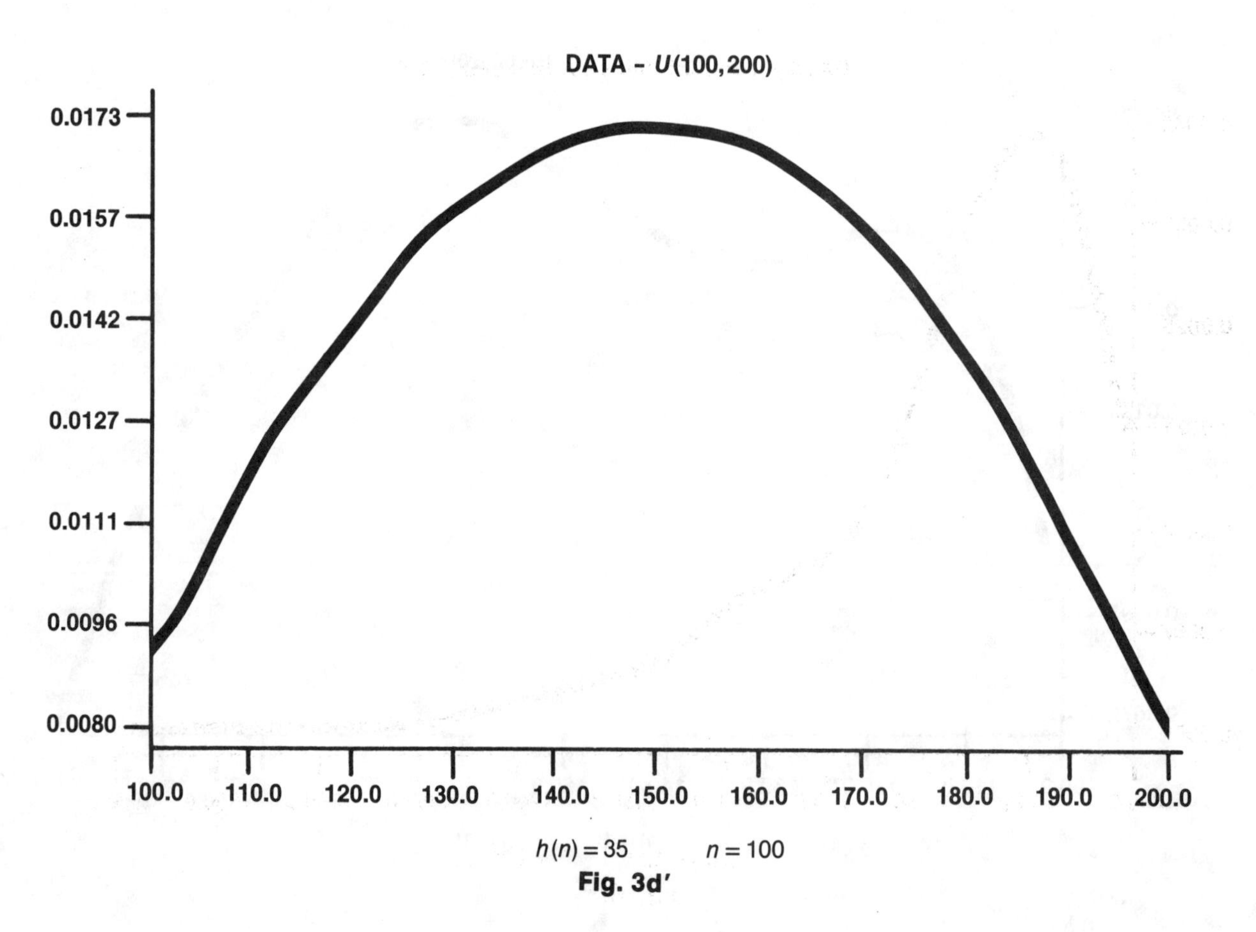

$h(n) = 35 \qquad n = 100$

Fig. 3d′

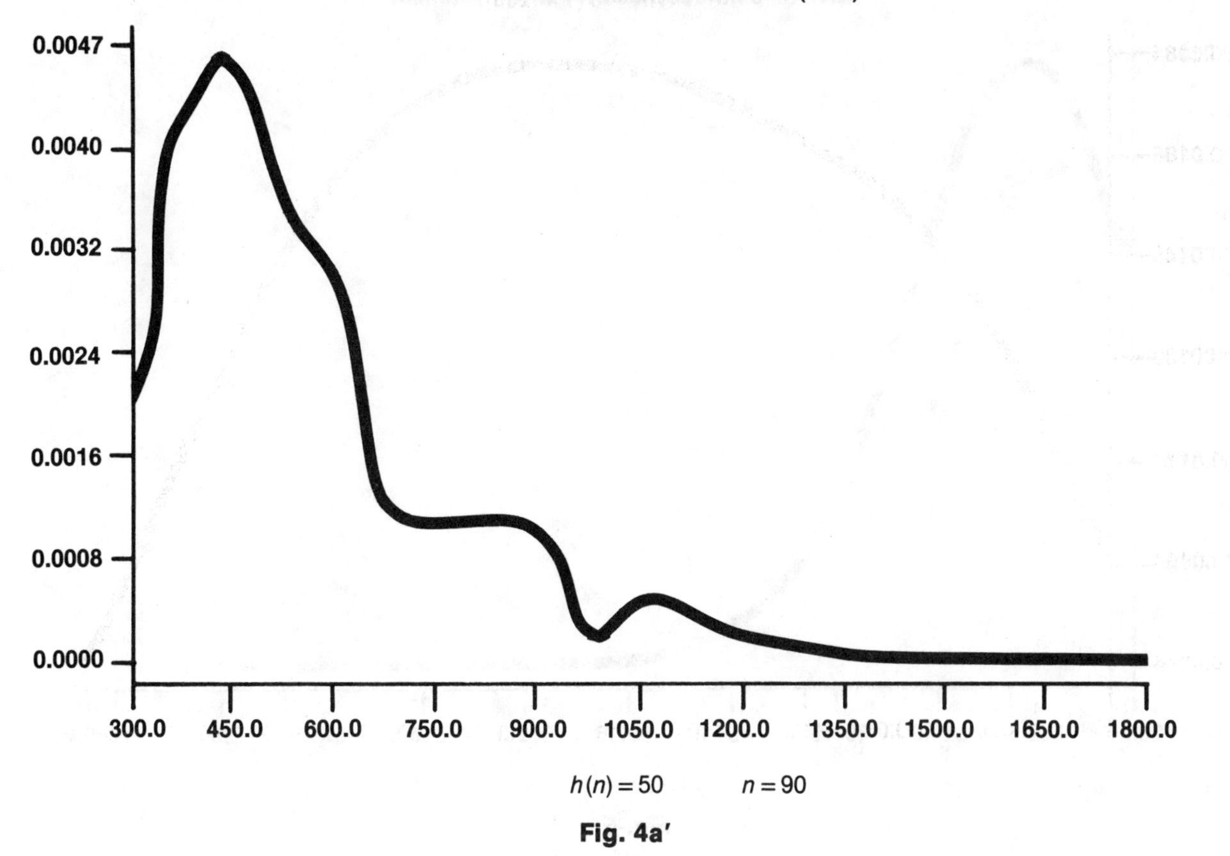

Fig. 4a′

DATA - connect times for IBM 370 (TSO)

0.0042
0.0035
0.0028
0.0021
0.0014
0.0007
0.0000

300.0 450.0 600.0 750.0 900.0 1050.0 1200.0 1350.0 1500.0 1650.0 1800.0

$h(n) = 100$ $n = 90$

Fig. 4b′

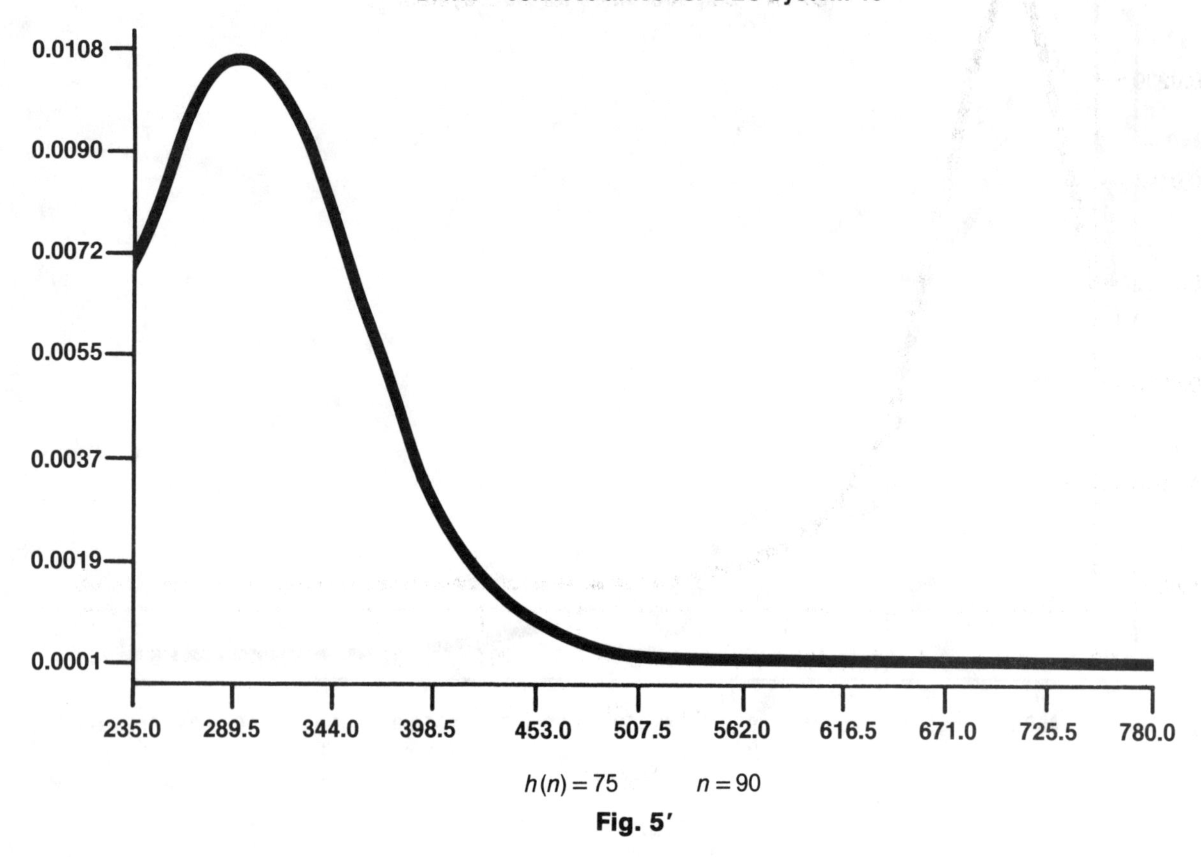
DATA - connect times for DEC system 10
0.0108
0.0090
0.0072
0.0055
0.0037
0.0019
0.0001
235.0
289.5
344.0
398.5
453.0
507.5
562.0
616.5
671.0
725.5
780.0
$h(n) = 75$ $n = 90$

Fig. 5′

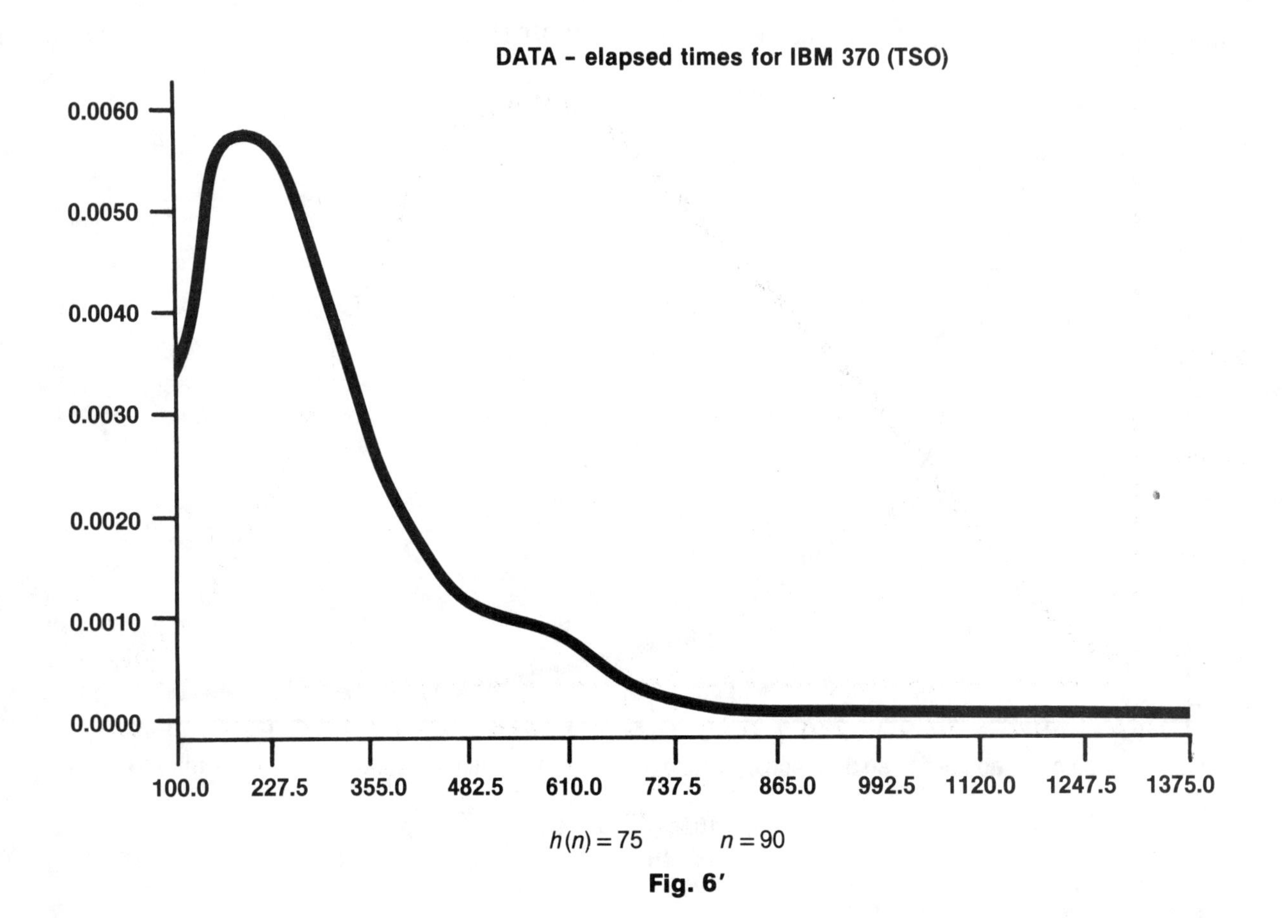
DATA - elapsed times for IBM 370 (TSO)
0.0060
0.0050
0.0040
0.0030
0.0020
0.0010
0.0000
100.0
227.5
355.0
482.5
610.0
737.5
865.0
992.5
1120.0
1247.5
1375.0
$h(n) = 75$ $n = 90$

Fig. 6′

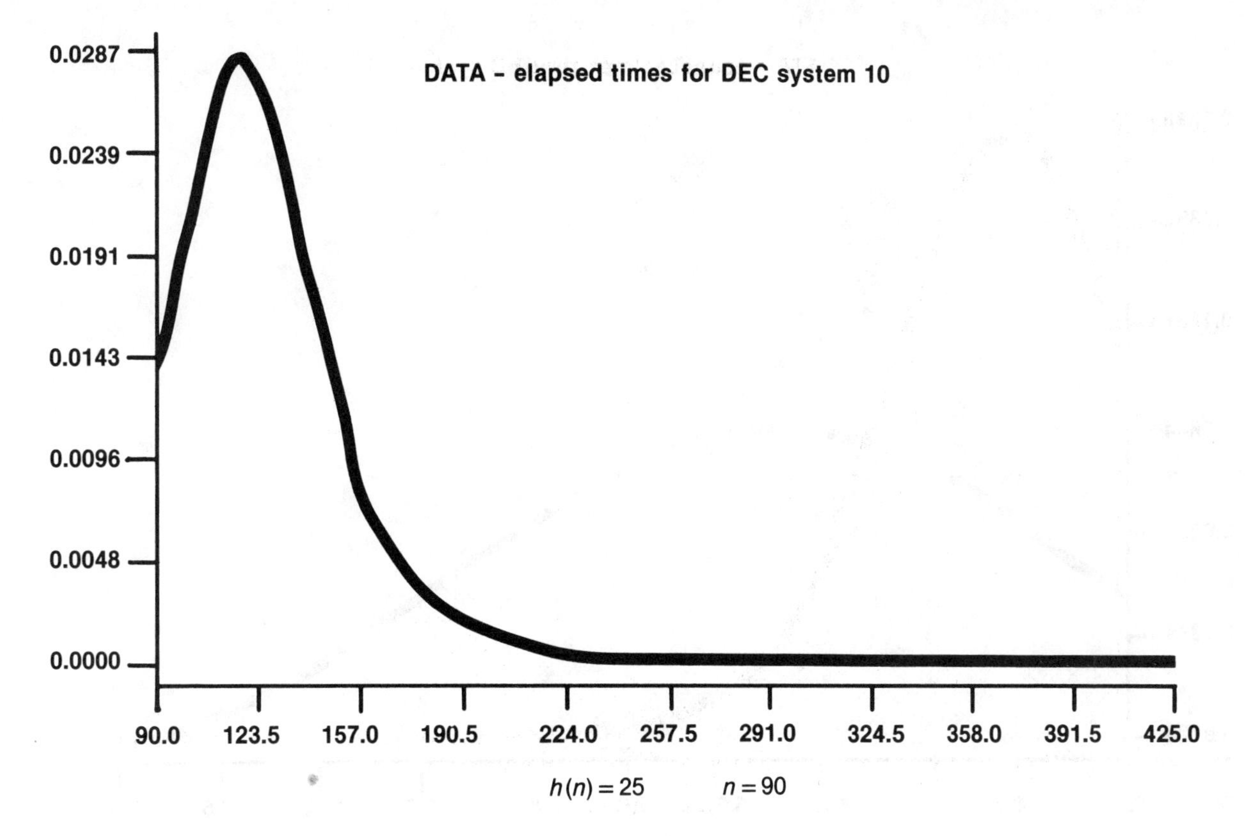
DATA - elapsed times for DEC system 10
0.0287
0.0239
0.0191
0.0143
0.0096
0.0048
0.0000
90.0
123.5
157.0
190.5
224.0
257.5
291.0
324.5
358.0
391.5
425.0
$h(n) = 25$
$n = 90$

Fig. 7′

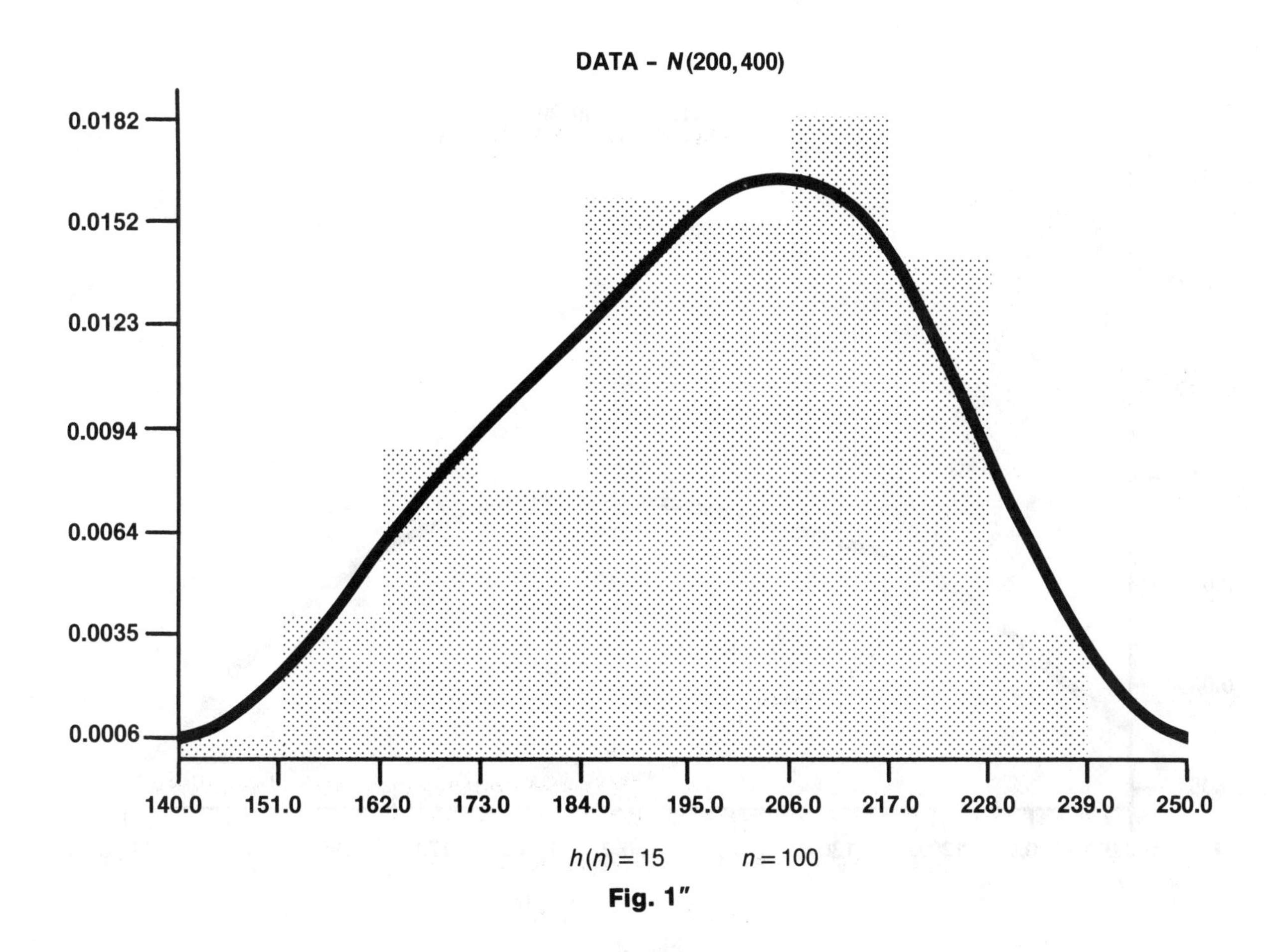
DATA - $N(200, 400)$
0.0182
0.0152
0.0123
0.0094
0.0064
0.0035
0.0006
140.0
151.0
162.0
173.0
184.0
195.0
206.0
217.0
228.0
239.0
250.0
$h(n) = 15$
$n = 100$

Fig. 1″

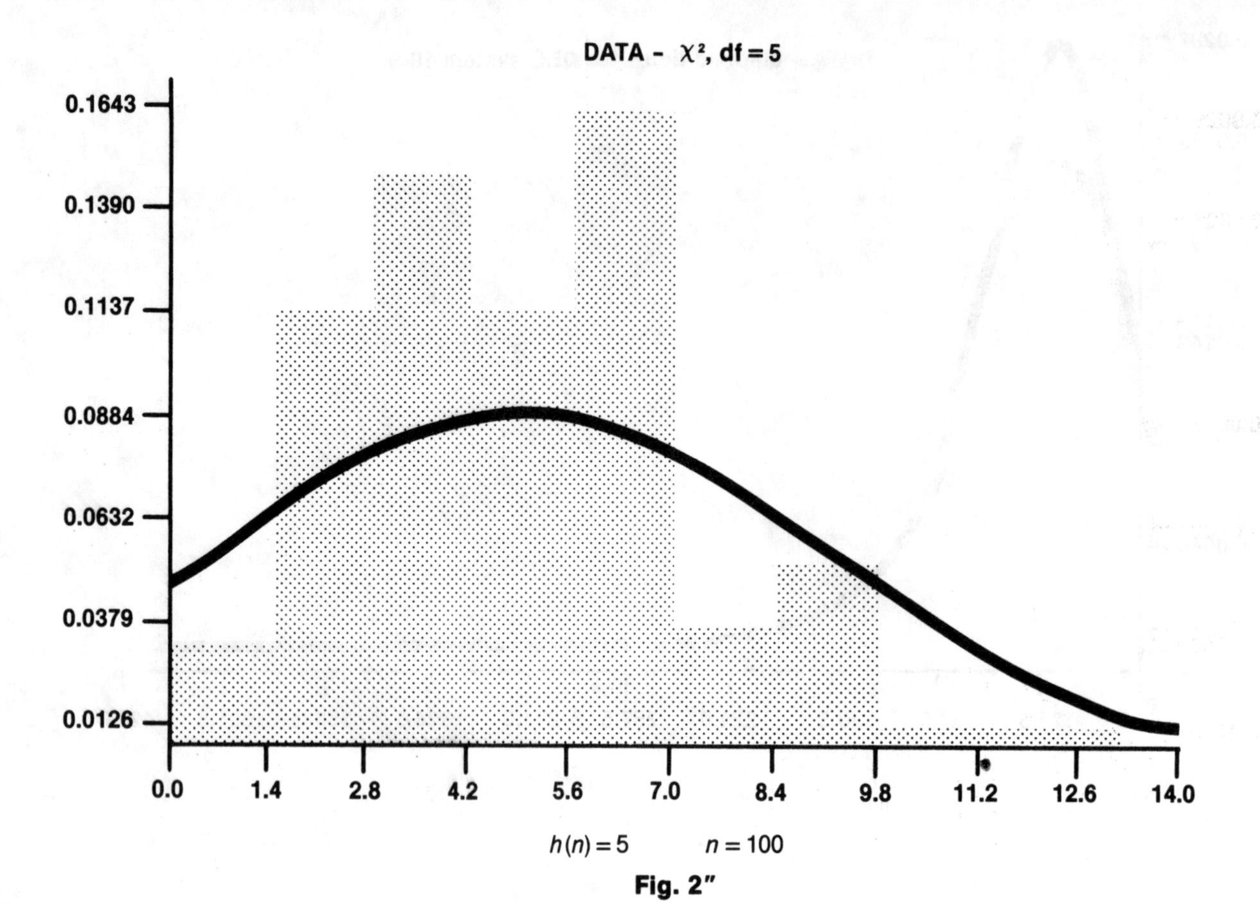
DATA - χ^2, df = 5
0.1643
0.1390
0.1137
0.0884
0.0632
0.0379
0.0126
0.0
1.4
2.8
4.2
5.6
7.0
8.4
9.8
11.2
12.6
14.0
$h(n) = 5$
$n = 100$

Fig. 2″

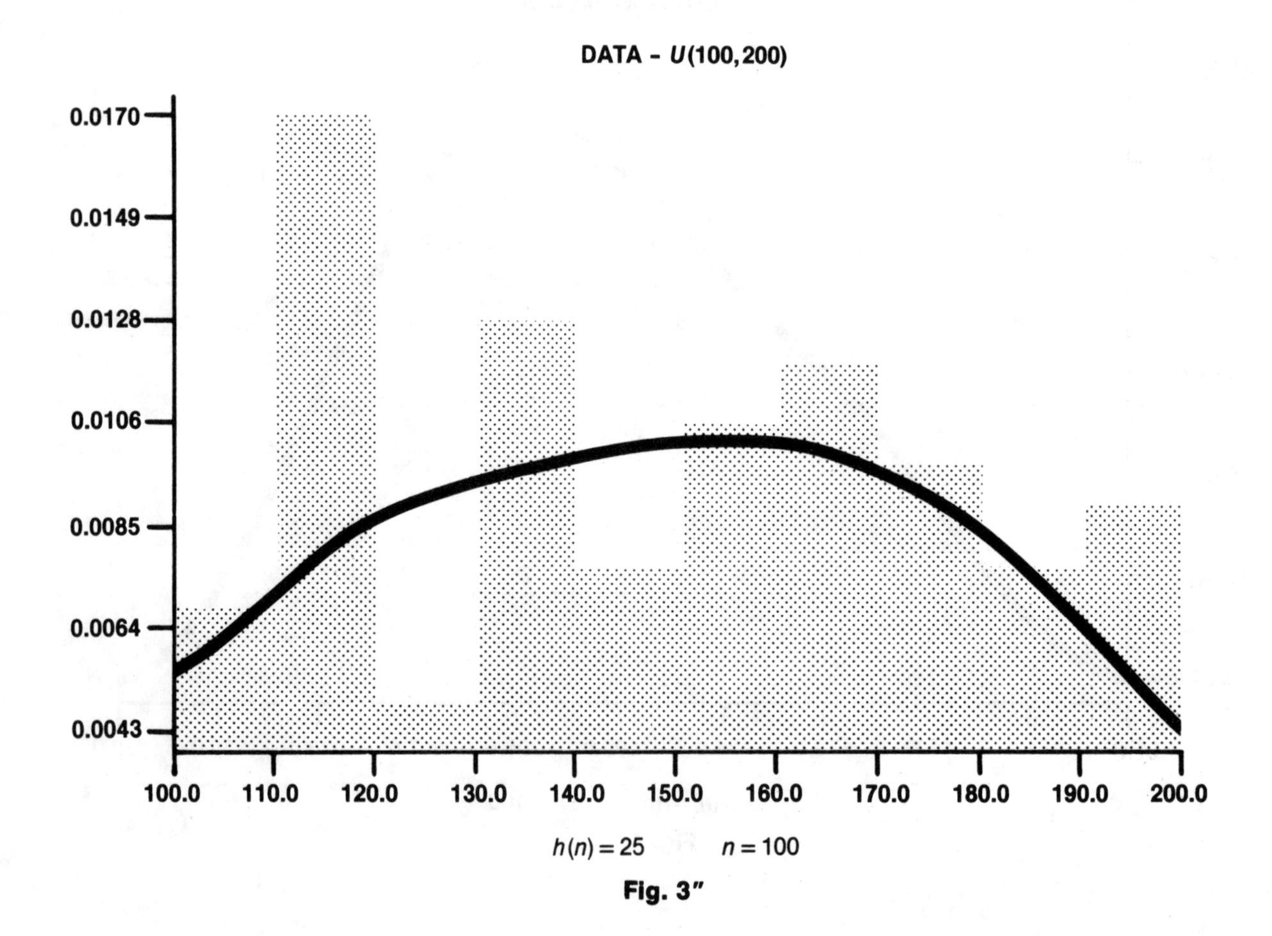
DATA - $U(100, 200)$
0.0170
0.0149
0.0128
0.0106
0.0085
0.0064
0.0043
100.0
110.0
120.0
130.0
140.0
150.0
160.0
170.0
180.0
190.0
200.0
$h(n) = 25$
$n = 100$

Fig. 3″

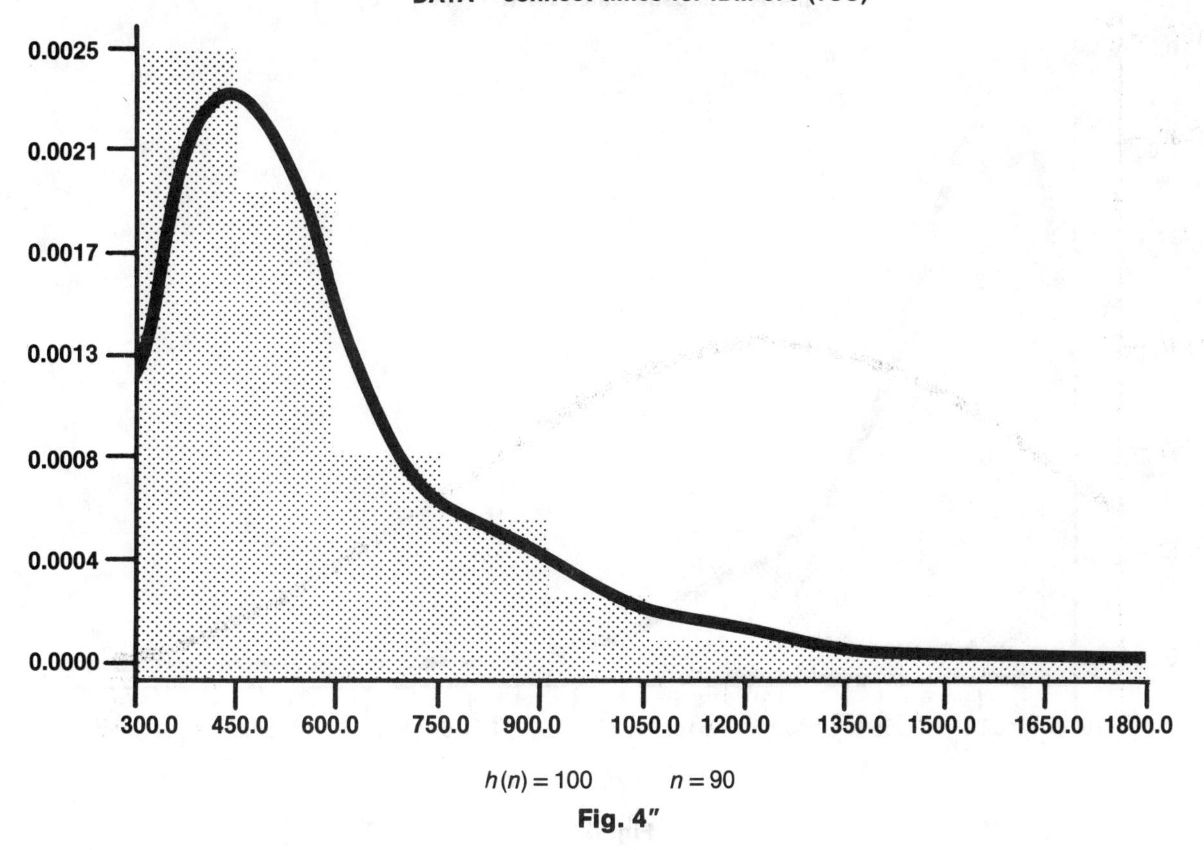
DATA - connect times for IBM 370 (TSO)
0.0025
0.0021
0.0017
0.0013
0.0008
0.0004
0.0000
300.0
450.0
600.0
750.0
900.0
1050.0
1200.0
1350.0
1500.0
1650.0
1800.0
$h(n) = 100$
$n = 90$

Fig. 4″

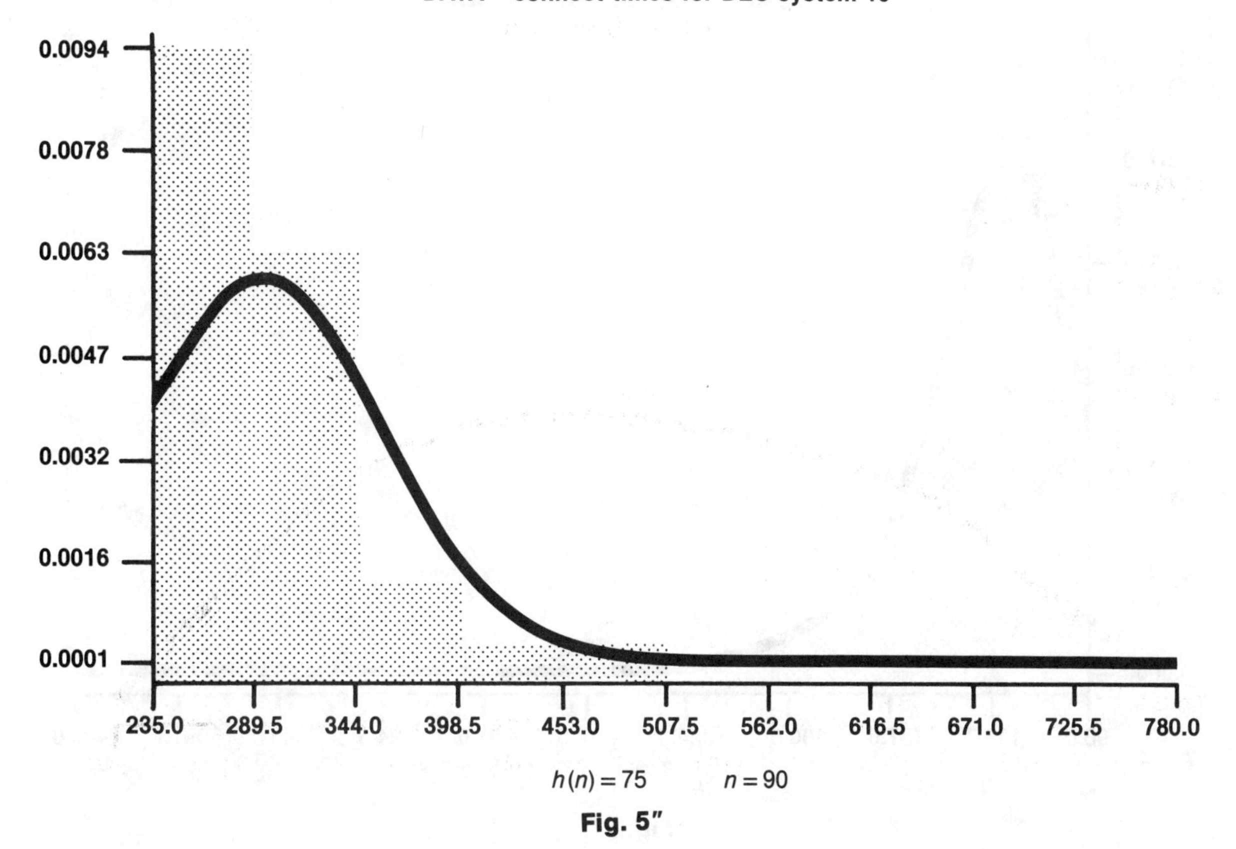
DATA - connect times for DEC system 10
0.0094
0.0078
0.0063
0.0047
0.0032
0.0016
0.0001
235.0
289.5
344.0
398.5
453.0
507.5
562.0
616.5
671.0
725.5
780.0
$h(n) = 75$
$n = 90$

Fig. 5″

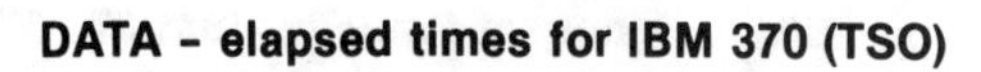

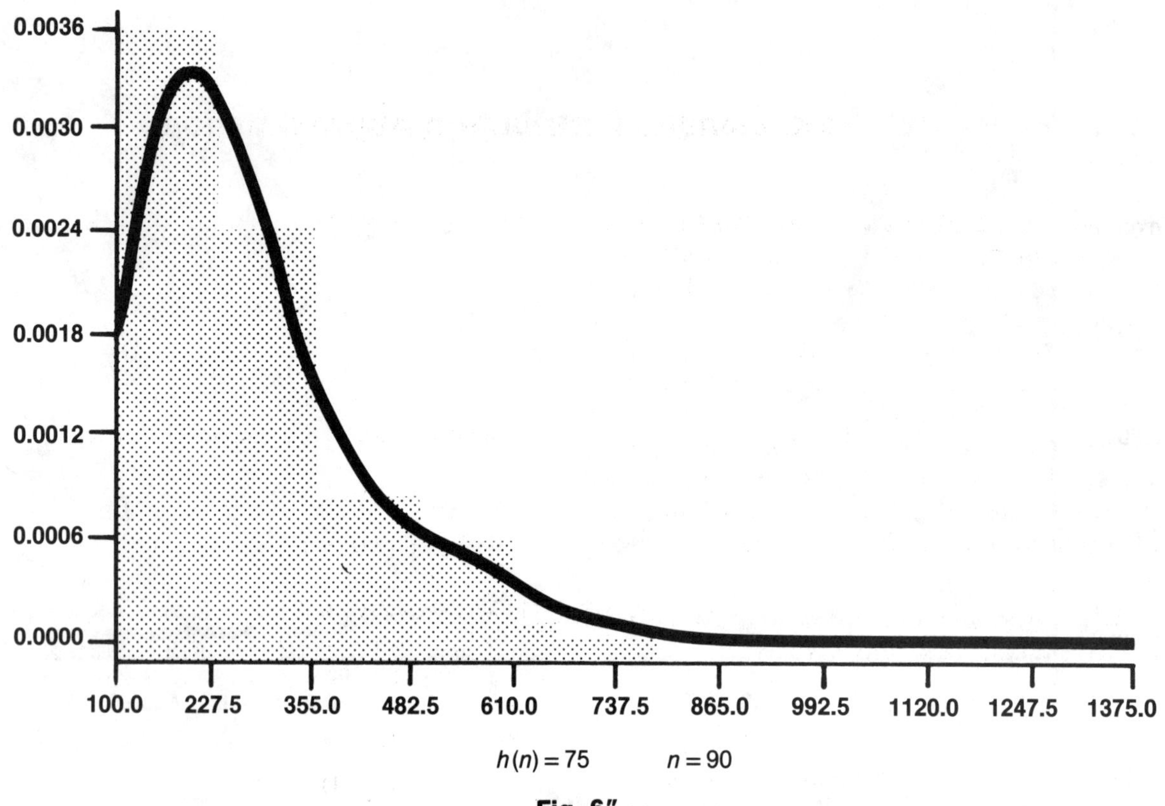

$h(n) = 75$ $n = 90$

Fig. 6″

DATA - elapsed times for DEC system 10

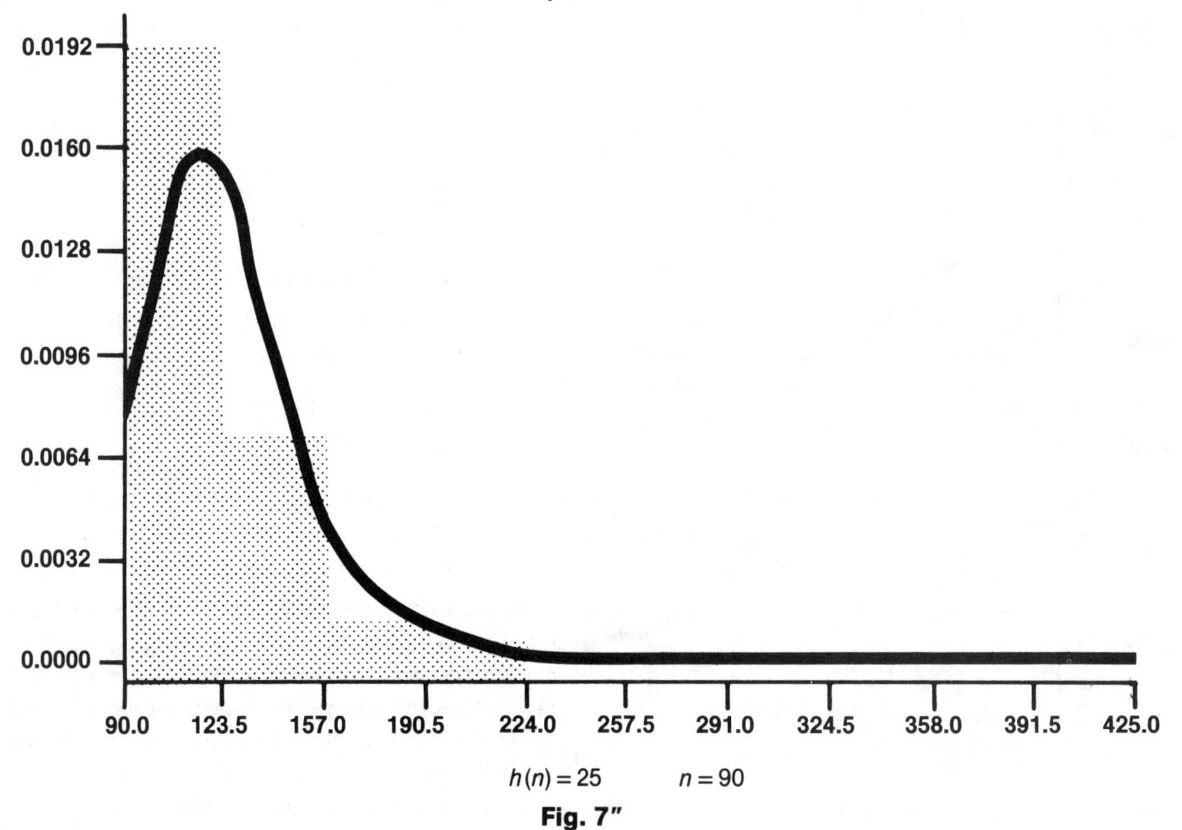

$h(n) = 25$ $n = 90$

Fig. 7″

Generalized Lambda Distribution Approximations

The generalized lambda distribution (GLD) was initially proposed by Ramberg and Schmeiser [4] and is used for the approximation of unimodal distributions. In addition to developing a basic algorithm for the approximation, Ramberg and Schmeiser obtained good GLD approximations of the exponential and gamma distributions. Dudewicz, Ramberg, and Tadikamalla [1] further elaborated on the original results and in a more recent paper Ramberg, Tadikamalla, Dudewicz, and Mykytka [5] have modified and extended the table of L or lambda values (to be discussed shortly) and have applied the GLD approximation to empirical data.

The inverse distribution function of a GLD with the parameters L_1, L_2, L_3, and L_4 is given by

$$x = F^{-1}(p) = R(p) = L_1 + [p^{L_3} - (1-p)^{L_4}]/L_2 \quad (9)$$

where p varies from 0 to 1. In order for (9) to be a legitimate inverse distribution function, its corresponding density function, $f(x)$, must satisfy

$$\int_{\mathbb{R}} f(x)dx = \int_{\mathbb{R}} f(R(p))R'(p) = 1 \quad (10)$$

and $f(x)$, or equivalently, $R'(p)$ must be non-negative.

There are four regions in which (9) describes a legitimate distribution; the four regions, together with the upper and lower bounds of the GLD, are given in Table 1 below.

Table 1

Region	Value of L_2	Value of L_3	Value of L_4	Lower Bound	Upper Bound
1	<0	<-1	>1	$-\infty$	L_1+1/L_2
2	<0	>1	<-1	L_1-1/L_2	∞
3	>0	>0	>0	L_1-1/L_2	L_1+1/L_2
	>0	$=0$	>0	L_1	L_1+1/L_2
	>0	>0	$=0$	L_1-1/L_2	L_1
4	<0	<0	<0	$-\infty$	∞
	<0	$=0$	<0	L_1	∞
	<0	<0	$=0$	$-\infty$	L_1

Ramberg and Schmeiser [4] show that for $L_1 = 0$, $E(X^k)$, **the expected value of X^k is given by**

$$E(X^k) = (1/L_2)^k \sum_i \binom{k}{i}(-1)^i B(L_3(k-i) + 1, L_4 i + 1) \quad (11)$$

where B denotes the beta function defined by

$$B(a,b) = \int_0^1 x^{a-1}(1-x)^{b-1}dx. \quad (12)$$

The mean, μ, of the GLD can be obtained by

$$\mu = L_1 + E(X) = (1/L_2)[B(L_3+1,1) - B(1,L_4+1)]$$
$$= L_1 + [1/(L_3+1) - 1/(L_4+1)]/L_2. \quad (13)$$

The second, third, and fourth moments designated by σ^2, μ_3, and μ_4 respectively are obtained by using (11) to compute $E((X-\mu)^k)$. The results are as follows:

$$\sigma^2 = \{[1/(2L_3+1) - 2B(L_3+1,L_4+1) + 1/(2L_4+1)] - [1/(L_3+1) - 1/(L_4+1)]^2\}/L_2^2, \quad (14)$$

$$\mu_3 = \{[1/(3L_3+1) - 3B(2L_3+1,L_4+1) + 3B(L_3+1,2L_4+1) - 1/(3L_4+1)] - 3[1/(2L_3+1) - 2B(L_3+1,L_4+1) + 1/(2L_4+1)][1/L_3+1) - 1/(L_4+1)] + 2[1/(L_3+1) - 1/(L_4+1)]^3\}/L_2^3, \quad (15)$$

$$\mu_4 = \{[1/(4L_3+1) - 4B(3L_3+1,L_4+1) + 6B(2L_3+1,2L_4+1) - 4B(L_3+1,3L_4+1) + 1/(4L_4+1)] - 4[1/(3L_3+1) - 3B(2L_3+1,L_4+1) + 3B(L_3+1,2L_4+1) - 1/(3L_4+1)][1/(L_3+1) - 1/(L_4+1)] + 6[1/(2L_3+1) - 2B(L_3+1,L_4+1) + 1/(2L_4+1)] \times [1/(L_3+1) - 1/(L_4+1)] - 3[1/(L_3+1) - 1/(L_4+1)]^4\}/L_2^4 \quad (16)$$

To determine a suitable GLD distribution for a given set of observations $x_1, x_2, \ldots, x_n$, the L_1, L_2, L_3, and L_4 are determined in such a way that the first four moments of the distribution are set equal to the corresponding sample moments. This is accomplished through the following steps.

Step 1. Compute the sample moments $\bar{x}$, m_2, m_3, and m_4 for $x_1, x_2, \ldots, x_n$. The appropriate formulas are

EH0227-9/85/0000/0265$01.00 © 1985 IEEE

$$\bar{x} = (1/n) \sum_i x_i, \tag{17}$$

$$m_k = (1/n) \sum_i (x_i - \bar{x})^k \tag{18}$$

Step 2. Solve the simultaneous equations

$$\mu_3/\sigma^3 = m_3/m_2^{3/2} \quad , \quad \mu_4/\sigma^4 = m_4/m_2^2 \tag{19}$$

from L_3 and L_4.

Step 3. Compute L_2 by setting σ^2 (14) equal to m_2. Note that L_2 will be the only unknown parameter in the equation at this point.

Step 4. Compute L_1, by setting μ (13) equal to $\bar{x}$.

With the exception of the second step, the procedure outlined above involves only direct computations and has no inherent complications. The solution of the simultaneous equations of step two, however, is quite another matter. To begin with, the evaluations of the beta function which are required will involve numerical methods since the beta function has no "closed form" representation. Thus, the equations given in the second step cannot be inverted to produce expressions for L_3 and L_4. An algorithm for searching optimal values on a surface, such as that of Nelder and Mead (discussed by Olsson and Nelson [3]) is needed to get approximate solutions to L_3 and L_4. **To reduce the difficulty of Step 2**, the tables provided by Ramberg and Schmeiser [4] for the determination of L_3 and L_4 have been expanded and improved by others. **The best tables to date have been compiled by Ramberg, Tadikamalla, Dudewicz, and Mykytka [5].**

In the following examples, the general algorithm described above, together with the available tables, **is used on the data sets in Examples 1–4** of the previous article on kernel estimation. Unfortunately, there are no tables extensive enough to take care of data sets in Examples 5–7. **The density functions seem to fit the data at least as well as the ones considered by the kernel method.**

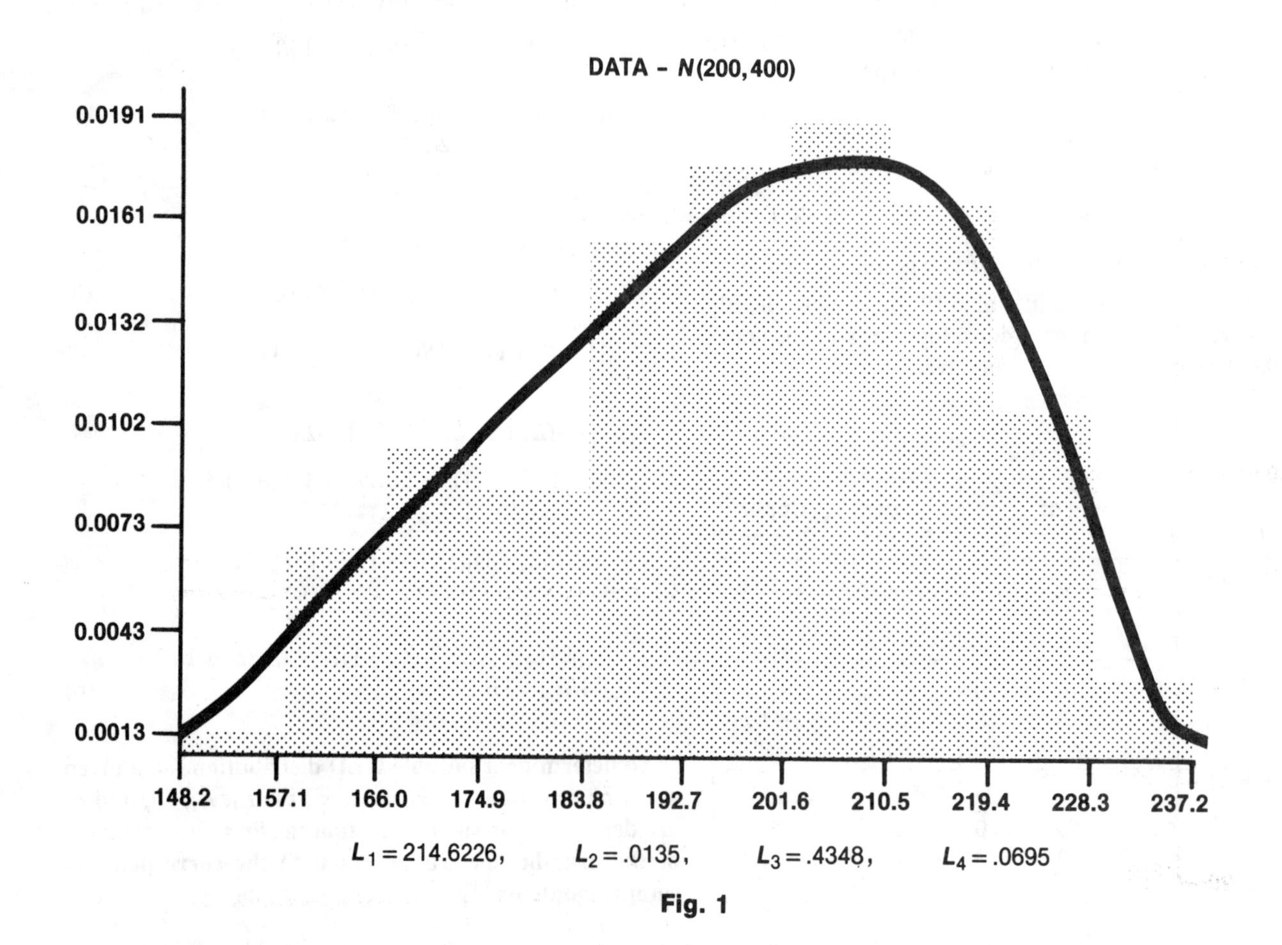

Fig. 1

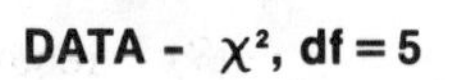

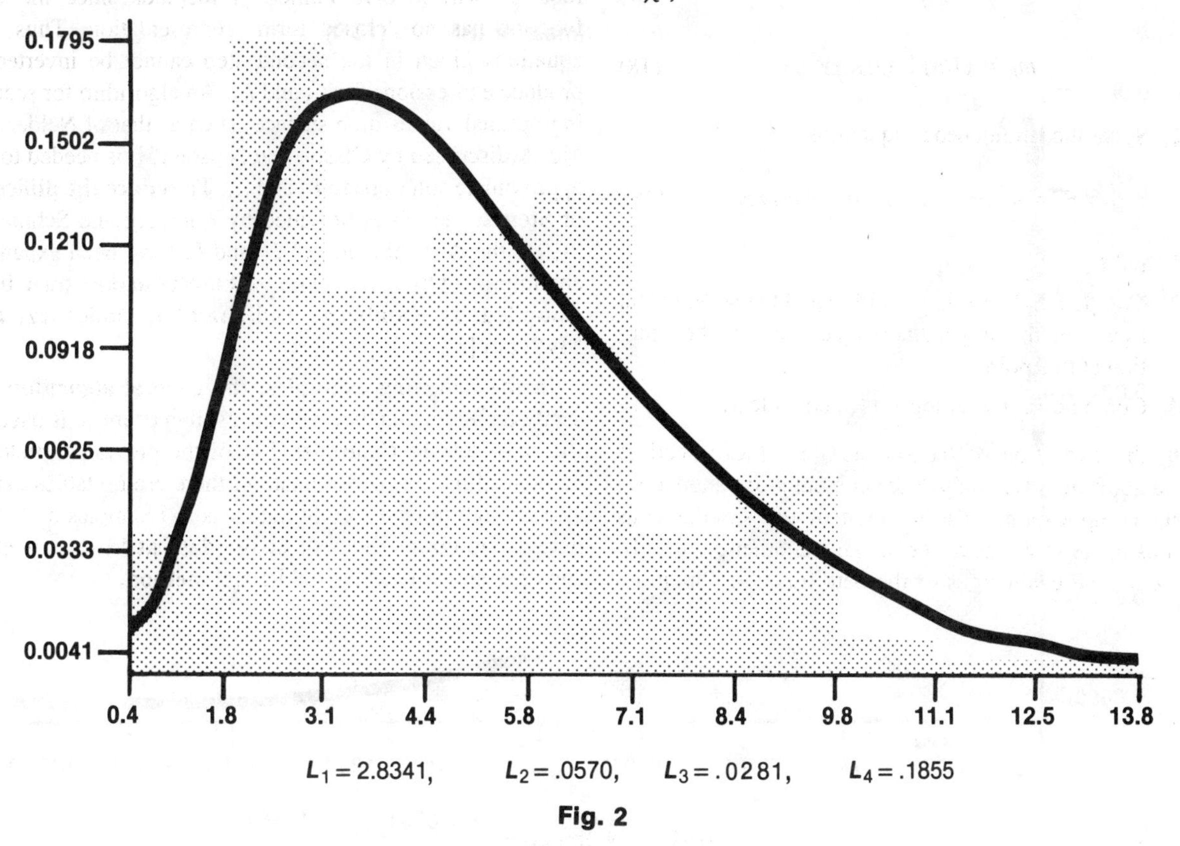

$L_1 = 2.8341, \quad L_2 = .0570, \quad L_3 = .0281, \quad L_4 = .1855$

Fig. 2

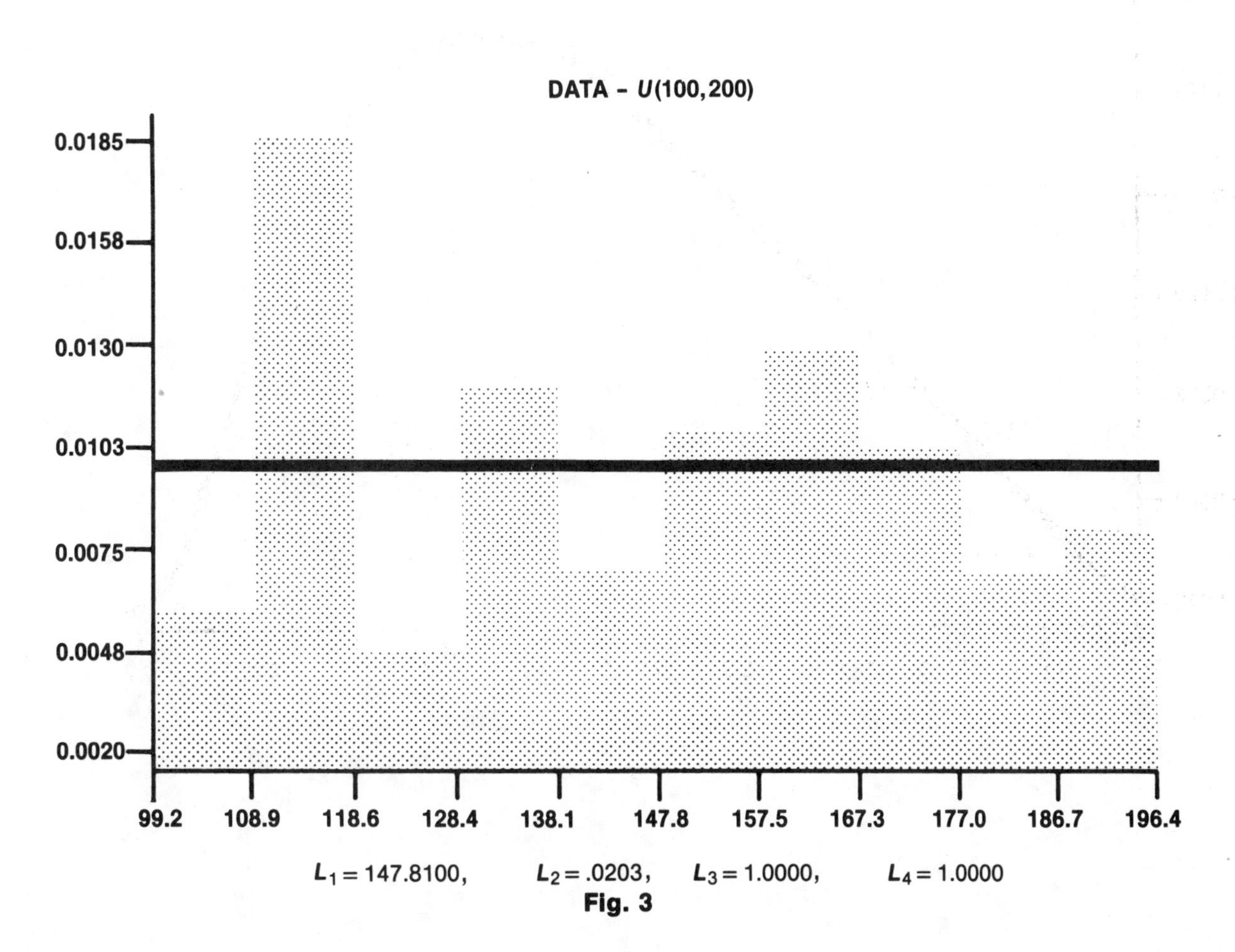

$L_1 = 147.8100, \quad L_2 = .0203, \quad L_3 = 1.0000, \quad L_4 = 1.0000$

Fig. 3

DATA - connect times for IBM 370 (TSO)

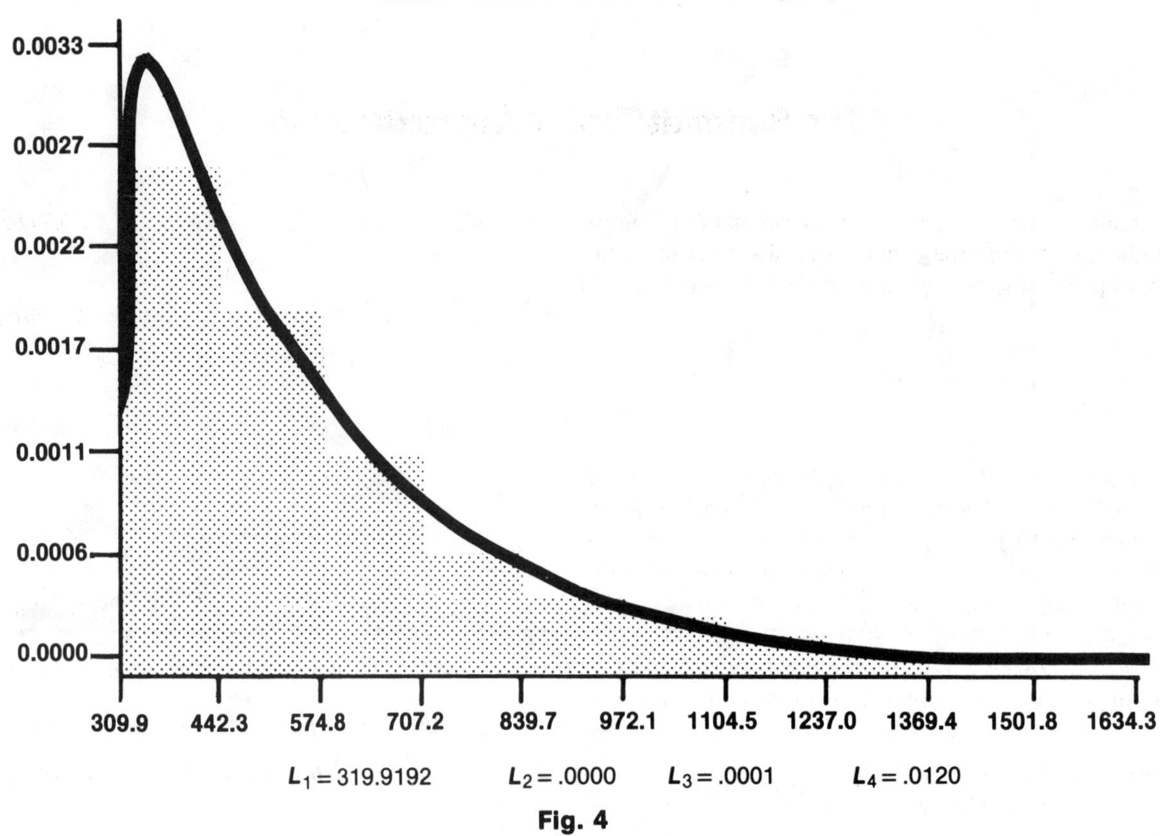
0.0033
0.0027
0.0022
0.0017
0.0011
0.0006
0.0000
309.9
442.3
574.8
707.2
839.7
972.1
1104.5
1237.0
1369.4
1501.8
1634.3
L_1 = 319.9192
L_2 = .0000
L_3 = .0001
L_4 = .0120

Fig. 4

The Schmidt-Taylor Approximation

Schmidt and Taylor [6] consider **polynomial and nonpolynomial approximations to inverse distribution functions. In particular** they assert [6, p. 288] that **the function**

$$x = F^{-1}(p) = R(p)$$
$$= a + bp + cp^2 + d(1-p)^2 \ln(p) + hp^2 \ln(1-p) \quad (20)$$

"has proven most useful in approximating inverse distribution functions . . ." There seems to be **no justification for the choice of (20)** other than the observation that it seems to have worked well often. Unlike the GLD approximation, this method does not try to produce distributions with some of the same statistical properties as were present the original data; in fact, **there is no assurance that the resulting function will be a legitimate inverse distribution function.**

Starting with a set $x_1, x_2, \ldots, x_n$ of empirical observations, a relative cumulative frequency table such as the one below is constructed.

Interval	Midpoint	Rel. Cum. Freq.
$a<x<b_1$	x_1	p_1
$a_2<x<b_2$	x_2	p_2
.	.	.
.	.	.
.	.	.
$a_k<x<b_k$	x_k	p_k

Since $(p_1,x_1), \ldots, (p_k,x_k)$ are points on the inverse distribution function (at least the data suggests that they should be), a, b, c, d, and h are determined in such a way that (20) is in good agreement with these data points. The least squares method is used for fitting (20) to the data. This is done by minimizing the sum of the squared error terms at each of the points. The expression to be minimized is

$$E(a,b,c,d,h) = \sum_i [x_i - a - bp_i - cp_i^2 - d(1-p_i)^2 \ln(p_i) - hp^2 \ln(1-p_i)]^2.$$

To minimize E, the first partial derivatives with respect to a, b, c, d, and h are set equal to zero yielding the following:

$$ka + b\Sigma p_i + c\Sigma p_i^2 + d\Sigma(1-p_i)^2 \ln(p_i) + h\,\Sigma\, p_i^2 \ln(1-p_i) = \Sigma x_i,$$

$$a\,\Sigma p_i + b\,\Sigma\, p_i^2 + c\,\Sigma\, p_i^3 + d\,\Sigma\, p_i(1-p)^2 \ln(p_i) + h\,\Sigma\, p_i^3 \ln(1-p_i) = \Sigma\, p_i\, x_i$$

$$a\,\Sigma p_i^2 + b\,\Sigma\, p_i^3 + c\,\Sigma\, p_i^4 + d\,\Sigma\, p_i^2(1-p_i)\ln(p_i) + h\,\Sigma\, p_i^4 \ln(1-p_i) = \Sigma\, p_i^2\, x_i$$

$$a\,\Sigma(1-p_i)^2 \ln(p_i) + b\,\Sigma\, p_i(1-p_i)^2 \ln(p_i) + c\,\Sigma\, p_i^2\,(1-p_i)^2 \ln(p_i) + d\,\Sigma\,(1-p_i)^4 \ln^2(p_i) + h\,\Sigma\, p_i^2(1-p_i)^2 \ln(p_i) \ln(1-p_i) = \Sigma\,(1-p_i)^2 \ln(p_i) x_i, \text{ and}$$

$$a\,\Sigma\, p_i^2 \ln(1-p_i) + b\,\Sigma\, p_i^3 \ln(1-p_i) + c\,\Sigma\, p_i^4 \ln(1-p_i) + d\,\Sigma\, p_i^2\,(1-p_i)^2 \ln(p_i) \ln(1-p_i) + h\,\Sigma\, p_i^4 \ln^2(1-p_i) = \Sigma\, p_i^2 \ln(1-p_i) x_i.$$

The five equations are linear in a, b, c, d, and h; therefore, they can be represented in matrix form by

$$AX = Y$$

where A is the 5×5 coefficient matrix of the linear system, $X^T = (a,b,c,d,h)$, and Y represents the right-hand side of the equations. Once the entries of A and Y are computed, a, b, c, d, and h can be obtained from

$$X = A^{-1}Y.$$

The data sets in Examples 1–4 which were used in the discussions of the kernel estimates and the generalized lambda distribution approximations in the previous articles are used here to illustrate this procedure. The figures below indicate that **this method seems to be less effective than the two previous methods. The approximation is especially poor in the case of the data with a rectangular distribution.** It can be seen in Figure 3a that the "density function" attains negative values for x in the vicinity of 105. A magnification of this portion of the graph given in Figure 3b shows that the function goes through a damped oscillation before stabilizing at larger values of x. Another problem, particularly apparent in Figure 4, is that this approximation will generate values well outside the data range.

DATA - $N(200,400)$

CUM. REL. FREQ.:
0.0100 / 0.1100 / 0.2200 / 0.3600 / 0.5000 / 0.7000 / 0.8700 / 1.0000 / 1.0000

CLASS MIDPOINTS:
140.00 / 160.00 / 175.00 / 185.00 / 195.00 / 205.00 / 215.00 / 227.50 / 245.00

CLASS LIMITS:
130.0 / 150.0 / 170.0 / 180.0 / 190.0 / 200.0 / 210.0 / 220.0 / 235.0 / 255.0

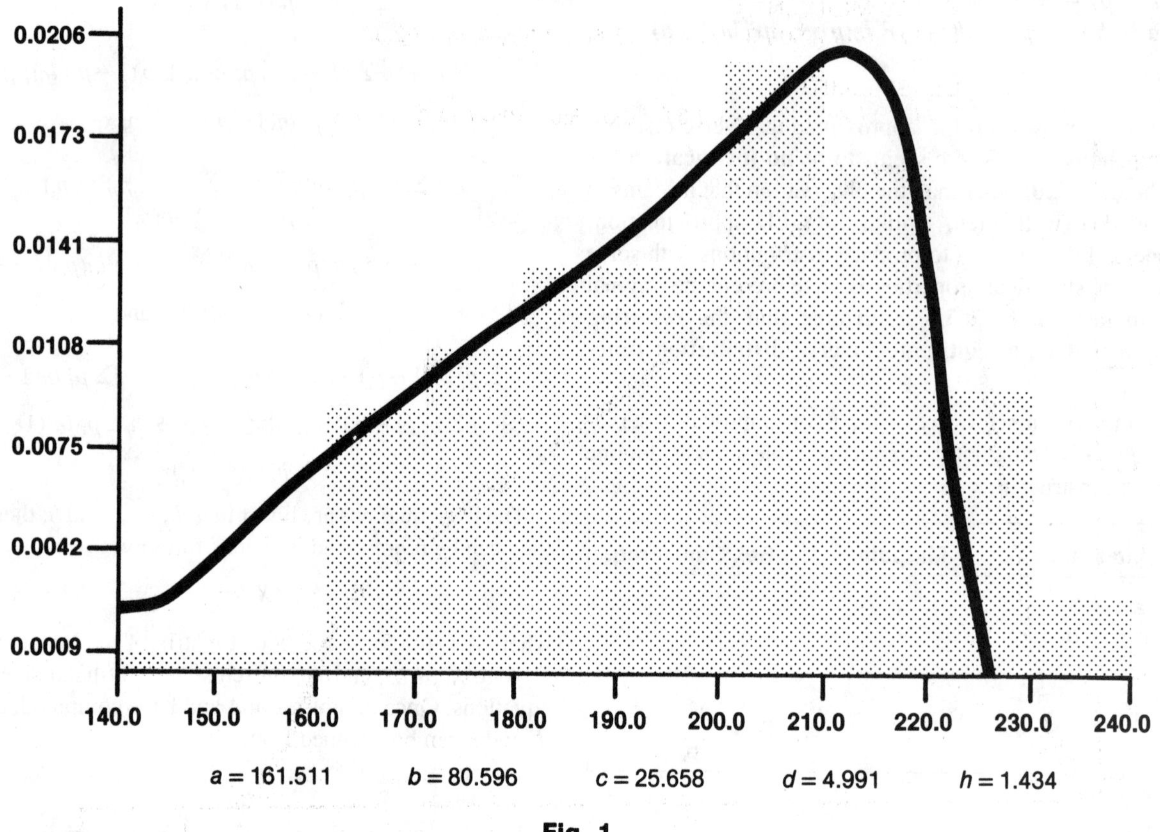

Fig. 1

DATA - χ^2, df = 5

CUM. REL. FREQ.:
0.0100 / 0.0800 / 0.2600 / 0.4000 / 0.5400 / 0.6500 / 0.8300 / 0.9600 / 1.0000

CLASS MIDPOINTS:
0.50 / 1.50 / 2.50 / 3.50 / 4.50 / 5.50 / 7.00 / 9.00 / 12.00

CLASS LIMITS:
0.0 / 1.0 / 2.0 / 3.0 / 4.0 / 5.0 / 6.0 / 8.0 / 10.0 / 14.0

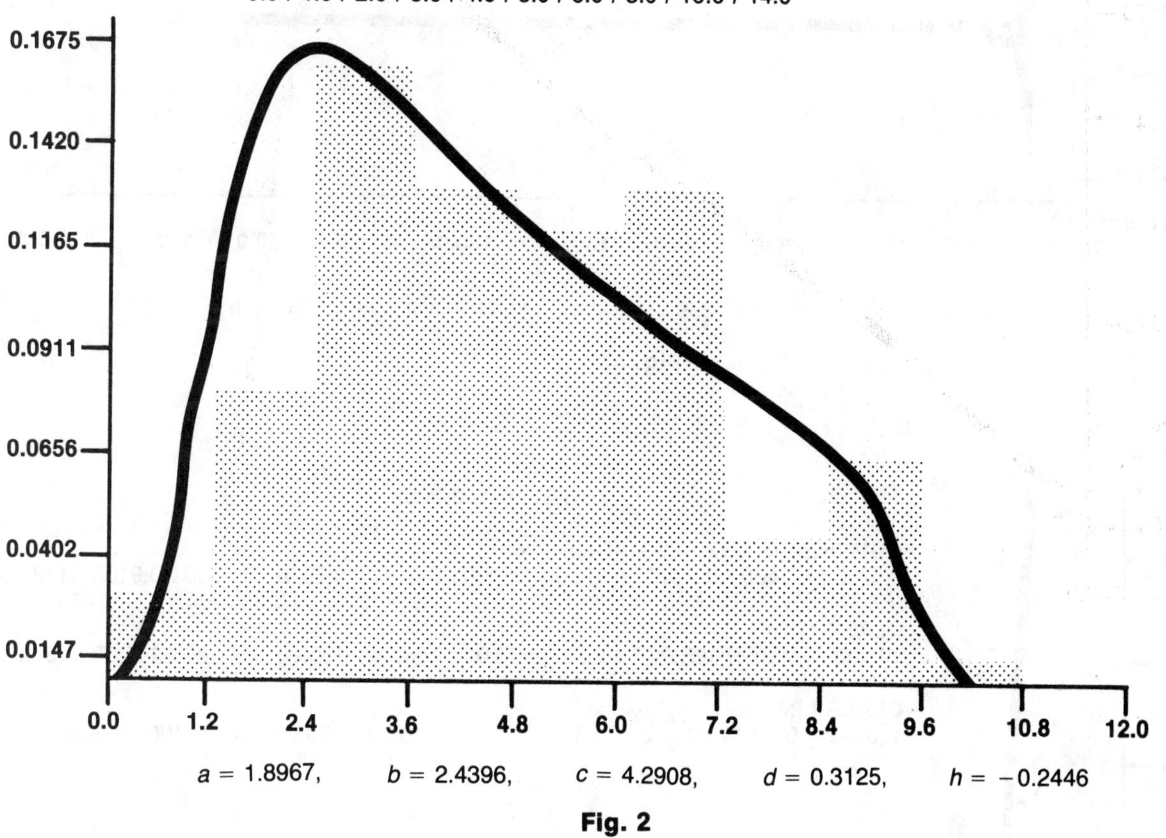

$a = 1.8967$, $b = 2.4396$, $c = 4.2908$, $d = 0.3125$, $h = -0.2446$

Fig. 2

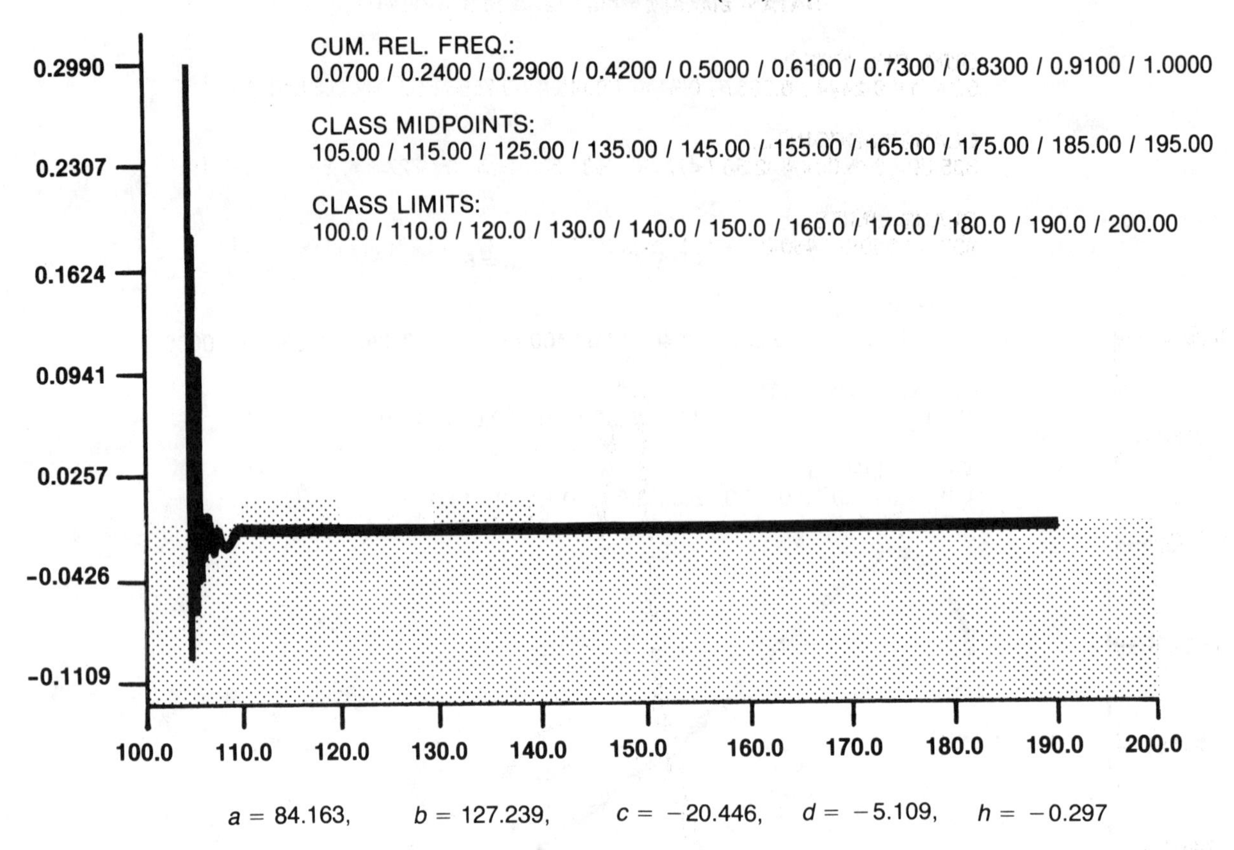

$a = 84.163$, $b = 127.239$, $c = -20.446$, $d = -5.109$, $h = -0.297$

Fig. 3a

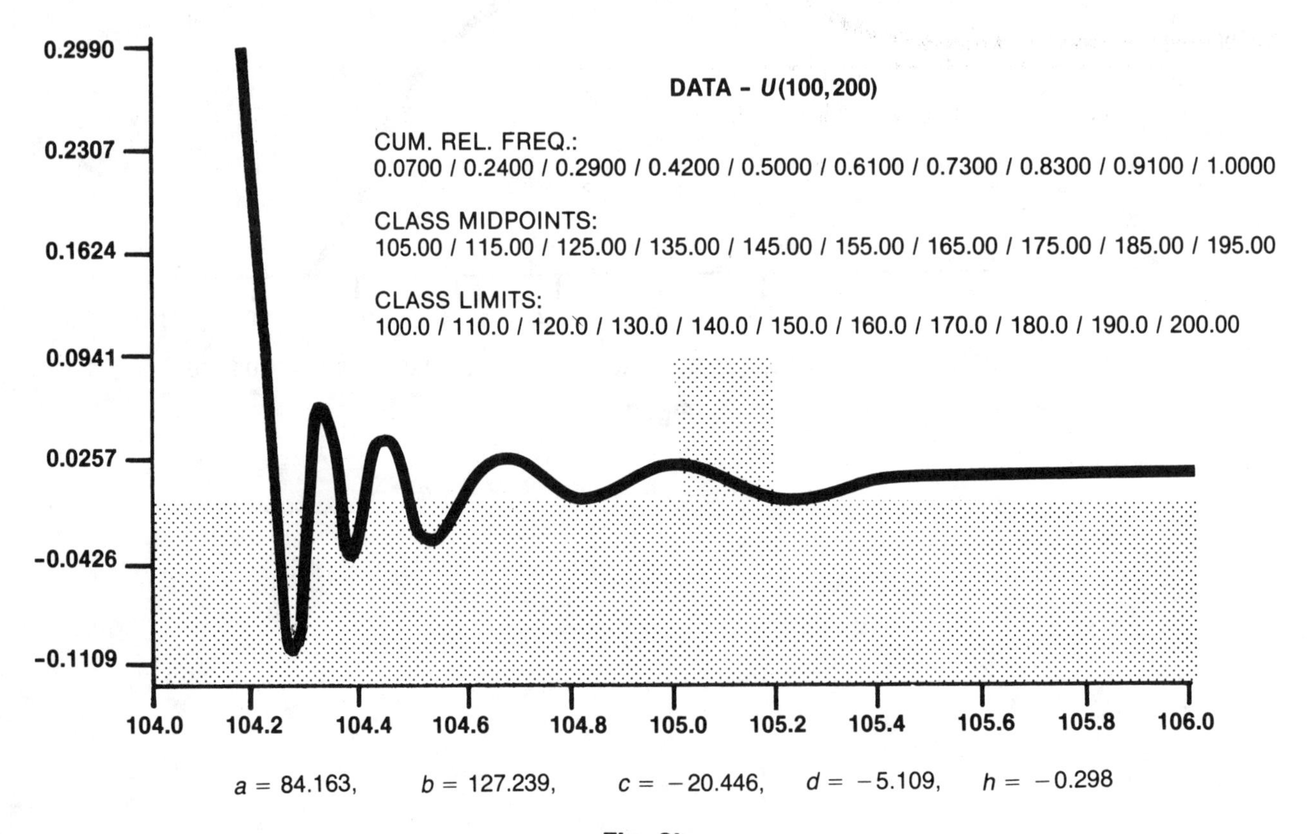

$a = 84.163$, $b = 127.239$, $c = -20.446$, $d = -5.109$, $h = -0.298$

Fig. 3b

DATA - elapsed times for IBM 370 (TSO)

CUM. REL. FREQ.:
0.1111 / 0.2444 / 0.3667 / 0.5111 / 0.6222 / 0.7556 / 0.8222 / 0.9222 / 1.0000

CLASS MIDPOINTS:
325.00 / 375.00 / 422.50 / 472.50 / 537.50 / 625.00 / 725.00 / 887.50 / 1425

CLASS LIMITS:
300.0 / 350.0 / 400.0 / 445.0 / 500.0 / 575.0 / 675.0 / 775.0 / 1000 / 1850

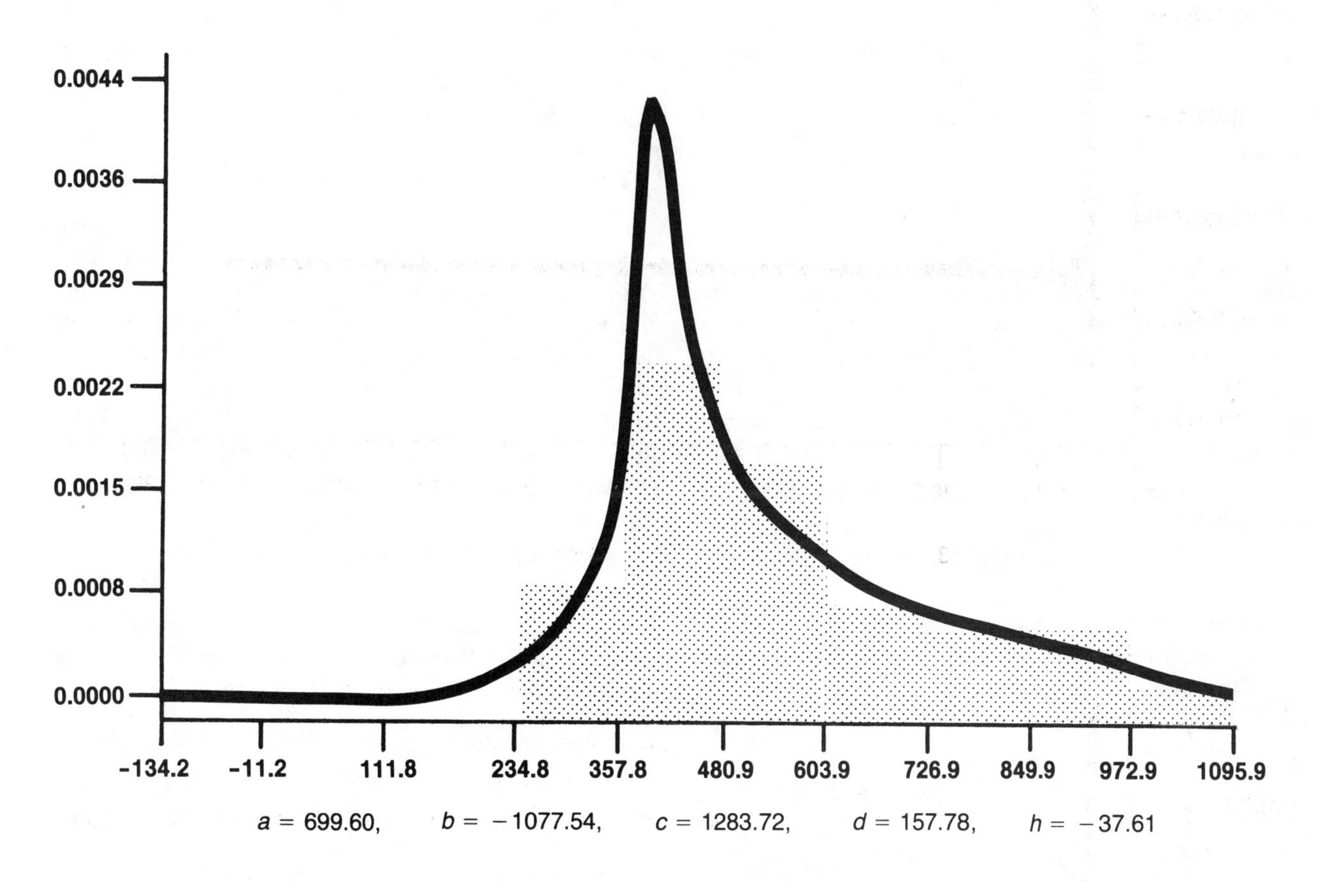

Fig. 4

Comparisons and Conclusions

This article will consider the relative merits of the three methods already discussed. Advantages and disadvantages of each method will be considered both from a practitioner's point of view as well as from the limited perspective of the first four data sets used in the illustration of the kernel method.

1. The Kernel Method

As the examples of kernel estimation indicate, **this method seems to work well in most cases; however, the third data set, $U(100, 200)$, is a notable exception to this observation.** The kernel method has **two serious disadvantages.** The *first* involves the ambiguity of the choice of $h(n)$. A practitioner will never really be certain that the choice of $h(n)$ which makes the graph of $\hat{f}$ smooth cannot be substantially improved. A *second* serious problem involves the definition of $\hat{f}$,

$$\hat{f}(x) = C \sum_j K((x - x_j)/h(n)),$$

which makes the computation of probabilities and the generation of observations rather difficult. It should be noted that both of these difficulties can be overcome since by numerical integration one can always obtain

$$P\,(a \leq x \leq b) = \int_a^b f(x)dx$$

for any given a and b.

Another observation that can be made from the examples of kernel estimation is that the modified version of the kernel method represents a significant improvement. As will be shown in "Comparison of Methods" below, this turns out to be a misleading observation; in fact, both methods seem to yield estimators of density functions of about the same quality.

The most salient features of the kernel method are its ability to produce estimators $\hat{f}$, of high quality and its independence from previously compiled tables. The kernel method is a good procedure with general applicability.

2. The Generalized Lambda Distribution (GLD) Method

Of the three methods considered, the GLD approximation gives the greatest accuracy and should be used whenever possible. There are no ambiguities in the choice of parameters as in the case of the kernel method, and once the four parameters L_1, L_2, L_3, and L_4 are determined, additional observations can be generated without difficulty. It is clear from the applications that were considered that the GLD method, whenever able to be applied (i.e., if tables were available), did at least as well as the other methods. Furthermore, it provided the only reasonable approximation to the uniformly distributed data of Example 3.

3. The Schmidt-Taylor Method

It was already observed that the Schmidt-Taylor approximation was **not guaranteed to produce legitimate distributions.** Furthermore, this method will, on occasion, produce values of x which are substantially removed from the interval from which the data was chosen. The generation of values of x for which $f(x)$ is negative is also a possibility. The illustrations of this method give the intuitive indication (this will be substantiated in the following section) of a comparatively poor fit.

4. Comparison of Methods

Figures 1 through 4 below compare the two reasonable methods (Kernel and GLD) and suggest, in a qualitative way, that **the GLD method generally provides a better approximation.**

To get a more **quantitative judgment of the relative performance of the methods,** the chi-square statistic and the Kolmogorov-Smirnov statistic are used. In the case of the GLD and Schmidt-Taylor methods, the percentiles $x_0, x_1, \ldots, x_{10}$ are computed in such a way that

$$P\,(x_i < x < x_{i+1}) = 0.1.$$

$x_0, x_1, \ldots, x_{10}$ are then used as interval endpoints to compute expected frequencies, E_i ($E_i = n/10$ where n is the sample size), and observed frequencies, O_i, for each interval. **The chi-square statistic** is now obtained from

$$\chi^2 = \Sigma\,[(E_i - O_i)^2/E_i].$$

To get the Kolmogorov-Smirnov statistic, the expected cumulative relative frequencies C_i ($C_i = i/10$; $1 \leq i \leq 10$) and the observed cumulative relative frequencies R_i are computed for each interval. **The Kolmogorov-Smirnov statistic** can then be obtained from

$$KS = \max |C_i - R_i| \cdot .$$

These results are summarized in Table 1 below. The entries for the kernel and modified kernel methods were derived by using the intervals obtained in the GLD method.

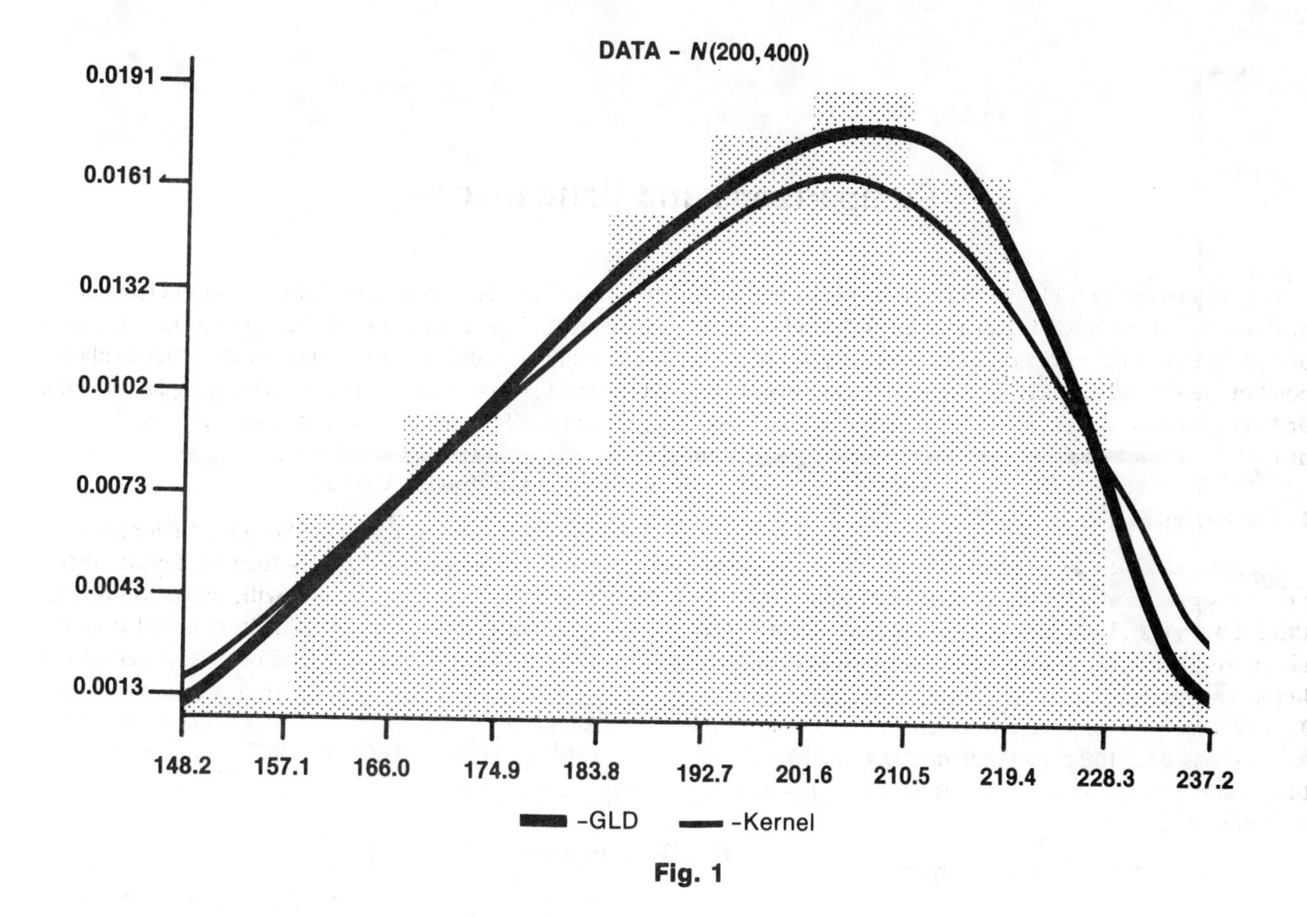
DATA - N(200,400)
0.0191
0.0161
0.0132
0.0102
0.0073
0.0043
0.0013
148.2
157.1
166.0
174.9
183.8
192.7
201.6
210.5
219.4
228.3
237.2
-GLD
-Kernel

Fig. 1

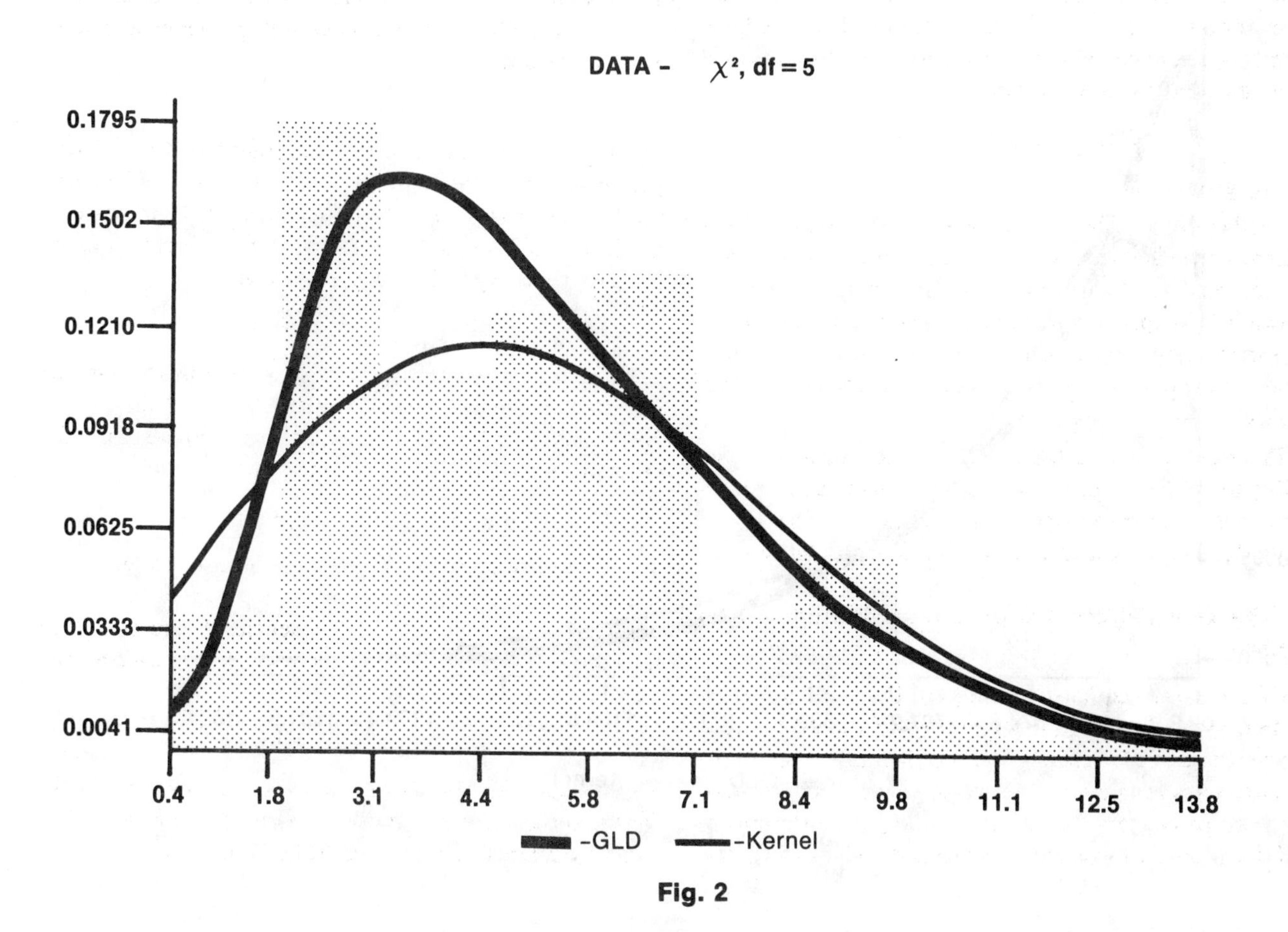
DATA - χ², df = 5
0.1795
0.1502
0.1210
0.0918
0.0625
0.0333
0.0041
0.4
1.8
3.1
4.4
5.8
7.1
8.4
9.8
11.1
12.5
13.8
-GLD
-Kernel

Fig. 2

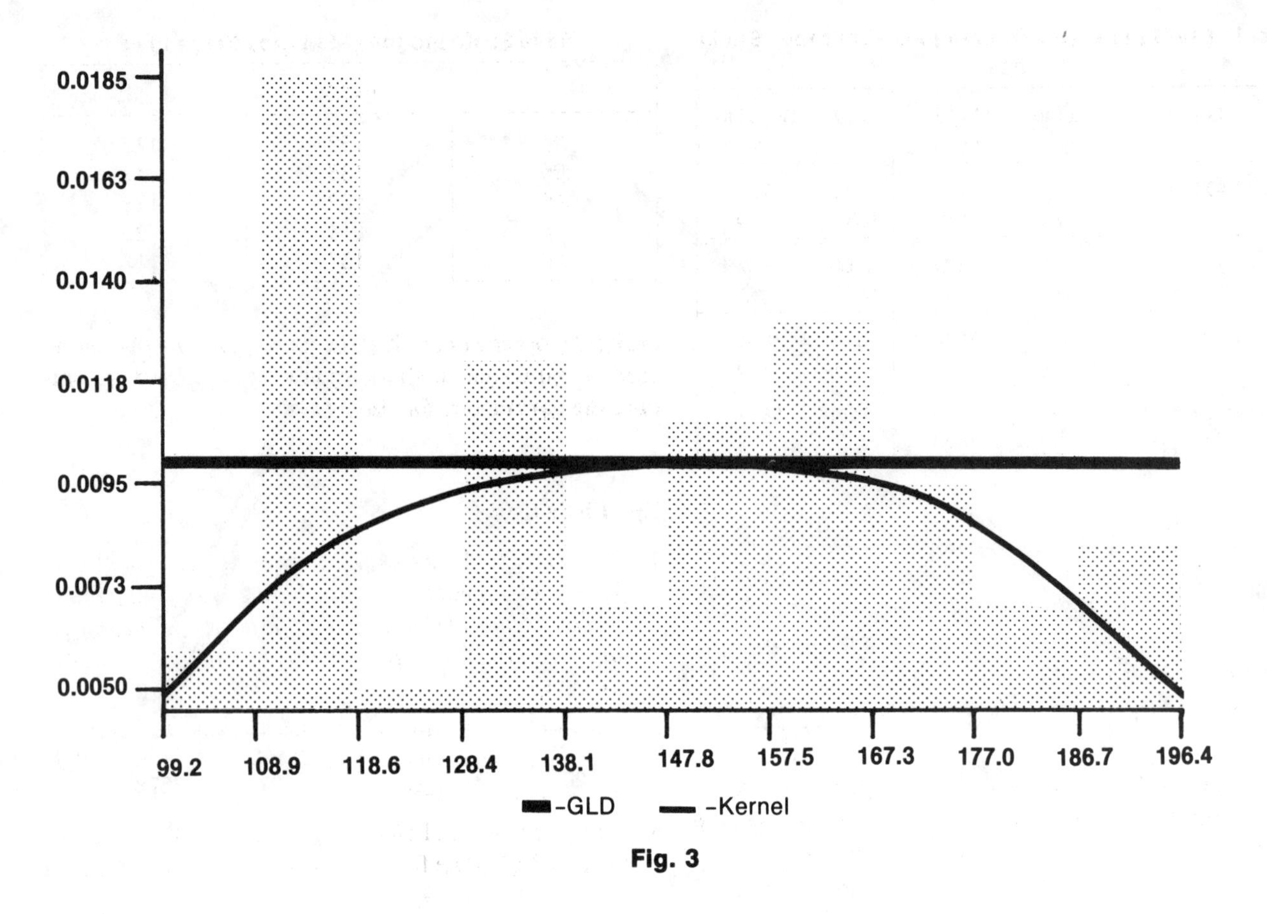

Fig. 3

DATA - connect times for IBM 370 (TSO)

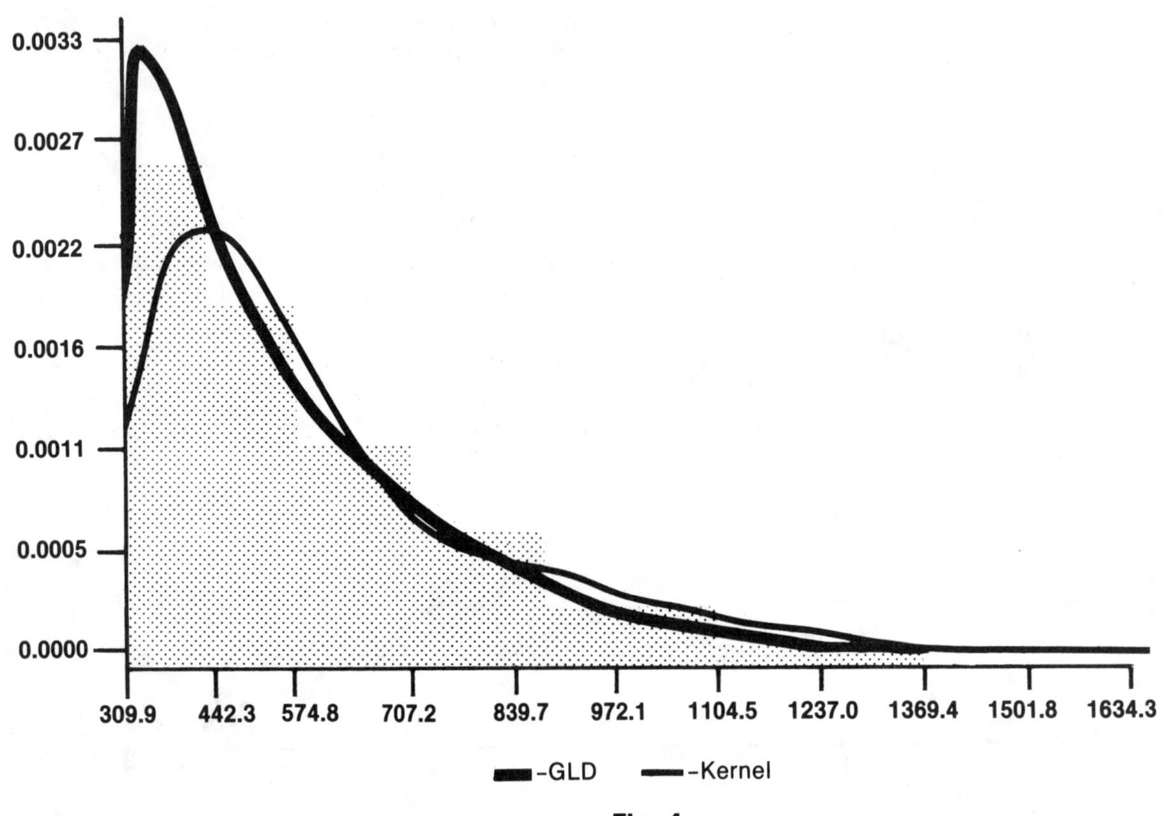

Fig. 4

Table 1: Chi-Square and (Kolmogorov-Smirnov) Statistics

Data	Kernel	Modified Kernel	GLD	Schmidt-Taylor
$N(200,400)$	1.27 (.02)	2.05 (.03)	2.20 (.04)	17.80 (.11)
χ^2, df=5	11.89 (.08)	13.35 (.08)	2.60 (.03)	17.80 (.08)
$U(100,200)$	20.28 (.06)	20.48 (.07)	12.90 (.04)	8.40 (.07)
Connect times for IBM 370 (TSO)	7.67 (.04)	8.35 (.04)	7.20 (.07)	28.00 (.14)

The χ^2 and the Kolmogorov-Smirnov statistics indicate the level of discrepency between the density function and the data; thus, the χ^2 and Kolmogorov-Smirnov values are small when there is a good fit. For purposes of comparison, the percentiles of the chi-square distribution (with 9 degrees of freedom) and the critical values for the Kolmogorov-Smirnov statistic are given in Tables 2 and 3, respectively.

The earlier observation that the GLD method is best is certainly substantiated. The observation that the modified kernel method improves the quality of the density function estimator is fallacious; the two kernel methods do not give significantly different results in spite of the fact that the modification produces a much smoother function. The Schmidt-Taylor method, in both the absolute and comparative senses, does very poorly and is not likely to be of any help.

A reasonable approach for a practitioner may be to **use the GLD method if tabled values of the parameters are available; otherwise, either the kernel method should be used or numerical methods should be employed to compute the parameters for the GLD.**

Table 2: Chi-Square (DF=9) Percentiles

.005	1.73		.750	11.39
.010	2.09		.900	14.68
.025	2.70		.950	16.92
.050	3.33		.975	19.02
.100	4.17		.990	21.67
.250	5.90		.995	23.59
.500	8.34			

Table 3: Kolmogorov-Smirnov Percentiles

	$n=100$	$n=90$
.01	.163	.172
.05	.136	.143
.10	.122	.128
.15	.114	.120
.20	.107	.113

REFERENCES

1. Dudewicz, E. J., Ramberg, J. S., and Tadikamalla, P. R., "A distribution for Data Fitting and Simulation," *Annual Technical Conference Transactions of the American Society of Quality Control,* Vol. 28, (1974), 407–418.
2. Mamrak, S. A. and DeRuyter, P. A., "Statistical Methods for Comparing Computer Services," *Computer,* Vol. 11 (1977), 32–39.
3. Olsson, D. M. and Nelson, L. S., "The Nelder-Mead Simplex Procedure for Function Minimization," *Technometrics,* Vol. 17 (1975), 45–51.
4. Ramberg, J. S. and Schmeiser, B. W., "An Approximate Method for Generating Asymmetric Random Variables," *Communications of the ACM,* Vol. 15 (1972), 78–82.
5. Ramberg, J. S., Tadikamalla, P. R., Dudewicz, E. J., and Mykytka, E. F., "A Probability Distribution for Fitting Empirical Data which Simplifies Applications," *Technometrics,* Vol. 21 (1979), 201–214.
6. Schmidt, J. W. and Taylor, R. E., *Simulation and Analysis of Industrial Systems,* Richard D. Irwin, Inc., Homewood, Ill., 1970.
7. Tarter, M. E. and Kronmal, R. A., "An Introduction to the Implementation and the Theory of Nonparametric Density Estimation," *The American Statistician,* Vol. 30 (1976), 105–111.

PART V: Computer Sorting Methods and Their Use in Simulation

The importance of sorting for simulation revolves around the fact that **in simulation, one often wishes to sort a relatively lengthy list numerous times.** The authors have encountered many situations where a simulation seemed too costly to run; it has not infrequently been the case that this was due to the fact that a "bubble sort" algorithm (an intuitively simple sort, but also one of the very least efficient) was incorporated into the simulation code. When simulating the genes of animal herds numerous times to find the best herd management techniques or when simulating nuclear attack strategies using the (over 30,000) airfields in the United States, **the difference between a bubble sort and a relatively efficient sort, such as the heapsort, can be measured in hours of computer time.** We have often found that this change alone made a simulation which previously timed out, run well.

Sorting algorithms typically are applied to **files of records** stored externally to the sorting program. There may be any number of **subfields** associated with each record within the file. For example, an employee record may contain employee name, job classification, social security number, wages, etc., as subfields. The particular subfields which are relevant in a specific application are called **keys**. If, in a particular application, we need to resequence the records of a file so that they would be in nondecreasing (or nonincreasing) order of hourly wages, then we would sort the file on the hourly wage key.

This part considers six sorting algorithms (insertion sort, bubble sort, shellsort, quicksort, mergesort, and heapsort). A descriptive illustration, an actual Pascal subprogram, and an analysis are given for each algorithm. The last article discusses the comparative performance of the six algorithms in actual computer runs. Knuth [4] is the most complete reference on sorting algorithms. Aho, Hopcroft, and Ullman [1], Tanenbaum and Augenstein [9], Horowitz and Sahni [3], and Reingold and Hansen [5] all contain segments on various sorting algorithms.

In the following articles, we will assume that the records $R_1, \ldots, R_N$ are to be sorted in ascending order on a single key with values $K_1, \ldots, K_N$, respectively. The keys will be implemented as an array $K[1], \ldots, K[N]$ in the Pascal subprograms. To avoid extraneous problems associated with file input and output, it will further be assumed that the records $R_1, \ldots, R_N$ are located in main memory.

EH0227-9/85/0000/0279$01.00 © 1985 IEEE

Insertion Sort

This algorithm is one of the simplest approaches to sorting. It sorts the keys $K[1], \ldots, K[N]$ by first considering $K[1]$ by itself as a sorted array. Next $K[2]$ is placed at its proper location relative to $K[1]$ (this may require moving $K[2]$ ahead of $K[1]$ if $K[2] < K[1]$) so that now the keys $K[1]$ and $K[2]$ form a sorted array. Now $K[3]$ is positioned in its proper location relative to $K[1]$ and $K[2]$ (moving $K[3]$ ahead of $K[1]$ or $K[2]$ if necessary) so that $K[1]$, $K[2]$, and $K[3]$ form a sorted array. This process is repeated for $K[4], \ldots, K[N]$.

Note that whenever $K[J]$ is considered, $K[1], \ldots, K[J-1]$ are already in ascending order and the expansion of the sorted portion of the array to include $K[J]$ required two actions: the determination of the correct location of $K[J]$ within the sorted subarray and the movement of $K[J]$ to that position. Figure 1 illustrates the application of the insertion sort to the keys:

42, 12, 17, 98, 56, 63, 34, 72, 25, 83

At each step the first key past the horizontal bar is moved to the left and positioned in its proper place within the sorted subarray. The underlined keys are the ones being positioned at each step.

A subprogram, written as a Pascal procedure, for executing an insertion sort is given in Figure 2. The procedure INSERT (lines 32 through 58) is a subprogram of the sorting routine INSERTSORT (lines 61 through 68). INSERTSORT does the sorting by calling upon INSERT to place each item in its correct location. Note that at line 63 an artificially small key value is added to the array to help simplify the loop at lines 65 and 66.

To develop some insight into the time requirements of this algorithm, we observe that in the process of sorting N keys, INSERT is invoked $N-1$ times to insert $K[2], \ldots, K[N]$. To insert $K[J]$ into the proper location among $K[1], \ldots, K[J-1]$, the subprocedure INSERT must make anywhere from 1 to J comparisons at line 49 and anywhere from 2 to $2J$ assignments at lines 51 and 52. On the average, if the keys are in random order, the insertion of $K[J]$ will require $J/2$ comparisons. Thus the computation time for the insertion itself will be proportional to J, and the entire computation time to sort the array will be proportional to

$$\sum_{J=2}^{N} J = \frac{N(N+1)}{2} - 1 = \tfrac{1}{2}N^2 + \tfrac{1}{2}N - 1$$

or simply proportional to N^2.

1–initial	42	12	17	98	56	63	34	72	25	83
2	12	42	17	98	56	63	34	72	25	83
3	12	17	42	98	56	63	34	72	25	83
4	12	17	42	98	56	63	34	72	25	83
5	12	17	42	56	98	63	34	72	25	83
6	12	17	42	56	63	98	34	72	25	83
7	12	17	34	42	56	63	98	72	25	83
8	12	17	34	42	56	63	72	98	25	83
9	12	17	25	34	42	56	63	72	98	83
10	12	17	25	34	42	56	63	72	83	98

Figure 1

```
 LINE         LEVEL
NUMBERS   PROC STMT  STATEMENT.        .        .        .        .        .        .

    23     1    0
    24     2    0    PROCEDURE INSERTSORT   (VAR K          : DATARRAY; {array to be sorted      }
    25     2    0                            N              : INTEGER); {# of items in the array}
    26     2    0
    27                {This procedure sorts the array K (of size N) using the INSERTION SORT
    28                method.  INSERTSORT only traverses the array; it calls the subprocedure
    29     2    0     INSERT to inserts specific items into the array.}
    30     2    0
    31     2    0            VAR     CTR      : INTEGER;        {used to step through the array       }
    32     2    0     {.........................................................................}
    33     2    0
    34     3    0            PROCEDURE INSERT(VAR SORTARRAY  : DATARRAY;
    35     3    0                                 LOC        : INTEGER);  {of next item in file     }
    36     3    0
    37                       {This procedure is a subprocedure to INSERTSORT for inserting the
    38                       LOCth entry of the array SORTARRAY into its proper position among
    39     3    0            SORTARRAY [1],  SORTARRAY [2], ... , SORTARRAY [LOC - 1]. }
    40     3    0
    41     3    0                 VAR     CNT      : INTEGER;
    42     3    0                         RECERD   : INTEGER;      {variable for array entry      }
    43     3    0
    44     3    0            BEGIN {of INSERT}
    45     3    1
    46     3    1               RECERD := SORTARRAY [LOC];          {hold the value at the loc     }
    47     3    1               CNT := LOC - 1;                     {start at previous item        }
    48     3    1
    49     3    1               WHILE RECERD < SORTARRAY [CNT] DO   {compare to find place         }
    50     3    1                  BEGIN {of WHILE}
    51     3    2                    SORTARRAY [CNT + 1] := SORTARRAY [CNT];{move next item up  }
    52     3    2                    CNT := CNT - 1                 {go to previous item           }
    53     3    2               END; {of WHILE}
    54     3    1
    55     3    1               SORTARRAY [CNT + 1] := RECERD       {put item in correct place     }
    56     3    1
    57     3    1            END; {of INSERT}
    58     3    0     {.........................................................................}
    59     3    0
    60     3    0
    61     2    0     BEGIN {of INSERTSORT}
    62     2    1
    63     2    1            K [0] := - MAXINT;                     {initialize to small value     }
    64     2    1
    65     2    1            FOR CTR := 2 TO N DO                   {call on INSERT to locate      }
    66     2    1                INSERT (K, CTR);                   {K[2],..., K [ N ] into K      }
    67     2    1
    68     2    1     END; {of INSERTSORT}
```

Figure 2

Bubble Sort

This sorting method scans successive pairs of keys $K[1]$ and $K[2]$, $K[2]$ and $K[3]$, and so on to $K[N-1]$ and $K[N]$, and it interchanges the records associated with the pair of keys if they are out of order. It is unlikely that after one sweep or pass of interchanges the keys will be sorted. In fact, all that can be assured is that the record with the largest key will "bubble up" to the end of the sequence. Repeated passes (possibly up to $N-1$ passes) are necessary to sort the entire sequence. The application of the bubble sort to the keys

42   12   17   98   56   63   34   72   25   83

is illustrated in Figure 3.

During pass 1, the first exchange switches $K[1]$ and $K[2]$ (42 and 12). The next exchange switches $K[2]$ and $K[3]$ (the new $K[2]-42$ is exchanged with 17), etc. The keys are sorted after six passes. Since additional passes cannot contribute to the sorting effort, a minimum number of passes should be executed. An additional seventh pass is needed to determine that no additional exchanges are necessary, and hence the keys are already sorted. The Pascal procedure given in Figure 2 uses the Boolean variable EXCH (in lines 106, 109, and 115) to detect a pass with no exchanges.

Further efficiency could be attained by observing that fewer pairs of keys need to be scanned for possible interchange on each successive pass because each pass moves a record with next largest key value into position. The upper limit of N-PASS is used at line 110 of the procedure for this reason. The SWAP subprocedure, which is invoked at line 114, simply interchanges K[CNT] and K[CNT + 1].

Since pass 1 will scan $N-1$ successive pairs of key values, its computation time will be proportional to $N-1$. More generally, the computation time for pass J will be proportional to $N-J$. The derivation of the computation time for the bubble sort is dependent on the number of passes that will be required for the sort. An expression for the number of passes in the average case (i.e., when the keys are in random order) is developed in Knuth [4] and requires a good deal of mathematical background. The overall computation time for the bubble sort when applied to random keys is proportional to N^2.

	K[1]	K[2]	K[3]	K[4]	K[5]	K[6]	K[7]	K[8]	K[9]	K[10]
	42	12	17	98	56	63	34	72	25	83
PASS 1	12	42	42	56	98	98	98	98	98	98
		17			63	34	72	25	83	
	12	17	42	56	63	34	72	25	83	98
PASS 2					34	63	25	72		
	12	17	42	56	34	63	25	72	83	98
PASS 3				34	56	25	63			
	12	17	42	34	56	25	63	72	83	98
PASS 4			34	42	25	56				
	12	17	34	42	25	56	68	72	83	98
PASS 5				25	42					
	12	17	34	25	42	56	63	72	83	98
PASS 6			25	34						
	12	17	25	34	42	56	63	72	83	98
PASS 7										
	12	17	25	34	42	56	63	72	83	98

Figure 3

Shellsort

A serious disadvantage of the bubble sort is that it always compares successive pairs of keys. Thus to move a record with a large key value to the end of the sequence, a large number of comparisons and exchanges will be needed. To speed up the movement of records with large key values to the back of the sequence, Shell [8] suggests that a positive integer, GAP, be chosen, and the pairs of keys $K[1]$ and $K[1+GAP]$, $K[2]$ and $K[2+GAP]$, etc. are compared and exchanged if necessary. If the value of GAP is taken to be 1, we obviously have the bubble sort.

Clearly, the larger the value of GAP, the fewer the number of comparisons needed. However, if GAP is chosen larger than 1, there will be no assurance that the records will eventually be sorted. This problem is resolved by starting with $GAP = N/2$ (the integer portion of $N/2$) and, after making all the exchanges that can be made, dividing GAP by 2 (again using integer division) until GAP assumes value 0. The sense of this algorithm can be described by the following:

1. Let $GAP = N/2$;
2. If $GAP > 0$, do a modified bubble sort comparing keys $K[J]$ and $K[J+GAP]$; redefine GAP to be half its current value; go to 2;
3. Sort is complete.

For the keys

42 12 17 98 56 63 34 72 25 83,

we take $GAP = N/2 = 5$ and compare the pairs of keys 42 and 63, 12 and 34, 17 and 72, 98 and 25 (this will require an exchange), and 56 and 83. No additional exchanges are

```
 LINE        LEVEL
NUMBERS   PROC STMT  STATEMENT.        .         .         .         .         .         .         .

   69      2    0
   70      2    0    PROCEDURE SWAP       (VAR K             : DATARRAY;
   71      2    0                  LOC               : INTEGER;      {loc of 1st item to be swapped}
   72      2    0                  GAP               : INTEGER);     {distance between 1st & 2nd   }
   73      2    0
   74      2    0    {This procedure swaps the values in an array at locations LOC and LOC + GAP.}
   75      2    0
   76      2    0         VAR     TEMP    :INTEGER;                   {temporary value holder       }
   77      2    0
   78      2    0      BEGIN {of SWAP}
   79      2    1
   80      2    1         TEMP := K [ LOC + GAP ];                    {temp set equal to second     }
   81      2    1         K [LOC + GAP] := K [LOC];                   {second equal to first        }
   82      2    1         K [LOC] := TEMP                             {first equal to temp          }
   83      2    1
   84      2    1      END; {of SWAP}
   85      2    0
   86      2    0    {---------------------------------------------------------------------------}
   87      2    0
   88      2    0    PROCEDURE BUBBLESORT    (VAR K           : DATARRAY; {array to be sorted      }
   89      2    0                              N               : INTEGER); {# of items to be sorted}
   90      2    0
   91                {This procedure sorts the array using the BUBBLE SORT method.  It scans the
   92                array and compares pairs of items, and swithces them if the pair is out of
   93                order.  After the first pass the last item in the file will automatically be
   94                the largest.  Therefore for the second pass it is only necessary to compare
   95      2    0    n-1 items; for the third, n-2; etc. }
   96      2    0
   97      2    0         VAR     EXCH    : BOOLEAN;                  {flag for swaps that are made}
   98      2    0                 PASS    : INTEGER;                  {counts the number of passes }
   99      2    0                 CNT     : INTEGER;                  {loop counter                }
  100      2    0
  101      2    0    BEGIN {of BUBBLESORT}
  102      2    1
  103      2    1      EXCH := TRUE;                                 {initialize EXCH             }
  104      2    1      PASS := 1;                                    {initialize pass             }
  105      2    1
  106      2    1      WHILE EXCH = TRUE DO                          {continue until no exchanges }
  107      2    1         BEGIN {of WHILE}
  108      2    2
  109      2    2            EXCH := FALSE;                          {no exchanges made yet       }
  110      2    2            FOR CNT := 1 TO ( N - PASS) DO          {compare 1 less item per pass}
  111      2    2               IF K [CNT] > K [CNT + 1] THEN
  112      2    2
  113      2    2               BEGIN {of IF/THEN}                   {first is larger than second }
  114      2    3                  SWAP (K, CNT, 1);
  115      2    3                  EXCH := TRUE                      {exchng made--file not sorted}
  116      2    3               END; {of IF/THEN}
  117      2    2
  118      2    2            PASS := PASS + 1                        {next pass--compare one less }
  119      2    2
  120      2    2         END {of WHILE}
  121      2    2
  122      2    2      END; {of BUBBLESORT}
```

Figure 4

possible with *GAP* = 5; therefore, the keys are now rearranged to

42 12 17 25 56 63 34 72 98 83

and *GAP* is redefined to be 2. Now, 17 and 42, and 34 and 56 are exchanged. Note, however, that after these exchanges, 42 and 34 will have to be exchanged at the next pass with *GAP* still equal to 2. The ordering of keys thus becomes

17 12 34 25 42 63 56 72 98 83.

GAP is set equal to 1, and a single pass will exchange 17 and 12, 31 and 25, 63 and 56, and 83 and 98 to complete the sorting.

The inner loop (lines 146 through 156) of the *SHELLSORT* procedure given in Figure 5 executes the modified bubble sort with comparisons and possible exchanges between *K*[*CNT*] and *K*[*CNT*+*GAP*] (lines 149 through 154). As in the bubble sort, the subprocedure *SWAP* is used to interchange the keys. The outer loop (lines 140 through 160) controls the value of *GAP* (lines 140 and 158) and executes the inner loop until *GAP* becomes 0. A *FORTRAN* program for Shellsort is given in Figure 6.

The analysis of the performance, in the sense of computer time, of Shellsort is very difficult. Experimental results, such as those given in the last article of this section show that Shellsort certainly does better than bubble-sort.

```
LINE       LEVEL
NUMBERS  PROC STMT  STATEMENT.      .       .       .       .       .       .

   123    2    0
   124    2    0    PROCEDURE SHELLSORT    (VAR K          : DATARRAY; {array to be sorted       }
   125    2    0                            N              : INTEGER); {# of items to be sorted  }
   126    2    0
   127              {This procedure sorts the array using the SHELLSORT method.  This method
   128              differs from the BUBBLE SORT in that it compares items that are seperated
   129    2    0    by a gap.  If the items ar out of order, they are switched. }
   130    2    0
   131    2    0          VAR     GAP     : INTEGER;               {distance between items     }
   132    2    0                  EXCH    : BOOLEAN;               {signal for continuing      }
   133    2    0                  CNT     : INTEGER;               {loop counter               }
   134    2    0
   135    2    0
   136    2    0    BEGIN {of SHELLSORT}
   137    2    1
   138    2    1      GAP := TRUNC (N/2);                          {initialize the gap         }
   139    2    1
   140    2    1      WHILE GAP <> 0 DO                            {continue until compare pairs}
   141    2    1
   142    2    1         BEGIN {of WHILE}
   143    2    2            EXCH := TRUE;                          {initialize EXCH            }
   144                      WHILE EXCH = TRUE DO                  {continue until no exchanges
   145    2    2                                                  are made -- list is sorted  }
   146    2    2               BEGIN {of WHILE}
   147    2    3                  EXCH := FALSE;                  {no exchanges made yet      }
   148    2    3                  FOR CNT := 1 TO (N - GAP) DO    {compare fewer items each pass}
   149    2    3                  IF K [CNT] > K [CNT + GAP] THEN
   150    2    3
   151    2    3                     BEGIN {of IF/THEN}          {first is larger than second }
   152    2    4                        SWAP (K, CNT, GAP);
   153    2    4                        EXCH := TRUE              {exchng made--file not sorted}
   154    2    4                     END {of IF/THEN}
   155    2    4
   156    2    4               END; {of WHILE}
   157    2    2
   158    2    2            GAP := TRUNC (GAP/2)                   {make gap smaller           }
   159    2    2
   160    2    2         END; {of WHILE}
   161    2    1
   162    2    1    END; {of SHELLSORT}
   163    2    0
```

Figure 5.

```
C
C     SUBROUTINE SORT(R,N)
C
C***********************************************************************
C
C PURPOSE
C    TO SORT A GIVEN ARRAY.
C
C USAGE
C    "CALL SORT(R,N)"
C
C DESCRIPTION OF PARAMETERS
C    R...IS THE NAME OF THE ONE-DIMENSIONAL ARRAY TO BE SORTED.
C    N...IS THE DIMENSION OF THE ARRAY TO BE SORTED.
C
C REMARKS
C    THIS ALGORITHM IS MORE EFFICIENT THAN EITHER THE "BUBBLE SORT"
C    OR THE "INTERCHANGE SORT" METHODS.  WHILE NOT OPTIMAL, IT
C    PROVIDES A SIMPLE AND RELATIVELY EFFICIENT SORTING TECHNIQUE
C    WHICH IS EASILY PROGRAMMED.
C
C SUBROUTINES AND FUNCTION SUBPROGRAMS REQUIRED
C    NONE
C
C METHOD
C    THE METHOD USED IS CALLED THE "SHELL SORT" AND WAS INTRODUCED BY
C    D.L. SHELL IN REFERENCE 1.  FOR MORE RECENT DEVELOPMENTS, SEE
C    ITEM 2. AND ITS REFERENCES:
C    1. SHELL,D.L.: "A HIGH-SPEED SORTING PROCEDURE," COMMUNICATIONS OF
C       THE ASSOCIATION FOR COMPUTING MACHINERY, VOL.2, NO.7(JULY,
C       1959), PP.30-32.
C    2. GHOSHDASTIDAR,D. AND ROY,M.K.: "A STUDY ON THE EVALUATION OF
C       SHELL'S SORTING TECHNIQUE," THE COMPUTER JOURNAL, VOL.18(1975),
C       PP.234-235.
C
C***********************************************************************
C
      SUBROUTINE SORT(R,N)
      DIMENSION R(1)
      M=N
    1 M=M/2
      IF(M.EQ.0) GOTO 5
      J=1
    2 I=J
    4 IF(R(I).LE.R(I+M)) GOTO 3
      SAV=R(I)
      R(I)=R(I+M)
      R(I+M)=SAV
      I=I-M
      IF(I.GE.1) GOTO 4
    3 J=J+1
      IF(J.GT.N-M) GOTO 1
      GOTO 2
    5 RETURN
      END
```

Figure 6.

Quicksort

Following the original quicksort method proposed by Hoare [2], a number of variations to the basic algorithm have appeared in the literature (e.g., Sedgewick [6,7]). In the original version, which will be considered here, the array of keys $K[1], \ldots, K[N]$ is rearranged in such a way that

1. All $K[J]$ with $K[J] < K[1]$ are moved to the left of $K[1]$;
2. All $K[J]$ with $K[J] > K[1]$ are moved to the right of $K[1]$; and
3. $K[1]$ is moved into its correct position within the array.

No stipulations are made regarding keys with values equal to $K[1]$; such keys may end up on either side of $K[1]$.

Rearrangement of keys in this way, of course, does not produce a sorted array; it simply subdivides the original array into two subarrays, one consisting of key values less than or equal to $K[1]$, and another with key values greater than or equal to $K[1]$. If this type of subdivision is now applied to each of the two subarrays, more and smaller subarrays will result.

Since with repeated applications of this process the subarrays get smaller, subarrays of size one will eventually be obtained and the subdivision process terminated.

To illustrate the quicksort algorithm in greater detail, consider again the sequence of keys:

42 12 17 98 56 63 34 72 25 83

To collect the keys with the values less than $K[1] = 42$ to the left and the keys with values greater than $K[1]$ to the right, the keys are scanned from the left until a key with value greater or equal to 42 is encountered. This occurs at $K[I] = 98$ ($I = 4$). The keys are also scanned from the right until a key value less than or equal to 42 is encountered; this happens at $K[J] = 25$ ($J=9$). If $I < J$, then we exchange $K[I]$ and $K[J]$ to get

42 12 17 25 56 63 34 72 98 83.

The scanning of the keys from the left and the right continues (from where it was left off—not from the beginning—to find $K[I] = 56 > 42$ ($I=5$) and $K[J] = 34 < 42$ ($J=7$). These are now exchanged to give

42 12 17 25 34 63 56 72 98 83.

Continuing, we find $K[I]=63>42$ with $I=6$, and $K[J]=34 < 42$ with $J=5$. However, in this case, since $I > J$, instead of exchanging $K[I]$ and $K[J]$, we exchange $K[1]$ and $K[J]$ to get

34 12 17 25 42 63 56 72 98 83.

(34 12 17 25: left subarray; 63 56 72 98 83: right subarray)

The key with value 42 has moved to its correct position; the keys to its left have values less than or equal to 42 and those to its right have values greater than or equal to 42.

This method is now applied to the subarrays

{ 34 12 17 25 } and { 63 56 72 98 83 }

to produce the subarrays

{ 25 12 17 }, empty subarray, {56}, and { 72 98 83 }.

Repeated applications eventually sort the entire array.

The Pascal procedure given in Figure 7 is an example of a recursive procedure (i.e., a procedure which calls itself repeatedly until certain conditions are met). The recursive calls for sorting the left and right subarrays are made at lines 210 and 211.

The performance of quicksort depends on how rapidly the array is divided into smaller subarrays. The best results are obtained if on each call to quicksort, the array to be sorted is

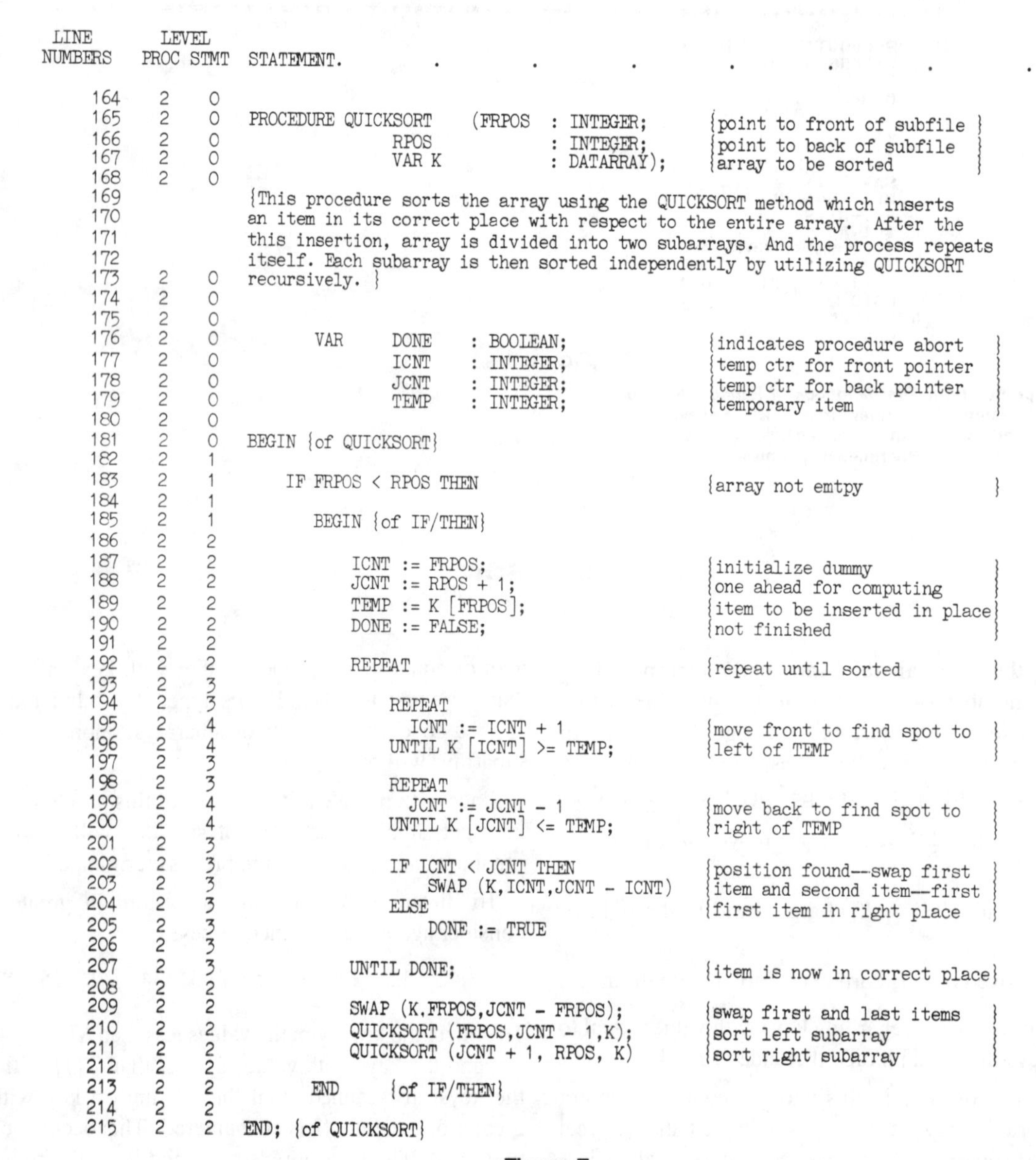

```
  LINE      LEVEL
NUMBERS  PROC STMT  STATEMENT.

  164     2    0
  165     2    0    PROCEDURE QUICKSORT    (FRPOS  : INTEGER;      {point to front of subfile }
  166     2    0                    RPOS   : INTEGER;      {point to back of subfile  }
  167     2    0                    VAR K  : DATARRAY);    {array to be sorted        }
  168     2    0
  169               {This procedure sorts the array using the QUICKSORT method which inserts
  170               an item in its correct place with respect to the entire array.  After the
  171               this insertion, array is divided into two subarrays. And the process repeats
  172               itself. Each subarray is then sorted independently by utilizing QUICKSORT
  173     2    0    recursively. }
  174     2    0
  175     2    0
  176     2    0        VAR   DONE    : BOOLEAN;               {indicates procedure abort  }
  177     2    0              ICNT    : INTEGER;               {temp ctr for front pointer }
  178     2    0              JCNT    : INTEGER;               {temp ctr for back pointer  }
  179     2    0              TEMP    : INTEGER;               {temporary item             }
  180     2    0
  181     2    0    BEGIN {of QUICKSORT}
  182     2    1
  183     2    1       IF FRPOS < RPOS THEN                     {array not emtpy            }
  184     2    1
  185     2    1          BEGIN {of IF/THEN}
  186     2    2
  187     2    2             ICNT := FRPOS;                    {initialize dummy           }
  188     2    2             JCNT := RPOS + 1;                 {one ahead for computing    }
  189     2    2             TEMP := K [FRPOS];                {item to be inserted in place}
  190     2    2             DONE := FALSE;                    {not finished               }
  191     2    2
  192     2    2             REPEAT                            {repeat until sorted        }
  193     2    3
  194     2    3                REPEAT
  195     2    4                  ICNT := ICNT + 1             {move front to find spot to }
  196     2    4                UNTIL K [ICNT] >= TEMP;        {left of TEMP               }
  197     2    3
  198     2    3                REPEAT
  199     2    4                  JCNT := JCNT - 1             {move back to find spot to  }
  200     2    4                UNTIL K [JCNT] <= TEMP;        {right of TEMP              }
  201     2    3
  202     2    3                IF ICNT < JCNT THEN            {position found--swap first }
  203     2    3                   SWAP (K,ICNT,JCNT - ICNT)   {item and second item--first}
  204     2    3                ELSE                           {first item in right place  }
  205     2    3                   DONE := TRUE
  206     2    3
  207     2    3             UNTIL DONE;                       {item is now in correct place}
  208     2    2
  209     2    2             SWAP (K,FRPOS,JCNT - FRPOS);      {swap first and last items  }
  210     2    2             QUICKSORT (FRPOS,JCNT - 1,K);     {sort left subarray         }
  211     2    2             QUICKSORT (JCNT + 1, RPOS, K)     {sort right subarray        }
  212     2    2
  213     2    2          END     {of IF/THEN}
  214     2    2
  215     2    2    END; {of QUICKSORT}
```

Figure 7

divided into two arrays of equal or almost equal size. The worst performance of quicksort occurs when arrays of size K are subdivided into subarrays of size 0 and $K-1$. If the original arrangement of keys is random, then quicksort will have a low computation time; if, on the other hand, the original array is in nearly sorted order (a frequent situation in many applications), the computation time of quicksort could be quite large.

To analyze the performance of quicksort with the "best case" assumption, let $CT(N)$ be the computation time required for sorting N keys. The computation time required to scan the array from the left and the right is independent of the particular ordering of the keys within the array. The number of exchanges required, however, is dependent on the key arrangement. Since at most $N/2$ exchanges will be necessary, we can assert that for some constant k, the computation time for scanning and exchanging keys is bounded by kN. Thus

$$CT(N) < kN + 2CT(N/2)$$

where $2CT(N/2)$ is the computation time for sorting the two subarrays, each of size $N/2$. If this argument is applied repeatedly, we would get

$$\begin{aligned} CT(N) &< kN + 2CT(N/2) \\ &< kN + 2(kN/2 + 2CT(N/4)) = 2kN + 4CT(N/4) \\ &< 2kN + 4(kN/4 + 2CT(N/8)) = 3kN + 8CT(N/8) \\ &\quad\vdots \\ &= tkN + 2^tCT(N/2^t) \\ &\quad\vdots \\ &= kN\log_2(N) + N\,CT(1). \end{aligned}$$

This makes $CT(N)$ proportional to $N\log_2(N)$ rather than N^2 as was the case in the insertion and bubble sorts. This substantial improvement of computation time is dependent on the optimal subdivision of the array of keys into equal-sized subarrays. Fortunately, for the average case (i.e., when the keys are in random order in the original array), the computation time is still proportional to $N\log_2(N)$.

When the subdivision pattern moves away from the optimal configuration, such as when an array of keys is in almost sorted order, the computation time of quicksort becomes proportional to N^2. In the "worst case," arrays of size K are subdivided into subarrays of size 0 and $K-1$. This gives

$$\begin{aligned} CT(N) &< kN + CT(N-1) \\ &< kN + k(N-1) + CT(N-2) \\ &\quad\vdots \\ &< kN + \cdots + k(2) + CT(1) \\ &= k\sum_{i=2}^{N} i + CT(1) \\ &= k\{N(N+1)/2 - 1\} + CT(1). \end{aligned}$$

Mergesort

The mergesort algorithm considers the initial array of keys $K[1], \ldots, K[N]$ as N sorted subarrays, each with a single element. It then merges pairs of these one-element arrays to form $N/2$ sorted subarrays, each with two elements. The first of these subarrays will be $K[1]$ and $K[2]$ arranged in ascending order. The merging process is repeated to form $N/4$ subarrays of size 4, etc. until the last two subarrays are merged into one sorted array. Figure 8 illustrates the application of the mergesort to the keys.

42 12 17 98 56 63 34 72 25 83.

Since N may not be a power of 2, on certain passes the number of subarrays to be merged may not be even. This happens on passes 2 and 3 in our example. In such cases the leftmost subarrays are merged in pairs, and the single rightmost subarray remains intact. The merging algorithm to be used should be flexible enough to allow the merges of subarrays of arbitrary sizes, x and y, into an array of size $x + y$.

During the merging process as the reordering of the keys is determined, the merged keys will have to be stored in a different array. Creation of additional arrays of size N at each pass of mergesort could be very costly. If sufficient memory could be allocated for one additional array, L, of size N, then pass 1 could merge the entries of the original array, K into L; pass 2 would then merge the entries of L into K; and so on. To make sure that eventually the array K contains all the keys in ascending order, an even number of passes should be executed—even if the last pass is unnecessary.

In the Pascal Program shown in Figure 9, the sorting is done at lines 342 through 348 of the MERGESORT procedure. Each entry into the while-loop of MERGESORT invokes MERGEPASS twice: the first time entries of K are merged into AUXARRAY; the second time the entries of AUXARRAY are merged into *K*. The subprocedure MERGEPASS uses the while-loop (lines 316 through 323) to call its own subprocedure, MERGE, to merge pairs of subarrays. Lines 325 and 326 take care of the situation where a rightmost segment of a different size needs to be merged, and lines 328 and 329 take care of the situation where a rightmost segment cannot be paired-up for merging and must be copied.

For the keys of the illustration in Figure 8, four passes will be necessary. Hence, the loop in MERGESORT will execute twice; taking two passes each time through the loop. MERGEPASS will be invoked a total of four times. MERGEPASS will call MERGE five times during the first pass, twice during the second pass, once during the third pass, and once during the fourth pass.

To analyze the computation time required for executing mergesort, we observe that since the sizes of the arrays to be merged double with each pass, approximately *t* passes will be needed where $N = 2^t$. During each of the *t* or $\log_2 N$ passes, at most *N* comparisons and assignments will be made; therefore the computation time for each pass is proportional to *N*. The computation time for the mergesort will be proportional to $N \log_2 N$.

```
Initial  {42}  {12}  {17}  {98}  {56}  {63}  {34}  {72}  {25}  {83}
Pass 1
         {12    42}  {17    98}  {56    63}  {34    72}  {25    83}
Pass 2
         {12    17    42    98}  {34    56    63    72}  {25    83}
Pass 3
         {12    17    34    42    56    63    72    98}  {25    83}
Pass 4
         {12    17    25    34    42    56    63    72    83    98}
```

Figure 8

```
 LINE       LEVEL
NUMBERS   PROC STMT  STATEMENT.      .      .      .      .      .      .      .

  216      2    0
  217      2    0    PROCEDURE MERGESORT     (VAR K            : DATARRAY;
  218      2    0                             N                : INTEGER);
  219      2    0
  220                {This procedure utilizes the MERGE SORT method.  It contains a subprocedure,
  221                MERGEPASS, which in turn contains the subprocedure MERGE.  This method first
  222                considers the entire array as N subarrays each containing one item.  These are
  223                merged to form N/2 arrays each of which contain 2 items.  The process is
  224      2    0    repeated recursively until there is only one array which is sorted.  }
  225      2    0
  226      2    0
  227      2    0          VAR     AUXARRAY         : DATARRAY;      {copy of array to be sorted }
  228      2    0                  MSSIZE           : INTEGER;       {size of subarray           }
  229      2    0
  230      2    0
  231      2    0    {**************************************************************************}
  232      2    0
  233      2    0
  234      2    0          PROCEDURE MERGEPASS
  235      3    0                        (VAR PASSARRAY   : DATARRAY; {input array for sorting }
  236      3    0                         VAR EXTRARRAY   : DATARRAY; {output array for sorting}
  237      3    0                         DATA            : INTEGER;
  238      3    0                         MPSIZE          : INTEGER); {size of subarray        }
  239      3    0
  240                      {This procedure merges ajdacent pairs of subfiles and transfers them
  241      3    0          to the dummy array.}
  242      3    0
  243      3    0
  244      3    0                  VAR     DEXCNT  : INTEGER;        {sub-subarray size           }
  245      3    0                          INDEX   : INTEGER;
  246      3    0
  247      3    0
  248      3    0    {............................................................................}
  249      3    0
  250      3    0
  251      4    0          PROCEDURE MERGE (VAR MERGEARRAY : DATARRAY;
  252      4    0                  MSIZE              : INTEGER;          {front of suarray          }
  253      4    0                  MLEN               : INTEGER;          {divider of subarrays      }
  254      4    0                  MNUM               : INTEGER;          {end of subarrays          }
  255      4    0                  VAR NEWARRAY       : DATARRAY);        {array for sorting         }
  256      4    0
  257                      {This procedure merges two sorted arrays. These arrays are subarrays of the
  258                      original array.  (Xmsize,...Xmlen) and (Xmlen+1,...Xmnum) = (Xmsize,...Xmnum)
  259      4    0          Each subarray is sorted and merged form NEWARRAY [MSIZE..MNUM].}
```

Figure 9

```
  LINE        LEVEL
NUMBERS    PROC STMT  STATEMENT.      .       .       .       .       .       .       .
  260       4    0
  261       4    0          VAR    NEWPOS  : INTEGER;              {new front of array        }
  262       4    0                 OPOS    : INTEGER;              {temporary to find MLEN    }
  263       4    0                 OLDPOS  : INTEGER;              {temporary (MLEN+1)        }
  264       4    0                 INDEX   : INTEGER;
  265       4    0                 CTR :INTEGER;
  266       4    0
  267       4    0    BEGIN {of MERGE}
  268       4    1        NEWPOS := MSIZE;                           {front of new subarray     }
  269       4    1        OPOS := NEWPOS;                            {front of first subarray   }
  270       4    1        OLDPOS := MLEN + 1;                        {front of second subarray  }
  271       4    1
  272       4    1        WHILE (OPOS <= MLEN) AND (OLDPOS <= MNUM) DO       {traverse subarray  }
  273       4    1           BEGIN {of WHILE}
  274       4    2
  275       4    2              IF MERGEARRAY [OPOS] <= MERGEARRAY [OLDPOS] THEN
  276       4    2
  277       4    2                 BEGIN {of IF/THEN}             {item in 1st <= item in 2nd }
  278       4    3                    NEWARRAY [NEWPOS] := MERGEARRAY [OPOS];
  279       4    3                    OPOS := OPOS + 1            {increment first subarray   }
  280       4    3                 END {of IF/THEN}
  281       4    3
  282       4    3              ELSE                              {item in 1st > item in 2nd  }
  283       4    2                 BEGIN {of ELSE}
  284       4    3                    NEWARRAY [NEWPOS] := MERGEARRAY [OLDPOS];
  285       4    3                    OLDPOS := OLDPOS + 1        {increment second subarray  }
  286       4    3                 END; {of ELSE}
  287       4    2
  288       4    2              NEWPOS := NEWPOS + 1              {increment new subarray     }
  289       4    2           END; {of WHILE}
  290       4    1
  291       4    1       IF OPOS > MLEN THEN                      {items in 1st subarray<second}
  292                        FOR INDEX := NEWPOS TO MNUM DO        {remaining 2nd subarray items
  293       4    1                                              in new subfile after 1st    }
  294       4    1              BEGIN {of IF/THEN}
  295       4    2                 NEWARRAY [INDEX] := MERGEARRAY [OLDPOS];
  296       4    2                 OLDPOS := OLDPOS + 1
  297       4    2              END {of IF/THEN}
  298       4    2
  299       4    2       ELSE                                     {items in 2nd subarray<first}
  300                        FOR INDEX := NEWPOS TO MNUM DO        {remaining 1st subfile items
  301       4    1                                               new subarray after first   }
  302       4    1              BEGIN {of ELSE}
  303       4    2                 NEWARRAY [INDEX] := MERGEARRAY [OPOS];
  304       4    2                 OPOS := OPOS + 1
  305       4    2              END; {of ELSE}
  306       4    1
  307       4    1    END; {of MERGE}
  308       4    0
  309       4    0    {.........................................................................}
  310       4    0
  311       4    0
  312       3    0    BEGIN {of MERGEPASS}
  313       3    1
  314       3    1        DEXCNT := 1;                                       {loc of first item in subarray}
  315       3    1
  316                     WHILE DEXCNT <= (DATA - (2 * MPSIZE) + 1) DO       {location  of 1st item in the
  317                                                                         first subarray <= loc of first
  318       3    1                                                            item in second of first half}
  319       3    1           BEGIN {of WHILE}
  320       3    2              MERGE (PASSARRAY, DEXCNT, DEXCNT + MPSIZE - 1, DEXCNT + (2 * MPSIZE) - 1,
  321       3    2                 EXTRARRAY);
  322       3    2              DEXCNT := DEXCNT + (2 * MPSIZE)              {subarray is larger now     }
  323       3    2           END; {of WHILE}
  324       3    1
  325       3    1        IF (DEXCNT + MPSIZE - 1) < DATA THEN               {2nd half has two subarrays }
  326       3    1           MERGE (PASSARRAY, DEXCNT, DEXCNT + MPSIZE - 1, DATA, EXTRARRAY)
  327       3    1
  328       3    1        ELSE                                               {merge second half          }
  329       3    1           FOR INDEX := DEXCNT TO DATA DO
  330       3    1              EXTRARRAY [INDEX] := PASSARRAY [INDEX]
  331       3    1
  332       3    1    END; {of MERGEPASS}
  333       3    0
  334       3    0
  335       3    0    {******************************************************************************}
  336       3    0
  337       3    0
  338       2    0        BEGIN {of MERGESORT}
  339       2    1
  340       2    1           MSSIZE := 1;                             {arrays are size one        }
  341       2    1
  342       2    1           WHILE MSSIZE < N DO                      {merge subarrays            }
```

Figure 9 (cont.)

```
343   2   1            BEGIN {of WHILE}
344   2   2               MERGEPASS (K, AUXARRAY, N, MSSIZE);
345   2   2               MSSIZE := 2 * MSSIZE;          {subarrays are larger      }
346   2   2               MERGEPASS (AUXARRAY, K, N, MSSIZE);
347   2   2               MSSIZE := 2 * MSSIZE           {subarrays are larger      }
348   2   2         END; {of WHILE}
349   2   1
350   2   1
351   2   1      END; {of MERGESORT}
```

Figure 9 (cont.)

Heapsort

This sorting method interprets the keys $K[1], \ldots, K[N]$ to be organized in the form of a binary tree as in Figure 10.

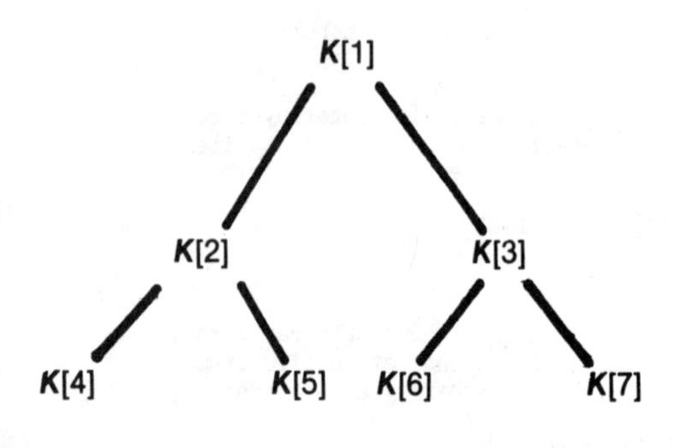

Figure 10

In this organization, $K[2J]$ and $K[2j+1]$ "emanate" from $K[J]$ (alternately $K[R]$ "emanates" from $K[S]$ where S is the integer portion of $R/2$). In standard terminology, $K[2J]$ and $K[2j+1]$ are called **children** of $K[J]$, or $K[J]$ is called the **parent** of $K[2J]$ and $K[2j+1]$. $K[J]$ is also called the **root** of the subtree of all descendants of $K[J]$. This type of arrangement of keys is called a **heap** if all key values are as great as the values of their children (if such children exist). For the keys we have considered previously, that is,

42 12 17 98 56 63 34 72 25 83,

the initial parent/child relationships are indicated in Figure 11.

This does not constitute a heap since the key value $K[2] = 12$ is smaller than the key values of its children, $K[4] = 98$ and $K[5] = 56$.

The first step in heapsort is to convert the original array of keys into a heap. This is done by starting with $K[N/2]$ or $K[5]$ in our case and converting the substructure emanating from $K[5]$ into a heap. This will be accomplished by exchanging $K[5]$ and $K[10]$. Next the substructure emanating from $K[4]$ is considered; since this is already a heap, nothing needs to be done. To convert the substructure emanating from $K[3] = 17$ into a heap, $K[3]$ is exchanged with its child with maximum value (i.e., $K[6] = 63$). Making a heap out of the substructure descending from $K[2]$ is more complicated because after exchanging $K[2] = 12$ with its child of maximum value, $K[4] = 98$, we will have $K[4] = 12$, $K[8] = 72$, and $K[9] = 25$, which violates the heap property that the value of a parent be at least as great as the values of its children. This forces us to move down along the path of exchanges and to restructure subportions (exchanging $K[4] = 12$ with $K[8] = 72$ in this case) into heaps. At this point the array has been rearranged as illustrated in Figure 12.

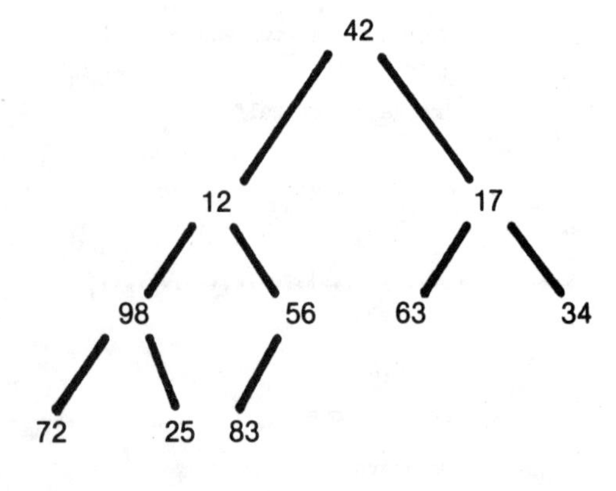

Figure 11

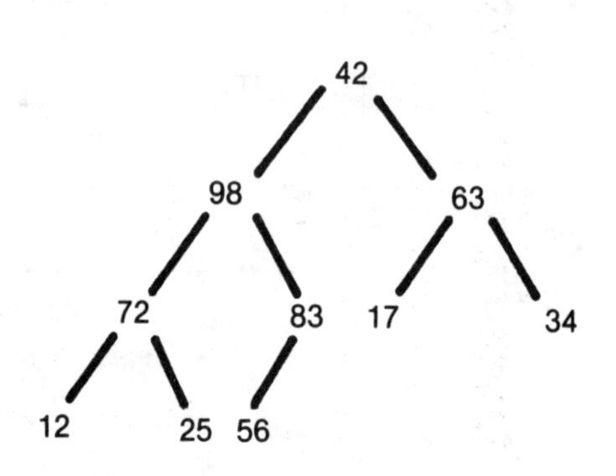

Figure 12

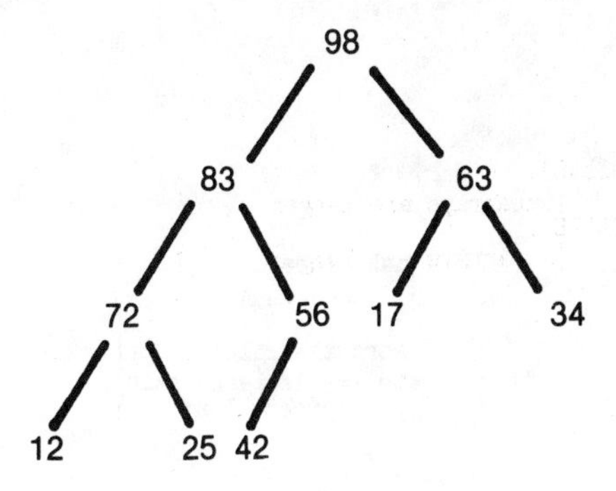

Figure 13

To make the entire array into a heap, $K[1] = 42$ and $K[2] = 98$ are exchanged, then $K[2] = 42$ and $K[5] = 83$ are exchanged, and finally $K[5] = 42$ and $K[10] = 56$ are exchanged. The final result is given in Figure 13:

After the array is converted to a heap, an exchange of $K[1] = 98$ and $K[N] = 42$ will place 98 in its correct position. To continue the sort, the subarray, $K[1], \ldots, K[N-1]$ is converted into a heap (this takes relatively little computer time since all the subtrees are already heaps), and $K[1]$ is exchanged with $K[N-1]$. Continuing this process with the subarrays $K[1], \ldots, K[N-2]$, etc. we will complete the sorting. The MAKEHEAP subprocedure (lines 361 through 403 of the Pascal procedures given in Figure 14), accepts an array whose left and right substructures are heaps and converts the entire array into a heap. In the main procedure, MAKEHEAP is repeatedly invoked at lines 408 and 409 to rearrange the keys of the original array into a heap. The loop (lines 411 through 415) exchanges $K[1]$ and $K[J]$ and rearranges the subarray $K[1], \ldots, K[J-1]$ into a heap for values of J from N to 1.

To analyze the computation time requirements of heapsort

```
  LINE          LEVEL
NUMBERS      PROC STMT   STATEMENT.          .          .          .          .          .          .

    352      2    0
    353      2    0      PROCEDURE HEAPSORT       (VAR K              : DATARRAY; {array to be sorted    }
    354      2    0                                N                  : INTEGER); {# of items in array   }
    355      2    0
    356                  {This is a procedure for sorting using the HEAPSORT algorithm.  HEAPSORT
    357                  uses the procedure MAKEHEAP to convert the original array into a heap and
    358      2    0      to reorganize the array into a heap when necessary. }
    359      2    0
    360      2    0             VAR    COUNT   : INTEGER;                {heap counter                     }
    361      2    0      {..................................................................................}
    362      2    0
    363      3    0             PROCEDURE MAKEHEAP        (POINT              : INTEGER;
    364      3    0                                        NODES              : INTEGER;
    365      3    0                                        VAR ADJARRAY       : DATARRAY);
    366      3    0
    367                         {This procedure organizes a tree into a heap.  It compares parents
    368                         to their children if the child is larger it switches the parent and
    369      3    0             child.}
    370      3    0
    371      3    0             VAR      JPOINT  : INTEGER;               {pointer to locations          }
    372      3    0                      TEMP    : INTEGER;               {temporary value               }
    373      3    0                      DONE    : BOOLEAN;               {continuancy variable          }
    374      3    0
    375      3    0             BEGIN
    376      3    1                 TEMP := ADJARRAY [POINT];             {initialize TEMP = parent      }
    377      3    1                 JPOINT := 2 * POINT;                  {points to left child          }
    378      3    1                 DONE := FALSE;                        {not a heap                    }
    379      3    1
    380      3    1                 WHILE (JPOINT <= NODES) AND (DONE = FALSE) DO
    381      3    1                     BEGIN {of WHILE}
    382      3    2
    383      3    2                         IF (JPOINT < NODES) AND
    384      3    2                             (ADJARRAY [JPOINT] < ADJARRAY [JPOINT + 1]) THEN
    385      3    2                             JPOINT := JPOINT + 1;    {right child is greater       }
    386      3    2
    387      3    2                         IF TEMP > ADJARRAY [JPOINT] THEN      {item > child          }
    388      3    2                             BEGIN {of IF/THEN}                {set parent =to item}
    389      3    3                                 ADJARRAY [TRUNC (JPOINT/2)] := TEMP;
    390      3    3                                 DONE := TRUE                  {tree is now a heap }
    391      3    3                             END; {of IF/THEN}
    392      3    2
    393      3    2                         IF NOT DONE THEN              {switch parent and child       }
    394      3    2                             BEGIN {of IF/THEN}
    395      3    3                                 ADJARRAY [TRUNC (JPOINT/2)] := ADJARRAY [JPOINT];
    396      3    3                                 JPOINT := 2 * JPOINT
    397      3    3                             END {of IF/THEN}
    398      3    3                     END; {of WHILE}
    399      3    1
    400      3    1                 IF NOT DONE THEN                              {set parent = item   }
    401      3    1                     ADJARRAY [TRUNC (JPOINT/2)] := TEMP
    402      3    1             END; {of MAKEHEAP}
    403      3    0      {..................................................................................}
```

Figure 14

```
LINE        LEVEL
NUMBERS  PROC STMT  STATEMENT.   .   .   .   .   .   .

  404   3   0
  405   3   0
  406   2   0        BEGIN {of HEAPSORT}
  407   2   1
  408   2   1           FOR COUNT := TRUNC (N/2) DOWNTO 1 DO           {make tree a heap    }
  409   2   1              MAKEHEAP (COUNT, N, K);
  410   2   1
  411   2   1           FOR COUNT := (N - 1) DOWNTO 1 DO               {sort the file        }
  412   2   1              BEGIN {of FOR}                         {array gets smaller each pass}
  413   2   2                 SWAP (K, COUNT + 1, - COUNT);          {put largest last     }
  414   2   2                 MAKEHEAP (1, COUNT, K)
  415   2   2              END; {of FOR}
  416   2   1
  417   2   1        END; {of HEAPSORT}
```

Figure 14 (cont.)

we observe that a binary tree starts with 1 key as its root which has 2 children; the 2 children have 4 children; and in general there are 2^i keys at level i of the tree (the root being considered level zero). The computation time required at line 409 is essentially the same for all keys at the same level, i. Furthermore, for any key at level i, the computation time required by line 409 is proportional to the number of levels available below that key. Since the total number of levels is approximately $\log_2 N$, the total computation time for the loop which constructs the initial heap (lines 408 and 409) is proportional to

$$\sum_{i=0}^{\log_2 N-1} 2^i (\log_2 N - i) = \sum_{j=1}^{\log_2 N} j\, 2^{(\log_2 N - j)}$$

$$= \sum_{j=1}^{\log_2 N} j\, 2^{-j}\, 2^{(\log_2 N)} = N \sum_{j=1}^{\log_2 N} j\, 2^{-j} \leq 2N.$$

In the sorting loop (lines 411 through 415), the SWAP and MAKEHEAP procedure calls at lines 413 and 414 require constant time, k, and at most time proportional to $log_2 N$ respectively. Hence the computing time used by the sorting loop will be proportional to $Nlog_2 N$. With the domination of the time requirement of the sorting loop, the total computation time will be proportional to $Nlog_2 N$.

Comparison of Sorting Methods

The analysis of sorting algorithms given in the preceding articles show that **the computation times of the insertion and bubble sorts are proportional to N^2** and the **computation times for the quicksort (the average case), mergesort, and heapsort are proportional to $N\log_2 N$**. Whatever the proportionality constants of the individual algorithms may be, for large values of N, sorting methods with $N\log_2 N$ computation times will significantly outperform those with N^2 times. The disparity between N^2 and $Nlog_2 N$ computation times (with proportionality constants equal to 1) is illustrated in Figure 15. For small values of N, a sort requiring $k_1 N^2$ computation time may be preferred to one requiring $k_2 Nlog_2 N$ time if k_1 is significantly smaller than k_2.

To obtain experimental results, the Pascal procedures given in the preceding articles were executed on a VAX-11/780 under the VMS operating system. The programs were run during off-peak hours so that variations of computing times due to the operating system would be within 5% of the computation times that were recorded. Each sorting method was tested on each of three data sets: DATA I consisting of 1000 random integers; DATA II consisting of the same integers arranged in ascending order, and DATA III consist-

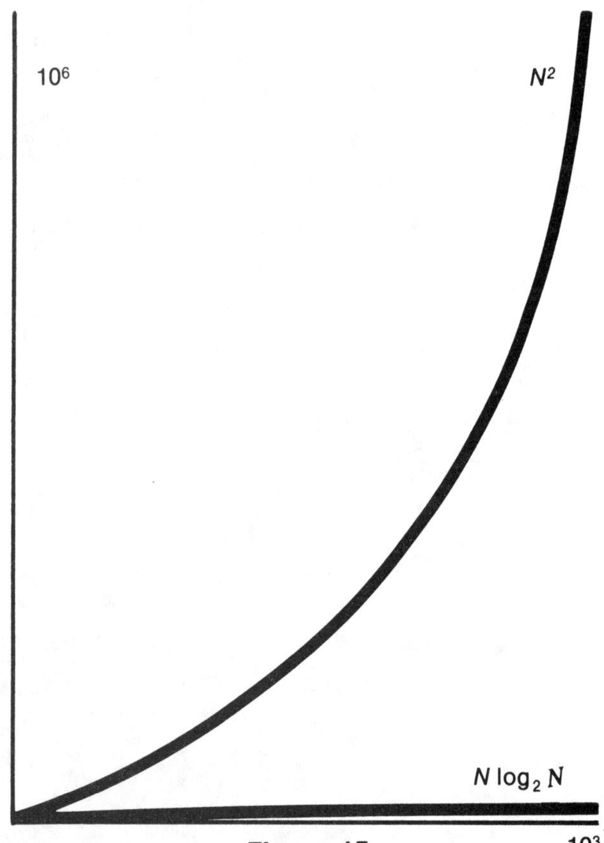

Figure 15

ing of the same integers in descending order. The results are summarized in Figure 16.

		TIMES IN MILLISECONDS		
		DATA I	DATA II	DATA III
SORT METHODS	INSERTION SORT	2,500	80	5,105
	BUBBLE SORT	14,745	4,165	25,145
	SHELLSORT	1,000	120	360
	QUICKSORT	255	3,275	3,325
	MERGESORT	300	270	260
	HEAPSORT	460	470	430

Figure 16

Mergesort and heapsort show little variation in computation time for sorting DATA I, DATA II, and DATA III. The computation time for quicksort, however, goes from 255 ms for DATA I to 3275 ms and 3325 ms for DATA II and DATA III, respectively. This illustrates the negative feature of quicksort when it is applied to data which are nearly in ascending or descending order. For data which are nearly in ascending order, insertion sort does very well. Mergesort, which does well on all types of data sets, has the disadvantage of requiring extra memory for a copy of the data during a pass. Heapsort, which performs nearly as well as mergesort, does not require a replica of the data for its execution.

REFERENCES

1. Aho, A. V., Hopcroft, J. E. and Ullman, J. D., *Data Structures and Algorithms*. Addison-Wesley 1983.
2. Hoare, C. A. R., "Quicksort," *Computer J.*, *5* (1962): 10–15.
3. Horowitz, E. and Sahni, S., *Fundamentals of Data Structures*. Computer Science Press, 1976.
4. Knuth, D. E., *The Art of Computer Programming, 3, Sorting and Searching*. Addison-Wesley, 1973.
5. Reingold, E. M. and Hansen, W. J., *Data Structures*. Little-Brown, 1983.
6. Sedgewick, R., "The Analysis of Quicksort Programs," *Acta Informatica, 7* (1977): 327–355.
7. Sedgewick, R., "Implementing Quicksort Programs," *Comm. ACM*, *21* (1978):847–857.
8. Shell, D. L., "A High Speed Sorting Procedure," *Comm. ACM*, 2 (1959), 30–32.
9. Tanenbaum, A. M. and Augenstein, M. J., *Data Structures Using Pascal*. Prentice-Hall, 1981.
10. Williams, J. W. J., "Algorithm 232: Heapsort," *Comm. ACM*, 2 (1959) 347—348.

Part VI: Applications of Simulations

Simulation has become the most widely used management science/operations research technique. Managers, researchers, and professionals in nearly all areas today recognize their need to understand and be able to apply this modern technique for research and decision-making. **In this part we give an overview of applications** by *citing a number of areas* in which we have personally been involved in simulations, *including papers* of simulations that, overall, are done well, *and including a large list* of simulations that have been recently reported in a large number of fields. **This should allow many practitioners to tie into the relevant literature**, which usually contains valuable insights on the systems being simulated. **If then approached with the efficient design techniques covered above, state-of-the-art simulation should result.**

Simulation Projects List

Titles of Projects

- The SIMSCRIPT II.5 Language and a Complex Example
- Ranking and Selection Procedures with Correlated Observations and Simulation Applications
- Variance-Reduction Techniques in Process Simulation
- A New Simulation Game for Hockey
- A DYNAMO Simulation and Its Relation to Forrester's Work
- Blood Bank Distribution and Simulation
- Generation of Correlated Random Sequences
- Configuration of Rochester's Off-Track Betting via GPSS Simulation
- Subset Selection Procedures: Tables and Applications
- A DYNAMO-PL/I Simulation of Engineering Employment and Enrollment Effects
- Random Number Generation: A Survey and Bibliographic Study
- An m-Dimensional Monte Carlo Integrator
- Parking Problems and Simulation
- Simulation of the ODLC Inventory Reordering System
- GPSS Simulation of a Production Line
- Cybernetic System Simulation: The University
- Random Number Generation: Theory, Tests, & Applications
- Capital Budgeting via Simulation
- Error Distributions in Munitions Ground Tests
- Simulation in Political Science: A Review and An Example (Welfare Services)
- Generalized Lambda Distribution Fitting Feller's Alpha-Distributions
- Pi, Data-Fudging, and Simulation
- Reliability of Estimates of Heritabilities and Genetic Correlations via Monte Carlo
- Simulation of a Computer Multi-Programming System
- Comparative Reliability of Estimates of Size of Treatment Effects in ANOVA via Monte Carlo
- International Relations Simulation
- The University Hospital Maternity Ward: A Simulation to Optimize Its Size
- GPSS and SIMSCRIPT: Examples and Comparisons
- Nonparametric vs. 2-Stage Procedures
- On Excess Over the Boundary: Simulation Results
- Personnel Management and Simulation
- Stochastic Epidemic Simulations
- DYNAMO and a Communication Simulation
- Effects of Pessimism/Realism/Optimism on Sales and Profit
- Simulation: Comparisons of GPSS/PLI/FORTRAN Language Effects
- Simulation of an Elevator System
- Response Time in a Multi On-Line System
- Simulation in Population Genetics
- Confidence Intervals with Correlated Data: Comparing Several Approaches
- Genetic Simulation with Inbreeding
- Simulation of Effects of Freight Consolidation on Costs and Delivery Times
- Simulation of Blackjack Shuffling and Strategy
- Simulation of Population Size Estimation
- GPSS vs. GASP vs. SIMSCRIPT Comparisons
- Simulation for Population Projection
- The GFSR Random Number Generator
- Simulation of Solar In-Ground Homes
- Simulation in Psycho-Acoustical Research
- Monte Carlo Evaluation of a New Test of Normality
- Baseball Season Simulation
- Airport Traffic Simulation: Port Columbus
- Supermarket Counter Strategy Simulation
- Library Elevator System Simulation
- Entropy Test of Random Number Generators
- Simulation of Information System Choice
- Inventory Model Error Effects
- NBU Prediction Lower-Bound Intervals
- Performance of a Correlated-Observations Ranking & Selection Procedure
- Football Simulation
- Monte Carlo Comparison of Tests of Bivariate Symmetry
- Swine Mating—Random vs. Selective
- Job-Shop Simulation: FORTRAN vs. SIMSCRIPT

EH0227-9/85/0000/0296$01.00 © 1985 IEEE

- Repairman Problem Simulation
- Ranking & Selection in Factorial Experiments
- Statistics for Tumor Studies
- Estimation Procedures for Mean Height: A Simulation Study
- GPSS Simulation of a Firm
- Two Data Structures: A Comparison
- Batch Arrivals in Queueing: A Simulation Study
- Bin Sort vs. Others via Simulation
- Optimum Markup in Competitive Bidding
- The Runs Test for Randomness
- PERT vs. CPM vs. Simulation for Construction Project Time
- CAS Transport Subsystem in GPSS
- Simulation Evaluation of CONJOINT
- Motel Cleaning Optimization in SLAM
- Optimal Grocery Store Peak Configuration
- Simulation Languages
- Pollutant Concentrations in Air
- Terrain Correlation Navigation Simulation
- Optimal Baseball Batting Order via Simulation
- What Are the Limits to Growth?
- Error Rates in Binary Communication Systems
- Human Factors Simulation: Nuclear Safety
- Testing Random Number Generators Using Spectral Analysis
- Serial Correlation Effects on Estimation of Var $(\overline{X}_T)$

AMERICAN JOURNAL OF MATHEMATICAL AND MANAGEMENT SCIENCES

OPTIMIZATION OF PRIORITY CLASS QUEUES, WITH A COMPUTER CENTER CASE STUDY

F. Keyzer, J. Kleijnen, E. Mullenders, and A. van Reeken
Tilburg University
Tilburg, Netherlands

SYNOPTIC ABSTRACT.

This paper discusses a case-study concerning key-punching in a computer center. This system is modeled as a queue with two servers (operators) and three priority classes (small, medium, and large jobs). Simulation is used to estimate the 90th quantiles of the waiting times per job class, for different borderlines c_1 and c_2 between the three job classes. Tradeoffs among the waiting times of each job class are quantified, and provide an overall citerion function. Several regression models are investigated, expressing the quantiles as functions in c_1 and c_2. Optimal c_1 and c_2 are determined by a numerical search. The resulting limits have been implemented. The methodology is more generally applicable.

1. INTRODUCTION

This paper illustrates methodology for fitting regression equations to simulation output data. The form of the regression models is based on qualitative knowledge of the queueing system. The resulting regression model is optimized. The illustration is based on a real-world problem. The paper is organized as follows: Introduction including a description of the case-study (Section 1); previous research including confidence intervals for estimated quantiles, and preliminary regression models (Section 2); reformulated problem, i.e., quantifying tradeoffs among job classes (Section 3); theoretical basis, the specification of the regression models based on queueing theory (Section 4); empirical models, a series of tentative regression equations (Section 5); optimization of the criterion, using a numerical search algorithm (Section 6); conclusions (Section 7).

Key Words and Phrases: case-study; methodology; computer center; priorities; scheduling; quantile; criterion selection; optimization; search algorithm.

1981, VOL. 1, NO. 4, 341-358
0196-6324/81/040341-18 $5.20

The methodology illustrated is relatively easy to master (e.g. two of the authors were gratude students in the Department of Econometrics). The results were immediately implemented since the manager of the computer center (Van Reeken) actively participated in the study.

The case-study of this paper arises from the following practical problem. The computer center of Tilburg University provides a key-punching service. There are two key-punching operators (servers) available. It is known from queueing theory that expected waiting time is minimized if jobs are served in the order of their service times, i.e. shortest-jobs-first (see Cobham (1954), Hofri (1980), and Jaiswal (1968)). However, the computer center's management decided that a complete ranking of jobs would be impractical. Instead three priority classes were established: small jobs (S-jobs) have priority over medium (M-)jobs, and M-jobs are key-punched before large (L-)jobs. Within each of the the three priority classes a first-come-first-served rule applies. The priorities are not preemptive. Classification of jobs is possible since key-punching times can be predicted accurately from the code sheets which are to be key-punched. (A consulting company, Berenschot Inc., provided a table with two dimensions, namely the number of symbols per line, and the percentage of numerical symbols. Multiplying the relevant entry in this table with the number of lines per sheet yields the predicted key-punching time.) Originally, management used intuition to choose both the borderline c_1 between S- and M-jobs, and c_2 between M- and L-jobs. The purpose of this study was to derive "optimum" c_1 and c_2 values. Similar priority rules can be used in many other practical queueing situations such as computer operating systems and automobile repair shops. (Also see Tsay and Ng (1980), who studied five priority rules, but not the one of the present paper.)

From the above problem statement, it follows that the system was modeled as a queueing system with two servers and three priority classes, priorities being based on the lengths of the service times. This model has not yet been solved analytically, even though in the present case the interarrival and service times are exponentially distributed. More precisely, the interarrival times form a Poisson process with parameter $\lambda = 0.033$ arrivals per minute. The service times come from truncating an exponential distribution with parameter $\mu = 0.021$ (mean = 1/.021 = 47.62 minutes). Truncation occurs at the lower end "a" because jobs requiring less than a = 10 minutes are key-punched by the users themselves. Truncation at the upper end "b" occurs because jobs requiring more than b = 900 minutes are served outside the computer center. (To simulate this, times are generated by sampling from the exponential distribution with parameter μ; if the generated value is within the limits $[a,b]$, then the value is accepted; otherwise sampling is repeated.) The probability density function of the truncated exponential variable X is

$$f_X(x) = \begin{cases} \mu e^{-\mu x}/c \;, & a \le x \le b \\ 0 \;, & \text{otherwise,} \end{cases} \tag{1}$$

where

$$c = \int_a^b \mu e^{-\mu x} dx = e^{-\mu a} - e^{-\mu b} = 0.811. \tag{2}$$

During the investigation it soon became evident that management is not interested so much in the average queueing time, as in helping "as many people as fast as possible". Management agreed that therefore the study should use the 90th quantile; if T denotes queueing (waiting) time, then the 90th quantile q satisfies

$$P(T \le q) = 0.90 \tag{3}$$

(There is only a 10% chance that customers have to wait longer than q.) Note that T is defined as waiting time excluding key-punching itself. Simulation is used to estimate the quantile q for various c_1 and c_2 values. (The present study takes the quantiles as criteria, whereas Hofri (1980) and Tsay and Ng (1980) studied the mean and coefficient of variation.)

2. PREVIOUS RESEARCH

In Coppus, van Dongen, and Kleijnen (1977), q was estimated together with a 90% confidence interval on q. To compute the confidence interval, the simulation was analyzed exploiting the "regenerative" property, that when the system becomes empty (both servers idle) a new history starts independently of the past simulated history (see Crane and Lemoine (1977) for a general exposition on renewal analysis in simulation). Each (c_1,c_2) combination was simulated until 500 regeneration cycles were obtained. In total, 19 (c_1,c_2) combinations were simulated, each combination yielding one observation (Q) for the 90th quantiles of the S-, M-, and L-jobs: Q^S, Q^M, Q^L respectively. For each of these three quantiles a quadratic function was fitted to its 19 observations; e.g. for S-jobs the 19 predictions were

$$\hat{Q}_i^S = \beta_0 + \beta_1 c_{1i} + \beta_2 c_{2i} + \beta_{11} c_{1i}^2 + \beta_{12} c_{1i} c_{2i} + \beta_{22} c_{2i}^2 \tag{4}$$

($i = 1, \ldots, 19$). The statistical tests applied to the resulting regression equations showed that c_1 and c_2 did affect the quantiles, but that the quadratic model did not correctly specify these effects. (When testing the individual regression parameters β, the resulting t-statistics were not significant at the 5% level. The F-statistic for testing all regression parameters simultaneously was significant, i.e., c_1 and c_2 did have effects. The F-test for lack-of-fit was significant, i.e. the quadratic model was not adequate. See Coppus, Van Dongen, and Kleijnen (1977), and Kleijnen (1981) for details.)

Van den Bogaard and Kleijnen (1977) increased the number of (c_1,c_2) combinations from 19 to 71. The following alternative regression models were investigated:

(i) Replace c_2 by (c_2-c_1) in (4); unfortunately, the individual regression effects remained not significant.

(ii) Replace c_1 and c_2 by p_1 and p_2, the probability of a job falling in classes 1 (S-jobs) and 2 (M-jobs) respectively; again the regression effects were not significant.

(iii) Next, some models were tried with the ratio c_1/c_2 as the explanatory variable; the results remained unsatisfactory.

(iv) Finally, some linear models in c_1 and c_2 were investigated, such as

$$\hat{Q}^S = 49.44 + 0.63c_1 - 0.03c_2 , \tag{5}$$

which had t-values of 10.79. 9.73,- 1.48 for the individual regression effects, and $R^2 = 0.86$; this model suggested that Q^S was sensitive to c_1 but not to c_2. Indeed the model with a single explanatory variable yielded

$$\hat{Q}^S = 43.66 + 0.63c_1 \tag{6}$$

with t-values of 17.70, 9.36 and $R^2 = 0.84$.
(This result will be useful later in the present paper.)

Van den Bogaard and Kleijnen (1977) continued with an approach different from that of the present paper, namely response surface methodology (RSM), in which locally a linear model such as (5) is fitted. The signs of the regression parameters tell whether c_1 and c_2 should be increased or decreased in order to minimize Q. The relative changes in c_1 and c_2 depend on the ratio of the regression parameters in accordance with the so-called steepest descent method. At the "bottom of the valley", the linear model becomes a bad guide for determining the c_1 and c_2 effects. At that stage of experimentation, a quadratic model for local search is introduced. Such a model also reveals the nature of the optimum region, e.g. whether there is a unique minimum, a saddle point, or a ridge. (For details of RSM, see Meyers (1971) and Kleijnen (1975).) In the present application, a problem was encountered with the path of steepest descent for large jobs in that it differed greatly from the paths for medium and small jobs, so that it was impossible to determine c_1 and c_2 such that all three job classes were optimized simultaneously.

3. REFORMULATED PROBLEM

Thus (see Section 2), previous research demonstrated that three job classes have different optimal c_1 and c_2 values. If only S-jobs are considered, then waiting times of these jobs decrease as c_1 decreases (irrespective of c_2), since as c_1 becomes lower fewer competing jobs remain in the S-class; also see the sign of the regression parameter in (6). Overall optimization of waiting time in a system with separate job classes requires tradeoffs among the waiting times per class. Therefore it was decided to reformulate the problem as follows. Let W^S denote the fraction of S-jobs. Since longer waiting times are acceptable if the service takes longer, the waiting time quantile q^S is utilized as a proportion of the mean service time per class (e.g. $\bar{B}^S$ for class S) and our goal is now to minimize

$$W^S\ (q^S/\bar{B}^S) + W^M\ (q^W/\bar{B}^M) + W^L\ (q^L/\bar{B}^L). \qquad (7)$$

(While other criteria are possible, since this one has some defects, e.g., if W^L is small one can have $q^L/\bar{B}^L$ very large and still tolerated and hence management desire a criterion like $q^L/\bar{B}^L \leq 2.0$, this criterion was imposed by the manager and has to be accepted by the analysts. Multi-attribute decision theory is another possible approach.)

Simulation is needed to estimate the quantiles q^S, q^M, q^L, but not the weights W nor the conditional mean service times $\bar{B}$. This is so because the fraction of S-jobs is (see (1))

$$W^S = P(X < c_1) = \frac{1}{c}\int_a^{c_1} \mu e^{-\mu x}\, dx = (e^{-\mu a} - e^{-\mu c_1})/0.811 \qquad (8)$$

and

$$\bar{B}^S = E(X|X < c_1) = \{(a + \frac{1}{\mu})e^{-\mu a} - (c_1 + \frac{1}{\mu})e^{-\mu c_1}\}.^{-1} \qquad (9)$$

Similarly for M- and L-jobs:

$$W^M = P(c_1 < X < c_2) = (e^{-\mu c_1} - e^{-\mu c_2})/0.811, \qquad (10)$$

$$W^L = P(X > c_2) = (e^{-\mu c_2} - e^{-\mu b})/0.811, \qquad (11)$$

$$\bar{B}^M = \{(c_1 + \frac{1}{\mu})e^{-\mu c_1} - (c_2 + \frac{1}{\mu})\}[e^{-\mu c_1} - e^{-\mu c_2}]^{-1} \qquad (12)$$

$$\bar{B}^L = \int_{c_2}^{b} x\mu e^{-\mu x}\, dx / (\int_{c_2}^{b} \mu e^{-\mu x}\, dx)$$

$$= \{(c_2 + \frac{1}{\mu})e^{-\mu c_2} - (b + \frac{1}{\mu})e^{-\mu b})\}\ [e^{-\mu c_2} - e^{-\mu b}]^{-1}. \qquad (13)$$

4. THEORETICAL BASIS

The reports by Coppus, van Dongen and Kleijnen (1977) and Van den Bogaard and Kleijnen (1977) - and experience in general - strongly suggest that regression analysis without a theoretical basis leads (at best) to very questionable results. Therefore, a qualitative theoretical analysis was made before processing the simulation results via regression.

The regression model explains how the quantiles q of the queue react to changes in c_1 and c_2. The model should explain the simulation model parsimoniously, i.e., with a minimal number of regression parameters. The selection of explanatory variables and functional form for the model should be based on theory and intuition (see Kleijnen (1979)). Without such a regression model, the simulation may result in a confusing number of tables, as in Tsay and Ng (1980).

Assumption 1: q^S depends on c_1 but not on c_2. The waiting times of S-jobs are affected by the borderline c_1 between S- and M-jobs, but not by the limit c_2 between M- and L-jobs because S-jobs have priority over all the other jobs, and are not influenced by the priority rules among these remaining jobs. [This assumption neglects the non-preemtive character of the priority rules, but is corroborated by various regression results such as (5) and (6).]

Assumption 2: q^L depends on c_2 but not on c_1. L-jobs have to wait until all S- and M-jobs have been served; the waiting times of L-jobs are not influenced by the subdivision into S- and M-jobs. [This assumption is again corroborated by various regression models (not shown here; see Van den Bogaard and Kleijnen (1977)).]

Assumption 3: q^M takes over the role of q^S when c_1 approaches zero. For M-jobs, waiting times depend on both c_1 and c_2. However, as c_1 approaches zero (or more precisely c_1 approaches the lower limit for service times, a = 10 minutes), no S-jobs remain and M-jobs acquire the highest priority. So, if $c_1 \downarrow a$ then M-jobs depend on their upper-limit c_2 just as S-jobs depended on their upper-limit c_1. In general,

$$q^S \equiv g^S(c_1) \quad (0 \leq c_1) \tag{14}$$

$$q^M \equiv g^M(c_1,c_2) \quad (0 \leq c_1 \leq c_2). \tag{15}$$

A special case is $c_1 = a$, and then (15) becomes

$$q^M = g^M(a,c_2) \equiv h^M(c_2) \quad (0 \leq c_2). \tag{16}$$

Equations (14) and (16) both specify how the waiting times of the top-priority jobs depend on their upper-limit. Hence assume

$$g^S(c_1) = h^M(c_2) \text{ for } c_1 = c_2.$$

Assumption 4: q^M takes over the role of q^L when c_2 approaches infinity. The reasoning leading to Assumption 3 can also be made for M- and L-jobs. If c_2 becomes large, say $c_2 = b$, then (15) becomes

$$q^M = g^M(c_1, b) \equiv k^M(c_1) \quad (0 \le c_1). \tag{18}$$

The equation

$$q^L = g^L(c_2) \quad (0 \le c_2) \tag{19}$$

and (18) both specify how the waiting times of the lowest priority jobs depend on their lower limit. Hence assume

$$g^L(c_2) = k^M(c_1) \text{ for } c_1 = c_2. \tag{20}$$

Assumption 5: The functional relationships between the quantiles and the class limits may be approximated by low-degree polynomials. It is well-known that a function may be represented as a Taylor series. Here it is assumed that this series may be cut off after the first or second derivatives. Hence the true functions are approximated by first- or second-degree polynomial regression models. For instance, (15) is approximated by

$$\hat{q}^M = \beta_0 + \beta_1 c_1 + \beta_2 c_2 + \beta_{11} c_1^2 + \beta_{12} c_1 c_2 + \beta_{22} c_2^2. \tag{21}$$

Likewise assumptions 1, 2 and 5 yield

$$\hat{q}^S = \alpha_0 + \alpha_1 c_1 + \alpha_2 c_1^2 \tag{22}$$

and

$$\hat{q}^L = \gamma_0 + \gamma_1 c_2 + \gamma_2 c_2^2. \tag{23}$$

The parameters are estimated via regression analysis. The resulting estimates are tested for significance using traditional (two-sided) t-statistics. The regression models are based on Ordinary Least Squares (OLS) assuming a common unknown variance σ^2, and the regression parameters' significance is measured through t-statistics using an estimate of σ^2 based on the Mean Square for Error; see regression textbooks such as Draper and Smith (1966). OLS remains unbiased even if the observations (quantiles estimated from simulation runs, Q) have non-constant variance σ^2, or if the observations are non-normal. The t-statistic is well-known for its robustness. (Nevertheless, it would have been better to estimate var(Q) and use Generalized Least Squares (GLS); see Kleijnen (1981). It would also have been interesting to investigate approximations based not on polynomials but on exponentials since exponential functions show a behavior comparable to the second-degree polynomials used in this report, exponential behavior is often found in queueing theory, and exponentials might lead to an explicit solution for the optimal c_1 and c_2 upon differentiation of an overall criterion (7) which comprises a number of exponentials (also see Section 6).)

Note that if first-order models like (6) were used instead of second-order models like (22), then one-sided instead of two-sided t-tests would be appropriate, since the sign of the regression coefficient of the first-order model would be tested. For instance, the null-hypothesis for q^S would be $\alpha_1 > 0$ (or $\partial q^S/\partial c_1 > 0$), since we know that if c_1 increases then more S-jobs result so their waiting times increase.

5. EMPIRICAL REGRESSION MODELS

This section explains successive specifications of empirical regression models. An estimated regression equation is accepted if it meets the following criteria (also see Kleijnen (1980 , pp. 178-181)): (i) The regression provides a good explanation of the changes in q as c_1 and/or c_2 vary ("high" R^2); and (ii) The estimated individual regression parameters are significantly different from zero at the 5% significance level. Readers interested only in the final form of the regression model can proceed to (31). (They may skip the "stepping stones" explained in the next paragraphs, with the exception of the three additional Assumptions 6, 7, and 8.)

One empirical result is that $\hat{\beta}_{12}$ in (21) is not significant. In the absence of interaction, response curves are parallel. Indeed the scatter diagrams show that, for a fixed c_1-value, parallel curves in the q^M, c_2 plane result. (For some c_1, c_2, q combinations, however, the patterns are very irregular.)

Originally, very low R^2 values were found for L-jobs, e.g. $R^2 = 0.63$ for (23). On inspection of the scatter diagram, this low R^2 was explained by the presence of some wild observations: outliers. These outliers occurred because waiting times of L-jobs have high variances. (The relative lengths of the confidence intervals around Q varied between 3% and 6% for S-jobs, between 5% and 25% for M-jobs, and between 9% and 84% for L-jobs, in the 19 c_1, c_2 combinations studied by Coppus, van Dongen and Kleijnen (1977). The higher c_2 becomes, the fewer L-jobs remain, and the higher their corresponding variance becomes.) Fortunately, in simulation experiments, it is easy to check whether an outlier is indeed caused by pure chance: simply repeat the c_1, c_2 combination with a new stream of random numbers; also see Kleijnen (1975, p. 358). In this way a number of suspicious observations q^L were checked. All new observations fell within the "cloud" of observations. Obvious outliers were eliminated. Consequently the total number of observations increase from 71 to 91. (More precisely, the first five suspected observations were simulated again, each replicated twice. These ten new observations showed that the suspected observations were outliers indeed. Next, fifteen more suspected observations were simulated again. However, these observations needed not be rejected.)

The requirements of high R^2 and significant parameters together with Assumptions 1, 2, and 5 of the preceding section, yield purely quadratic models, i.e. in (21) through (23) the first-degree effects are not significant, so that they are eliminated. This yields:

$$\hat{q}^S = 56.07 + 0.0087\, c_1^2 \tag{24}$$

with t-values of 89.1 , 17.43 and $R^2 = 0.82$.Further

$$\hat{q}^M = 30.5 + 0.0323\, c_1^2 + 0.0092\, c_2^2 \tag{25}$$

with t-values 3.9 , 6.75 , 31.19 and $R^2 = 0.95$.Finally

$$\hat{q}^L = 548.9 + 0.1564\, c_2^2 \tag{26}$$

with t-values 2.3 , 16.0 and $R^2 = 0.74$. This type of model also agrees with Assumption 3: $\hat{\beta}_{22} = 0.0092$ and $\hat{\alpha}_2 = 0.0087$ do not differ significantly so that the null-hypothesis $\beta_{22} = \alpha_2$ is not rejected. Alternative models fail at this point. Note that the best linear unbiased estimate (BLUE) of, say, α_1 in (22) is $\hat{\alpha}_1$. So, even though $\hat{\alpha}_1$ is not significant, this value might be retained in (22). This is a moot issue in statistics; see Box (1954, p. 57). In the present case, however, Assumption 3 forces the issue, i.e. α_1 is eliminated.

Assumption 3 further implies the hypothesis $\alpha_0 = \beta_0$ but the intercepts $\hat{\alpha}_0$ and $\hat{\beta}_0$ of (24) and (25) differ significantly. However, q^S is estimated for small c_1 values, whereas q^M is estimated for large c_2 (which take over the c_1 role; see (17)). In other words, $\hat{\beta}_0$ is based on an extrapolation beyond the range of observed values; see Figure 1. Consequently $\hat{\alpha}_0$ is more accurate than $\hat{\beta}_0$ when estimating the intercept. (Indeed the t-statistic for $\hat{\alpha}_0$ is 89.1 and only 3.9 for $\hat{\beta}_0$.) Also note that the coefficients $\hat{\beta}_{11}$ and $\hat{\gamma}_2$ of (25) and (26) differ significantly, and so do $\hat{\beta}_0$ and $\hat{\gamma}_0$, violating Assumption 4.

The dual role of (25) - model for q^M and for q^S when substituting $c_1 = a$ - leads to the following idea: When fitting the q^M model, take into account the q^S observations, or more generally, fit the q^S, q^M, and q^L models simultaneously. This can be formalized as

Assumption 6: Per job-class, the quantile q depends on its left and right class limits ℓ and r. The variable c_1 plays the role of right-hand class limit for q^S but c_1 is also the left-hand limit for q^M. Similarly, c_2 is the right class limit for q^M and also the left limit for q^L. The minimum service time "a" (see (1) and (2)) is the left limit for q^S. The maximum service time "b" forms the right limit for q^L.

Assumption 6 together with the results of this section (especially (24) through (26)) yields

$$q = \delta_0 + \delta_1 \ell^2 + \delta_2 r^2 \qquad (27)$$

or in more elaborated format the assumption results in Table 1.

TABLE 1: Single, Simultaneous Model.

Left	Right	$\hat{q} = \hat{g}(\ell, r)$
a	c_1	$\hat{q}^S = \delta_0 + \delta_1 a^2 + \delta_2 c_1^2$
c_1	c_2	$\hat{q}^M = \delta_0 + \delta_1 c_1^2 + \delta_2 c_2^2$
c_2	b	$\hat{q}^L = \delta_0 + \delta_1 c_2^2 + \delta_2 b^2$

Comparison of Table 1 with (22) and (23) shows the identities $\alpha_0 \equiv \delta_0 + \delta_1 a^2$, $\alpha_2 \equiv \delta_2$, $\gamma_0 \equiv \delta_0 + b^2\delta_2$, etc. The coefficients δ are estimated using all observations on q^S, q^M and q^L simultaneously:

$$q = \delta_0 \underline{I} + \delta_1 \underline{z}_1 + \delta_2 \underline{z}_2 \qquad (28)$$

with the vectors

$$\begin{aligned} \underline{q}' &= (q_1^S, \ldots, q_{71}^S, q_1^M, \ldots, q_{71}^M, q_1^L, \ldots, q_{71}^L) \\ \underline{z}_1' &= (a^2, \ldots, a^2, c_{1,1}^2, \ldots, c_{1,71}^2, c_{2,1}^2, \ldots, c_{2,71}^2) \qquad (29) \\ \underline{z}_2' &= (c_{1,1}^2, \ldots, c_{1,71}^2, c_{2,1}^2, \ldots, c_{2,71}^2, b^2, \ldots, b^2) \end{aligned}$$

$\underline{I}$ being a vector with 3 times 71 elements equal to one. For a = 10 and b = 900 (28) results in:

$$\begin{bmatrix} q^S \\ q^M \\ q^L \end{bmatrix} = 67 \begin{bmatrix} 1 \\ 1 \\ 1 \end{bmatrix} + 0.15 \begin{bmatrix} 10^2 \\ c_1^2 \\ c_2^2 \end{bmatrix} + 0.00076 \begin{bmatrix} c_1^2 \\ c_2^2 \\ 900^2 \end{bmatrix} \qquad (30)$$

with t-values 1.3, 23.75, 3.73 and $R^2 = 0.90$.

To the comment below (23) concerning the use of OLS should now be added that (28) raises one more statistical complication; namely, the observations are multivariate, i.e., q_1^S, q_1^M, q_1^L are correlated, and so are q_2^S, q_2^M, q_2^L, etc. Hence Generalized Least Squares would have been better. Nevertheless, OLS still yields unbiased estimators of δ_0, δ_1, δ_2. Also R^2 is not affected by the covariance structure of the observations. Although the R^2 of (30) is high, inspection of the residuals $q-\hat{q}$ shows that the model systematically over-estimates near the origin. (Also compare $\hat{\alpha}_0 = 56.07$ in (24) with $\hat{\delta}_0 + \hat{\delta}_1 a^2 = 67 + (0.15)(100) = 82$.) In the estimation of the intercept, the influence of q^M and q^L observations far away from the origin presumably creates this over-estimation. Therefore, we proceed as follows. First, introduce Assumption 7.

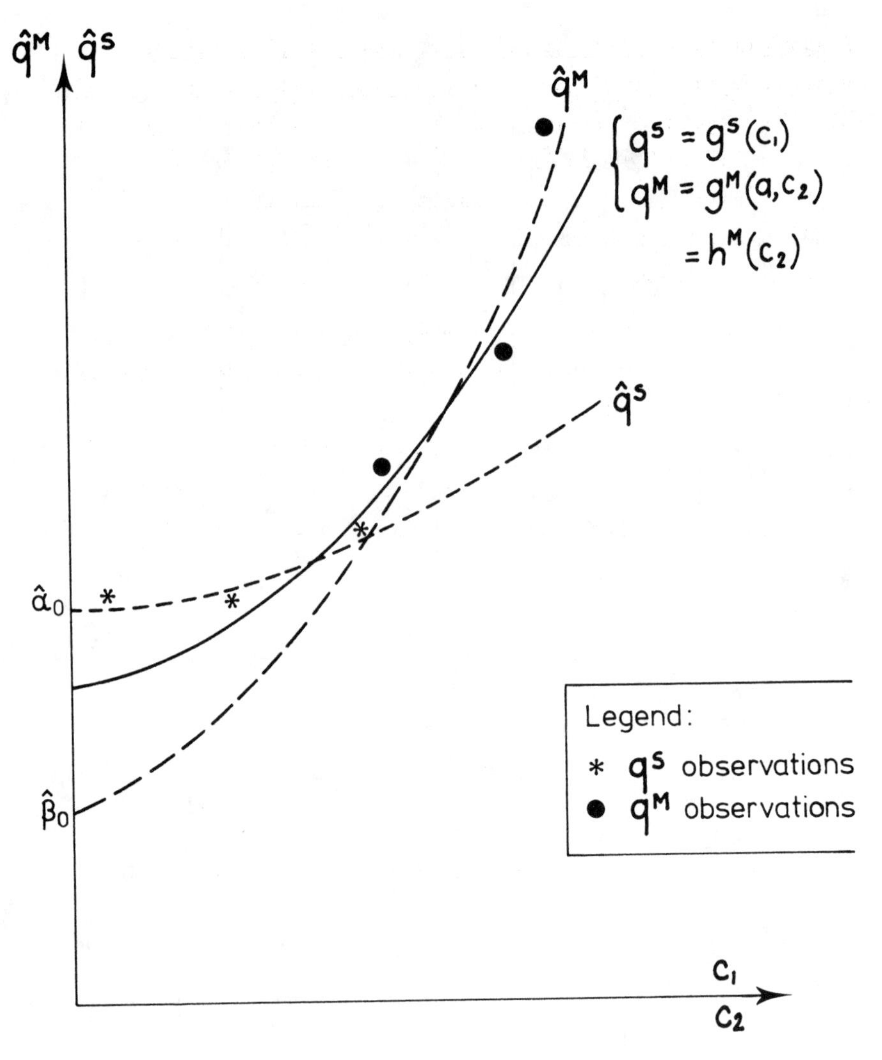

FIGURE 1. The Intercept Estimates in $\hat{q}^S$ and $\hat{q}^M$.

Assumption 7: The $\hat{q}^M$-model (25) provides the best estimates of the reaction coefficients δ_1 and δ_2 in (28); the $\hat{q}^S$-model provides the best estimate of the intercept δ_0. Here the starting point is the $\hat{q}^M$ model (25) (with R^2 as high as 0.95) where $\hat{\delta}_1 = 0.0323$ and $\hat{\delta}_2 = 0.0092$. Substitute $\hat{\delta}_1$ and $\hat{\delta}_2$ into the $\hat{q}^S$ model of Table 1. Each of the 71 q^S-values yields an estimated intercept. Their average $\bar{\delta}_0$ yields $\bar{\delta}_0 = 52.33$. (This approach could have been improved by an iterative procedure: re-estimate $\hat{q}^M$ under the condition $\delta_0 = 52.33$; use the resulting estimates of δ_1 and δ_2 in $\hat{q}^S$, etc.)

Next, apply a similar procedure to determine an "effective" upperbound for the $\hat{q}^L$ function. (The factor b = 900 in Table 1 is an "absolute" upperbound, i.e. jobs with service times longer than 900 are not accepted.) This reasoning yields the following assumption.

Assumption 8: The "effective" upperbound for q^L is not the absolute limit b = 900, but the upperbound under which virtually all jobs remain. To estimate the effective upperbound, substitute $\hat{\delta}_1$ and $\hat{\delta}_2$ (from $\hat{q}^M$) and $\hat{\delta}_0$ (from $\hat{q}^S$) into the $\hat{q}^L$ function of table 1. Each observation q^L corresponds with an effective upperbound, its average being 563.15. Note that the probability of service times higher than 563.15 is virtually zero.

Upon substitution of the rounded value b = 560 into (29), simultaneous estimation via (27) and (28) yields the final model:

$$\begin{bmatrix} \hat{q}^S \\ \hat{q}^M \\ \hat{q}^L \end{bmatrix} = 47.9 \begin{bmatrix} 1 \\ 1 \\ 1 \end{bmatrix} + 0.123 \begin{bmatrix} 10^2 \\ c_1^2 \\ c_2^2 \end{bmatrix} + 0.00400 \begin{bmatrix} c_1^2 \\ c_2^2 \\ 560^2 \end{bmatrix} \quad (31)$$

with t-values 0.61, 13.99, 5.43 and R^2 = 0.82. Now the residuals are acceptable, i.e., no systematic over- or underestimation occurs. (Also compare $\hat{\alpha}_0$ = 56.07 with $\hat{\delta}_0 + \hat{\delta}_1 a^2$ = 47.9 + 0.123 (100) = 60.2, a much better result than the value 82 obtained before Assumptions 7 and 8 were introduced).

6. OPTIMIZATION OF THE CRITERION

The criterion is defined by (7). The quantiles q^S, q^M, q^L were approximated by quadratic functions in c_1 and c_2 in (31). Substituting these functions into (7), and substituting the functions in c_1 and c_2 for the weights W and the conditional mean service times $\bar{B}$((8) through (13)) results in a criterion that is an explicit function in c_1 and c_2. Since via derivates we were not able to derive an explicit solution for the estimated optimal values, a computerized iterative search procedure was used, starting at c_1 = 30 and c_2 = 180, the values in current use. The individual $\hat{q}$ models of (24) through (26) yielded $\hat{c}_1^*$ = 44.83 and $\hat{c}_2^*$ = 177.19. The simultaneous model (31) resulted in $\hat{c}_1^*$ = 47.46 and $\hat{c}_2^*$ = 183.65. These results were implemented by the computer center: c_2 was maintained at 180 minutes, and c_1 was increased to 45 minutes. (On hindsight, it would have been prudent to double-check the estimated optimal values of c_1 and c_2 by simulating a few c_1, c_2 combinations around $\hat{c}_1^*$, $\hat{c}_2^*$.)

Note that the effects of changes in the arrival intensity λ and the service intensity μ (sensitivity studies) were not investigated. The effects of μ on the weights W and on the mean conditional service times $\bar{B}$ follow from (8) through (13). The effects of λ and μ on the quantiles q would require additional simulation experimentation and statistical analysis.

7. CONCLUSIONS

In many practical queueing systems priorities are introduced so that small jobs (short service times) are served first. The resulting model cannot be solved analytically; simulation is often used. The interpretation of the (voluminous) simulation output data can be based on regression analysis. Choice of an appropriate regression model required a qualitative theoretical analysis. This has been illustrated by a case study for which practical results were derived and implemented.

REFERENCES

Box, G.E.P. (1954). The exploration and exploitation of response surfaces, some general considerations and examples. Biometrics, 10, 16-60.

Cobham, A. (1954). Priority assignment in waiting line problems. Journal of the Operations Research Society of America, 2, 70-76.

Coppus, G., van Dongen, M., and Kleijnen J.P.C., (1977). Quantile estimation in regenerative simulation: a case study. Performance Evaluation Review, 5, 5-15. (Reprinted in Simuletter, 8 (1977), 38-47).

Crane, A. and Lemoine, J. (1977). An Introduction to the Regenerative Method for Simulation Analysis. Springer-Verlag, Berlin.

Draper, N.R. and Smith, H. (1966). Applied Regression Analysis. John Wiley & Sons, Inc., New York.

Hofri, M. (1980). Disk scheduling: FCFS vs. SSTF revisited. Communications of the ACM, 23, 645-653.

Jaiswal, N.K. (1968). Priority Queues. Academic Press, New York.

Kleijnen, J.P.C. (1974/1975). Statistical Techniques in Simulation, Vol. I/II. Marcel Dekker, Inc., New York.

Kleijnen, J.P.C. (1979) The role of statistical methodology in simulation. Methodology in Systems Modelling and Simulation (B. Zeigler, M.S. Elzas, G.J. Klir and T.I. Ören, eds.). North-Holland Publishing Company, Amsterdam.

Kleijnen, J.P.C. (1980). Computers and Profits; Quantifying Financial Benefits of Information. Addison - Wesley Publishing Company, Reading, Massachusetts.

Kleijnen, J.P.C. (1981). Regression analysis for simulation practitioners. Journal Operational Research Society, 32, 35-43.

Meyers, R.H. (1971). Response Surface Methodology. Allyn and Bacon, Boston.

Scheffé, H. (1959). The Analysis of Variance. John Wiley & Sons, Inc., New York.

Tsay, O. and Ng, P.A. (1980). A simulation approach to the comparative analysis of various job scheduling methods. Information Science, 21, 31-58.

Van den Bogaard, W. and Kleijnen, J.P.C. (1977). Minimizing waiting times using priority classes: a case study in response surface methodology. Report 77.056, Department of Business and Economics, Katholieke Hogeschool Tilburg, Netherlands.

Received 8/5/80; Revised 2/5/82.

AMERICAN JOURNAL OF MATHEMATICAL AND MANAGEMENT SCIENCES

COMPUTER SIMULATION PROGRAM FOR FOOD SCIENTISTS: PARAMETER SELECTION FOR CANNED FOOD THERMAL PROCESSING SYSTEMS

Jack Hachigian
Department of Mathematical Sciences
Hunter College
695 Park Avenue
New York, New York 10021

SYNOPTIC ABSTRACT

Food scientists conduct studies aimed at selecting parameters for a thermal (heat) processing system for canned food which will guarantee commercial sterility, economic shelf-life, and a wholesome, appetizing product. The computer simulation program given in this paper is based on the probabalistic nature of such studies and is a realistic model of the process in that it includes lag-time, cooling-time, and thermal transfer considerations. It allows for the inexpensive reproduction of these studies and can be used in many different ways by food scientists, medical and public health researchers, and microbiologists.

1.INTRODUCTION

Many researchers have contributed to the understanding of heat sterilization of canned food, beginning with Esty and Meyer (1922) and coming to the works of Ball and Olson (1957), Stumbo (1966), Charm (1971), and Potter (1973). Based on these results and methods, food scientists conduct studies aimed at selecting the parameters for a thermal process which will yield commercial sterility, economic shelf-life, and a wholesome, appetizing canned food product.

Hachigian (1978) described a probabilistic computer simulation model which was designed primarily as an additional tool for use by food scientists in their quest to understand the factors involved in destroying the target bacterium C. botulinum by irradiation. That simulation model provides partial spoilage data in an inexpensive way so that researchers can evaluate and analyze the kinetic parameters and/or kinetic processes involved in such irradiation studies. It would also be useful in simulating microwave sterilization processes. The purpose of this

Key Words and Phrases: simulation; partial spoilage data; thermal processing; sterilization of food; inoculated pack studies.

1982, VOL. 2, NO. 1, 13-58
0196-6324/82/010013-46 $9.40

paper is to provide a simulation program for heat processes which includes "lag-time," "cooling-time", and thermal transfer considerations. The algorithm and flow chart described below (in Sections 2 and 3, respectively) give researchers a basis for a computer program for their own computing facility. The program listing provided in Appendix A-3 is written in APL and is interactive in its design. The algorithm and program provide considerable flexibility, and can be used in many different ways by food scientists, medical and public health researchers, and microbiologists.

A detailed step-by-step description of the methods employed in the design of the simulation is given below. So as to facilitate the understanding of that description the following paragraphs present an overview of the program using as a model a canned food sterilization study.

An inoculated pack study consists of cans which have been inoculated with a microbiological contaminant of interest to a researcher. A researcher selects a fixed temperature for the experiment and then heats N cans for various predetermined lengths of time. We allow for an initial choice of heating time and an incremental change in time of exposure. The simulation is conducted at multiples of this exposure time increment until termination.

The number of cans N_i heated at each time point t_i may (but need not) be equal in number, as the user chooses. After the application of heat, the cans are examined to determine whether any of them contain at least one remaining viable microbiological contaminant of interest. Estimates of the remaining number of microbiological contaminants in each can are assumed to not be possible, either because of potential hazards or expense. These partial spoilage data are then recorded to be further analyzed, e.g. to determine the "D factor" (i.e. the reciprocal of the death rate constant).

Prior to heating, an inoculant is placed into each can. This inoculation procedure may take one of three forms: a surface placement of the bacterium before sealing, a center placement of the bacterium, or mixing of the inoculant throughout the media. Our algorithm allows for each of these procedures. The procedure desired by the experimenter is selected prior to a computer run. The inoculation distribution is based on probabilistic considerations as a matter of spatial distribution. Specifically, consider the case when the inoculant is mixed throughout the media in a can which has been conceptually divided into concentric shells of equal wall thickness. Each shell occupies a different proportion of the total volume of the can. As a consequence there is a natural probability distribution function (determined by the different proportions of volume) by which to distribute the inoculant through the can. We assume this natural distribution function when an experiment is to be conducted with an inoculant mixed throughout the media, and each shell of the can has probability of receiving inoculant equal to the proportion of the total volume occupied by that shell. Thermal processes involve "lag-time" and "cooling-time" (Charm (1971) and Potter (1973)). "Lag-time" is defined to be the time it takes for the center of the can to reach the desired temperature and "cooling-time" is defined as the time it takes for the center to reach the termperature of a cooling bath. These considerations are included in the algorithm and are under user control. Their choice will of course influence the output partial spoilage data. Both "lag-time" and "cooling-time" are assumed to be linear in this paper for purposes of demonstration; modifications for other assumptions can be made by an experienced user. Another reason for assuming linearity, for the purpose of this paper, is the ease it provides in explaining the algorithm rather than dealing with the complication of heat transfer through a specific medium. Since this is an important simplifying assumption it is recommended that a careful effort be made to modify the simulation so as to conform to the process to be simulated before using the simulation in practice. (In particular the specific heat of the media, the heat transfer function, and other considerations must be taken into account.) It is essential to note that any such modification must not change the underlying probabilistic nature of the process by insuring that the various probabilities involved are appropriately calculated especially with regard to the transfer function (see Figure 7 below).

2. METHODS

This computer simulation algorithm is based on inoculated pack studies. It can also be used to simulate model systems studies (i.e, studies carried out on a small scale in a laboratory in test tubes) by making appropriate use of available options. The algorithm given below is based on that of Hachigian (1978) but is specifically designed for a thermal process. It provides the options found in Hachigian (1978) plus other options especially useful for simulating thermal processes, and considerable flexibility in use.

Each step of the algorithm is followed by an explanation, example, or discussion to clarify the point-of-view and reasons behind the development.

Step 1. Select the initial length of time the pack is exposed to heat (called INITIALTIME, T). Specifically, this means the length of time the user wishes the center of each can to be at process temperature. (Of course this is not the actual time since cooling considerations must be taken into account. See Step 10 below.)

We assume that the simulation is conducted at a fixed temperature and we are interested in determining the death kinetics of, say, Purtrefactive Anerobe 3679 over time at that temperature. Since only part of the time scale may be of interest, we allow for an optional initial heating time length. For example, T may be chosen to be twenty minutes. The temperature at which the experiment is to be conducted is manifested in the death rate constant B (see Step 10 below).

<u>Step 2</u>. Select a time length increment (called TIMEINC, ΔT).

Obviously an experiment would not be conducted in time increments which are very small, but the possibility of this choice does not alter the algorithm. The simulated experiment proceeds in multiples of the choice of T, i.e. T, $T+\Delta T$, $T+2\Delta T$, ..., until termination. Termination will be described as the last step (Step 12) of the algorithm. Data for a single time length T can be obtained by choosing $\Delta T=0$. This is a useful computer-time-saving option if an experimenter is interested in only one time length, or the repetition of the experiment at one given time length.

<u>Step 3</u>. Select the number of cans to be thermally processed (called N) at each time length.

An experimenter using this simulation has been provided the option of choosing the number of replicates as he would in an actual experimental situation. One may change the number N of cans for different time lengths of thermal processing by using the flexibility of steps 1 and 2 above (see Hachigian (1978)).

<u>Step 4</u>. "Inoculate" each of the N cans by one of three optional methods described below. The choice of method is left to the user of the simulation model. The justification for each option is contained in Hachigian (1978).

<u>4. (A)</u> Inoculate each of the cans with a fixed number LAMBDA of spores.

<u>4. (B)</u> Inoculate each of the N cans with an average number LAMBDA of spores. The inoculation procedure follows the Poisson distribution having an average inoculum count of LAMBDA (L). Choose LAMBDA the average number of bacteria.

<u>4. (C)</u> Inoculate each of the N cans with an average number of spores LAMBDA where the inoculation procedure follows the truncated Poisson distribution. Choose LAMBDA the average number of spores, and the percent of LAMBDA truncation.

Options 4.(A) and 4.(C) above are provided for in the program listed in Appendix A-3, as well as in the flow-chart of Section 3 below. They are included in the program primarily because experimenters often believe they have greater control over the inoculation procedure and inoculation is not strictly governed by the Poisson. We will not dwell on these reasons or how they affect the simulation. (The explanatory paragraphs in 4A.

and C. in Hachigian (1978) apply as well for thermal processing and it will be assumed that the user will reference that paper.)

We suppose henceforth that inoculation occurs according to Step 4.(B), the Poisson probability distribution with average inoculum level equal to LAMBDA. To obtain this average inoculum for each of the N cans we computer exp (-L), generate uniform random variates U[J], and compute $\prod_{J=1}^{r} U[J]$ repetitively until for the first time

$\prod_{J=1}^{K+1} U[J] \leq e^{-L}$. Then K(1) = K is the inoculum level used for the first can. We proceed to the second can, generate new uniform random variates, and again use the smallest K for which $\prod_{J=1}^{K+1} U[J] \leq e^{-L}$. Then K(2) = K is the inoculum level for the second can. We continue in this way until all N cans receive their inoculi K(1), K(2), ..., K(N).

Up to this point we have followed the outline set forth in Hachigian (1978), which was designed primarily for radioactive sterilization of the food contaminants. However, we must now alter that algorithm for use in thermal processes, and although the description hereafter applies only under hypothesized conditions, modifications can accommodate other models. Specifically, we will assume the package (can) is cylindrical in form (see Figure 1) and that thermal transfer occurs directly from the exterior of the can toward the center. We do not deal with convection transfer or combination transfers of heat, nor do we consider non-cylindrical shapes; note again that a modification of the algorithm can be made to accommodate these specific other processes.

Step 5. Choose M, the number of concentric shells.

This choice divides the can into M-1 parts, each in the shape of a closed cylinder with equal wall thickness, R, and in addition, there is the center which will have the shape of a solid cylinder (see Figure 2 for the size and proportions of this solid cylinder). The choice M=1 implies the can is not subdivided and is useful for model systems studies.

Step 6. Choose G, the ratio of can height to can diameter. There are three cases: $G < 1$, $G = 1$, $G > 1$.

In particular the user must decide the type of can to be used in the simulation of partial spoilage data. A critical aspect of a can in a thermal process is the height to diameter ratio, for this determines how heat penetrates through the flat top and bottom relative to the sides of the cylinder. This is so since the simulation is based on the probabilistic nature of the process, i.e., the probability of destroying a spore in a given shell must be constant throughout that shell which requires that the shell thickness be constant.

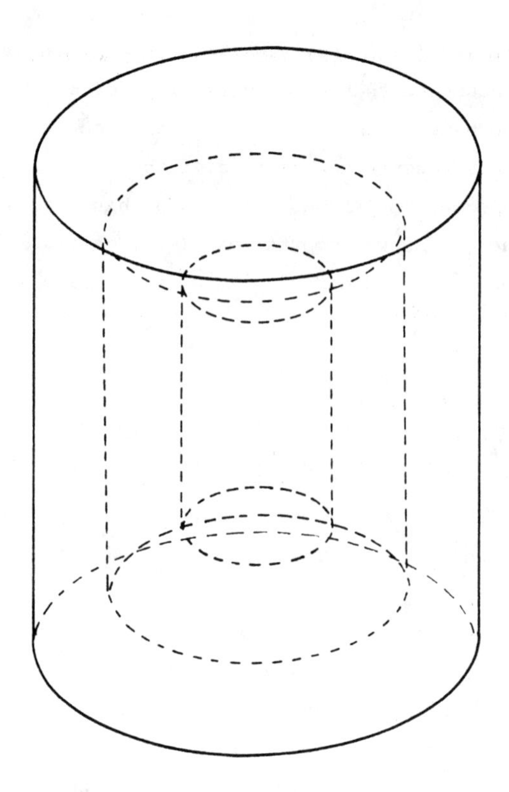

Figure 1. View of Concentric Shells in Can.

Since this simulation may be used for a variety of purposes, an additional option (Step 7) allows the user to choose how the contaminant is distributed in the can. In particular, the inoculate may be placed in the geometric center of the can, or at the surface of the contents, or distributed (mixed) throughout the M shells. Since each shell is assumed to have the same thickness (see Figure 2) it will have a different volume; consequently any distribution of contaminant (uniform or other) in the can must take into account these differing volumes. We use a uniform distribution of spores throughout the can, proportional to the proportion of volume, for demonstration purposes. Other distributions such as the spatial Poisson can be used by modifying the program.

We account for these differing volumes by computing the proportion of volume accounted for by each shell. This computation is as follows (refer to Figure 2 below). The total volume TV_n of each concentric solid sub-can in terms of shell wall thickness and height-to-diameter ratios is given by the formulas

$$TV_n = \begin{cases} 2\pi n^2 R^3 \, [M(G-1)+n] , & \text{if } G>1 \\ 2\pi n^3 R^3 , & \text{if } G=1 \\ 2\pi n R^3 \quad \frac{M(1-G)}{G} + n^2 , & \text{if } G<1 \end{cases} \qquad (1)$$

where n=1,2,3,......,M, G=height-to-diameter ratio and R=radius divided by M when $G \geq 1$, and R=radius times G divided by M when

$G < 1$ (see Figure 2). Using these formulas we obtain the volume $V_n = TV_n - TV_{n-1}$ of each of the M shells.

$$V_n = \begin{cases} 2\pi R^3 \,[3n^2 - 3n + 1 + (2n - 1)M(G-1)] & \text{, if } G>1 \\ 2\pi R^3 \,[3n^2 - 3n + 1] & \text{, if } G=1 \\ 2\pi R^3 \,[M^2(1-G)^2 + 2(2n-1)M(1-G) + 3n^2 + 3n + 1] & \text{, if } G<1 \end{cases} \quad (2)$$

where n=1, 2, ..., M. As an example, if we choose G = 2 and M = 3 then $V_1 = 8\pi R^3$, $V_2 = 32\pi R^3$, and $V_3 = 68\pi R^3$, where V_1 is the innermost solid volume, V_2 is the volume of the middle shell, and V_3 is the outermost volume.

The total volume, in each case, can be found by using (1) or by taking the sum of the constituent parts, i.e.

$$TV_M = \begin{cases} 2\pi M^3 R^3 \,(G) & \text{, if } G > 1 \\ 2\pi M^3 R^3 & \text{, if } G = 1 \\ 2\pi M^3 R^3 \,(\frac{1}{G})^2 & \text{, if } G < 1. \end{cases}$$

The proportion $SP_n = V_n/TV_n$ of the total volume that each shell number n occupies is

$$SP_n = \frac{2\pi R^3[3n^2 - 3n + 1 + (2n - 1)M(G-1)]}{2\pi R^3 M^3 G}$$

$$= \frac{3n^2 - 3n + 1 + (2n - 1)M(G-1)}{M^3 G}, \qquad \text{if } G > 1 \quad (3)$$

$$SP_n = \frac{2\pi R^3[3n^2 - 3n + 1]}{2\pi R^3 M^3} = \frac{3n^2 - 3n + 1}{M^3}, \qquad \text{if } G = 1 \quad (4)$$

$$SP_n = \frac{2\pi R^3 \left[\frac{M^2(1-G)^2}{G^2} + \frac{2(2n-1)M(1-G)}{G} + 3n^2 - 3n + 1 \right]}{2\pi R^3 \; M^3 \left(\frac{1}{G}\right)^2}$$

$$= \frac{G^2}{M^3}\left[\frac{M^2(1-G)^2}{G^2} + \frac{2(2n-1)M(1-G)}{G} + 3n^2 - 3n + 1\right], \text{ if } G < 1. \quad (5)$$

Continuing the example with M = 3 initiated after (2), we obtain (using (3)) the following proportions of total volume upon choosing G = 2:

$$SP_1 = \frac{V_1}{TV_3} = \frac{4}{54}$$

$$SP_2 = \frac{V_2}{TV_3} = \frac{16}{54}$$

$$SP_3 = \frac{V_3}{TV_3} = \frac{34}{54}$$

which add to one as they should.

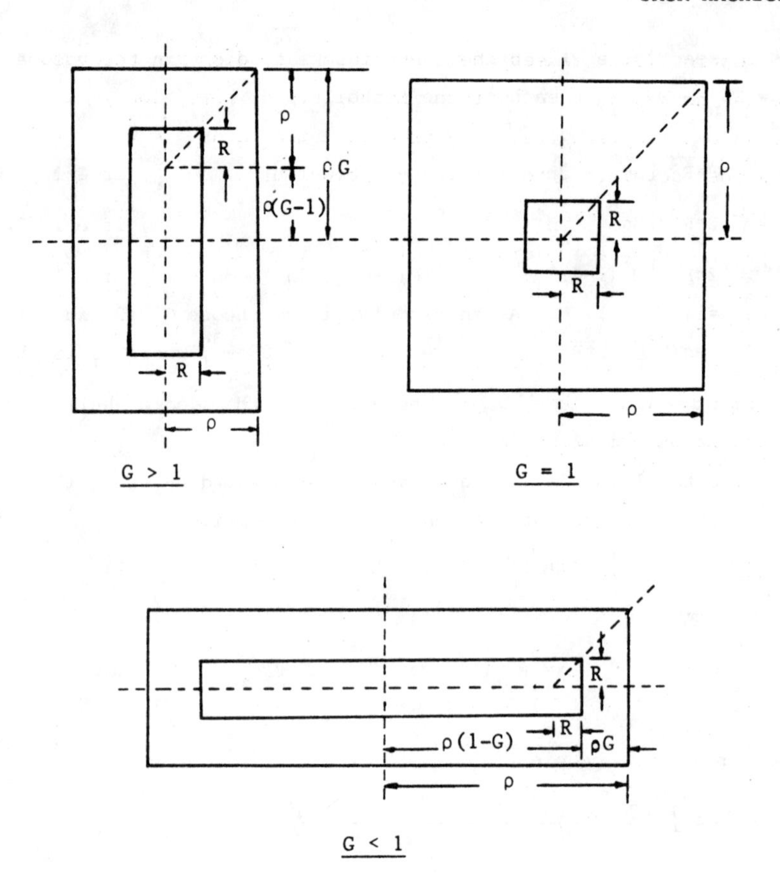

Figure 2. Cross-Section Views of Cans When $G<1$, $G=1$, and $G>1$ ($R = \rho/M$, $R = \rho/M$ and $R = \rho G/M$, Respectively).

Step 7. Distribute the inoculate according to one of three options (a) In the outershell, (b) In the center, or (c) Throughout the can.

One of these options is chosen as an initial condition in the simulation. Option 7.(a) automatically puts the inoculate in the Mth shell (outer shell, which is the equivalent of the top surface) by using the distribution function shown in Figure 3.

Shell No.	M	M-1	...	3	2	1
Probability	1	0		0	0	0

Figure 3. Distribution for Surface Placement of Inoculate.

Option 7.(b) inserts all the inoculate into the center (shell #1) by using the distribution function shown in Figure 4.

Shell No.	M	M-1	...	3	2	1
Probability	0	0	...	0	0	1

Figure 4. Distribution for Center Placement of Inoculate.

Option 7.(c) causes the contaminants to disperse throughout the can according to a uniform probability distribution. Since such a process is necessarily probabilistic in nature we distribute the inoculate into the various concentric shells according to their proportionate volume, i.e. SP_n. Using formula (3), (4), or (5) we compute the fraction of the total volume, SP_n, occupied by each of the M shells (given the choice of G). Note that it is not necessary to choose or determine R, as it cancels in (3), (4), and (5). These proportions form a probability distribution across the M shells as shown in Figure 5.

Shell No.	M	...	2	1
Probability	SP_M	...	SP_2	SP_1

Figure 5. Distribution for Mixed Placement of Inoculate.

In the example with M=3 and G=2 this distribution is

Shell No.	3	2	1
Probability	34/54	16/54	4/54

To distribute the spores according to the probabilities $\{SP_n\}$, in can no. 1 we generate a uniform random number W[1,1] on (0,1) and use it to determine in which shell the spore lies. (W[I,J] is a uniform random number used to place the Jth spore in the Ith can.) To specify the placement process, let us denote $SP[1] = 0$, $SP[2] = SP_1$, $SP[3] = SP_1 + SP_2, \ldots, SP[M] = \sum_{n=1}^{M-1} SP_n$, and $SP[M+1] = 1$. If $SP[2] < W[1,1] \leq SP[3]$ then the first spore (of the random number K(1) of spores in can no. 1) is inserted into shell number 2. We now generate W[1,2] and correspondingly insert another spore into the appropriate shell, shell J if $SP[J] < W[1,2] \leq SP[J+1]$. We continue until all K(1) spores are distributed probabilistically throughout the M shells of can no. 1. The process proceeds similarly for can no. 2, using random numbers W[2,1],I=1, ... , K(2). We proceed in this way until all N cans have been "mixed", and the spores are distributed throughout each individual can according to the distribution in Figure 5.

Figure 6. shows a possible resulting configuration after "mixing": N=5, M=3, and LAMBDA = 25.

		Shell No. 3	2	1	Total No. Spores K(I)
Can No.	1	17	7	3	27
	2	19	7	0	26
	3	15	6	2	23
	4	16	5	3	24
	5	19	6	1	26

Figure 6. Example of "Mixed" Inoculate Placement.

Step 8. Choose LAG, the time necessary for the center of the can to reach the desired temperature.

Step 9. Choose COOL, the time it takes for the center of the can to reach process temperature after cooling bath.

Steps 8 and 9 use variables pre-chosen from experimental efforts using a specific food and cans equipped with thermocouples which, when preset to a desired temperature, can signal when the center of the can is at that temperature. This allows for measuring the elapsed time from the beginning of the heating or cooling process. (Alternatively, LAG and COOL may be chosen by an experimenter using this simulation to obtain desired parameter information.)

Step 10. "Heat and cool" the cans. Cans having at least one survivor are recorded.

We assume that sterilization is governed by a probability distribution whose form is left to the user who may enter any justifiable extinction time distribution function by modifying the program. We otherwise choose the exponential distribution since there is a great deal of literature and experimental evidence supporting its use for thermal processes (e.g., see Stumbo (1966)). This choice also facilitates our explanation of the simulation. The exponential distribution is used in the form

$$F(T) = 1 - \exp(-BT)$$

where B is the death rate constant.

"Destruction" of the spores is performed probabilistically. In Step 1 of the algorithm, we choose the initial length of time the center shell of the can is to be heated, called INITIALTIME or T. We have also chosen LAG, COOL, and the number of shells M.

Recall that in heat sterilization of a canned product the object is to raise the temperature of the center of the can to a given temperature for a specified (sterilizing) length of time. It is therefore apparent, even without taking note of the cooling process, that each shell is at "the given temperature" for a different length of time: the exterior shell, M, is at temperature for $T+(M-1)\frac{LAG}{M}$; the (M-1)st shell is at temperature for $T+(M-2)\frac{LAG}{M}$; ...; shell number 2 is at temperature for $T+\frac{LAG}{M}$; and the center (shell number one) is at temperature for time length T.

The cans are also to be cooled. The cooling bath reduces the time at temperature, hence shell M is at temperature for $T+(M-1)\frac{LAG}{M}$, shell M-1 is at temperature for time $T+(M-2)\frac{LAG}{M}+\frac{COOL}{M-1}$, ..., the second shell is at temperature $T+\frac{LAG}{M}+(M-2)\frac{COOL}{M-1}$, and the center is at temperature T+COOL time units. Figure 7 summarizes these considerations.

SHELL NO.	HEATING TIME	HEATING AND COOLING
M	$T + \frac{(M-1)\,LAG}{M}$	$T + \frac{M-1}{M}\,LAG$
M-1	$T + \frac{(M-2)\,LAG}{M}$	$T + \frac{(M-2)\,LAG}{M} + \frac{1}{M-1}\,COOL$
M-2	$T + \frac{(M-3)\,LAG}{M}$	$T + \frac{(M-3)\,LAG}{M} + \frac{2}{M-1}\,COOL$
⋮	⋮	⋮
2	$T + \frac{LAG}{M}$	$T + \frac{LAG}{M} + \frac{M-2}{M-1}\,COOL$
1	T	T + COOL

Figure 7. Heat Transfer Function Table, Listing the Length of Time That Each Shell is At Process Temperature in the Two Cases Indicated.

In both the heating and cooling cycle we have assumed that the outer shell is at temperature instantaneously. Refinements must be made in the above considerations before applying the simulation in practice (see Charm (1971)). Alternatively, if the number of shells is chosen to be large then this assumption has little effect in that each shell (the outer one in particular) is of smaller thickness and therefore this assumption is reasonable.

As previously mentioned we have also assumed that heat is transferred or removed through the shells linearly. Other transfer functions for heat and for cooling can be incorporated into the simulation by modifying the coefficients of LAG and/or COOL in Figure 7.

The simulated destruction of spores by heat is conducted as follows: Generate a random number V(1,1,1) for can #1, (center) shell #1, where V(I,J,K) is a uniform (0,1) random number to be used in deciding if the Kth spore in the Jth shell of the Ith can be destroyed. If $V(1,1,1) \leq F(T+COOL) = 1-\exp[-B(T+COOL)]$, then the number of spores in shell #1 is decreased by one. Generate another random number V(1,1,2) and test to determine whether $V(1,1,2) \leq F(T+COOL)$. If so, we decrease the number of remaining spores in shell #1 of can #1 by one. Continue in this way until all the spores in shell #1 are tested. The destruction of spores continues in shell #2 of can #1 in a similar fashion. Generate a random number V(1,2,1); if $V(1,2,1) \leq$

$F(T + \frac{LAG}{M} + \frac{M-2}{M-1}\,COOL)$ the number of spores in shell #2 is decreased by one. Continuing in this way, all the spores in shell #2 of can #1 are tested. Note that F is evaluated at the time-length of exposure for shell #2 having accounted for LAG and COOL (see Figure 7).

The computer processing proceeds in this way until all M

shells of can #1 have been "heated" and "cooled". Upon completing the tests for can #1, can #2 is "heated" and "cooled" by generating random numbers V(2,J,P), where J ranges over the shells and P over the spores. When all N cans have been tested, each can is checked to determine whether any "viable spores" remain, that is, if any shell of any can has a non-zero entry. When all the shells in a can have zeros, that can is considered "<u>sterile</u>". The number of "sterile" cans is recorded as the numerator of the fraction Q/N and the fraction (N-Q)/N is printed out as LIVE/N.

<u>Step 11</u>. The process is repeated at time length $T + \Delta T$.

<u>Step 12</u>. The program stops.

The stopping procedure provides optional points at which the program stops. It may be set to stop when, for the first time, all cans are sterile. Or, it may be set to stop when all cans are sterile for a predetermined number of increments of ΔT (see Hachigian (1978)). (See for example, Example 2.) The object of this optional stopping procedure is to provide data on "skips" and "tailings", that is, data of the form where all cans are sterile at one point in time and, at a subsequent time or times, not all cans are sterile (see Hachigian (1978)).

3.ALGORITHM FLOWCHART

An algorithm flowchart (showing the twelve steps in the algorithm) is given in Figure 8. A more detailed flowchart is included in Appendix A-2 to aid readers in implementing the simulation or modifying the program for their computer facility.

Note that, in addition to the partial spoilage data which is the primary thrust of this simulation, additional information is printed out. Specifically, estimates of F (which contains the parameter B) and of B itself are printed out and are similar to those in Hachigian (1978), except that at each increment of time length T, F is evaluated at T+COOL. Consequently, $\hat{F}(T+COOL) = 1 + \frac{1}{L} \ln Q/N$ when using option 4.(B), the Poisson inoculation procedure; while $\hat{F}(T+COOL) = (Q/N)^{1/L}$ when 4.(A) is used. The death rate constant B is estimated, in all three cases, by $\hat{B} = \frac{-1}{T+COOL} \ln [\frac{-1}{L} \ln \frac{Q}{N}]$.

The sequence of estimates F resulting from heating at temperature for time lengths T_1, T_2, ..., T_s will result in a Maximum Likelihood Estimate (MLE) of F provided the partial spoilage data are non-increasing. If a reversal occurs, one may use the "pooled-adjacent" algorithm (Barlow, Bartholomew, Bremner, and Brunk (1972), chapter 1) to obtain an MLE. The computer program listed in Appendix A.3. does not compute the pooled-adjacent estimates.

It is appropriate at this point to explain the <u>reasons for</u>

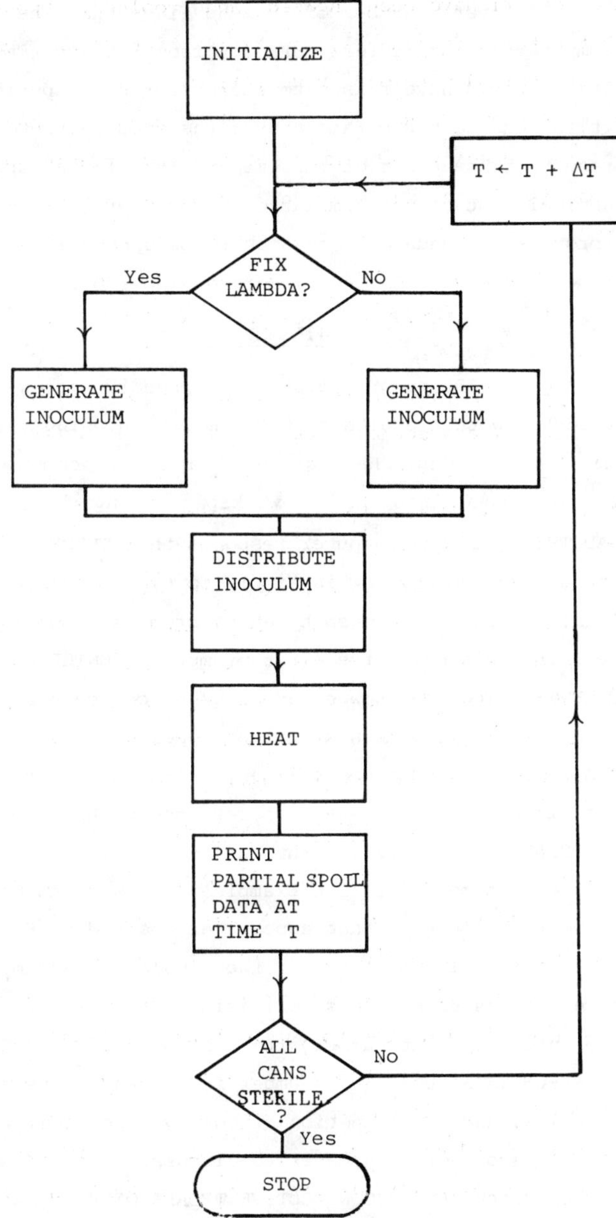

Figure 8. Algorithm Flowchart.

computing estimates of F and/or B when in fact both are known before generating data. Although our discussion of this simulation technique is in the context of food sterilization, it is clear that this model is useful in other experimental and public health situations (i.e. wherever the lethal agent is heat). In these cases (see Koch and Tolley (1975) and Cornell and Speckman (1967), for example) it is not known apriori what F is and the object of the research is to determine F. The utility of this simulation in these situations is to provide a means by which statistical methodologies for estimating F can be tested.

Moreover, even in the situation described herein, an experimenter having performed an experiment and plotted his results can now turn to this simulation to "repeat" the experiment and if after a number of runs he finds that the "shape" of the computer output differs markedly from his experimental results he would have reason to reconsider his experimental effort, or change the parameters of the simulation to confirm his results. The merit of this methodology is the speed, ease, and nominal cost of the simulation. Moreover the simulation will provide a standard upon which to base comparisons from experiment to experiment and laboratory to laboratory.

4.EXAMPLES

Using the APL program listed in Appendix A-3, we have reproduced below the results of the simulation using various combinations of options. Having signed on to the computer one loads the program into the APL workspace using the command) LOAD THERMAL. The computer responds that THERMAL is loaded. All that is necessary now is simply to type in RUN which as you will note appears before each set of prompts shown in the examples. At this point the first prompt is "ENTER LENGTH OF TIME INCREMENT:" The user now enters his/her choice and types in CR (carriage return). The next prompt appears, etc.

A review of the prompts in any of the examples indicates some additional options. A Y in response to DO YOU WANT TO PRINT DETAILS? Y/N: will print out the details of the simulation at each time point (see Example 2 below). An N simply prints out the summary that appears at the end of Example 2.

If a user wishes to repeat the simulation a number of different times using the same initializations just entered, he/she enters in that number after the prompt: NUMBER OF REPEATS FOR THIS RUN: This number is chosen by the user just as in practice, and is a function of cost vs. accuracy which it is beyond the scope of this paper to discuss.

The prompt "ENTER CONSECUTIVE TRIALS OF VOID CANS:" is explained in the algorithm after Step 12.

It should be noted that another time-saving feature is available for users. After familiarization with the order of the requests all the entries may be entered as a response to the first prompt provided each entry is separated by a slash. For example the responses in Example 1 below may be entered as follows:

ENTER LENGTH OF THE INCREMENT: .5/1/1.2/1.8/10/F/20/S/4.5/0/Y/0/2/1.7/CR

Example 1. (This example demonstrates a surface placement of a fixed number of spores.)

```
      RUN
ENTER LENGTH OF TIME INCREMENT: .5
ENTER INITIAL TIME: 1
ENTER LAG TIME: 1.2
ENTER COOL TIME: 1.8
ENTER NUMBER OF REPLICATES: 10
```

```
ENTER FIXED, POISSON, TRUNCATED: F
ENTER AVERAGE SAMPLE SIZE: 20
INOCULATE: SURFACE, CENTER, MIXED: S
ENTER DEATH RATE: 4.5
ENTER CONSECUTIVE TRIALS OF VOID CANS: 0
DO YOU WANT TO PRINT THE DETAILS? Y/N: Y
NUMBER OF REPEATS FOR THIS RUN: 0
NUMBER OF SHELLS: 2
RATIO G (HEIGHT÷DIAMETER): 1.7

*** TIME IN THE DISPLAYS BELOW IS ***
*** TIME+COOL FOR THE CENTER SHELL OF CAN ***

*** TIME =        2.8000 ***
INITIAL CELLS:
     20 20 20 20 20 20 20 20 20 20
----------------------------------
  1|  0  0  0  0  0  0  0  0  0  0
  2| 20 20 20 20 20 20 20 20 20 20
AVERAGE INOCULANT COUNT: 20

 DESTRUCTION
 PROBABILITY
 .9999966280
 .9992534142

LIVE/N = 0/10  FHAT:    -
AVERAGE RESIDUAL COUNT: 0


     N  LAMDA     T    LIVE  LIVE/N
    10   20     2.80     0    .000

0.735 CPU SECONDS
```

It should be observed that the left-hand column in the output of Example 1 indicates the shell numbers where 1 is the inner solid cylinder and 2 is the outermost (surface) shell (where 2 was the number of shells selected). The columns are the N (in this case 10) replicates with a constant (in this case 20) input. ***TIME = 2.8000*** at the top of the table is a result of "INITIALTIME" being chosen to be 1 and "COOL TIME" chosen to be 1.8 which sums to be 2.8. DESTRUCTION PROBABILITY is what results by entering the appropriate numbers in Table 7 for shells 1 and 2 and calculating F(·). Here they turn out to be .9999966280 etc.

Example 2. (This example demonstrates a variety of features of the simulation.) The inoculant is a Poisson variate with mean LAMBDA chosen to be 22. The inoculate is mixed throughout the can as can be seen in the tabled format under TIME. The replicates (N) are the columns and the shells are the rows. INITIAL CELLS are before heating and SURVIVING CELLS are after heating and cooling. The Summary is at the end of the run with columns labeled accordingly. Note that CONSECUTIVE TRIALS OF VOID CANS was chosen to be 2, and that three zeros appeared in the summary under LIVE before termination. Note also that after the summary 22.12 CPU SECONDS indicates the end of the first run. Since REPEATS FOR THIS RUN was chosen as 2 the program continues and will repeat the simulation from the beginning. We have only shown the first (i.e. TIME = 2.0000) part of that repeat to conserve space.

A word about REPEATS FOR THIS RUN. The default is that the simulation is conducted once, i.e. if nothing is entered after the prompt. If a zero or one is entered the simulation is also conducted once. Thereafter, the simulation is repeated according to what is selected, i.e. 2, 3, 4, etc.

```
    RUN
ENTER LENGTH OF TIME INCREMENT: .5
ENTER INITIAL TIME: 1
ENTER LAG TIME: 1
ENTER COOL TIME: 1
ENTER NUMBER OF REPLICATES: 7
ENTER FIXED, POISSON, TRUNCATED: P
ENTER AVERAGE SAMPLE SIZE: 22
INOCULATE: SURFACE, CENTER, MIXED: M
ENTER DEATH RATE: 1.23
ENTER CONSECUTIVE TRIALS OF VOID CANS: 2
DO YOU WANT TO PRINT THE DETAILS? Y/N: Y
NUMBER OF REPEATS FOR THIS RUN: 2
NUMBER OF SHELLS: 5
RATIO G (HEIGHT÷DIAMETER): .5
*** TIME IN THE DISPLAYS BELOW IS ***
*** TIME+COOL FOR THE CENTER SHELL OF CAN ***

*** TIME =        2.0000 ***
INITIAL CELLS:
     27 23 20 27 20 17 24
-------------------------
  1|  1  2  1  1  3  0  1
  2|  6  4  3  2  3  2  0
  3|  3  9  6  5  2  2  7
  4| 11  5  3  6  3  3  7
  5|  6  3  7 13  9 10  9
AVERAGE INOCULANT COUNT: 22.57142857

SURVIVING CELLS:            DESTRUCTION
      3  1  2  4  3  0  3   PROBABILITY
-------------------------
  1|  0  1  0  0  0  0  0   .9145650490
  2|  0  0  0  0  0  0  0   .9091458677
  3|  1  0  1  0  1  0  2   .9033829449
  4|  2  0  1  1  0  0  0   .8972544772
  5|  0  0  0  3  2  0  1   .8907372776
LIVE/N = 6/7  FHAT:   0.9115495387
AVERAGE RESIDUAL COUNT: 2.285714286
BHAT: 1.212656321

*** TIME =        2.5000 ***
INITIAL CELLS:
     13 27 28 30 24 19 25
-------------------------
  1|  1  4  2  2  2  0  4
  2|  1  3  2  4  2  2  3
  3|  2  7  6  4  3  3  3
  4|  3  5  8 10 12  5  4
  5|  6  8 10 10  5  9 11
AVERAGE INOCULANT COUNT: 23.71428571
```

```
SURVIVING CELLS:                  DESTRUCTION
      0  2  1  1  0  2  1         PROBABILITY
   -------------------------
   1|  0  0  0  0  0  0  0   .9538103716
   2|  0  0  0  0  0  0  0   .9508805406
   3|  0  0  0  1  0  0  0   .9477648689
   4|  0  0  1  0  0  2  0   .9444515685
   5|  0  2  0  0  0  0  1   .9409281040
LIVE/N = 5/7  FHAT:   0.9430562287
AVERAGE RESIDUAL COUNT: 1
BHAT: 1.146276387

*** TIME =        3.0000 ***
INITIAL CELLS:
     24 19 19 21 24 19 15
   -------------------------
   1|  4  0  1  4  1  4  1
   2|  2  5  1  1  3  2  1
   3|  5  3  2  2  6  4  1
   4|  3  5  6  7  8  3  7
   5| 10  6  9  7  6  6  5
AVERAGE INOCULANT COUNT: 20.14285714

SURVIVING CELLS:                  DESTRUCTION
      0  2  1  1  0  0  0         PROBABILITY
   -------------------------
   1|  0  0  0  0  0  0  0   .9750279980
   2|  0  0  0  0  0  0  0   .9734440115
   3|  0  1  0  0  0  0  0   .9717595519
   4|  0  0  0  0  0  0  0   .9699682463
   5|  0  1  1  1  0  0  0   .9680633172
LIVE/N = 3/7  FHAT:   0.9745629187
AVERAGE RESIDUAL COUNT: 0.5714285714
BHAT: 1.223849092


*** TIME =        3.5000 ***
INITIAL CELLS:
     21 20 26 10 23 21 17
   -------------------------
   1|  3  2  1  0  2  1  6
   2|  2  3  3  0  5  2  2
   3|  3  3  3  2  3  4  0
   4|  6  5 16  2  6  5  5
   5|  7  7  3  6  7  9  4
AVERAGE INOCULANT COUNT: 19.71428571

SURVIVING CELLS:                  DESTRUCTION
      0  0  0  0  2  0  0         PROBABILITY
   -------------------------
   1|  0  0  0  0  0  0  0   .9864991145
   2|  0  0  0  0  1  0  0   .9856427466
   3|  0  0  0  0  0  0  0   .9847320589
   4|  0  0  0  0  0  0  0   .9837636058
   5|  0  0  0  0  1  0  0   .9827337232
LIVE/N = 1/7  FHAT:   0.9929931509
AVERAGE RESIDUAL COUNT: 0.2857142857
BHAT: 1.417390619

*** TIME =        4.0000 ***
INITIAL CELLS:
     19 18 20 18 19 33 13
   -------------------------
   1|  1  2  0  1  0  1  0
   2|  2  4  2  6  3  3  0
   3|  4  2  3  1  1  8  1
   4| 10  4  7  4  6  5  2
   5|  2  6  8  6  9 16 10
AVERAGE INOCULANT COUNT: 20
```

```
SURVIVING CELLS:                       DESTRUCTION
      0  0  1  0  0  1  0              PROBABILITY
-------------------------
   1|  0  0  0  0  0  0  0             .9927008692
   2|  0  0  0  0  0  0  0             .9922378817
   3|  0  0  0  0  0  1  0             .9917455266
   4|  0  0  1  0  0  0  0             .9912219413
   5|  0  0  0  0  0  0  0             .9906651447
LIVE/N = 2/7  FHAT:     0.9847058074
AVERAGE RESIDUAL COUNT: 0.2857142857
BHAT: 1.045070523


*** TIME =         4.5000 ***
INITIAL CELLS:
     19 34 31 17 22 23 26
-------------------------
   1|  0  2  2  1  3  2  0
   2|  1  2  6  3  4  2  3
   3|  3 10  2  3  3  6  5
   4|  6 10 10  5  6  7 11
   5|  9 10 11  5  6  6  7
AVERAGE INOCULANT COUNT: 24.57142857

SURVIVING CELLS:                       DESTRUCTION
      2  0  0  0  0  0  0              PROBABILITY
-------------------------
   1|  0  0  0  0  0  0  0             .9960537914
   2|  0  0  0  0  0  0  0             .9958034814
   3|  0  0  0  0  0  0  0             .9955372941
   4|  1  0  0  0  0  0  0             .9952542225
   5|  1  0  0  0  0  0  0             .9949531955
LIVE/N = 1/7  FHAT:     0.9929931509
AVERAGE RESIDUAL COUNT: 0.2857142857
BHAT: 1.102414926
*** TIME =         5.0000 ***
INITIAL CELLS:
     13 26 29 19 27 20 21
-------------------------
   1|  0  0  3  1  1  2  1
   2|  1  7  4  3  5  3  3
   3|  3  2  6  3  9  4  5
   4|  6  8  5  5  6  4  7
   5|  3  9 11  7  6  7  5
AVERAGE INOCULANT COUNT: 22.14285714

SURVIVING CELLS:                       DESTRUCTION
      0  0  0  0  1  0  0              PROBABILITY
-------------------------
   1|  0  0  0  0  0  0  0             .9978665182
   2|  0  0  0  0  0  0  0             .9977311904
   3|  0  0  0  0  0  0  0             .9975872787
   4|  0  0  0  0  0  0  0             .9974342386
   5|  0  0  0  0  1  0  0             .9972714911
LIVE/N = 1/7  FHAT:     0.9929931509
AVERAGE RESIDUAL COUNT: 0.1428571429
BHAT: 0.9921734336


*** TIME =         5.5000 ***
INITIAL CELLS:
     18 17 19 17 17 28 27
-------------------------
   1|  1  1  1  1  1  1  2
   2|  1  2  1  0  2  3  6
   3|  1  4  3  2  2  6  5
   4|  4  7  5  9  5  6  7
   5| 11  3  9  5  7 12  7
AVERAGE INOCULANT COUNT: 20.42857143
```

```
 DESTRUCTION
 PROBABILITY
 .9988465525
 .9987733888
 .9986955842
 .9986128445
 .9985248565

LIVE/N = 0/7  FHAT:    1
AVERAGE RESIDUAL COUNT: 0

*** TIME =         6.0000 ***
INITIAL CELLS:
     23 19 17 23 20 29 18
-------------------------
  1|  3  3  1  0  1  1  2
  2|  3  2  5  3  4  7  3
  3|  4  3  3  3  2  1  2
  4|  7  6  5  4  4  6  6
  5|  6  5  3 13  9 14  5
AVERAGE INOCULANT COUNT: 21.28571429

 DESTRUCTION
 PROBABILITY
 .9993763991
 .9993368438
 .9992947795
 .9992500470
 .9992024771

LIVE/N = 0/7  FHAT:    1
AVERAGE RESIDUAL COUNT: 0


*** TIME =         6.5000 ***
INITIAL CELLS:
     20 20 25 18 30 24 23
-------------------------
  1|  1  2  1  0  1  1  3
  2|  2  1  2  1  4  3  1
  3|  5  2  7  2  4  6  4
  4|  5  6  3  7  7  6  8
  5|  7  9 12  8 14  8  7
AVERAGE INOCULANT COUNT: 22.85714286

 DESTRUCTION
 PROBABILITY
 .9996628559
 .9996414706
 .9996187289
 .9995945447
 .9995688265
 N  LAMDA     T     LIVE  LIVE/N
 7    22     2.00     6     .857
 7    22     2.50     5     .714
 7    22     3.00     3     .429
 7    22     3.50     1     .143
 7    22     4.00     2     .286
 7    22     4.50     1     .143
 7    22     5.00     1     .143
 7    22     5.50     0     .000
 7    22     6.00     0     .000
 7    22     6.50     0     .000

22.12 CPU SECONDS      (Indicates the end of first run.)
```

```
*** TIME =        2.0000 ***
INITIAL CELLS:
     24 24 22 18 25 24 28
-------------------------
  1|  2  0  3  0  2  3  0
  2|  4  5  1  4  1  2  7
  3|  2  6  3  2  3  6  5
  4|  8  9  5  3  9  5  8
  5|  8  4 10  9 10  8  8
AVERAGE INOCULANT COUNT: 23.57142857

SURVIVING CELLS:               DESTRUCTION
      1  2  3  4  3  3  1      PROBABILITY
-------------------------
  1|  1  0  0  0  1  0  0     .9145650490
  2|  0  1  0  0  0  0  0     .9091458677
  3|  0  0  0  1  0  0  0     .9033829449
  4|  0  1  0  0  0  1  0     .8972544772
  5|  0  0  3  3  2  2  1     .8907372776
LIVE/N = 7/7  FHAT:    -
AVERAGE RESIDUAL COUNT: 2.428571429


*** TIME =        2.5000 ***
INITIAL CELLS:
     23 28 25 24 25 15 27
-------------------------
  1|  1  2  0  2  2  2  3
  2|  4  4  3  2  2  2  4
  3|  5  7  4  4  7  1  6
  4|  5  7  7 10  6  4  2
  5|  8  8 11  6  8  6 12
AVERAGE INOCULANT COUNT: 23.85714286
```

BREAK! (Author induced to conserve space.)

5. DISCUSSION

The simulation of partial spoilage data of radio-sterilization processes using computer and probabilistic techniques was introduced by Hachigian (1978). The algorithm contained herein is designed for the thermal processing of food and, therefore, should find wide applicability when modified for specific foods.

The simulation algorithm has a variety of potential uses for food scientists when the experiment basically follows the methodology of an inoculated pack or model system in its broad aspects. Among the uses forseen are: prediction of spoilage rates, assistance in the design and testing of new packages, and economic generation of test data for evaluation of new procedures or products. Most important among these potential applications, however, is the generation of partial data on an economical basis so that comprehensive statistical analyses can be conducted. By their very nature inoculated pack studies are very expensive and are not often reproduced. Valid statistical analyses consequently, are difficult, if not impossible, to perform at acceptable confidence levels.

In addition, such simulations have educational uses. For example, an instructor can use such a model to demonstrate how micro-organisms can be distributed throughout a can and how that distribution can effect the outcomes realized with various parameters for a process. The simulation can also produce data which students might analyze.

There are a number of streams of random numbers called for in this simulation, and it is important that these streams be independently generated. Moreover a poor random number generator or dependence among generators will result in invalid data. A useful reference in these regards is Dudewicz and Ralley (1981).

An interesting observation in this simulation is that can size per se does not enter into the considerations. Can size is absorbed into the specification of M (number of concentric shells), G(height to diameter ratio), and thermal rate of transfer through the shells. This observation simplifies the simulation considerably, making it possible to handle all sizes.

In describing the algorithm it was noted that we assumed that heat penetrates the can to its interior by conduction through the media. Heat transfer by convection or a combination of conduction and convection was not described. However, modification of the algorithm to allow for the incorporation of such considerations can be accomplished by modifying Figure 7 (which is, in effect, the heat transfer function across shells). In particular, convection may transfer heat to certain inner shells before shells near the outside. Suppose, for example, there are M shells, the outermost being designated as M. Convection transfer of heat may then heat shell M-4 faster than $M-j$, $1<j\leq 3$. In this case shell M-4 will be at temperature for a different length of time than if heat were transferred by conduction alone. Moreover, if convection occurs it is possible that spores may travel from shell to shell. Therefore, the simulation should given different spoilage data as output. On the other hand, when heat is transferred directly by conduction, but in a non-linear way, as is primarily the case with solid foods or products, the output data will also be different. These situations can be incorporated into the simulation by appropriately modifying Figure 7.

One last observation, with regard to M (the number of concentric shells): the larger M is, the greater the time required to complete the simulation, hence the greater the computer cost. However, the larger M, the more precise the simulation becomes, specifically regarding the assumption that the outer shell is at temperature instantaneously both during the heating and cooling cycle.

ACKNOWLEDGMENTS

I wish to acknowledge the assistance of Mr. J. C. Lincoln while writing the APL program. In addition I wish to thank the referees for a careful reading of the paper, and also the Editor for a painstaking effort to improve the presentation and clarity.

APPENDIX A.

This appendix contains three basic subdivisions useful in user implementation of the algorithm using APL. These subdivisions are: A-1, a description of the computer configuration used to program the simulation; A-2, a detailed flowchart based on the 12-step algorithm, and A-3, APL program documentation (i.e. complete listing of the APL program used at CUNY). The program documentation includes a listing of all functions in the workspace called THERMAL, a calling sequence of those functions, a listing of the variables in the workspace, and a listing of the APL code with appropriate comments in the listings.

A-1.COMPUTER CONFIGURATION

Hunter College's Laboratory of Data Analysis (Jack Hachigian, Director) consists of 11 ADDS 980 Consuls and a Tektronix 4013 with APL, connected through a Timplex multiplexer and modem via leased line to the CUNY/ Computer Center, which has the following equipment configuration on which APL is supported. the CPU is an AMDAHL 470/V6-II with a controller IBM 3705, IBM DISK DRIVES and TAPE DRIVES. The OPERATING SYSTEM is MVS/JE53.

A-2. DETAILED FLOWCHART

DETAILED FLOWCHART

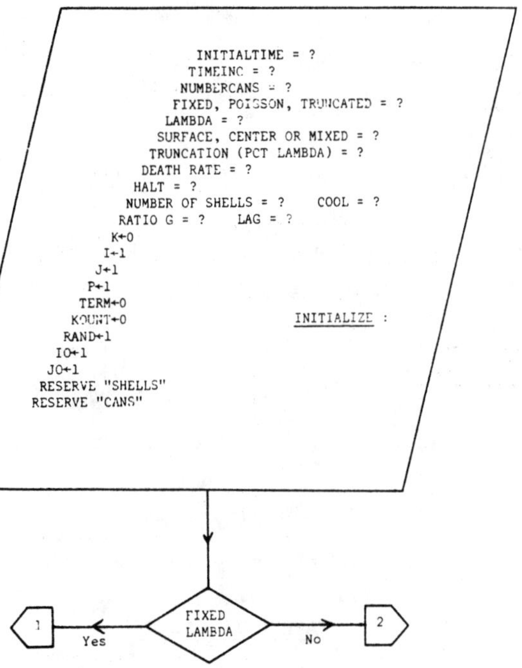

INITIALTIME = ?
TIMEINC = ?
NUMBERCANS = ?
FIXED, POISSON, TRUNCATED = ?
LAMBDA = ?
SURFACE, CENTER OR MIXED = ?
TRUNCATION (PCT LAMBDA) = ?
DEATH RATE = ?
HALT = ?
NUMBER OF SHELLS = ? COOL = ?
RATIO G = ? LAG = ?
K←0
I←1
J←1
P←1
TERM←0
KOUNT←0
RAND←1
IO←1
JO←1
RESERVE "SHELLS"
RESERVE "CANS"

INITIALIZE :

FIXED LAMBDA

Yes → 1

No → 2

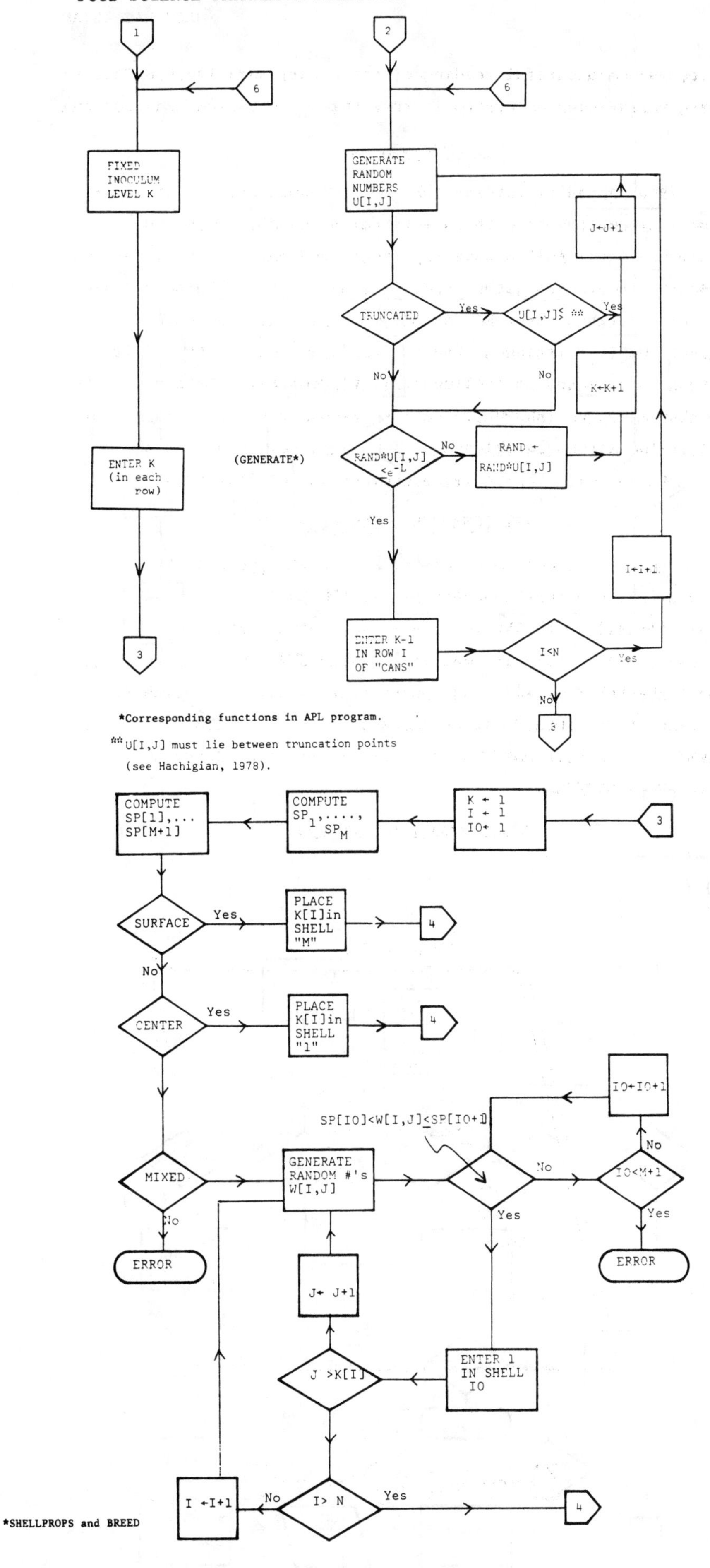
1
6
FIXED INOCULUM LEVEL K
ENTER K (in each row)
3
2
6
GENERATE RANDOM NUMBERS U[I,J]
J←J+1
TRUNCATED
Yes
U[I,J]≤ **
Yes
No
No
K←K+1
(GENERATE*)
RAND*U[I,J] ≤e-L
No
RAND ← RAND*U[I,J]
Yes
I←I+1
ENTER K-1 IN ROW I OF "CANS"
I<N
Yes
No
3
*Corresponding functions in APL program.
**U[I,J] must lie between truncation points (see Hachigian, 1978).
COMPUTE SP[1],... SP[M+1]
COMPUTE SP1,....,SPM
K ← 1
I ← 1
IO← 1
3
SURFACE
Yes
PLACE K[I]in SHELL "M"
4
No
CENTER
Yes
PLACE K[I]in SHELL "1"
4
IO←IO+1
SP[IO]<W[I,J]≤SP[IO+1]
MIXED
GENERATE RANDOM #'s W[I,J]
No
IO<M+1
No
Yes
Yes
No
ERROR
ERROR
J← J+1
ENTER 1 IN SHELL IO
J >K[I]
I ←I+1
No
I> N
Yes
4
*SHELLPROPS and BREED

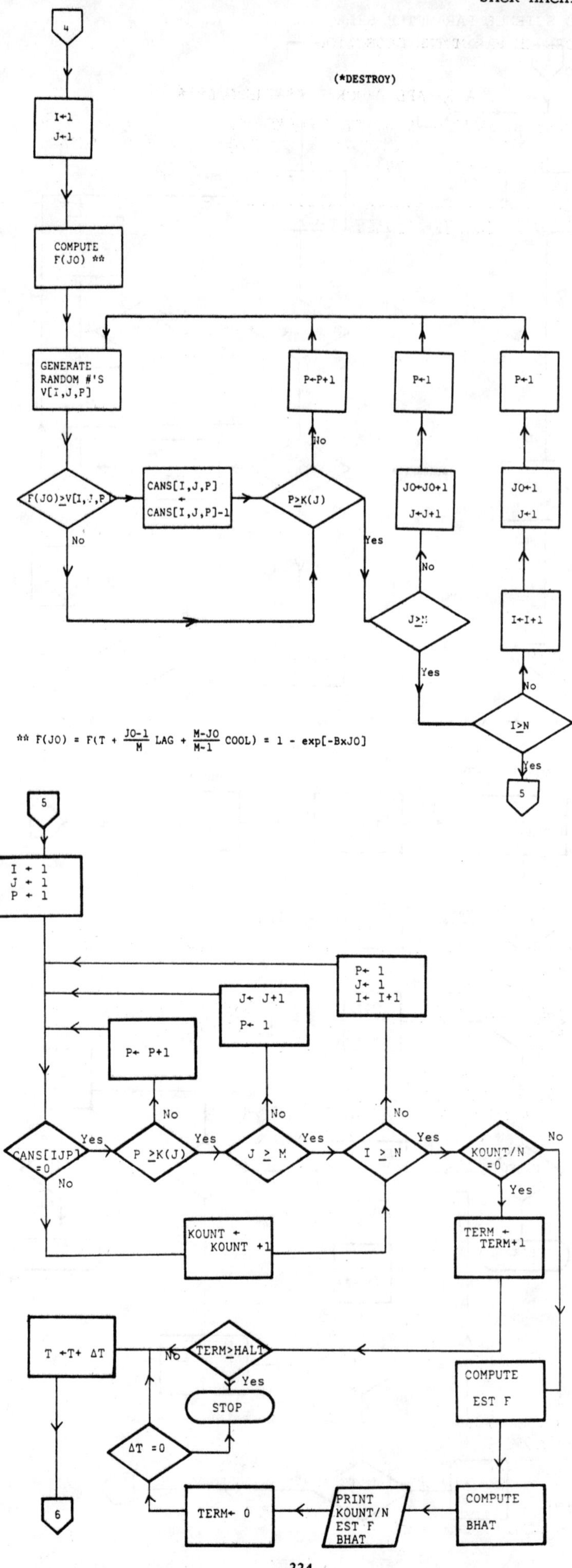
4
I←1
J←1
COMPUTE
F(JO) **
(*DESTROY)
GENERATE
RANDOM #'S
V[I,J,P]
P←P+1
P←1
P←1
F(JO)≥V[I,J,P]
CANS[I,J,P]
←
CANS[I,J,P]-1
P≥K(J)
JO←JO+1
J←J+1
JO←1
J←1
No
No
Yes
No
J≥M
I←I+1
Yes
No
I≥N
Yes
5
** F(JO) = F(T + (JO-1)/M LAG + (M-JO)/(M-1) COOL) = 1 - exp[-BxJO]
5
I ← 1
J ← 1
P ← 1
P← 1
J← 1
I← I+1
J← J+1
P← 1
P← P+1
No
No
No
CANS[IJP]
=0
Yes
P ≥K(J)
Yes
J ≥ M
Yes
I ≥ N
Yes
KOUNT/N
=0
No
No
Yes
KOUNT ←
KOUNT +1
TERM ←
TERM+1
T ←T+ ΔT
TERM≥HALT
No
Yes
STOP
COMPUTE
EST F
ΔT =0
6
TERM← 0
PRINT
KOUNT/N
EST F
BHAT
COMPUTE
BHAT

A-3. APL PROGRAM ***THERMAL***

```
THERE ARE 13 FUNCTIONS IN THIS WORKSPACE:
BREED      CPU      DESTROY    GENERATE   RUN    SETUP
SHELLPROPS SUMMARY  YN         II         IJ     JI
JJ
--------------------------------------------------------------
TOPOLOGICAL CALLING SEQUENCE


RUN
  SETUP
     II

       IJ
       JI
     JJ
       IJ

       JI
     YN
       JJ

          IJ

          JI

       SHELLPROPS
    BREED
       GENERATE
    DESTROY
       GENERATE
    CPU
     YN
      JJ

        IJ
        JI
TS: TIMESTAMP,SIGNATURE
DS: DESCRIPTION
RT: EXPLICIT RESULT
LA: LEFT ARGUMENT
RA: RIGHT ARGUMENT
GU: GLOBAL UPDATE
GR: GLOBAL REFERENCE
     ∇ RUN;CELLM;CELLS;INSCOUNT;KOUNT;REPCOUNT;TIME;CPU
[1]  ⍝TS: 12/9/80  JCL
[2]  ⍝DS: TOPLEVEL PROGRAM
[3]  ⍝GU: ALLCELLS,ALLLIVE,ALLTIMES,BHATS,FHATS,TIMES
[4]  ⍝GR: COOL,NOPRINT,REPEATS,TIME0,VOIDS,ΔT,MAXCOUNT,PASS
[5]   SETUP
[6]   REPCOUNT←0
[7]  B0:CPU←⎕AI[2]
[8]   TIMES←FHATS←BHATS←⍳KOUNT←0
[9]   ALLTIMES←ALLCELLS←ALLLIVE←⍳0
[10]  TIME←TIME0+COOL
[11] ⍝ VOID COUNTER
[12] B1:INSCOUNT←0
[13] B2:ALLTIMES←ALLTIMES,TIME←TIME+ΔT
[14]  →NOPRINT/B2A
[15]  '*** TIME = ',(¯12↑4⍕TIME),' ***'
[16] B2A:BREED
[17]  DESTROY
[18] ⍝ AUSTERITY CHECK
[19]  →(MAXCOUNT≤KOUNT←KOUNT+1)/B4
[20] ⍝ IF ΔT=0, THEN ONE SHOT DEAL
[21]  →(~×ΔT)/0
[22] ⍝ SOME NON-STERILE?
[23] B5:→(×+/CELLS)/B1
[24] ⍝ MORE TERMINAL SKIPS?
[25]  →(VOIDS≥INSCOUNT←INSCOUNT+1)/B2
[26] B3:SUMMARY
[27]  ''
[28] ⍝ WHAT'S THE DAMAGE?
```

```
[29]  CPU
[30]  2 1 ρ' '
[31]  →(REPEATS>REPCOUNT←REPCOUNT+1)/B0
[32]  →0
[33] ⍝ DON'T SPEND TOO MUCH
[34] B4:'THAT''S ',(⍕MAXCOUNT),' RUNS'
[35]  →(~'TRY FOR MORE' YN 'MORE')/B3
[36]  →(~∧/(4↑PASS)=4↑'WHAT''S THE SECRET WORD' JJ 'SECRET')/B3
[37] ⍝ RESET AUSTERITY COUNTER
[38]  KOUNT←0
[39]  →B5
    ∇
**** THERMAL ****  DISPLAYED  12/11/80 11:54 AM

    ∇ SETUP;LAG
[1]  ⍝TS: 12/9/80  JCL
[2]  ⍝DS: PROMPT USER FOR SETUP PARAMETERS
[3]  ⍝GU: ALL PARAMETERS
[4]   VOIDS←V←W←0×A←1
[5]   ΔT←1↑'ΔT' II 'ENTER LENGTH OF TIME INCREMENT'
[6]   TIME0←1↑(-ΔT)+'TIME' II 'ENTER INITIAL TIME'
[7]   LAG←1↑'LAG' II 'ENTER LAG TIME'
[8]   COOL←1↑'COOL' II 'ENTER COOL TIME'
[9]  ⍝ ΔT=0 IMPLIES ONE SHOT ONLY
[10]  →(~×ΔT)/B5
[11]  N←1↑'N' II 'ENTER NUMBER OF REPLICATES'
[12] B0:→(B3,B3,B3,B0)[DIST←'FPT'⍳1↑'F,P,T' JJ 'ENTER FIXED, POISS
ON, TRUNCATED']
[13] B3:LAMDA←1↑'LAMDA' II 'ENTER AVERAGE SAMPLE SIZE'
[14] B6:→(3<INOC←'SCM'⍳'S,C,M' JJ 'INOCULATE: SURFACE, CENTER, MIX
ED')/B6
[15]  →(3≠DIST)/B4
[16] B1:PCT←100⌊1↑'TRUNC' II 'ENTER TRUNCATION (PCT LAMDA)'
[17]  V←LAMDA+W←⌊0.5+LAMDA×PCT÷100
[18]  W←LAMDA-W
[19] ⍝ TOPCUT AND BOTCUT ARE TRUNCATION LIMITS FOR RANDOM NUMBERS
[20]  TOPCUT←+/((*-LAMDA)×LAMDA*0,⍳V)÷!0,⍳V
[21]  BOTCUT←+/((*-LAMDA)×LAMDA*0,⍳W-1)÷!0,⍳W-1
[22] B4:B←1↑'B' II 'ENTER DEATH RATE'
[23]  VOIDS←1↑'VOIDS' II 'ENTER CONSECUTIVE TRIALS OF VOID CANS'
[24] B5:NOPRINT←~'DETAIL' YN 'DO YOU WANT TO PRINT THE DETAILS'
[25]  REPEATS←1↑'REPEATS' II 'NUMBER OF REPEATS FOR THIS RUN'
[26]  MM←1⌈1↑'M' II 'NUMBER OF SHELLS'
[27]  COMCOOL←⌽(LAG×(MM-⍳MM)÷MM)+COOL×(0,⍳MM-1)÷MM-1
[28]  G←'G' II 'RATIO G (HEIGHT÷DIAMETER)'
[29]  SP←MM SHELLPROPS G
[30]  ''
[31]  '*** TIME IN THE DISPLAYS BELOW IS ***'
[32]  '*** TIME+COOL FOR THE CENTER SHELL OF CAN ***'
[33]  ''
    ∇
```

THE FUNCTIONS II, JJ, IJ, AND JI REQUEST TERMINAL INPUT

REQUESTS NUMERIC INPUT

```
    ∇ IOZ←IOS II IOL;IOA;IOB;IOC
[1]   →IJ/7,IOZ←⍳0
[2]   ⍞←(⍎'IO','SL'[SW+1]),': ',IOA←''
[3]   IOB[(IOC←CL=¯1↑IOB)/⍴IOB←,⍞]←' '
[4]   IOA←IOA,IOB
[5]   →IOC/3
[6]   JI
[7]   →3××⍴IOA←(')'≠1↑IOA)/IOA←(∨\IOA≠' ')/IOA←(~IOA∊'⎕→')/IOA
[8]   →9+∧/('X'=IOA),1=⍴IOA
[9]   →2×0≠1↑0↑IOZ←,⍎IOA
[10]  →1⌈SW←~SW
    ∇
```

TEST INPUT BUFFER FOR STACKED INPUT.

```
     ∇ Z←IJ;A
[1]   →2×Z←×ρIJD
[2]   →3×Z←×ρIJM
[3]   IOA←(¯1+A←IJM⍳IJD)↑IJM
[4]   IJM←A↓IJM
     ∇
```

UPDATE BUFFER AFTER EXTRACTING MOST CURRENT RESPONSE.

```
     ∇ JI;A
[1]   →2××ρIJD
[2]   IJM←(A←IOA⍳IJD)↓IOA
[3]   IOA←(¯1+A)↑IOA
     ∇
```

REQUESTS CHARACTER INPUT.

```
     ∇ IOA←S JJ L;B;C
[1]   →IJ/9,IOA←''
[2]   B←⍞←(⍎'SL'[SW+I]),': '
[3]   C←(M←(⌽B)⍳CR)↓⍞
[4]   C[(M↑B←(M↑C)=CR,CL)/ρC]←' '
[5]   IOA←IOA,C
[6]   →8-∨/B
[7]   →4,ρC←,⍞
[8]   JI
[9]   →B/I,SW←T|SW+B←∧/('X'=IOA),I=ρIOA
     ∇

     ∇ Z←S YN L
[1] ⍝TS: 12/10/80  JCL
[2] ⍝DS: ASK A YES OR NO QUESTION
[3] ⍝RT: Z ↔ 1 IS YES, 0 IS NOT YES
[4] ⍝LA: S ↔ SHORT PROMPT
[5] ⍝RA: L ↔ LONG PROMPT
[6]  Z←'Y'=I↑(S,'? Y/N') JJ L,'? Y/N'
     ∇



     ∇  Z←MM SHELLPROPS G;I
[1]   ⍝TS: 12/9/80  JCL
[2]   ⍝DS: DEFINE SHELL VOLUMETRIC PROPORTIONS
[3]   ⍝RT: Z ↔ VECTOR OF PROPORTIONS
[4]   ⍝LA: MM ↔ NUMBER OF SHELLS
[5]   ⍝RA: G ↔ HEIGHT÷DIAMETER OF CAN
[6]    Z←⍳I←0
[7]    →(G≤1)/B2
[8]   B1:→(MM<I←I+1)/B5
[9]    Z←Z,(((¯1+2×I)×MM×(G-1))+1+3×I×I-1)÷G×MM*3
[10]   →B1
[11] B2:→(G<1)/B4
[12] B3:→(MM<I←I+1)/B5
[13]   Z←Z,(1+3×I×I-1)÷MM*3
[14]   →B3
[15] B4:→(MM<I←I+1)/B5
[16]   Z←Z,(G×G÷MM*3)×((MM×MM×(1-G)*2)÷G×G)+(2×(¯1+2×I)×(MM÷G)×1-G)
+1+3×I×I-1
[17]   →B4
[18] B5:Z←+\Z
     ∇
```

```
**** THERMAL ****  DISPLAYED  12/11/80 11:54 AM

     ∇  BREED;C;COL;CULTURE;EML;NN;PROD;RANDOM
[1]   ⍝TS: 12/9/80  JCL
[2]   ⍝DS: BREED CYCLE
[3]   ⍝GU: ACTUAL,CELLM,CELLS,CWID
[4]   ⍝GR: BOTCUT,INOC,LAMDA,MM,N,NOPRINT,SP,TOPCUT
[5]    CELLS←⍳0
[6]    →(FIXED≠DIST)/B0
[7]   ⍝ FIXED DISTRIBUTION
[8]    CELLS←NρLAMDA
[9]    →B4
[10] B0:EML←*-LAMDA
[11] B1:CULTURE←¯1
```

```
[12]  PROD←1
[13] B2:RANDOM←GENERATE
[14]  →(DIST≠TRUNCATED)/B3
[15] ⍝ IF THIS RANDOM OUTSIDE TRUNCATION POINTS, GET ANOTHER
[16]  →((RANDOM>TOPCUT)∨RANDOM<BOTCUT)/B2
[17] B3:CULTURE←CULTURE+1
[18] ⍝ DECREASE PROD
[19]  PROD←PROD×RANDOM
[20] ⍝ ARE WE DOWN TO EML YET?
[21]  →(PROD>EML)/B2
[22] ⍝ BREED ANOTHER CELL, IF NOT DONE
[23]  →(N>⍴CELLS←CELLS,CULTURE)/B1
[24] B4:CELLM←(MM,N)⍴0
[25]  →(B5,B6,B7)[INOC]
[26] B5:CELLM[MM;]←CELLS
[27]  →B8
[28] B6:CELLM[1;]←CELLS
[29]  →B8
[30] B7:COL←C←1
[31] B7A:CELLM[NN;COL]←1+CELLM[NN←1+MM-+/SP≥GENERATE;COL]
[32]  →(CELLS[COL]≥C←C+1)/B7A
[33]  →(N≥COL←COL+C←1)/B7A
[34] ⍝ CWID IS STANDARD FORMAT VALUES
[35] B8:CWID←2↑2+⌊10⍟1.1⌈⌈/CELLS
[36]  →NOPRINT/0
[37]  'INITIAL CELLS:'
[38]  '     ',CWID⍕CELLS
[39]  (4+N×1↑CWID)⍴'-'
[40]  (3 0 ⍕(MM,1)⍴⍳MM),'|',CWID⍕CELLM
[41]  'AVERAGE INOCULANT COUNT: ',⍕ACTUAL←(+/CELLS)÷N
    ∇

    ∇ RANDOM←GENERATE
[1] ⍝TS: 12/9/80  JCL
[2] ⍝DS: GENERATE A RANDOM NUMBER FROM 0 TO 1
[3] ⍝RT: RANDOM ↔ THE RANDOM NUMBER
[4] ⍝GR: TWOBILLS ↔ ¯1+2*31
[5]  RANDOM←(?TWOBILLS)÷TWOBILLS
    ∇

    ∇ DESTROY;CAN;COUNT;CURRENT;SUBCAN;TIMEPROBE;WID
[1]  ⍝TS: 12/18/80  JCL
[2]  ⍝DS: DESTROY CYCLE
[3]  ⍝GU: BHAT,BHATS,CELLM,CELLS,EST,FHAT,FHATS,RATIO,TIMES
[4]  ⍝GR: B,COMCOOL,COOL,CR,CWID,DIST,LAMDA,MM,NOPRINT
[5]  ⍝GR: POISSON,TRUNCATED
[6]  ⍝ COMPUTE CURRENT DESTRUCTION PROBABILITY
[7]  B1:TIMEPROBE←1-*-B×TIME+COMCOOL-COOL
[8]   →NOPRINT/B2
[9]  B2:CAN←1
[10] B2A:SUBCAN←1
[11] B2B:→(~×CURRENT←COUNT←CELLM[SUBCAN;CAN])/B4
[12] ⍝ TEST ONE
[13] B3:CURRENT←CURRENT-TIMEPROBE[SUBCAN]≥GENERATE
[14] ⍝ MORE THIS CELL?
[15]  →(×COUNT←COUNT-1)/B3
[16]  CELLM[SUBCAN;CAN]←CURRENT
[17] ⍝ UPDATE SUBCAN
[18] B4:→(MM≥SUBCAN←SUBCAN+1)/B2B
[19]  →(N≥CAN←CAN+1)/B2A
[20]  CELLS←+⌿CELLM
[21]  BHAT←FHAT←¯1
[22]  EST←' -'
[23]  →(DIST=TRUNCATED)/B8
[24] ⍝ COMPUTE RATIO FOR HAT VALUES
[25]  RATIO←(+/0=CELLS)÷⍴CELLS
[26]  →(DIST=POISSON)/B5
[27]  FHAT←RATIO*÷LAMDA
[28]  →B7
[29] ⍝ NO ESTIMATE IF TRUE
[30] B5:→(RATIO<*-LAMDA)/B7
```

```
[31]  FHAT←1+(⍟RATIO)÷LAMDA
[32]  →((RATIO∊ 0 1)∨A≠1)/B6
[33]  BHAT←(-÷TIME)×⍟(-÷LAMDA)×⍟RATIO
[34] B6:EST←⍕FHAT
[35] ⍝ TIMES, FHATS, AND BHATS ARE USED IN SUMMARY AND PLOT
[36] B7:TIMES←TIMES,TIME
[37]  FHATS←FHATS,FHAT
[38]  BHATS←BHATS,BHAT
[39] B8:ALLCELLS←ALLCELLS,(+/0≠CELLS)÷ρCELLS
[40]  ALLLIVE←ALLLIVE,+/0≠CELLS
[41]  →NOPRINT/0
[42]  →((INOC≠3)∨∧/0=CELLS)/B8A
[43]  CR,'SURVIVING CELLS:',(WID≥12)/((0⌈¯12+WID←N×1↑
CWID)ρ' '),'   DESTRUCTION'
[44]  '     ',(CWID⍕CELLS),'   PROBABILITY'
[45]  (4+N×1↑CWID)ρ'-'
[46]  (3 0 ⍕(MM,1)ρ⍳MM),'|',(((N×2)ρCWID), 14 10)⍕CELLM
,TIMEPROBE
[47]  →B8B
[48] B8A:CR,' DESTRUCTION'
[49]  ' PROBABILITY'
[50]  12 10 ⍕((ρTIMEPROBE),1)ρTIMEPROBE
[51]  ''
[52] B8B:'LIVE/N = ',(⍕+/0≠CELLS),'/',(⍕ρCELLS),'  FHAT
:  ',EST
[53]  'AVERAGE RESIDUAL COUNT: ',⍕(+/CELLS)÷ρCELLS
[54]  →(BHAT=¯1)/B9
[55]  'BHAT: ',⍕BHAT
[56] B9:2ρCR
    ∇


    ∇ SUMMARY;A;MAT
[1] ⍝TS: 12/9/80  JCL
[2] ⍝DS: PRINT OUT A SUMMARY OF THIS RUN
[3] ⍝GR: ALLCELLS,ALLLIVE,ALLTIMES,LAMDA,N
[4]  A←ρALLTIMES
[5]  MAT←(AρN),(AρLAMDA),ALLTIMES,ALLLIVE,[1.5]
ALLCELLS
[6]  '      N  LAMDA     T     LIVE  LIVE/N'
[7]  6 0 6 0 8 2 6 0 8 3 ⍕MAT
    ∇

    ∇ CPU;⎕IO;A
[1] ⍝TS: 12/8/80  JCL
[2] ⍝DS: PRINT OUT ACCOUNTING INFORMATION
[3] ⍝GU: CPU
[4]  ⎕IO←1
[5]  →(0≠⎕NC 'CPU')/B1
[6]  CPU←⎕AI[2]
[7]  →0
[8] B1:A←0.001×⎕AI[2]-CPU
[9]  (⍕A),' CPU SECONDS'
    ∇
```

THERE ARE 17 VARIABLES IN THIS WORKSPACE:

CR	I	M	O	SW	T
CL	CR	FIXED	IJD	IJM	IJX
MAXCOUNT	PASS	POISSON	TRUNCATED	TWOBILLS	

```
CR          [LITERAL SCALAR] ρ=
  (CARRIAGE RETURN)                     INPUT CR

I           [ BINARY SCALAR] ρ=         INPUT GLOBAL
1

M           [INTEGER SCALAR] ρ=           "       "
¯1
```

O 0	[*BINARY SCALAR*] ρ=	" "
SW 1	[*INTEGER SCALAR*] ρ=	" "
T 2	[*INTEGER SCALAR*] ρ=	" "
CL ←	[*LITERAL SCALAR*] ρ=	" "
CR →	[*LITERAL SCALAR*] ρ=	" "
FIXED 1	[*BINARY SCALAR*] ρ=	DISTRIBUTION TYPE VALUE ARBITRARY
IJD /	[*LITERAL VECTOR*] ρ=1	INPUT GLOBAL (DELIMITER)
IJM	[*LITERAL VECTOR*] ρ=0	INPUT GLOBAL (BUFFER)
IJX ↔	[*LITERAL VECTOR*] ρ=3	INPUT GLOBAL (CONTINUATION)
MAXCOUNT 30	[*INTEGER SCALAR*] ρ=	AUSTERITY LIMIT
PASS *HIHO*	[*LITERAL VECTOR*] ρ=4	PASWORD TO CONTINUE BYPASSING AUSTERITY
POISSON 2	[*INTEGER VECTOR*] ρ=1	DISTRIBUTION TYPE
TRUNCATED 3	[*INTEGER SCALAR*] ρ=	DISTRIBUTION TYPE
TWOBILLS 2147483647	[8-*BYTE SCALAR*] ρ=	RANDOM NUMBER BASE/DIVISOR

REFERENCES

Barlow, R.E., Bartholomew, D.J., Bremner, J.M., and Brunk, H.D. (1972). Statistical Inference Under Order Restrictions, John Wiley & Sons, Inc., New York.

Ball, C.O. and Olson, F.C.W. (1957). Sterilization Food Technology, McGraw-Hill Book Company, New York.

Charm, S.E. (1971). The Fundamentals of Food Engineering, AVI Publishing Company, Westport, Connecticut.

Cornell, R.G. and Speckman, J.A. (1967). Estimation for a simple exponential. Biometrics, 23, 717-37.

Dudewicz, E.J. and Ralley, T.G. (1981). The Handbook of Random Number Generation and Testing with TEST-RAND Computer Code. American Sciences Press, Inc., Columbus, Ohio.

Esty, J.R. and Meyer, K.F. (1922). The head resistance of the spores of B. Botuluius and allied anaerobes. Journal of Infectious Diseases, 31, 650.

Hachigian, J. (1978). Computer simulation of partial spoilage data. Journal of Food Sciences, 43, 1741-1748.

Koch, G.G. and Tolley, H.D. (1975). A generalized modified-χ^2 analysis of categorical bacteria survival data from a complex dilution experiment. Biometrics, 31, 59-92.

Potter, N.N. (1973). Food Science. AVI Publishing Company, Westport, Connecticut.

Stumbo, C. (1966). Thermobacteriology. Academic Press, New York.

Received 4/7/80; Revised 3/1/82.

Recent Simulations

Acoustics

Docherty, R. and Corlett, E.N.: "A Laboratory and Simulation Comparison of the Effectiveness of Open Entry Rooms for Protection from High Noise-Levels," *Applied Acoustics*, Vol. 16, 1983, No. 6, pages 409 ff.

Singh, R. and Kung, C.H.: "Simulating the Response to a Sonic-Boom," *Simulation* (Simulation Councils, Inc.), Vol. 39, 1982, No. 2, pages 43 ff.

Aeronautics

Clements, R.R., Wright, J.H., and Tonkin, S.W.: "Digital-Simulation of Counterspun Nutation Damper Operation under Non-Ideal Operating Conditions," *Aeronautical Journal*, Vol. 87, 1983, No. 867, pages 277 ff.

Air Service, Airport Access, and Future Technology, Washington, D.C.: National Academy of Sciences, 1981, iv + 52 pages.

Harrison, J.M.: "A Multiple-Aircraft Radar Simulation," *Simulation* (Simulation Councils, Inc.), Vol. 40, 1983, No. 3, pages 77 ff.

Stirling, R.: "Simulation of a Digital Aircraft Flight Control-System," *Simulation* (Simulation Councils, Inc.), Vol. 40, 1983, No. 5, pages 171 ff.

Agriculture, Food, and Nutrition

Crawford, E.W. and Milligan, R.A.: "A Multi-Year, Stochastic, Farm Simulation-Model for Northern Nigeria—An Experimental-Design Approach," *American Journal of Agricultural Economics*, Vol. 64, 1982, No. 4, pages 728 ff.

Freeman, R. and Hay, R.: "Food Shortage Simulations—An Application of the Food Accounting Matrix with Proportionality Assumptions," *Ecology of Food and Nutrition*, Vol. 12, 1982, No. 3, pages 155 ff.

Kornher, A. and Torssell, B.W.: "Simulation of Weather × Management Interactions in Temporary Grasslands in Sweden," *Swedish Journal of Agricultural Research,* Vol. 13, 1983, No. 3, pages 145 ff.

Kumar, R. and Goss, J.R.: "Computer-Simulation of Harvesting Alfalfa Seed," *Transactions of the American Society of Agricultural Engineers* (ASAE), Vol. 24, 1981, No. 5, pages 1135 ff.

Russell, N.P., Milligan, R.A., and Ladue, E.L.: "A Stochastic Simulation-Model for Evaluating Forage Machinery Performance," *Agricultural Systems*, Vol. 10, 1983, No. 1, pages 39 ff.

Air Quality

Gross, M.: "Urban Simulation-Models and Air-Quality Control—Overview of Major Issues," *Resource Management & Optimization*, Vol. 2, 1983, No. 4, pages 371 ff.

Turner, D.B. and Irwin, J.S.: "Extreme Value Statistics Related to Performance of a Standard Air-Quality Simulation-Model Using Data at 7 Power-Plant Sites," *Atmospheric Environment*, Vol. 16, 1982, No. 8, pages 1907 ff.

Astronomy and Astrophysics

Kornilov, V.G. and Lipunov, V.M.: "Neutron Stars in Massive Binary-Systems. 2. Computer-Simulation" (in Russian), *Astronomicheskii Zhurnal*, Vol. 60, 1983, No. 3, pages 574 ff.

Macgilli, H.T. and Dodd, R.J.: "Monte-Carlo Simulations of Galaxy Systems. 4. Static Properties for Galaxies in Supercluster Cells," *Astrophysics and Space Science*, Vol. 86, 1982, No. 2, pages 419 ff.

Macgilli, H.T. and Dodd, R.J.: "Monte-Carlo Simulations of Galaxy Systems. 5. Computer-Models of Galaxy Fields," *Astrophysics and Space Science*, Vol. 86, 1982, No. 2, pages 437 ff.

Sobolev, A.M. and Strelnit, V.S.: "Monte-Carlo Simulation of Methanol Masers," *Soviet Astronomy Letters*, Vol. 9, 1983, No. 1, pages 12 ff.

Automata

Menzel W., and Sperschn, V.: "Universal Automata with Uniform Bounds on Simulation Time," *Information And Control*, Vol. 52, 1982, No. 1, pages 19 ff.

Ballistics, Military Applications

Brennan K., Hess, K., and Iafrate, G.J.: "Monte-Carlo Simulation of Reflecting Contact-Behavior on Ballistic Device Speed," *IEEE Electron Device Letters* (was *Elec. Dev.* 6), Vol. 4, 1983, No. 9, pages 332 ff.

Bryant, R.M.: "Discrete System Simulation in ADA," *Simulation* (Simulation Councils, Inc.), Vol. 39, 1982, No. 4, pages 111 ff.

Elphick, A.: "Perfecting Aerial Combat Systems with Simulation Testing," *Simulation* (Simulation Councils, Inc.), Vol. 39, 1982, No. 2, pages 70 ff.

Lee, W.C.Y. and Smith, H.L.: "A Computer-Simulation Model for the Evaluation of Mobile Radio Systems in the

Military Tactical Environment," *IEEE Transactions on Vehicular Technology*, Vol. 32, 1983, No. 2, pages 177 ff.

Liberg, D.A. and Morgan, G.B.: "Computer-Simulation of the Use of Barker Phase Codes for the Detection of Overlapping Targets in Clutter," *IEEE Proceedings-F* (continuation of *IEEE Reviews*), Vol. 130, 1983, No. 6, pages 557 ff.

Morawski, P.: "A Computer-Simulation of Launch Procedures for a Squadron of Minuteman-II ICBMs," *Simulation* (Simulation Councils, Inc.), Vol. 41, 1983, No. 1, pages 16 ff.

Sisle, M.E., and McCarthy, E.D.: "Hardware-in-the-Loop Simulation for an Active Missile," *Simulation* (Simulation Councils, Inc.), Vol. 39, 1982, No. 5, pages 159 ff.

Biology

Busico, V. and Vacatell, M.: "Lipid Bilayers in the Fluid State—Computer-Simulation and Comparison with Model Compounds," *Molecular Crystals and Liquid Crystals*, Vol. 97, 1983, Nos. 1–4, pages 195 ff.

Harte, C. and Lindenma, A.: "Mitotic Index in Growing Cell-Populations—Mathematical-Models and Computer-Simulations," *Biologisches Zentralblatt*, Vol. 102, 1983, No. 5, pages 509 ff.

Groner, G.F. and Clark, R.L.: *BIOMOD: An Interactive Graphics System for Analysis Through Simulation*. Santa Monica, Calif.: Rand Corp., 1971, 11 pages

Lookman, T., Pink, D.A., Grundke, E.W., Zuckerman, M.J., and Dev: "Phase-Separation in Lipid Bilayers Containing Integral Proteins—Computer-Simulation Studies," *Biochemistry*, Vol. 21, 1982, No. 22, pages 5593 ff.

Lopezsae, J.F., Calvo, A., Cruz, J.L., Gutierre, C., and Carmo: "Cell-Proliferation in File Meristems—A General-Theory and Its Analysis by Computer-Simulation," *Environmental and Experimental Botany*, Vol. 23, 1983, No. 1, pages 59 ff.

Lucas, P.W. and Luke, D.A.: "Computer-Simulation of the Breakdown of Carrot Particles during Human Mastication," *Archives of Oral Biology*, Vol. 28, 1983, No. 9, pages 821 ff.

Matela, R.J., Ransom, R., and Bowles, M.A.: "Computer-Simulation of Compartment Maintenance in the *Drosophila* Wing Imaginal Disk," *Journal of Theoretical Biology*, Vol. 103, 1983, No. 3, pages 357 ff.

Munson, P.J.: "Experimental Artifacts and the Analysis of Ligand-Binding Data—Results of a Computer-Simulation," *Journal of Receptor Research*, Vol. 3, 1983, Nos. 1–2, pages 249 ff.

Saroff, H.A. and Kutyna, F.A.: "The Uniqueness of Protein Sequences—A Monte-Carlo Analysis," *Bulletin of Mathematical Biology*, Vol. 43, 1981, No. 6, pages 619 ff.

Brownian Motion

Ciccotti, G., Ferrario, M., and Ryckaert, J.P.: "Computer-Simulation of the Generalized Brownian-Motion. 2. An Argon Particle in Argon Fluid," *Molecular Physics*, Vol. 46, 1982, No. 4, pages 875 ff.

Nowakows, B. and Sitarski, M.: "Brownian Coagulation of Aerosol-Particles by Monte-Carlo Simulation," *Journal of Colloid and Interface Science*, Vol. 83, 1981, No. 2, pages 614 ff.

Bus Systems

Maze, T.H., Jackson, G.C., and Dutta, U.: "Bus Maintenance Planning with Computer-Simulation," *J. Transp. E.*, Vol. 109, 1983, No. 3, pages 389 ff.

Tsugawa, S. and Shigeta, K.: "Simulation of the Demand Bus System for Tsukuba Science City. 1.," (in Japanese), *Journal of Mechanical Engineering Laboratory*, Vol. 36, 1982, No. 5, pages 216 ff.

Chemical Engineering

Botev, T., Boyadjie, C., and Stateva, R.: "On Simulation of Chemical-Engineering Systems Having Controlled Output Variables," *Hungarian Journal of Industrial Chemistry*, Vol. 10, 1982, No. 2, pages 175 ff.

Dohnal, M. and Ulmanova, M.: "Simulation of Chemical Processes with Uncertain Parameters," *Collection Of Czechoslovak Chemical Communications*, Vol. 48, 1983, No. 6, pages 1588 ff.

Felder, R.M.: "Simulation—A Tool for Optimizing Batch-Process Production," *Chemical Engineering*, Vol. 90, 1983, No. 8, pages 79 ff.

Felder, R.M., Kester, P.M., and McConney, J.M.: "Simulation Optimization of a Specialties Plant," *Chemical Engineering Progress*, Vol. 79, 1983, No. 6, pages 84 ff.

Kurinny, A.I.: "Digital-Simulation of a Complex Chemical-Engineering Process Flow Diagram" (in Russian), *Avtomatika*, Vol. 1982, 1982, No. 6, pages 29 ff.

Chemistry

Adams, D.J.: "On the Use of the Ewald Summation in Computer-Simulation," *Journal of Chemical Physics*, Vol. 78, 1983, No. 5, pages 2585 ff.

Anacker, E.W., Westwell, A.E., and Anacker, D.C.: "Computer-Simulation of the Brice-Phoenix Differential Refractometer," *Journal of Colloid and Interface Science*, Vol. 90, 1982, No. 2, pages 352 ff.

Brender, C. and Lax, M.: "A Monte-Carlo Off-Lattice Method—The Slithering Snake in a Continuum," *Journal of Chemical Physics*, Vol. 79, 1983, No. 5, pages 2423 ff.

Duben, A.J. and Bush, C.A.: "Monte-Carlo Calculations on the Conformations of Models for the Glycopeptide Linkage of Glycoproteins," *Archives of Biochemistry and Biophysics*, Vol. 225, 1983, No. 1, pages 1 ff.

Evans, G.J. and Evans, M.W.: "Computer-Simulation of Some Structural and Spectral Properties of Liquid and Rotator Phase Tert-Butyl Chloride," *Journal of the Chemical Society. Faraday Transactions II*, Vol. 79, 1983, No. 6, pages 767 ff.

Garland, G.E. and Dufty, J.W.: "Computer-Simulation of the Velocity Auto-Correlation Function at Low-Density," *International Journal of Quantum Chemistry*, Vol. 1982, 1982, No. 16, pages 91 ff.

Goodnick, S.E., Porod, W., Grondin, R.O., and Goodnick, S.M.: "A Monte-Carlo Study of SI(111) Surface Oxidation," *Journal of Vacuum Science & Technology Processing & Phenomena*, Vol. 1, 1983, No. 3, pages 767 ff.

Le S.Y. and Jiang, S.P.: "Monte-Carlo Simulation of the Nucleic-Acid Base (Cytosine) Hydration," *Kexue Tong*, Vol. 27, 1982, No. 11, pages 1218 ff.

Meirovit, H.: "Computer-Simulation of Self-Avoiding Walks—Testing the Scanning Method," *Journal of Chemical Physics*, Vol. 79, 1983, No. 1, pages 502 ff.

Mouritse, O.G., Boothroy, A., Harris, R., Jan, N., and Lookma: "Computer-Simulation of the Main Gel Fluid Phase–Transition of Lipid Bilayers," *Journal of Chemical Physics*, Vol. 79, 1983, No. 4, pages 2027 ff.

Paskal, Y.I. and Domrache, V.Y.: "Statistical Simulation of Isomorphous Precipitation" (in Russian), *Fizika Metallov I Metallovedenie*, Vol. 55, 1983, No. 6, pages 1051 ff.

Simkin, V.Y. and Sheikhet, I.I.: "Characteristic Features of Calculations of the Structure of Solutions by the Monte-Carlo Method," *Journal of Structural Chemistry*, Vol. 24, 1983, No. 1, pages 65 ff.

Sitarski, M. and Nowakows, B.: "Condensation Kinetics of Trace Lennard-Jones Vapor on Sub-Micron Aerosols by Monte-Carlo Simulation," *Polish Journal of Chemistry* (see *Roczniki Chemii*), Vol. 55, 1981, No. 5, pages 1075 ff.

Stoop, L.C.A.: "A Monte-Carlo Calculation of Lateral Interactions in Silver Monolayers on W(110)," *Thin Solid Films*, Vol. 103, 1983, No. 4, pages 375 ff.

Termonia, Y.: "Fluctuations and Transitions at Chemical Instabilities. 1. A Computer-Simulation Approach," *Journal of Chemical Physics*, Vol. 79, 1983, No. 8, pages 3778 ff.

Ciphers

Gabriel, R.: "Enciphering Mappings with Pseudoinverses, Random Number Generators and Partitions" (in German), *Kybernetika*, Vol. 18, 1982, No. 6, pages 485 ff.

Climatology, Meteorology, and Solar Energy

Howells, P.B. and Marshall, R.H.: "An Improved Computer Code for the Simulation of Solar Heating-Systems," *Solar Energy*, Vol. 30, 1983, No. 2, pages 99 ff.

Larsen, G.A. and Pense, R.B.: "Stochastic Simulation of Daily Climatic Data for Agronomic Models," *Agronomy Journal*, Vol. 74, 1982, No. 3, pages 510 ff.

Livezey, R.E. and Chen, W.Y.: "Statistical Field Significance and Its Determination by Monte-Carlo Techniques," *Monthly Weather Review*, Vol. 111, 1983, No. 1, pages 46 ff.

Rachner, M.: "An Attempt to Estimate the Depth of a Snow Cover Using Simulation-Models" (in German), *Zeitschrift für Meteorologie*, Vol. 33, 1983, No. 4, pages 234 ff.

Small, M.J. and Samson, P.J.: "Stochastic Simulation of Atmospheric Trajectories," *J. Clim. Appl.* Vol. 22, 1983, No. 2, pages 266 ff.

Terjung, W.H. and Orourke, P.A.: "Winter Energy Budget Simulation for a Latitudinal Transect Along the East Coast of the America," *Archives for Meteorology, Geophysics & Bioclimatology, Series B-Climatology Env.*, Vol. 32, 1983, Nos. 2–3, pages 187 ff.

Communications

Ahamed, S.V.: "Simulation and Design Studies of Digital Subscriber Lines," *Bell System Technical Journal*, Vol. 61, 1982, No. 6, pages 1003 ff.

Baets, W.: "Model for the Simulation of the Internal Telephone Traffic Based on Segmentation of the Market " (in Dutch), *Cahiers Economiques de Bruxelles*, Vol. 1982, 1982, No. 94, pages 283 ff.

Fillmer, J.L. and Mellicha, J.M.: "Simulation—An Operational Planning Device for the Bell System," *Interfaces*, Vol. 12, 1982, No. 3, pages 54 ff.

Gomtsyan, O.A. and Rizkin, I.K.: "The Estimation of Signal and Noise Digital-Simulation Error in Radio Systems" (in Russian), *Radiotekhnika I Elektronika*, Vol. 27, 1982, No. 7, pages 1428 ff.

Lam, S.S. and Lien, Y.C.L.: "Congestion Control of Packet Communication Networks by Input Buffer Limits—A Simulation Study," *IEEE Transactions on Computers*, Vol. 30, 1981, No. 10, pages 733 ff.

McDonnel, M. and Georgana, N.D.: "Simulation Study of a Signaling Protocol for Mobile Radio Systems," *IEEE Proceedings-F* (continuation of *IEE Reviews*), Vol. 129, 1982, No. 6, pages 459 ff.

Warfield, R.E.: "Bayesian-Analysis of Teletraffic Measurements and Simulation Results," *Australian Telecommunication Research*, Vol. 15, 1981, No. 2, pages 43 ff.

Computer Devices

Akers, L.A. and Wang, T.M.: "Simulation of a 3-Dimensional Semiconductor-Device," *Simulation* (Simulation Councils, Inc.), Vol. 40, 1983, No. 2, pages 43 ff.

Bowes, S.R. and Clements, R.R.: "Digital-Computer Simulation of Variable-Speed PWM Inverter-Machine Drives," *IEE Proceedings-B* (continuation of *IEE Reviews*), Vol. 130, 1983, No. 3, pages 149 ff.

Groner, G.F. and Clark, R.L. *An Interactive Computer Graphics System for Simulating Dynamic Systems*, Santa Monica, Calif., Rand Corp., 1973, v + 14 pages.

Loveluck, J.M. and Balcar, E.: "Computer-Simulation Calculations for One-Dimensional Magnetic Systems, with Some Remarks on Optimization on the Cray Computer," *Computer Physics Communications*, Vol. 27, 1982, No. 4, pages 335 ff.

Oh, S.Y.: "MOS Device and Process Design Using Computer-Simulations," *Hewlett-Packard Journal*, Vol. 33, 1982, No. 10, pages 288 ff.

Computer Networks

Chlamtac, I. and Franta, W.R.: "A Generalized Simulator for Computer-Networks," *Simulation* (Simulation Councils, Inc.), Vol. 39, 1982, No. 4, pages 123 ff.

Gomaa, H.: "The Design and Calibration of a Simulation-Model of a Star Computer Network ," *Software: Practice and Experience*, Vol. 12, 1982, No. 7, pages 599 ff.

Ramshaw, L.A. and Amer, P.D.: "Generating Artificial Traffic over a Local Area Network Using Random Number Generators," *Computer Networks*, Vol. 7, 1983, No. 4, pages 233 ff.

Schoemaker, S.: *Computer Networks and Simulation II*, Amsterdam: North-Holland Pub. Co., 1982, xv + 326 pages.

Unger, B.W. and Bidulock, D.S.: "The Design and Simulation of a Multicomputer Network Message Processor," *Computer Networks*, Vol. 6, 1982, No. 4, pges 263 ff.

Correlation in Simulation

Parisi, G.: "Correlation-Functions and Computer-Simulations. 2.," *Nuclear Physics, Part B*, Vol. 205, 1982, No. 3, pages 337 ff.

Crystallography

Doster, J.M. and Gardner, R.P.: "The Complete Spectral Response for EDXRF Systems—Calculation by Monte-Carlo and Analysis Applications. 2. Heterogeneous Samples," *X-Ray Spectrometry*, Vol. 11, 1982, No. 4, pages 181 ff.

Kim, K.M., Kran, A., Smetana, P., and Schwuttk, G.H.: "Computer-Simulation and Controlled Growth of Large Diameter Czochaalski Silicon-Crystals," *Journal of the Electrochemical Society*, Vol. 130, 1983, No. 5, pages 1156 ff.

Luckhurs, G.R., Romano, S., and Simpson, P.: "Computer-Simulation Studies of Anisotropic Systems. 9. The Maier-Saupe Theory for Nematic Liquid-Crystals and the Molecular-Field," *Chemical Physics*, Vol. 73, 1982, No. 3, pages 337 ff.

Rikvold, P.A.: "Simulations of a Stochastic-Model for Cluster Growth on a Square Lattice," *Physical Review A. General Physics*, Vol. 26, 1982, No. 1, pges 647 ff.

Yamamoto, T.: "An Application of Monte-Carlo Method to Simulate Disorders in Polymer Crystals," *Polymer*, Vol. 24, 1983, No. 8, pages 943 ff.

Dosimetry

Beck, J.W., Dunn, W.L., and Ofoghlud, F.: "A Monte-Carlo Model for Absorbed Dose Calculations in Computed-Tomography," *Medical Physics*, Vol. 10, 1983, No. 3, pages 314 ff.

Kulkarni, R.N.: "Monte-Carlo Calculation of the Dose Distributions across a Plane Bone-Marrow Interface During Diagnostic-X-Ray Examinations," *British Journal of Radiology*, Vol. 54, 1981, No. 646, pages 875 ff.

Tory, E.M., Jodrey, W.S., and Pickard, D.K.: "Simulation of Random Sequential Absorption—Efficient Methods and Resolution of Conflicting Results," *Journal of Theoretical Biology*, Vol. 102, 1983, No. 3, pages 439 ff.

Williams, J.F., Morin, R.L., and Khan, F.M.: "Monte-Carlo Evaluation of the Sievert Integral for Brachytherapy Dosimetry," *Physics in Medicine and Biology*, Vol. 28, 1983, No. 9, pages 1021 ff.

Electronics

Batz, O. and Fischer, C.P.: "Optimization of Transponders by Computer-Simulation" (in German), *NTZ Archiv*, Vol. 5, 1983, No. 6, pages 171 ff.

Edward, L.N.M.: "LINSIM, a Linear Electrical Network Simulation and Optimization Program," *International Journal of Electrical Engineering Education*, Vol. 20, 1983, No. 2, pages 151 ff.

Moltgen, G.: "Digital-Simulation of a Current-Source Inverter with Interphase Commutation" (in German), *Siemens Forschungs-Und Entwicklungs Berichie/Search and Development Reports*, Vol. 12, 1983, No. 3, pages 166 ff.

Morris, A.S.: "Simple-Model Order Reduction by Truncation with Performance Verification by Fast Digital-Simulation," *Electronics Letters*, Vol. 19, 1983, No. 9, pages 343 ff.

Reller, H., Kirowaei, E., and Gileadi, E.: "Esembles of Microelectrodes—A Digital-Simulation," *Journal of Electroanalytical Chemistry and Interfacial Electrochemistry*, Vol. 138, 1982, No. 1, pages 65 ff.

Energy

Bergendahl, P.A. and Bergstrom, C.: *Long Term Energy Options for Sweden: The IEA Model and Some Simulation Results*. Stockholm: Energy Research and Development Commission (DFE), 1981, 261 pages.

Corum, K.R. and VanDyke, J.: "Stimulating Energy-Conservation in Commercial Buildings—A Simulation Analysis of Alternative Policies," *Energy Policy*, Vol. 11, 1983, No. 1, pages 52 ff.

Thomas, M.A.: *An Energy Crisis Management Simulation for the State of California*. Santa Monica, Calif.: Rand Corp., 1982, xvii + 60 pages.

Entropy

Meirovit, H.: "Methods for Estimating Entropy with Computer-Simulation—The Simple Cubic Ising Lattice," *Journal of Physics, Mathematical and General*, Vol. 16, 1983, No. 4, pages 839 ff.

Fermentation

Lee, J.M., Pollard, J.F., and Coulman, G.A.: "Ethanol Fermentation with Cell Recycling—Computer-Simulation," *Biotechnology and Bioengineering*, Vol. 25, 1983, No. 2, pages 497 ff.

Finance

Andersen, D.F.: "A System Dynamics Simulation of Educational Finance Policies," *Simulation* (Simulation Councils, Inc.), Vol. 40, 1983, No. 6, pages 227 ff.

Cassidy, H.J.: "Monte-Carlo Simulation Estimates of the Expected Value of the Due-on-Sale Clause in Home Mortgages," *Hous. Finan.*, Vol. 2, 1983, No. 1, pages 33 ff.

Hawkins, C.A. and Leggett, D.N.: "The Investment Tax Credit in Capital Replacement—A Simulation," *Journal of Accounting & Public Policy*, Vol. 2, 1983, No. 3, pages 167 ff.

Tugcu, S.: "A Simulation Study on the Determination of the Best Investment Plan for Istanbul Seaport," *Journal of the Operational Research Society* (cont. *Operational Research Quarterly*), Vol. 34, 1983, No. 6, pages 479 ff.

Witt, U. and Perske, J.: "SMS—A Program Package for Simulation and Gaming of Stochastic Market Processes and Learning-Behavior," *Lect. in Econ.*, Vol. 202, 1982, pages 1 ff.

Fire Science

Fahy, R.: "Building Fire Simulation-Model," *Fire J.*, Vol. 77, 1983, No. 4, pages 93 ff.

Hansenta, E. and Baunan, T.: "Fire Risk Assessment by Simulation—FIRESIM," *Fire Safety Journal* (was *Fire Research*), Vol. 5, 1983, Nos. 3–4, pages 205 ff.

Nakaya, I. and Akita, K.: "A Simulation-Model for Compartment Fires," *Fire Safety Journal* (was *Fire Research*), Vol. 5, 1983, No. 2, pages 157 ff.

Fisheries

Jovellan, C.L. and Gaskin, D.E.: "Predicting the Movements of Juvenile Atlantic Herring (Clupea-Harengus-Harengus) in the S.W. Bay of Fundy Using Computer-Simulation Techniques," *Canadian Journal of Fisheries & Aquatic Sciences* (continuation of *Journal of The Fisheries Research*), Vol. 40, 1983, No. 2, pages 139 ff.

Kleinstr, C. and Logan, B.E.: "Generalized Computer-Simulation Model for the Impact Assessment of Industrial Water-Use on Fish Populations," *Water Science & Technology* (continuation of *Progress in Water Technology*), Vol. 13, 1981, No. 7, pages 363 ff.

Kuipers, H.: "The Role of a Game-Simulation in a Project of Change—A Case for Deep-Sea Fishing," *Simulation and Games*, Vol. 14, 1983, No. 3, pages 275 ff.

Forestry

Allen, T.F.H. and Shugart, H.H.: "Ordination of Simulated Complex Forest Succession—A New Test of Ordination Methods," *Vegetatio; Acta Geobotanica*, Vol. 51, 1983, No. 3, pages 141 ff.

Barber, R.L.: "HARVEST, an Interactive Harvest Scheduling Simulator for Even-Aged Forests," *Journal of Forestry*, Vol. 81, 1983, No. 8, pages 532 ff.

Cropper, W.P. and Ewel, K.C.: "Computer-Simulation of Long-Term Carbon Storage Patterns in Florida Slash Pine Plantations," *Forest Ecology*, Vol. 6, 1983, No. 2, pages 101 ff.

Hines, G.S., Liu, C.Y., Webster, D.B., and Sirois, D.L.: "Features of a Skidding Module in a Timber Harvesting Computer-Simulation Model," *Forest Products Journal*, Vol. 33, 1983, Nos. 7–8, pages 11 ff.

Miller, P.C., Kendall, R., and Oechel, W.C.: "Simulating Carbon Accumulation in Northern Ecosystems," *Simulation* (Simulation Councils, Inc.), Vol. 40, 1983, No. 4, pages 119 ff.

Schmidt, J.S. and Tedder, P.L.: "A Comprehensive Examination of Economic Harvest Optimization Simulation Methods," *Forest Science*, Vol. 27, 1981, No. 3, pages 523 ff.

Webster, D.B., Padgett, M.L., and Sirois, D.L.: "Features of a Felling Module Using a Feller-Buncher in a Timber Harvesting Computer-Simulation Model," *Forest Products Journal*, Vol. 33, 1983, No. 6, pages 11 ff.

West, P.W.: "Simulation of Diameter Growth and Mortality in Regrowth Eucalypt Forest of Southern Tasmania," *Forest Science*, Vol. 27, 1981, No. 3, pages 603 ff.

Gaming

Ellington, H., Addinall, E., and Percival, F.: *Games and Simulations in Science Education*. London: Kogan Page, 1981, 216 pages.

Perspectives on Academic Gaming & Simulation. London: Kogan Page, 1978–1981, 6 V. in 5.

Ellington, H., Adinall, E., and Percival, F.: "Games and Simulations Teach Social Relevance of Science," *Impact of Science On Society*, Vol. 32, 1982, No. 4, pages 481 ff.

Greenblat, C.S., Duke, R.D., and Greenblat: *Principles and Practices of Gaming-Simulation*. Beverly Hills: Sage Publications, 1981, 283 pages.

Johnson, W.B. and Rouse, W.B.: "Training Maintenance Technicians for Troubleshooting—2 Experiments with Computer-Simulations," *Human Factors*, Vol. 24, 1982, No. 3, pages 271 ff.

Ruben, B.D. and Lederman, L.C.: "Instructional Simulation Gaming—Validity, Reliability, and Utility," *Simulation and Games*, Vol. 13, 1982, No. 2, pages 233 ff.

Yefimov, V.M. and Komarov, V.F.: "Developing Management Simulation Games," *Simulation and Games*, Vol. 13, 1982, No. 2, pages 145 ff.

Health Systems

Friedman, B.A., Abbott, R.D., and Williams, G.W.: "A Blood Ordering Strategy for Hospital Blood Banks Derived from a Computer-Simulation," *American Journal of Clinical Pathology*, Vol. 78, 1982, No. 2, pages 154 ff.

Lambo, E.: "An Optimization-Simulation Model of a Rural Health-Center in Nigeria," *Interfaces*, Vol. 13, 1983, No. 3, pages 29 ff.

Herd Management

Allen, M.A. and Stewart, T.S.: "A Simulation-Model for a Swine Breeding Unit Producing Feeder Pigs," *Agricultural Systems*, Vol. 10, 1983, No. 4, pages 193 ff.

Bennett, G.L., Tess, W.M., Dickerso, G.E., and Johnson, R.K.: "Simulation of Breed and Crossbreeding Effects on Costs of Pork Production," *Journal of Animal Science*, Vol. 56, 1983, No. 4, pages 801 ff.

Davis, M.E. and Brinks, J.S.: "Selection and Concurrent Inbreeding in Simulated Beef Herds," *Journal of Animal Science*, Vol. 56, 1983, No. 1, pages 40 ff.

Holography

Bianco, B., Beltrame, F., and Chiabrer, A.: "Computer-Simulation of 3D Imaging in Holography—A Preliminary-Study for Automated Holographic Microscopy," *Proceedings of the Society of Photo-Optical Instrumentation Engineers*, Vol. 375, 1982, pages 32 ff.

Information Theory

Bucy, R.S., Moura, J.M.F., and Mallinck, A.J.: "A Monte-Carlo Study of Absolute Phase Determination," *IEEE Transactions on Information Theory*, Vol. 29, 1983, No. 4, pages 509 ff.

Insurance

Feldman, R.D. and Dowd, B.E.: "Simulation of a Health-Insurance Market with Adverse Selection," *Operations Research*, Vol. 30, 1982, No. 6, pages 1027 ff.

Inventory Management and Policies

Gaither, N.: "Using Computer-Simulation to Develop Optimal Inventory Policies," *Simulation* (Simulation Councils, Inc.), Vol. 39, 1982, No. 3, pages 81 ff.

Pope, J.A.: "An Inventory Management Simulation," *Journal of Experiential Learning & Simulation*, Vol. 3, 1981, Nos. 3–4, pages 261 ff.

Irrigation

Kundu, S.S., Skogerbo, G.V., and Walker, W.R.: "Using a Crop Growth Simulation-Model for Evaluating Irrigation Practices," *Agricultural Water Management*, Vol. 5, 1982, No. 3, pages 253 ff.

Job Shops, Queueing

Browne, J. and Davies, B.J.: "A Preliminary Review of Batch Sizes in a Job Shop Using a Digital-Simulation Model," *Simulation* (Simulation Councils, Inc.), Vol. 41, 1983, No. 4, pages 149 ff.

Gupta, S.M.: "A Simulation-Model for Automated Planning and Optimization of Machining Conditions for Multi-Station Synchronous Machines," *Journal of Engineering Sciences*, Vol. 8, 1982, No. 1, pages 21 ff.

Korovyan, E.I.: "Studying the Productivity of Automatic Lines by Computer-Simulation," *Soviet Engineering Research*, Vol. 1, 1981, No. 4, pages 60 ff.

Lavenber, S.S., Moeller, T.L., and Welch, P.D.: "Statistical Results on Control Variables with Application to Queueing Network Simulation," *Operations Research*, Vol. 30, 1982, No. 1, pages 182 ff.

Ovuworie, G.C.: "Some Simulation Results of a Model of an Unreliable Series Production System," *International Journal of Production Research*, Vol. 20, 1982, No. 5, pages 619 ff.

Shedler, G.S. and Southard, J.: "Regenerative-Simulation of Networks of Queues with General Service Times—Passage through Sub-Networks," *IBM Journal of Research and Development*, Vol. 26, 1982, No. 5, pages 625 ff.

Maintenance

Barry, D.M. and Garty, R.J.F.: "A Motorway Lighting System Maintenance Model Based on Simulation," *Reliability Engineering*, Vol. 6, 1983, No. 2, pages 63 ff.

Management, Planning, and Decision-Making

Binkley, J.K. and McKinzie, L.: "Alternative Methods of Estimating Export Demand—A Monte-Carlo Comparison," *Can. J. Ag. Ec.*, Vol. 29, 1981, No. 2, pages 187 ff.

Burns, A.C. and Sherrell, D.L.: "A Microcomputer Simulation for Teaching Retail Location Strategy," *Journal of Experiential Learning & Simulation*, Vol. 3, 1981, Nos. 3–4, pages 239 ff.

Cooley, J.W. and Cooley, B.J.: "Internal Accounting Control-Systems—A Simulation Program for Assessing Their Reliabilities," *Simulation and Games*, Vol: 13, 1982, No. 2, pages 211 ff.

Dawson, W., Lakshmin, S., Landry, A.A., and McLeod, J.B.: "The Acceptance of a Simulation-Model for Planning Decisions at the St.-Lawrence-Seaway-Authority," *Infor: Canadian Journal of Operational Research and Information Processing*, Vol. 20, 1982, No. 1, pages 16 ff.

Gray, L. and Waitt, I.: *Simulation in Management & Business Education: The Proceedings of the 1981 Conference Of SAGSET, The Society For Academic Gaming and Simulation*. London, England: Kogan Page, 1982, 192 pages.

Jaffe, J.M.: "Decentralized Simulation of Resource Managers," *Journal of the Association for Computing Machinery*, Vol. 30, 1983, No. 2, pages 300 ff.

Miles, R.H., and Randolph, W.A.: *The Organization Game: A Simulation in Organizational Behavior, Design, Change, and Development*. Glenview, Illinois: Scott, Foresman, 1981, x + 325 pages.

Naylor, T.H. and Mann, M.H.: *Computer Based Planning Systems*. Oxford, Ohio: Planning Executives Institute, 1982, 199 pages.

Picardi, A.C. and Shorb, A.M.: "Simulating Saudi-Arabia Oil Export Strategy," *Simulation* (Simulation Councils, Inc.), Vol. 40, 1983, No. 1, pages 20 ff.

Pritsker, A.A.B. and Sigal, C.E.: *Management Decision Making: A Network Simulation Approach*. Englewood Cliffs, N.J.: Prentice-Hall, 1983, xxi + 521 pages.

Shelden, R.A.: "Computer-Simulation and the Optimal Management of Non-Renewable Resources—Taxation, Profit Maximization, and Negotiation," *Journal of Environmental Management*, Vol. 15, 1982, No. 2, pages 131 ff.

Sherman, M.D.: "MANAGE—An Interactive Simulation-Model for Evaluating the Effects of Management Structure on Organization Performance," *Simulation* (Simulation Councils, Inc.), Vol. 41, 1983, No. 2, pages 49 ff.

Sule, D.R., Tolbert, D., and Gomez, G.: "Simulation in Decision-Making—A Case-Study," *Computers and Industrial Engineering*, Vol. 5, 1981, No. 4, pages 227 ff.

Taylor, B.W., Keown, A.J., and Barrett, R.T.: "Analyzing Court System Congestion with Q-GERT Network Modeling and Simulation," *Computer and Operations Research*, Vol. 9, 1982, No. 3, pages 163 ff.

Theil, H., Finke, R., and Rosalsky, M.C.: "Verifying a Demand System by Simulation," *Economics Letters*, Vol. 13, 1983, No. 1, pages 15 ff.

Warren, A., Leigh, M., and Black, P.: "Simulation Testing Zips Datacomm Products to Market," *Computer Design*, Vol. 22, 1983, No. 4, pages 169 ff.

Manufacturing, Production, and Distribution Systems

Garciadi, A., Hogg, G.L., and Phillips, D.T.: "Combined Simulation and Network Optimization Analysis of a Production Distribution-System," *Simulation* (Simulation Councils, Inc.), Vol. 40, 1983, No. 2, pages 59 ff.

Klitz, J.K.: "Simulation of an Automated Logistics and Manufacturing System," *European Journal of Operational Research*, Vol. 14, 1983, No. 1, pages 36 ff.

Lenz, J.E.: "MAST—A Simulation Tool for Designing Computerized Metalworking Factories," *Simulation* (Simulation Councils, Inc.), Vol. 40, 1983, No. 2, pages 51 ff.

Mohanty, R.P. and Chandras, V.: "Computer-Simulation Study for a Production Distribution-System," *International Journal of Physical Distribution & Materials Management*, Vol. 13, 1983, No. 3, pages 51 ff.

Newman, K.W.: "A Monte-Carlo Approach to Linkage Clustering," *Taxon: Official News Bulletin of the International Association For Plant Technology*, Vol. 31, 1982, No. 4, pages 662 ff.

Ohmi, T. and Kimura, Y.: "Effects of System Factors in Flexible Manufacturing. 1. Model and Simulation Program" (in Japanese), *Journal of Mechanical Engineering Laboratory*, Vol. 37, 1983, No. 2, pages 87 ff.

Rushton, J.D.: "Management Benefits from Process Simulation—Meaningful, Complete Information," *Tappi J.*, Vol. 65, 1982, No. 10, pages 53 ff.

Schonber, R.J.: "Why Projects Are Always Late—A Rationale Based on Manual Simulation of a PERT-CPM Network," *Interfaces*, Vol. 11, 1981, No. 5, pages 66 ff.

Shah, S.A., Okos, M.R., and Reklaiti, G.V.: "Simulation Modeling of a Sausage Manufacturing Plant," *Transactions of the American Society of Agricultural Engineers (ASAE)*, Vol. 26, 1983, No. 2, pages 635 ff.

Talavage, J.: "Simulating Manufacturing Systems," *IEEE Spectrum*, Vol. 20, 1983, No. 5, pages 53 ff.

Taylor, B.W., and Russell, R.S.: "A Simulation Approach for Adapting a Production Line Balancing Procedure to a Probabilistic Environment," *International Journal of Production Research*, Vol. 20, 1982, No. 6, pages 787 ff.

Wemmerlo, U.: "Statistical Aspects of the Evaluation of Lot-Sizing Techniques by the Use of Simulation," *International Journal of Production Research*, Vol. 20, 1982, No. 4, pages 461 ff.

Medical Curriculum

Verhagen, P.: "Computer-Simulation and Mathematical-Models in the Medical Curriculum—MIN, FBM," *EC&TJ-Educational Communication & Technology Journal*, Vol. 31, 1983, No. 1, pages 56 ff.

Medicine

Bert, J.L. and Pinder, K.L.: "Analog Simulation of the Human Micro-Vascular Exchange System," *Simulation* (Simulation Councils, Inc.), Vol. 39, 1982, No. 3, pages 89 ff.

Kasmia, A.-H.: *Computer Simulation of the Cardiovascular System Mean Circulatory Filling Pressure*. 101 pages.

Rattanar, S., Grossman, M., Fernando, R.L., and Shanks, R.D.: "A Monte-Carlo Comparison of Estimators of Average Daily Gain in Body-Weight," *Journal of Animal Science*, Vol. 57, 1983, No. 4, pages 885 ff.

Sorahan, T. and Waterhou, J.A.: "The Generation of Simulated Data in the Form of Industrial Cohorts," *Computers and Biomedical Research*, Vol. 16, 1983, No. 3, pages 260 ff.

Vaughan, C.L.: "Simulation of a Sprinter. 2. Implementation on a Programmable Calculator," *International Journal of Bio-Medical Computing*, Vol. 14, 1983, No. 1, pages 75 ff.

Microcomputers

Dunlop, D.L. and Sigmund, T.F.: *Problem-Solving with the Programmable Calculator: Puzzles, Games & Simulations with Math & Science Applications*. Englewood Cliffs, N.J.: Prentice-Hall, 1983, xi + 227 pages.

Hollocks, B.: "Simulation and the Micro," *Journal of the Operational Research Society* (continuation of *Operational Research Quarterly*), Vol. 34, 1983, No. 4, pages 331 ff.

Leventhal, L.A.: *Modeling and Simulation on Microcomputers*. La Jolla, Calif. (P.O. Box 2228, La Jolla 92038): Society for Computer Simulation, 1982, ix + 119 pages.

Pimentel, J.R.: "Real-Time Simulation Using Multiple Microcomputers," *Simulation* (Simulation Councils, Inc.), Vol. 40, 1983, No. 3, pages 93 ff.

Migration

Stoddard, P.K., Marsden, J.E., and Williams, T.C.: "Computer-Simulation of Autumnal Bird Migration over the Western North-Atlantic," *Animal Behaviour*, Vol. 31, 1983, No. EB, pages 173 ff.

Mining

Adams, T.F.: "Computer-Simulation of Explosive Fracture of Oil-Shale," *American Chemical Society Symposium Series*, Vol. 1981, 1981, No. 163, pages 13 ff.

Malmqvis, K. and Malmqvis, L.: "Simulation as a Tool for Planning of Exploration for Deep Seated Base-Metal Deposits," *European Journal of Operational Research*, Vol. 14, 1983, No. 2, page 163.

Modeling

A Directory of Computer Software Applications: Administration and Management. 1970–July 1978, Springfield, Va. (5285 Port Royal Road, Springfield 22161): U.S. Dept. Of Commerce, vii + 98 pages.

Adequate Modeling of Systems: Proceedings of the International Working Conference On Model Realism, held in Bad Honnef, Federal Republic of Germany. Berlin: Springer-Verlag, 1983, xi + 335 pages.

Alpaugh, G.L.: "Simulation Modeling—The System Dynamics Way," *Simulation* (Simulation Councils, Inc.), Vol. 41, 1983, No. 4, pages 155 ff.

Armstrong, J.R.: "Chip Level Modeling and Simulation," *Simulation* (Simulation Councils, Inc.), Vol. 41, 1983, No. 4, pages 141 ff.

Johnson, T.G.: "The Use of Simulation Techniques in Dynamic Input-Output Modeling," *Simulation*, (Simulation Councils, Inc.), Vol. 41, 1983, No. 3, pages 93 ff.

Landwehr, C.E.: "Applying Software Engineering to Protocol Simulation," *Simulation* (Simulation Councils, Inc.), Vol. 37, 1981, No. 5, pages 157 ff.

Levendel, Y.H., Menon, P.R., and Patel, S.H.: "Special-Purpose Computer for Logic Simulation Using Distributed-Processing," *Bell System Technical Journal*, Vol. 61, 1982, No. 10, pages 2873 ff.

Lotwick, H.W.: "Simulation of Some Spatial Hard-Core Models, and the Complete Packing Problem," *J. Stat. Comp.*, Vol. 15, 1982, No. 4, pages 295 ff.

Sheppard, S.: "Applying Software Engineering to Simulation," *Simulation* (Simulation Councils, Inc.), Vol. 40, 1983, No. 1, pages 13 ff.

Spriet, J.A. and Vansteenkiste, G.C.: *Computer-Aided Modelling and Simulation*. London: Academic Press, 1982, x + 490 pages.

Molecular Science

Decoster, D., Constant, E., and Constant, M.: "Computer-Simulation of Molecular-Dynamics of Anisotropic Fluids," *Molecular Crystals and Liquid Crystals*, Vol. 97, 1983, Nos. 1–4, pages 263 ff.

Evans, M.W.: "Computer-Simulation of Some Field-Induced Phenomena of Molecular Liquids," *Journal of Molecular Liquids*, Vol. 26, 1983, No. 1, pages 49 ff.

Mattice, W.L.: "Subchain Expansion in Generator Matrix and Monte-Carlo Treatments of Simple Chains with Excluded Volume," *Macromolecules*, Vol. 14, 1981, No. 5, pages 1491 ff.

Meirovit, H.: "Improved Computer-Simulation Method for Estimating the Entropy of Macromolecules with Hard-Core Potential," *Macromolecules*, Vol. 16, 1983, No. 10, pages 1628 ff.

Pletneva, S.G. and Khalatur, P.G.: "Study by the Monte-Carlo Method of the Liquid Equilibrium Structure, Formed by Chain Molecules" (in Russian), *Zhurnal Fizicheskoi Khimii*, Vol. 57, 1983, No. 9, pages 2236 ff.

Monte Carlo Methods

Abramov, A.A.: "The Calculation of Macroscopic Values in the Monte-Carlo Direct Statistical Simulation Method" (in Russian), *Akademiia Nauk SSSR. Doklady*, Vol. 271, 1983, No. 2, pages 315 ff.

Bell, R.C.: "Monte-Carlo Debugging—A Brief Tutorial," *Communications Of The ACM*, Vol. 26, 1983, No. 2, pages 126 ff.

Blomquis, R.N. and Gelbard, E.M.: "An Assessment of Existing Klein-Nishina Monte-Carlo Sampling Methods," *Nuclear Science And Engineering*, Vol. 83, 1983, No. 3, pages 380 ff.

Dubi, A., Elperin, T., and Rief, H.: "On Confidence-Limits and Statistical Convergence of Monte-Carlo Point-Flux Estimators with Unbounded Variance," *Annals of Nuclear Energy* (formerly *Annals of Nuclear Science and Engineering*), Vol. 9, 1982, No. 11, pages 675 ff.

Finney, J.L.: "Monte-Carlo Techniques," *Biochemical Society Transactions*, Vol. 10, 1982, No. 5, pages 305 ff.

Goldman, S.: "A Simple New Way to Help Speed up Monte-Carlo Convergence-Rates—Energy-Scaled Displacement Monte-Carlo," *Journal of Chemical Physics*, Vol. 79, 1983, No. 8, pages 3938 ff.

Gupta, H.C.: "A Class of Zero-Variance Biasing Schemes for Monte-Carlo Reaction-Rate Estimators," *Nuclear Science and Engineering*, Vol. 83, 1983, No. 2, pages 187 ff.

Johnson, M.E., Ramberg, J.S., and Wang, C.: "The Johnson Translation System in Monte-Carlo Studies," *Communications in Statistics Part B—Simulation & Computation*, Vol. 11, 1982, No. 5, pages 521 ff.

Larsen, U.: "The Time Scale of Monte-Carlo Experiments," *Physics Letters A*, Vol. 97, 1983, No. 4, pages 147 ff.

Mikhailo, G.A.: "Uniform Optimization of the Monte-Carlo Weight Methods (a Minimax Approach)" (in Russian), *Akademiia Nauk SSSR. Doklady*, Vol. 270, 1983, No. 5, pages 1054 ff.

Murthy, K.P.N.: "Monte-Carlo Methods—An Introduction," *Lect. N. Phys.*, Vol. 184, 1983, pages 116 ff.

Pearson, R.B., Richards, J.L., and Toussain, D.: "A Fast Processor for Monte-Carlo Simulation," *Journal of Computational Physics*, Vol. 51, 1983, No. 2, pages 241 ff.

Ragheb, M.M.H.: "Optimal Multistage Sequential Monte-Carlo," *Atomkernenergie-Kerntechnik* (cont. *Atomkernenergie & Kerntechnik*), Vol. 43, 1983, No. 2, pages 127 ff.

Suri, R.: "Implementation of Sensitivity Calculations on a Monte-Carlo Experiment," *Journal of Optimization Theory and Applications*, Vol. 40, 1983, No. 4, pages 625 ff.

Windeban, E.: "A Monte-Carlo Simulation Method versus a General Analytical Method for Determining Reliability-Measures of Repairable Systems," *Reliability Engineering*, Vol. 5, 1983, No. 2, pages 73 ff.

Monte Carlo Methods—Applications

Brender, C., Benavrah, D., and Havlin, S.: "Study of Monte-Carlo Methods for Generating Self-Avoiding Walks," *Journal of Statistical Physics*, Vol. 31, 1983, No. 3, pages 661 ff.

Seriozhi, V.G.: "Solution of a Non-Linear Equation by the Monte-Carlo Method" (in Russian), *Leningrad Universitet, Vestnik, Seriya Matematiki, Mekhaniki I Astronomii*, Vol. 1983, 1983, No. 3, pages 104 ff.

Stewart, L.: "Bayesian-Analysis Using Monte-Carlo Integration—A Powerful Methodology for Handling Some Difficult Problems," *The Statistician*, Vol. 32, 1983, Nos. 1–2, pages 195 ff.

Yang, Y.S., Howell, J.R., and Klein, D.E.: "Radiative Heat-Transfer through a Randomly Packed-Bed of Spheres by the Monte-Carlo Method," *Journal of Heat Transfer*, Vol. 105, 1983, No. 2, pages 325 ff.

Monte Carlo Methods—Variance Reduction

Arvidsen, N.I., and Johnsson, T.: "Variance Reduction through Negative Correlation, a Simulation Study," *J. Stat. Comp.*, Vol. 15, 1982, Nos. 2–3, pages 119 ff.

Dwivedi, S.R.: "A New Importance Biasing Scheme for Deep-Penetration Monte-Carlo," *Annals of Nuclear Energy* (Formerly *Annals of Nuclear Science and Engineering*), Vol. 9, 1982, No. 7, pages 359 ff.

Lux, I.: "A Handy Method for Approximate Optimization of Splitting in Monte-Carlo," *Nuclear Science and Engineering*, Vol. 83, 1983, No. 2, pages 198 ff.

Lux, I.: "On Zero Variance Monte-Carlo Path-Stretching Schemes," *Nuclear Science and Engineering*, Vol. 84, 1983, No. 4, pages 388 ff.

Tadikamalla, P.R.: *Modern Digital Simulation Methodology: Input, Modeling, and Output*. Columbus, Ohio: American Sciences Press, Inc., 1984, ii + 208 pages.

Natural Resource Planning

Lonergan, S.C.: "A Simulation Optimization Model for Natural-Resource Planning—The Chesapeake Bay Experience," *Resource Management & Optimization*, Vol. 2, 1983, No. 4, pages 293 ff.

Navigation

Davis, P.V., Dove, M.J., and Stockel, C.T.: "A Computer-Simulation of Multi-Ship Encounters," *Journal of Navigation*, Vol. 35, 1982, No. 2, pages 347 ff.

Nuclear Physics

Caldi, D.G.: "How Well Do Monte-Carlo Simulations Distinguish between U(1) and SU(2)—A Study of the String Tension in U(1) Lattice Gauge-Theory," *Nuclear Physics, Part B*, Vol. 220, 1983, No. 1, pages 48 ff.

Hashikur, H., Fukumoto, H., Oka, Y., Akiyama, M., and An, S.: "Neutron Streaming through a Slit and Duct in Concrete Shields and Comparison with a Monte-Carlo Analysis," *Nuclear Science and Engineering*, Vol. 84, 1983, No. 4, pages 337 ff.

Kristiak, J., Tomek, S., and Hnatowic, V.: "Timing Characteristics of Pulses from GE(LI) Detector and Their Simulation by a Monte-Carlo Method," *Nuclear Instruments & Methods in Physics Research* (was *Nucl. Instr.*), Vol. 211, 1983, Nos. 2–3, pages 413 ff.

Latta, B.M.: "Comparison of Blaugrund and Monte-Carlo Centroid-Shift Calculations," *Nuclear Instruments & Methods in Physics Research* (was *Nucl. Instr.*), Vol. 211, 1983, Nos. 2–3, pages 447 ff.

Lillie, R.A., White, T.L., Gabriel, T.A., and Alsmille, R.G.: "Microwave Transport in EBT Distribution Manifolds Using Monte-Carlo Ray Tracing Techniques," *Nuclear Technology-Fusion*, Vol. 4, 1983, No. 2, pages 1436 ff.

Noma, H., Avignone, F.T., Moltz, D.M., and Toth, K.S.: "A Monte-Carlo Study of the Response of a Germanium Detector to Electrons and Positrons," *Nuclear Instruments & Methods in Physics Research* (was *Nucl. Instr.*), Vol. 211, 1983, Nos. 2–3, pages 391.

Radi, H.M.A., Rasmusse, J.O., Donangel, R., Canto, L.F., and Oli: "Monte-Carlo Studies of Alpha-Accompanied Fission," *Physical Review C. Nuclear Physics*, Vol. 26, 1982, No. 5, pages 2049 ff.

Reiter, D. and Nicolai, A.: "Monte-Carlo Simulation of the Neutral Gas-Density and Temperature Distribution Due to the Recycling Processes at a Poloidal, Toroida," *Journal of Nuclear Materials*, Vol. 111, 1982, pages 434 ff.

Sanders, L.G.: "The Application of Monte-Carlo Computations to Formation Analysis by Neutron Interactions," *International Journal of Applied Radiation and Isotopes*, Vol. 34, 1983, No. 1, pages 173 ff.

Santoro, R.T. and Barnes, J.M.: "Monte-Carlo Calculations of Neutron and Gamma-Ray Energy-Spectra for Fusion-Reactor Shield Design—Comparison with Experiment," *Nuclear Technology-Fusion*, Vol. 4, 1983, No. 2, pages 367 ff.

Ueki, K., Inoue, M., and Maki, Y.: "Validity of the Monte-Carlo Method for Shielding Analysis of a Spent-Fuel Shipping Cask—Comparison with Experiment," *Nuclear Science and Engineering*, Vol. 84, 1983, No. 3, pages 271 ff.

Ueki, K. and Ryufuku, H.: "Analysis of Shielding Problems with Monte-Carlo Methods" (in Japanese), *Journal of the Atomic Energy Society of Japan*, Vol. 23, 1981, No. 9, pages 632 ff.

Zaider, M., Brenner, D.J., and Wilson, W.E.: "The Applications of Track Calculations to Radiobiology. 1. Monte-Carlo Simulation of Proton Tracks," *Radiation Research 1*, Vol. 95, 1983, No. 2, pages 231 ff.

Optics

Beck, J.W., Jaszczak, R.J., and Starmer, C.F.: "Analysis of Spect Using Monte-Carlo Simulation," *Proceedings of the Society of Photo-Optical Instrumentation Engineers (1979)*, Vol. 372, 1982, pages 32 ff.

Bruscagl, P., Ismaelli, A., Zaccanti, G., and Pantani, L.: "Modified Monte-Carlo Method to Evaluate Multiple-Scattering Effects on Lightbeam Transmission through a Turbid Atmosphere," *Proceedings of the Society of Photo-Optical Instrumentation Engineers (1979)*, Vol. 369, 1983, pages 164 ff.

Carrasco, M., Unamuno, S., and Agullolo, F.: "Monte-Carlo Simulation of the Performance of PMMA Luminescent Solar Collectors," *Applied Optics*, Vol. 22, 1983, No. 20, pages 3236 ff.

Hilmes, G.L., Harris, H.H., and Riehl, J.P.: "Time-Dependent Concentration Depolarization—A Monte-Carlo Study," *Journal of Luminescence*, Vol. 28, 1983, No. 2, pages 135 ff.

Ohtsubo, J.: "Non-Gaussian Speckle—A Computer-Simulation," *Applied Optics*, Vol. 21, 1982, No. 22, pages 4167 ff.

Tang, T.T.: "Monte-Carlo Simulation of Discrete Space-Charge Effects in Electron-Beam Lithography Systems," *Optik*, Vol. 64, 1983, No. 3, pages 237 ff.

Tomiyasu, K.: "Computer-Simulation of Speckle in a Synthetic Aperture Radar Image Pixel," *IEEE Transactions on Geoscience & Remote Sensing*, Vol. 21, 1983, No. 3, pages 357 ff.

Optometry

Nussenblatt, H.: "Computer-Simulation of an Optometric Practice," *American Journal of Optometry and Physiological Optics*, Vol. 60, 1983, No. 9, pages 754 ff.

Paper and Pulp; Textiles

Abbott, R.D., Edwards, L.L., and Boyle, T.J.: "Economic-Benefits from Paper-Machine Simulation," *Appita*, Vol. 36, 1983, No. 5, pages 377 ff.

Arthur, D.F. and Nield, R.: "Computer-Simulation of the Deposition of Particles in the Rotor Groove During Open-End Spinning. 1. The Development of a Model," *Journal of the Textile Institute*, Vol. 73, 1982, No. 5, pages 201 ff.

Bouchard, D.C., Roche, A.A., and Nobleza, G.C.: "Process Simulation in an Industrial-Research Context," *Pulp and Paper Magazine of Canada*, Vol. 83, 1982, No. 12, pages 49 ff.

Mahig, J.: "Improvements to Multivariable Paper Plant Simulations Obtained through Observer-Based Estimators," *I.S.A. Transactions*, Vol. 21, 1982, No. 4, pages 57 ff.

Peters, N.: "A User's Perception of Computer Process Simulation in the Pulp and Paper-Industry," *Pulp and Paper Magazine of Canada*, Vol. 83, 1982, No. 12, pages 66 ff.

Rounsley, R.R.: "Discrete Event Simulation of a Broke System," *Tappi Journal*, Vol. 66, 1983, No. 9, pages 69 ff.

Shewchuk, C.F.: "Process Simulation Technology for the Pulp and Paper-Industry," *Pulp and Paper Magazine of Canada*, Vol. 83, 1982, No. 12, pages 60 ff.

Strauss, R.W. and Rushton, J.D.: "Process Simulation—

Why Do It—A Corporate Viewpoint," *Tappi Journal*, Vol. 66, 1983, No. 3, pages 67 ff.

Parasitology

Smith, R.D.: "Babesia-Bovis—Computer-Simulation of the Relationship between the Tick Vector, Parasite, and Bovine Host," *Experimental Parasitology*, Vol. 56, 1983, No. 1, pages 27 ff.

Pharmacokinetics

Binder, H.: "Monte-Carlo Simulation of One-Component and 2-Component Systems—Physical-Properties of Model Membranes," *Studia Biophysica*, Vol. 93, 1983, No. 3, pages 217 ff.

Laskarze, P.M., Weiner, D.L., and Ott, L.: "A Simulation Study of Parameter-Estimation in the One and 2 Compartment Models," *Journal of Pharmacokinetics and Biopharmaceutics*, Vol. 10, 1982, No. 3, pages 317 ff.

Sundqvis, T., Tagesson, C., and Magnusso, K.E.: "Simulation of a Multicompartment Model for the Intestinal Permeability to Low-Molecular-Weight Probes (Polyethyleneglycol-400)," *Mathematical Biosciences*, Vol. 56, 1981, Nos. 3–4, pages 287 ff.

Photographic Science

Xia, P.J.: "A Computer-Simulation of Latent Image-Formation in Photographic-Emulsion by a Monte-Carlo Method," *Journal of Photographic Science*, Vol. 30, 1982, No. 5, pages 142 ff.

Physics

Aleksand, I.V. and Kaybyshe, D.A.: "Computer-Simulation of Texture Formation During Plastical Deformation in Metals with FCC and HCP Lattices" (in Russian), *Fizika Metallov I Metallovedenie*, Vol. 54, 1982, No. 4, pages 818 ff.

Anisimov, S.I., Berezovs, M.A., Ivanov, M.F., and Petrov, I.V.: "Computer-Simulation of the Langmuir Collapse," *Physics Letters A*, Vol. 92, 1982, No. 1, pages 32 ff.

Azetsu, A.: "The Analysis of Performance of Stirling Engines. 1. Computer-Simulation Model," *Japan Society of Mechanical Engineers. Bulletin of the JSME*, Vol. 25, 1982, No. 210, pages 1953 ff.

Bhanot, G., Heller, U.M., and Stamates, I.O.: "A New Method for Fermion Monte-Carlo," *Physics Letters B*, Vol. 129, 1983, No. 6, pages 440 ff.

Boissona, J., Barreau, F., and Carmona, F.: "The Percolation of Fibers with Random Orientations—A Monte-Carlo Study," *Journal of Physics, Mathematical and General*, Vol. 16, 1983, No. 12, pages 2777 ff.

Brant, E.H.: "Computer-Simulation of Vortex Pinning in Type-II Superconductors. 2. Random Point Pins," *Journal of Low Temperature Physics*, Vol. 53, 1983, Nos. 1–2, pages 71 ff.

Bucy, R.S., Moura, J.M.F., and Mallinck, A.J.: "A Monte-Carlo Study of Absolute Phase Determination," *IEEE Transactions on Information Theory*, Vol. 29, 1983, No. 4, pages 509 ff.

Ceperley, D.: "The Simulation of Quantum-Systems with Random-Walks—A New Algorithm for Charged Systems," *Journal of Computational Physics*, Vol. 51, 1983, No. 3, pages 404 ff.

Deraedt, H. and Lagendij, A.: "Monte-Carlo Methods for Quantum-Lattice Models," *Helvetica Physica Acta*, Vol. 56, 1983, Nos. 1–3, pages 593 ff.

Eguchi, T. and Fukugita, M.: "Monte-Carlo Simulation of Quantum String Theory," *Physics Letters B*, Vol. 117, 1982, Nos. 3–4, pages 223 ff.

Farantos, S.C.: "Evaluation of an Upper Bound of the Maximal Lyapunov Characteristic Number by Monte-Carlo Integration in the Chaotic Regions of Phase-Space," *Chemical Physics*, Vol. 71, 1982, No. 2, pages 157 ff.

Felstein, J.: "Monte-Carlo Study of the Effect of Multiple-Scattering on Electron-Energy Loss in Aluminum," *Journal Of Physics: F. Metal Physics*, Vol. 13, 1983, No. 11, pages 2229 ff.

Filipo, W.A.S. and Hohmann, G.W.: "Computer-Simulation of Low-Frequency Electromagnetic Data Acquisition," *Geophysics*, Vol. 48, 1983, No. 9, pages 1219 ff.

Fujiki, S., Shutoh, K., Abe, Y., and Katsura, S.: "Monte-Carlo Simulation of the Anti-Ferromagnetic Ising-Model on the Triangular Lattice with the 1st and 2nd Neighbor Interactions," *Journal of the Physical Society of Japan and Supplement*, Vol. 52, 1983, No. 5, pages 1531 ff.

Gerlach, P., Prandl, W., and Lefebvre, J.: "The Plastic State of C2CL6—A Comparison of a Monte-Carlo Simulation of the Molecular-Distribution with Single-Crystal Neutron Data," *Molecular Physics*, Vol. 49, 1983, No. 4, pages 991 ff.

Gharadje, F.: "Computer-Simulation of a Nematic Guest-Host Display," *Journal of Applied Physics*, Vol. 54, 1983, No. 9, pages 4989 ff.

Hafner, J. and Punz, G.: "Static Lattice-Distortions in Substitutional Alloys—A Computer-Simulation," *Journal of Physics: F. Metal Physics*, Vol. 13, 1983, No. 7, pages 1393 ff.

Hale, B.N. and Ward, R.C.: "A Monte-Carlo Method for Approximating Critical Cluster Size in the Nucleation of Model Systems," *Journal of Statistical Physics*, Vol. 28, 1982, No. 3, pages 487 ff.

Harris, R. and Lewis, L.J.: "Chemical Short-Range Order in Computer-Simulated Metallic Glasses," *Journal of Physics: F. Metal Physics*, Vol. 13, 1983, No. 7, pages 1359 ff.

Hoogland, A., Spaa, J., Selman, B., and Compagne, A.: "A Special-Purpose Processor for the Monte-Carlo Simulation of Ising Spin Systems," *Journal of Computational Physics,* Vol. 51, 1983, No. 2, pages 250 ff.

Jacucci, G.: "Linear and Non-Linear Response in Computer-Simulation Experiments," *Physica A*, Vol. 118, 1983, Nos. 1–3, pages 157 ff.

Jacucci, G. and Omerti, E.: "Monte-Carlo Calculation of the Radial-Distribution Function of Quantum Hard-Spheres at Finite Temperatures Using Path-Integrals," *Journal of Chemical Physics*, Vol. 79, 1983, No. 6, pages 3051 ff.

Kang, N.K. and Swanson, L.W.: "Computer-Simulation of Liquid-Metal Ion-Source Optics," *Applied Physics A-Solids & Surfaces* (continuation of *Appl. Phy.*, see also *Appl. Phys. B*), Vol. 30, 1983, No. 2, pages 95 ff.

Koch, S.W. and Abraham, F.F.: "Freezing Transition of Xenon on Graphite—A Computer-Simulation Study," *Physical Review: B. Solid State*, Vol. 27, 1983, No. 5, pages 2964 ff.

Koiwa, M. and Ishioka, S.: "Random-Walks and Correlation Factor in Diffusion in a 3-Dimensional Lattice with Coordination Number-8," *Philosophical Magazine A-Electronic, Optical & Magnetic Properties* (cont. *Philos. M.*), Vol. 48, 1983, No. 1, pages 1 ff.

Koonin, S.E., Sugiyama, G., and Friedric, H.: "Mean-Field Monte-Carlo Method for Many-Body Ground-States," *Lect. N. Phys.*, Vol. 171, 1982, pages 214 ff.

Kushner, M.J.: "Monte-Carlo Simulation of Electron Properties in RF Parallel Plate Capacitively Coupled Discharges," *Journal of Applied Physics*, Vol. 54, 1983, No. 9, pages 4958 ff.

Li, P. and Strieder, W.: "Monte-Carlo Simulation of the Conductivity of the 2-Dimensional Triangular Site Network," *Journal of Physics: C. Solid State Physics and Supplement*, Vol. 15, 1982, No. 32, pages 6591 ff.

Markzon, G.: *The Monte-Carlo Method in Three-Dimensional Problems of Elasticity*. 34 pages.

Matsumot, H. and Omura, Y.: "Computer-Simulation Studies of VLF Triggered Emissions Deformation of Distribution Function by Trapping and Detrapping," *Geophysical Research Letters,* Vol. 10, 1983, No. 8, pages 607 ff.

Meirovit, H.: "An Approximate Stochastic-Process for Computer-Simulation of the Ising-Model at Equilibrium," *Journal of Physics, Mathematical and General*, Vol. 15, 1982, No. 7, pages 2063 ff.

Mezei, M.: "Virial-Bias Monte-Carlo Methods Efficient Sampling in the (T,P,N) Ensemble," *Molecular Physics*, Vol. 48, 1983, No. 5, pages 1075 ff.

Murch, G.E.: "Computer-Simulation of the Vacancy-Wind Effect in a Concentrated Alloy with Long-Range and Short-Range Order," *Philosophical Magazine A-Electronic, Optical & Magnetic Properties* (continuation of *Philos. M.*), Vol. 46, 1982, No. 4, pages 575 ff.

Muto, K. and Horie, H.: "Monte-Carlo Calculation of Gamow-Teller Transition Strength Function," *Physics Letters B*, Vol. 127, 1983, No. 5, pages 291 ff.

Mutter, K.H. and Schillin, K.: "Glueball Mass from Variant Actions in Lattice Monte-Carlo Simulations," *Physics Letters B*, Vol. 121, 1983, No. 4, pages 267 ff.

Noack, K.: "Variance Analysis of the Monte-Carlo Perturbation Source Method in Inhomogeneous Linear Particle-Transport Problems. 2.," *Kernenergie*, Vol. 26, 1983, No. 7, pages 282 ff.

Ogielski, A.T.: "Monte-Carlo Study of Scale-Covariant Field-Theories," *Physical Review D. Particles and Fields*, Vol. 28, 1983, No. 6, pages 1461 ff.

Ohuchi, M. and Kubota, T.: "Monte-Carlo Simulation of Electronics in the Cathode Region of the Glow-Discharge in Helium," *Journal of Physics: D. Applied Physics*, Vol. 16, 1983, No. 9, pages 1705 ff.

Okeefe, M.A., Fryer, J.R., and Smith, D.J.: "Image Interpretation for Aromatic-Hydrocarbons by Computer-Simulation," *Inst. Phys. C.*, Vol. 1982, 1982, No. 61, pages 337 ff.

Park, Y.J., Tang, T.W., and Navon, D.H.: "Monte-Carlo Surface Scattering Simulation in Mosfet Structures," *IEEE Transactions on Electron Devices,* Vol. 30, 1983, No. 9, pages 1110 ff.

Ramachan, V. and Vaya, P.R.: "Computer-Aided Study of Hot Wall Epitaxy System Using a Monte-Carlo Technique," *Journal of Applied Physics,* Vol. 54, 1983, No. 9, pages 5385 ff.

Ravishan, G., Mezei, M., and Beveridg, D.L.: "Monte-Carlo Computer-Simulation Study of the Hydrophobic Effect—Potential of Mean Force for <(CH4)2>AQ at 25-Degrees-C and 50-Degrees-F," *Faraday Sym.*, Vol. 1982, 1982, No. 17, pages 79 ff.

Retterer, J.M., Chang, T., and Jasperse, J.R.: "Ion-Acceleration in the Suprauroral Region—A Monte-Carlo Model," *Geophysical Research Letters*, Vol. 10, 1983, No. 7, pages 583 ff.

Ries, B., Schonher, G., Bassler, H., and Silver, M.: "Monte-Carlo Simulations of Geminate-Pair Dissociation in Discrete Anisotropic Lattices," *Philosophical Magazine B-Defects & Mechanical Properties* (cont. *Philos. Mag.*), Vol. 48, 1983, No. 1, pages 87 ff.

Rizzoli, V. and Someda, C.G.: "Bandwidth of Multimode Fibers with Random Fluctuation in the Index Profile—A Monte-Carlo Approach," *Optical and Quantum Electronics*, Vol. 15, 1983, No. 5, pages 407 ff.

Roman, E. and Majlis, N.: "Computer-Simulation Model of the Structure of Ion-Implanted Impurities in Semiconductors," *Solid State Communications*, Vol. 47, 1983, No. 4, pages 259 ff.

Rotzoll, G.: "Computer-Simulation of Different Designs of Pseudo-Random Time-of-Flight Velocity Analyzers for Molecular-Beam Scattering Experiments," *Journal of Scientific*

Instruments. Journal of Physics E, Vol. 15, 1982, No. 7, pages 708 ff.

Samoto, N. and Shimizu, R.: "Theoretical-Study of the Ultimate Resolution in Electron-Beam Lithography by Monte-Carlo Simulation, Including Secondary-Electron Generation," *Journal of Applied Physics*, Vol. 54, 1983, No. 7, pages 3855 ff.

Singh, J. and Madhukar, A.: "Prediction of Kinetically Controlled Surface Roughening—A Monte-Carlo Computer-Simulation Study," *Physical Review Letters*, Vol. 51, 1983, No. 9, pages 794 ff.

Strandbu, K.J., Solla, S.A., and Chester, G.V.: "Monte-Carlo Studies of a Laplacian Roughening Model for 2-Dimensional Melting," *Physical Review: B. Solid State*, Vol. 28, 1983, No. 5, pages 2717 ff.

Stratt, R.M.: "Monte-Carlo Evaluation of Path-Integrals—Quantal Intramolecular Degrees of Freedom in Solution," *Journal of Chemical Physics*, Vol. 77, 1982, No. 4, pages 2108 ff.

Szekely, J.G., Perry, K.A., and Petkau, A.: "Simulated Response to Log-Normally Distributed Continuous Low Radiation-Doses," *Health Physics*, Vol. 45, 1983, No. 3, pages 699 ff.

Tsong, T.T.: "Monte-Carlo Simulation of Atomic Processes on the Solid-Surfaces," *Surface Science*, Vol. 122, 1982, No. 1, pages 99 ff.

Vangunst, W.F., Berendse, H.J., Hermans, J., and Hol, W.G.J.: "Computer-Simulation of the Dynamics of Hydrated Protein Crystals and Its Comparison with X-Ray Data," *Proceedings of the National Academy of Sciences of the United States of America—Biology*, Vol. 80, 1983, No. 14, pages 4315 ff.

Vanhemme, J.L. and Morgenst, I.: "On Sampling of Monte-Carlo Data and Disordered-Systems,"*Journal of Physics: C. Solid State Physics and Supplement*, Vol. 15, 1982, No. 20, pages 4353 ff.

Wada, K. and Ishikawa, T.: "Monte-Carlo Study of a Triangular Ising Lattice. 1.," *Journal of the Physical Society of Japan and Supplement*, Vol. 52, 1983, No. 5, pages 1774 ff.

Weaire, D. and Kermode, J.P.: "Computer-Simulation of a 2-Dimensinal Soap Froth. 1. Method and Motivation," *Philosophical Magazine B-Defects & Mechanical Properties* (continuation of *Philos. Mag.*), Vol. 48, 1983, No. 3, pages 245 ff.

Zinenko, V.I. and Tomshina, N.G.: "Monte-Carlo Investigation of the Mitsui Model," *Journal of Physics: C. Solid State Physics and Supplement*, Vol. 16, 1983, No. 25, pages 4997 ff.

Police Patrols

Saladin, B.A.: "Simulation of a Police Patrol Activity," *Omega—International Journal of Management Science*, Vol. 11, 1983, No. 4, pages 377 ff.

Politics and Elections

Bordley, R.F.: "A Pragmatic Method for Evaluating Election Schemes through Simulation," *American Political Science Review*, Vol. 77, 1983, No. 1, pages 123 ff.

Woodworth, J.R. and Gump, W.R.: *CAMELOT, a Role Playing Simulation for Political Decision-Making*,. Homewood, Ill.: Dorsey Press, 1982, xiv + 137 pages.

Polymers

Guttman, C.M. and Dimarzio, E.A.: "Monte-Carlo Modeling of Kinetics of Polymer Crystal-Growth—Regime-III and Its Implications on Chain Morphology," *Journal of Applied Physics*, Vol. 54, 1983, No. 10, pages 5541 ff.

Khalatur, P.G. Klyushni, B.N., and Pakhomov, P.M.: "Study of Conformational Elasticity of Polymer-Chains by the Monte-Carlo Method" (in Rusian), *Vysokomolekularnye Soedineniya, Seriya A*, Vol. 25, 1983, No. 7, pages 1510 ff.

Khalatur, P.G. and Pletneva, S.G.: "Concentrational Dependence of the Radius of Correlation—Results of the Simulation of Polymer-Solution by Monte-Carlo Method" (in Russian), *Vysokomolekularnye Soedineniya, Seriya B*, Vol. 24, 1982, No. 6, pages 449 ff.

Odriscol, K.F.: "Compositional Heterogeneity in Low-Molecular Weight Co-Polymers as Revealed by Monte-Carlo Simulations," *Journal of Coatings Technology*, Vol. 55, 1983, No. 705, pages 57 ff.

Weber, T.A. and Helfand, E.: "Time-Correlation Functions from Computer-Simulations of Polymers," *Journal of Physical Chemistry*, Vol. 87, 1983, No. 15, pages 2881 ff.

Webman, I., Lebowitz, J.L., and Kalos, M.H.: "A Monte-Carlo Study of the Collapse of a Polymer-Chain," *Macromolecules*, Vol. 14, 1981, No. 5, pages 1495 ff.

Population

Hogeweg, P. and Richter, A.F.: "INSTAR, a Discrete Event Model for Simulating Zooplankton Population-Dynamics," *Hydrobiologia*, Vol. 95, 1982 No. EP, pages 275 ff.

Patil, M.K., Janahanl, P.S., and Ghista, D.N.: "Mathematical Simulation of Impact of Birth-Control Policies on Indian Population System," *Simulation* (Simulation Councils, Inc.), Vol. 41, 1983, No. 3, pages 103 ff.

Relethfo, J.H.: "Simulation of the Island Model of Population-Structure under Conditions of Population-Growth," *Human Biology*, Vol. 53, 1981, No. 3, pages 295 ff.

Sale, P.F.: "Stock-Recruit Relationships and Regional Coexistence in a Lottery Competitive System—A Simulation Study," *American Naturalist*, Vol. 120, 1982, No. 2, pages 139 ff.

Samuels, M.L.: "POPREG-I—A Simulation of Population Regulation among the Maring of New-Guinea," *Human Ecology*, Vol. 10, 1982, No. 1, pages 1 ff.

Weinstei, D.A., Shugart, H.H., and Brandt, C.C.: "Energy-Flow and the Persistence of a Human-Population—A Simulation Analysis," *Human Ecology*, Vol. 11, 1983, No. 2, pages 210 ff.

Population Ecology and Wildlife Management

Clark, W.R. and Innis, G.S.: "Forage Interactions and Black-Tailed Jack Rabbit-Population Dynamics—A Simulation-Model," *Journal of Wildlife Management*, Vol. 46, 1982, No. 4, pages 1018 ff.

Nakamura, K.: "Prey Capture Tactics of Spiders—An Analysis Based on a Simulation-Model for Spider's Growth," *Researches on Population Ecology*, Vol. 24, 1982, No. 2, pages 302 ff.

Rollo, C.D., Vertinsk, I.B., Wellingt, W.G., and Thompson, W.A.: "Description and Testing of a Comprehensive Simulation-Model of the Ecology of Terrestrial Gastropods in Unstable Environments," *Researches on Population Ecology*, Vol. 25, 1983, No. 1, pages 150 ff.

Troester, S.J.: "Damage and Yield Reduction in Field Corn Due to Black Cutworm (Lepidoptera, Noctuidae) Feeding—Results of a Computer-Simulation Study," *Journal of Economic Entomology*, Vol. 75, 1982, No. 6, pages 1125 ff.

Steiner, A.J.: "COMTRAP—A Trapping Simulation Interactive Computer-Program," *Journal of Wildlife Management*, Vol. 47, 1983, No. 2, pages 561 ff.

Power Systems and Apparatus

Blacksto, J.H., Hogg, G.L., and Patton, A.D.: "A Simulation-Model for Assessment of Large-Scale Power-System Reliability," *IIE Trans.*, Vol. 14, 1982, No. 1, pages 60 ff.

Elkady, M.A.: "Probabilistic Short-Circuit Analysis by Monte-Carlo Simulation," *IEEE Transactions on Power Apparatus and Systems*, Vol. 102, 1983, No. 5, pages 1308 ff.

Ford, A.: "Using Simulation for Policy Evaluation in the Electric Utility Industry," *Simulation* (Simulation Councils, Inc.), Vol. 40, 1983, No. 3, pages 85 ff.

Juves, J.A.: "Quality Assurance Comes to Public Utilities in Simulation Software," *Industrial Research & Development* (was *Indust. Research & Research Development*), Vol. 24, 1982, No. 11, pages 132 ff.

Lasseter, R.H. and Lee, S.Y.: "Digital-Simulation of Static Var System Transients," *IEEE Transactions on Power Apparatus and Systems*, Vol. 101, 1982, No. 10, pages 4171.

Schmidt, K. and Leonhard, W.: "Simulation of Electric-Power Systems by Parallel Computation," *I.S.A. Transactions*, Vol. 21, 1982, No. 4, pages 15 ff.

Timko, K.J., Bose, A., and Anderson, P.M.: "Monte-Carlo Simulation of Power-System Stability," *IEEE Transactions on Power Apparatus and Systems*, Vol. 102, 1983, No. 10, pages 3453 ff.

Thum, P.C., Liew, A.C., and Wong, C.M.: "Computer-Simulation of the Initial-Stages of the Lightning Protection Mechanism," *IEEE Transactions on Power Apparatus and Systems*, Vol. 101, 1982, No. 11, pages 4370 ff.

Undrill, J.M. and Laskowsk, T.F.: "Model Selection and Data Assembly for Power-System Simulations," *IEEE Transactions on Power Apparatus and Systems*, Vol. 101, 1982, No. 9, pages 3333 ff.

Psychiatry

Colby, K.M.: *Artificial Paranoia: A Computer Simulation of Paranoid Processes*. New York: Pergamon Press, 1974, 900 pages.

Random Figures

Kellerer, A.M.: "On the Number of Clumps Resulting from the Overlap of Randomly Placed Figures in a Plane," *Journal of Applied Probability*, Vol. 20, 1983, No. 1, pages 126 ff.

McDiarmi, "On the Chromatic Forcing Number of a Random Graph," *Discrete Applied Mathematics*, Vol. 5, 1983, No. 1, pages 123 ff.

Random Numbers and Psychical Research

Radin, D.I.: "Experimental Attempts to Influence Pseudo-Random Number Sequences," *Journal of the American Society for Psychical Research*, Vol. 76, 1982, No. 4, pages 359 ff.

Random Number Generation

Fushimi, M. and Tezuka, S.: "The *K*-Distribution of Generalized Feedback Shift Register Pseudo-Random Numbers," *Communications of the ACM*, Vol. 26, 1983, No. 7, pages 516 ff.

Havel, J., Morozevi, A.N., and Jarmolik, V.M.: "Randomizing Generator of Pseudo-Random Number Sequences" (in Russian), *Kybernetika*, Vol. 19, 1983, No. 1, pages 58 ff.

Pearson, R.B.: "An Algorithm for Pseudo Random Number Generation Suitable for Large-Scale Integration," *Journal of Computational Physics*, Vol. 49, 1983, No. 3, pages 478 ff.

Ripley, B.D.: "The Lattice Structure of Pseudo-Random Number Generators," *Proceedings of the Royal Society, Series A, Mathematical and Physical Science*, Vol. 389, 1983, No. 179, pages 197 ff.

Roberts, C.S.: "Implementing and Testing New Versions of a Good, 48-Bit, Pseudo-Random Number Generator," *Bell System Technical Journal*, Vol. 61, 1982, No. 8, pages 2053 ff.

Wichmann, B.A. and Hill, I.D.: "An Efficient and Portable Pseudo-Random Number Generator," *Appl. Stat.*, Vol. 31, 1982, No. 2, pages 188 ff.

Yegorov, V.A. and Yegorov, Y.A.: "An Effective Technique of Pseudo-Random Numbers Having Uniform-Distribution Pattern" (in Russian), *Izvestiya Vysshikh Uchebnykhika Zavedenii Radioelektronika*, Vol. 25, 1982, No. 7, pages 58 ff.

Random Number Generation by Physical Devices

Inoue, H., Kumahora, H., and Yoshizaw, Y.: "Random Numbers Generated by a Physical Device," *Applied Statistics—Journal of the Royal Statistical Society, Series C*, Vol. 32, 1983, No. 2, pages 115 ff.

Random Number Generation On Microcomputers

Jennergren, L.P.: "Another Method for Random Number Generation on Microcomputers," *Simulation* (Simulation Councils, Inc.), Vol. 41, 1983, No. 2, pages 79 ff.

Random Process Generation

Erber, T., Rynne, T.M., Darsow, W.F. and Frank, M.J.: "The Simulation of Random-Processes on Digital-Computers—Unavoidable Order," *Journal of Computational Physics*, Vol. 49, 1983, No. 3, pages 394 ff.

Random Variate Generation

Atkinson, A.C.: "The Simulation of Generalized Inverse Gaussian and Hyperbolic Random-Variables," *SIAM Journal on Scientific & Statistical Computing*, Vol. 3, 1982, No. 4, pages 502 ff.

Bondesso, L.: "On Simulation from Infinitely Divisible Distributions," *Advances in Applied Probability*, Vol. 14, 1982, No. 4, pages 855 ff.

Leutjen, P. and Hartman, P.: "Simulation with the Restricted Erlang Distribution," *Proceedings, Annual Reliability and Maintainability Symposium*, Vol. 1982, 1982, pages 233 ff.

Popova, V.I. and Tyurin, S.V.: "Random Event Simulator with Selectable Probability Characteristics," *Instruments and Experimental Techniques*, Vol. 25, 1982, No. 2, pages 355 ff.

Sakasega, H.: "Stratified Rejection and Squeeze Method for Generating Beta-Random Numbers," *Annals of the Institute of Statistical Mathematics*, Vol. 35, 1983, No. 2, pages 291 ff.

Reliability

Depuy, K.M., Hobbs, J.R., Moore, A.H., and Johnston, J.W.; "Accuracy of Univariate, Bivariate, and a Modified Double Monte-Carlo Technique for Finding Lower Confidence-Limits of System Reliability," *IEEE Transactions on Reliability*, Vol. 31, 1982, No. 5, pages 474 ff.

Hauptman, U. and Yllera, J.: "Fault-Tree Evaluation by Monte-Carlo Simulation," *Chemical Engineering*, Vol. 90, 1983, No. 1, pages 91 ff.

Robotics and Automation

Cheng, F.T.: *Computer Simulation of the Dynamics and Control of an Energy-Efficient Robot Leg*. 241 pages.

Fonarev, A.S. and Sherstyu, A.V.: "Algorithms and Methods for Computer-Simulation of Transonic Flow," *Automation and Remote Control, USSR*, Vol. 43, 1982, No. 7, pages 843 ff.

Medeiros, D.J. and Sadowski, R.P.: "Simulation of Robotic Manufacturing Cells—A Modular Approach," *Simulation* (Simulation Councils, Inc.), Vol. 40, 1983, No. 1, pages 3 ff.

Yong, Y. and Bonner M.: "Simulation—Preventing Some Nasty Snarl-Ups," *Ind. Robot*, Vol. 1983, 1983, pages 30 ff.

Scientific Discovery

Bradshaw, G.F., Langley, P.W., and Simon, H.A.: "Studying Scientific Discovery by Computer-Simulation," *Science*, Vol. 22, 1983, No. 462, pages 971 ff.

Simulation Languages

Grant, F.H., Polito, J., and Sabuda, J.: "SNAP SOS—A Package for Simulating and Analyzing Safeguards Systems," *Simulation* (Simulation Councils, Inc.), Vol. 41, 1983, No. 2, pages 65 ff.

Hac, A.: "Computer-System Simulation in Pascal," *Software: Practice and Experience*, Vol. 12, 1982, No. 8, pages 777 ff.

Hidinger, R.M.: "Digital-Computer Benchmarks of a Continuous System Simulation," *Simulation* (Simulation Councils, Inc.), Vol. 39, 1982, No. 1, pages 27 ff.

Huang, P.Y. and Ghandfor, P.: "Simulation Language Selection," *Journal of Systems Management*, Vol. 34, 1983, No. 4, pages 10 ff.

Hunt, E.: "Simulation in Pascal," *Behavior Research Methods and Instrumentation*, Vol. 15, 1983, No. 2, pages 305 ff.

Korn, G.A.: "A Wish List for Simulation-Language Specifications," *Simulation* (Simulation Councils, Inc.), Vol. 40, 1983, No. 1, pages 30 ff.

McCormick, S.F. and Ruge, J.W.: " Unigrid for Multigrid Simulation, " *Mathematics of Computation*, Vol. 41, 1983, No. 163, pages 43 ff.

Payne, J.A.: *Introduction to Simulation: Programming Techniques and Methods of Analysis*. New York: McGraw-Hill, 1982, xi + 413 pages.

Richardson, G.P. and Pugh, A.: *Introduction to System Dynamics Modeling with DYNAMO*. Cambridge, Mass.: MIT Press, 1981, xi + 413 pages.

Roberts, N.: *Introduction to Computer Simulation: The System Dynamics Approach*. Reading, Mass.: Addison-Wesley, 1983, xiii + 562 pages.

User Guide and Reference Manual for MICRO-DYNAMO: System Dynamics Modeling Language. Reading, Mass.: Addison-Wesley, 1982, x + 113 pages.

Withers, S.J.: "Voice Control of an Interactive Simulation," *Simulation* (Simulation Councils, Inc.), Vol. 40, 1983, No. 1, pages 28 ff.

Zimmerma, B.G. "Model-S, a Sampled-Data Simulation Language," *Simulation* (Simulation Councils, Inc.), Vol. 40, 1983, No. 5, pages 183 ff.

Simulation Theory and Conferences

Bronson, R.: "Spreading the Gospel of Simulation," *Simulation* (Simulation Councils, Inc.), Vol. 39, 1982, No. 3, pages 100 ff.

Cellier, F.E.: *Progress in Modelling and Simulation*. London: Academic Press, 1982, viii + 466 pages.

Curry G.L. and Hartfiel, D.J.: "A Simulation-Optimization Method—Its Convergence and Utility," *Naval Research Logistics Quarterly*, Vol. 30, 1983, No. 2, pages 227 ff.

Discrete Simulation and Related Fields: Proceedings of the IMACS European Simulation Meeting on Discrete Simulation and Related Fields. Amsterdam: North-Holland Pub. Co., 1982, vii + 246 pages.

Drenick, R.F. and Kozin, F.: *System Modeling and Optimization: Proceedings of the 10th IFIP Conference , New York City, USA, August 31—September 4, 1981*. Berlin: Springer-Verlag, 1982, xi + 893 pages.

Innis, G. and Rexstad, E.: "Simulation-Model Simplification Techniques," *Simulation* (Simulation Councils, Inc.), Vol. 41, 1983, No. 1, pages 7 ff.

Lundqvis, S.O. and Brattber, T.: "How Shall It Be Shorted —Simulation Helps You Make the Correct Choice" (in Swedish), *Svensk Papperstidning*, Vol. 86, 1983, No. 2, pages 26 ff.

Mitrani, I.: *Simulation Techniques for Discrete Event Systems*. Cambridge: Cambridge University Press, 1982, ix + 185 pages.

Pappas, I.A., Maniatop, K., Protosig, S., and Vakalopo, A.: "A Tool for Teaching Monte-Carlo Simulation Without Really Meaning It," *European Journal of Operational Research*, Vol. 11, 1982, No. 3, pages 217 ff.

Paul, W.: "Online Simulation of $K+1$ Tapes by K-Tapes Requires Non-Linear Time," *Information and Control*, Vol. 53, 1982, Nos. 1–2, pages 1 ff.

Schwefel, H.: *Numerical Optimization of Computer Models*. Chichester, England: Wiley, 1981, vii + 389 pages.

Smith, H.W., Winer, J.L., and George, C.E.: "The Relative Efficacy of Simulation Experiments," *Journal of Vocational Behavior*, Vol. 22, 1983, No. 1, pages 96 ff.

Sol, H.G.: Review of *Simulation Techniques for Discrete Event Systems* by I. Mitrani, *European Journal of Operational Research*, Vol. 14, 1983, No. 2, pages 197 ff.

Spivey, J.E. and Jackson, P.: "Computer-Simulation—A Case of Conflict," *Behavior Research Methods and Instrumentation*, Vol. 15, 1983, No. 2, pages 187 ff.

Sugarman, R. and Wallich, P.: "The Limits to Simulation," *IEEE Spectrum*, Vol. 20, 1983, No. 4, pages 36 ff.

UKSC 81: Proceedings of the 1981 UKSC Conference On Computer Simulation 13–15 May 1981, Old Swan Hotel, Harrogate, England. Guildford, Surrey, England: Westbury House, 1981.

Winter Simulation Conference. Association for Computing Machinery (1133 Avenue of the Americas, New York 10036), New York, 1976.

1981 Winter Simulation Conference: Proceedings, December 9–11, 1981, Peachtree Plaza, Atlanta, Georgia. New York, N.Y. (345 East 47th St., New York 10017): Institute of Electrical and Electronic Engineers.

Yen, K. and Cook, G.: "Digital-Simulation Algorithms Using Parallel Processing," *IEEE Ind. E.*, Vol. 29, 1982, No. 3, pages 217 ff.

Social Systems and Public Policy

Botman, J.J.: *Dynamics of Housing and Planning: A Regional Simulation Model*. Delft, Holland: Delft University Press, 1981, 270 pages.

Cohen, C.F.: "The Impact on Women of Proposed Changes in the Private Pension System—A Simulation," *Industrial and Labor Relations Review*, Vol. 36, 1983, No. 2, pages 258 ff.

Henize, J.: "Can a Shorter Workweek Reduce Unemployment—A German Simulation Study," *Simulation* (Simulation Councils, Inc.), Vol. 37, 1981, No. 5, pages 145 ff.

Johnson, W.R. and Browning, E.K.: "The Distributional and Efficiency Effects of Increasing the Minimum-Wage—A Simulation," *American Economic Review*, Vol. 73, 1983, No. 1, pages 204 ff.

Kroeck, K.G., Barrett, G.V., and Alexande, R.A.: "Imposed Quotas and Personnel-Selection—A Computer-Simulation Study," *Journal of Applied Psychology*, Vol. 68, 1983, No. 1, pages 123 ff.

Pike, M: "A Simulation of the Impact of Introducing Equality for Males and Females into Social-Security," *Economica*, Vol. 50, 1983, No. 199, pages 351 ff.

Sullivan, W.G. and Orr, R.G.: "Monte-Carlo Simulation Analyzes Alternatives in Uncertain Economy," *Industrial Engineering*, Vol. 14, 1982, No. 11, pages 42 ff.

Space Flight

Hsing, J.C.: "Digital Simulations of SBS Spacecraft Despin Operation," *COMSAT Technical Review*, Vol. 12, 1982, No. 1, pages 213 ff.

Jayarama, T.S., Menon, K.A.P., and Ganeshan, A.S.: "Frames of Reference for Launch-Vehicle Trajectory Simulation," *Simulation* (Simulation Councils, Inc.), Vol. 39, 1982, No. 1, pages 3 ff.

Landauer, J.P.: "Using a Multi-Processing Hybrid Computer for Flight Simulation ," *Simulation* (Simulation Councils, Inc.), Vol. 39, 1982, No. 1, pages 23 ff.

Szatkows, G.P. and Nelander, H.C.: "Real-Time Microcomputer Simulation for Space-Shuttle Centaur Avionics," *Simulation* (Simulation Councils, Inc.), Vol. 39, 1982, No.5, pages 169 ff.

Whitsid, R.: "Microprocessors Appear in Flight Simulation," *Simulation* (Simulation Councils, Inc.), Vol. 39, 1982, No. 3, pages 96 ff.

Statistical Procedures' Properties via Simulation

Bhansali, R.J.: "A Simulation Study of Autoregressive and Window Estimators of the Inverse Correlation-Function," *Applied Statistics—Journal of the Royal Statistical Society, Series C*, Vol. 32, 1983, No. 2, pages 141 ff.

Burguill, F.J., Wright, A.J., and Bardsley, W.G.: "Use of the *F*-Test for Determining the Degree of Enzyme-Kinetic and Ligand-Binding Data—A Monte-Carlo Simulation Study," *Biochemical Journal*, Vol. 211, 1983, No. 1, pages 23 ff.

Chen, W.: "Simulation on Probability Points for Testing of Lognormal or Weibull Distribution with a Small Sample," *J. Stat. Comp.*, Vol. 15, 1982, Nos. 2–3, pages 201 ff.

Evans, D.J. and Dunbar, R.C.: "The Parallel Quicksort Algorithm. 2. Simulation," *International Journal of Computer Mathematics*, Vol. 12, 1982, No. 2, pages 125 ff.

Gaver, D.P. and Miller, R.G.: "Jackknifing the Kaplan-Meier Survival Estimator for Censored-Data—Simulation Results and Asymptotic Analysis," *Communications in Statistics—Theory and Methods*, Vol. 12, 1983, No. 15, pages 1701 ff.

Hoffman, K.N. and Kalnay, E.: "Lagged Average Forecasting, an Alternative to Monte-Carlo Forecasting," *Tellus A*, Vol. 35, 1983, No. 2, pages 100 ff.

Iglehart, D.L. and Shedler, G.S.: "Simulation of Non-Markovian Systems," *IBM Journal of Research and Development*, Vol. 27, 1983, No. 5, pages 472 ff.

Johnson, D.R. and Creech, J.C.: "Ordinal Measures in Multiple Indicator Models—A Simulation Study of Categorization Error," *American Sociological Review*, Vol. 48, 1983, No. 3, pages 398 ff.

Ketellap, R.H.: "The Relevance of Large Sample Properties of Estimators for the Errors-in-Variables Model—A Monte-Carlo Study," *Communications in Statistics, Part B—Simulation & Computation*, Vol. 11, 1982, No. 5, pages 625 ff.

Nathan, G.: "A Simulation Comparison of Estimators for a Regression Coefficient Under Differential Non-Response," *Communications in Statistics—Theory and Methods*, Vol. 12, 1983, No. 6, pages 645 ff.

Obrien, P.N., Parente, F.J., and Schmitt, C.J.: "A Monte-Carlo Study on the Robustness of 4 MANOVA Criterion tests," *J. Stat. Comp.*, Vol. 15, 1982, Nos. 2–3, pages 183 ff.

Orton, C.: "Computer-Simulation Experiments to Assess the Performance of Measures of Quantity of Pottery," *World Archaeology*, Vol. 14, 1982, No. 1, pages 1 ff.

Rae, G.: "A Monte-Carlo Comparison of Small Sample Procedures for Testing the Hypothesis That 2 Variables Measure the Same Trait Except for Error," *British Journal of Mathematical and Statistical Psychology*, Vol. 35, 1982, pages 228 ff.

Recchia, M. and Rocchett, M.: "The Simulated Randomization Test," *Computer Programs in Biomedicine*, Vol. 15, 1982, No. 2, pages 111 ff.

Saxena, K.M.L. and Alam, K.: "Estimation of the Non-Centrality Parameter of a Chi-Squared Distribution," *Annals of Statistics*, Vol. 10, 1982, No. 3, pages 1012 ff.

Schneide, W. and Scheible, D.: "On the Evaluation of Clustering Algorithms—A Monte-Carlo Approach" (in German), *Psychologische Beitraege*, Vol. 25, 1983, Nos. 1–2, pages 238 ff.

Schwertm, N.C.: "A Monte-Carlo Study of the LN Statistic for the Multivariate Nonparametric Median and Rank Sum Tests for 2 Populations," *Communications in Statistics, Part B—Simulation & Computation*, Vol. 11, 1982, No. 6, pages 667 ff.

Starks, T.H.: "A Monte-Carlo Evaluation of Response Surface-Analysis Based on Paired-Comparison Data," *Communications in Statistics, Part B—Simulation & Computation*, Vol. 11, 1982, No. 5, pages 603 ff.

Stoodley, K.D.: "Adaptive Filtering—A Theoretical and Simulation Study of the Convergence of the Parameter Estimates," *Journal of the Operational Research Society* (continuation of *Operational Research Quarterly*), Vol. 33, 1982, No. 12, pages 1077 ff.

Turiel, T.P., Hahn, G.J., and Tucker, W.T.: "New Simulation Results for the Calibration and Inverse Median Estimation Problems," *Communications in Statistics, Part B—Simulation & Computation*, Vol. 11, 1982, No. 6, pages 677 ff.

Vandijk, H.K. and Kloek, T.: "Monte-Carlo Analysis of Skew Posterior Distributions—An Illustrative Econometric Example," *The Statistician*, Vol. 32, 1983, Nos. 1–2, pages 216 ff.

Velicer, W.F., Peacock, A.C., and Jackson, D.N.: "A Comparison of Component and Factor Patterns—A Monte-Carlo Approach," *Multivariate Behavioral Research*, Vol. 17, 1982, No. 3, pages 371 ff.

Wiegersm, S.: "Sequential Response Bias in Randomized-Response Sequences—A Computer-Simulation," *ACTA Psychologica*, Vol. 52, 1982, No. 3, pages 249 ff.

Yu, M.C. and Dunn, O.J.: "Robust-Tests for the Equality of 2 Correlation-Coefficients—A Monte-Carlo Study," *Educa-*

tional and Psychological Measurement, Vol. 42, 1982, No. 4, pages 987 ff.

Yunker, J.A.: "Optimal Redistribution with Interdependent Utility-Functions—A Simulation Study," *Public Finance*, Vol. 38, 1983, No. 1, pages 132 ff.

Statistics in Simulation

Adam, N.R.: "Achieving a Confidence-Interval for Parameters Estimated by Simulation," *Management Science*, Vol. 29, 1983, No. 7, pages 856 ff.

Chandra, M., Singpurwalla, N.D., and Stephens, M.A.: "Some Problems in Simulating the Quantiles of the Maxima and Other Functionals of Gaussian-Processes," *Journal of Statistical Computation & Simulation*, Vol. 18, 1983, No. 1, pages 45 ff.

Clayton, E.R., Weber, W.E., and Taylor, B.W.: A Goal Programming Approach to the Optimization of Multiresponse Simulation-Models," *IIE Trans.*, Vol. 14, 1982, No. 4, pages 282 ff.

Dorea, C.C.Y.: "Expected Number of Steps of a Random Optimization Method," *Journal of Optimization Theory and Applications*, Vol. 39, 1983, No. 2, pages 165 ff.

Fagerstr, R.M. and Meeter, D.A.: "The Design and Analysis of a Simulation—Tolerance Quantiles and Random-Walk Designs," *Communications in Statistics-Simulation*, Vol. 12, 1983, No. 5, pages 541 ff.

Feoli, E. and Lagone, G.R.M.: "A Resemblance Function Based on Probability—Applications to Field and Simulated Data," *Vegetatio; Acta Geobotanica*, Vol. 53, 1983, No. 1, pages 3 ff.

Fishman, G.S.: "Accelerated Accuracy in the Simulation of Markov-Chains," *Operations Research*, Vol. 31, 1983, No. 3, pages 466 ff.

Golden, B.L., Assad, A.A., and Zanakis, S.H.: *Statistics and Optimization*. Columbus, Ohio: American Sciences Press, Inc., 1984, 208 pages.

Gordon, H.L., Rothstei, S.M., and Proctor, T.R.: "Efficient Variance-Reduction Transformations for the Simulation of a Ratio of 2 Means—Application to Quantum Monte-Carlo Simulations," *Journal of Computational Physics*, Vol. 47, 1982, No. 3, pages 375 ff.

Iglehart, D.L. and Shedler, G.S.: "Statistical Efficiency of Regenerative-Simulation Methods for Networks of Queues," *Advances in Applied Probability*, Vol. 15, 1983, No. 1, pages 183 ff.

Kleijnen, J.P.: "Cross-Validation Using the t-Statistic," *European Journal of Operational Research*, Vol. 13, 1983, No. 2, pages 133 ff.

Kottas, J.F. and Lau, H.S.: "A 4-Moments Alternative to Simulation for a Class of Stochastic Management Models," *Management Science*, Vol. 28, 1982, No. 7, pages 749 ff.

Reynolds, M.R. and Deaton, M.L. "Comparisons of Some Tests for Validation of Stochastic Simulation-Models," *Communications in Statistics, Part B— Simulation & Computation*, Vol. 11, 1982, No. 6, pages 769 ff.

Rothery, P.: "The Use of Control Variates in Monte-Carlo Estimation of Power," *Appl. Stat.*, Vol. 31, 1982, No. 2, pages 125 ff.

Rukhin, A.L.: "Adaptive Procedures in Multiple Decision-Problems and Hypothesis-Testing," *Annals of Statistics*, Vol. 10, 1982, No. 4, pages 1148 ff.

Traffic Engineering

Bayat-Mokhtari, J.B.: *Computer Simulation of Real World Pedestrian Accidents*. 149 pages.

The Application of Traffic Simulation Models: Proceedings of a Conference on the Application of Traffic Simulation Models. Washington, D.C.: National Academy of Sciences, 1981, v + 114 pages.

Elgindy, M. and Ilosvai, L.: "Computer-Simulation Study on a Vehicle's Directional Response in Some Severe Maneuvers. 1. Rapid Lane-Change Maneuvers," *International Journal of Vehicle Design*, Vol. 4, 1983, No. 4, pages 386 ff.

Elgindy, M. and Ilosvai, L.: "Computer-Simulation Study on a Vehicle's Directional Response in Some Severe Maneuvers. 2. Steering and Braking Maneuvers," *International Journal of Vehicle Design*, Vol. 4, 1983, No. 5, pages 501 ff.

Transportation

Hammesfa, R.D. and Clayton, E.R.: "A Computer-Simulation Model to Assist Intermodal Terminal Managers in Operations Analysis," *Transportation Journal*, Vol. 22, 1983, No. 4, pages 55 ff.

Petersen, E.R. and Taylor, A.J.: "Line Block Prevention in Rail Line Dispatch and Simulation-Models ," *Infor: Canadian Journal of Operational Research and Information Processing*, Vol. 21, 1983, No. 1, pages 46 ff.

Tumor Growth

Duchting, W. and Vogelsae, T.: "3-Dimensional Simulation of Tumor-Growth," *Simulation* (Simulation Councils, Inc.), Vol. 40, 1983, No. 5, pages 163 ff.

Vehicle Design

Elimam, Y.A.: "Simulation of a High-Gain Adaptive Speed Regulating System for an Internal-Combustion Engine," *Simulation* (Simulation Councils, Inc.), Vol. 39, 1982, No. 4, pages 133 ff.

Ferber, R. and Stotter, A.: "Simulation for Optimization of Vehicle Powertrain Performance," *Israel Journal of Technology*, Vol. 20, 1982, Nos. 1–2, pages 81 ff.

Jernstro, C.: "Computer-Simulation of a Motor-Vehicle Crash Dummy and Use of Simulation in the Design Analysis

Process," *International Journal of Vehicle Design*, Vol. 4, 1983, No. 2, pages 136 ff.

Khandelw, R.S. and Nigam, N.C: "Digital-Simulation of the Dynamic-Response of a Vehicle Carrying Liquid Cargo on a Random Uneven Surface," *Vehicle System Dynamics*, Vol. 11, 1982, No. 4, pages 195 ff.

Shung, J.B., Tomizuka, M., Auslander, D.M., and Stout, G.: "Feedback-Control and Simulation of a Wheelchair," *Journal of Dynamic Systems Measurement and Control—Transactions of the ASME Series*, Vol. 105, No. 2, pages 96 ff.

Water

Jorgense, W.L.: "Convergence of Monte-Carlo Simulations of Liquid Water in the NPT Ensemble," *Chemical Physics Letters*, Vol. 92, 1982, No. 4, pages 405 ff.

Kataoka, Y., Hamada, H., Nose, S., and Yamamoto, T.: "Studies of Liquid Water by Computer-Simulations. 2. Static Properties of a 3D Model," *Journal of Chemical Physics*, Vol. 77, 1982, No. 11, pages 5699 ff.

Romano, S.: "Monte-Carlo Simulation of Water Solvent with Biomolecules—Serines with Reaction-Field Correction," *International Journal of Quantum Chemistry*, Vol. 20, 1981, No. 4, pages 921 ff.

Tani, A.: "Non-Polar Solute Water Pair Correlation-Functions—A Comparison Between Computer-Simulation and Theoretical Results," *Molecular Physics*, Vol. 48, 1983, No. 6, pages 1229 ff.

Water Systems

Brizkish, B.H. and Avadhanu, R.V.: "An Efficient Procedure in the Digital-Simulation of Aquifer Systems," *Journal of Hydrology*, Vol. 64, 1983, Nos. 1–4, pages 159 ff.

Fewkes, A. and Ferris, S.A.: "The Recycling of Domestic Waste-Water—A Study of the Factors Influencing the Storage Capacity and the Simulation of the Usage Pattern," *Building and Environment*, Vol. 17, 1982, No. 3, pages 209 ff.

Latouche, M.C. and Atkin, R.C.: "Application of a Simulation-Model Computer-Program in Planning a Multipurpose Water Scheme in Peru," *Proceedings of the Institution of Civil Engineers, Part I—Design and Construction*, Vol. 72, 1982, pages 611 ff.

Mote, C.R. and Pote, J.W.: "Computer Design and Simulation of Systems for Pressurized Distribution of Septic-Tank Effluent in Sloping Filter Fields," *Arkansas, Agricultural Experiment Station, Report Series*, Vol. 1982, 1982, No. 273, pages 1 ff.

Skaggs, R.W.: "Field Evaluation of a Water Management Simulation-Model," *Transactions of the American Society of Agricultural Engineers (ASAE)*, Vol. 25, 1982, No. 3, pages 666 ff.

Part VII: Appendix—Review of Statistical Concepts and Modeling

Statistical Principles On Which Experimentation Is Based

Specification of **modern statistical design and analysis of experiments is based on statistical principles** which, as necessary background material, are given a review in this Appendix at a level such that only some familiarity with basic statistics (such as from an introductory course or through use of statistical methods at work) is necessary as a prerequisite. These basics include frequency distributions, distribution characteristics (mean, variance), normal distribution, estimates from samples (mean, standard deviation), distributions of sample mean, significance tests, confidence intervals, and prediction intervals; in addition some elementary mathematical notation is covered first.

MATHEMATICAL NOTATIONS

$\{x: \mathscr{P}(x)\}$ the set of all x such that $\mathscr{P}(x)$ holds; for example,

$$\{x: x \text{ is an integer}\} = \{\ldots, -3, -2, -1, 0, 1, 2, 3, \ldots\}$$

$\in$ is a member of (belongs to)

$\notin$ is not a member of (does not belong to)

=> implies

≠> does not imply

$\ni$ such that

iff if and only if

<=> if and only if

$\geq$ greater than or equal to

$>$ greater than

$\leq$ less than or equal to

$<$ less than

$>>$ much larger than

$\equiv$ identically equal to

$\emptyset$ the empty set

GREEK ALPHABET

A	α	alpha	N	ν	nu
B	β	beta	Ξ	ξ	xi
Γ	γ	gamma	O	o	omicron
Δ	δ	delta	Π	π	pi
E	ϵ	epsilon	P	ρ	rho
Z	ζ	zeta	Σ	σ	sigma
H	η	eta	T	τ	tau
Θ,ϴ	θ	theta	Υ	υ	upsilon
I	ι	iota	Φ	ϕ,φ	phi
K	κ	kappa	X	χ	chi
Λ	λ	lambda	Ψ	ψ	psi
M	μ	mu	Ω	ω	omega

Summation Notation

If $x_1, x_2, \ldots, x_n$ are n numbers, then we write their sum as

$$\sum_{i=1}^{n} x_i = x_1 + x_2 + \ldots + x_n,$$

and the sum of their squares as

$$\sum_{i=1}^{n} x_i^2 = x_1^2 + \ldots + x_n^2.$$

For example, if $x_1 = 3.2$, $x_2 = 5.7$, $x_3 = -5.0$, and $x_4 = .76$, then

$$\sum_{i=1}^{3} x_i = 3.2 + 5.7 - 5.0 = 3.9, \sum_{i=1}^{4} x_i = 4.66, \text{ and}$$

$$\sum_{i=1}^{4} \frac{1}{x_i} = \frac{1}{3.2} + \frac{1}{5.7} - \frac{1}{5.0} + \frac{1}{.76} = 1.6037.$$

Note that $\dfrac{1}{\sum_{i=1}^{4} x_i} = .2146 \neq \sum_{i=1}^{4} \dfrac{1}{x_i} = 1.6037.$

Functional Notation

For each x, $f(x)$ represents a number, and f or $f(\cdot)$ or $f(x)$ is called a function of x. See Example 1.

Properties of Summation and Functional Notation

(1) $\sum_{i=1}^{n} c = nc$

(2) $\sum_{i=1}^{n} cf(x_i) = c \sum_{i=1}^{n} f(x_i)$

(3) $\sum_{i=1}^{n} (f(x_i) + g(x_i)) = \sum_{i=1}^{n} f(x_i) + \sum_{i=1}^{n} g(x_i)$

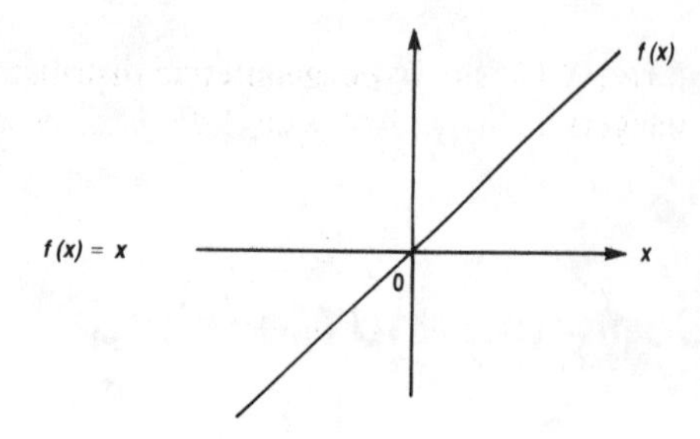

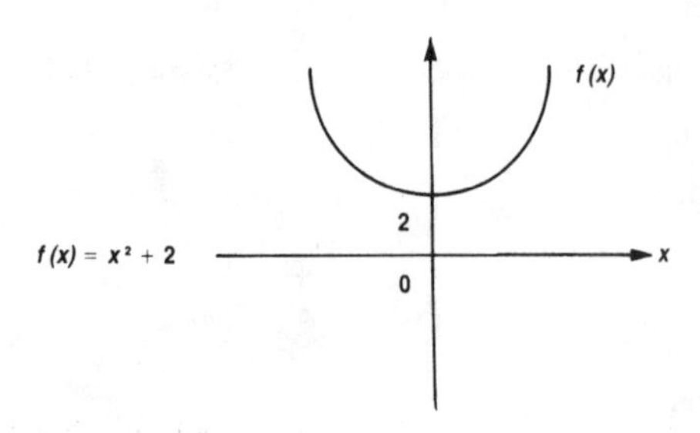

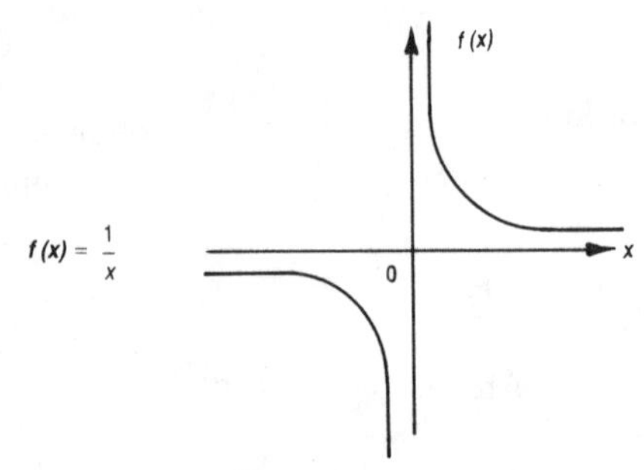

Example 1

(4) $\sum_{i=1}^{n} f(x_i) \leq \sum_{i=1}^{n} g(x_i)$ if $f(x) \leq g(x)$ for all x

(5) $\left| \sum_{i=1}^{n} f(x_i) \right| \leq \sum_{i=1}^{n} |f(x_i)|$.

For example, if $f(x) = x^2$, $g(x) = x$, $x_1 = .38$, $x_2 = 2.5$, $x_3 = -1.1$, then

$$\sum_{i=1}^{4} x_i^2 = 7.6044 + x_4^2, \sum_{i=1}^{3} (f(x_i) - g(x_i)) = 5.8244.$$

Probability Concepts

Suppose a bin contains 1,000,000 computer chips and among them the following two types are present: defective (30% are of this type) and good (70% are of this type), distinguishable only by an electronic analysis. If the bin is well mixed and we select a chip at random, what are the chances (**probability**) that the chip is defective?

If N = No. of chips in bin, then
No. of defective chips = $.30N$
No. of total chips = N

and

$$\textbf{Probability} \text{ (Draw a defective chip)} \equiv P \text{ (defective)}$$
$$= \frac{\textbf{No. of ways favorable to defective}}{\textbf{Total No. of ways to draw}}$$
$$= \frac{.30N}{N} = .30.$$

If this experiment is repeated a large number of times (with a thoroughly remixed bin), we expect that

$$\frac{\text{No. of draws of defective}}{\text{No. of draws made}} \rightarrow .30.$$

Relative Frequency Definition of Probability

How is such a probability function $P(\cdot)$ obtained in general? One general method of conceptualizing obtaining such a probability function $P(\cdot)$ (which is actually a conceptual connection between mathematical probability theory and the real world) is **frequency probability**. This says that **if an event A occurs $n(A)$ times in n (unrelated) trials of an appropriate experiment, then we should define a frequency function (which depends on the number of trials n) by**

$$R(A) = \frac{n(A)}{n}$$

and define a probability function by

$$P(A) = \lim_{n \to \infty} R(A) = \lim_{n \to \infty} \frac{n(A)}{n}.$$

In practice, n is finite, which is but one of the practical difficulties with such empirical estimates of probability (others including the problem of *having* unrelated trials). One often simply takes $P(A) = R(A)$ for all A, and it is easily verified that this *is* a probability function [that is, $P(\cdot)$ so defined satisfies the following definition:

Definition. *$P(\cdot)$ is a* **probability function** *if to each event $A \subseteq \Omega$ it assigns a number $P(A)$ such that*

$$P(A) \geq 0,$$
$$P(\Omega) = 1,$$

and (**Countable Additivity**) *if $A_1, A_2, \ldots$ are events with $A_i \cap A_j = \emptyset$ $(i \neq j)$, then*

$$P\left(\bigcup_{n=1}^{\infty} A_n\right) = \sum_{n=1}^{\infty} P(A_n).\ ]$$

Here Ω is a set (called the **sample space**). The points $\omega \in \Omega$ are called **elementary events**. Ω corresponds to the collection of distinct outcomes that may result when we perform some particular experiment of interest. Each $A \subseteq \Omega$ is called an **event**. (For instance, in our example, Ω consisted of 2 possible outcomes, defective and good.) We have formalized the notion of the "chance" of obtaining an outcome lying in a specified subset A of Ω into a definition of probability; we will wish to talk of the probability of each event $A \subseteq \Omega$, and will hence need a function which assigns a unique number (called **probability**) to each such A.

Suppose boxes are filled from the bin of computer chips mentioned earlier. If a box contains 5 chips, what are the chances that 3 will be defective?

In order to answer this question, we need a **probability model** for how a sample of 5 should behave when drawn from a bin as specified. ("On the average" we expect 30% defective chips in the sample, i.e. $(.30)(5) = 1.5$ defective chips. But, what are the chances of 3? Are they so small that, if we find 3, 4, or 5 of 5 are defective, we should doubt the "30%" figure?) Two models are commonly used for this situation:

Definition. An r.v. X has the **Binomial** distribution if (for some positive integer n, and some p with $0 \leq p \leq 1$)

$$P[X=x] = p_X(x) = \begin{cases} \binom{n}{x} p^x(1-p)^{n-x}, & x = 0, 1, \ldots, n \\ 0, & \text{otherwise.} \end{cases}$$

Definition. An r.v. X has the **hypergeometric** distribution if (for some integers n, a, and N with $1 \leq n \leq N$ and $0 \leq a \leq N$)

$$P[X=x] = p_X(x) = \begin{cases} \dfrac{\binom{a}{x}\binom{N-a}{n-x}}{\binom{N}{n}} \end{cases}$$

if $x = \max(0, n-(N-a)), \ldots, \min(n,a)$, and
0 otherwise,
where

Definition. $(n)_0 \equiv 1$. If n, r are positive integers,

$$(n)_r = \begin{cases} n(n-1)\ldots(n-r+1), & \text{if } r \leq n, \\ 0, & \text{if } r > n. \end{cases}$$

Definition. **(n-factorial) $0! \equiv 1$.** If n is a positive integer,

$$n! = n(n-1)\cdot\ldots\cdot 2\cdot 1.$$

Definition. **(Binomial Coefficients)** If n and r are nonnegative integers,

Table 1

						p					
n	x	.05	.10	.15	.20	.25	.30	.35	.40	.45	.50
1	0	.9500	.9000	.8500	.8000	.7500	.7000	.6500	.6000	.5500	.5000
	1	.0500	.1000	.1500	.2000	.2500	.3000	.3500	.4000	.4500	.5000
2	0	.9025	.8100	.7225	.6400	.5625	.4900	.4225	.3600	.3025	.2500
	1	.0950	.1800	.2550	.3200	.3750	.4200	.4550	.4800	.4950	.5000
	2	.0025	.0100	.0225	.0400	.0625	.0900	.1225	.1600	.2025	.2500
3	0	.8574	.7290	.6141	.5120	.4219	.3430	.2746	.2160	.1664	.1250
	1	.1354	.2430	.3251	.3840	.4219	.4410	.4436	.4320	.4084	.3750
	2	.0071	.0270	.0574	.0960	.1406	.1890	.2389	.2880	.3341	.3750
	3	.0001	.0010	.0034	.0080	.0156	.0270	.0429	.0640	.0911	.1250
4	0	.8145	.6561	.5220	.4096	.3164	.2401	.1785	.1296	.0915	.0625
	1	.1715	.2916	.3685	.4096	.4219	.4116	.3845	.3456	.2995	.2500
	2	.0135	.0486	.0975	.1536	.2109	.2646	.3105	.3456	.3675	.3750
	3	.0005	.0036	.0115	.0256	.0469	.0756	.1115	.1536	.2005	.2500
	4	.0000	.0001	.0005	.0016	.0039	.0081	.0150	.0256	.0410	.0625
5	0	.7738	.5905	.4437	.3277	.2373	.1681	1160	.0778	.0503	.0312
	1	.2036	.3280	.3915	.4096	.3955	.3602	.3124	.2592	.2059	.1562
	2	.0214	.0729	.1382	2048	.2637	.3087	.3364	.3456	.3369	.3125
	3	.0011	.0081	.0244	.0512	.0879	.1323	.1811	.2304	.2757	.3125
	4	.0000	.0004	.0022	.0064	.0146	.0284	.0488	.0768	.1128	.1562
	5	.0000	.0000	.0001	.0003	.0010	.0024	.0053	.0102	.0185	.0312
6	0	.7351	.5314	.3771	.2621	.1780	.1176	.0754	.0467	.0277	.0156
	1	.2321	.3543	.3993	.3932	.3560	.3025	.2437	.1866	.1359	.0938
	2	.0305	.0984	.1762	.2458	.2966	.3241	.3280	.3110	.2780	.2344
	3	.0021	.0146	.0415	.0819	.1318	.1852	.2355	.2765	.3032	.3125
	4	.0001	.0012	.0055	.0154	.0330	.0595	.0951	.1382	.1861	.2344
	5	.0000	.0001	.0004	.0015	.0044	.0102	.0205	.0369	.0609	.0938
	6	.0000	.0000	.0000	.0001	.0002	.0007	.0018	.0041	.0083	.0156
7	0	.6983	.4783	.3206	.2097	.1335	.0824	.0490	.0280	.0152	.0078
	1	.2573	.3720	.3960	.3670	.3115	.2471	.1848	.1306	.0872	.0547
	2	.0406	.1240	.2097	.2753	.3115	.3177	.2985	.2613	.2140	.1641
	3	.0036	.0230	.0617	.1147	.1730	.2269	.2679	.2903	.2918	.2734
	4	.0002	.0026	.0109	.0287	.0577	.0972	.1442	.1935	.2388	.2734
	5	.0000	.0002	.0012	.0043	.0115	.0250	.0466	.0774	.1172	.1641
	6	.0000	.0000	.0001	.0004	.0013	.0036	.0084	.0172	.0320	.0547
	7	.0000	.0000	.0000	.0000	.0001	.0002	.0006	.0016	.0037	.0078
8	0	.6634	.4305	.2725	.1678	.1001	.0576	.0319	.0168	.0084	.0039
	1	.2793	.3826	.3847	.3355	.2670	.1977	.1373	.0896	.0548	.0312
	2	.0515	.1488	.2376	.2936	.3115	.2965	.2587	.2090	.1569	.1094
	3	.0054	.0331	.0839	.1468	.2076	.2541	.2786	.2787	.2568	.2188
	4	.0004	.0046	.0185	.0459	.0865	.1361	.1875	.2322	.2627	.2734
	5	.0000	.0004	.0026	.0092	.0231	.0467	.0808	.1239	.1719	.2188
	6	.0000	.0000	.0002	.0011	.0038	.0100	.0217	.0413	.0703	.1094
	7	.0000	.0000	.0000	.0001	.0004	.0012	.0033	.0079	.0164	.0312
	8	.0000	.0000	.0000	.0000	.0000	.0001	.0002	.0007	.0017	.0039

$$\binom{n}{r} \equiv n_{C_r} \equiv \frac{(n)_r}{r!}.$$

(Thus, $\binom{n}{0} \equiv 1$.)

and a **random variable (r.v.)** X is a number we can calculate after the experiment has been performed, such as

— number of defective chips among 5 drawn
— mg. of vitamin C in a 5 oz. glass of drink
— no. of offspring of a fruitfly.

With the binomial model, in our case *we find*

$$P(x \text{ of 5 are defective}) = \binom{5}{x} (.30)^x(1-.30)^{5-x}$$

so

$$P[3 \text{ defective}] = \binom{5}{3}(.3)^3(.7)^2$$

$$= \frac{5!}{3!2!}(.027)(.49)$$

$$= (10)(.01323) = .1323.$$

Thus, in the long run we expect 13.23% of the boxes to have 3 defective and 2 good chips. Similarly,

$$P[X=0] = .1681$$
$$P[X=1] = .3602$$
$$P[X=2] = .3087$$
$$P[X=3] = .1323$$
$$P[X=4] = .0284$$
$$P[X=5] = .0024,$$

and tables are widely available, such as Table 1. (With the hypergeometric distribution, the same results are obtained when N is large, as is the case here.)

Model Building: Probability Models vs. Statistics

Probability deals with situations where the underlying experimental structure is fully specified, as with our computer chip bin; we can then compute the probabilities of outcomes of interest.

Statistics deals with situations where the structure of the experiment is only incompletely specified, and with it we can make inferences based on data obtained from real-world experiments. (We are concerned with designing such experiments so that maximal information may be extracted from their data.)

The sample space Ω (set of possible outcomes of our experiment) can be finite, denumerable, or **continuous** (non-denumerably infinite). If it is not continuous, then it is called **discrete**. The outcome of an experiment (for example, the outcome of playing a game of chance) is completely described by a sample space Ω with probability function $P(\cdot)$ on its events, and by observing an $\omega \in \Omega$ selected according to $P(\cdot)$ one can answer questions concerning the experiment. However, usually we need only observe some *function* of ω (for historical reasons such functions are called **random variables**) in order to answer the questions of interest. [For example, we may toss a pair of dice and, instead of observing the pair (No. on first die, No. on second die), simply report the resulting sum; this sum is a *function* of (No. on first die, No. on second die), namely the sum of the two numbers, and it is often sufficient to answer the questions of practical import.]

Definition. A **random variable** X is a real-valued function with domain Ω [that is, for each $\omega \in \Omega$, $X(\omega) \in \mathbf{R} = \{y: -\infty < y < +\infty\}$]. An ***n*-dimensional random variable** (or n-dimensional random vector, or vector random variable), $\mathbf{X} = (X_1, \ldots, X_n)$, is a function with domain Ω and range in Euclidean n-space R^n [that is, for each $\omega \in \Omega$, $X(\omega) \in \mathbf{R}^n = \{(y_1, \ldots, y_n): -\infty < y_i < +\infty \ (1 \leq i \leq n)\}$].

We may picture the situation as shown in Figure 1

We can associate probabilities with values of X (or $\mathbf{X}$) because probabilities are already associated with points $\omega \in \Omega$ by the probability function $P(\omega)$. For example, suppose we have the situation shown in Example 2. Then $P[X=2] = P(\{\omega: \omega \in \Omega, X(\omega) = 2\}) = P(\{\omega_1, \omega_2\}) = .4$, $P[X=0] =$

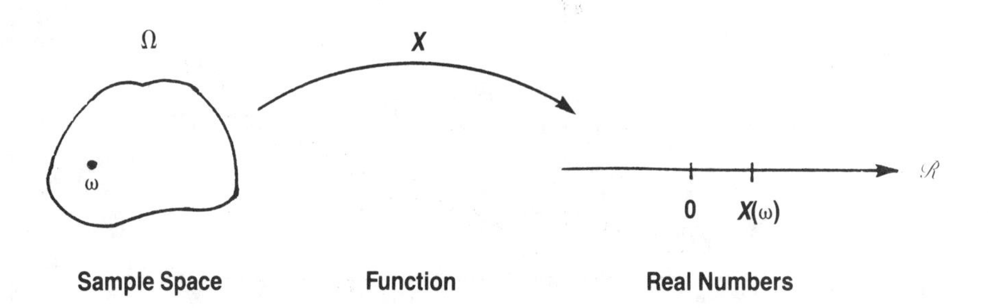

Figure 1

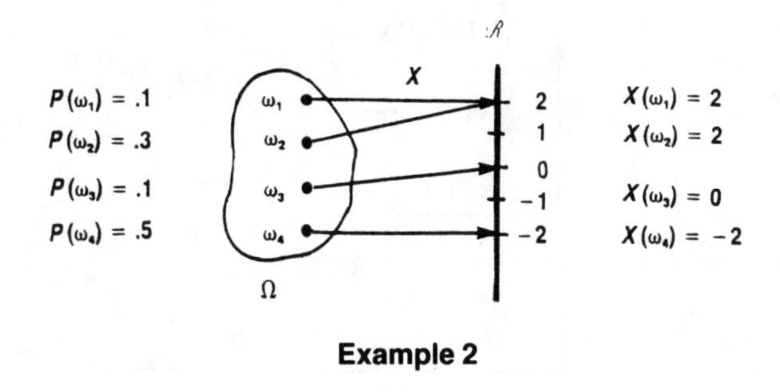

Example 2

.1, $P[X=-2] = .5$. Our discussion will now be for a one-dimensional random variable X.

Definition. If X is a random variable, its **distribution function** (or **cumulative distribution function**, or **d.f.**, or **c.d.f.**) is defined by

$$F_X(x) = P[X \leq x], \text{ for all } x \in (-\infty, +\infty).$$

Definition. X is called a (one-dimensional) **discrete random variable** if it is a random variable which assumes only a finite or denumerable number of values.

Definition. The **probability function** of a discrete r.v. X which assumes values $x_1, x_2, \ldots$ is $p(x_1), p(x_2), \ldots,$ where $p(x_i) = P[X=x_i]$ $(i = 1, 2, \ldots)$.

For example, a typical discrete probability function and its associated distribution function are as shown in Example 3.

In the particular case $P(\omega_1) = .1$, $P(\omega_2) = .3$, $P(\omega_3) = .1$, $P(\omega_4) = .5$ with $X(\omega_1) = 2$, $X(\omega_2) = 2$, $X(\omega_3) = 0$, $X(\omega_4) = -2$, we obtain the discrete probability function and distribution function as shown in Figure 2.

Test Problem

A machine pits prunes with an average of 1 percent of its output being defective. What is the probability function of the number of defectives in a sample of 60 consecutive prunes if

(1) each "pitted" prune has probability 1/100 of being defective and is defective or good independently of the other prunes;

(2) the machine turns out cycle after cycle of prunes with 99 good followed by 1 bad, and the "time" we started to sample in a cycle was random.

Solution

In case **(1)** the number of defectives X in a sample of size 60 is binomial with $n=60$ and $p=.01$. Hence

$$p_X(x) = \begin{cases} \binom{60}{x} (.01)^x(.99)^{60-x}, & x=0, 1, \ldots, 60 \\ 0, & \text{otherwise.} \end{cases}$$

In case **(2)** the number of defectives X in a sample of 60 consecutive prunes is either 1 or 0. It is 1 (i.e., $X=1$) iff our sample overlaps the 100th prune in the cycle, i.e. iff we start sampling with a prune other than prune 1, or prune 2, or . . . , or prune 40; and it is 0 (i.e. $X = 0$) otherwise. Hence

$$p_X(x) = \begin{cases} 40/100, & x = 0 \\ 60/100, & x = 1 \\ 0, & \text{otherwise.} \end{cases}$$

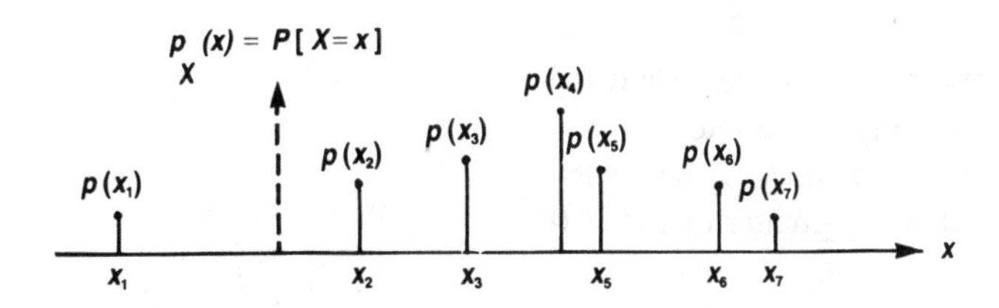

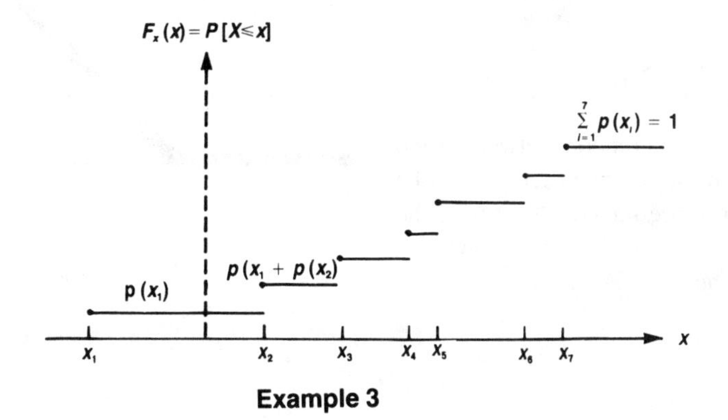

Example 3

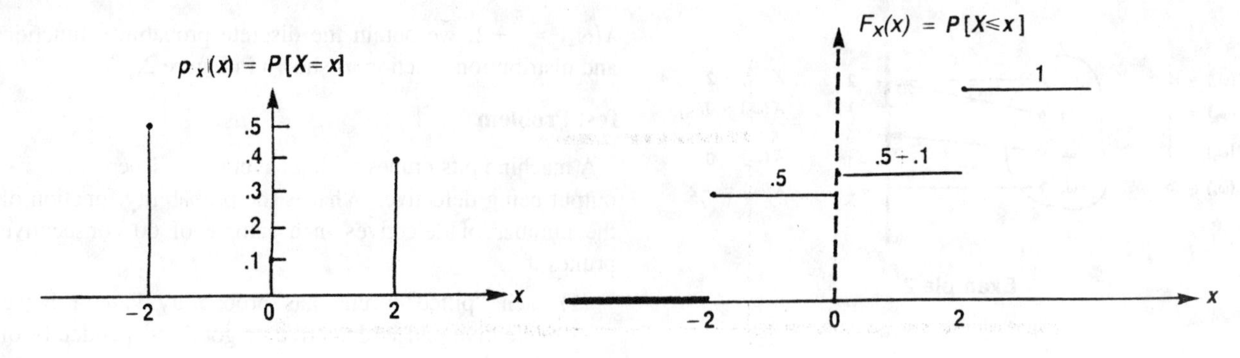

Figure 2

Definition. X is called a (one-dimensional) **(absolutely) continuous random variable** if its d.f. $F_X(x)$ may be represented as

$$F_X(x) = \int_{-\infty}^{x} f_X(y)dy \qquad (-\infty < x < +\infty),$$

where $f_X(y)$ is a **probability density function (p.d.f.)**; that is by definition,

$$f_X(y) \geq 0 \text{ for all } y, \text{ and}$$

$$\int_{-\infty}^{\infty} f_X(y) \geq 0 \text{ for all } y.$$

In this case $F_X(x)$ is called an **(absolutely) continuous d.f.**

Example. Suppose that we pick a number "at random" between 0 and 1, say X (a random variable). What is the d.f. of X, and is X absolutely continuous or discrete (or as is sometimes possible, neither)?

Since the number is picked at random between 0 and 1, it should have equal probability of being in the interval $(0, \frac{1}{2})$ and the interval $(\frac{1}{2}, 1)$, while its probability of being less than 0 is 0, and its probability of being greater than 1 is 0. Thus,

$$F_X(x) = 0, \text{ if } x \leq 0,$$

$$F_X(\tfrac{1}{2}) = \tfrac{1}{2},$$

$$F_X(x) = 1, \text{ if } x \geq 1.$$

However, since the point is chosen at random between 0 and 1, the probability that it lies in the interval $(0,x)$ should be the proportion that the length of the interval $(0,x)$ is of the interval $(0,1)$; that is, $x/1 = x$ if $0 \leq x \leq 1$. Thus $F_X(x) = x$ if $0 \leq x \leq 1$, and $F_X(x)$ is as shown in Figure 3.

Now let us define $f(x) = 1$ if $0 \leq x \leq 1$ and $f(x) = 0$ otherwise. Then $f(y) \geq 0$ for all y and $\int_{-\infty}^{\infty} f(y)dy = 1$, so by definition $f(y)$ is a p.d.f. But, for all x,

$$F_X(x) = \int_{-\infty}^{x} f(y)dy,$$

so X is an (absolutely) continuous r.v. with p.d.f. $f(x)$ as shown in Figure 4. Note that the probability $P[X \leq 3]$, in the continuous r.v. case, is the area under the p.d.f. $f(x)$ from $-\infty$ to 3 (see Figure 5). It follows that by plotting a **histogram** of your data you can obtain at least a rough approximation of the p.d.f. $f(x)$. For example, suppose 100 batteries are tested on hours-to-failure in continuous operation and we find that 58 fail by 1 hour, 22 fail between 1 and 2 hours, 13 fail between 2 and 3 hours, and 7 fail between 3 and 4 hours. Then a plot as shown in Figure 6 suggests a possible form (see Figure 7).

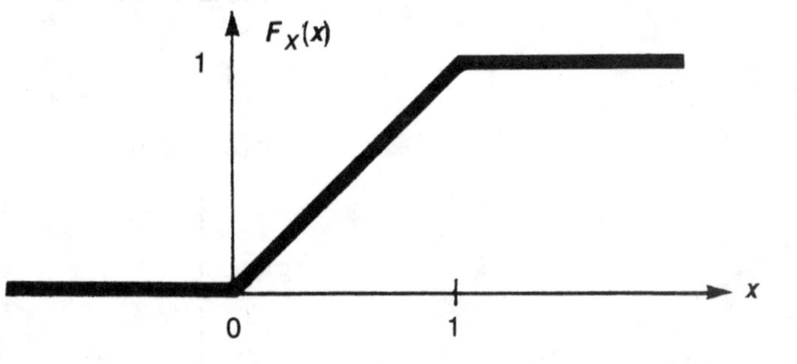

Figure 3

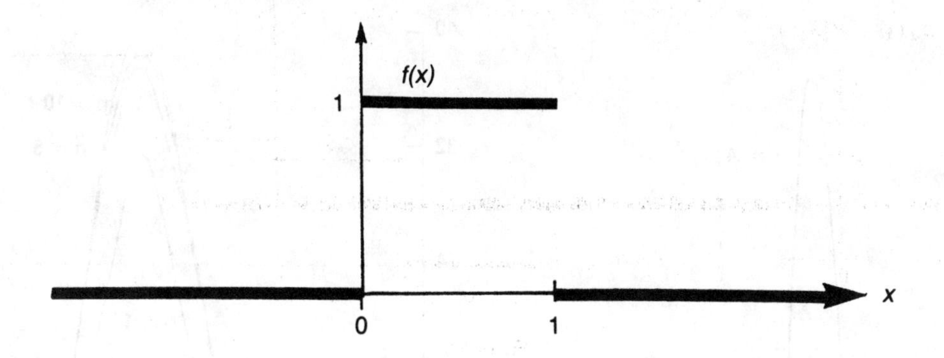

Figure 4

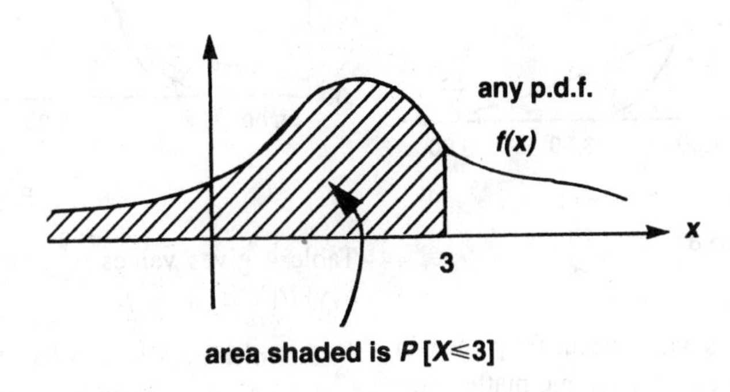

Figure 5

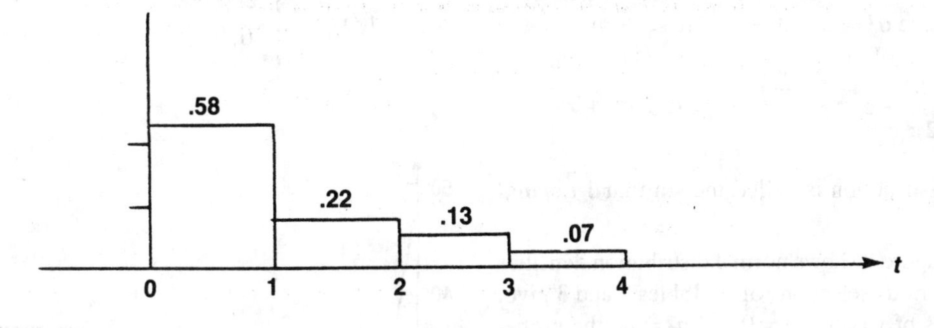

Figure 6

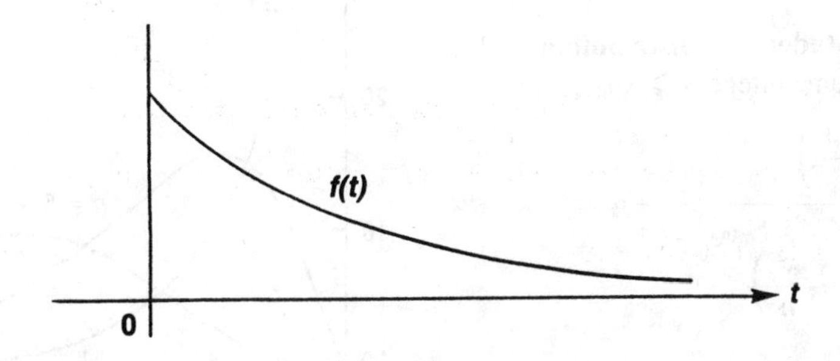

Figure 7

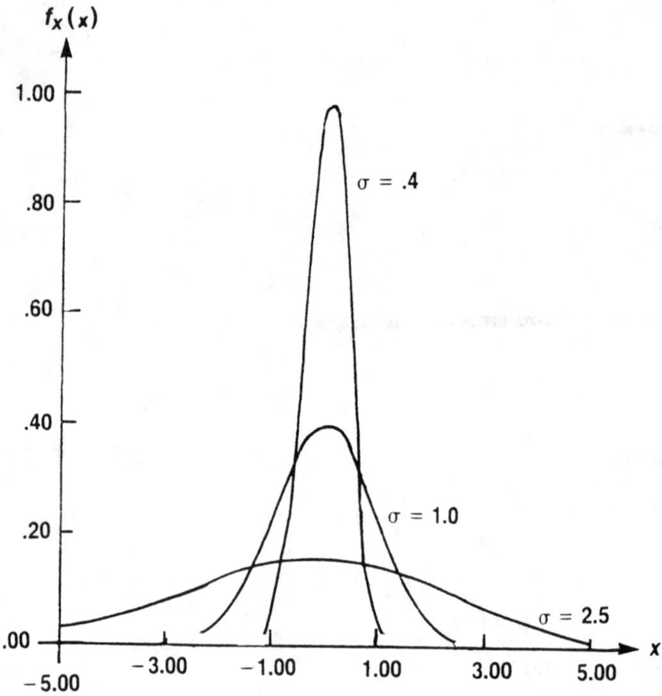

Figure 8

for the underlying p.d.f. P.d.f.'s which occur frequently in statistics, with their special names, graphs, and mathematical formulas and tables, are:

Definition. An r.v. X has the **normal** distribution $N(\mu,\sigma^2)$ if (for some $\sigma^2 > 0$ and $-\infty < \mu < +\infty$)

$$f_X(x) = \frac{1}{\sqrt{2\pi}\,\sigma} e^{-\frac{1}{2}\left(\frac{x-\mu}{\sigma}\right)^2}, \quad -\infty < x < +\infty.$$

The $N(0,1)$ distribution is called the **standard normal** distribution.

The graph of Figure 8 shows normal distribution densities with $\mu = 0$ and various selections of σ. Tables 2 and 3 give, for various values of y (when $\mu = 0$ and $\sigma^2 = 1$), the probability $P[X \leq y]$.

Definition. A r.v. X has **Student's *t*-distribution** with n degrees of freedom if (for some integer $n > 0$)

$$f_X(x) = \frac{\Gamma\left(\frac{n+1}{2}\right)}{\sqrt{n\pi}\,\Gamma\left(\frac{n}{2}\right)\left(1 + \frac{x^2}{n}\right)^{(n+1)/2}}, \quad -\infty < x < +\infty.$$

The following graph (Figure 9) shows t-distribution densities with several selections of n, and the $N(0,1)$ density (labeled $n = \infty$).

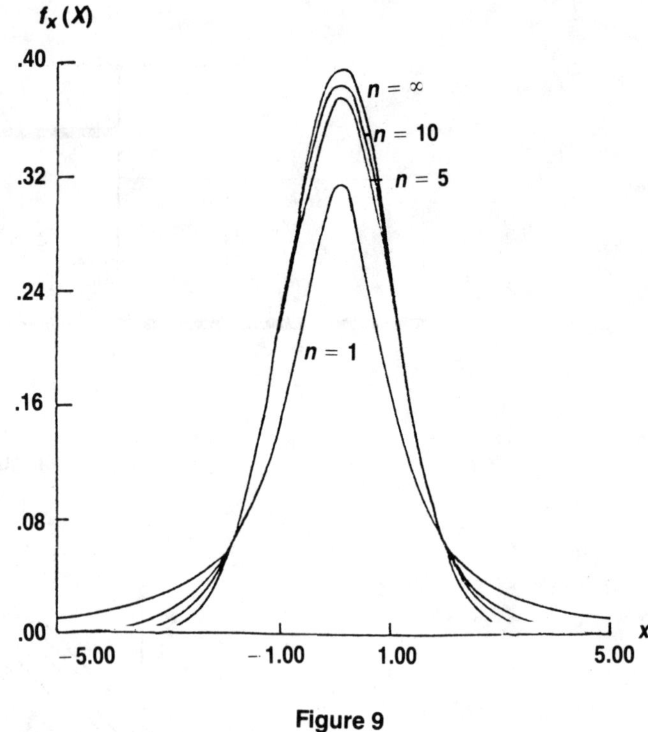

Figure 9

Table 4 gives values of y such that (for various values of n,γ) $P[X \leq y] = \gamma$.

Definition. An r.v. X has the **chi-square** distribution with n degrees of freedom, say $X_n^2(0)$, if (for some integer $n > 0$)

$$f_X(x) = \begin{cases} \frac{1}{2^{n/2}\Gamma(n/2)} x^{(n/2)-1}e^{-x/2}, & 0 \leq x < \infty \\ 0, & \text{otherwise.} \end{cases}$$

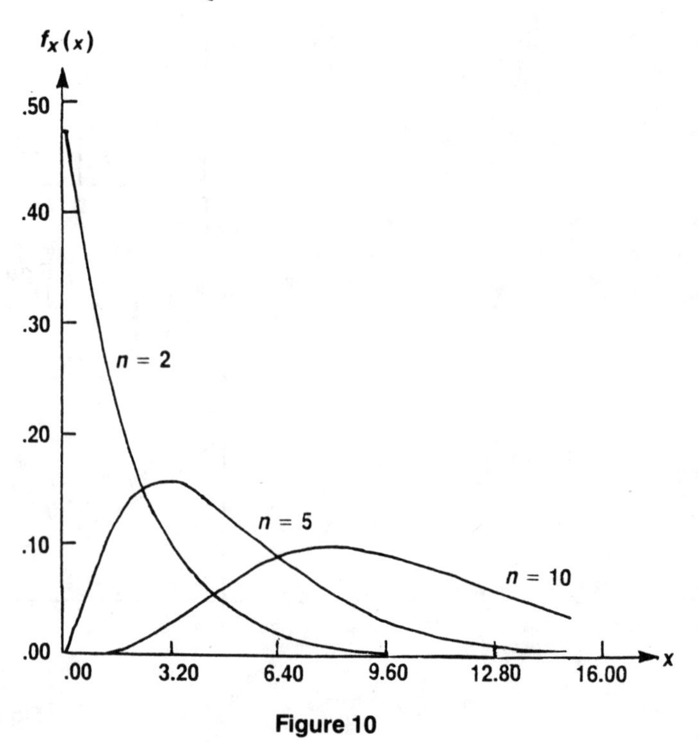

Figure 10

Table 2: $P[X \leq y] = \int_{-\infty}^{y} f_X(x)dx$

y	0	1	2	3	4	5	6	7	8	9
−3.0	.0013	.0010	.0007	.0005	.0003	.0002	.0002	.0001	.0001	.0000
−2.9	.0019	.0018	.0017	.0017	.0016	.0016	.0015	.0015	.0014	.0014
−2.8	.0026	.0025	.0024	.0023	.0023	.0022	.0021	.0021	.0020	.0019
−2.7	.0035	.0034	.0033	.0032	.0031	.0030	.0029	.0028	.0027	.0026
−2.6	.0047	.0045	.0044	.0043	.0041	.0040	.0039	.0038	.0037	.0036
−2.5	.0062	.0060	.0059	.0057	.0055	.0054	.0052	.0051	.0049	.0048
−2.4	.0082	.0080	.0078	.0075	.0073	.0071	.0069	.0068	.0066	.0064
−2.3	.0107	.0104	.0102	.0099	.0096	.0094	.0091	.0089	.0087	.0084
−2.2	.0139	.0136	.0132	.0129	.0126	.0122	.0119	.0116	.0113	.0110
−2.1	.0179	.0174	.0170	.0166	.0162	.0158	.0154	.0150	.0146	.0143
−2.0	.0228	.0222	.0217	.0212	.0207	.0202	.0197	.0192	.0188	.0183
−1.9	.0287	.0281	.0274	.0268	.0262	.0256	.0250	.0244	.0238	.0233
−1.8	.0359	.0352	.0344	.0336	.0329	.0322	.0314	.0307	.0300	.0294
−1.7	.0446	.0436	.0427	.0418	.0409	.0401	.0392	.0384	.0375	.0367
−1.6	.0548	.0537	.0526	.0516	.0505	.0495	.0485	.0475	.0465	.0455
−1.5	.0668	.0655	.0643	.0630	.0618	.0606	.0594	.0582	.0570	.0559
−1.4	.0808	.0793	.0778	.0764	.0749	.0735	.0722	.0708	.0694	.0681
−1.3	.0968	.0951	.0934	.0918	.0901	.0885	.0869	.0853	.0838	.0823
−1.2	.1151	.1131	.1112	.1093	.1075	.1056	.1038	.1020	.1003	.0985
−1.1	.1357	.1335	.1314	.1292	.1271	.1251	.1230	.1210	.1190	.1170
−1.0	.1587	.1562	.1539	.1515	.1492	.1469	.1446	.1423	.1401	.1379
− .9	.1841	.1814	.1788	.1762	.1736	.1711	.1685	.1660	.1635	.1611
− .8	.2119	.2090	.2061	.2033	.2005	.1977	.1949	.1922	.1894	.1867
− .7	.2420	.2389	.2358	.2327	.2297	.2266	.2236	.2206	.2177	.2148
− .6	.2743	.2709	.2676	.2643	.2611	.2578	.2546	.2514	.2483	.2451
− .5	.3085	.3050	.3015	.2981	.2946	.2912	.2877	.2843	.2810	.2776
− .4	.3446	.3409	.3372	.3336	.3300	.3264	.3228	.3192	.3156	.3121
− .3	.3821	.3783	.3745	.3707	.3669	.3632	.3594	.3557	.3520	.3483
− .2	.4207	.4168	.4129	.4090	.4052	.4013	.3974	.3936	.3897	.3859
− .1	.4602	.4562	.4522	.4483	.4443	.4404	.4364	.4325	.4286	.4247
− 0	.5000	.4960	.4920	.4880	.4840	.4801	.4761	.4721	.4681	.4641

The graph given in Figure 10 shows chi-square distribution densities with various selections of n.

Table 5 gives values of y such that (for various values of n, γ) $P[X \leq y] = \gamma$.

Definition. A r.v. X has the ***F*-distribution** $F(n_1, n_2)$ if (for some positive integers n_1, n_2)

$$f_x(x) = \begin{cases} \dfrac{\Gamma\left(\dfrac{n_1+n_2}{2}\right)\left(\dfrac{n_1}{n_2}\right)^{n_1/2}}{\Gamma\left(\dfrac{n_1}{2}\right)\Gamma\left(\dfrac{n_2}{2}\right)} \dfrac{x^{(1/2)(n_1-2)}}{\left(1+\dfrac{n_1}{n_2}x\right)^{(1/2)(n_1+n_2)}}, & 0 < x < \infty \\ 0, & \text{otherwise.} \end{cases}$$

The following graphs (Figures 11, 12, 13) show F-distribution densities with various selections of n_1 and n_2.

Table 6 gives values of y such that (for various values of n_1, n_2) $P[X \leq y] = .95$.

Note that in these pdf's we have used the gamma function, a generalization of the factorial function on positive integers:

Definition. (Gamma Functions) For $n > 0$ (not necessarily an integer),

Table 3: $P[X \leq y] = \int_{-\infty}^{y} f_x(x)dx$ for various y

y	0	1	2	3	4	5	6	7	8	9
.0	.5000	.5040	.5080	.5120	.5160	.5199	.5239	.5279	.5319	.5359
.1	.5398	.5438	.5478	.5517	.5557	.5596	.5636	.5675	.5714	.5753
.2	.5793	.5832	.5871	.5910	.5948	.5987	.6026	.6064	.6103	.6141
.3	.6179	.6217	.6255	.6293	.6331	.6368	.6406	.6443	.6480	.6517
.4	.6554	.6591	.6628	.6664	.6700	.6736	.6772	.6808	.6844	.6879
.5	.6915	.6950	.6985	.7019	.7054	.7088	.7123	.7157	.7190	.7224
.6	.7257	.7291	.7324	.7357	.7389	.7422	.7454	.7486	.7517	.7549
.7	.7580	.7611	.7642	.7673	.7703	.7734	.7764	.7794	.7823	.7852
.8	.7881	.7910	.7939	.7967	.7995	.8023	.8051	.8078	.8106	.8133
.9	.8159	.8186	.8212	.8238	.8264	.8289	.8315	.8340	.8365	.8389
1.0	.8413	.8438	.8461	.8485	.8508	.8531	.8554	.8577	.8599	.8621
1.1	.8643	.8665	.8686	.8708	.8729	.8749	.8770	.8790	.8810	.8830
1.2	.8849	.8869	.8888	.8907	.8925	.8944	.8962	.8980	.8997	.9015
1.3	.9032	.9049	.9066	.9082	.9099	.9115	.9131	.9147	.9162	.9177
1.4	.9192	.9207	.9222	.9236	.9251	.9265	.9278	.9292	.9306	.9319
1.5	.9332	.9345	.9357	.9370	.9382	.9394	.9406	.9418	.9430	.9441
1.6	.9452	.9463	.9474	.9484	.9495	.9505	.9515	.9525	.9535	.9545
1.7	.9554	.9564	.9573	.9582	.9591	.9599	.9608	.9616	.9625	.9633
1.8	.9641	.9648	.9656	.9664	.9671	.9678	.9686	.9693	.9700	.9706
1.9	.9713	.9719	.9726	.9732	.9738	.9744	.9750	.9756	.9762	.9767
2.0	.9772	.9778	.9783	.9788	.9793	.9798	.9803	.9808	.9812	.9817
2.1	.9821	.9826	.9830	.9834	.9838	.9842	.9846	.9850	.9854	.9857
2.2	.9861	.9864	.9868	.9871	.9874	.9878	.9881	.9884	.9887	.9890
2.3	.9893	.9896	.9898	.9901	.9904	.9906	.9909	.9911	.9913	.9916
2.4	.9918	.9920	.9922	.9925	.9927	.9929	.9931	.9932	.9934	.9936
2.5	.9938	.9940	.9941	.9943	.9945	.9946	.9948	.9949	.9951	.9952
2.6	.9953	.9955	.9956	.9957	.9959	.9960	.9961	.9962	.9963	.9964
2.7	.9965	.9966	.9967	.9968	.9969	.9970	.9971	.9972	.9973	.9974
2.8	.9974	.9975	.9976	.9977	.9977	.9978	.9979	.9979	.9980	.9981
2.9	.9981	.9982	.9982	.9983	.9984	.9984	.9985	.9985	.9986	.9986
3.	.9987	.9990	.9993	.9995	.9997	.9998	.9998	.9999	.9999	1.0000

Table 4: $P[X \leq y] = \int_{-\mu}^{y} f_x(x)dx = \gamma$

n \ γ	·60	·75	·90	·95	·975	·99	·995	·9975	·999	·9995
1	·325	1·000	3·078	6·314	12·706	31·821	63·657	127·32	318·31	636·62
2	·289	·816	1·886	2·920	4·303	6·965	9·925	14·089	22·327	31·598
3	·277	·765	1·638	2·353	3·182	4·541	5·841	7·453	10·214	12·924
4	·271	·741	1·533	2·132	2·776	3·747	4·604	5·598	7·173	8·610
5	·267	·727	1·476	2·015	2·571	3·365	4·032	4·773	5·893	6·869
6	·265	·718	1·440	1·943	2·447	3·143	3·707	4·317	5·208	5·959
7	·263	·711	1·415	1·895	2·365	2·998	3·499	4·029	4·785	5·408
8	·262	·706	1·397	1·860	2·306	2·896	3·355	3·833	4·501	5·041
9	·261	·703	1·383	1·833	2·262	2·821	3·250	3·690	4·297	4·781
10	·260	·700	1·372	1·812	2·228	2·764	3·169	3·581	4·144	4·587
11	·260	·697	1·363	1·796	2·201	2·718	3·106	3·497	4·025	4·437
12	·259	·695	1·356	1·782	2·179	2·681	3·055	3·428	3·930	4·318
13	·259	·694	1·350	1·771	2·160	2·650	3·012	3·372	3·852	4·221
14	·258	·692	1·345	1·761	2·145	2·624	2·977	3·326	3·787	4·140
15	·258	·691	1·341	1·753	2·131	2·602	2·947	3·286	3·733	4·073
16	·258	·690	1·337	1·746	2·120	2·583	2·921	3·252	3·686	4·015
17	·257	·689	1·333	1·740	2·110	2·567	2·898	3·222	3·646	3·965
18	·257	·688	1·330	1·734	2·101	2·552	2·878	3·197	3·610	3·922
19	·257	·688	1·328	1·729	2·093	2·539	2·861	3·174	3·579	3·883
20	·257	·687	1·325	1·725	2·086	2·528	2·845	3·153	3·552	3·850
21	·257	·686	1·323	1·721	2·080	2·518	2·831	3·135	3·527	3·819
22	·256	·686	1·321	1·717	2·074	2·508	2·819	3·119	3·505	3·792
23	·256	·685	1·319	1·714	2·069	2·500	2·807	3·104	3·485	3·767
24	·256	·685	1·318	1·711	2·064	2·492	2·797	3·091	3·467	3·745
25	·256	·684	1·316	1·708	2·060	2·485	2·787	3·078	3·450	3·725
26	·256	·684	1·315	1·706	2·056	2·479	2·779	3·067	3·435	3·707
27	·256	·684	1·314	1·703	2·052	2·473	2·771	3·057	3·421	3·690
28	·256	·683	1·313	1·701	2·048	2·467	2·763	3·047	3·408	3·674
29	·256	·683	1·311	1·699	2·045	2·462	2·756	3·038	3·396	3·659
30	·256	·683	1·310	1·697	2·042	2·457	2·750	3·030	3·385	3·646
40	·255	·681	1·303	1·684	2·021	2·423	2·704	2·971	3·307	3·551
60	·254	·679	1·296	1·671	2·000	2·390	2·660	2·915	3·232	3·460
120	·254	·677	1·289	1·658	1·980	2·358	2·617	2·860	3·160	3·373
∞	·253	·674	1·282	1·645	1·960	2·326	2·576	2·807	3·090	3·291

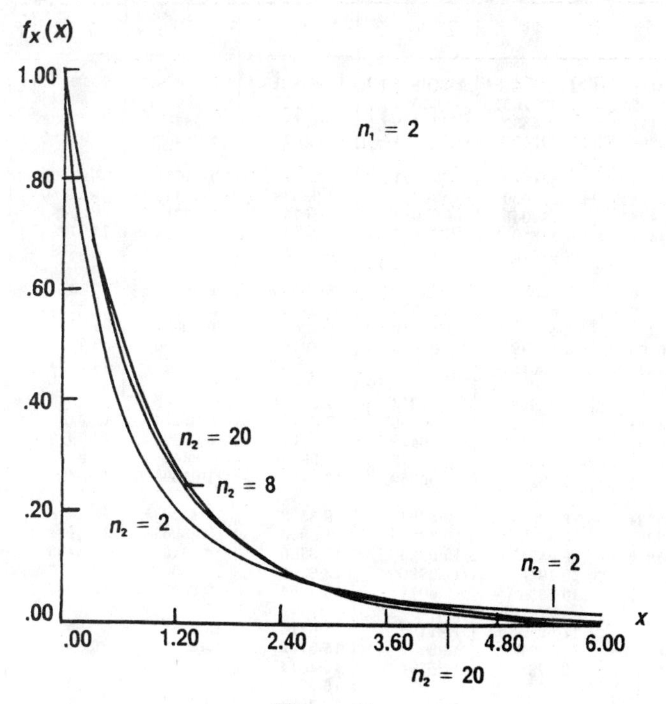

Figure 11

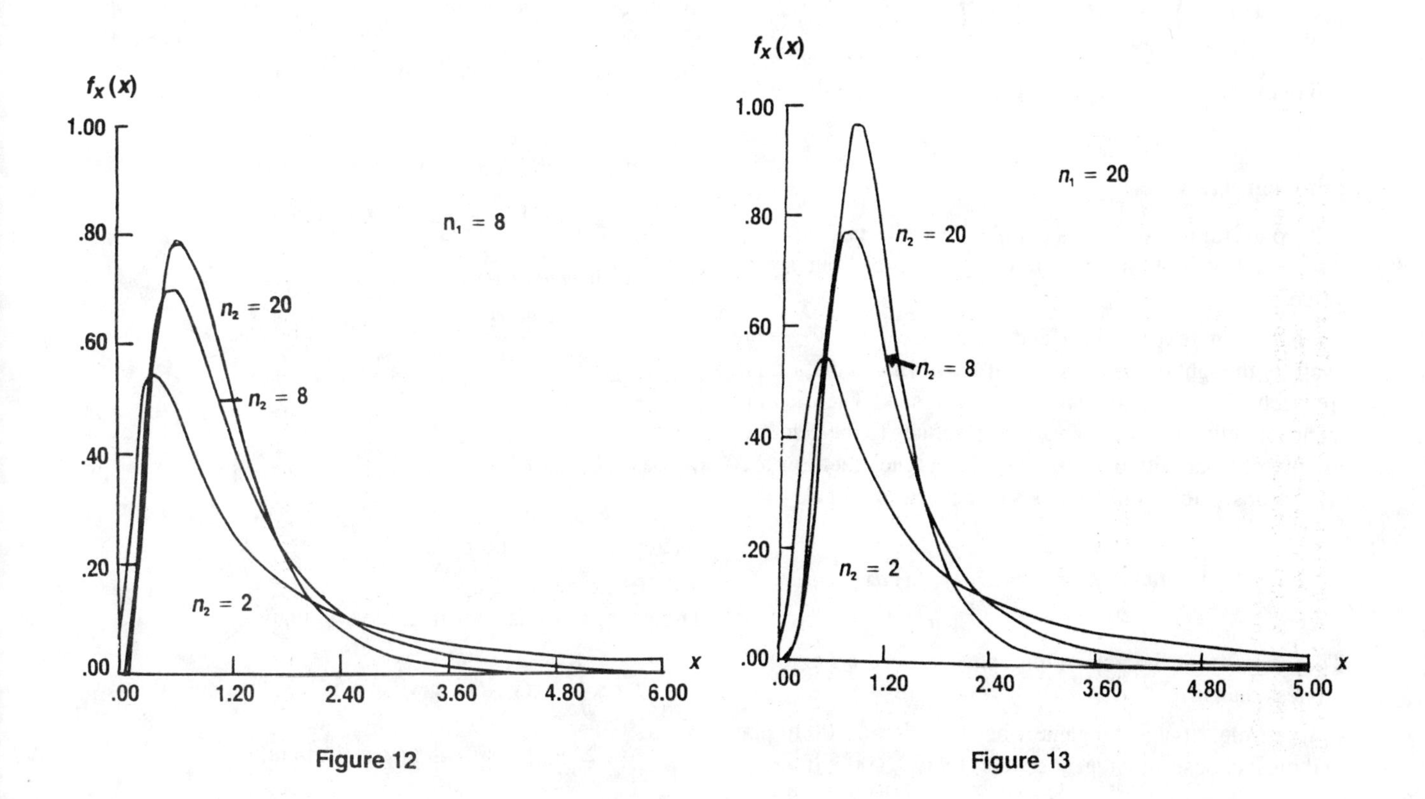

Figure 12

Figure 13

Table 5: $P[X \leq y] = \int_0^y f_x(x)\,dx = \gamma$

n \ γ	·005	·01	·025	·05	·10	·25	·50
1	392704.10^{-10}	157088.10^{-9}	982069.10^{-9}	393214.10^{-8}	·0157908	·1015308	·454936
2	·0100251	·0201007	·0506356	·102587	·210721	·575364	1·38629
3	·0717218	·114832	·215795	·351846	·584374	1·212534	2·36597
4	·206989	·297109	·484419	·710723	1·063623	1·92256	3·35669
5	·411742	·554298	·831212	1·145476	1·61031	2·67460	4·35146
6	·675727	·872090	1·23734	1·63538	2·20413	3·45460	5·34812
7	·989256	1·239043	1·68987	2·16735	2·83311	4·25485	6·34581
8	1·34441	1·64650	2·17973	2·73264	3·48954	5·07064	7·34412
9	1·73493	2·08790	2·70039	3·32511	4·16816	5·89883	8·34283
10	2·15586	2·55821	3·24697	3·94030	4·86518	6·73720	9·34182
11	2·60322	3·05348	3·81575	4·57481	5·57778	7·58414	10·3410
12	3·07382	3·57057	4·40379	5·22603	6·30380	8·43842	11·3403
13	3·56503	4·10692	5·00875	5·89186	7·04150	9·29907	12·3398
14	4·07467	4·66043	5·62873	6·57063	7·78953	10·1653	13·3393
15	4·60092	5·22935	6·26214	7·26094	8·54676	11·0365	14·3389
16	5·14221	5·81221	6·90766	7·96165	9·31224	11·9122	15·3385
17	5·69722	6·40776	7·56419	8·67176	10·0852	12·7919	16·3382
18	6·26480	7·01491	8·23075	9·39046	10·8649	13·6753	17·3379
19	6·84397	7·63273	8·90652	10·1170	11·6509	14·5620	18·3377
20	7·43384	8·26040	9·59078	10·8508	12·4426	15·4518	19·3374
21	8·03365	8·89720	10·28293	11·5913	13·2396	16·3444	20·3372
22	8·64272	9·54249	10·9823	12·3380	14·0415	17·2396	21·3370
23	9·26043	10·19567	11·6886	13·0905	14·8480	18·1373	22·3369
24	9·88623	10·8564	12·4012	13·8484	15·6587	19·0373	23·3367
25	10·5197	11·5240	13·1197	14·6114	16·4734	19·9393	24·3366
26	11·1602	12·1981	13·8439	15·3792	17·2919	20·8434	25·3365
27	11·8076	12·8785	14·5734	16·1514	18·1139	21·7494	26·3363
28	12·4613	13·5647	15·3079	16·9279	18·9392	22·6572	27·3362
29	13·1211	14·2565	16·0471	17·7084	19·7677	23·5666	28·3361
30	13·7867	14·9535	16·7908	18·4927	20·5992	24·4776	29·3360
40	20·7065	22·1643	24·4330	26·5093	29·0505	33·6603	39·3353
50	27·9907	29·7067	32·3574	34·7643	37·6886	42·9421	49·3349
60	35·5345	37·4849	40·4817	43·1880	46·4589	52·2938	59·3347
70	43·2752	45·4417	48·7576	51·7393	55·3289	61·6983	69·3345
80	51·1719	53·5401	57·1532	60·3915	64·2778	71·1445	79·3343
90	59·1963	61·7541	65·6466	69·1260	73·2911	80·6247	89·3342
100	67·3276	70·0649	74·2219	77·9295	82·3581	90·1332	99·3341

$$\Gamma(n) \equiv \int_0^\infty x^{n-1} e^{-x}\,dx = 2\int_0^\infty y^{2n-1} e^{-y^2}\,dy.$$

Theorem. $\Gamma(1) = 1$. If n is a positive integer, then $\Gamma(n) = (n-1)!$. $\Gamma(\frac{1}{2}) = \sqrt{\pi}$.

Pdf Characteristics

Two useful measures of a population are its **center** (e.g., its average value) and **dispersion** (how spread-out are its values).

The mean (expectation) and variance of a r.v. X are usefully thought of as **measures of location** and **dispersion** (respectively) of the distribution of the r.v. X. The **mean** of an absolutely continuous r.v. X corresponds to the center of gravity of a bar with density $f_X(x)$. Then "the" **median** of X (if X is absolutely continuous) **is any point m where**

$$\int_{-\infty}^{m} f_X(x)\,dx = \tfrac{1}{2} = \int_{m}^{\infty} f_X(x)\,dx.$$

That is, 50% of the population values exceed m, and 50% are less than m.

Example. Imagine a game where you pay \$1.00 to play and must choose an integer between 1 and 10,000; if you do not choose the integer that was previously written on a slip of paper, you lose your dollar, while if you guess correctly you win back your entry fee plus \$10,000. In this game you win X and

$$P[X = -1] = 9{,}999/10{,}000$$
$$P[X = 10{,}000] = 1/10{,}000.$$

Hence your *mean (net) winnings* are

$$(-1)\frac{9{,}999}{10{,}000} + (10{,}000)\frac{1}{10{,}000} = \frac{1}{10{,}000}.$$

In general,

Definition. For a r.v. X we define its **expectation** [EX or $E(X)$] as $EX = \int_{-\infty}^{\infty} x f_X(x)\,dx$ [$EX = \sum x_i p_X(x_i)$] if X is absolutely continuous with density function $f_X(x)$ [if X is discrete with probability function $p_X(x)$].

Other expected values:

Theorem. If X is a r.v., then the expectation

$$Eg(X) = \int_{-\infty}^{\infty} g(x) f_X(x)\,dx \text{ when } X \text{ is absolutely continuous}$$
$$= \sum g(x_i)\, p_X(x_i) \text{ when } X \text{ is discrete.}$$

Table 5: The Chi-Square Distribution (Continued)

y such that $P[X \leq y] = \int_0^y f_x(x)\,dx = \gamma$

n \ γ	·75	·90	·95	·975	·99	·995	·999
1	1·32330	2·70554	3·84146	5·02389	6·63490	7·87944	10·828
2	2·77259	4·60517	5·99146	7·37776	9·21034	10·5966	13·816
3	4·10834	6·25139	7·81473	9·34840	11·3449	12·8382	16·266
4	5·38527	7·77944	9·48773	11·1433	13·2767	14·8603	18·467
5	6·62568	9·23636	11·0705	12·8325	15·0863	16·7496	20·515
6	7·84080	10·6446	12·5916	14·4494	16·8119	18·5476	22·458
7	9·03715	12·0170	14·0671	16·0128	18·4753	20·2777	24·322
8	10·2189	13·3616	15·5073	17·5345	20·0902	21·9550	26·125
9	11·3888	14·6837	16·9190	19·0228	21·6660	23·5894	27·877
10	12·5489	15·9872	18·3070	20·4832	23·2093	25·1882	29·588
11	13·7007	17·2750	19·6751	21·9200	24·7250	26·7568	31·264
12	14·8454	18·5493	21·0261	23·3367	26·2170	28·2995	32·909
13	15·9839	19·8119	22·3620	24·7356	27·6882	29·8195	34·528
14	17·1169	21·0641	23·6848	26·1189	29·1412	31·3194	36·123
15	18·2451	22·3071	24·9958	27·4884	30·5779	32·8013	37·697
16	19·3689	23·5418	26·2962	28·8454	31·9999	34·2672	39·252
17	20·4887	24·7690	27·5871	30·1910	33·4087	35·7185	40·790
18	21·6049	25·9894	28·8693	31·5264	34·8053	37·1565	42·312
19	22·7178	27·2036	30·1435	32·8523	36·1909	38·5823	43·820
20	23·8277	28·4120	31·4104	34·1696	37·5662	39·9968	45·315
21	24·9348	29·6151	32·6706	35·4789	38·9322	41·4011	46·797
22	26·0393	30·8133	33·9244	36·7807	40·2894	42·7957	48·268
23	27·1413	32·0069	35·1725	38·0756	41·6384	44·1813	49·728
24	28·2412	33·1962	36·4150	39·3641	42·9798	45·5585	51·179
25	29·3389	34·3816	37·6525	40·6465	44·3141	46·9279	52·618
26	30·4346	35·5632	38·8851	41·9232	45·6417	48·2899	54·052
27	31·5284	36·7412	40·1133	43·1945	46·9629	49·6449	55·476
28	32·6205	37·9159	41·3371	44·4608	48·2782	50·9934	56·892
29	33·7109	39·0875	42·5570	45·7223	49·5879	52·3356	58·301
30	34·7997	40·2560	43·7730	46·9792	50·8922	53·6720	59·703
40	45·6160	51·8051	55·7585	59·3417	63·6907	66·7660	73·402
50	56·3336	63·1671	67·5048	71·4202	76·1539	79·4900	86·661
60	66·9815	74·3970	79·0819	83·2977	88·3794	91·9517	99·607
70	77·5767	85·5270	90·5312	95·0232	100·425	104·215	112·317
80	88·1303	96·5782	101·879	106·629	112·329	116·321	124·839
90	98·6499	107·565	113·145	118·136	124·116	128·299	137·208
100	109·141	118·498	124·342	129·561	135·807	140·169	149·449

Table 6: $P[X \leq y] = \int_0^y f_x(x)\,dx = .95$

n_2 \ n_1	1	2	3	4	5	6	7	8	9	10	12	15	20	24	30	40	60	120	∞
1	161·4	199·5	215·7	224·6	230·2	234·0	236·8	238·9	240·5	241·9	243·9	245·9	248·0	249·1	250·1	251·1	252·2	253·3	254·3
2	18·51	19·00	19·16	19·25	19·30	19·33	19·35	19·37	19·38	19·40	19·41	19·43	19·45	19·45	19·46	19·47	19·48	19·49	19·50
3	10·13	9·55	9·28	9·12	9·01	8·94	8·89	8·85	8·81	8·79	8·74	8·70	8·66	8·64	8·62	8·59	8·57	8·55	8·53
4	7·71	6·94	6·59	6·39	6·26	6·16	6·09	6·04	6·00	5·96	5·91	5·86	5·80	5·77	5·75	5·72	5·69	5·66	5·63
5	6·61	5·79	5·41	5·19	5·05	4·95	4·88	4·82	4·77	4·74	4·68	4·62	4·56	4·53	4·50	4·46	4·43	4·40	4·36
6	5·99	5·14	4·76	4·53	4·39	4·28	4·21	4·15	4·10	4·06	4·00	3·94	3·87	3·84	3·81	3·77	3·74	3·70	3·67
7	5·59	4·74	4·35	4·12	3·97	3·87	3·79	3·73	3·68	3·64	3·57	3·51	3·44	3·41	3·38	3·34	3·30	3·27	3·23
8	5·32	4·46	4·07	3·84	3·69	3·58	3·50	3·44	3·39	3·35	3·28	3·22	3·15	3·12	3·08	3·04	3·01	2·97	2·93
9	5·12	4·26	3·86	3·63	3·48	3·37	3·29	3·23	3·18	3·14	3·07	3·01	2·94	2·90	2·86	2·83	2·79	2·75	2·71
10	4·96	4·10	3·71	3·48	3·33	3·22	3·14	3·07	3·02	2·98	2·91	2·85	2·77	2·74	2·70	2·66	2·62	2·58	2·54
11	4·84	3·98	3·59	3·36	3·20	3·09	3·01	2·95	2·90	2·85	2·79	2·72	2·65	2·61	2·57	2·53	2·49	2·45	2·40
12	4·75	3·89	3·49	3·26	3·11	3·00	2·91	2·85	2·80	2·75	2·69	2·62	2·54	2·51	2·47	2·43	2·38	2·34	2·30
13	4·67	3·81	3·41	3·18	3·03	2·92	2·83	2·77	2·71	2·67	2·60	2·53	2·46	2·42	2·38	2·34	2·30	2·25	2·21
14	4·60	3·74	3·34	3·11	2·96	2·85	2·76	2·70	2·65	2·60	2·53	2·46	2·39	2·35	2·31	2·27	2·22	2·18	2·13
15	4·54	3·68	3·29	3·06	2·90	2·79	2·71	2·64	2·59	2·54	2·48	2·40	2·33	2·29	2·25	2·20	2·16	2·11	2·07
16	4·49	3·63	3·24	3·01	2·85	2·74	2·66	2·59	2·54	2·49	2·42	2·35	2·28	2·24	2·19	2·15	2·11	2·06	2·01
17	4·45	3·59	3·20	2·96	2·81	2·70	2·61	2·55	2·49	2·45	2·38	2·31	2·23	2·19	2·15	2·10	2·06	2·01	1·96
18	4·41	3·55	3·16	2·93	2·77	2·66	2·58	2·51	2·46	2·41	2·34	2·27	2·19	2·15	2·11	2·06	2·02	1·97	1·92
19	4·38	3·52	3·13	2·90	2·74	2·63	2·54	2·48	2·42	2·38	2·31	2·23	2·16	2·11	2·07	2·03	1·98	1·93	1·88
20	4·35	3·49	3·10	2·87	2·71	2·60	2·51	2·45	2·39	2·35	2·28	2·20	2·12	2·08	2·04	1·99	1·95	1·90	1·84
21	4·32	3·47	3·07	2·84	2·68	2·57	2·49	2·42	2·37	2·32	2·25	2·18	2·10	2·05	2·01	1·96	1·92	1·87	1·81
22	4·30	3·44	3·05	2·82	2·66	2·55	2·46	2·40	2·34	2·30	2·23	2·15	2·07	2·03	1·98	1·94	1·89	1·84	1·78
23	4·28	3·42	3·03	2·80	2·64	2·53	2·44	2·37	2·32	2·27	2·20	2·13	2·05	2·01	1·96	1·91	1·86	1·81	1·76
24	4·26	3·40	3·01	2·78	2·62	2·51	2·42	2·36	2·30	2·25	2·18	2·11	2·03	1·98	1·94	1·89	1·84	1·79	1·73
25	4·24	3·39	2·99	2·76	2·60	2·49	2·40	2·34	2·28	2·24	2·16	2·09	2·01	1·96	1·92	1·87	1·82	1·77	1·71
26	4·23	3·37	2·98	2·74	2·59	2·47	2·39	2·32	2·27	2·22	2·15	2·07	1·99	1·95	1·90	1·85	1·80	1·75	1·69
27	4·21	3·35	2·96	2·73	2·57	2·46	2·37	2·31	2·25	2·20	2·13	2·06	1·97	1·93	1·88	1·84	1·79	1·73	1·67
28	4·20	3·34	2·95	2·71	2·56	2·45	2·36	2·29	2·24	2·19	2·12	2·04	1·96	1·91	1·87	1·82	1·77	1·71	1·65
29	4·18	3·33	2·93	2·70	2·55	2·43	2·35	2·28	2·22	2·18	2·10	2·03	1·94	1·90	1·85	1·81	1·75	1·70	1·64
30	4·17	3·32	2·92	2·69	2·53	2·42	2·33	2·27	2·21	2·16	2·09	2·01	1·93	1·89	1·84	1·79	1·74	1·68	1·62
40	4·08	3·23	2·84	2·61	2·45	2·34	2·25	2·18	2·12	2·08	2·00	1·92	1·84	1·79	1·74	1·69	1·64	1·58	1·51
60	4·00	3·15	2·76	2·53	2·37	2·25	2·17	2·10	2·04	1·99	1·92	1·84	1·75	1·70	1·65	1·59	1·53	1·47	1·39
120	3·92	3·07	2·68	2·45	2·29	2·17	2·09	2·02	1·96	1·91	1·83	1·75	1·66	1·61	1·55	1·50	1·43	1·35	1·25
∞	3·84	3·00	2·60	2·37	2·21	2·10	2·01	1·94	1·88	1·83	1·75	1·67	1·57	1·52	1·46	1·39	1·32	1·22	1·00

Theorem. (Properties of Expectation) If c is a constant and $g(X)$, $g_1(X)$, and $g_2(X)$ are functions whose expectations exist, then

(1) $E(c) = c$;

(2) $E(cg(X)) = cEg(X)$;

(3) $E(g_1(X) + g_2(X)) = Eg_1(X) + Eg_2(X)$;

(4) $Eg_1(X) \leq Eg_2(X)$ if $g_1(x) \leq g_2(x)$ for all x;

(5) $|Eg(X)| \leq E|g(X)|$.

Dispersion is often measured by averages of deviations from the mean:

Definition. Let X be an r.v. The **n^{th} central moment** of X is $\mu_n \equiv E(X - EX)^n$. The **variance** of X, denoted **Var(X)** or $\sigma^2(X)$, is μ_2 (the second central moment of X).

It turns out, as we shall shortly see, that the square root of Var(X) is most "natural" in statistical uses, so it is specially named:

Definition. Let us denote $\sigma^2(X)$ as, simply, σ^2. Then σ (the positive square root of σ^2) is called the **standard deviation** of X and is often written $\sigma(X)$.

Lemma. $\sigma^2(X) \geq 0$ with equality iff X is a constant with probability 1.

Lemma. $\text{Var}(X) = EX^2 - (EX)^2$.

Theorem. $\text{Var}(aX + b) = a^2\,\text{Var}(X)$.

Theorem. $X^* = \dfrac{(X - EX)}{\sigma(X)}$ has $EX^* = 0$, $\text{Var}(X^*) = 1$.

(*Note* that X^* is called the **standardized** X.)

One instance where σ (rather than σ^2) plays a natural role is **Chebyshev's Inequality**:

Corollary. Let $\delta > 0$ be fixed. Let X be an r.v. with mean μ and variance $\sigma^2 > 0$. Then

$$P\left[\left|\frac{X-\mu}{\sigma}\right| \geq \delta\right] \leq \frac{1}{\delta^2}.$$

Thus we may find

P [X is more than 1 standard
deviation away from its mean] $\leq$ 1.00

P [. . . . . 2 . . . .
. . . . s . . . .] $\leq$ .25

P [. . . . . 3. . . . .
. . . . s . . . .] $\leq$ .11

no matter what distribution (pdf) X has.

If X has a normal pdf, we can make stronger statements. Since

Theorem. If X has the normal distribution

$$f_X(x) = \frac{1}{\sqrt{2\pi}\,\sigma}\, e^{-\frac{1}{2}\left(\frac{x-\mu}{\sigma}\right)^2}, \quad -\infty < x < +\infty,$$

then $EX = \mu$, $\text{Var}(X) = \sigma^2$, and from our Tables 2 and 3, we find that **for a normal r.v. X**

$$P[\,|X - \mu| \geq \sigma\,] = .32$$
$$P[\,|X - \mu| \geq 2\sigma\,] = .05$$
$$P[\,|X - \mu| \geq 3\sigma\,] = .003.$$

The **sample mean $\overline{X}$**, i.e.,

$$\overline{\mathbf{X}} \equiv \frac{X_1 + X_2 + \ldots + X_n}{n},$$

is often used to summarize the results of n experiments. Here note that (if $X_1, \ldots, X_n$ are not dependent on each other, and have the same distribution)

$$E(\overline{X}) = \mu,\ \text{Var}(\overline{X}) = \frac{\text{Var}(X_1)}{n} = \frac{\sigma^2}{n}.$$

A good estimator of σ^2: it is used often and is usually denoted by

$$s^2 \equiv \frac{\sum_{i=1}^{n}(X_i - \overline{X})^2}{n-1}.$$

s^2 is called **the sample variance**, and $E(s^2) = \sigma^2$. Note that algebraic manipulation shows

$$s^2 = \frac{1}{n-1}\left\{\sum_{i=1}^{n} X_i^2 - \frac{1}{n}\left(\sum_{i=1}^{n} X_i\right)^2\right\}.$$

Example. A newspaper stand has observed that, over the course of 10 days, the numbers of newspapers demanded were $X_1, X_2, \ldots, X_{10}$, respectively. If we assume that these are independent and identically distributed observations of demand, what bound can be put on the probability that their average $\overline{X} \equiv (X_1 + \ldots + X_{10})/10$ is more than 3σ newspapers from the (unknown) average demand μ? We have $E\overline{X} = \mu$, $\text{Var}(\overline{X}) = \sigma^2/10$, so by Chebyshev's Inequality

$$P[\,|\overline{X} - \mu| \geq 3\sigma\,]$$
$$= P\left[\left|\frac{\overline{X}-\mu}{\sigma/\sqrt{10}}\right| \geq 3\sqrt{10}\right] \leq \frac{1}{(3\sqrt{10})^2} = \frac{1}{90}.$$

$\overline{X}$ becomes a closer estimate of μ as the sample size n increases.

Coding of Data.

It is sometimes easier to work with, and maintain accuracy in, $Y_i \equiv a(X_i + b)$ instead of X_i $(i=1, \ldots, n)$. If

$$\overline{X} = \frac{1}{n}\sum_{i=1}^{n} X_i,\ s_X^2 = \frac{1}{n-1}\sum_{i=1}^{n} (X_i - \overline{X})^2$$

$$\overline{Y} = \frac{1}{n}\sum_{i=1}^{n} Y_i,\ s_Y^2 = \frac{1}{n-1}\sum_{i=1}^{n} (Y_i - \overline{Y})^2,$$

then

$$\overline{X} = \frac{1}{a}\overline{Y} - b,\ s_X^2 = \frac{1}{a^2}s_Y^2.$$

Example. For
$$X_1 = 9869.00013$$
$$X_2 = 9869.00007$$
$$X_3 = 9869.00015$$
$$X_4 = 9869.00008$$
$$X_5 = 9869.00009$$

find $\overline{X}$ and s_X^2. If we calculate directly on a TI-55 (or IBM-370 in single precision), we find

$$\sum_{i=1}^{5} X_i = 49345.\ ,\ \sum_{i=1}^{5} X_i^2 = 4.8698581 \times 10^8.$$

$$\overline{X} = 9869.\quad ,\quad s_x^2 = .985.$$

Coding with $Y_i = a(X_i + b)$, $b = -9869.$, and $a = 10^5$, we have

$$Y_1 = 13$$
$$Y_2 = 7$$
$$Y_3 = 15$$
$$Y_4 = 8$$
$$Y_5 = 9,$$

$\sum_{i=1}^{5} Y_i = 52.$, $\sum_{i=1}^{5} Y_i^2 = 588.$, $\overline{Y} = 10.4$, and $s_Y^2 = 11.8$ (all exact), hence (exactly)

$$\overline{X} = 10.4 \times 10^{-5} + 9869. = 9869.000104,$$

$$s_X^2 = 11.8 \times 10^{-10}.$$

As previously noted, in **statistics** one **seeks to make inferences about population characteristics from sample data.** In this process one often takes **random samples** $X_1, X_2, \ldots, X_n$ (at random to eliminate selection biases and other systematic effects) and uses $\overline{X}$ to estimate the population mean μ. How good is this estimate?

Theorem. If $X_1, X_2, \ldots, X_n$ is a random sample from a population with mean μ and variance σ^2, then $\overline{X}$ has a distribution with mean μ and variance σ^2/n. (So, as n increases, $\overline{X}$ is more tightly clustered about μ.)

Theorem. (Central Limit Theorem) Let $X_1, X_2, \ldots$ be independent and identically distributed r.v.'s with $EX_1 = \mu$ and $\text{Var}(X_1) = \sigma^2 > 0$ (both finite). Then (for all z, $-\infty < z < +\infty$) as $n \to \infty$,

$$P\left[\frac{(X_1 - \mu) + \ldots + (X_n - \mu)}{\sqrt{n}\,\sigma} < z\right] \to \frac{1}{\sqrt{2\pi}}\int_{-\infty}^{z} e^{-\frac{1}{2}y^2}\,dy.$$

Note that, if X is normal with mean μ and variance σ^2, then the standardized X, $X^* = (X-\mu)/\sigma$, is *normal* with mean 0 and variance 1.

Thus, by the Central Limit Theorem, $\dfrac{\overline{X}-\mu}{\sigma/\sqrt{n}}$ has approximately a standard normal distribution (see Tables 2 and 3).

Example. A casino has a coin and wishes to estimate p, the probability that a Head will appear on any toss. The owners wish to do this in such a way that they can be 99 percent confident that the estimate ($\hat{p}$, say) is within .01 of p.

They take n independent tosses, which yield $X_1, \ldots, X_n$ with (for $i = 1, \ldots, n$)

$$X = \begin{cases} 1, & \text{if a head occurs on the } i\text{th toss} \\ 0, & \text{otherwise.} \end{cases}$$

If they decide to use $\hat{p} = \overline{X} \equiv (X_1 + \ldots + X_n)/n$ to estimate p, then how large must n be? We have

$$\begin{aligned}
&P[\,|\overline{X} - p| < .01\,] \\
&= P[n(p-.01) < X_1 + \cdots + X_n < n(p+.01)] \\
&= P[-.01n < X_1 + \cdots + X_n - np < .01n] \\
&= P\left[\frac{-.01n}{\sqrt{npq}} < \frac{X_1 + \cdots + X_n - np}{\sqrt{npq}} < \frac{.01n}{\sqrt{npq}}\right] \\
&= P\left[-.01\sqrt{\frac{n}{pq}} < \frac{X_1 + \cdots + X_n - np}{\sqrt{npq}} < .01\sqrt{\frac{n}{pq}}\right] \\
&\approx P\left[-.01\sqrt{\frac{n}{pq}} < Y < .01\sqrt{\frac{n}{pq}}\right],
\end{aligned}$$

where Y is a $N(0,1)$ r.v. (This follows by the CLT; here $EX_1 = p$, $\text{Var}(X_1) = pq$.) Now this approximation is a function of p, and is clearly minimized when $pq = p(1-p)$ is maximized, which occurs at $p = \frac{1}{2}$ (so $pq = \frac{1}{4}$). We will now find n such that

$$P [-.02 \sqrt{n} < Y < .02 \sqrt{n}] \approx .99.$$

Let $\Phi(z) \equiv P [Y \leq z]$, $-\infty < z < +\infty$ (the area under the normal density to the left of z). Of course $\Phi(0) = \frac{1}{2}$. Then we wish to set n by

$$\begin{aligned} .99 &= P [-.02 \sqrt{n} < Y < .02 \sqrt{n}] \\ &= \Phi (.02 \sqrt{n}) - \Phi (-.02 \sqrt{n}) \\ &= \Phi (.02 \sqrt{n}) - [1 - \Phi (-.02 \sqrt{n})] \\ &= 2\Phi (.02 \sqrt{n}) - 1 \end{aligned}$$

so $\Phi (.02 \sqrt{n}) = .995$ and $.02 \sqrt{n} = 2.58$, or $n = (129)^2 \approx (130)^2 = 16{,}900$. So our approximation to n_0 is $n_0 = 16{,}900$.

Example. A standard process is known to yield about 80 percent acceptable items. A new process has been found to yield 85 acceptable items out of its first 100. Is the superiority of the new process well established?

If the new process' probability of producing an acceptable item (p_2, say) is less than or equal to the old process' probability of producing an acceptable item (p_1, say), how large can the probability of obtaining 85 acceptables in a sample of 100 from the new process be? Denote the 100 items by $X_1, \ldots, X_{100}$, where

$$X_i = \begin{cases} 1, & \text{if the } i\text{th unit is acceptable} \\ 0, & \text{otherwise.} \end{cases}$$

Then $X_1, \ldots, X_{100}$ will be assumed independent with $P [X_i = 1] = p_2$ $(i = 1, \ldots, 100)$. We wish to study

$$\begin{aligned} &P_{p_2} [X_1 + \ldots + X_{100} \geq 85] \\ &= P_{p_2} \left[\frac{(X_1 + \ldots + X_{100}) - 100p_2}{\sqrt{100} \sqrt{p_2 (1 - p_2)}} \geq \frac{85 - 100p_2}{\sqrt{100} \sqrt{p_2 (1 - p_2)}} \right] \\ &\approx 1 - \Phi \left(\frac{85 - 100p_2}{10 \sqrt{p_2 (1 - p_2)}} \right) \text{ by the CLT.} \end{aligned}$$

For $p_2 \leq p_1 = .80$, this is maximized when

$\Phi \left(\frac{85 - 100p_2}{10 \sqrt{p_2 (1 - p_2)}} \right)$ is a minimum, that is, when $\frac{85 - 100p_2}{10 \sqrt{p_2 (1 - p_2)}}$ is a minimum. Since the derivative of this expression with respect to p_2 is negative for $0 \leq p_2 \leq 1$, the minimum for $p_2 \leq .80$ occurs at $p_2 = .80$.

Now

$$\begin{aligned} 1 - \Phi \left(\frac{85 - 80}{10 \sqrt{(.8)(.2)}} \right) &= 1 - \Phi (1.25) \\ &= 1 - .8944 = .1056 \approx .11. \end{aligned}$$

So, there is at most 1 chance in 10 of obtaining 85 or more acceptable items out of 100 if, in fact, $p_2 \leq p_1 = .80$.

We could decide to switch to the new process based on this evidence, or we could decide to wait for more evidence, or we could decide to reject the new process. (For example, we might have decided in advance to reject upon observing 75 or fewer acceptable items out of 100, wait for more evidence upon observing 76, . . . , 90 acceptable out of 100, and accept upon observing 91 or more acceptable out of 100.) Reaching a realistic decision rule involves consideration of cost of observations, loss suffered by not switching to a better process, loss suffered by switching to an inferior process, and so on, and is studied in Statistical Decision Theory. For further details, see Chapter 12 of E.J. Dudewicz' *Introduction to Statistics and Probability*, 1976, American Sciences Press, Inc., Columbus, Ohio.

In both of these examples, we made use of the fact that the number of successes $\sum_{i=1}^{n} X_i$ in independent trials has a binomial distribution with mean $\mu = np$ and variance $\sigma^2 = np(1-p)$ where p is the (constant) probability of a success on any trial. In many other settings it is necessary to estimate σ^2 via s^2.

Test Problem. Let $X_1, \ldots, X_n$ be independent random variables, each $N(\mu, \sigma^2)$ where μ is unknown $(-\infty < \mu < +\infty)$ and $\sigma^2 > 0$ is known. Let $\bar{X} = (X_1 + \ldots + X_n)/n$.

Only some pairs of real numbers (d_1, d_2) will be such that (irrespective of the value of μ)

$$P [\bar{X} - d_1 \leq \mu \leq \bar{X} + d_2] = .99; \qquad (*)$$

if (d_1, d_2) is such a pair, then the interval of real numbers $(\bar{X} - d_1, \bar{X} + d_2)$ is called a **confidence interval** for μ with **confidence coefficient** .99.

Specify precisely which pairs (d_1, d_2) satisfy (*).

Solution.

The pair (d_1, d_2) will satisfy (*) iff

$$\begin{aligned} .99 &\equiv P_\mu [\bar{X} - d_1 \leq \mu \leq \bar{X} + d_2] \\ &\equiv P_\mu [-d_2 \leq \bar{X} - \mu \leq d_1] \\ &\equiv P_\mu \left[\frac{-d_2}{\sigma/\sqrt{n}} \leq \frac{\bar{X} - \mu}{\sigma/\sqrt{n}} \leq \frac{d_1}{\sigma/\sqrt{n}} \right] \\ &\equiv \Phi \left(\frac{d_1}{\sigma/\sqrt{n}} \right) - \Phi \left(\frac{-d_2}{\sigma/\sqrt{n}} \right) \\ &\equiv \Phi \left(\frac{\sqrt{n}}{\sigma} d_1 \right) + \Phi \left(\frac{\sqrt{n}}{\sigma} d_2 \right) - 1, \end{aligned}$$

i.e., iff

$$\Phi \left(\frac{\sqrt{n}}{\sigma} d_1 \right) + \Phi \left(\frac{\sqrt{n}}{\sigma} d_2 \right) = 1.99.$$

Solutions (d_1, d_2) clearly exist for any fixed n, σ $(n = 1, 2, \ldots; \sigma > 0)$.

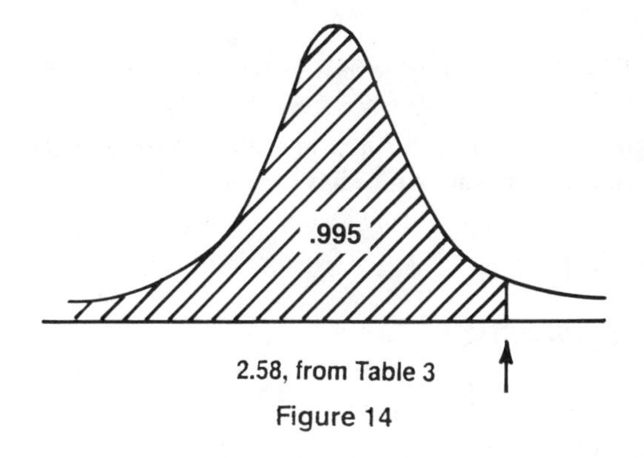

Figure 14

The interval has the *smallest length* $d_1 + d_2$ when

$$d_1 = d_2 = 2.58\,\frac{\sigma}{\sqrt{n}}.$$

Why? With $d_2 = d_1$,

$$\Phi\left(\frac{\sqrt{n}}{\sigma}\,d_1\right) + \Phi\left(\frac{\sqrt{n}}{\sigma}\,d^1\right) = 1.99$$

$$\Phi\left(\frac{\sqrt{n}}{\sigma}\,d_1\right) = .995$$

$$\frac{\sqrt{n}}{\sigma}\,d_1 = 2.58$$

$$d_1 = 2.58\,\frac{\sigma}{\sqrt{n}}.$$

What if σ^2 is unknown and estimated by s^2? Then
Theorem. If Y_1 has the $N(0,1)$ distribution (see Figure 8), if Y_2 has the $X^2_m(0)$ distribution (see Table 5), and if Y_1 and Y_2 are independent r.v.'s, then

$$Z = \frac{Y_1}{\sqrt{Y_2/m}}$$

has the **Student's *t*-distribution** with m degrees of freedom (see Figure 9).

Lemma. If $X_1, \ldots, X_n$ are independent r.v.'s, each of which is $N(\mu_0,\sigma^2)$, then $\overline{X} = (X_1 + \ldots + X_n)/n$ **and** $s^2 = \sum_{i=1}^{n} (X_i - \overline{X})^2/(n-1)$ **are independent r.v.'s.** It follows that

$$\frac{\overline{X} - \mu_0}{s\sqrt{n}}$$

has the Student's t-distribution with $m = n - 1$ degrees of freedom. Note that $(n-1)s^2/\sigma^2$ has a $\chi^2_{n-1}(0)$ distribution. Now, from our t-distribution as shown in Table 4, we find we use, instead of 2.58,

Example. A chemical analysis device claims to measure concentration of a specific chemical, in the 0 to 5% range, with a standard deviation of at most .01%. A standard solution of .7% concentration is analyzed, yielding

$$X_1 = .70,\ X_2 = .70,\ X_3 = .69,\ X_4 = .66, \text{ and } X_5 = .63.$$

Are these measurements consistent with the accuracy claim?

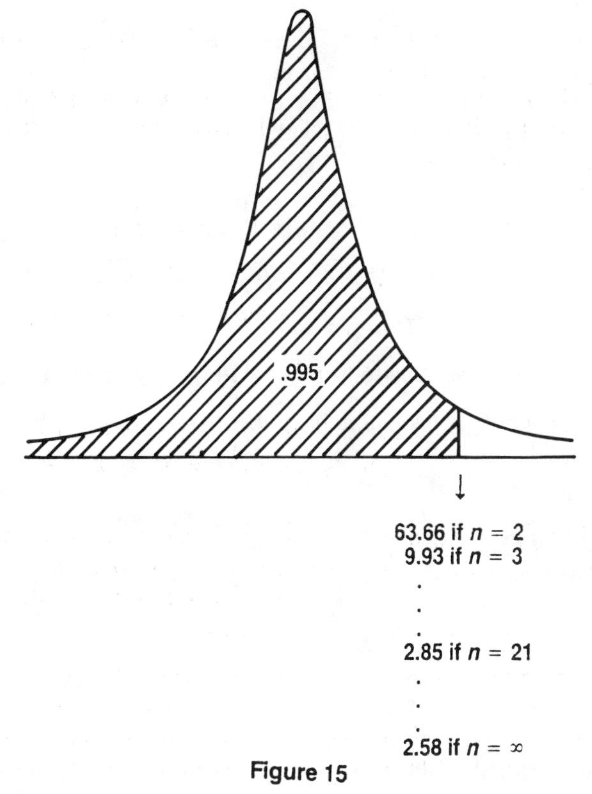

Figure 15

Assuming that $X_1, \ldots, X_n$ have (approximately) a normal distribution, we know (p. 382) that $\frac{(n-1)s^2}{\sigma^2}$ has a $\chi^2_{n-1}(0)$ distribution. Thus, from Table 5,

$$P\left[\ .207 \le \frac{(5-1)s^2}{\sigma^2} \le 14.86\ \right] = .99$$

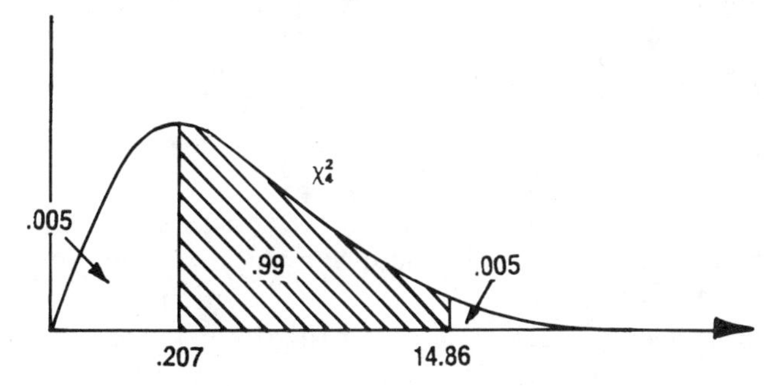

or

$$P\left[\ \frac{4s^2}{14.86} \le \sigma^2 \le \frac{4s^2}{.207}\ \right] = .99.$$

Since in 99% of the experiments we would find an s^2 such that σ^2 lies between these two numbers, we say we are 99% confident that (since $s^2 = .00093$ and $s = .0305$)

$$.00025 \le \sigma^2 \le .01797$$

or

$$.016 \le \sigma \le .134.$$

Thus the claim $\sigma \le .01$ is not consistent with the data.

Now if Y_1 has the χ_n^2 (0) distribution, Y_2 has the χ_m^2 (0) distribution, and Y_1 and Y_2 are independent r.v.'s (with $n \geq 2$, $m \geq 2$), then

$$Y = \frac{Y_1/n}{Y_2/m}$$

has the F-distribution $F(n,m)$. This would allow us to compare two variances. This is often called a **hypothesis testing problem**, but can also conveniently and intuitively be considered an **interval estimation problem.**

TESTS OF HYPOTHESES

Simple and Composite Null and Alternative Hypotheses

A hypothesis-testing problem usually results from questions like the following: Does smoking cause cancer? Do seat belts reduce car accident injuries? Do atomic power plants increase radiation levels? or Does adding putative catalyst A to a certain chemical process increase the yield? Thus, we have an underlying parameter θ (in our four examples this might be a cancer rate, an injury rate, a radiation level, or a process yield, respectively) and we wish to determine whether it changes in specified ways (for example, does it increase?) when an element of a system is changed (for example, when smoking is added, seat belts are added, atomic power plants are added, or a putative catalyst is added).

Formally, let us make the following assumption (which represents no restriction, but gives us a precise notation).

Assumption 1. Suppose that *X* is a r.v. with *df* $F(x|\theta)$ where θ is unknown, $\theta \in \Theta$. (Here X and θ may be multivariate.)

Let us now talk in terms of the question, Do atomic power plants increase radiation levels? Then it may be that a long history of precise study has shown that to date the average level of radiation per year per person is $\theta_0 = 1.2$ units on some standard scale in the area within a 25-mile radius of a proposed model atomic power plant site. In fact, study has shown that the level of radiation per year in this area is random and normally distributed with a mean level of $\theta_0 = 1.2$ and a variance of $\sigma^2 = .09$; thus, a person chosen at random has a level X_1 which is an $N(\theta_0,\sigma^2)$ r.v. If n persons' radiation exposure is measured in this area next year, we will observe independent r.v.'s $X_1, \ldots, X_n$ which are each $N(\theta,\sigma^2)$ where: $\theta = \theta_0$ if an atomic power plant does not operate in the area; and θ is unknown, $\theta \in \Theta = \{x: -\infty < x < +\infty\}$ if an atomic power plant does operate in the area.

The hypothesis that the atomic power plant has no effect (or a null effect) is called the "null hypothesis," denoted H_0, and is the assertion that $\theta = \theta_0$. The "alternative hypothesis," denoted H_1, that it has an effect is the assertion that $\theta \neq \theta_0$. Note that each "hypothesis" is a subset of Θ. It is also useful to distinguish whether, for any specified hypothesis $H \subseteq \Theta$, the distribution of X is completely determined or not.

Definition 2. Suppose Assumption 1 holds. Any $H \subseteq \Theta$ is called a **hypothesis**. Often one hypothesis H_0 is called the **null hypothesis**, and another hypothesis H_1 is called the **alternative hypothesis**.

Definition 3. A hypothesis $H \subseteq \Theta$ is called **simple** if knowing $\theta \in H$ completely specifies $F(x|\theta)$. Otherwise, H is called a **composite hypothesis**.

Thus, in our atomic power plant example, H_0 is a simple null hypothesis, while H_1 is a composite alternative hypothesis. (Note that it is *not* the case that H_0 and H_1 are unique for a particular problem. Thus, the alternative hypothesis in the atomic power plant example could just as well be taken to be $H_1^* = \{\theta: \theta \in \Theta, \theta > \theta_0\}$. This would be an alternative of increase, while H_1 was an alternative of change (+ or −).)

It is important to note that it is not as simple to "test" a hypothesis as the above simplified discussion might imply. Thus, we talked of sampling n people during a year of atomic power plant operation and comparing their radiation levels to θ_0. Yet the n people we choose will usually only be a sample of the N people living within 25 miles of the atomic power plant. For the N people, the level θ may vary not only randomly (being a $N(\theta,\sigma^2)$ r.v.) but also by class of person (for example, it might be: higher for dentists and X-ray technicians; different—higher or lower—for new residents, people who were away part of the year on business or vacation, and babies; lower as distance from the atomic power plant increases; higher downwind from the plant; lower for people who dwell in basement apartments; and so on). It could also vary in general with wind velocities and directions, and with precipitation (for example, precipitation might tend to take radioactive particles to ground level, while wind would tend to take them to certain sections of the area or out of the area). It could also vary with the correlation experienced between peak load periods (when the most radiation might be emitted) and wind velocity. (Other factors which we would like to control include the level of background radiation, which varies as sunspots vary, and so on.)

These **problems of experimental validity** also occur in other problem areas. If the question is, *Does smoking cause cancer?* and we just compare smokers and nonsmokers, the retort is that *smokers may tend to be people with a predisposition to cancer*. If one "smokes up" nonsmokers, the retort is, *those who let you do it are different*; if one "desmokes" smokers, the same retort calls the experimental results' validity into question. (Cigarette companies are experts at such retorts!) (To effectively, totally answer such questions, we really need to find the causal mechanism. Nonetheless, an experiment was performed on dogs[1] which seems to obviate all retorts except this one: *Humans are not dogs*. To this we can say: *Some are—especially those who, for economic gain, lead others to smoke*. However, dogs, as well as guinea pigs, and so on, are customarily used to test drugs, heart disease remedies and theories, and so on, and are felt to be sufficiently comparable to humans to validate the results. Only in a Hitlerite society could many of the important medical hypotheses be tested on human populations.)

Figuring out how to conduct an experiment so that factors such as those outlined above do not nullify its validity is called **designing an experiment** (or **experimental design**) and is a specialty of statisticians. It is for this reason that a statistician is usually employed on good research projects before any observations are made: in conjunction with experts in the subject matter of the specific area where the question arose, the statistician formulates an experimental design which will yield data valid for the study of the question.

Tests; Errors of Type I and Type II; Power

Suppose that X is a r.v. with *df* $F(x|\theta)$ where θ is unknown, $\theta \in \Theta$. Suppose we have a null hypothesis H_0 and an alternative hypothesis H_1. Our inference as to whether H_0 or H_1 contains the true θ will be based on observing X (whose *d.f.* $F(x|\theta)$ varies as $\theta \in \Theta$ varies). (In terms of the atomic power plant example, we can think of $\Theta = R$, $H_0 = \{\theta_0\}$, and $H_1 = \{\theta : \theta \in \Theta, \theta \neq \theta_0\}$. We then wish to decide whether the true $\theta = \theta_0$ or if the true $\theta \neq \theta_0$.) So, for some outcomes $X = x$ we will decide $\theta \in H_0$, while for the other outcomes $X = x$ we will decide $\theta \in H_1$. A "test" is a specification of these subsets of the set of possible outcomes $\mathcal{X}$ for some specification of H_0, H_1, Θ, and X:

Definition. Suppose that X is a r.v. with *df* $F(x|\theta)$ where θ is unknown, $\theta \in \Theta$, and that a null hypothesis H_0 and an alternative hypothesis H_1 have been specified. The problem of deciding (after observing X) either that $\theta \in H_0$ (called **accepting the null hypothesis**) or that $\theta \in H_1$ (called **rejecting the null hypothesis**) is called a **hypothesis-testing problem.** H_0 is called **true** if $\theta \in H_0$ and is called **false** if $\theta \notin H_0$.

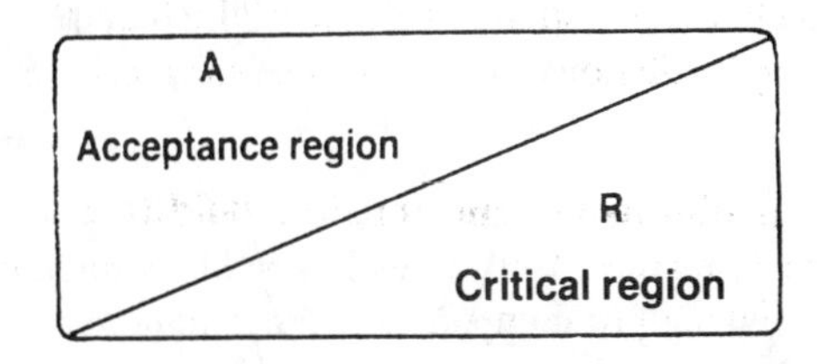

Figure 16: A nonrandomized test

[1]See E.C. Hammond, O. Auerbach, D. Kirman, and L. Garfinkel, "Effects of cigarette smoking on dogs," *Archives of Environmental Health, 21* (1970): 740–768.

If the set of possible outcomes of X is $\mathcal{X}$, then a division of $\mathcal{X}$ into two disjoint and exhaustive subsets A (called the **acceptance region**, or **region of acceptance of H_0**) and R (called the **critical region**, or **region of rejection of H_0**), such that if we find $X \in A$ we accept H_0 and if we find $X \in R$ we reject H_0, is called a **(nonrandomized) test of H_0 versus H_1**. (See Figure 16.)

Now when we conduct a test in a hypothesis-testing problem, **two errors are possible**: we can reject H_0 when it is true, and we can accept H_0 when it is false. These errors, and their probabilities of occurrence, are of central importance and hence have special names associated with them:

Definition. In a hypothesis-testing problem, rejecting H_0 when it is true is called an **error of Type I**. [The probability of rejecting a true H_0 depends on the test $\phi(\cdot)$ employed and the true $\theta \in H_0$, and is called the **level (of significance)** of the test or the **size of the critical region**.]

Accepting H_0 when it is false is called an **error of Type II**. [The probability of rejecting a false H_0 depends on the test $\phi(\cdot)$ employed and the true $\theta \in \Theta \cap \overline{H_0}$, and is called the **power** of the test.]

Example. Let us now illustrate these concepts with the use of the atomic power plant example. There we observe that $X = (X_1, \ldots, X_n) \in \mathcal{X} = R^n$ where $X_1, \ldots, X_n$ are independent r.v.'s, each $N(\theta, \sigma^2)$ with θ unknown, $\theta \in \Theta = R$, and $\sigma^2 = .09$. (The case of unknown σ^2 will be considered later, since it is more complicated.) Our null hypothesis is $H_0 = \{\theta_0\}$ where $\theta_0 = 1.2$, and our alternative hypothesis is $H_1 = \{\theta : \theta \in R, \theta \neq \theta_0\}$. (As we noted, H_0 is a simple hypothesis while H_1 is a composite hypothesis.) In our decision to accept or reject H_0 we will (providing we restrict ourselves to nonrandomized tests; this restriction is possible here without loss) need to decide, for each possible value $x = (x_1, \ldots, x_n)$ of $X = (X_1, \ldots, X_n)$, whether $x \in A$ or whether $x \in R$. A reasonable way to do this would be to base our decision on a statistic for the problem that retains all information about θ which is possessed by $X = (X_1, \ldots, X_n)$. $X_1 + \ldots + X_n$ is such a statistic. But we know that in this situation $\overline{X} = (X_1 + \ldots + X_n)/n$ is a good estimator of θ, so it may make sense to compare $\overline{X}$ with θ_0 in order to make our acceptance or rejection decision.

Suppose we desire to have the level of our test be $\alpha = .05$. Two possible tests are the test which rejects H_0 iff $|\overline{X} - \theta_0| > a$, that is, the test with **critical function** (we flip a coin with probability $\phi(X)$ of heads and reject (accept) H_0 if heads (tails) occurs)

$$\phi_1(x) = \begin{cases} 1, & \text{if } |\overline{x} - \theta_0| > a \\ 0, & \text{otherwise;} \end{cases}$$

and the test which rejects H_0 iff $\overline{X} > \theta_0 + b$, that is, the test with critical function

$$\phi_2(x) = \begin{cases} 1, & \text{if } \overline{x} > \theta_0 + b \\ 0, & \text{otherwise.} \end{cases}$$

In both cases a and b are to be set to give level $\alpha = .05$. What a and b are these, and **how do the tests compare with regard to power?**

For the test with critical function $\phi_1(\cdot)$, the level is

$$
\begin{aligned}
&P_{\theta_0}[\,|\bar{X}-\theta_0|>a] \\
&= 1 - P_{\theta_0}[\,|\bar{X}-\theta_0|\le a] \\
&= 1 - P_{\theta_0}[-a \le \bar{X}-\theta_0 \le a] \\
&= 1 - P_{\theta_0}\left[-\frac{a}{\sigma/\sqrt{n}} \le \frac{\bar{X}-\theta_0}{\sigma/\sqrt{n}} \le \frac{a}{\sigma/\sqrt{n}}\right] \\
&= 1 - P\left[-a\frac{\sqrt{n}}{\sigma} \le Y \le a\frac{\sqrt{n}}{\sigma}\right] \\
&= 1 - \left[\Phi\left(a\frac{\sqrt{n}}{\sigma}\right) - \Phi\left(-a\frac{\sqrt{n}}{\sigma}\right)\right] \\
&= 1 - \left[\Phi\left(a\frac{\sqrt{n}}{\sigma}\right) - \left\{1 - \Phi\left(a\frac{\sqrt{n}}{\sigma}\right)\right\}\right] \\
&= 2\left[1 - \Phi\left(a\frac{\sqrt{n}}{\sigma}\right)\right],
\end{aligned}
$$

where: Y is a $N(0,1)$ r.v. since $\bar{X}$ is $N(\theta_0,\sigma^2/n)$, and $\Phi(z) = P[Y \le z]$. To obtain level $\alpha = .05$ for our test, we set

$$
\left\{
\begin{aligned}
.05 &= 2\left[1 - \Phi\left(a\frac{\sqrt{n}}{\sigma}\right)\right], \\
.025 &= 1 - \Phi\left(a\frac{\sqrt{n}}{\sigma}\right), \\
\Phi\left(a\frac{\sqrt{n}}{\sigma}\right) &= .975, \\
a\frac{\sqrt{n}}{\sigma} &= \Phi^{-1}(.975) = 1.96, \\
a &= 1.96\frac{\sigma}{\sqrt{n}} = \frac{.588}{\sqrt{n}},
\end{aligned}
\right.
$$

where $\Phi^{-1}(.975) = 1.96$ (that is, $\Phi(1.96) = .975$) was determined using Tables 2 and 3 and we used the fact that $\sigma^2 = .09$ (so $\sigma = .3$) in this example. So if, say, $n = 100$, then $a = .0588$; that is, we reject H_0 iff $|\bar{X} - \theta_0| > .0588$.

The power of the test with critical function $\phi_1(\cdot)$ is (for $\theta \in \Theta \cap \bar{H}_0$, that is, for $\theta \neq \theta_0$) the probability

$$
\begin{aligned}
&P_{\theta_0}[\,|\bar{X}-\theta_0|>a] \\
&= 1 - P_{\theta}[\,|\bar{X}-\theta_0|\le a] \\
&= 1 - P_{\theta}[-a \le \bar{X}-\theta_0 \le a] \\
&= 1 - P_{\theta}[-a+\theta_0-\theta \le \bar{X}-\theta_0 \le a+\theta_0-\theta] \\
&= 1 - P_{\theta}\left[\frac{-a+\theta_0-\theta}{\sigma/\sqrt{n}} \le \frac{\bar{X}-\theta}{\sigma/\sqrt{n}} \le \frac{a+\theta_0-\theta}{\sigma/\sqrt{n}}\right] \\
&= 1 - P_{\theta}\left[\frac{-a+\theta_0-\theta}{\sigma/\sqrt{n}} \le Y \le \frac{a+\theta_0-\theta}{\sigma/\sqrt{n}}\right] \\
&= 1 - \left[\Phi\left(\frac{a+\theta_0-\theta}{\sigma/\sqrt{n}}\right) - \Phi\left(\frac{-a+\theta_0-\theta}{\sigma/\sqrt{n}}\right)\right] \\
&= 1 - \left[\Phi\left(1.96 + \frac{\theta_0-\theta}{\sigma}\sqrt{n}\right) - \Phi\left(-1.96 + \frac{\theta_0-\theta}{\sigma}\sqrt{n}\right)\right],
\end{aligned}
$$

where Y is a $N(0,1)$ r.v. and $a = 1.96\sigma/\sqrt{n}$ was substituted. As a function of θ, say $f(\theta)$, the rejection probability has the properties: $f(\theta_0) = \alpha = .05$; $f(\theta_0+x) = f(\theta_0-x)$ for all x; $f(\theta_2) > f(\theta_1)$ for all $\theta_2 > \theta_1 \ge \theta_0$; and $\lim_{\theta\to+\infty} f(\theta) = 1$.

Now, **for the test with critical function $\phi_2(\cdot)$ the probability of rejection of H_0 is**

$$
\begin{aligned}
P_{\theta_0}[\bar{X} > \theta_0 + b] &= 1 - P_{\theta}[\bar{X} \le \theta_0 + b] \\
&= 1 - P_{\theta}\left[\frac{\bar{X}-\theta}{\sigma/\sqrt{n}} \le \frac{b+\theta_0-\theta}{\sigma/\sqrt{n}}\right] \\
&= 1 - P_{\theta}\left[Y \le \frac{b+\theta_0-\theta}{\sigma/\sqrt{n}}\right] \\
&= 1 - \Phi\left(\frac{b+\theta_0-\theta}{\sigma/\sqrt{n}}\right),
\end{aligned}
$$

where Y is a $N(0,1)$ r.v. To set the level to $\alpha = .05$ requires that

$$
\left\{
\begin{aligned}
.05 &= P_{\theta_0}[\bar{X} > \theta_0 + b]. \\
.05 &= 1 - \Phi\left(\frac{b+\theta_0-\theta_0}{\sigma/\sqrt{n}}\right) \\
.05 &= 1 - \Phi\left(b\frac{\sqrt{n}}{\sigma}\right), \\
\Phi\left(b\frac{\sqrt{n}}{\sigma}\right) &= .95, \\
b\frac{\sqrt{n}}{\sigma} &= 1.64 \\
b &= 1.64\frac{\sigma}{\sqrt{n}} = \frac{.492}{\sqrt{n}}.
\end{aligned}
\right.
$$

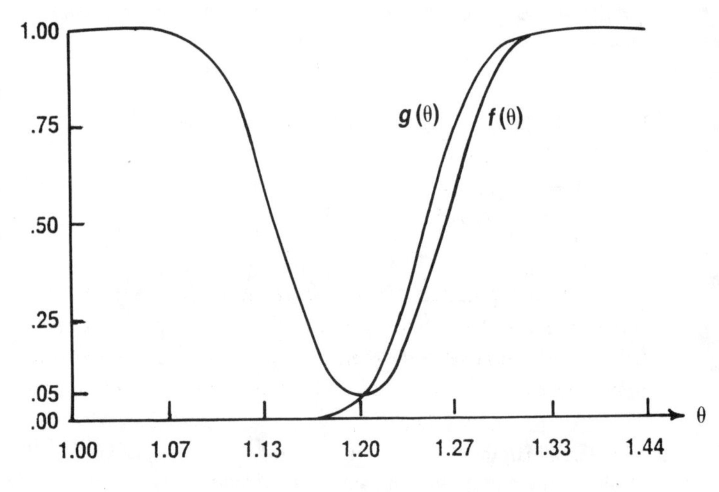

Figure 17: Rejection probabilities of two tests

So if, say, $n = 100$, then $b = .0492$; that is, we reject H_0 iff $\overline{X} - \theta_0, > .0492$. As a function of θ, say $g(\theta)$, the rejection probability has the properties: $g(\theta_0) = .05$; $g(\theta_2) > g(\theta_1)$ for all $\theta_2 > \theta_1$; and $\lim_{\theta \to +\infty} g(\theta) = 1$.

The rejection probabilities of the tests with critical functions $\phi_1(\cdot)$ and $\phi_2(\cdot)$ are plotted for $n = 100$. From this graph (Figure 17) we see that **both tests have the same level;** and $g(\theta) > f(\theta)$ for $\theta > \theta_0$, while $g(\theta) < f(\theta)$ for $\theta < \theta_0$. Thus if we are sure that either $\theta = \theta_0$ or $\theta > \theta_0$, we should use $\phi_2(\cdot)$ (it yields the same level as $\phi_1(\cdot)$ and more power for all $\theta > \theta_0$); while if we a priori think we may encounter $\theta < \theta_0$ also, we should use $\phi_1(\cdot)$ in order to have a suitably high probability of rejection when θ is less than θ_0. The graphs would be useful in the situation where we could only afford $n = 100$ and wanted to see what **power** this would yield for the two tests considered. **Often n is set so that the power at a specified θ has at least a certain specified value;** for example, if for the test using $\phi_1(\cdot)$ we wanted to have power $\geq .95$ when $\theta = 1.23$, we would set n as follows:

$$\begin{cases} .95 \leqslant P_{\theta=1.23}\ [\ |\ \overline{X} - \theta_0\ | > a\ ]. \\ .95 \leqslant 1 - \left[\Phi\left(1.96 + \dfrac{1.2-1.23}{.3} \sqrt{n} \right) \right. \\ \qquad \left. - \Phi\left(-1.96 + \dfrac{1.2-1.23}{.3} \sqrt{n} \right) \right]. \\ \Phi\,(1.96 - .1\sqrt{n}) - \Phi\,(-1.96 - .1\sqrt{n}) \leqslant .05. \end{cases}$$

Solving by trial and error (that is, choose an n, calculate $\Phi\,(1.96 - .1\sqrt{n}) - \Phi\,(-1.96 - .1\sqrt{n})$, and so on), we find that the smallest n which will satisfy is $n = 1296$.

Now suppose our hypothesis-testing problem is $X = (X_1, \ldots, X_n)$ where $X_1, \ldots, X_n$ are independent r.v.'s, each $N(\mu,\sigma^2)$ where μ **is unknown and σ^2 is unknown.**

For the σ^2 unknown case, our intuition leads us (by analogy to the σ^2 known case) to consider the test which rejects when

$$\frac{\overline{X} - \mu_0}{s/\sqrt{n}} \tag{1}$$

is "too large." In order to obtain level α for this test, we need to set c so that

$$P_{\mu_0}\left[\frac{\overline{X} - \mu_0}{s/\sqrt{n}} > c \right] = \alpha.$$

We are in a position to provide a c, since we know that (1) has the Student's t-distribution with $n-1$ degrees of freedom. First we need some notation. [For the $N(0,1)$ distribution, we also needed a notation for that number c such that if X is an $N(0,1)$ r.v., then $P\,[X \leq c] = x$; namely, $\Phi\,(c) = x$ or $c = \Phi^{-1}(x)$.]

Definition. If an r.v. X has the Student's t-distribution with m degrees of freedom, we will say ***X has the t_m distribution*** and will (for $0 < \alpha < 1$) denote the solution c of the equation

$$P\,[\,X \leq c\,] = \alpha$$

by writing $t_m\,(c) = \alpha$, or $c = t_m^{-1}\,(\alpha)$. Thus, we have the following.

Theorem (the t-test). For the hypothesis-testing problem with $X = (X_1, \ldots, X_n)$ where $X_1, \ldots, X_n$ are independent r.v.'s, each $N(\mu,\sigma^2)$ where μ $(-\infty < \mu < +\infty)$ and σ^2 ($\sigma^2 > 0$) are both unknown, $H_0 = \{\mu_0\}$, and $H_1 = \{\mu_1\}$ with $\mu_1 > \mu_0$, the test which rejects iff

$$\frac{\overline{X} - \mu_0}{s/\sqrt{n}} > t_{n-1}^{-1}\,(1 - \alpha)$$

has level α. Similar tests are easily obtained for other alternatives H_1.

Testing Normal Variances (One Population)

The hypothesis-testing problem of concern in this section is the one with $X = (X_1, \ldots, X_n)$ where $X_1, \ldots, X_n$ are independent r.v.'s, each $N(\mu,\sigma^2)$ where μ is known ($-\infty < \mu < +\infty$), σ^2 is unknown with $\sigma^2 \in \Theta = \{x: x > 0\}$, $H_0 = \{\sigma_0^2\}$, and H_1 may have several forms. **First we will study the case where $H_1 = \{\sigma_1^2\}$ with $\sigma_1^2 > \sigma_0^2$,** and we will reason intuitively to a test as follows. For this problem we know that

$$\hat{\sigma}^2 = \frac{\sum_{i=1}^{n} (X_1 - \mu)^2}{n}$$

is a good estimator of σ^2. Thus, it seems intuitively reasonable to reject H_0 iff

$$\hat{\sigma}^2 > k,$$

where k is a constant chosen (if such a choice is possible) to give level α. Now

$$P_{\sigma_0^2}\,[\hat{\sigma}^2 > k] = P_{\sigma_0^2}\left[\frac{\sum_{i=1}^{n} (X_i - \mu)^2}{n} > k \right]$$

$$= P_{\sigma_0^2}\left[\sum_{i=1}^{n} \left(\frac{X_i - \mu}{\sigma_0} \right)^2 > \frac{nk}{\sigma_0^2} \right],$$

and since $\sum_{i=1}^{n} ((X_i - \mu)/\sigma_0)^2$ has the $\chi_n^2\,(0)$ distribution, a table can be used to find k. Thus, we have the following:

Theorem. For the hypothesis-testing problem with $X = (X_1, \ldots, X_n)$ where $X_1, \ldots, X_n$ are independent r.v.'s, each $N(\mu,\sigma^2)$ where μ is known ($-\infty < \mu < +\infty$), σ^2 is unknown ($\sigma^2 > 0$), $H_0 = \{\sigma_0^2\}$, and $H_1 = \{\sigma_1^2\}$ with $\sigma_1^2 > \sigma_0^2$, the test which rejects iff

$$\frac{\sum_{i=1}^{n} (X_i - \mu)^2}{n} > k$$

(where k satisfies $P\ [Y > nk/\sigma_0^2] = \alpha$ for Y a $\chi_n^2\ (0)$ r.v.) has level α. Similar tests are easily obtained for other alternatives H_1. **Now if μ ($-\infty < \mu < +\infty$) is unknown, what do we do?** We replace μ by the good estimator $\overline{X}$ and reject iff

$$\frac{\sum_{i=1}^{n} (X_i - \overline{X})^2}{n} > k'.$$

Since $(n-1)\ s^2/\sigma^2$ has a $\chi_{n-1}^2\ (0)$ distribution in this case, we may (by choosing k' to satisfy

$$P\left[\ Y > \frac{nk'}{\sigma_0^2}\ \right] = \alpha$$

for Y *a* $\chi_{n-1}^2\ (0)$ r.v.) obtain our level α test. [This justifies the **"loss of 1 degree of freedom"** nomenclature.]

Testing Normal Variances (Two Populations)

Previously we considered testing whether the observations from one source (each such source is called a **population**) have a specified variance ($\sigma^2 = \sigma_0^2$). In the present section we consider **the problem of testing whether observations from two sources (or populations) have the same variance.** (For example, we might be concerned with whether the variability in height of people who smoked before their twenty-first birthday is equal to the variability in height of people who did not smoke until after their twenty-first birthday.)

Formally the hypothesis-testing problem of concern is one with $X = (X_1, \ldots, X_n, Y_1, \ldots, Y_m)$ where $X_1, \ldots, X_n, Y_1, \ldots, Y_m$ are $n + m$ independent r.v.'s, each of $X_1, \ldots, X_n$ is $N\ (\mu_1, \sigma_1^2)$, each of $Y_1, \ldots, Y_m$ is $N\ (\mu_2, \sigma_2^2)$, μ_1 and μ_2 are known ($-\infty < \mu_1, \mu_2 < +\infty$), σ_1^2 and σ_2^2 are unknown ($\sigma_1^2, \sigma_2^2 \in \Theta = \{ (x,y) : x > 0, y > 0 \}$), $H_0 = \{ (\sigma_1^2, \sigma_2^2) : \sigma_1^2 = \sigma_2^2 > 0 \}$, and H_1 may have several forms. **Let us consider first the case where $H_1 = \{ (\sigma_1^2, \sigma_2^2) : \sigma_1^2 > \sigma_2^2 \}$.** We can reason to a test intuitively as follows. We know that $\hat{\sigma}_i^2$ is a good estimator of $\sigma_i^2\ (i = 1,2)$ where

$$\hat{\sigma}_1^2 = \frac{\sum_{j=1}^{n} (X_j - \mu_1)^2}{n}, \quad \hat{\sigma}_2^2 = \frac{\sum_{j=1}^{m} (Y_j - \mu_2)^2}{m}.$$

So, we can consider rejecting H_0 if $\hat{\sigma}_1^2/\hat{\sigma}_2^2$ is much larger than 1; that is, reject H_0 iff

$$\frac{\hat{\sigma}_1^2}{\hat{\sigma}_2^2} > k,$$

where k is a constant chosen (if such a choice is possible) to give level α. Now

$$P_{\sigma_1^2,\sigma_2^2}\left[\frac{\hat{\sigma}_1^2}{\hat{\sigma}_2^2} > k\right]$$

$$= P_{\sigma_1^2,\sigma_2^2}\left[\frac{\sum_{j=1}^{n} (X_j - \mu_1)^2/n}{\sum_{j=1}^{m} (Y_j - \mu_2)^2/m} > k\right]$$

$$= P_{\sigma_1^2,\sigma_2^2}\left[\frac{\sum_{j=1}^{n} \left(\frac{X_j - \mu_1}{\sigma_1}\right)^2/n}{\sum_{j=1}^{m} \left(\frac{Y_j - \mu_2}{\sigma_2}\right)^2/m} > \frac{\sigma_2^2}{\sigma_1^2}\,k\right]$$

$$= P\left[\ Y > \frac{\sigma_2^2}{\sigma_1^2}\,k\ \right],$$

where Y has the $F(n,m)$ distribution, so Table 6 can be used to find a k which gives level σ. Thus, we have the following:

Theorem. For the hypothesis-testing problem with $X = (X_1, \ldots, X_n, Y_1, \ldots, Y_m)$ where $X_1, \ldots, X_n, Y_1, \ldots, Y_m$ are $n + m$ independent r.v.'s, each of $X_1, \ldots, X_n$ is $N\ (\mu_1, \sigma_1^2)$, each of $Y_1, \ldots, Y_m$ is $N\ (\mu_2, \sigma_2^2)$, μ_1 and μ_2 are known ($-\infty < \mu_1, \mu_2 < +\infty$), σ_1^2 and σ_2^2 are unknown ($\sigma_1^2, \sigma_2^2 > 0$), $H_0 = \{\sigma_1^2 = \sigma_2^2\}$, and $H_1 = \{\sigma_1^2 > \sigma_2^2\}$, **the test which rejects iff**

$$\frac{\sum_{j=1}^{n} (X_j - \mu_1)^2/n}{\sum_{i=1}^{m} (Y_j - \mu_2)^2/m} > k$$

(where k satisfies $P\ [Y > \sigma_2^2 k/\sigma_1^2\ ] = \alpha$ for Y an $F(n,m)$ r.v.) **has level α.** Similar tests are easily obtained for other alternatives H_1.

Now **if μ_1 and μ_2 are unknown**, we may consider estimating them by $\overline{X}$ and $\overline{Y}$ respectively, then rejecting iff

$$\frac{\sum_{j=1}^{n} (X_j - \overline{X})^2/(n - 1)}{\sum_{i=1}^{m} (Y_j - \overline{Y})^2/(m - 1)} > k'.$$

Since the left-hand side has the $F(n-1,m-1)$ distribution when multiplied by σ_2^2/σ_1^2, we may (by choosing k' to satisfy

$$P\left[\ Y > \frac{\sigma_2^2\,k'}{\sigma_1^2}\ \right] = \alpha$$

for Y an $F(n-1,m-1)$ r.v.) obtain our level α test. [Note that when μ_1 and μ_2 are unknown we must have $n \geq 2$ and $m \geq 2$ (otherwise we do not obtain a proper F-distribution).]

Note that in these cases $k[k']$ is chosen so the quantity in question equals α *when H_0 is true*, that is, when $\sigma_1^2 = \sigma_2^2$. Hence, our solution does not involve the (unknown) quantities σ_1^2 and σ_2^2.

Testing Normal Means (Two Populations with Equal Variances)

We considered testing whether the observations from one source have a specified mean ($\mu = \mu_0$). In the present section we consider the problem of testing whether observations from two sources (or populations) have the same mean when the sources have equal variability. (The case of unequal variabilities is also considered.) In terms of the example, we are concerned here with whether the "early smokers" have the same mean height as the "late and nonsmokers," and are assuming the variabilities of the heights to be equal.

Formally, the hypothesis-testing problem of concern is one with $X = (X_1, \ldots, X_n, Y_1, \ldots, Y_m)$ where $X_1, \ldots, X_n, Y_1, \ldots, Y_m$ are $n + m$ independent r.v.'s, each of $X_1, \ldots, X_n$ is $N(\mu_1, \sigma^2)$, each of $Y_1, \ldots, Y_m$ is $N(\mu_2, \sigma^2)$, μ_1 and μ_2 are unknown $(-\infty < \mu_1, \mu_2 < +\infty)$, σ^2 is known $(\sigma^2 > 0)$, $H_0 = \{\mu_1 = \mu_2\}$, and H_1 may have several forms. **Let us consider first the case where $H_1 = \{\mu_1 > \mu_2\}$.** We can reason to a test intuitively as follows. We know that $\overline{X}$ and $\overline{Y}$ are respectively good estimators of μ_1 and μ_2. So consider rejecting H_0 if $\overline{X}$ is much larger than $\overline{Y}$; that is, reject H_0 iff

$$\overline{X} - \overline{Y} > c,$$

where c is a constant to be chosen (if such a choice is possible) to give level α. Now

$$P_{\mu_1=\mu_2}[\overline{X} - \overline{Y} > c] = \alpha,$$

$$P_{\mu_1=\mu_2}\left\{\frac{\overline{X}-\overline{Y}}{\sqrt{\sigma^2\left(\frac{1}{n}+\frac{1}{m}\right)}} > \frac{c}{\sqrt{\sigma^2\left(\frac{1}{n}+\frac{1}{m}\right)}}\right\} = \alpha$$

$$1 - \Phi\left(\frac{c}{\sqrt{\sigma^2\left(\frac{1}{n}+\frac{1}{m}\right)}}\right) = \alpha$$

$$c = \sigma\sqrt{\frac{1}{n}+\frac{1}{m}}\,\Phi^{-1}(1-\alpha).$$

So c can be found as desired. Similar tests are easily obtained for other alternatives H_1. Note that **we can rewrite our test as one which rejects iff**

$$\frac{\overline{X} - \overline{Y}}{\sqrt{\sigma^2\left(\frac{1}{n}+\frac{1}{m}\right)}} > \Phi^{-1}(1 - \alpha).$$

(This will help us in the case of unknown σ^2.)

Now **if σ^2 is unknown**, we would consider estimating it by

$$s_1^2 = \frac{\sum_{j=1}^{n}(X_j - \overline{X})^2}{n - 1}$$

or by

$$s_2^2 = \frac{\sum_{j=1}^{m}(Y_j - \overline{Y})^2}{m - 1},$$

and our estimators would have reasonable properties.

Can we somehow combine s_1^2 and s_2^2 to obtain an even better estimator of σ^2? If we combine them according to their respective *df* of $n-1$ and $m-1$ into a weighted average, **we obtain a third estimator**

$$s_P^2 = \frac{\sum_{j=1}^{n}(X_j - \overline{X})^2 + \sum_{j=1}^{m}(Y_j - \overline{Y})^2}{n + m - 2}.$$

So, what about the test that rejects iff

$$\frac{\overline{X} - \overline{Y}}{\sqrt{\frac{\sum_{j=1}^{n}(X_j - \overline{X})^2 + \sum_{j=1}^{m}(Y_j - \overline{Y})^2}{n + m - 2}\left(\frac{1}{n}+\frac{1}{m}\right)}} > k'?$$

It follows that when H_0 is true, the left-hand side has Student's t-distribution with $n + m - 2$ *df* [this is because the left-hand side is the ratio of the $N(0,1)$ r.v.

$$\frac{\overline{X} - \overline{Y}}{\sigma\sqrt{\frac{1}{n}+\frac{1}{m}}}$$

and the square root of the χ^2_{n+m-2} r.v.

$$\sum_{j=1}^{n}\left(\frac{X_j - \overline{X}}{\sigma}\right)^2 + \sum_{j=1}^{m}\left(\frac{Y_j - \overline{Y}}{\sigma}\right)^2$$

divided by its *df*, and these two r.v.'s are independent]. Hence, a proper choice for k' is

$$k' = t^{-1}_{n+m-2}(1 - \alpha).$$

Similar tests are easily obtained for other alternatives H_1.

Testing Normal Means (Two Populations with Unequal Variances; The Behrens-Fisher Problem)

We considered hypothesis-testing problems involving the

means of two populations which have the same variance (which may be known or unknown). In this section we consider such problems **when we cannot assume the variabilities of the two populations** (for example, the populations "early smokers" and "late smokers and nonsmokers") **to be the same.**

Formally, the hypothesis-testing problem of concern is one with $X = (X_1, \ldots, X_n, Y_1, \ldots, Y_m)$ where $X_1, \ldots, X_n, Y_1, \ldots, Y_m$ are $n+m$ independent r.v.'s, each of $X_1, \ldots, X_n$ is $N(\mu_1,\sigma_1^2)$, each of $Y_1, \ldots, Y_m$ is $N(\mu_2,\sigma_2^2)$, μ_1 and μ_2 are unknown $(-\infty < \mu_1, \mu_2 < +\infty)$, σ_1^2 and σ_2^2 are known $(\sigma_1^2, \sigma_2^2 > 0$ but $\sigma_1^2 \neq \sigma_2^2)$, $H_0 = \{\mu_1 = \mu_2\}$, and H_1 may have several forms. **Let us consider first the case where $H_1 = \{\mu_1 > \mu_2\}$.** Consider rejecting H_0 iff

$$\bar{X} - \bar{Y} > c,$$

where c is a constant to be chosen (if such a choice is possible) to give level α. We can find such a c:

$$P_{\mu_1=\mu_2}[\bar{X} - \bar{Y} > c] = \alpha,$$

$$P_{\mu_1=\mu_2}\left[\frac{\bar{X} - \bar{Y}}{\sqrt{\frac{\sigma_1^2}{n} + \frac{\sigma_2^2}{m}}} > \frac{c}{\sqrt{\frac{\sigma_1^2}{n} + \frac{\sigma_2^2}{m}}}\right] = \alpha,$$

$$1 - \Phi\left[\frac{c}{\sqrt{\frac{\sigma_1^2}{n} + \frac{\sigma_2^2}{m}}}\right] = \alpha,$$

$$c = \sqrt{\frac{\sigma_1^2}{n} + \frac{\sigma_2^2}{m}}\,\Phi^{-1}(1 - \alpha).$$

Similar tests are easily obtained for other alternatives H_1.

Now **if σ_1^2 and σ_2^2 are unknown**, we would consider estimating them by s_1^2 and s_2^2, respectively, and rejecting iff

$$\frac{\bar{X} - \bar{Y}}{\sqrt{\frac{s_1^2}{n} + \frac{s_2^2}{m}}} > k'.$$

However, the distribution of the left-hand side is not Student's t-distribution when H_0 is true; in fact, the distribution of the left-hand side depends on n, m, and σ_1^2/σ_2^2. Hence, it is not possible to choose a k' such that the test has level α (since the probability of rejection varies as σ_1^2/σ_2^2 varies, instead of remaining constant at α). **The distribution of the quantity**

$$\frac{\bar{X} - \bar{Y}}{\sqrt{\frac{s_1^2}{n} + \frac{s_2^2}{m}}}$$

(when H_0 is true) is called the Behrens-Fisher distribution with parameters n, m, and σ_1^2/σ_2^2. **The hypothesis-testing problem we are considering (with σ_1^2 and σ_2^2 unknown) is called the Behrens-Fisher problem** and represents the first time we have encountered a situation where we could not obtain a level α test by "substituting a good estimator of an unknown parameter in the procedure used when that parameter is known."

One proposed solution to the Behrens-Fisher problem, which we will call the **Neyman-Bartlett solution**, proceeds as follows. Either $n < m$ or $n \geq m$, so let us suppose without any loss of generality that $n \geq m$. Define new r.v.'s $Z_1, \ldots, Z_m$ by

$$Z_i = X_i - Y_i \ (i = 1, \ldots, m).$$

Then $Z_1, \ldots, Z_m$ are independent $N(\mu_1 - \mu_2, \sigma_1^2 + \sigma_2^2)$ r.v.'s, and we can solve the problem of testing H_0: $\mu_1 - \mu_2 = 0$ versus H_1: $\mu_1 - \mu_2 > 0$ by rejecting H_0 iff

$$\frac{\bar{Z} - 0}{\sqrt{\dfrac{\sum_{i=1}^{m}(Z_i - \bar{Z})^2}{(m-1)\,m}}} > t_{m-1}^{-1}(1-\alpha),$$

that is, iff (letting $\bar{X}(m) = (X_1 + \ldots + X_m)/m$, which is to be distinguished from $\bar{X} = (X_1 + \ldots + X_m + X_{m+1} + \ldots X_n)/n$)

$$\frac{\bar{X}(m) - \bar{Y}}{\sqrt{\dfrac{\sum_{i=1}^{m}((X_i - Y_i) - (\bar{X}(m) - \bar{Y}))^2}{(m-1)\,m}}} > t_{m-1}^{-1}(1-\alpha).$$

As Scheffé has pointed out, *the Neyman-Bartlett solution* essentially *has two drawbacks*. *First,* it throws away the $n-m$ observations $X_{m+1}, \ldots, X_n$ completely. *Second,* the outcome of the test depends on the order of the observations in the samples. Scheffé has proposed another solution (which we will call **the Scheffé solution**) as follows: Reject H_0 iff the equation on the top of p. 385 holds.

When H_0 is true, the left-hand side has Student's t-distribution with $m - 1$, df, so the test is of level α. Note that this test uses $\bar{X}$ (not $\bar{X}(m)$) in the numerator and hence does not suffer from one of the maladies of the Neyman-Bartlett solution. However, *the order of the observations still affects the outcome of the test*. Scheffé has shown that, in a certain

$$\frac{\bar{X} - \bar{Y}}{\sqrt{\dfrac{\sum_{i=1}^{m}\left(\left(X_i - \sqrt{\frac{m}{n}}\,Y_i\right) - \left(\bar{X}(m) - \sqrt{\frac{m}{n}}\,\bar{Y}\right)\right)^2}{(m-1)m}}} > t_{m-1}^{-1}(1-\alpha).$$

framework, this dependence on order cannot be eliminated, and has recommended that because of this dependence his solution should *not* be used.

A third solution is to return to the Behrens-Fisher statistic and attempt to use it. What we will call **the Hsu solution** takes $k' = t_{f_0-1}^{-1}(1-\alpha)$ where $f_0 = \min(m,n)$, and it **has level $\leq \alpha$.** Other authors have suggested replacing $f_0 - 1$ with other (larger) quantities in order to obtain higher power (and also, of course, higher level, hopefully $\approx \alpha$ and hopefully never $>> \alpha$). For example, **Welch suggested**

$$\mathcal{F} = \frac{\left(\dfrac{s_1^2}{n} + \dfrac{s_2^2}{m}\right)^2}{\dfrac{(s_1^2 n)^2}{n-1} + \dfrac{(s_2^2 m)^2}{m-1}}$$

and we can show that

$$f_0 - 1 = \min(n-1, m-1) \leq \mathcal{F} \leq (n-1) + (m-1)\ .$$

Such tests do not guarantee the stated level of test α, and for reasonably large $f_0 - 1$ (for example, 10 or more) do not reduce the $t_f^{-1}(1-\alpha)$ value appreciably (and hence do not greatly affect the probabilities of acceptance and rejection). Hence, **we would usually choose** $f = f_0 - 1$ rather than $f = \mathcal{F}$ or some other quantity.

Test Problem. For the hypothesis-testing problem discussed, a third test sets $\phi_3(x) = .05$ for all $x \in \mathcal{X}$. (Thus, if this test is to be utilized, then we can just as well take no observations since this test ignores them; it simply flips a coin with probability .05 of heads.) Show that this test has level $\alpha = .05$.

Thus, **obtaining a test with level α is trivial in general. Much more important is obtaining a test of level α which has "good" properties with regard to power**; this is often overlooked in "cookbook" statistics courses and texts. Graph the power of the test utilizing $\phi_3(\cdot)$ as a function of θ, say $h(\theta)$. Compare it with $f(\theta)$ and $g(\theta)$ as shown.

Test Problem. We have seen that in cases of two-sided alternatives we usually obtain a class of tests rather than one "best" test.

That one of these tests which devotes $\alpha/2$ of α as a probability of rejecting due to "large" values of the test statistic (and $\alpha/2$ as a probability of rejecting due to "small" values of the test statistic) when H_0 is true, is called a **two-sided equal-tail test.**

Let $X = (X_1, \ldots, X_n)$ where $X_1, \ldots, X_n$ are independent $N(\mu,\sigma^2)$ r.v.'s and μ is known. For $n = 10$ and $\alpha = .10$, compute the power curve of the two-sided equal-tail test of H_0: $\sigma^2 = 1$ versus $H_1: \sigma^2 \neq 1$. For $n = 10$ and $\alpha = .10$, compute the power curve of the one-sided test of $H_0: \sigma^2 = 1$ versus H_1: $\sigma^2 > 1$. Plot the two power curves on the same graph.

Solution.

The two-sided equal-tail test rejects if either

$$\sum_{i=1}^{10}(X_i - \mu)^2 < 3.94 \text{ or } \sum_{i=1}^{10}(X_i - \mu)^2 > 18.3$$

(since $\sum_{i=1}^{10}(X_i - \mu)^2$ has a $\chi_{10}^2(0)$ distribution and we know that $P[Y < 3.94] = P[Y > 18.3] = .05$ if Y has a $\chi_{10}^2(0)$ distribution). The one-sided test rejects if

$$\sum_{i=1}^{10} X_i - \mu)^2 > 16.0$$

(note that 16.0 is 15.9872 rounded to one decimal place). The respective power functions are, for the two-sided equal-tail test,

$$1 - P_{\sigma^2}\left[3.94 \leq \sum_{i=1}^{10}(X_i - \mu)^2 \leq 18.3\right]$$
$$= 1 - P\left[\frac{3.94}{\sigma^2} \leq Y \leq \frac{18.3}{\sigma^2}\right]$$

(where Y is a $\chi_{10}^2(0)$ r.v.) and, for the one-sided test,

$$1 - P_{\sigma^2}\left[\sum_{i=1}^{10}(X_i - \mu)^2 \leq 16.0\right] = 1 - P\left[Y \leq \frac{16}{\sigma^2}\right]$$

(where Y is a $\chi_{10}^2(0)$ r.v.) The following points can be plotted to yield the power curves:

σ^2	0	.25	.50	.75	1.0
P [One-sided rejects]	0	.00	.00	.02	.10
P [Two-sided rejects]	1	.89	.35	.13	.10

σ^2	1.5	2.0	2.5	3.0
P [One-sided rejects]	.40	.63	.78	.88
P [Two-sided rejects]	.30	.51	.70	.82

Test Problem. Man A (a track star) has run the mile twice in competitions, with times of 3.8 and 3.4. Man B (another track star) has run the mile thrice in competitions, with times of 3.7, 3.6, and 4.1. Assume that man A and man B ran independently of each other and that each one's times are independent observations of his running ability. Let $\mu_A(\mu_B)$ denote the average time it takes man $A(B)$ to run the mile in competitions. Similarly, let his variance be σ_A^2 (σ_B^2).

(a) Based on these results perform a two-sided, equal-tail test of $\mu_A = \mu_B$ at level $\alpha = .05$. Assume $\sigma_A^2 = \sigma_B^2 = .04$. State precisely the statistical assumptions that you are making.

(b) As in part (a), but with $\sigma_A^2 = .04$, $\sigma_B^2 = .09$.

(c) As in part (a), but with $\sigma_A^2 = \sigma_B^2$ unknown.

(d) For the problem as in part (a), but with σ_A^2 and σ_B^2 both unknown, what can be said?

Solution.

From the given data, we easily find that $\bar{X} = 3.6$, $\Sigma(X_i - \bar{X})^2 = .08$, $\bar{Y} = 3.8$, $\Sigma(Y_i - \bar{Y})^2 = .14$.

(a) We will calculate

$$T_1 = \frac{(\bar{X} - \bar{Y})}{\sqrt{\frac{.04}{2} + \frac{.04}{3}}},$$

which has an $N(0,1)$ distribution under H_0: $\mu_A = \mu_B$, and reject if either $T_1 < -a$ or $T_1 > a$ where a is set so that

$$P_{H_0}[T_1 < -a] = P_{H_0}[T_1 > a] = .025,$$

namely (check this) $a = 1.96$. Since we find that $T_1 = -\sqrt{1.2} = -1.095$, we accept H_0.

(b) Here we will calculate

$$T_2 = \frac{\bar{X} - \bar{Y}}{\sqrt{\frac{.04}{2} + \frac{.09}{3}}},$$

and (check this), for a two-sided equal-tail test, will reject if either $T_2 < -1.96$ or $T_2 > 1.96$. We find $T_2 = -\sqrt{.80} = -.8944$ and accept H_0.

(c) Here, we will calculate

$$T_3 = \frac{\bar{X} - \bar{Y}}{\sqrt{\dfrac{\sum_{i=1}^{2}(X_i - \bar{X})^2 + \sum_{i=1}^{3}(Y_i - \bar{Y})^2}{(2-1) + (3-1)}\left(\frac{1}{2} + \frac{1}{3}\right)}}$$

and (check this), for a two-sided equal-tail test, will reject if either $T_3 < -3.18$ or $T_3 > 3.18$. [Note that, under H_0: $\mu_A = \mu_B$, T_3 has Student's t-distribution with 3 degrees of freedom.] We find $T_3 = -\sqrt{.654545} = -.81$ and accept H_0.

(d) Under these assumptions, we have the Behrens-Fisher problem. Using the Scheffé solution will mean calculating the statistic T_4, and (since we have a Student's t-distribution with 1 df under H_0: $\mu_A = \mu_B$) rejecting if either $T_4 < -12.7$ or $T_4 > 12.7$. [Note that a t-distribution with one degree of freedom is undesirable, as $t_1^{-1}(1-\alpha)$ is much larger than $t_r^{-1}(1-\alpha)$ for $r \geq 2$. This is accounted for by the fact that the t-distribution with one degree of freedom is a Cauchy distribution, for which the mean does not exist.] We find

$$T_4 = \frac{-.2}{\sqrt{\dfrac{\left(\left(3.8 - \sqrt{\frac{2}{3}}\,3.7\right) - \left(3.6 - \sqrt{\frac{2}{3}}\,3.8\right)\right)^2 + \left(\left(3.4 - \sqrt{\frac{2}{3}}\,3.6\right) - \left(3.6 - \sqrt{\frac{2}{3}}\,3.8\right)\right)^2}{(2)(1)}}}$$

$$= \frac{-.2\sqrt{2}}{\sqrt{(1.333 - 1.067)^2 + (1.000 - 1.067)^2}}$$

$$= \frac{-.283}{.275} = -1.029$$

Table 7: Seasonal Snowfall (in Inches) at Rochester, New York

Winter	*Amount*	*Winter*	*Amount*	*Winter*	*Amount*
1884-85	90.3	1914-15	60.8	1944-45	94.7
1885-86	64.7	1915-16	106.9	1945-46	49.5
1886-87	72.6	1916-17	92.9	1946-47	75.5
1887-88	50.7	1917-18	79.3	1947-48	63.4
1888-89	82.2	1918-19	36.0	1948-49	50.9
1889-90	65.8	1919-20	97.5	1949-50	81.7
1890-91	84.6	1920-21	61.3	1950-51	75.8
1891-92	77.1	1921-22	59.7	1951-52	75.8
1892-93	96.6	1922-23	102.5	1952-53	41.7
1893-94	102.4	1923-24	65.6	1953-54	77.5
1894-95	93.4	1924-25	60.8	1954-55	69.2
1895-96	123.8	1925-26	55.3	1955-56	121.4
1896-97	66.5	1926-27	80.6	1956-57	79.2
1897-98	91.0	1927-28	68.5	1957-58	130.8
1898-99	96.0	1928-29	55.1	1958-59	140.6
1899-1900	131.3	1929-30	59.5	1959-60	161.7
1900-01	141.5	1930-31	77.3	1960-61	89.4
1901-02	79.0	1931-32	75.7	1961-62	65.6
1902-03	68.3	1932-33	29.2	1962-63	76.4
1903-04	95.7	1933-34	77.1	1963-64	92.0
1904-05	68.3	1934-35	49.5	1964-65	71.1
1905-06	44.3	1935-36	90.0	1965-66	103.2
1906-07	63.7	1936-37	65.9	1966-67	74.0
1907-08	67.3	1937-38	54.7	1967-68	76.7
1908-09	66.5	1938-39	79.1	1968-69	79.8
1909-10	84.9	1939-40	54.5	1969-70	119.6
1910-11	107.4	1940-41	73.7	1970-71	142.7
1911-12	62.0	1941-42	66.3	1971-72	97.8
1912-13	59.2	1942-43	70.6		
1913-14	86.7	1943-44	46.1		

and again accept H_0. [One could also use the Neyman-Bartlett or Hsu solutions here. The differences among all of the solutions proposed lie to a large extent in their power (not their level).]

Test Problem. Suppose that $X_1, \ldots, X_n$ are independent r.v.'s, each $N(\mu,\sigma^2)$ with both μ and σ^2 unknown, $(\mu,\sigma^2) \in \Theta = \{(x,y): -\infty < x < +\infty, y > 0\}$. Suppose that $n = 88$ and that $X_1, \ldots, X_{88}$ are taken to be the seasonal snowfalls observed at Rochester, New York, since the 1884–85 winter.

Test Problem (a)

Estimate μ and σ^2.

Solution to (a)

The Rochester snowfall data yields (by "lengthy and trivial" calculations)

$$\hat{\mu} = 79.6989, \qquad \hat{\sigma}^2 = 619.6809$$

Since this author was in Rochester for the 1967–68 through 1971–72, winters, this means that 4 of the 5 winters he observed had snowfall greater than the estimated average (of more than 6 feet!). [Of course a person in Rochester for only the 5 winters of 1949–50 through 1953–54 would observe 4 of the 5 had less snowfall than the estimated average. These two observations together call the independence assumption (i.e., the assumption that $X_1, \ldots, X_{88}$ are independent r.v.'s) into question, and an interesting project would be to apply statistical tests of independence to this data.]

Note that calculations like these can be rendered simple by **coding the data**. Namely, work with $y_i = (x_i - a)/b$ $(i = 1, \ldots, 88)$ [instead of the basic $x_1, \ldots, x_{88}$] where a and b are rough guesses (rounded to be easy to work with) of $\hat{\mu}$ and $\hat{\sigma}$. From $\overline{Y}$ and $\sum_{i=1}^{n} (Y_i - \overline{Y})^2/n$ it is easy to find $\overline{X}$ and $\sum_{i=1}^{n} (X_i - \overline{X})^2/n$, since (using $x_i = by_i + a$)

$$\overline{X} = b\overline{Y} + a, \sum_{i=1}^{n} (X_i - \overline{X})^2/n = b^2 \sum_{i=1}^{n} (Y_i - \overline{Y})^2.$$

The Rochester, New York snowfall data was taken from newspaper articles published in Rochester. For data about other cities (as well as for Rochester data), one can write to the Environmental Data Service, National Oceanic and Atmospheric Administration, U.S. Department of Commerce (write to it at the National Climatic Center, Federal Building, Asheville, North Carolina 28801), asking for "Local Climatological Data, Annual Summary with Comparative Data" for the city in question. They will respond with a 4-page publication which contains a wealth of data (for your area if you asked for it) including temperature, precipitation, and snowfall by year and month for many years.

The Environmental Data Service (EDS) brochure dated 1974 gives Rochester data for 1935–36 through 1973–74 (with incomplete figures on 1974–75 with regard to snowfall. Comparing those figures with ours, we find that the EDS lists 50.6 (not 50.9) for 1948–49 and that the EDS lists 105.1 (not 97.8) for 1971–72; these are the only discrepancies. The first discrepancy is minor and seems likely to be due to an error by the newspaper's typesetter (not unexpected in a table of this size!). The second discrepancy seems difficult to attribute to typesetting, and a closer look at the monthly data reveals that the difference of 105.1 − 97.8 = 7.3 inches was exactly the April 1972 snowfall. An interesting assignment might be to recompute this problem with corrected 1948–49 and 1971–72 data (as one would expect, the effect of the discrepancies is negligible). Note these discrepancies and the problems in dealing with really large data sets (this is only a moderate sized data set) should increase respect for data and its handling.

Other data which may be useful for projects is the 1972–73 and 1973–74 snowfalls (73.0 and 99.1 inches, respectively). As a final note, in examining the data it seems odd that 1950–51 and 1951–52 *both* had 75.5 inches of snowfall exactly. Until the EDS confirms such odd coincidences one should suspect (for example) a typing error. (In fact these data are confirmed, however, since the different monthly snowfalls for those two seasons really do each add to 75.8 inches.) Nevertheless, such oddities should always be suspect until confirmed.

Test Problem (b)

Now it turns out that the snowfalls for 1939–40 and prior years were measured at the Federal Building at Church and North Fitzhugh Streets in downtown Rochester, but that from 1940–41 onward the snowfalls have been measured by the National Weather Service at the Rochester-Monroe County Airport. This poses an interesting question: Should snowfalls from 1939–40 and previous years be considered in a statement about the average Rochester snowfall?

If $X_1, \ldots, X_{88}$ are independent $N(\mu,\sigma^2)$ r.v.'s, the answer is clearly yes; this is so because our estimator will be more concentrated about μ if it is based on 88 observations than if it is based on only 32 observations. Yet if $X_1, \ldots, X_{88}$ are independent r.v.'s, but $X_1, \ldots, X_{56}$ are $N(\mu_1,\sigma^2)$ while $X_{57}, \ldots, X_{88}$ are $N(\mu_2,\sigma^2)$ where $\mu_1 \neq \mu_2$, then the observations would be measuring different phenomena and should probably not be combined. Thus, test (at level $\alpha = .05$) H_0: $\mu_1 = \mu_2$ versus H_1: $\mu_1 \neq \mu_2$, assuming σ^2 is unknown. What is the **significance probability** of the data? [Use a two-sided equal-tail test.]

Now a skeptical person (for example, a scientist) might question the assumption of equal variances. If $X_1, \ldots, X_{56}$ are $N(\mu_1,\sigma_1^2)$ while $X_{57}, \ldots, X_{88}$ are $N(\mu_2,\sigma_2^2)$, test (via a two-sided equal-tail test) H_0: $\sigma_1^2 = \sigma_2^2$ versus H_1: $\sigma_1^2 \neq \sigma_2^2$. Use level $\alpha = .05$. Find the significance probability of the data.

Now, what is your estimate of the average seasonal snowfall in Rochester, and why have you chosen a particular estimator?

[Reference: *The Times-Union*, Rochester, New York, March 10, 1972, p. 1B.]

Suppose we are to observe X and have a hypothesis-testing problem of H_0 versus H_1, where H_0 and/or H_1 may be composite. Suppose we have a test of level α for each α $(0 \leq \alpha \leq 1)$, say $\phi_\alpha(X)$. Our intention is to use test $\phi_{\alpha^*}(X)$ if we decide to test the hypothesis at level $\alpha = \alpha^*$. If we observe $X = x$, then **the significance probability of** $\boldsymbol{x}$, say $\alpha_{SP}(x,\phi_\alpha)$, **is defined as the smallest** α **for which** $\phi_\alpha(x) = 1$ [the smallest α for which we would surely reject if we observed $X = x$]. Often an experimenter will report $\alpha_{SP}(x,\phi_\alpha)$ in his work. Then an experimenter with a possibly different desired level, say α', will often know that he would also decide to reject if the reported $\alpha_{SP}(x,\phi_\alpha) \leq \alpha'$ and that he would not decide to reject if the reported $\alpha_{SP}(x,\phi_\alpha) > \alpha'$. (We say often, not always, because the assumption that the tests $\phi_\alpha(X)$ have the property

$$\phi_{\alpha_1}(x) = 1 \Rightarrow \phi_{\alpha_2}(x) = 1, \text{ for all } \alpha_2 > \alpha_1, \text{ for all } x \quad (1)$$

is needed, but it is not always satisfied.)

Solution (b).

Here we may calculate

$$\sum_{i=1}^{56} X_i = 4279.1, \quad \sum_{i=1}^{56} X_i^2 = 353838.71$$

$$\sum_{i=57}^{88} X_i = 2734.4, \quad \sum_{i=57}^{88} X_i^2 = 259661.7$$

hence

$$\overline{X}_1 = \sum_{i=1}^{56} X_i/56 = 76.4125, \quad \overline{X}_2 = \sum_{i=57}^{88} X_i/32 = 85.45$$

$$\sum_{i=1}^{56} (X_i - \overline{X}_1)^2 = 26861.98, \quad \sum_{i=57}^{88} (X_i - \overline{X}_2)^2 = 26007.22$$

$$s_1^2 = \sum_{i=1}^{56} (X_i - \overline{X}_1)^2/55 = 488.40,\ s_2^2 = \sum_{i=57}^{88} (X_i - \overline{X}_2)^2/31 = 838.94.$$

To test H_0: $\mu_1 = \mu_2$ versus H_1: $\mu_1 \neq \mu_2$ (assuming a common unknown variance σ^2), we calculate the statistic T_1 and, for a two-sided equal-tail test of level $\alpha = .05$, reject if

$$T_1 > t_{86}^{-1}(.985) \text{ or } T_1 < -t_{86}^{-1}(.975).$$

(Note that $t_{86}^{-1}(.975) = 1.99$, approximately.) Since we calculate

$$T_1 = \frac{-9.0375}{\sqrt{(614.76)\left(\frac{1}{56}+\frac{1}{32}\right)}} = -1.64,$$

we do not reject H_0. The significance probability of the data (for a two-sided equal-tail test) is approximately .10.

For testing H_0: $\sigma_1^2 = \sigma_2^2$ versus H_1: $\sigma_1^2 \neq \sigma_2^2$, we use the statistic T_2 and, for a two-sided equal-tail test of level $\alpha = .10$, reject if

$$T_2 > c_U \text{ or } T_2 < c_L$$

where c_U and c_L are chosen so that if Y has the $F(55,31)$ distribution, then $P[Y > c_U] = P[Y < c_L] = .05$, namely approximately

$$c_U = 1.75 \text{ and } c_L = 1/1.66 = .60.$$

Since

$$T_2 = \frac{s_1^2}{s_2^2} = \frac{488.40}{838.94} = 0.58,$$

we do not reject H_0: $\sigma_1^2 = \sigma_2^2$ *at level .10*, hence (why?) we will not reject H_0 at level .05 (for which our table 7 is not appropriate, but rather tables from the literature must be used; we need not examine those tables here, as variance differences of the type seen or larger would be expected in about 10% of all experiments even if H_0 were true).

Ignoring possible time trends (which we were unable to demonstrate when we tested H_0: $\mu_1 = \mu_2$) and the question of which location is of interest to us (and why),

$$\sum_{i=1}^{88} X_i/88 = 79.7$$

inches seems an adequate estimate.

INTERVAL ESTIMATION

We consider "interval estimation"; that is, how to estimate θ by an interval of values $\{x: \theta_L < x < \theta_U\}$ (abbreviated (θ_L,θ_U)) which has high probability of including θ but which also has, say, small average length $E(\theta_U - \theta_L)$. This subject is intimately related to tests of hypotheses.

Confidence Sets and Tests of Hypotheses

Some situations in which hypothesis-testing problems arise were described. Generally, these were of the type "Does making a certain modification to some system or process change a particular parameter θ?" or "Is my parameter θ equal to θ_0?" **In interval estimation we** are concerned with the same situations, but instead **ask "How much does making a certain modification to some system or process change a particular parameter θ?" or "How large is my parameter θ?"** (Another approach to these latter questions is that of point estimation.)

Let us consider a special case: the results here will be helpful to our intuition. Suppose X is a univariate r.v. with d.f. $F(x|\theta)$ where θ is an unknown univariate parameter, $\theta \in \Theta$. Suppose we have available a test of the simple hypothesis H_0: $\theta = \theta_0$ versus the composite alternative H_1: $\theta \neq \theta_0$, and that test rejects iff $X < a$ or $X > b$, where a and b are specified numbers such that

$$P_{\theta_0}\ [X<a, \text{ or } X>b] = \alpha;$$

that is, such that the test has level α. Then usually a and b will be functions of θ_0, say $a = a(\theta_0)$ and $b = b(\theta_0)$. Consider plotting $a(\cdot)$ and $b(\cdot)$ for every θ. Then for a specified $\theta = \theta_0$, we have heights $a(\theta_0) < b(\theta_0)$, and

$$P_{\theta_0}\ [a(\theta_0) \leq X \leq b(\theta_0)] = 1 - \alpha.$$

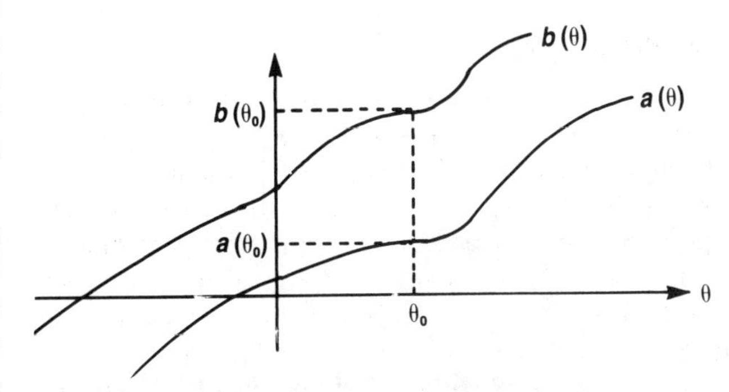

Figure 18: The functions $a(\cdot)$ and $b(\cdot)$

EH0227-9/85/0000/0389$01.00

Now plot the r.v. X on the vertical axis and draw a horizontal dotted line through it. Let $L(X)$ denote the value of θ at the intersection of the dotted line with the graph of $b(\theta)$, and let $U(X)$ denote the value of θ at the intersection with the graph of $a(\theta)$. (We assume for now that $a(\theta)$ and $b(\theta)$ are such that there is, for each possible value of X, exactly one intersection of each type. More general cases are considered in the context of confidence sets.) Then

$$P_\theta\,[L(X) \leq \theta \leq U(X)] \stackrel{\theta}{\equiv} 1 - \alpha; \tag{2}$$

that is, θ **has probability** $\mathbf{1-\alpha}$ (for example, .95 if α = .05) **of lying between** $\boldsymbol{L(X)}$ **and** $\boldsymbol{U(X)}$. The **proof** is as follows: If θ^* is the true value of θ, then $P_{\theta^*}\,[a(\theta^*) \leq X \leq b(\theta^*)] = 1-\alpha$. But $a(\theta^*) \leq X \leq b(\theta^*)$ iff $L(X) \leq \theta^* \leq U(X)$, which proves the result.

The interval $\boldsymbol{(L(X), U(X))}$ **is then called a "confidence interval" for** θ **with "confidence coefficient"** $\mathbf{1-\alpha}$. The concept extends to the multivariate case (Figure 19).

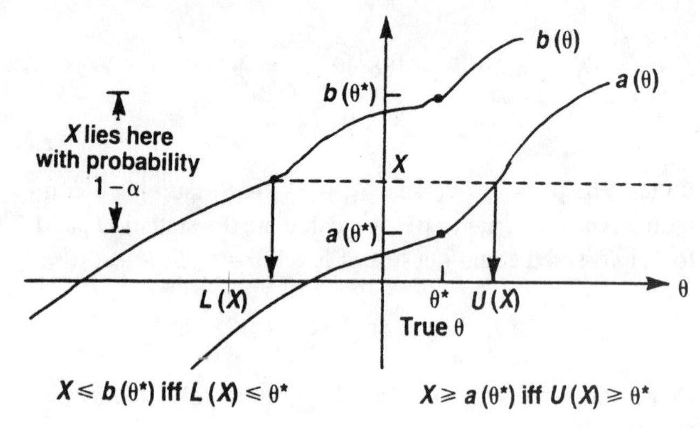

Figure 20: Proof of (2)

Definition. Let X be a r.v. with d.f. $F(x|\theta)$ where θ is unknown, $\theta \in \Theta$. [Here X and θ may be multivariate.] If (2) holds, then the interval $(L(X), U(X))$ is called a **confidence interval** for θ with **confidence coefficient** $1-\alpha$.

Confidence Intervals for One Normal Mean

We consider confidence intervals (and make some remarks about confidence sets) for one normal mean, first in the case where σ^2 is known, and then in the case where σ^2 is unknown.

Example. let $X_1, \ldots, X_n$ be independent $N(\mu,\sigma^2)$ r.v.'s where μ is unknown with $\mu \in \Theta = \{x: -\infty < x < +\infty\}$ and where $\sigma^2 > 0$ is known. Find a confidence interval for μ with confidence coefficient $1-\alpha = .95$.

Here $\overline{X} = (X_1 + \ldots + X_n)/n$, and we should let our confidence interval for μ depend on $X_1, \ldots, X_n$ solely through $\overline{X}$. In this case good tests of the hypothesis H_0: $\mu = \mu_0$ were studied and a typical member of a reasonable class rejects H_0: $\mu = \mu_0$ iff

$$\begin{aligned} &\overline{X} - \mu_0 < \frac{\sigma}{\sqrt{n}}\,\Phi^{-1}(\alpha p) \text{ or} \\ &\overline{X} - \mu_0 > \frac{\sigma}{\sqrt{n}}\,\Phi^{-1}(1 - \alpha(1-p)), \end{aligned} \tag{3}$$

where p $(0 \leq p \leq 1)$ is fixed but arbitrary. Since when H_0 is true (3) occurs with probability α, its complement occurs with probability $1-\alpha$:

$$P_{\mu_0}\left[\mu_0 + \frac{\sigma}{\sqrt{n}}\,\Phi^{-1}(\alpha p) \leqslant \overline{X} \leqslant \mu_0 + \frac{\sigma}{\sqrt{n}}\,\Phi^{-1}(1-\alpha(1-p))\right] = 1 - \alpha.$$

Thus here our functions $a(\mu)$ and $b(\mu)$ are

$$\begin{cases} a(\mu) = \mu + \dfrac{\sigma}{\sqrt{n}}\,\Phi^{-1}(\alpha p), \\ b(\mu) = \mu + \dfrac{\sigma}{\sqrt{n}}\,\Phi^{-1}(1-\alpha(1-p)). \end{cases}$$

To construct our confidence interval $(L(\overline{X}), U(\overline{X}))$, we proceed obtaining the solution via solving $\overline{X} = b(\mu)$ for $L(\overline{X})$ and $\overline{X} = a(\mu)$ for $U(\overline{X})$:

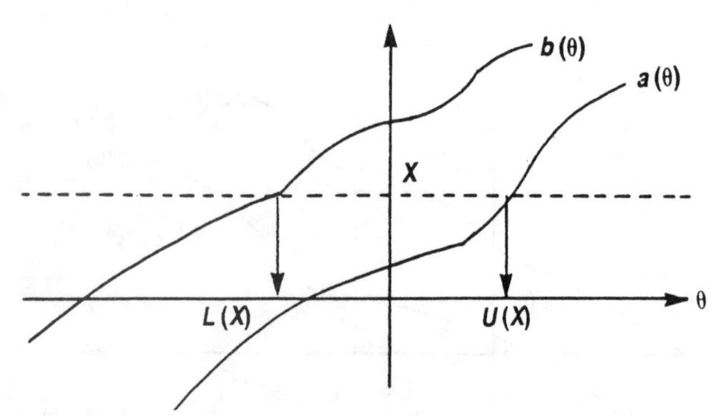

Figure 19: Construction of the Interval (L(X), U(X))

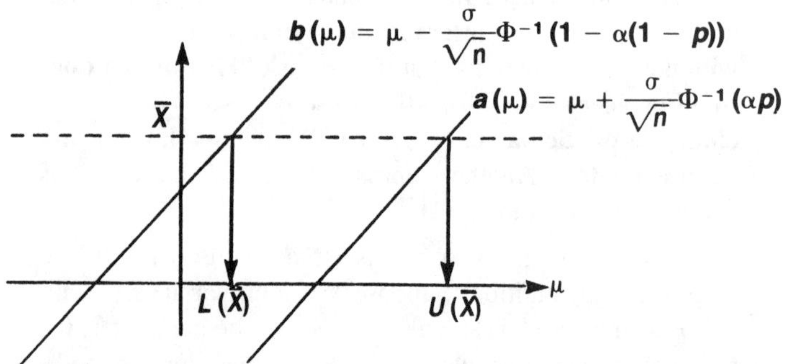

Figure 21: Construction of (L(X̄), U(X̄))

$$\begin{cases} L(\overline{X}) = \overline{X} - \dfrac{\sigma}{\sqrt{n}}\Phi^{-1}(1-\alpha(1-p)) \\ U(\overline{X}) = \overline{X} - \dfrac{\sigma}{\sqrt{n}}\Phi^{-1}(\alpha p), \end{cases}$$

Thus

$$P_\mu\left[\overline{X} - \frac{\sigma}{\sqrt{n}}\Phi^{-1}(1-\alpha(1-p)) \leq \mu \leq \overline{X} - \frac{\sigma}{\sqrt{n}}\Phi^{-1}(\alpha p)\right]$$
$$\overset{\mu}{\equiv} 1 - \alpha.$$

In the desired case of $1-\alpha = .95$, if we choose $p = .5$ [one of many possible choices], we find our interval is

$$\left(\overline{X} - 1.96\frac{\sigma}{\sqrt{n}}, \overline{X} + 1.96\frac{\sigma}{\sqrt{n}}\right)$$

In fact, $p = .5$ is a reasonable choice since of all p $(0 \leq p \leq 1)$ it yields the one with the **smallest length**, namely length equal to

$$\frac{\sigma}{\sqrt{n}}\left\{\Phi^{-1}\left(1 - \frac{\alpha}{2}\right) - \Phi^{-1}\left(\frac{\alpha}{2}\right)\right\} = 2\frac{\sigma}{\sqrt{n}}\Phi^{-1}\left(1 - \frac{\alpha}{2}\right).$$

Note that if we were concerned with **a confidence interval which would put an upper bound on** μ, we could choose $p = 1$ and obtain the interval

$$\left(-\infty, \overline{X} + \frac{\sigma}{\sqrt{n}}\Phi^{-1}(1-\alpha)\right).$$

We can similarly obtain a **lower bound** on μ.

Also *note* that we obtained parallel lines $a(\mu)$ and $b(\mu)$ due to our holding p fixed independent of μ. If we chose different p's for various μ values, we could obtain other curves.

We next consider the case where σ^2 is unknown. Example. Let $X_1, \ldots, X_n$ be independent $N(\mu,\sigma^2)$ r.v.'s where μ is unknown $(-\infty < \mu < +\infty)$ and σ^2 is unknown $(\sigma^2 > 0)$. Find a confidence interval for μ with confidence coefficient $1-\alpha = .95$.

In this case, good tests of the hypothesis H_0: $\mu = \mu_0$ were obtained and a typical test rejects H_0: $\mu = \mu_0$ iff

$$\overline{X} - \mu_0 < \frac{s}{\sqrt{n}}t_{n-1}^{-1}(\alpha p) \text{ or } \overline{X} - \mu_0 > \frac{s}{\sqrt{n}}t_{n-1}^{-1}(1-\alpha(1-p)).$$

where p $(0 \leq p \leq 1)$ is fixed but arbitrary. By a development similar to that above, we find

$$\begin{cases} L = \overline{X} - \dfrac{s}{\sqrt{n}}t_{n-1}^{-1}(1-\alpha(1-p)), \\ U = \overline{X} - \dfrac{s}{\sqrt{n}}t_{n-1}^{-1}(\alpha p). \end{cases}$$

In the desired case of $1-\alpha = .95$, if we choose $p = .5$ [one of many possible choices], we find our interval is

$$\left(\overline{X} - 2.57\frac{s}{\sqrt{6}}, \overline{X} + 2.57\frac{s}{\sqrt{6}}\right) \text{ if } n = 6.$$

Here we needed to assume a value of n (we took $n = 6$ for purposes of illustration) since the quantity $t_{n-1}^{-1}(\cdot)$ is a function of n.

In fact, $p = .5$ is a reasonable choice here also since, as was the case in the instance of a known $\sigma^2 > 0$, it yields the **shortest length**

$$2\frac{s}{\sqrt{n}}t_{n-1}^{-1}\left(1 - \frac{\alpha}{2}\right)$$

Note that in the case of known σ^2 the intervals had fixed length, so by an appropriate choice of n we could specify the length of the interval on μ. (for example, if we have $p = .5$, we simply take n to be the smallest integer greater than or equal to the solution x of the equation

$$2\frac{\sigma}{\sqrt{x}}\Phi^{-1}\left(1 - \frac{\alpha}{2}\right) = L_0$$

and obtain an interval of length L_0 [or slightly less].) In this case the length is a random variable since s is a r.v., and hence cannot be specified in advance of the experiment unless we have some idea of how large s will be (which we might if we had an idea of the size of σ^2).

Above we have obtained confidence intervals from tests of hypotheses. However, another method of obtaining confidence intervals (which is not always feasible, but when it is often yields answers quickly and neatly) **is available**, and will now be explained.

Lemma. Let X be a r.v. with d.f. $F(x|\theta)$ where θ is unknown, $\theta \in \Theta$. [Here X and θ may be multivariate.] Suppose there is a function $g(X,\theta)$ whose distribution is known and which does not depend on θ. Then it is easy to find sets S of possible values of $g(X,\theta)$ such that

$$P\,[g(X,\theta) \in S] \equiv 1 - \alpha.$$

However, each such set S yields us a confidence set for θ since

$$P_\theta\,[\theta \in \{\theta\colon g(X,\theta) \in S\}] \overset{\theta}{\equiv} 1 - \alpha.$$

We now demonstrate the use of this new method of finding confidence intervals.

Example. Let $X_1, \ldots, X_n$ be independent $N(\mu,\sigma^2)$ r.v.'s with μ unknown and σ^2 known. Then

$$g\,(X_1, \ldots, X_n, \mu) = \frac{\overline{X} - \mu}{\sigma/\sqrt{n}}$$

has the $N(0,1)$ distribution, which does not depend on μ. One possible choice of S, $S = \{x: -1.96 \leq x \leq 1.96\}$, yields

$$P[g(X_1, \ldots, X_n, \mu) \in S]$$

$$= .95$$

$$= P\left[-1.96 \leq \frac{\overline{X} - \mu}{\sigma/\sqrt{n}} \leq 1.96 \right]$$

$$= P\left[-1.96 \frac{\sigma}{\sqrt{n}} \leq \overline{X} - \mu \leq 1.96 \frac{\sigma}{\sqrt{n}} \right]$$

$$= P\left[\overline{X} - 1.96 \frac{\sigma}{\sqrt{n}} \leq \mu \leq \overline{X} + 1.96 \frac{\sigma}{\sqrt{n}} \right].$$

This is the same interval as before.

Confidence Intervals for One Normal Variance

We consider confidence intervals for one normal variance in the case where μ is known. The case where μ is unknown is handled by a simple modification.

Example. Let $X_1, \ldots, X_n$ be independent $N(\mu,\sigma^2)$ r.v.'s where μ is known $(-\infty < \mu < +\infty)$ and $\sigma^2 > 0$ is unknown. Find a confidence interval for σ^2 with confidence coefficient $1-\alpha = .95$ when $n = 10$.

We know that

$$g(X_1, \ldots, X_n, \sigma) = \sum_{i=1}^{n} \left(\frac{X_i - \mu}{\sigma} \right)^2$$

has the $\chi_n^2\ (0)$ distribution. To find the desired confidence interval,

$$P\left[3.25 \leq \frac{\sum_{i=1}^{n}(X_i - \mu)^2}{\sigma^2} \leq 20.48 \right] = .95$$

$$= P\left[\frac{\sum_{i=1}^{n}(X_i - \mu)^2}{20.48} \leq \sigma^2 \leq \frac{\sum_{i=1}^{n}(X_i - \mu)^2}{3.25} \right]$$

In the case of unknown μ we know that

$$g^*(X_1, \ldots, X_n, \sigma) = \sum_{i=1}^{n} \left(\frac{X_i - \overline{X}}{\sigma} \right)^2 \qquad (4)$$

has the $\chi_{n-1}^2(0)$ distribution, so the desired result is easily obtained.

In the case of unknown μ, (4) requires $n \geq 2$ in order to yield a confidence interval. However, **even in the case of $n = 1$, an interval for σ which has probability at least $1-\alpha$ of including σ can be obtained,** as we will now see.

Example. Let X be a $N(\mu,\sigma^2)$ r.v. with μ $(-\infty < \mu < +\infty)$ and $\sigma^2 > 0$ both unknown. Let $c > 0$ be chosen to satisfy

$$\Phi\left(\frac{1}{c}\right) - \Phi\left(-\frac{1}{c}\right) = \alpha$$

Then

$$P[\sigma \in (0,c|X|)] = P\left[\frac{\sigma}{c} \leq |X|\right] = 1 - P\left[\frac{\sigma}{c} > |X|\right]$$

$$= 1 - P\left[-\frac{\sigma}{c} < X \frac{\sigma}{c}\right]$$

$$= 1 - P\left[-\frac{1}{c} - \frac{\mu}{\sigma} < \frac{\overline{X}-\mu}{\sigma} < \frac{1}{c} - \frac{\mu}{\sigma} \right]$$

$$= 1 - \left\{ \Phi\left(\frac{1}{c} - \frac{\mu}{\sigma}\right) - \Phi\left(-\frac{1}{c} - \frac{\mu}{\sigma}\right) \right\}.$$

Since $\Phi(a-b) - \Phi(-a-b)$ is largest when $b = 0$, we find

$$P\,[\sigma \in (0,c|X|)] \geq 1 - \left\{ \Phi\left(\frac{1}{c}\right) - \Phi\left(-\frac{1}{c}\right) \right\} = 1 - \alpha$$

Hence, the interval $(0,c|X|)$ has probability at least $1-\alpha$ of covering σ.

Test Problem. In the example of seasonal snowfall in Rochester, New York what confidence interval on μ do we obtain if we assume that:

(a) $X_1, \ldots, X_{88}$ are independent $N(\mu,\sigma^2)$ r.v.'s with μ and σ^2 unknown;

(b) $X_1, \ldots, X_{56}$ are independent $N(\mu_1,\sigma^2)$ r.v.'s, $X_{57}, \ldots, X_{88}$ are independent $N(\mu,\sigma^2)$ r.v.'s, and μ_1, μ, σ^2 are unknown;

(c) $X_1, \ldots, X_{56}$ are independent $N(\mu_1,\sigma_1^2)$ r.v.'s, $X_{57}, \ldots, X_{88}$ are independent $N(\mu,\sigma_2^2)$ r.v.'s, and μ_1, μ, σ_1^2, σ_2^2 are unknown.

[Hint: Your interval in (*b*) should be "better" than your interval in (*c*) since in (*b*) more is known about σ^2.] Which interval do you recommend should be used to guide the Highway Department in its purchases of road salt for the next season?

Solution.

(a) Here we calculate

$$\overline{X} = \sum_{i=1}^{88} X_i/88 = 7013.5/88 = 79.7,$$

$$s^2 = \sum_{i=1}^{88} (X_i - \overline{X})^2/87 = 54532.43/87 = 626.81.$$

Hence, using a two-sided equal-tail interval with confidence coefficient .95, we find

$$\bar{X} - t_{n-1}^{-1}(.975)\frac{s}{\sqrt{n}} \le \mu \le \bar{X} + t_{n-1}^{-1}(.975)\frac{s}{\sqrt{n}},$$

that is

$$74.39 \le \mu \le 85.01.$$

(b) Here $X_{57}, \ldots, X_{88}$ are used to estimate μ, and s_1^2 and s_2^2 are used to estimate σ^2 and our interval is

$$\bar{X}_2 - t_{86}^{-1}(.975)\frac{\sqrt{614.76}}{\sqrt{32}} \le \mu \le \bar{X}_2 + t_{86}^{-1}(.975)\frac{\sqrt{614.76}}{\sqrt{32}},$$

that is

$$76.73 \le \mu \le 94.17.$$

(c) Here only $X_{57}, \ldots, X_{88}$ can be used to estimate μ and σ^2, and (as in part (a)) our interval is

$$\bar{X}_2 - t_{31}^{-1}(.975)\frac{s_2}{\sqrt{32}} \le \mu \le \bar{X}_2 + t_{31}^{-1}(.975)\frac{s_2}{\sqrt{32}},$$

that is

$$75.01 \le \mu \le 95.89.$$

It would seem reasonable for total *mean* snowfall of 96 inches (8 feet) to be allowed for in road salt purchases (subject to further investigation, e.g. using nonparametric methods).

(However, for next year alone we more appropriately need a prediction interval on the next snowfall to guide us. Also, the relation between snowfall and road salt needs may bear investigation via regression methods.)

STATISTICAL PRINCIPLES — TOPICS, USES, A LOOK AHEAD

At this point we wish to look in some detail at applications of some important statistical analyses, namely

- paired and unpaired *t*-tests (Student's *t*-tests)
- correlation coefficients
- linear regression
- parallel regressions
- all possible regressions
- two-way analysis of variance.

These analyses embody the basic principles already reviewed, incorporate real-data examples, and furnish an introductory entry into some of the important topics of design and analysis of experiments, which will come in full detail later.

1. PAIRED AND UNPAIRED t-TESTS (STUDENT'S t-TESTS)

Statistical Theory

Suppose we are given a **data** matrix $X^{m \times n}$, i.e.,

$$X^{m \times n} = \begin{pmatrix} X_{11} & X_{12} & \ldots & X_{1n} \\ X_{21} & X_{22} & \ldots & X_{2n} \\ \cdot & & & \\ \cdot & & & \\ \cdot & & & \\ X_{m1} & X_{m2} & \ldots & X_{mn} \end{pmatrix}.$$

Each of the m rows of X will often in practice represent an experiment, with each of the n columns representing some measured (or derived) variable.

Zeros in X are often taken to mean no observation was present, (e.g., a certain measurement may not be available on a certain patient; this is called a case of **"incomplete data"** or **"missing observations,"** and occurs often in research).

The statistical summary measures "mean," "variance," "standard deviation," "standard error of the mean," and "100 $(1-\alpha)$% confidence interval for the mean" are often of interest for each column. For column i ($1 \le i \le n$) of the data matrix $X^{m \times n}$ they are:

$\bar{X}_i = \Sigma X_{ij}/n_i$ --(**mean**)

$s_i^2 = \Sigma (X_{ij} - \bar{X}_i)^2/(n_i - 1)$ -------------------- (**variance**)

s_i -------------------------------------- (**standard deviation**)

$s_i/\sqrt{n_i}$ --------------------- (**standard error of the mean**)

$$\bar{X}_i - t_\alpha(n_i - 1)\frac{s_i}{\sqrt{n_i}} \leq \text{MEAN} \leq \bar{X}_i + t_\alpha(n_i - 1)\frac{s_i}{\sqrt{n_i}}$$

--------- **(100(1 − α)% confidence interval for the mean)**.

Here n_i is the number of non-zero items in column i, and $t_\alpha(n_i - 1)$ is the two-tail α-point of the Student's t-distribution with $n_i - 1$ degrees of freedom. [These values may be found on p. 464 of Dixon and Massey (1969) and references given therein; e.g., $t_{.05}(1) = 12.706$, $t_{.05}(2) = 4.303$, $t_{.05}(3) = 3.182$, $t_{.05}(4) = 2.776$, $t_{.05}(5) = 2.571$, etc.]

For an unpaired t-test of the hypothesis that columns i and j ($i \neq j$) have the same mean value, we are interested in "numbers of observations on each variable, t-statistic, significance probability, are means significantly different at level α, and 100(1 − α)% confidence interval for difference of means." For columns i and j of the data matrix $X^{m \times n}$, they are:

(n_i, n_j) -------- (number of observations on each variable)

$$t = \frac{\bar{X}_i - \bar{X}_j}{\sqrt{\frac{s_i^2}{n_i} + \frac{s_j^2}{n_j}}} \text{ --------------------------------- } (t\text{-statistic})$$

$2P[T \geq |t|]$ --------------------- (significance probability)

$\begin{cases} \text{NO if } \alpha < 2P[T \geq |t|] \\ \text{YES if } \alpha \geq 2P[T \geq |t|] \end{cases}$ --------- (are means significantly different at level α)

$$(\bar{X}_i - \bar{X}_j) - t_\alpha(f_{ij} - 1)\sqrt{\frac{s_i^2}{n_i} + \frac{s_j^2}{n_j}} \leq \text{MEAN}(I)$$

$$- \text{MEAN}(J) \leq (\bar{X}_i - \bar{X}_j) + t_\alpha(f_{ij} - 1)\sqrt{\frac{s_i^2}{n_i} + \frac{s_j^2}{n_j}}$$

------------------------ (100(1 − α)% confidence interval for difference of means)

where $f_{ij} = \min(n_i, n_j)$ and T is a Student's t random variable with $f_{ij} - 1$ degrees of freedom.

(Note on unpaired t-tests.) The above confidence interval is that given on p. 1502, equation (2.1), of Scheffé (1970), which reviews the Behrens-Fisher problem. As is shown there, it has at least 100(1 − α)% confidence, and can be called **the Hsu solution**. Others have suggested replacing $f_{ij} - 1$ with other (larger) quantities in order to obtain shorter confidence intervals (and hence more results judged "significant"); e.g., **Welch suggested**

$$\mathcal{F} = \frac{\left(\frac{s_i^2}{n_i} + \frac{s_j^2}{n_j}\right)^2}{\frac{(s_i^2/n_i)^2}{n_i - 1} + \frac{(s_j^2/n_j)^2}{n_j - 1}}$$

(e.g., see Dixon and Massey (1969), p. 119, where $n_i - 1$ and $n_j - 1$ are replaced by n_i and n_j). One can show

$$f_{ij} - 1 = \min(n_i - 1, n_j - 1) \leq \mathcal{F} \leq (n_i - 1) + (n_j - 1).$$

Such tests do not necessarily guarantee the stated level of test α, and for reasonably large $f_{ij} - 1$ (e.g. 10 or more), they do not reduce the $t_\alpha(f_{ij} - 1)$ value appreciably. Hence we have chosen not to use them.

(Note on significance probability.) The significance probability of a t-statistic is the lowest significance level α for which the results can be called "significant." In the past it has not been possible to report the significance probability precisely, and people often say, for example, "p < .01." Dudewicz and Dalal (1972) use a formula of Zelen and Severo (1964) to allow exact statements, for example, "p = .003." The value "p" is calculated as $2(1 - F_\upsilon(|t|))$, where $|t|$ is the absolute value of the t-statistic obtained, υ is the number of degrees of freedom, and

$$F_\upsilon(z) = \begin{cases} \frac{1}{2} + \frac{\theta}{\pi}, \upsilon = 1 \\ \frac{1}{2} + \frac{1}{\pi}\{\theta + \sin\theta[\cos\theta + \frac{2}{3}\cos^3\theta \\ \quad + \ldots + \frac{2 \cdot 4 \cdots (\upsilon - 3)}{1 \cdot 3 \cdots (\upsilon - 2)}\cos^{\upsilon - 2}\theta\}, \upsilon > 1 \text{ odd} \\ \frac{1}{2} + \frac{\sin\theta}{2}\{1 + \frac{1}{2}\cos^2\theta + \frac{1 \cdot 3}{2 \cdot 4}\cos^4\theta \\ \quad + \ldots + \frac{1 \cdot 3 \cdot 5 \cdots (\upsilon - 3)}{2 \cdot 4 \cdot 6 \cdots (\upsilon - 2)}\cos^{\upsilon - 2}\theta\}, \upsilon \text{ even,} \end{cases}$$

where $\theta = \arctan(z/\sqrt{\upsilon})$. This allows much more accurate reporting of the significance of scientific experiments.

For a paired t-test of the hypothesis that columns i and j ($i \neq j$) have the same mean value, we are interested in "number of observation pairs, t-statistic, significance probability, are means significantly different at level α, and 100(1 − α)% confidence interval for difference of means." For columns i and j of the data matrix $X^{m \times n}$, they are:

(n_{ij}) ------------------------- (number of observation pairs)

$$t = \frac{\bar{Y}_{ij}}{\sqrt{\frac{s_{ij}^2}{n_{ij}}}} \text{ -- } (t\text{-statistic})$$

$2P[T \geq |t|]$ --------------------- (significance probability)

$\begin{cases} \text{NO if } \alpha < 2P[T \geq |t|] \\ \text{YES if } \alpha \geq 2P[T \geq |t|] \end{cases}$ --------- (are means significantly different at level α)

$$\bar{Y}_{ij} - t_\alpha(n_{ij} - 1)\sqrt{\frac{s_{ij}^2}{n_{ij}}} \leq \text{MEAN}(I) - \text{MEAN}(J) \leq (\bar{Y}_{ij} + t_\alpha(n_{ij} - 1)\sqrt{\frac{s_{ij}^2}{n_{ij}}}$$

-------- (100(1 − α)% confidence interval for difference of means)

where n_{ij} is the number of pairs in the data matrix, $Y_{ij1}, \ldots, Y_{ijn_{ij}}$ are the differences $X_{il} - X_{jl}$ of X_{il} and X_{jl} in these pairs, $\bar{Y}_{ij} = \Sigma Y_{ijl}/n_{ij}$, $s_{ij}^2 = \Sigma(Y_{ijl} - \bar{Y}_{ij})^2/(n_{ij} - 1)$, and T is a Stu-

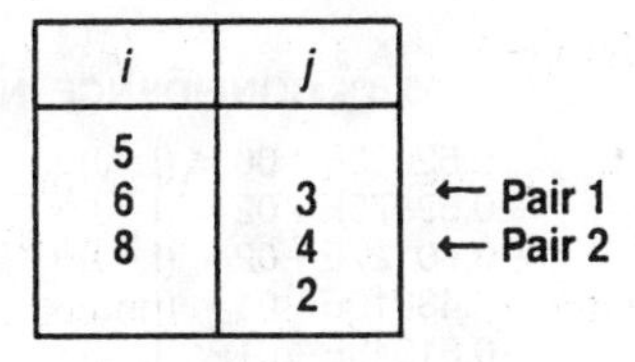

Figure 22

dent's t random variable with $n_{ij}-1$ degrees of freedom. For example, if columns i and j are as shown in Figure 22, then there are $n_{ij}=2$ pairs, and $Y_1=6-3=3$, $Y_2=8-4=4$. (We must have n_{ij} at least 2 or this statistical analysis is not possible.)

(Note on assumptions.) The above analyses make the (usual) assumptions of independence and identical normal distribution of data.

1. Paired and Unpaired *t*-Tests (Student's *t*-Tests): References

Dixon, W.J. and Massey, F.J., Jr. (1969): *Introduction to*

First Example

Dixon and Massey, *Intro. Stat. Ana.*, 1969, pp. 11–22.

T Level = .050

\# Cases = 12 # Variables = 8

Data by cases (0.0 is considered as no obs. present)

RATN A	RATN B	PAINT1	PAINT2	BOYS	GIRLS	BEFORE	AFTER
31.00	26.00	85.00	89.00	28.00	19.00	120.00	128.00
34.00	24.00	87.00	89.00	18.00	38.00	124.00	131.00
29.00	28.00	92.00	90.00	22.00	42.00	130.00	131.00
26.00	29.00	80.00	84.00	27.00	25.00	118.00	127.00
32.00	30.00	84.00	88.00	25.00	15.00	140.00	132.00
35.00	29.00	.00	.00	30.00	31.00	128.00	125.00
38.00	32.00	.00	.00	21.00	22.00	140.00	141.00
34.00	26.00	.00	.00	21.00	37.00	135.00	137.00
30.00	31.00	.00	.00	20.00	30.00	126.00	118.00
29.00	29.00	.00	.00	27.00	24.00	130.00	132.00
32.00	32.00	.00	.00	.00	.00	126.00	129.00
31.00	28.00	.00	.00	.00	.00	127.00	135.00

# OBSERVATIONS		MEAN	VARIANCE	STANDARD DEVIATION	STANDARD ERROR OF THE MEAN
RATN A	12	0.3175E+02	0.1020E+02	0.3194E+01	0.9222E+00
RATN B	12	0.2867E+02	0.6061E+01	0.2462E+01	0.7107E+00
PAINT1	5	0.8560E+02	0.1930E+02	0.4393E+01	0.1965E+01
PAINT 2	5	0.8800E+02	0.5500E+01	0.2345E+01	0.1049E+01
BOYS	10	0.2390E+02	0.1610E+02	0.4012E+01	0.1269E+01
GIRLS	10	0.2830E+02	0.7779E+02	0.8820E+01	0.2789E+01
BEFORE	12	0.1287E+03	0.4806E+02	0.6933E+01	0.2001E+01
AFTER	12	0.1305E+03	0.3500E+02	0.5916E+01	0.1708E+01

	95.0% CONFIDENCE INTERVALS
RATN A	0.2972E+02 < MEAN < 0.3378E+02
RATN B	0.2710E+02 < MEAN < 0.3023E+02
PAINT 1	0.8015E+02 < MEAN < 0.9105E+02
PAINT 2	0.8509E+02 < MEAN < 0.9091E+02
BOYS	0.2103E+02 < MEAN < 0.2677E+02
GIRLS	0.2199E+02 < MEAN < 0.3461E+02
BEFORE	0.1243E+03 < MEAN < 0.1331E+03
AFTER	0.1267E+03 < MEAN < 0.1343E+03

UNPAIRED T TEST

I		J	T	SIGPROB	SIG?	N(I),N(J)	95.0% CONFIDENCE INTERVALS
RATN A	−	RATN B	2.64838	.02265	YES	(12, 12)	0.52085E+00 < (I−J) < 0.56458E+01
RATN A	−	PAINT1	−24.81189	.00002	YES	(12, 5)	−0.59875E+02 < (I−J) < −0.47825E+02
RATN A	−	PAINT2	−40.27756	.00000	YES	(12, 5)	−0.60127E+02 < (I−J) < −0.52373E+02
RATN A	−	BOYS	5.00461	.00073	YES	(12, 10)	0.43019E+01 < (I−J) < 0.11398E+02
RATN A	−	GIRLS	1.17444	.27036	NO	(12, 10)	−0.31948E+01 < (I−J) < 0.10095E+02
RATN A	−	BEFORE	−43.98227	.00000	YES	(12, 12)	−0.10177E+03 < (I−J) < −0.92067E+02
RATN A	−	AFTER	−50.87880	.00000	YES	(12, 12)	−0.10302E+03 < (I−J) < −0.94478E+02
RATN B	−	PAINT1	−27.25044	.00001	YES	(12, 5)	−0.62733E+02 < (I−J) < −0.51134E+02
RATN B	−	PAINT2	−46.83313	.00000	YES	(12, 5)	−0.62850E+02 < (I−J) < −0.555816+02
RATN B	−	BOYS	3.27760	.00957	YES	(12, 10)	0.14770E+01 < (I−J) < 0.80563E+01
RATN B		GIRLS	12740	.90143	NO	(12, 10)	−0.61438E+01 < (I−J) < 0.68771E+01
RATN B	−	BEFORE	−47.08678	.00000	YES	(12, 12)	−0.10467E+03 < (I−J) < 0.95326E+02
RATN B	−	AFTER	−55,05125	.00000	YES	(12, 12)	−0.10590E+03 < (I−J) < −0.97762E+02
PAINT1	−	PAINT2	−1.07763	.34185	NO	(5, 5)	−0.85824E+01 < (I−J) < 0.37824E+01
PAINT1	−	BOYS	26.38116	.00001	YES	(5, 10)	0.55208E+02 < (I−J) < 0.68192E+02
PAINT1	−	GIRLS	16.79575	.00007	YES	(5, 10)	0.47829E+02 < (I−J) < 0.66771E+02
PAINT1	−	BEFORE	−15.35632	.00011	YES	(5, 12)	−0.50852E+02 < (I−J) < −0.35281E+02
PAINT1	−	AFTER	−17.24805	.00007	YES	(5, 12)	−0.52126E+02 < (I−J) < −0.37674E+02
PAINT2	−	BOYS	38.93810	.00000	YES	(5, 10)	0.59530E+02 < (I−J) < 0.68670E+02
PAINT2	−	GIRLS	20.03525	.00004	YES	(5, 10)	0.51428E+02 < (I−J) < 0.67972E+02
PAINT2	−	BEFORE	−17.99832	.00006	YES	(5, 12)	−0.46939E+02 < (I−J) < −0.34394E+02
PAINT2	−	AFTER	−21.20587	.00003	YES	(5, 12)	−0.48064E+02 < (I−J) < −0.36936E+02
BOYS	−	GIRLS	−1.43597	.18484	NO	(10, 10)	−0.11331E+02 < (I−J) < 0.25311E+01
BOYS	−	BEFORE	−44.21208	.00000	YES	(10, 12)	−0.11013E+03 < (I−J) < −0.99407E+02
BOYS	−	AFTER	−50.10361	.00000	YES	(10, 12)	−0.11141E+03 < (I−J) < −0.10179E+03
GIRLS	−	BEFORE	−29.23756	.00000	YES	(10, 12)	−0.10813E+03 < (I−J) < −0.92602E+02
GIRLS	−	AFTER	−31.24992	.00000	YES	(10, 12)	−0.10960E+03 < (I−J) < −0.94802E+02
BEFORE	−	AFTER	−.69683	.50037	NO	(12, 12)	−0.76240E+01 < (I−J) < 0.39574E+01

PAIRED T TEST

I		J	T	SIGPROB	SIG?	N(IJ)	95.0% CONFIDENCE INTERVALS
RATN A	−	RATN B	2.70966	.02030	YES	(12)	0.57881E+00 < (I−J) < 0.55879E+01
RATN A	−	PAINT1	−27.80923	.00001	YES	(5)	−0.60710E+02 < (I−J) < −0.49690E+02
RATN A	−	PAINT2	−55.94713	.00000	YES	(5)	−0.60458E+02 < (I−J) < −0.54742E+02
RATN A	−	BOYS	4.14516	.00250	YES	(10)	0.35890E+01 < (I−J) < 0.12211E+02
RATN A	−	GIRLS	1.17211	.27125	NO	(10)	−0.32545E+01 < (I−J) < 0.10255E+02
RATN A	−	BEFORE	−58.77277	.00000	YES	(12)	−0.10055E+03 < (I−J) < −0.93287E+02
RATN A	−	AFTER	−65.47609	.00000	YES	(12)	−0.10207E+03 < (I−J) < −0.95430E+02
RATN B	−	PAINT1	−23.11411	.00002	YES	(5)	−0.65190E+02 < (I−J) < −0.51210E+02
RATN B	−	PAINT2	−33.56345	.00000	YES	(5)	−0.65612E+02 < (I−J) < −0.55588E+02
RATN B	−	BOYS	3.23174	.01029	YES	(10)	0.13503E+01 < (I−J) < 0.76497E+01
RATN B	−	GIRLS	.03106	.97590	NO	(10)	−0.71826E+01 < (I−J) < 0.73826E+01
RATN B	−	BEFORE	−52.99539	.00000	YES	(12)	−0.10415E+03 < (I−J) < −0.95847E+02
RATN B	−	AFTER	−53.19791	.00000	YES	(12)	−0.10605E+03 < (I−J) < −0.97620E+02
PAINT1	−	PAINT2	−2.05798	.10870	NO	(5)	−0.56373E+01 < (I−J) < 0.83734E+00
PAINT1	−	BOYS	18.27647	.00005	YES	(5)	0.52244E+02 < (I−J) < 0.70956E+02
PAINT1	−	GIRLS	14.04335	.00015	YES	(5)	0.46374E+02 < (I−J) < 0.69226E+02
PAINT1	−	BEFORE	−10.62700	.00044	YES	(5)	−0.51458E+02 < (I−J) < −0.30142E+02
PAINT1	−	AFTER	−27.73334	.00001	YES	(5)	−0.48624E+02 < (I−J) < −0.39776E+02
PAINT2	−	BOYS	25.70303	.00001	YES	(5)	0.57088E+02 < (I−J) < 0.70912E+02
PAINT2	−	GIRLS	12.10309	.00027	YES	(5)	0.46392E+02 < (I−J) < 0.74008E+02
PAINT2	−	BEFORE	−10.38978	.00049	YES	(5)	−0.48660E+02 < (I−J) < −0.28140E+02
PAINT2	−	AFTER	−48.59024	.00000	YES	(5)	−0.44188E+02 < (I−J) < −0.39412E+02
BOYS	−	GIRLS	−1.22644	.25115	NO	(10)	−0.12515E+02 < (I−J) < 0.37152E+01
BOYS	−	BEFORE	−35.16176	.00000	YES	(10)	−0.11197E+03 < (I−J) < −0.98432E+02
BOYS	−	AFTER	−40.76003	.00000	YES	(10)	−0.11220E+03 < (I−J) < −0.10040E+03
GIRLS	−	BEFORE	−25.57918	.00000	YES	(10)	−0.10971E+03 < (I−J) < −0.91886E+02
GIRLS	−	AFTER	−28.97826	.00000	YES	(10)	−0.10985E+03 < (I−J) < −0.93946E+02
BEFORE	−	AFTER	−1.08965	.29916	NO	(12)	−0.55365E+01 < (I−J) < 0.18698E+01

Note: SIGPROB = Lowest level for which T is significant (P value)

OUTPUT FOR SECOND EXAMPLE

PAD Measurements from Patients of Class II or III

T Level = .050
\# Cases = 7 # Variables = 2

Data by Cases (0.0 is considered as no. obs. present)

Day1	Day 2
.00	.00
12.00	8.00
10.00	10.00
15.00	13.00
12.00	12.00
22.00	20.00
30.00	.00

# Observations		**Mean**	**Variance**	**Standard Deviation**	**Standard Error of the Mean**	**95.0% Confidence Intervals**
Day 1	6	0.1683E+02	0.5937E+02	0.7705E+01	0.3146E+01	0.8746E+01 < MEAN < 0.2492E+02
Day 2	5	0.1260E+02	0.2080E+02	0.4561E+01	0.2040E+01	0.6938E+01 < MEAN < 0.1826E+02

PAIRED T TEST

I		**J**	**T**	**SIGPROB**	**SIG?**	**N(IJ)**	**95.0% CONFIDENCE INTERVALS**
DAY 1	–	DAY 2	2.13809	.09930	NO	(5)	−0.47737E+00 < (I−J) < 0.36774E+01

Note: SIGPROB = Lowest level for which T is significant (P Value)

Statistical Analysis (Third Edition), McGraw-Hill Book Company, New York.

Dudewicz, E.J. and Dalal, S.R. (1972): "On approximations to the *t*-distribution," *Journal of Quality Technology*, Vol. 4, pp. 196–198.

Scheffé, H. (1970): "Practical solutions of the Behrens-Fisher problem," *Journal of the American Statistical Association*, Vol. 65, pp. 1501–1508.

Zelen, M. and Severo, N.C. (1964): "Probability functions," Chapter 26, pp. 925–995 of *Handbook of Mathematical Functions with Formulas, Graphs, and Mathematical Tables* edited by M. Abramowitz and I.A. Stegun, Vol. 55, Applied Mathematics Series, National Bureau of Standards, United States Department of Commerce, Washington, D.C.

1. Paired and Unpaired t-Tests (Student's *t*-Tests): Examples

For a **first example**, we have taken data from Dixon and Massey (1969), pp. 117-122, dealing with two rations (p. 117), two paints (p. 118), boys and girls (p. 121), and blood pressure before and after a stimulus (p. 122). Our numerical answers agree with those obtained by Dixon and Massey, but are much more detailed. We have analyzed all pairs of columns (e.g., paint one vs. girls) in order to demonstrate how to handle columns of differing length.

For a **second example**, we have data from a group of patients of "Class II or III," and we wish to test whether their pulmonary arterial diastolic pressure (PAD) changed from "Day 1" to "Day 2." The data are shown in Table 8.

Table 8

PAD Measurements		
Patient	Day 1	Day 2
1	0	0
2	12	8
3	10	10
4	15	13
5	12	12
6	22	20
7	30	0

We see that a paired *t*-test is appropriate (since we will thus eliminate differences due to patients rather than to days); 5 pairs are available. The difference observed is not judged to be significant at level 0.05, but would be judged to be significant at any level 0.09930 or greater (e.g., at level 0.10 we would judge that PAD had decreased over time).

We can be 95% confident that the mean difference between Day 1 and Day 2 readings is between −0.48 and 3.68.

2. CORRELATION COEFFICIENTS

Statistical Theory

Suppose we are given a **data** matrix $X^{m\times n}$, i.e.,

$$X^{m\times n} = \begin{pmatrix} X_{11} & X_{12} & \cdots & X_{1n} \\ X_{21} & X_{22} & \cdots & X_{2n} \\ \cdot & & & \\ \cdot & & & \\ \cdot & & & \\ X_{m1} & X_{m2} & \cdots & X_{mn} \end{pmatrix}$$

each of the m rows of X will often in practice represent an experiment, with each of the n columns representing some measured (or derived) variable.

The **statistical summary measures** "mean," "variance," "standard deviation," and "standard error of the mean" are often of interest for each column. For column i ($1 \le i \le n$) of the data matrix $X^{m\times n}$ they are

$\bar{X}_i = \Sigma X_{ij}/n_i$ --(mean)

$s_i^2 = \Sigma (X_{ij} - \bar{X}_i)^2/(n_i - 1)$ -------------------- (variance)

s_i -- (standard deviation)

$s_i/\sqrt{n_i}$ ------------------------ (standard error of the mean),

where n_i is the number of non-zero items in column i. The **sample correlation coefficient r_{ij}** between columns i and j is:

$$r_{ij} = \frac{\Sigma (X_{i\ell} - \bar{X}_{i(j)})(X_{j\ell} - \bar{X}_{j(i)})/(n_{ij} - 1)}{\sqrt{\dfrac{\Sigma (X_{i\ell} - \bar{X}_{i(j)})^2}{n_{ij}-1}}\sqrt{\dfrac{\Sigma (X_{j\ell} - \bar{X}_{j(i)})^2}{n_{ij}-1}}}$$

$$= \frac{\Sigma X_{i\ell} X_{j\ell} - \dfrac{(\Sigma X_{i\ell})(\Sigma X_{j\ell})}{n_{ij}}}{\sqrt{\Sigma (X_{i\ell} - \bar{X}_{i(j)})^2}\sqrt{\Sigma (X_{j\ell} - \bar{X}_{j(i)})^2}}$$

where n_{ij} is the number of (row) pairs of observations on i and j in the data matrix, $\bar{X}_{i(j)}$ is the mean of variable i values which are paired, $\bar{X}_{j(i)}$ is the mean of variable j values which are paired, and all summations are over indices ℓ for which a pair of observations (one on variable i and one on variable j) is present in the data matrix.

The correlation coefficient r_{ij} based on n_{ij} pairs will be judged significantly different from zero at level 0.05 using a two-tailed test if

$$|r_{ij}| \ge R_{n_{ij}-2}\,(2Q = 0.05).$$

(We underscore any correlation r_{ij} which is found to be significantly different from zero.) The quantities $R_\upsilon(2Q=0.05)$ were obtained from Table 13 on p. 146 of Pearson and Hartley (1970) for υ = 1(1)20(5)50(10)100. **A confidence interval** (with confidence coefficient 0.95 or 0.99, whichever is desired) may be obtained from r_{ij} and $n_{ij} - 2$ by using the charts of Table 15 of Pearson and Hartley (1970), pp. 148–149.

(Note on assumptions.) The above analysis makes the (usual) assumptions of independent pairs of observations, each with the same bivariate normal distribution.

2. Correlation Coefficients References

Pearson, E.S. and Hartley, H.O. (Editors) (1970): *Biometrika Tables for Statisticians*, Volume I (Third Edition Reprinted with additions), Cambridge University Press, London.

Snedecor, G.W. and Cochran, W.G. (1967): *Statistical Methods* (Sixth Edition), The Iowa State University Press, Ames, Iowa.

2. Correlation Coefficients Examples (input and output)

For a **first example**, we have taken data from Snedecor and Cochran (1967), pp. 172–174, and wish to analyze all pairs of columns (even though only special pairs are of interest).

Table 9

Patient	CBV	PEV	PCW	PAD	PAM
1	569	80	9	16	22
2	723	121	24	24	28
3	729	114	26	27	35
4	750	134	12	12	22
5	599	100	20	22	30
6	458	66	—	15	15
7	605	155	—	25	31
8	633	179	—	40	50
9	753	115	19	22	32
10	1140	310	—	27	35
11	816	141	6	7	14
12	634	189	26	28	34
13	552	103	9	11	18
14	530	155	27	27	32
15	522	93	3	7	13
16	513	77	14	12	21
17	805	113	9	10	14
18	688	144	16	19	21
19	663	54	6	11	18
20	570	60	9	11	16

For a **second example**, we have data from a group of patients and wish to correlate central blood volume (CBV), pulmonary extravascular volume (PEV), pulmonary capillary wedge mean pressure (PCW), pulmonary artery diastolic pressure (PAD), and pulmonary artery mean pressure (PAM). The data are shown in Table 9.

Note that the output matrices of cases per correlation, and of correlation coefficients, give some data which is (strictly speaking) duplicative since $n_{ij} = n_{ji}$ and $r_{ij} = r_{ji}$. This is done for ease of visual interpretation; if half of each matrix were deleted, one would no longer be able to simply glance across a row and see a specified variable's correlations with all other variables.

First Example

Snedecor & Cochran, Stat. Meth., 6th Ed., 1967, pp. 172–4.

Cases = 10   # Variables = 8

Variable Format = (8F4.0)

DATA   (0.0 is considered as no. obs. present)

VAR 1	VAR 2	VAR 3	VAR 4	VAR 5	VAR 6	VAR 7	VAR 8
1.00	3.00	2.00	1.00	3.00	11.00	.00	11.00
2.00	5.00	5.00	2.00	5.00	5.00	4.00	13.00
3.00	7.00	6.00	2.00	8.00	6.00	6.00	8.00
4.00	9.00	8.00	3.00	11.00	8.00	8.00	4.00
5.00	11.00	10.00	2.00	12.00	7.00	12.00	7.00
.00	.00	12.00	4.00	12.00	18.00	14.00	6.00
.00	.00	14.00	3.00	17.00	9.00	16.00	3.00
.00	.00	15.00	4.00	.00	.00	22.00	2.00
.00	.00	18.00	4.00	.00	.00	26.00	.00
.00	.00	20.00	5.00	.00	.00	.00	.00

	# OBSERVATIONS	MEAN	VARIANCE	STANDARD DEVIATION	STANDARD ERROR OF THE MEAN
VAR 1	5	0.3000E+01	0.2500E+01	0.1581E+01	0.7071E+00
VAR 2	5	0.7000E+01	0.1000E+02	0.3162E+01	0.1414E+01
VAR 3	10	0.1100E+02	0.3422E+02	0.5850E+01	0.1850E+01
VAR 4	10	0.3000E+01	0.1556E+01	0.1247E+01	0.3944E+00
VAR 5	7	0.9714E+01	0.2257E+02	0.4751E+01	0.1796E+01
VAR 6	7	0.9143E+01	0.1914E+02	0.4375E+01	0.1654E+01
VAR 7	8	0.1350E+02	0.5914E+02	0.7690E+01	0.2719E+01
VAR 8	8	0.6750E+01	0.1479E+02	0.3845E+01	0.1359E+01

MATRIX OF # OF CASES/CORRELATION

	VAR 1	VAR 2	VAR 3	VAR 4	VAR 5	VAR 6	VAR 7	VAR 8
VAR 1	5	5	5	5	5	5	4	5
VAR 2	5	5	5	5	5	5	4	5
VAR 3	5	5	10	10	7	7	8	8
VAR 4	5	5	10	10	7	7	8	8
VAR 5	5	5	7	7	7	7	6	7
VAR 6	5	5	7	7	7	7	6	7
VAR 7	4	4	8	8	6	6	8	7
VAR 8	5	5	8	8	7	7	7	8

Confidence Intervals can be obtained from: E.S. Pearson & H.O. Hartley (Editors), *Biometrika Tables for Statisticians*, Volume 1, Cambridge University Press, London, 1970, Table 15.

CORRELATION COEFFICIENTS

	VAR 1	VAR 2	VAR 3	VAR 4	VAR 5	VAR 6	VAR 7	VAR 8
VAR 1	1.00000	1.00000	.99044	.67082	.98974	−.34340	.98271	−.76642
	-----	-----	-----		-----		-----	
VAR 2	1.00000	1.00000	.99044	.67082	.98974	−.34340	.98271	−.76642
	-----	-----	-----		-----		-----	
VAR 3	.99044	.99044	1.00000	.91372	.96743	.34496	.98433	−.85417
	-----	-----	-----	-----	-----		-----	-----
VAR 4	.67082	.67082	.91372	1.00000	.71381	.56878	.80257	−.76184
			-----	-----			-----	-----
VAR 5	.98974	.98974	.96743	.71381	1.00000	.20274	.93440	−.89816
	-----	-----	-----		-----		-----	-----
VAR 6	−.34340	−.34340	.34496	.56878	.20274	1.00000	.62828	−.26915

VAR 7	.98271	.98271	.98433	.80257	.93440	.62828	1.00000	−.79986
	-----	-----	-----	-----	-----		-----	-----
VAR 8	−.76642	−.76642	−.85417	−.76184	−.89816	−.26915	−.79986	1.00000
			-----	-----	-----		-----	-----

----- Indicates significantly different from 0.0 with a 2-tailed test at level of sig. 0.05

Second Example

Data from a group of MIRU patients.

CASES = 20 # VARIABLES = 5

VARIABLE FORMAT = (5F6.0)

DATA

(0.0 is considered as no. obs. present)

CBV	PEV	PCW	PAD	PAM
569.00	80.00	9.00	16.00	22.00
723.00	121.00	24.00	24.00	28.00
729.00	114.00	26.00	27.00	35.00
750.00	134.00	12.00	12.00	22.00
599.00	100.00	20.00	22.00	30.00
458.00	66.00	.00	15.00	15.00
605.00	155.00	.00	25.00	31.00
633.00	179.00	.00	40.00	50.00
753.00	115.00	19.00	22.00	32.00
1140.00	310.00	.00	27.00	35.00
816.00	141.00	6.00	7.00	14.00
634.00	189.00	26.00	28.00	34.00
552.00	103.00	9.00	11.00	18.00
530.00	155.00	27.00	27.00	32.00
522.00	93.00	3.00	7.00	13.00
513.00	77.00	14.00	12.00	21.00
805.00	113.00	9.00	10.00	14.00
688.00	144.00	16.00	19.00	21.00
663.00	54.00	6.00	11.00	18.00
570.00	60.00	9.00	11.00	16.00

	# OBSERVATIONS	MEAN	VARIANCE	STANDARD DEVIATION	STANDARD ERROR OF THE MEAN
CBV	20	0.6626E+03	0.2297E+05	0.1516E+03	0.3389E+02
PEV	20	0.1251E+03	0.3333E+04	0.5773E+02	0.1291E+02
PCW	16	0.1469E+02	0.6450E+02	0.8031E+01	0.2008E+01
PAD	20	0.1865E+02	0.7761E+02	0.8810E+01	0.1970E+01
PAM	20	0.2505E+02	0.9416E+02	0.9703E+01	0.2170E+01

MATRIX OF # OF CASES/CORRELATION

	CBV	PEV	PCW	PAD	PAM
CBV	20	20	16	20	20
PEV	20	20	16	20	20
PCW	16	16	16	16	16
PAD	20	20	16	20	20
PAM	20	20	16	20	20

Confidence Intervals can be obtained from: E.S. Pearson & H.O. Hartley (Editors), *Biometrika Tables for Statisticians*, Volume 1, Cambridge University Press, London, 1970, Table 15.

CORRELATION COEFFICIENTS

	CBV	PEV	PCW	PAD	PAM
CBV	1.00000	.70558	.03983	.16434	.22412
	-----	-----			
PEV	.70558	1.00000	.57198	.58015	.58616
	-----	-----	-----	-----	-----
PCW	.03983	.57198	1.00000	.96336	.93497
		-----	-----	-----	-----
PAD	.16434	.58015	.96336	1.00000	.96441
		-----	-----	-----	-----
PAM	.22412	.58616	.93497	.96441	1.00000
		-----	-----	-----	-----

----- Indicates significantly different from 0.0 with a 2-tailed test at level of sig. 0.05

3. LINEAR REGRESSION

Statistical Theory

Suppose we are given **data**

$$\begin{pmatrix} X_1 & Y_1 \\ X_2 & Y_2 \\ \cdot & \\ \cdot & \\ \cdot & \\ X_n & Y_n \end{pmatrix}$$

each row will often in practice represent an experiment; with each of the two columns representing some measured (or derived) variable.

For those m pairs of observations (X_i, Y_i) actually present,

$\overline{X} = \Sigma X_i / m$ ----------------------------------- (mean of X)

$\overline{Y} = \Sigma Y_i / m$ ------------------------------------ (mean of Y)

$s_X^2 = \Sigma (X_i - \overline{X})^2/(m - 1)$ ---------------- (variance of X)

$s_Y^2 = \Sigma (Y_i - \overline{Y})^2/(m - 1)$ ---------------- (variance of Y)

$R_{XY} = \dfrac{\Sigma X_i Y_i - m\overline{X}\overline{Y}}{(m-1)\, S_X S_Y}$ ----------- (correlation coefficient).

The "best" (least-squares) a and b for predicting Y from X via a linear equation $Y = a + bX$, say $a = \alpha$ and $b = \beta$, has

$\alpha = \overline{Y} - r_{XY} \dfrac{S_Y}{S_X} \overline{X}$ --------------- **(least-squares intercept)**

$\beta = r_{XY} \dfrac{s_Y}{s_X}$ ------------------------- **(least-squares slope)**.

The residual variance, and the residual standard deviation, are measures of how well we can predict Y from X via $Y = \alpha + \beta X$:

$$s^2_{\text{RESID}} = \frac{\Sigma \left(\{Y_i - (\alpha + \beta X_i)\} - \dfrac{\Sigma\{Y_i - (\alpha + \beta X_i)\}}{m} \right)^2}{m-1}$$

--- **(residual variance)**

$s_{\text{RESID}} = \sqrt{s^2_{\text{RESID}}}$ -------- **(residual standard deviation)**

$Y_i - (\alpha + \beta X_i)$ is called a **residual**; if we had perfect prediction, all residuals would be zero.

Finally, a **plot** (Y on the vertical axis, X on the horizontal axis) containing all m data points (X_i, Y_i), the regression line $Y = \alpha + \beta X$, and prediction and confidence bands is useful. The **prediction bands are interpreted as follows:** at any value X, 95% of the Y-values will be between Y_L and Y_U given by

$$Y_U = (\alpha + \beta X) + t_{.05}(m-2) \sqrt{\frac{m\, s_Y^2(1-r_{XY}^2)}{m-2}} \sqrt{1 + \frac{1}{m} + \frac{(X-\bar{X})^2}{m\, s_X^2}}$$

$$Y_L = (\alpha + \beta X) - t_{.05}(m-2) \sqrt{\frac{m\, s_Y^2(1-r_{XY}^2)}{m-2}} \sqrt{1 + \frac{1}{m} + \frac{(X-\bar{X})^2}{m\, s_X^2}},$$

where $t_{.05}\,(m-2)$ is an appropriate Student's t percentage point.

The **confidence bands (for the regression line) are interpreted as follows:** at any value X, the interval (Y_L^*, Y_U^*) provides a 95% confidence interval for $\bar{Y}$, where

$$Y_U^* = (\alpha + \beta X) + t_{.05}(m-2) \sqrt{\frac{m\, s_Y^2(1-r_{XY}^2)}{m-2}} \sqrt{\frac{1}{m} + \frac{(X-\bar{X})^2}{m\, s_X^2}}$$

$$Y_L^* = (\alpha + \beta X) - t_{.05}(m-2) \sqrt{\frac{m\, s_Y^2(1-r_{XY}^2)}{m-2}} \sqrt{\frac{1}{m} + \frac{(X-\bar{X})^2}{m\, s_X^2}},$$

For a discussion see, for example, Hays and Winkler (1970), pp. 23–48 (pp. 23–42 for regression line; pp. 42–48 for confidence and prediction bands).

3. Linear Regression: References

Hays, W.L. and Winkler, R.L. (1970): *Statistics: Probability, Inference, and Decision: Volume II*, Holt, Rinehart and Winston, Inc., New York.

Snedecor, G.W. and Cochran, W.G. (1967): *Statistical Methods* (Sixth Edition), The Iowa State University Press, Ames, Iowa.

3. Linear Regression: Examples

For a **first example**, we have taken data from Snedecor and Cochran (1967), p. 136, and, for a **second example**, from Snedecor and Cochran (1967), p. 150.

First Example

Snedecor & Cochran, *Stat. Meth.*, 6th ed., 1967, p. 136.

LINEAR REGRESSION

OF CASES = 5 INDEX OF DEP VAR = 2
VARIABLE FORMAT = (2F4.0)

Y =	X =		
ESTIMATE OF BETA =	0.13800E+01	STANDARD ERROR BETA =	0.10263E+00
ESTIMATE OF ALPHA =	0.65100E+02	STANDARD ERROR ALPHA =	0.58284E+01
RESIDUAL VARIANCE =	0.10533E+02	RESIDUAL STANDARD DEVIATION =	0.32455E+01
MEAN X =	0.55000E+02	VARIANCE OF X =	0.25000E+03
MEAN Y =	0.14100E+03	VARIANCE OF Y =	0.48400E+03

EQUATION OF REGRESSION LINE IS Y = 0.651000E+02 + 0.138000E+01 X

CORRELATION COEFFICIENT = .9918053 (SIG. AT 0.05 LEVEL WITH 2-TAILED TEST)

SUM X =	0.27500E+03	SUM Y =	0.70500E+03	SUM CROSS PRODUCTS =	0.40155E+05
SUM SQUARES X =	0.16125E+05	SUM SQUARES Y =	0.10134E+06		
SUM SQUARES X CORRECTED FOR THE MEAN =	0.10000E+04	SUM SQUARES Y CORRECTED FOR THE MEAN =	0.19360E+04	SUM CROSS PRODUCTS CORRECTED FOR THE MEAN =	0.13800E+05

	X	Y	ESTIMATE OF Y	RESIDUAL
1	0.35000E+02	0.11400E+03	0.11340E+03	0.60001E+00
2	0.45000E+02	0.12400E+03	0.12720E+03	−0.32000E+01
3	0.55000E+02	0.14300E+03	0.14100E+03	0.20000E+01
4	0.65000E+02	0.15800E+03	0.15480E+03	0.32000E+01
5	0.75000E+02	0.16600E+03	0.16860E+03	−0.26000E+01

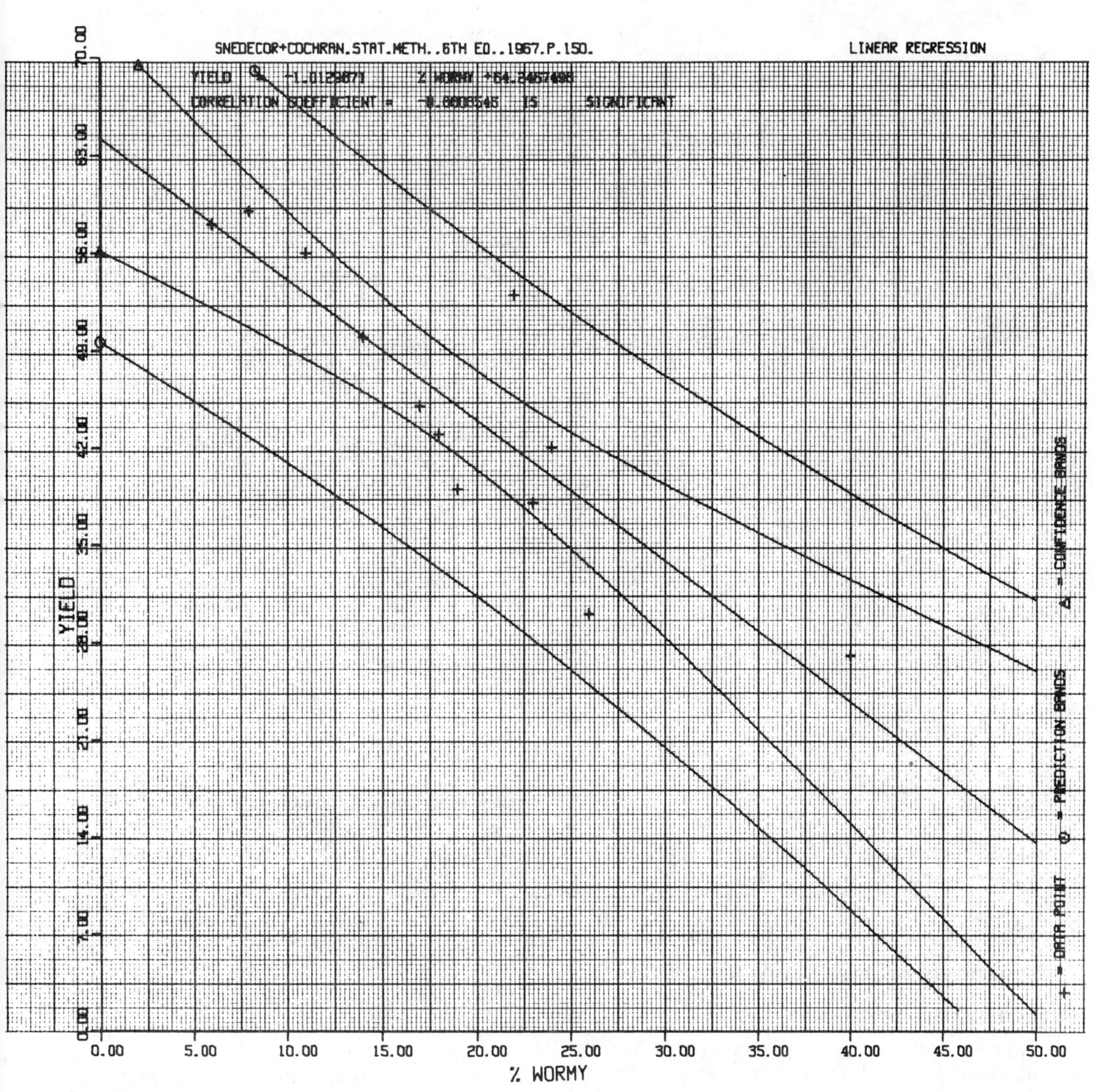

(Graph for Second Example.)

Second Example

Snedecor & Cochran, *Stat. Meth.*, 6th ed., 1967, p. 150.

LINEAR REGRESSION

\# OF CASES = 12 INDEX OF DEP VAR = 2
VARIABLE FORMAT = (8F3.0)

Y = YIELD	X = % WORMY		
ESTIMATE OF BETA =	− 0.10130E+01	STANDARD ERROR BETA =	0.17215E+00
ESTIMATE OF ALPHA =	0.64247E+02	STANDARD ERROR ALPHA =	0.36029E+01
RESIDUAL VARIANCE =	0.27384E+02	RESIDUAL STANDARD DEVIATION =	0.52330E+01
MEAN X =	0.19000E+02	VARIANCE OF X =	0.84000E+03
MEAN Y =	0.45000E+03	VARIANCE OF Y =	0.11109E+03

EQUATION OF REGRESSION LINE IS Y = 0.642467E+02 − 0.101299E+01 X

CORRELATION COEFFICIENT = −.8808546 (SIG. AT 0.05 LEVEL WITH 2-TAILED TEST)

SUM X =	0.22800E+03	SUM Y =	0.54000E+03	SUM CROSS PRODUCTS =	0.93240E+04
SUM SQUARES X =	0.52560E+04	SUM SQUARES Y =	0.25522E+05		
SUM SQUARES X CORRECTED FOR THE MEAN =	0.92400E+03	SUM SQUARES Y CORRECTED FOR THE MEAN =	0.12220E+04	SUM CROSS PRODUCTS CORRECTED FOR THE MEAN =	−0.93600E+03

	X	Y	ESTIMATE OF Y	RESIDUAL
1	0.80000E+01	0.59000E+02	0.56143E+02	0.28571E+01
2	0.60000E+01	0.58000E+02	0.58169E+02	−0.16882E+00
3	0.11000E+02	0.56000E+02	0.53104E+02	0.28961E+01
4	0.22000E+02	0.53000E+02	0.41961E+02	0.11039E+02
5	0.14000E+02	0.50000E+02	0.50065E+02	−0.64926E−01
6	0.17000E+02	0.45000E+02	0.47026E+02	−0.20260E+01
7	0.18000E+02	0.43000E+02	0.46013E+02	−0.30130E+01
8	0.24000E+02	0.42000E+02	0.39935E+02	0.20649E+01
9	0.19000E+02	0.39000E+02	0.45000E+02	−0.60000E+01
10	0.23000E+02	0.38000E+02	0.40948E+02	−0.29480E+01
11	0.26000E+02	0.30000E+02	0.37909E+02	−0.79091E+01
12	0.40000E+02	0.27000E+02	0.23727E+02	0.32727E+01

4. PARALLEL REGRESSIONS

Statistical Theory

Suppose we are given **data**

$$\begin{pmatrix} X_{11} & Y_{11} \\ X_{12} & Y_{12} \\ \cdot & \cdot \\ \cdot & \cdot \\ \cdot & \cdot \\ X_{1n_1} & Y_{1n_1} \end{pmatrix} \cdots \begin{pmatrix} X_{r1} & Y_{r1} \\ X_{r2} & Y_{r2} \\ \cdot & \cdot \\ \cdot & \cdot \\ \cdot & \cdot \\ X_{rn_r} & Y_{rn_r} \end{pmatrix}$$

each pair of columns represents the values of two variables (measured or derived) obtained under specified (perhaps different) circumstances. (For example, columns 1 and 2 contain n_1 measurements of age and serum-cholesterol concentration in a group of Iowa women; columns 3 and 4 contain similar measurements for n_2 Nebraska women; etc.)

Suppose we had ***R* pairs of columns**; we would then have ***R* collections of output (one for each individual regression,** giving regression, etc.). Suppose the regressions fitted are

$$Y_1 = \alpha_1 + \beta_1 X_1$$
$$Y_2 = \alpha_2 + \beta_2 X_2$$
$$\cdot$$
$$\cdot$$
$$\cdot$$
$$Y_R = \alpha_R + \beta_R X_R .$$

One is also often interested in **what regressions one would obtain if one assumed all lines were parallel** ($\beta_1 = \beta_2 = \ldots = \beta_R$), say

$$Y_1 = \alpha_1^* + \beta X_1$$
$$Y_2 = \alpha_2^* + \beta X_2$$
$$\cdot$$
$$\cdot$$
$$\cdot$$
$$Y_R = \alpha_R^* + \beta X_R$$

(the validity of this assumption is tested), and also **what regressions one would obtain if one assumed all lines were the same (i.e., had the same intercept $\alpha_1 = \alpha_2 = \ldots = \alpha_R = \alpha^{**}$ as well as the same slope β^{**})**

$$Y_i = \alpha^{**} + \beta^{**} X_i \; (i = 1, 2, \ldots, R)$$

(the validity of this assumption is also tested).

Following this, an **analysis of variance** table is given, and the hypotheses

$$H_1\colon \beta_1 = \beta_2 = \ldots = \beta_R$$
$$H_2\colon \beta_1 = \beta_2 = \ldots = \beta_R \text{ and } \alpha_1 = \alpha_2 = \ldots = \alpha_R$$

are tested. (If H_1 is true, then the lines are **parallel**; if H_2 is true, the lines **coincide**.)

These tests are done at level of significance 0.05, using $F(0.05, \upsilon_1, \upsilon_2)$ values obtained from pp. 426–427 of Scheffé (1959).

Finally, we give **simultaneous confidence intervals** (with confidence coefficient at least 0.95) **on all pairs of differences of slopes** $\beta_i - \beta_j$ $(1 \leq i < j \leq R)$. As given on pp. 68–72 of Scheffé (1959), these are, for example,

$$(\hat{\beta}_1 - \hat{\beta}_2) - c \leq \beta_1 - \beta_2 \leq (\hat{\beta}_1 - \hat{\beta}_2) + c,$$
$$c = \sqrt{(R-1)\,F\,(0.05, R-1, n-r)} \quad \sqrt{s^2(a_1^2 + a_2^2)}\,,$$
$$a_i^2 = \frac{1}{(n_i - 1)\, s_i^2}.$$

(For example, in the first example of Section 4.C.—Nebraska and Iowa women—$R = 2$, $n - r = 26$, $s^2 = 1861$, $a_1^2 = \frac{1}{18} \cdot \frac{1}{309.1}$, and $a_2^2 = \frac{1}{10} \cdot \frac{1}{182.8}$.)

4. Parallel Regressions: References

Scheffé, H. (1959): *The Analysis of Variance*, John Wiley & Sons, Inc., New York.

Snedecor, G.W. and Cochran, W.G. (1967): *Statistical Methods* (Sixth Edition), The Iowa State University Press, Ames, Iowa.

4. Parallel Regressions: Examples

For a **first example**, we have taken data from Snedecor and Cochran (1967), pp. 432–436. This compares the relation between age and serum-cholesterol level in Iowa and Nebraska women. ($R = 2$ regressions.)

For a **second example**, we compare $R = 5$ regressions involving digitalis decay in 5 patients.

First Example

Age & Conc. of Cholesterol in Serum of Iowa & Nebraska Women

PARALLEL REGRESSIONS

REGRESSION 1 N = 19 NEBRASKA

ESTIMATE OF BETA =	0.25204E+01	STANDARD ERROR BETA =	0.53297E+00
ESTIMATE OF ALPHA =	0.10130E+03	STANDARD ERROR ALPHA =	0.26132E+02
RESIDUAL VARIANCE =	0.15808E+04	RESIDUAL STANDARD DEVIATION =	0.39759E+02
MEAN X =	0.45947E+02	VARIANCE OF X =	0.30916E+03
MEAN Y =	0.21711E+03	VARIANCE OF Y =	0.34570E+04

EQUATION OF REGRESSION LINE IS Y = 0.10130E+03 + 0.25204E+01 X

CORRELATION COEFFICIENT = .7537426 (SIG. AT 0.05 LEVEL WITH 2-TAILED TEST)

SUM X =	0.87300E+03	SUM Y =	0.41250E+04	SUM CROSS PRODUCTS =	0.20356E+06
SUM SQUARES X =	0.45677E+05	SUM SQUARES Y =	0.95778E+06		
SUM SQUARES X CORRECTED FOR THE MEAN =	0.55649E+04	SUM SQUARES Y CORRECTED FOR THE MEAN =	0.62225E+05	SUM CROSS PRODUCTS CORRECTED FOR THE MEAN =	0.14026E+05

X	Y	ESTIMATE OF Y	RESIDUAL
0.18000E+02	0.13700E+03	0.14667E+03	−0.96655E+01
0.44000E+02	0.17300E+03	0.21220E+03	−0.39197E+02
0.33000E+02	0.17700E+03	0.18447E+03	−0.74722E+01
0.78000E+02	0.24100E+03	0.29789E+03	−0.56892E+02
0.51000E+02	0.22500E+03	0.22984E+03	−0.48401E+01
0.43000E+02	0.22300E+03	0.20968E+03	0.13323E+02
0.44000E+02	0.19000E+03	0.21220E+03	−0.22197E+02
0.58000E+02	0.25700E+03	0.24748E+03	0.95168E+01
0.63000E+02	0.33700E+03	0.26009E+03	0.76915E+02
0.19000E+02	0.18900E+03	0.14919E+03	0.39814E+02
0.30000E+02	0.14000E+03	0.17691E+03	−0.36911E+02
0.47000E+02	0.19600E+03	0.21976E+03	−0.23758E+02
0.58000E+02	0.26200E+03	0.24748E+03	0.14517E+02
0.70000E+02	0.26100E+03	0.27773E+03	−0.16729E+02
0.67000E+02	0.35600E+03	0.27017E+03	0.85833E+02
0.31000E+02	0.15900E+03	0.17943E+03	−0.20431E+02
0.21000E+02	0.19100E+03	0.15423E+03	0.36773E+02
0.56000E+02	0.19700E+03	0.24244E+03	−0.45442E+02
0.42000E+02	0.21400E+03	0.20716E+03	0.68438E+01

DATA POINTS ARE PLOT # 1
REGRESSION LINE IS PLOT # 2

REGRESSION 2 N = 11 IOWA

ESTIMATE OF BETA =	0.32381E+01	STANDARD ERROR BETA =	0.11434E+01
ESTIMATE OF ALPHA =	0.35812E+02	STANDARD ERROR ALPHA =	0.62471E+02
RESIDUAL VARIANCE =	0.23912E+04	RESIDUAL STANDARD DEVIATION =	0.48900E+02
MEAN X =	0.53091E+02	VARIANCE OF X =	0.18289E+03
MEAN Y =	0.20773E+03	VARIANCE OF Y =	0.40698E+04

EQUATION OF REGRESSION LINE IS Y = 0.35812E+02 + 0.32381E+01 X

CORRELATION COEFFICIENT = .6864406 (SIG. AT 0.05 LEVEL WITH 2-TAILED TEST)

SUM X =	0.58400E+03	SUM Y =	0.22850E+04	SUM CROSS PRODUCTS =	0.12723E+05
SUM SQUARES X =	0.32834E+05	SUM SQUARES Y =	0.51535E+06		
SUM SQUARES X CORRECTED FOR THE MEAN =	0.18289E+04	SUM SQUARES Y CORRECTED FOR THE MEAN =	0.40698E+05	SUM CROSS PRODUCTS CORRECTED FOR THE MEAN =	0.59222E+05

X	Y	ESTIMATE OF Y	RESIDUAL
0.46000E+02	0.18100E+03	0.18477E+03	−0.37660E+01
0.52000E+02	0.22800E+03	0.20419E+03	0.23805E+02
0.39000E+02	0.18200E+03	0.16210E+03	0.19901E+02
0.65000E+02	0.24900E+03	0.24629E+03	0.27095E+01
0.54000E+02	0.25900E+03	0.21067E+03	0.48329E+02
0.33000E+02	0.20100E+03	0.14267E+03	0.58330E+02
0.49000E+02	0.12100E+03	0.19448E+03	−0.73480E+02
0.76000E+02	0.33900E+03	0.28191E+03	0.57090E+02
0.71000E+02	0.22400E+03	0.26572E+03	−0.41719E+02
0.41000E+02	0.11200E+03	0.16858E+03	−0.56575E+02
0.58000E+02	0.18900E+03	0.22362E+03	−0.34624E+02

DATA POINTS ARE PLOT # 3
REGRESSION LINE IS PLOT # 4

COMBINED REGRESSION ASSUMING PARALLELISM

ESTIMATE OF BETA = 0.269797E+01
STANDARD ERROR BETA = 0.501737E+00

REGRESSION	ESTIMATE OF ALPHA	STANDARD ERROR OF ALPHA
1	0.931408E+02	0.283565E+02
2	0.644898E+02	0.551164E+02

OVERALL REGRESSION ASSUMING SAME SLOPE AND INTERCEPT

ESTIMATE OF BETA =	0.251397E+01	STANDARD ERROR BETA =	0.490093E+00
ESTIMATE OF ALPHA =	0.826355E+02	STANDARD ERROR ALPHA =	0.250716E+02
RESIDUAL VARIANCE =	0.186133E+04	RESIDUAL STANDARD DEVIATION =	0.431431E+02

ANALYSIS OF VARIANCE **R = 2**

SOURCE	DEGREES OF FREEDOM	SUM OF SQUARES	MEAN SQUARE	F
OVERALL REGRESSION	1	0.489764E+05		
DIFFERENCE OF POSITIONS (COMMON INTERCEPT)	1	0.616606E+04	0.616606+04	3.3127E+00
COMBINED REGRESSION (COMMON BETA)	1	0.538200E+05		
DIFFERENCE OF REGRESSION (PARALLELISM)	1	0.709027E+03	0.709027E+03	.3809E+05
INDIVIDUAL REGRESSIONS	2	0.545291E+05		
COMBINED RESIDUAL ERROR	26	0.483946E+05	0.186133E+04	
TOTAL WITHIN SETS (COMBINED)	28	0.102924E+06		
TOTAL OVERALL	29	0.103537E+06		

THE ANOVA TEST THAT ALL REGRESSION LINES HAVE SAME SLOPE REJECTS THE HYPOTHESIS IF AND ONLY IF:

$F > F(0.05, 1, 26) = 4.23$

$F = .380925$

THE ANOVA TEST THAT ALL REGRESSION LINES HAVE SAME SLOPE AND INTERCEPT REJECTS THE HYPOTHESIS IF AND ONLY IF:

$F > F(0.05, 2, 26) = 3.37$

$F = 3.312720$

CONFIDENCE INTERVALS

THE CONFIDENCE IS 0.95 THAT ALL OF THE FOLLOWING COVER THE TRUE POINTS

	COVERS ZERO?
$-0.31093E+01 < \text{BETA } 1 - \text{BETA } 2 < 0.16739E+01$	YES

NOTE . . . THE ANOVA (ANALYSIS OF VARIANCE) TEST OF THE HYPOTHESIS OF EQUAL SLOPES WILL REJECT THAT HYPOTHESIS . . . IF AND ONLY IF . . . SOME INTERVAL ABOVE ON A BETA I − BETA J (OR AN INTERVAL ON SOME OTHER LINEAR FUNCTION OF BETA 1, BETA 2, . . . WHOSE COEFFICIENTS SUM TO ZERO) DOES NOT INCLUDE ZERO.

Second Example

RADIOIMMUNOASSAY

PARALLEL REGRESSIONS

REGRESSION 1 N = 6 X = TIME Y = PATIENT 1 (DABA)

ESTIMATE OF BETA =	−0.95238E−02	STANDARD ERROR BETA =	0.28107E−02
ESTIMATE OF ALPHA =	0.92976E+00	STANDARD ERROR ALPHA =	0.10784E+00
RESIDUAL OF VARIANCE =	0.21235E−01	RESIDUAL STANDARD DEVIATION =	0.14572E+00
MEAN X =	0.32000E+02	VARIANCE OF X =	0.53760E+03
MEAN Y =	0.62500E+00	VARIANCE OF Y =	0.65750E−01

EQUATION OF REGRESSION LINE IS Y = 0.92976E+00 −0.95238E−02 X

CORRELATION COEFFICIENT = −.8611761 (SIG AT 0.05 LEVEL WITH 2-TAILED TEST)

SUM X =	0.19200E+03	SUM Y =	0.37500E+01	SUM CROSS PRODUCTS =	0.94400E+02
SUM SQUARES X =	0.88320E+04	SUM SQUARES Y =	0.26725E+01		
SUM SQUARES X CORRECTED FOR THE MEAN =	0.26880E+04	SUM SQUARES Y CORRECTED FOR THE MEAN =	0.32875E+00	SUM CROSS PRODUCTS CORRECTED FOR THE MEAN =	−0.25600E+02

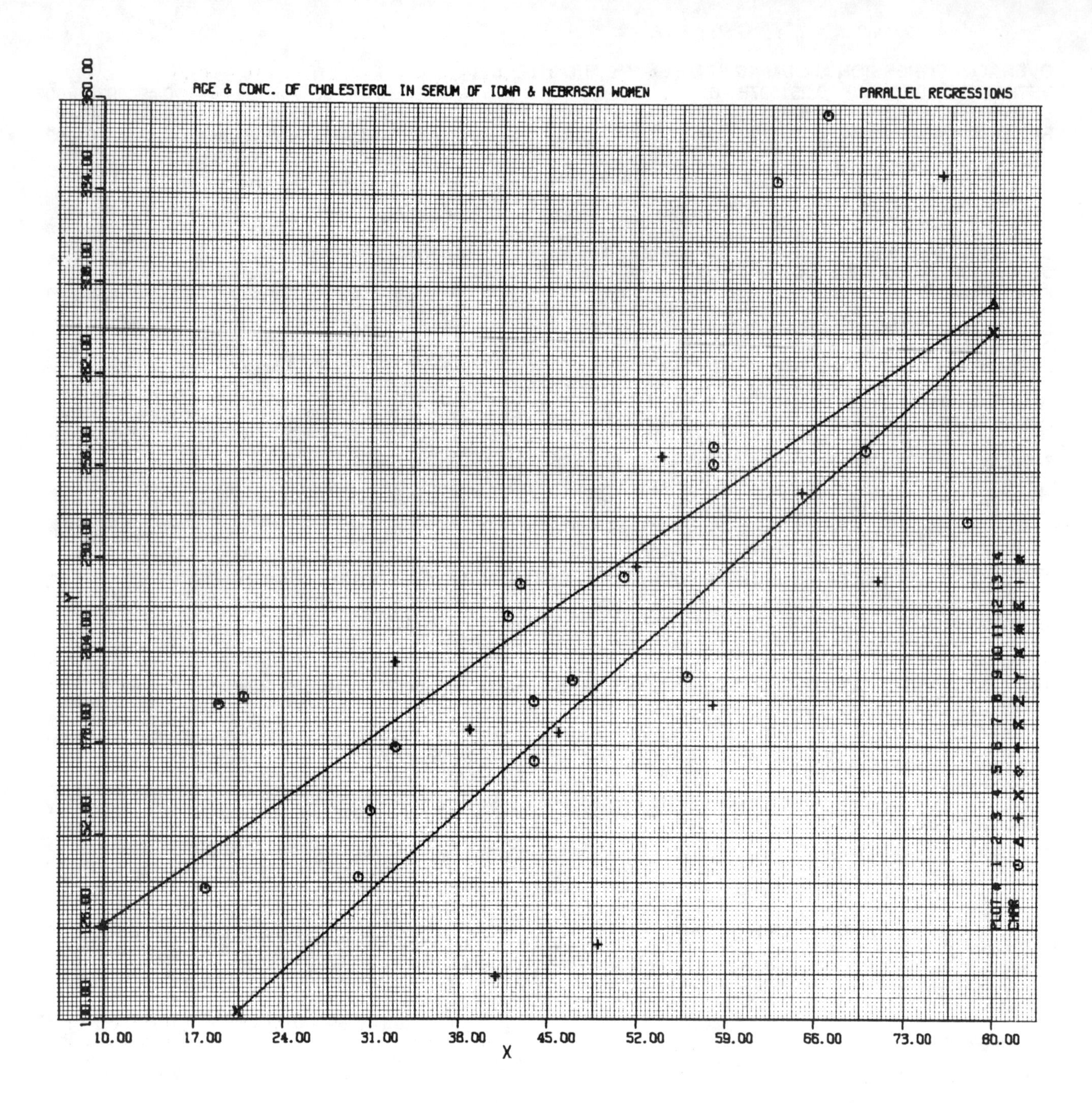

AGE & CONC. OF CHOLESTEROL IN SERUM OF IOWA & NEBRASKA WOMEN
PARALLEL REGRESSIONS
360.00
334.00
308.00
282.00
256.00
230.00
204.00
178.00
152.00
126.00
100.00
Y
10.00
17.00
24.00
31.00
38.00
45.00
52.00
59.00
66.00
73.00
80.00
X
PLOT # 1 2 3 4 5 6 7 8 9 10 11 12 13 14
CHAR

X	Y	ESTIMATE OF Y	RESIDUAL
0.00000E+00	0.10000E+01	0.92976E+00	0.70239E−01
0.16000E+02	0.90000E+00	0.77738E+00	0.12262E+00
0.24000E+02	0.50000E+00	0.70119E+00	−0.20119E+00
0.40000E+02	0.50000E+00	0.54881E+00	−0.48809E−01
0.48000E+02	0.40000E+00	0.47262E+00	−0.72619E−01
0.64000E+02	0.45000E+00	0.32024E+00	0.12976E+00

DATA POINTS ARE PLOT # 1
REGRESSION LINE IS PLOT # 2

REGRESSION 2 N = 6 X = TIME Y = PATIENT 2 (JB)

ESTIMATE OF BETA = −0.11815E−01
ESTIMATE OF ALPHA = 0.94143E+00
RESIDUAL VARIANCE = 0.21269E−01
MEAN X = 0.32000E+02
MEAN Y = 0.56333E+00

STANDARD ERROR BETA = 0.28129E−02
STANDARD ERROR ALPHA = 0.10792E+00
RESIDUAL STANDARD DEVIATION = 0.14584E+00
VARIANCE OF X = 0.53760E+03
VARIANCE OF Y = 0.92067E−01

EQUATION OF REGRESSION LINE IS Y = 0.94143E+00 −0.11815E−01 X

CORRELATION COEFFICIENT = −.9028777 (SIG. AT 0.05 LEVEL WITH 2-TAILED TEST)

SUM X = 0.19200E+03
SUM SQUARES X = 0.88320E+04
SUM SQUARES X CORRECTED FOR THE MEAN = 0.26880E+04

SUM Y = 0.33800E+01
SUM SQUARES Y = 0.23644E+01
SUM SQUARES Y CORRECTED FOR THE MEAN = 0.46033E+00

SUM CROSS PRODUCTS = 0.76400E+02
SUM CROSS PRODUCTS CORRECTED FOR THE MEAN = −0.31760E+02

X	Y	ESTIMATE OF Y	RESIDUAL
0.00000E+00	0.11300E+01	0.94143E+00	0.18857E+00
0.16000E+02	0.55000E+00	0.75238E+00	−0.20238E+00
0.24000E+02	0.60000E+00	0.65786E+00	−0.57857E−01
0.40000E+02	0.45000E+00	0.46881E+00	−0.18809E−01
0.48000E+02	0.40000E+00	0.37429E+00	0.25714E−01
0.64000E+02	0.25000E+00	0.18524E+00	0.64762E−01

DATA POINTS ARE PLOT # 3
REGRESSION LINE IS PLOT # 4

REGRESSION 3 N = 6 X = TIME Y = PATIENT 3 (DABR)

ESTIMATE OF BETA = −0.12202E−01
ESTIMATE OF ALPHA = 0.10155E+01
RESIDUAL VARIANCE = 0.58782E−02
MEAN X = 0.32000E+02
MEAN Y = 0.62500E+00

STANDARD ERROR BETA = 0.14788E−02
STANDARD ERROR ALPHA = 0.56737E−01
RESIDUAL STANDARD DEVIATION = 0.76670E−01
VARIANCE OF X = 0.53760E+03
VARIANCE OF Y = 0.84750E−01

EQUATION OF REGRESSION LINE IS Y = 0.101550E+01 −0.12202E−01 X

CORRELATION COEFFICIENT = −.9718602 (SIG. AT 0.05 LEVEL WITH 2-TAILED TEST)

SUM X = 0.19200E+03
SUM SQUARES X = 0.88320E+04
SUM SQUARES X CORRECTED FOR THE MEAN = 0.26880E+04

SUM Y = 0.37500E+01
SUM SQUARES Y = 0.27675E+01
SUM SQUARES Y CORRECTED FOR THE MEAN = 0.42375E+00

SUM CROSS PRODUCTS = 0.87200E+02
SUM CROSS PRODUCTS CORRECTED FOR THE MEAN = −0.32800E+02

X	Y	ESTIMATE OF Y	RESIDUAL
0.00000E+00	0.11000E+01	0.10155E+01	0.84525E−01
0.16000E+02	0.80000E+00	0.82024E+00	−0.20237E−01
0.24000E+02	0.65000E+00	0.72262E+00	−0.72619E−01
0.40000E+02	0.45000E+00	0.52738E+00	−0.77381E−01
0.48000E+02	0.45000E+00	0.42976E+00	0.20238E−01
0.64000E+02	0.30000E+00	0.23452E+00	0.65476E−01

DATA POINTS ARE PLOT # 5
REGRESSION LINE IS PLOT # 6

REGRESSION 4 N = 6 X = TIME Y = PATIENT 4 (EB)

ESTIMATE OF BETA =	−0.10119E−01	STANDARD ERROR BETA =	0.18512E−02
ESTIMATE OF ALPHA =	0.76548E+00	STANDARD ERROR ALPHA =	0.71023E−01
RESIDUAL VARIANCE =	0.92114E−02	RESIDUAL STANDARD DEVIATION =	0.95976E−01
MEAN X =	0.32000E+02	VARIANCE OF X =	0.53760E+03
MEAN Y =	0.44167E+00	VARIANCE OF Y =	0.62417E−01

EQUATION OF REGRESSION LINE IS Y = 0.76548E+00 −0.10119E−01 X

CORRELATION COEFFICIENT = −.9391149 (SIG. AT 0.05 LEVEL WITH 2-TAILED TEST)

SUM X = 0.19200E+03
SUM SQUARES X = 0.88320E+04
SUM SQUARES X CORRECTED FOR THE MEAN = 0.26880E+04

SUM Y = 0.26500E+01
SUM SQUARES Y = 0.14825E+01
SUM SQUARES Y CORRECTED FOR THE MEAN = 0.31208E+00

SUM CROSS PRODUCTS = 0.57600E+02
SUM CROSS PRODUCTS CORRECTED FOR THE MEAN = −0.27200E+02

X	Y	ESTIMATE OF Y	RESIDUAL
0.00000E+00	0.80000E+00	0.76548E+00	0.34524E−01
0.16000E+02	0.60000E+00	0.60357E+00	−0.35710E−02
0.24000E+02	0.40000E+00	0.52262E+00	−0.12262E+00
0.40000E+02	0.50000E+00	0.36071E+00	0.13929E+00
0.48000E+02	0.25000E+00	0.27976E+00	−0.29762E−01
0.64000E+02	0.10000E+00	0.11786E+00	−0.17857E−01

DATA POINTS ARE PLOT # 7
REGRESSION LINE IS PLOT # 8

REGRESSION 5 N = 6 X = TIME Y = PATIENT 5 (JG)

ESTIMATE OF BETA =	−0.10714E−01	STANDARD ERROR BETA =	0.39589E−02
ESTIMATE OF ALPHA =	0.93452E+00	STANDARD ERROR ALPHA =	0.15189E+00
RESIDUAL VARIANCE =	0.42129E−01	RESIDUAL STANDARD DEVIATION =	0.20525E+00
MEAN X =	0.32000E+02	VARIANCE OF X =	0.53760E+03
MEAN Y =	0.59167E+00	VARIANCE OF Y =	0.95417E−01

EQUATION OF REGRESSION LINE IS Y = 0.93452E+00 −0.10714E−01 X

CORRELATION COEFFICIENT = −.8042272 (NOT SIG. AT 0.05 LEVEL WITH 2-TAILED TEST)

SUM X = 0.19200E+03
SUM SQUARES X = 0.88320E+04
SUM SQUARES X CORRECTED FOR THE MEAN = 0.26880E+04

SUM Y = 0.35500E+01
SUM SQUARES Y = 0.25775E+01
SUM SQUARES Y CORRECTED FOR THE MEAN = 0.47709E+00

SUM CROSS PRODUCTS = 0.84800E+02
SUM CROSS PRODUCTS CORRECTED FOR THE MEAN = −0.28800E+02

X	Y	ESTIMATE OF Y	RESIDUAL
0.00000E+00	0.12000E+01	0.93452E+00	0.26548E+00
0.16000E+02	0.50000E+00	0.76309E+00	−0.26309E+00
0.24000E+02	0.55000E+00	0.67738E+00	−0.12738E+00
0.40000E+02	0.55000E+00	0.50595E+00	0.44048E−01
0.48000E+02	0.40000E+00	0.42024E+00	−0.20238E−01
0.64000E+02	0.35000E+00	0.24881E+00	0.10119E+00

DATA POINTS ARE PLOT # 9
REGRESSION LINE IS PLOT # 10

COMBINED REGRESSION ASSUMING PARALLELISM

ESTIMATE OF BETA = −0.108750E−01
STANDARD ERROR BETA = 0.121818E−02

REGRESSION	ESTIMATE OF ALPHA	STANDARD ERROR OF ALPHA
1	0.972999E+00	0.104508E+00
2	0.911332E+00	0.104508E+00
3	0.972999E+00	0.104508E+00
4	0.789666E+00	0.104508E+00
5	0.939666E+00	0.104508E+00

OVERALL REGRESSION ASSUMING SAME SLOPE AND INTERCEPT

ESTIMATE OF BETA =	−0.108750E−01	STANDARD ERROR BETA =	0.121818E−02
ESTIMATE OF ALPHA =	0.917332E+00	STANDARD ERROR ALPHA =	0.467376E−01
RESIDUAL VARIANCE =	0.199445E−01	RESIDUAL STANDARD DEVIATION =	0.141225E+00

ANALYSIS OF VARIANCE

SOURCE	DEGREES OF FREEDOM	SUM OF SQUARES	MEAN SQUARE	R = F
OVERALL REGRESSION	1	0.158949E+01		
DIFFERENCE OF POSITIONS (COMMON INTERCEPT)	4	0.151814E+00	0.379536E−01	1.90299
COMBINED REGRESSION (COMMON BETA)	1	0.158949E+01		
DIFFERENCE OF REGRESSION (PARALLELISM)	4	0.136261E−01	0.340652E−02	.17080
INDIVIDUAL REGRESSIONS	5	0.160311E+01		
COMBINED RESIDUAL ERROR	20	0.398890E+00	0.199445E−01	
TOTAL WITHIN SETS (COMBINED)	25	0.200200E+01		
TOTAL OVERALL	29	0.214020E+01		

THE ANOVA TEST THAT ALL REGRESSION LINES HAVE SAME SLOPE REJECTS THE HYPOTHESIS IF AND ONLY IF:

F > F(0.05, 4, 20) = 2.87

F = .170800

THE ANOVA TEST THAT ALL REGRESSION LINES HAVE SAME SLOPE AND INTERCEPT REJECTS THE HYPOTHESIS IF AND ONLY IF:

F > F(0.05, 8, 20) = 2.45

F = 1.902959

CONFIDENCE INTERVALS

THE CONFIDENCE IS 0.95 THAT ALL OF THE FOLLOWING COVER THE TRUE POINTS

	COVERS ZERO ?
−0.10761E−01 < BETA 1 − BETA 2 < 0.15344E−01	YES
−0.10374E−01 < BETA 1 − BETA 3 < 0.15731E−01	YES
−0.12457E−01 < BETA 1 − BETA 4 < 0.13647E−01	YES
−0.11862E−01 < BETA 1 − BETA 5 < 0.14243E−01	YES
−0.12665E−01 < BETA 2 − BETA 3 < 0.13439E−01	YES
−0.14749E−01 < BETA 2 − BETA 4 < 0.11356E−01	YES
−0.14153E−01 < BETA 2 − BETA 5 < 0.11951E−01	YES
−0.15136E−01 < BETA 3 − BETA 4 < 0.10969E−01	YES
−0.14540E−01 < BETA 3 − BETA 5 < 0.11564E−01	YES
−0.12457E−01 < BETA 4 − BETA 5 < 0.13647E−01	YES

NOTE . . . THE ANOVA (ANALYSIS OF VARIANCE) TEST OF THE HYPOTHESIS OF EQUAL SLOPES WILL REJECT THAT HYPOTHESIS . . . IF AND ONLY IF . . . SOME INTERVAL ABOVE ON A BETA I − BETA J (OR AN INTERVAL ON SOME OTHER LINEAR FUNCTION OF BETA 1, BETA 2, . . . WHOSE COEFFICIENTS SUM TO ZERO) DOES NOT INCLUDE ZERO.

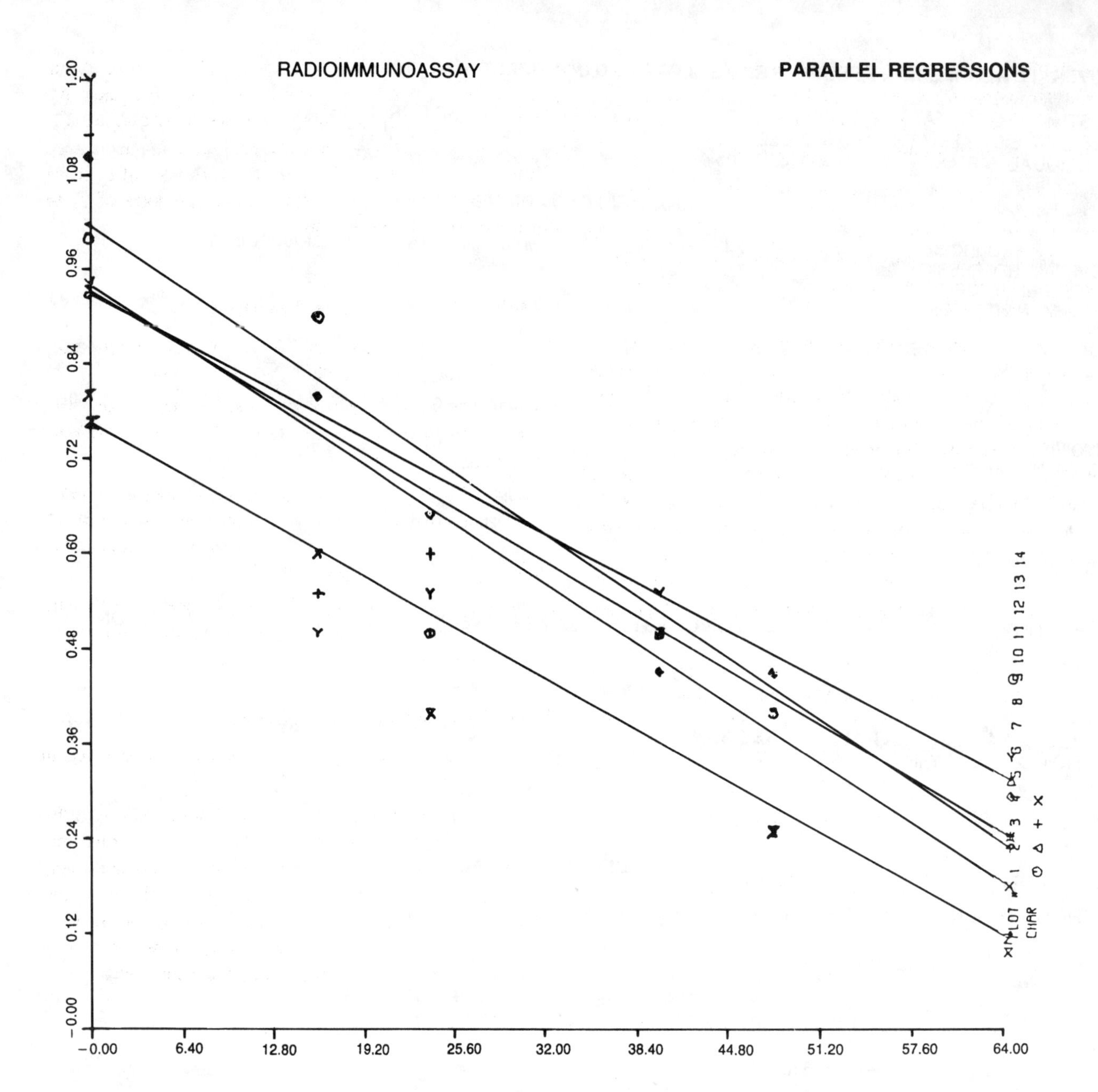

5. ALL POSSIBLE REGRESSIONS

Statistical Theory

Suppose we are given a **data** matrix $X^{m \times n}$, i.e.,

$$X^{m \times n} = \begin{pmatrix} X_{11} & X_{12} & \cdots & X_{1n} \\ X_{21} & X_{22} & \cdots & X_{2n} \\ \bullet & & & \\ \bullet & & & \\ \bullet & & & \\ X_{m1} & X_{m2} & \cdots & X_{mn} \end{pmatrix}$$

each of the m rows of X will often in practice represent an experiment, with each of the n columns representing some measured (or derived) variable.

In "Linear Regression" we discussed the problem of predicting a (dependent) variable Y from a (independent) variable X by a linear equation $Y = \alpha + \beta X$, where α and β are chosen so that for the n observed pairs of values (X_1, Y_1), . . ., (X_n, Y_n) the average squared error $\sum_{i=1}^{n} (\hat{Y}_i - Y_i)^2/n$ is minimized.

In the present section, we wish **to predict a (dependent)**

variable Y from (independent) variables $X_1, \ldots, X_{n-1}$ by a linear equation $\hat{Y} = \alpha + \beta_1 X_1 + \ldots + \beta_{n-1} X_{n-1}$, where $\alpha, \beta_1, \ldots, \beta_{n-1}$ are chosen so that for the m observed sets of values $(X_{11}, \ldots, X_{1,n-1}, Y_1), \ldots, (X_{m1}, \ldots, X_{m,n-1}, Y_n)$, the average squared error $\sum_{i=1}^{m} (\hat{Y}_i - Y_i)^2/m$ is minimized. (The details are generalizations of the details of "Linear Regression" and are fairly complex as well as being well explained in Draper and Smith (1966), so we shall not dwell on them here.)

Now although **we have available $n-1$ variables $X_1, \ldots, X_{n-1}$** to use in prediction of Y, **it may be that some of these have essentially no relation to Y; these will increase the variability of our predictor $\hat{Y}$ without any gain. Hence, it may be that using only some** one of $X_1, \ldots, X_{n-1}$ (or only some two, or only some three, . . ., or all) **would be best** (in terms of making $\hat{Y}$ "close" to Y in the future). For this reason we wish to calculate **all possible regressions** (Y on X_1, Y on X_2, . . .; Y on X_1 and X_2, . . .; . . .; Y on X_1 and . . . and X_{n-1}). (This is feasible using the computational method described by Schatzoff, Tsao, and Fienberg (1968).)

After calculating all possible regressions, one wishes **to choose the few best regressions** and examine them and their residuals $\hat{Y}_i - Y_i$ $(i = 1, \ldots, m)$. Now the total squared error for any regression $\hat{Y}$ containing p of $X_1, \ldots, X_{n-1}$ is

$$E\sum_{i=1}^{m} (\hat{Y}_i - Y_i)^2 = E\sum_{i=1}^{m} (\, (\hat{Y}_i - \mu_i) - (Y_i - \mu_i)\,)^2$$

$$= \sum_{i=1}^{m} E(Y_i - \mu_i)^2 + \sum_{i=1}^{m} E(\hat{Y}_i - \mu_i)^2 - 2\sum_{i=1}^{m} E(\hat{Y}_i - \mu_i)(Y_i - \mu_i)$$

$$= \sum_{i=1}^{m} E\,(Y_i - \mu_i)^2 + \sum_{i=1}^{m} \mathrm{Var}\,(\hat{Y}_i).$$

If we define the last quantity divided by the variance of our observations Y_i, say σ^2, to have the name Γ_p, it is clear we would be interested in a regression which minimizes Γ_p. However, a quantity C_p introduced by C. Mallows (see Gorman and Toman (1966)) is a good estimator of Γ_p if we have a good estimate of σ^2; hence **we use values of C_p to rank the possible regressions.**

5. All Possible Regressions: References

Draper, N.R. and Smith, H. (1966): *Applied Regression Analysis*, John Wiley & Sons, Inc., New York.

Gorman, J.W. and Toman, R.J. (1966): "Selection of Variables for Fitting Equations to Data," *Technometrics*, Vol. 8, pp. 27–51.

Longley, J.W. (1967): "An Appraisal of Least Squares Programs for the Electronic Computer from the Point of View of the User," *Journal of the American Statistical Association*, Vol. 62, pp. 819–841.

Schatzoff, M., Tsao, R., and Fienberg, S. (1968): "Efficient Calculation of All Possible Regressions," *Technometrics*, Vol. 10, pp. 769–779.

5. All Possible Regressions: Examples

For a **first example**, we consider some data analyzed in detail by Draper and Smith (1966).

For a **second example**, we consider some data used by Longley (1967) to evaluate various regression programs, which are commonly available, with regard to numerical accuracy. (Comparing our regression #42—the one containing all variables—with the true regression on all variables given by Longley, we find ours is very accurate—much better than almost all considered by Longley.)

First Example

Efficient Calculation of all Possible Regressions

DRAPER & SMITH, APPLIED REG. ANA., 1966, P.366.

INDEPENDENT VARIABLES 4

OF CASES 13

INDEX OF DEPENDENT VARIABLE 5

VARIABLE FORMAT (4F5.0, F6.1)

DATA WITH DEPENDENT VARIABLE IN LAST POSITION

7.	26.	6.	60.	78.5
1.	29.	15.	52.	74.3

11.	56.	8.	20.	104.3
11.	31.	8.	47.	87.6
7.	52.	6.	33.	95.9
11.	55.	9.	22.	109.2
3.	71.	17.	6.	102.7
1.	31.	22.	44.	72.5
2.	54.	18.	22.	93.1
21.	47.	4.	26.	115.9
1.	40.	23.	34.	83.8
11.	66.	9.	12.	113.3
10.	68.	8.	12.	109.4

	# OBS	MEAN
X1	13	0.7461538E+01
X2	13	0.4815385E+02
X3	13	0.1176923E+02
X4	13	0.3000000E+02
X5	13	0.9542308E+02

SUM OF SQUARES AND CROSS PRODUCTS MATRIX

	1	2	3	4	5
1	0.4152308E+03	0.2510769E+03	−0.3726154E+03	−0.2900000E+03	0.7759615E+03
2	0.2510769E+03	0.2905692E+04	−0.1665385E+03	−0.3041000E+04	0.2292954E+04
3	−0.3726154E+03	−0.1665385E+03	0.4923077E+03	0.3800000E+02	−0.6182308E+03
4	−0.2900000E+03	−0.3041000E+04	0.3800000E+02	0.3362000E+04	−0.2481700E+04
5	0.7759615E+03	0.2292954E+04	−0.6182308E+03	−0.2481700E+04	0.2715764E+04

- - - - - **BETAS** - - - - -

STEP	CONST	X1	X2	X3	X4
1	0.8147934E+02	0.1868748E+01			
2	0.5257735E+02	0.1468306E+01	0.6622505E+00		
3	0.5742368E+02		0.7891248E+00		
4	0.7207467E+02		0.7313296E+00	−0.1008386E+01	
5	0.4819363E+02	0.1695890E+01	0.6569149E+00	0.2500178E+00	
6	0.7234899E+02	0.2312469E+01		0.4944684E+00	
7	0.1102027E+03			−0.1255781E+01	
8	0.1312824E+03			−0.1199851E+01	−0.7246001E+00
9	0.1116844E+03	0.1051854E+01		−0.4100431E+00	−0.6427961E+00
10	0.6240442E+02	0.1551112E+01	0.5101774E+00	0.1019194E+00	−0.1440514E+00
11	0.2036419E+03		−0.9234149E+00	−0.1447971E+01	−0.1557044E+01
12	0.9416003E+02		0.3109053E+00		−0.4569414E+00
13	0.7164826E+02	0.1451938E+01	0.4161103E+00		−0.2365397E+00
14	0.1030974E+03	0.1439958E+01			−0.6139536E+00
15	0.1175679E+03				−0.7381618E+00

ALL INDEPENDENT VARIABLES ARE INCLUDED IN REGRESSION 10, WHICH IS USED TO CALCULATE THE CP STATISTICS

STEP	**SUM OF SQUARES**	**P**	**CP**
1	0.1265687E+04	1	0.2269906E+03
2	0.5790484E+02	2	0.1888003E+01
3	0.9063367E+03	1	0.1594209E+03
4	0.4154431E+03	2	0.6911689E+02
5	0.4811096E+02	3	0.2046433E+01
6	0.1227072E+04	2	0.2217297E+03
7	0.1939401E+04	1	0.3536708E+03
8	0.1757386E+03	2	0.2404459E+02
9	0.5083662E+02	3	0.2558946E+01
10	0.4786402E+02	4	0.4000000E+01
11	0.7381531E+02	3	0.6879692E+01
12	0.8688806E+03	2	0.1543780E+03
13	0.4797313E+02	3	0.2020517E+01
14	0.7476259E+02	2	0.5057810E+01
15	0.8838674E+03	1	0.1551960E+03

LOWEST 15 CPS

STEP	SUM OF SQUARES	P	CP
2	0.5790484E+02	2	0.1888003E+01
13	0.4797313E+02	3	0.2020517E+01
5	0.4811096E+02	3	0.2046433E+01
9	0.5083662E+02	3	0.2558946E+01
10	0.4786402E+02	4	0.4000000E+01
14	0.7476259E+02	2	0.5057810E+01
11	0.7381531E+02	3	0.6879692E+01
8	0.1757386E+03	2	0.2404459E+02
4	0.4154431E+03	2	0.6911689E+02
12	0.8688806E+03	2	0.1543780E+03
15	0.8838674E+03	1	0.1551960E+03
3	0.9063367E+03	1	0.1594209E+03
6	0.1227072E+04	2	0.2217297E+03
1	0.1265687E+04	1	0.2269906E+03
7	0.1939401E+04	1	0.3536708E+03

DRAPER & SMITH, APPLIED REG. ANA., 1966, P. 366.

REGRESSION # 2

X5 = 0.52577E+02
+ 0.14683E+01 * X1
+ 0.66225E+00 * X2

CASE	OBSERVED	PREDICTED	RESIDUAL
1	0.78500E+02	0.80074E+02	0.15740E+01
2	0.74300E+02	0.73251E+02	−0.10491E+01
3	0.10430E+03	0.10581E+03	0.15147E+01
4	0.87600E+02	0.89258E+02	0.16585E+01
5	0.95900E+02	0.97293E+02	0.13925E+01
6	0.10920E+03	0.10515E+03	−0.40475E+01
7	0.10270E+03	0.10400E+03	0.13021E+01
8	0.72500E+02	0.74575E+02	0.20754E+01
9	0.93100E+02	0.91275E+02	−0.18245E+01
10	0.11590E+03	0.11454E+03	−0.13625E+01
11	0.83800E+02	0.80536E+02	−0.32643E+01
12	0.11330E+03	0.11244E+03	−0.86276E+00
13	0.10940E+03	0.11229E+03	0.28934E+01

AVERAGE SQUARED DEVIATION = 0.4454191E+01

```
                              X = PREDICTED       Y = RESIDUALS
        -0.16515E+02            -0.82573E+01          -0.11176E-07            0.82573E+01            0.16515E+02
    -0.20643E+02        -0.12386E+02         -0.41287E+01          0.41287E+01           0.12386E+02          0.20643E+02
0.733E+02●●●●●● +●●●●●● +●●●●●● +●●●●●● +●●●●●● +●●●●●● +●●●●●● +●●●●●● +●●●●●● +●●●●●● Y•AXIS
         •                                              *          *
         •
0.774E+02
         •                                                        *
         •                                            *
0.815E+02
         •
         •
0.856E+02
         •
         •                                                          *
0.898E+02
         •
         •
0.939E+02
         •
         •                                                         *
0.980E+02
         •
         •
0.102E+03
         •                                                         *
         •                                             *           *
0.106E+03
         •
         •
0.110E+03
         •                                              *                  *
         •
0.115E+03                                             *
   X AXIS
```

DRAPER & SMITH, APPLIED REG. ANA., 1966, P.366.

REGRESSION # 13

```
X5     =     0.71648E+02
                  +
      0.14519E+01 *   X1
                  +
      0.41611E+00 *   X2
                  +
     -0.23654E+00 *   X4
```

CASE	OBSERVED	PREDICTED	RESIDUAL
1	0.78500E+02	0.78438E+02	-0.61684E-01
2	0.74300E+02	0.72867E+01	-0.14327E+01
3	0.10430E+03	0.10619E+03	0.18910E+01
4	0.87600E+02	0.89402E+02	0.18016E+01
5	0.95900E+02	0.95644E+02	-0.25624E+00
6	0.10920E+03	0.10530E+03	-0.38982E+01
7	0.10270E+03	0.10413E+03	0.14287E+01
8	0.72500E+02	0.75592E+02	0.30919E+01
9	0.93100E+02	0.91818E+02	-0.12818E+01
10	0.11590E+03	0.11555E+03	-0.35388E+00
11	0.83800E+02	0.81702E+02	-0.20977E+01
12	0.11330E+03	0.11224E+03	-0.10556E+01
13	0.10940E+03	0.11162E+03	0.22247E+01

```
AVERAGE SQUARED DEVIATION = 0.3690210E+01

                              X = PREDICTED      Y = RESIDUALS
       -0.17072E+02          -0.85358E+01         -0.74506E-08          0.85358E+01          0.17072E+02
   -0.21339E+02       -0.12804E+02        -0.42679E+01         0.42679E+01        0.12804E+02        0.21339E+02
0.729E+02●●●●●● +●●●●●● +●●●●●● +●●●●●● +●●●●●● +●●●●●● +●●●●●● +●●●●●● +●●●●●● +●●●●●● Y•AXIS
        •                                     *       *
        •
0.771E+02                                        *
        •
        •                                    *
0.814E+02
        •
        •
0.857E+02
        •
        •                                          *
0.899E+02
        •                                     *
        •
0.942E+02
        •                                        *
        •
0.985E+02
        •
        •
0.103E+03                                         *
        •                                 *
        •                                           *
0.107E+03
        •
        •
0.111E+03
        •                                        *     *
        •
0.116E+03                                          *
     X AXIS
```

Second Example

Efficient calculation of all possible regressions

DATA SET 2 FROM LONGLEY

INDEPENDENT VARIABLES 6

OF CASES 16

INDEX OF DEPENDENT VARIABLE 7

VARIABLE FORMAT (1X, F12.1, F13.0, F9.0, F10.0, F12.0, F8.0, F11.0)

DATA WITH DEPENDENT VARIABLE IN LAST POSITION

83.0	234289.	2356.	1590.	107608.	1947.	60323.
88.5	259426.	2325.	1456.	108632.	1948.	61122.
88.2	258054.	3682.	1616.	109773.	1949.	60171.
89.5	284599.	3351.	1650.	110929.	1950.	61187.
96.2	328975.	2099.	3099.	112075.	1951.	63221.
98.1	346999.	1932.	3594.	113270.	1952.	63639.
99.0	365385.	1870.	3547.	115094.	1953.	64989.
100.0	363112.	3578.	3350.	116219.	1954.	63761.
101.2	397469.	2904.	3048.	117388.	1955.	66019.
104.6	419180.	2822.	2857.	118734.	1956.	67857.
108.4	442769.	2936.	2798.	120445.	1957.	68169.
110.8	444546.	4681.	2637.	121950.	1958.	66513.

112.6	482704.	3813.	2552.	123366.	1959.	68655.
114.2	502601.	3931.	2514.	125368.	1960.	69564.
115.7	518173.	4806.	2572.	127852.	1961.	69331.
116.9	554894.	4007.	2827.	130081.	1962.	70551.

	# OBS	MEAN
IVAR 1	16	0.1016812E+03
IVAR 2	16	0.3876984E+06
IVAR 3	16	0.3193312E+04
IVAR 4	16	0.2606687E+04
IVAR 5	16	0.1174240E+06
IVAR 6	16	0.1954500E+04
DEPVAR	16	0.6531700E+05

SUM OF SQUARES AND CROSS PRODUCTS MATRIX

	1	2	3	4	5	6	7
1	0.1746865E+04	0.1595406E+08	0.9387999E+05	0.5235380E+05	0.1102545E+07	0.7638516E+03	0.5519499E+09
2	0.1595406E+08	0.1481903E+12	0.8418655E+09	0.4632064E+09	0.1027861E+11	0.7064656E+07	0.5149952E+11
3	0.9387999E+05	0.8418655E+09	0.1309835E+08	-0.1730681E+07	0.6694112E+08	0.4459550E+05	0.2473654E+09
4	0.5235380E+05	0.4632064E+09	-0.1730681E+07	0.7264561E+07	0.2646147E+08	0.2073650E+05	0.1676521E+09
5	0.1102545E+07	0.1027861E+11	0.6694112E+08	0.2646147E+08	0.7258099E+09	0.4937600E+06	0.3519292E+09
6	0.7638516E+03	0.7064656E+07	0.4459550E+05	0.2073650E+05	0.4937600E+06	0.3400000E+03	0.2436140E+07
7	0.5519499E+06	0.5149952E+10	0.2473654E+08	0.1676521E+08	0.3519292E+09	0.2436140E+06	0.1850087E+09

STEP

----BETAS----

STEP	CONST	IVAR 1	IVAR 2	IVAR 3	IVAR 4	IVAR 5	IVAR 6
1	0.3318918E+05	0.3159661E+03					
2	0.5694470E+05	−0.8510079E+02	0.4391418E−01				
3	0.5184359E+05		0.3475229E−01				
4	0.5238217E+05		0.3784032E−01	−0.5435739E+00			
5	0.5392685E+05	−0.2593690E+02	0.4057516E−01	−0.5334516E+00			
6	0.3179977E+05	0.3488421E+03		−0.6117374E+00			
7	0.5928636E+05			0.1888523E+01			
8	0.5066281E+05			0.2264743E+01	0.2847352E+01		
9	0.3039342E+05	0.3980757E+03		−0.1072498E+01	−0.8165279E+00		
10	0.5008324E+05	0.5626853E+02	0.3526266E−01	−0.8538066E+00	−0.5495464E+00		
11	0.5330646E+05		0.4078798E−01	−0.7968154E+00	−0.4827645E+00		
12	0.5168347E+05		0.3439346E−01		0.1147959E+00		
13	0.5752446E+05	−0.9842276E+02	0.4485814E−01		0.1568496E+00		
14	0.3320645E+05	0.3147916E+03			0.3918673E−01		
15	0.5930126E+05				0.2307808E+01		
16	0.9426420E+04				0.6245613E+00	0.4621078E+00	
17	0.2418175E+05	0.2008590E+03			0.2369445E+00	0.1711236E+00	
18	0.1203252E+06	−0.1363180E+03	0.9659914E−01		−0.4689414E+00	−0.6589470E+00	
19	0.1094733E+06		0.7992898E−01		−0.4979036E+00	−0.6288900E+00	
20	0.8261704E+05		0.6210455E−01	−0.5197647E+00	−0.5917338E+00	−0.3251093E+00	
21	0.9246629E+05	−0.4846761E+02	0.7200771E−01	−0.4038200E+00	−0.5605055E+00	−0.4035609E+00	
22	0.1378174E+05	0.2070514E+03		−0.1241233E+01	−0.5968439E+00	0.3065936E+00	
23	−0.1323061E+04			−0.1229179E+01	−0.1892958E+00	0.6051459E+00	
24	−0.1353047E+03			−0.1115148E+01		0.5877276E+00	
25	0.1152506E+05	0.1366942E+03		−0.9610262E+00		0.3658670E+00	
26	0.9421795E+05	−0.1268793E+03	−0.7554401E−01	0.1003855E+00		−0.3829493E+00	
27	0.6616096E+05		0.4761798E−01	−0.3853356E+00		−0.1539283E+00	
28	0.8894000E+05		0.6317334E−01			−0.4097562E+00	
29	0.1020314E+06	−0.1481440E+03	0.8234889E−01			−0.4562732E+00	
30	0.2685174E+05	0.2408464E+03				0.1190191E+00	
31	0.8380694E+04					0.4848779E+00	
32	−0.1911869E+07					−0.2119700E+00	0.1024342E+04
33	−0.1110111E+07	0.1255396E+03				−0.1152120E+00	0.6017866E+03
34	−0.2978593E+06	−0.1817028E+03	0.8089616E−01			−0.5282346E+00	0.2109573E+03
35	0.4162973E+06		0.6788354E−01			−0.3598436E+00	−0.1714220E+03
36	−0.1141570E+07		0.1199295E−01	−0.8470903E+00		−0.2734734E−01	0.6181394E+03
37	−0.1122431E+07	−0.1278769E+03	0.3986900E−01	−0.5633937E+00		−0.2572112E+00	0.6228167E+03
38	−0.1880549E+07	−0.6489812E+02		−0.1051931E+01		−0.2795996E−03	0.1000694E+03
39	−0.1485065E+07			−0.1007126E+01		0.3859266E−01	0.7925640E+03
40	−0.2445601E+07			−0.1500441E+01	−0.9342135E+00	−0.2284880E+00	0.1302110E+03
41	−0.2704733E+07	−0.4396420E+02		−0.1526284E+01	−0.9256726E+00	−0.2523796E+00	0.1438446E+03

42	−0.3475320E+07	0.1458437E+02	−0.3554047E−01	−0.2016329E+01	−0.1032120E+01	−0.5233285E−01	0.1825637E+03
43	−0.3444198E+07		−0.3181068E−01	−0.1969862E+01	−0.1019309E+01	−0.7791332E−01	0.1811176E+03
44	0.3023190E+06		0.8239755E−01		−0.4886410E+00	−0.5953517E+00	−0.1011845E+03
45	−0.4042531E+06	−0.1799876E+03	0.9517997E−01		−0.4850263E+00	−0.7604110E+00	0.2770658E+03
46	−0.1015142E+07	0.1042640E+03			0.1843882E+00	−0.5467166E−01	0.5504203E+03
47	−0.1594946E+07				0.2531425E+00	−0.1063431E+00	0.8555079E+03
48	−0.1298040E+07				0.3178860E+00		0.6971240E+03
49	−0.8332135E+06	0.1112752E+03			0.2109344E+00		0.4536537E+03
50	0.1168280E+07	−0.1976103E+02	0.6437298E−01		−0.1004744E−01		−0.5760474E+03
51	0.1217983E+07		0.6354349E−01		−0.2457025E−01		−0.6023217E+03
52	−0.3593508E+07		−0.4007422E−01	−0.2086581E+01	−0.1013918E+01		0.1884711E+03
53	−0.3559711E+07	0.2753840E+02	−0.4199398E−01	−0.2101962E+01	−0.1041449E+01		0.1866429E+03
54	−0.1829189E+07	−0.7345399E+01		−0.1473453E+01	−0.7680269E+00		0.9731187E+03
55	−0.1797222E+07			−0.1469672E+01	−0.7722828E+00		0.9563801E+03
56	−0.1587139E+07			−0.9955309E+00			0.8470888E+03
57	−0.1879567E+07	−0.6484448E+02		−0.1051969E+01			0.1000172E+03
58	−0.1444051E+07	−0.6843475E+02	0.1046012E−01	−0.9327913E+00			0.7752621E+03
59	−0.1198447E+07		0.9000885E−02	−0.8902905E+00			0.6462612E+03
60	0.1197990E+07		0.6297521E−01				−0.5920123E+03
61	0.1157235E+07	−0.2125742E+02	0.6422884E−01				−0.5703034E+03
62	−0.6882896E+06	0.1507960E+03					0.3777301E+03
63	−0.1335105E+07						0.7165118E+03

ALL INDEPENDENT VARIABLES ARE INCLUDED IN REGRESSION 42, WHICH IS USED TO CALCULATE THE CP STATISTICS

STEP	SUM OF SQUARES	P	**CP**
1	0.1061126E+08	1	0.1126197E+03
2	0.5824189E+07	2	0.5749765E+02
3	0.6036107E+07	1	0.5802638E+02
4	0.3579036E+07	2	0.3070716E+02
5	0.3560203E+07	3	0.3248243E+02
6	0.7597620E+07	2	0.7865928E+02
7	0.1382932E+09	1	0.1636196E+04
8	0.8125036E+08	2	0.9575271E+03
9	0.5509989E+07	3	0.5574843E+02
10	0.2683787E+07	4	0.2402452E+02
11	0.2756687E+07	3	0.2289441E+02
12	0.5959454E+07	2	0.5911171E+02
13	0.5686281E+07	3	0.5785206E+02
14	0.1060251E+08	2	0.1145154E+03
15	0.1463178E+09	1	0.1731950E+04
16	0.1190857E+08	2	0.1301001E+03
17	0.9948759E+07	3	0.1087145E+03
18	0.2533127E+07	4	0.2222676E+02
19	0.3050507E+07	3	0.2640045E+02
20	0.2366479E+07	4	0.2023823E+02
21	0.2335113E+07	5	0.2186394E+02
22	0.3537496E+07	4	0.3421149E+02
23	0.5619407E+07	3	0.5705407E+02
24	0.5755110E+07	2	0.5667336E+02
25	0.4573530E+07	3	0.4457406E+02
26	0.3245922E+07	4	0.3073226E+02
27	0.3482170E+07	3	0.3155131E+02
28	0.3874205E+07	2	0.3422930E+02
29	0.3259871E+07	3	0.2889870E+02
30	0.1018726E+08	2	0.1095604E+03
31	0.1436599E+08	1	0.1574235E+03
32	0.1006307E+08	2	0.1080785E+03
33	0.9659948E+07	3	0.1052682E+03
34	0.3197227E+07	4	0.3015120E+02
35	0.3812026E+07	3	0.3584733E+02
36	0.3236605E+07	4	0.3062107E+02
37	0.2996641E+07	5	0.2975768E+02
38	0.3165666E+07	4	0.2977459E+02
39	0.3259932E+07	3	0.2889942E+02

40	0.9860347E+06	4	0.3765947E+01
41	0.9429648E+06	5	0.5252011E+01
42	0.8380411E+06	6	0.6000000E+01
43	0.8407840E+06	5	0.4032730E+01
44	0.3029128E+07	4	0.2814534E+02
45	0.2425925E+07	5	0.2294756E+02
46	0.9519358E+07	4	0.1055906E+03
47	0.9776263E+07	3	0.1066561E+03
48	0.9850114E+07	2	0.1055374E+03
49	0.9537637E+07	3	0.1038087E+03
50	0.4899723E+07	4	0.5046637E+02
51	0.4908735E+07	3	0.4857392E+02
52	0.8603082E+06	4	0.2265703E+01
53	0.8430253E+06	5	0.4059474E+01
54	0.1321920E+07	4	0.7773922E+01
55	0.1323225E+07	3	0.5789496E+01
56	0.3271995E+07	2	0.2704338E+02
57	0.3165667E+07	3	0.2777460E+02
58	0.3121456E+07	4	0.2924705E+02
59	0.3239018E+07	3	0.2864987E+02
60	0.4911906E+07	2	0.4661175E+02
61	0.4900195E+07	3	0.4847201E+02
62	0.9756517E+07	2	0.1044205E+03
63	0.1045641E+08	1	0.1107720E+03

LOWEST 20 CPS

STEP	SUM OF SQUARES	P	**CP**
52	0.8603082E+06	4	0.2265703E+01
40	0.9860347E+06	4	0.3765947E+01
43	0.8407840E+06	5	0.4032730E+01
53	0.8430253E+06	5	0.4059474E+01
41	0.9429648E+06	5	0.5252011E+01
55	0.1323225E+07	3	0.5789496E+01
42	0.8380411E+06	6	0.6000000E+01
54	0.1321920E+07	4	0.7773922E+01
20	0.2366479E+07	4	0.2023823E+02
21	0.2335113E+07	5	0.2186394E+02
18	0.2533127E+07	4	0.2222676E+02
11	0.2756687E+07	3	0.2289441E+02
45	0.2425925E+07	5	0.2294756E+02
10	0.2683787E+07	4	0.2402452E+02
19	0.3050507E+07	3	0.2640045E+02
56	0.3271995E+07	2	0.2704338E+02
57	0.3165667E+07	3	0.2777460E+02
44	0.3029128E+07	4	0.2814534E+02
59	0.3239018E+07	3	0.2864987E+02
29	0.3259871E+07	3	0.2889870E+02

DATA SET 2 FROM LONGLEY

REGRESSION # 52

DEPVAR = −0.35935E+07
+ −0.40074E−01 * IVAR 2
+ −0.20866E+01 * IVAR 3
+ −0.10139E+01 * IVAR 4
+ 0.18847E+04 * IVAR 6

CASE	OBSERVED	PREDICTED	RESIDUAL
1	0.60323E+05	0.60107E+05	−0.21561E+03
2	0.61122E+05	0.61185E+05	0.63301E+02

3	0.60171E+05	0.60131E+05	−0.39725E+02
4	0.61187E+05	0.61608E+05	0.42140E+03
5	0.63221E+05	0.62858E+05	−0.36299E+03
6	0.63639E+05	0.63867E+05	0.22799E+03
7	0.64989E+05	0.65192E+05	0.20292E+03
8	0.63761E+05	0.63804E+05	0.42582E+02
9	0.66019E+05	0.66024E+05	0.50234E+01
10	0.67857E+05	0.67403E+05	−0.45356E+03
11	0.68169E+05	0.68165E+05	−0.42051E+01
12	0.66513E+05	0.66500E+05	−0.12549E+02
13	0.68655E+05	0.68753E+05	0.98346E+02
14	0.69564E+05	0.69633E+05	0.69012E+02
15	0.69331E+05	0.69009E+05	−0.32188E+03
16	0.70551E+05	0.70831E+05	0.27989E+03

AVERAGED SQUARED DEVIATION = 0.5366776E+05

```
                                X = PREDICTED      Y = RESIDUALS
       -0.42894E+04         -0.21447E+04        -0.28610E-05        0.21447E+04        0.42894E+04
   -0.53618E+04        -0.32171E+04        -0.10724E+04       0.10724E+04       0.32171E+04       0.53618E
0.601E+05........ +........ +........ +........ +........ +........ +........ +........ +........ +........ Y.A
        .                                       * *
        .
0.612E+05                                          *
        .                                            *
        .
0.623E+05
        .                                    *
        .
0.633E+05
        .                                          *  *
        .
0.644E+05
        .
        .                                           *
0.655E+05
        .                                          *
        .                                         *
0.665E+05
        .
        .                                   *
0.676E+05
        .                                          *
        .
0.687E+05                                     *    *
        .
        .                                           *
0.698E+05
        .
        .
0.708E+05                                            *
    X AXIS
```

DATA SET 2 FROM LONGLEY

REGRESSION # 40

DEPVAR = −0.24456E+07
+ −0.15004E+01 * IVAR 3
+ −0.93421E+00 * IVAR 4
+ −0.22849E+00 * IVAR 5
+ 0.13021E+04 * IVAR 6

CASE	OBSERVED	PREDICTED	RESIDUAL
1	0.60323E+05	0.60000E+05	−0.32285E+03
2	0.61122E+05	0.61240E+05	0.11799E+03
3	0.60171E+05	0.60096E+05	−0.75180E+02
4	0.61187E+05	0.61599E+05	0.41168E+03
5	0.63221E+05	0.63164E+05	−0.57178E+02
6	0.63639E+05	0.63981E+05	0.34203E+03
7	0.64989E+05	0.65003E+05	0.14311E+02
8	0.63761E+05	0.63670E+05	−0.91340E+02
9	0.66019E+05	0.65998E+05	−0.20904E+02
10	0.67857E+05	0.67294E+05	−0.56287E+03
11	0.68169E+05	0.68089E+05	−0.79633E+02
12	0.66513E+05	0.66580E+05	0.66744E+02
13	0.68655E+05	0.68940E+05	0.28511E+03
14	0.69564E+05	0.69643E+05	0.79230E+02
15	0.69331E+05	0.69011E+05	−0.32029E+03
16	0.70551E+05	0.70764E+05	0.21314E+03

AVERAGE SQUARED DEVIATION = 0.6160763E+05

```
                                   X = PREDICTED       Y = RESIDUALS
         -0.43056E+04         -0.21528E+04         -0.57220E-05         0.21528E+04         0.43056E+04
   -0.53820E+04        -0.32292E+04        -0.10764E+04        0.10764E+04        0.32292E+04        0.53820E+04
0.600E+05........ +........ +........ +........ +........ +........ +........ +........ +........ +........Y.AXIS
        .
        .                                   *  *
0.611E+05                                          *
        .                                             *
        .
0.622E+05
        .
        .                                       *
0.632E+05
        .                                       *
        .                                            *
0.643E+05
        .                                        *
        .
0.654E+05
        .                                       *
        .
0.665E+05                                        *
        .
        .                                 *
0.675E+05
        .                                        *
        .
0.686E+05                                           *
        .                                     *
        .                                          *
0.697E+05
        .
        .
0.708E+05
   X AXIS                                          *
```

DATA SET 2 FROM LONGLEY

REGRESSION # 43

DEPVAR = −0.34442E+07
+ −0.31811E−01 * IVAR 2
+ −0.19699E+01 * IVAR 3

```
                                        +
                     -0.10193E+01       *       IVAR 4
                                        +
                     -0.77913E-01       *       IVAR 5
                                        +
                      0.18112E+04       *       IVAR 6
```

CASE	OBSERVED	PREDICTED	RESIDUAL
1	0.60323E+05	0.60064E+05	-0.25926E+03
2	0.61122E+05	0.61193E+05	0.71160E+02
3	0.60171E+05	0.60123E+05	-0.48111E+02
4	0.61187E+05	0.61617E+05	0.42995E+03
5	0.63221E+05	0.62916E+05	-0.30450E+03
6	0.63639E+05	0.63886E+05	0.24662E+03
7	0.64989E+05	0.65140E+05	0.15085E+03
8	0.63761E+05	0.63772E+05	0.10961E+02
9	0.66019E+05	0.66035E+05	0.15652E+02
10	0.67857E+05	0.67407E+05	-0.45047E+03
11	0.68169E+05	0.68170E+05	0.59180E+00
12	0.66513E+05	0.66534E+05	0.20682E+02
13	0.68655E+05	0.68817E+05	0.16218E+03
14	0.69564E+05	0.69646E+05	0.81729E+02
15	0.69331E+05	0.68985E+05	-0.34574E+03
16	0.70551E+05	0.70769E+05	0.21765E+03

AVERAGE SQUARED DEVIATION = 0.5245964E+05

```
                         X = PREDICTED     Y = RESIDUALS
     -0.42820E+04        -0.21410E+04       -0.19073E-05       0.21410E+04        0.42820E+04
-0.53525E+04       -0.32115E+04        -0.10705E+04       0.10705E+04       0.32115E+04       0.53525E+04
0.601E+05........+........+........+........+........+........+........+........+........+........Y.AXIS
         .                                            **
         .
0.611E+05                                               *
         .                                                 *
         .
0.622E+05
         .                                            *
         .
0.633E+05
         .                                              * *
         .
0.643E+05
         .
         .                                               *
0.654E+05
         .                                              *
         .
0.665E+05                                                *
         .
         .                                           *
0.676E+05
         .                                               *
         .
0.686E+05                                                 *
         .                                            *
         .                                               *
0.697E+05
         .
         .
0.708E+05                                                   *.
  X AXIS
```

6. TWO-WAY ANALYSIS OF VARIANCE

Statistical Theory

Table 10

	Column 1	Column 2	...	Column n
Row 1	$X_{111}, \ldots, X_{11a_{11}}$	$X_{121}, \ldots, X_{12a_{12}}$	...	$X_{1n1}, \ldots, X_{1na_{1n}}$
2	$X_{211}, \ldots, X_{21a_{21}}$	$X_{221}, \ldots, X_{22a_{22}}$	...	$X_{2n1}, \ldots, X_{2na_{2n}}$
·	·			
·	·			
·	·			
m	$X_{m11}, \ldots, X_{m1a_{m1}}$	$X_{m21}, \ldots, X_{m2a_{m2}}$	...	$X_{mn1}, \ldots, X_{mna_{mn}}$

Suppose we are given an array of **data** as shown in Table 10. **Each row could represent a patient and each column a time period,** for example. The X's would then be (independent) measurements of a given patient's heart rate, for example, in a given time period; for patient 1 in time period 1 there was a_{11} (0 or 1 or 2 or . . .) such measurements, etc.

Besides some usual summary measures, we are interested in comparing days and patients individually with regard to heart rate, to see if heart rate varies from day to day, and to see if heart rate varies from patient to patient. If we have an **additive model** $E(HR) = \alpha_i + \beta_j$ (α_i due to patient i, β_j due to day j), we are then testing the hypothesis H: $\beta_1 = \ldots = \beta_n$ and the hypothesis H: $\alpha_1 = \ldots = \alpha_m$, respectively. **If we have interaction** present (i.e., different patients behave differently on different days), $E(HR) = \alpha_i + \beta_j + \gamma_{ij}$ and the situation becomes complex; a test of whether interaction is present is performed (i.e., a test of H: $\gamma_{ij} \equiv 0$). Computational details are given in Scheffé (1959), pp. 112–119, and in Bancroft (1968), pp. 16–29.

6. Two-Way Analysis of Variance: References

Bancroft, T.A. (1968): *Topics in Intermediate Statistical Methods, Vol. 1*, The Iowa State University Press, Ames, Iowa.

Scheffé, H. (1959): *The Analysis of Variance*, John Wiley & Sons, Inc., New York.

6. Two-Way Analysis of Variance: Examples (Input and Output)

The **first example** uses data from Bancroft (1968), and has empty cells. The **second example** has at least one observation in each cell.

First Example

T. A. Bancroft, *Topics In Int. Stat. Meth.*, Vol. 1, 1968, pp. 20-25.

TWO-WAY ANALYSIS OF VARIANCE WITH UNEQUAL SUBCLASS FREQUENCIES

```
2 22. 25.                 ( 1, 1)
3 -1. 40. 18.             ( 1, 2)
2 41. 41.                 ( 2, 1)
2 23. 13.                 ( 2, 2)
3 29. 20. 37.             ( 3, 1)
                          ( 3, 2)  EMPTY
2 49. 50.                 ( 4, 1)
1 61.                     ( 4, 2)
1 55.                     ( 5, 1)
                          (5, 2)   EMPTY
```

CELL MEANS

```
0.2350E+02     0.1900E+02
0.4100E+02     0.1800E+02
0.2867E+02     0.0000E+00
0.4950E+02     0.6100E+02
0.5500E+02     0.0000E+00
```

 # OBS / CELL

```
2   3
2   2
3   0
2   1
1   0
```

CELL TOTALS

```
0.4700E+02     0.5700E+02
0.8200E+02     0.3600E+02
0.8600E+02     0.0000E+00
0.9900E+02     0.6100E+02
0.5500E+02     0.0000E+00
```

OBS / ROW

5 4 3 3 1

OBS / COLUMN

10 6

ROW TOTALS

0.1040E+03 0.1180E+03 0.8600E+02 0.1600E+03 0.5500E+02

COLUMN TOTALS

0.3690E+03 0.1540E+03

NORMAL EQUATIONS MATRIX A

0.3100E+01	−0.1400E+01	−0.6000E+00	−0.9000E+00
−0.1400E+01	0.2933E+01	−0.6000E+00	−0.7333E+00
−0.6000E+00	−0.6000E+00	0.2100E+01	−0.6000E+00
−0.9000E+00	−0.7333E+00	−0.6000E+00	0.2433E+01

NORMAL EQUATIONS VECTOR FOR ALPHA

−0.4680E+02 −0.7133E+01 −0.2470E+02 0.6053E+02

NORMAL EQUATIONS MATRIX B

0.2867E+01

NORMAL EQUATIONS VECTOR FOR BETA

0.2073E+02

ANALYSIS OF VARIANCE TABLE

SOURCE OF VARIATION	DEGREES OF FREEDOM	SUM OF SQUARES	MEAN SQUARE	F
ROW	4	0.22490550E+04	0.56226376E+03	3.6673083
COLUMN	1	0.14995504E+03	0.14995504E+03	0.9780665
INTERACTION	2	0.49151164E+03	0.24575582E+03	1.8874047
ERROR UNDER ADDITIVE MODEL	10	0.15331783E+04	0.15331783E+03	
ERROR	8	0.10416667E+04	0.13020833E+03	
TOTAL (UNADJUSTED FOR THE MEAN)	16	0.21351000E+05		

MULTIPLE COMPARISONS

	ROW EFFECT	COLUMN EFFECT
1	−0.29860465E+02	0.72325582E+01
2	−0.21883721E+02	0.00000000E+00
3	−0.26333333E+02	
4	0.74418628E+00	
5	0.00000000E+00	

MULTIPLE COMPARISONS FOR ROW EFFECTS

BASED ON UPPER 5% POINT OF F ON 4 AND 10 DEGREES OF FREEDOM

F = 3.48 AND THE CRITICAL VALUE = 3.731

DIFFERENCE		VARIANCE OF DIFFERENCE	STANDARD ERROR	CRITICAL RATIO
(2−1) =	0.79767442E+01	0.59047965E+02	0.76842674E+01	1.03806177 NS
(3−1) =	0.35271318E+01	0.85796189E+02	0.92626232E+01	.38079189 NS
(3−2) =	−0.44496124E+01	0.87310239E+02	0.93439948E+01	−.47620022 NS
(4−1) =	0.30604651E+02	0.72674419E+02	0.85249292E+01	3.59001821 NS
(4−2) =	0.22627907E+02	0.77216570E+02	0.87872959E+01	2.57507055 NS
(4−3) =	0.27077519E+02	0.91852390E+02	0.95839653E+01	2.82529398 NS
(5−1) =	0.29860465E+02	0.17260174E+03	0.13137798E+02	2.27286675 NS
(5−2) =	0.21883721E+02	0.17411579E+03	0.13195294E+02	1.65844885 NS
(5−3) =	0.26333333E+02	0.17361111E+03	0.13176157E+02	1.99855947 NS
(5−4) =	−0.74418628E+00	0.17865794E+03	0.13366299E+02	−.05567632 NS

MULTIPLE COMPARISONS FOR COLUMN EFFECTS

BASED ON UPPER 5% POINT OF F ON 1 AND 10 DEGREES OF FREEDOM
F = 4.96 AND THE CRITICAL VALUE = 2.227

DIFFERENCE	VARIANCE OF DIFFERENCE	STANDARD ERROR	CRITICAL RATIO
(2 − 1) = −0.72325582E+01	0.45421512E+02	0.67395483E+01	−1.07315176 NS

ASSUMING INTERACTION - METHOD OF WEIGHTED SQUARES OF MEANS

CELL(3, 2) IS EMPTY, DIVISION BY # CELL VALUES IS NOT POSSIBLE. PROGRAM CANNOT DO ANALYSIS ON CELL MEANS AND CELL NUMBERS FOR ROW EFFECTS.

Second Example

Tensile Strength 1,2,4 wks rats no-fe, fe added, chow macon

TWO-WAY ANALYSIS OF VARIANCE WITH UNEQUAL SUBCLASS FREQUENCIES

```
2110.255.                              ( 1, 1)
5650.620.575.650.750.                  ( 1, 2)
6800.625.450.600.150.375.              ( 1, 3)
6 60. 95.350.100.130.120.              ( 2, 1)
6500.630.500.475.525.840.              ( 2, 2)
6550.100.150.550.750.850.              ( 2, 3)
7175.115.200.220.280.175.183.          ( 3, 1)
7940.900.500.650.660.875.925.          ( 3, 2)
6500.200.750.650.375.475.              ( 3, 3)
```

```
                                           CELL MEANS
 0.1825E+03          0.6490E+03              0.1833E+04
 0.1425E+03          0.5783E+03              0.1992E+04
 0.1924E+03          0.7786E+03              0.2492E+04
                                           # OBS / CELL
    2      5            6
    6      6            6
    7      7            6
                                           CELL TOTALS
 0.3650E+03          0.3245E+04              0.1100E+05
 0.8550E+03          0.3470E+04              0.1195E+05
 0.1347E+04          0.5450E+04              0.1495E+05
                                           # OBS / ROW
   13     18           20
                                           # OBS / COLUMN
   15     18           18
                                           ROW TOTALS
 0.1461E+05          0.1627E+05              0.2175E+05
                                           COLUMN TOTALS
 0.2567E+04          0.1216E+05              0.3790E+05
                                           NORMAL EQUATIONS MATRIX A
 0.9344E+01         -0.4467E+01
-0.4467E+01          0.1160E+02
                                           NORMAL EQUATIONS VECTOR
                                           FOR ALPHA
-0.1745E+04         -0.1440E+04
                                           NORMAL EQUATIONS MATRIX B
 0.1024E+02         -0.5219E+01
-0.5219E+01          0.1163E+02
                                           NORMAL EQUATIONS VECTOR
                                           FOR BETA
-0.1272E+05         -0.6491E+04
```

ANALYSIS OF VARIANCE TABLE

SOURCE OF VARIATION	DEGREES OF FREEDOM	SUM OF SQUARES	MEAN SQUARE	F
ROW	2	0.87218400E+06	0.43609200E+06	4.6504375
COLUMN	2	0.34552246E+08	0.17276123E+08	184.2306819
INTERACTION	4	0.68763594E+06	0.17190898E+06	1.9912310
ERROR UNDER ADDITIVE MODEL	46	0.43136227E+07	0.93774408E+05	
ERROR	42	0.36259868E+07	0.86333019E+05	
TOTAL (UNADJUSTED FOR THE MEAN)	51	0.93647174E+08		

MULTIPLE COMPARISONS

	ROW EFFECT	COLUMN EFFECT
1	−0.30156658E+03	−0.19787175E+04
2	−0.24026989E+03	−0.14464759E+04
3	0.00000000E+00	0.00000000E+00

MULTIPLE COMPARISONS FOR ROW EFFECTS

BASED ON UPPER 5% POINT OF F ON 2 AND 46
DEGREES OF FREEDOM
F = 3.21 AND THE CRITICAL VALUE = 2.532

DIFFERENCE	VARIANCE OF DIFFERENCE	STANDARD ERROR	CRITICAL RATIO
(2−1) = 0.61296692E+02	0.11724371E+05	0.10827913E+03	.56609883 NS
(3−1) = 0.30156658E+03	0.11323074E+05	0.10640994E+03	2.83400775 SIG
(3−2) = 0.24026989E+03	0.91213655E+04	0.95505840E+02	2.51576125 NS

MULTIPLE COMPARISONS FOR COLUMN EFFECTS

BASED ON UPPER 5% POINT OF F ON 2 AND 46 DEGREES OF FREEDOM
F = 3.21 AND THE CRITICAL VALUE = 2.532

DIFFERENCE	VARIANCE OF DIFFERENCE	STANDARD ERROR	CRITICAL RATIO
(2−1) = 0.53224162E+03	0.10744629E+05	0.10365630E+03	5.13467700 SIG
(3−1) = 0.19787175E+04	0.10929008E+05	0.10454190E+03	18.92750752 SIG
(3−2) = 0.14464759E+04	0.96275055E+04	0.98119852E+02	14.74192923 SIG

ASSUMING INTERACTION - METHOD OF WEIGHTED SQUARES OF MEANS

ANALYSIS OF VARIANCE TABLE

SOURCE OF VARIATION	DEGREES OF FREEDOM	SUM OF SQUARES	MEAN SQUARE	F
ROW	2	0.76742422E+06	0.38371211E+06	4.4445579
COLUMN	2	0.31051224E+08	0.15525612E+08	179.8339962
INTERACTION	4	0.68763594E+06	0.17190898E+06	
ERROR	42	0.36259868E+07	0.86333019E+05	

MULTIPLE COMPARISONS

	ROW EFFECT	COLUMN EFFECT
1	0.88827778E+03	0.17247619E+03
2	0.90416667E+03	0.66863492E+03
3	0.11542222E+04	0.21055556E+04

MULTIPLE COMPARISONS FOR ROW EFFECTS

BASED ON UPPER 5% POINT OF F ON 2 AND 42 DEGREES OF FREEDOM
F = 3.22 AND THE CRITICAL VALUE = 2.539

DIFFERENCE	VARIANCE OF DIFFERENCE	STANDARD ERROR	CRITICAL RATIO
(2−1) = 0.15888889E+02	0.13109829E+05	0.11449816E+03	.13876982 NS
(3−1) = 0.26594445E+03	0.12653040E+05	0.11248573E+03	2.36425044 NS
(3−2) = 0.25005556E+03	0.91357692E+04	0.95581218E+02	2.61615788 SIG

MULTIPLE COMPARISONS FOR COLUMN EFFECTS

BASED ON UPPER 5% POINT OF F ON 2 AND 42 DEGREES OF FREEDOM
F = 3.22 AND THE CRITICAL VALUE = 2.539

DIFFERENCE	VARIANCE OF DIFFERENCE	STANDARD ERROR	CRITICAL RATIO
(2−1) = 0.49615873E+03	0.12653040E+05	0.11248573E+03	4.41085916 SIG
(3−1) = 0.19330794E+04	0.12561683E+05	0.11207891E+03	17.24748503 SIG
(3−2) = 0.14369206E+04	0.96839154E+04	0.98406887E+02	14.60183003 SIG

STATISTICAL ANALYSIS TECHNIQUES

In experimental designs, it is common for the characteristics of the random quantity of interest to be related to other (random or non-random) quantities which are under the control (or partial control) of the experimenter, and which in many cases are able to be set without error. We will now detail the specification and analysis of models for this setting.

Example: If Y is the amount of seed (in lbs./1000 ft²) delivered by a drop spreader, this amount is a random variable: from experimental run to experimental run it will vary (even though we try to hold all experimental conditions constant). If we denote $E(Y)$ by μ and $Var(Y)$ by σ^2, we have already seen how to estimate μ and σ^2 based on a random sample $Y_1, Y_2, \ldots, Y_n$ using $\overline{Y}$ and s^2. (Recall that

$$\overline{Y} = \frac{Y_1 + Y_2 + \ldots + Y_n}{n} \text{ and } s^2 = \frac{\sum_{i=1}^{n} (Y_i - \overline{Y})^2}{n-1}.)$$

We also saw how to obtain confidence intervals for μ (the mean amount of seed delivered) and σ^2 (the variance of the amount of seed delivered), assuming a normal distribution for Y.

Now we realize, however, that the mean amount delivered, $E(Y)$, is related to the non-random dial setting x used during the experimental runs, so that $E(Y)$ is a function of the setting x:

$$E(Y) = f(x).$$

It is often appropriate to assume that, over the range of x values of interest (e.g., from $x = 0.50$ to $x = 3.00$ lbs./1000 ft²), **this function is linear (Figure 23):**

$$E(Y) = \alpha + \beta x$$

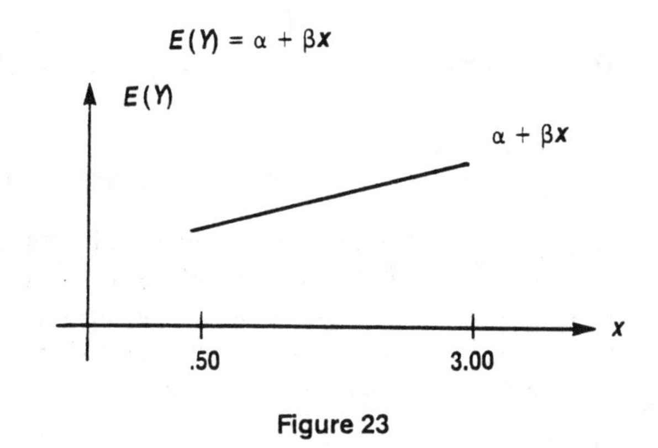

Figure 23

where of course α and β are unknown. On any experiment we observe

$$Y = \alpha + \beta x + \epsilon$$

where ϵ is an *error* with $E(\epsilon) = 0$. In this setting, x is called an **independent variable**, while Y is called a **dependent variable.**

In *other examples*, Y might be the amount of yield of a chemical process while x is the amount of catalyst used, Y might be turf density and x the amount of fertilizer used, and so on. We will touch on other instances of specific importance below. (x,Y) data such as we are discussing is often plotted as

and a line is fitted. This constitutes a **linear model**.

One of the first statistical questions which arises for such a linear model is: If *I* observe (x_1,Y_1), (x_2,Y_2), . . ., (x_n,Y_n), e.g., set x to x_1 and observe yield Y_1 and so on, **how**

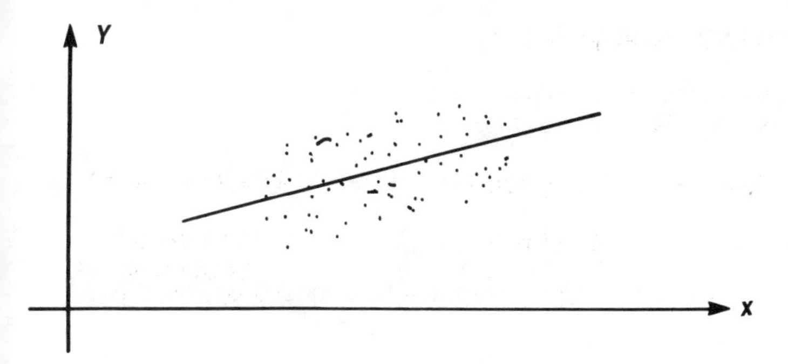

Figure 24

should α and β be estimated? The usefulness of this answer is that it provides us with a *model for future use*. If we estimate α by $\hat{\alpha}$ and β by $\hat{\beta}$, then to estimate what $E(Y)$ would equal at some new setting x^*, we will simply calculate $\hat{\alpha}+\hat{\beta}x^*$.

If we were to estimate α by a and β by b, we would predict $a+bx_1$ at setting x_1, while in fact we observed Y_1. Our estimate would thus differ from the observed value by

$$Y_1 - (a+bx_1).$$

Similarly, at x_2 we predict $a+bx_2$ but observe Y_2, which differs by

$$Y_2 - (a+bx_2).$$

And so on to $Y_n - (a+bx_n)$ at setting x_n. **It is reasonable to try to choose a and b to minimize the sum of these deviations (of observed from predicted) squared,** and **this choice is called the method of least squares of choosing estimates a and b** (for α and β). We choose a and b to minimize

$$\sum_{i=1}^{n} [Y_i - (a+bx_i)]^2.$$

It is easy to show that **the resulting estimators are**

$$\hat{\alpha} = \overline{Y} - r_{xY}\frac{s_Y}{s_X}\overline{x} \text{ and } \hat{\beta} = r_{xY}\frac{s_Y}{s_x}$$

where we use the *usual notation*

$$\begin{cases} \overline{x} = \dfrac{x_1+\ldots+x_n}{n}, & s_x^2 = \dfrac{n\sum_{i=1}^{n}x_i^2-\left(\sum_{i=1}^{n}x_i\right)^2}{(n-1)n} \\ \overline{Y} = \dfrac{Y_1+\ldots+Y_n}{n}, & s_Y^2 = \dfrac{n\sum_{i=1}^{n}Y_i^2-\left(\sum_{i=1}^{n}Y_i\right)^2}{(n-1)n} \end{cases}$$

and the *new notation* (to be considered further later, under correlation)

$$r_{xY} = \frac{n\sum_{i=1}^{n} x_iY_i - \sum_{i=1}^{n} x_i \sum_{i=1}^{n} Y_i}{\sqrt{n\sum_{i=1}^{n}x_i^2-\left(\sum_{i=1}^{n}x_i\right)^2}\sqrt{n\sum_{i=1}^{n}Y_i^2-\left(\sum_{i=1}^{n}Y_i\right)^2}}$$

Remark: These expressions for the estimators $\hat{\alpha}$ and $\hat{\beta}$ are those which make the most *intuitive sense* when we study *correlation* later. (That applies to cases where x itself is also random, for example the dissolution rate of a chemical powder which cannot be set exactly, but only within 5 or 10 percent of the target value.)

For actual calculations one would use algebraic simplifications of these formulas, namely,

$$\hat{\beta} = \frac{\sum_{i=1}^{n} x_iY_i - \dfrac{\left(\sum_{i=1}^{n}x_i\right)\left(\sum_{i=1}^{n}Y_i\right)}{n}}{\sum_{i=1}^{n} x_i^2 - \dfrac{\left(\sum_{i=1}^{n}x_i\right)^2}{n}},$$

$$\hat{\alpha} = \overline{Y} - \hat{\beta}\overline{x}.$$

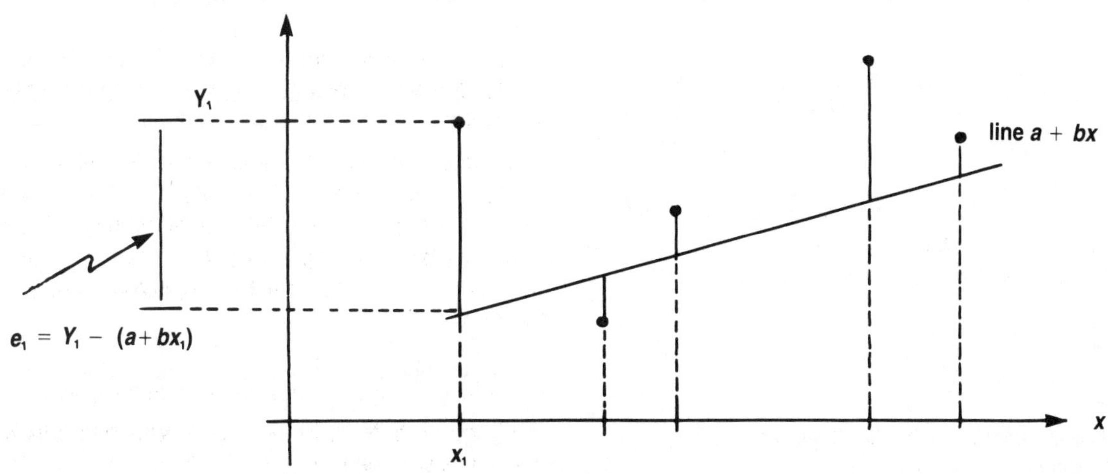

Figure 25

For the *relation with multiple linear regression* (useful in cases where Y, e.g. process yield, is a function of two or more variables which the experimenter sets, such as amount of catalyst *and* dissolution rate) to be clear later, it is useful to note that $\hat{\alpha}$ **and** $\hat{\beta}$ **are also expressible as the solution of the simultaneous linear equations**

$$\begin{cases} na + b\sum_{i=1}^{n} x_i = \sum_{i=1}^{n} Y_i \\ a\sum_{i=1}^{n} x_i + b\sum_{i=1}^{n} x_i^2 = \sum_{i=1}^{n} x_i Y_i . \end{cases}$$

Gauss-Markov Optimality Theorem: It can be seen that $\hat{\alpha}$ and $\hat{\beta}$ are linear functions of $Y_1, \ldots, Y_n$. Of all linear functions which are right on the average (i.e. for which $E(\hat{\alpha}) = \alpha$ and $E(\hat{\beta}) = \beta$), these least-squares estimators have the smallest variances. [This assumes $E(Y_i) = \alpha + \beta x_i$ with $Var(Y_i) = \sigma^2$ for $i = 1, 2, \ldots, n$. No specific distribution is assumed for $Y_1, Y_2, \ldots, Y_n$.] Thus, $\hat{\alpha}$ and $\hat{\beta}$ are *most reliable* (in the sense of having the smallest chance variation).

Graphical Interpretation (Figure 25):

$\hat{\alpha}$ and $\hat{\beta}$ *are* the a and b values which minimize $\sum_{i=1}^{n} e_i^2$.

Example (Analytical Chemistry). The following measurements of absorbance (Y) were made at various fixed micrograms (μg) x of a certain compound in a standard solution medium

x_i	1	3	5	7	10
Y_i	.032	.106	.179	.243	.333

Example 4

What is the best linear relationship $a + bx$ for estimating Y from x? Here we can calculate (with $n = 5$)

$$\sum_{i=1}^{5} x_i = 26 \quad , \bar{x} = 5.2$$

$$\sum_{i=1}^{5} Y_i = .893 \ , \bar{Y} = .1786$$

$$\sum_{i=1}^{5} x_i Y_i = 6.276$$

$$\sum_{i=1}^{5} x_i^2 = 184 \ .$$

Hence

$$\hat{\beta} = \frac{(6.276) - \dfrac{(26)(.893)}{5}}{184 - \dfrac{(26)^2}{5}} = .0334508196$$

$$\hat{\alpha} = .1786 - (.0334508196)(5.2) = .0046557377$$

and our fitted line predicts Y by

$$.0047 + .0335x.$$

Example 4 is a *good* one, in that the fit is "obviously good." However, it is a *largely trivial* one in that just about anyone who plotted the data carefully could put a ruler to the paper and come up with about the same line without using any statistical techniques. The *reason* for this is that the variability of Y's at any fixed x value here is very small (though it is present).

Remark. In actual use, this line is "read inversely" for what is called the inverse linear regression problem, or the calibration problem: We extract a sample from a lot of submitted chemical, calculate the absorbance, and solve to find the unknown percent of iron x in the lot:

$$x = \frac{Y - \alpha}{\beta} = \frac{Y - .0047}{.0335} .$$

$$\hat{\alpha} \quad \hat{\beta}$$

This is the **classical solution** to estimating x. If we find $Y = .20$ for a certain batch of chemical, we will say its μg of iron is $x = 5.840$.

We should note, however, that it has been shown [by Krutchkoff in *Technometrics* in 1967] that for this inverse problem, the **inverse solution** is often superior* (it has a smaller squared error on the average, i.e., smaller $E((\text{est. of } x) - x)^2)$. This method chooses c and d to minimize

*However, for most uses the direct solution is preferable, as it allows one to obtain a confidence interval for x at Y, while the inverse solution does not.

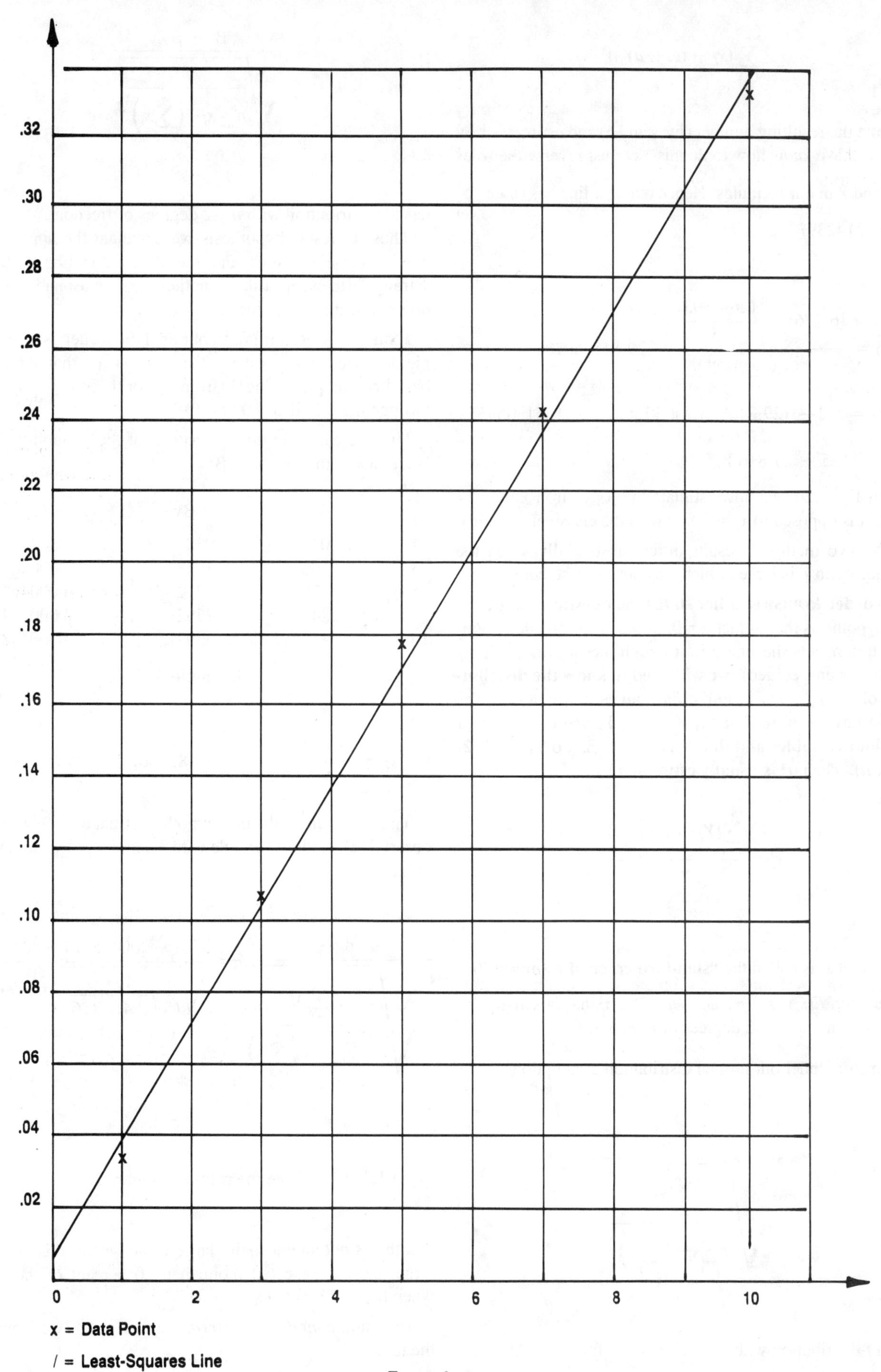
.32
.30
.28
.26
.24
.22
.20
.18
.16
.14
.12
.10
.08
.06
.04
.02
0
2
4
6
8
10
x = Data Point
/ = Least-Squares Line
Example 4

$$\sum_{i=1}^{n} (x_i - (c + dY_i))^2$$

(denote the resulting numbers by $\hat{\gamma}$ and $\hat{\delta}$) and estimates x by $\hat{\gamma} + \hat{\delta} Y$. We *know* how to do this . . . just reverse the roles of x and Y in our formulas. Hence our new line has (note: $\sum_{i=1}^{5} Y_i^2 = .214239$)

$$\begin{cases} \hat{\delta} = \dfrac{(6.276) - \dfrac{(26)(.893)}{5}}{.214239 - \dfrac{(.893)^2}{5}} = 29.81596078, \\ \hat{\gamma} = 5.2 - (29.81596078)(.1786) = -.1251305955 \end{cases}$$

i.e., $-.125 + 29.816\, Y$.

When $Y = .20$, we now estimate the μg of iron x by $x = 5.838$ (as opposed to $x = 5.840$ with the classical method).

The two methods' results differ substantially when the variability in Y is large, which was not the case here.

In order to answer other statistical questions (e.g., test the hypothesis that in fact $\alpha = 0$; in the drop spreader example, this means the line goes through the origin, while its slope β is unspecified), **we will need to know the distribution of $Y_1, Y_2, \ldots, Y_n$ and obtain an estimate of σ^2.**

We now assume that $Y_1, Y_2, \ldots, Y_n$ are independent random variables and that Y_i is $N(\alpha + \beta x_i, \sigma^2)$ ($i = 1, 2, \ldots, n$). Then **σ^2 is usually estimated by**

$$s_e^2 = \frac{\sum_{i=1}^{n} [Y_i - (\hat{\alpha} + \hat{\beta} x_i)]^2}{n-2}.$$

(Note that s_e is called the **"standard error of estimate."**)

Theorem: $E(s_e^2) = \sigma^2$, and $(n-2)s_e^2/\sigma^2$ has a chi-square distribution with $n-2$ degrees of freedom.

Theorem: Under our normal distribution assumptions,

$$\frac{\hat{\alpha} - \alpha}{\sqrt{\dfrac{\sum_{i=1}^{n} x_i^2}{n\sum_{i=1}^{n} x_i^2 - \left(\sum_{x=1}^{n} x_i\right)^2}}\, s_e}$$

has a t-distribution with $n-2$ degrees of freedom. Also,

$$\frac{\hat{\beta} - \beta}{\sqrt{\dfrac{n}{n\sum_{i=1}^{n} x_i^2 - \left(\sum_{x=1}^{n} x_i\right)^2}}\, s_e}$$

has a t-distribution with $n-2$ degrees of freedom.

Thus, to test a hypothesis we calculate the appropriate one of these statistics and see if the resulting value is "strange" (then reject the hypothesis) or "reasonable" (then do not reject the hypothesis).

Example 5. Returning to our analytical chemistry example, let us test the hypothesis that $\alpha = 0$ (i.e., that in fact the best line for predicting Y from x, for $1 \leq x \leq 10$, goes through the origin ($x = 0$, $Y = 0$)).

First we calculate our estimate s_e^2 of the variability (variance) about the line $\alpha + \beta x$.

X_i	Y_i	$\hat{\alpha} + \hat{\beta} x_i$	$[Y_i - (\hat{\alpha} + \hat{\beta} x_i)]^2$
1	.032	.0382	.00003844
3	.106	.1052	.00000064
5	.179	.1722	.00004624
7	.243	.2392	.00001444
10	.333	.3397	.00004489

Example 5

$$s_e^2 = \frac{.00014465}{5-2} = .0000482166,\ s_e = .00694382$$

Thus, here under the null hypothesis (that $\alpha = 0$) we have observed a t random variable with 3 degrees of freedom with value

$$\frac{\hat{\alpha} - \alpha}{\sqrt{\dfrac{\sum_{i=1}^{5} x_i^2}{5\sum_{i=1}^{5} x_i^2 - \left(\sum_{i=1}^{5} x_i\right)^2}}\, s_e} = \frac{.00465574 - 0}{\sqrt{\dfrac{184}{(5)(184) - (26)^2}}(.00694382)}$$

$$= .7721.$$

From Table 4, we see the results in Figure 26.

that this is not an unusually large value: values this large or larger can be expected more than 20 percent of the time when the null hypothesis is true.

To obtain *confidence intervals* (e.g., for α), we "invert" the tests:

$$P\left[-t_{.025} \leq \frac{\hat{\alpha} - \alpha}{\sqrt{\dfrac{\sum_{i=1}^{n} x_i^2}{n \sum_{i=1}^{n} x_i^2 - \left(\sum_{i=1}^{n} x_i\right)^2}}\, s_e} \leq t_{.025}\right] = .95$$

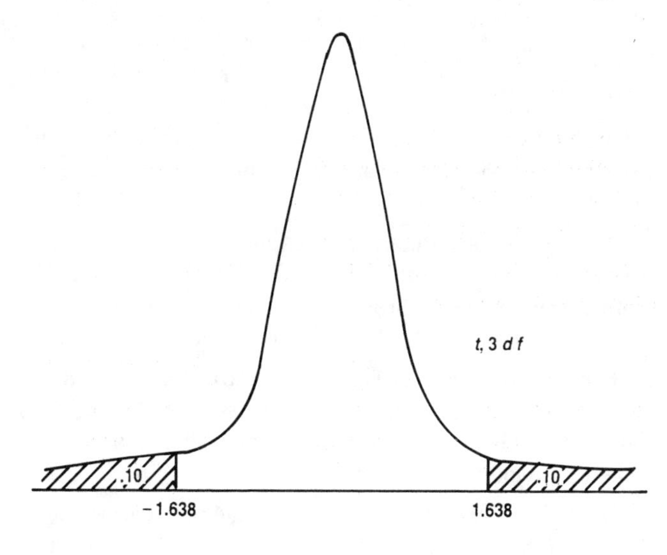

Figure 26

means

$$P\left[\hat{\alpha} - t_{.025} \sqrt{\frac{\sum_{i=1}^{n} x_i^2}{n \sum_{i=1}^{n} x_i^2 - \left(\sum_{i=1}^{n} x_i\right)^2}}\, s_e \leq \alpha \leq \hat{\alpha} + t_{.025} \sqrt{\frac{\sum_{i=1}^{n} x_i^2}{n \sum_{i=1}^{n} x_i^2 - \left(\sum_{i=1}^{n} x_i\right)^2}}\, s_e\right] = .91$$

In the analytical chemistry example, our 95 percent confidence interval on α is

$$.0046 \pm (3.182)\,(.0060299325) = .0046 \pm .0192,$$

i.e. $-\ .0146 \leq \alpha \leq .0238$.

Often, we will want **a confidence interval for** $\alpha + \beta x^*$ (at a specific value $x = x^*$). Since $\hat{\alpha} + \hat{\beta}x^*$ is an unbiased estimator for $\alpha + \beta x^*$ and

$$\text{Var}\,(\hat{\alpha} + \hat{\beta}x^*) = \sigma^2 \left(\frac{1}{n} + \frac{(x^* - \bar{x})^2}{(n-1)s_x^2}\right),$$

$$\hat{\alpha} + \hat{\beta}x^* \pm t_{.025}\, s_e \sqrt{\frac{1}{n} + \frac{(x^* - \bar{x})^2}{(n-1)s_x^2}}$$

is a 95 percent confidence interval for $\alpha + \beta x^*$. (The $t_{.025}$ used here is for $n-2$ degrees of freedom.)

If we plot the confidence interval for $\alpha + \beta x^*$ for all x^*, the result is called **a confidence band (for the regression line)** (see Figure 27).

For any x you choose, from such a graph it is immediately easy to find a 95 percent confidence interval for $E(Y)$ at that x. Note that *as n (the number of observations we have made) grows, the length of this confidence interval on $\alpha + \beta x$ goes to zero.*

A related, but different, interval is sometimes desired. For example, we may wish to know: **If we set $x = x^*$, within what limits can we be 95 percent sure Y will lie?** (The confidence interval on $\alpha + \beta x^*$ answers this question for $E(Y)$, *not* for Y.) **The resulting answer is called a *prediction interval:***

$$\hat{\alpha} + \hat{\beta}x^* \pm t_{.025}\, s_e \sqrt{1 + \frac{1}{n} + \frac{(x^* - \bar{x})^2}{(n-1)s_x^2}}$$

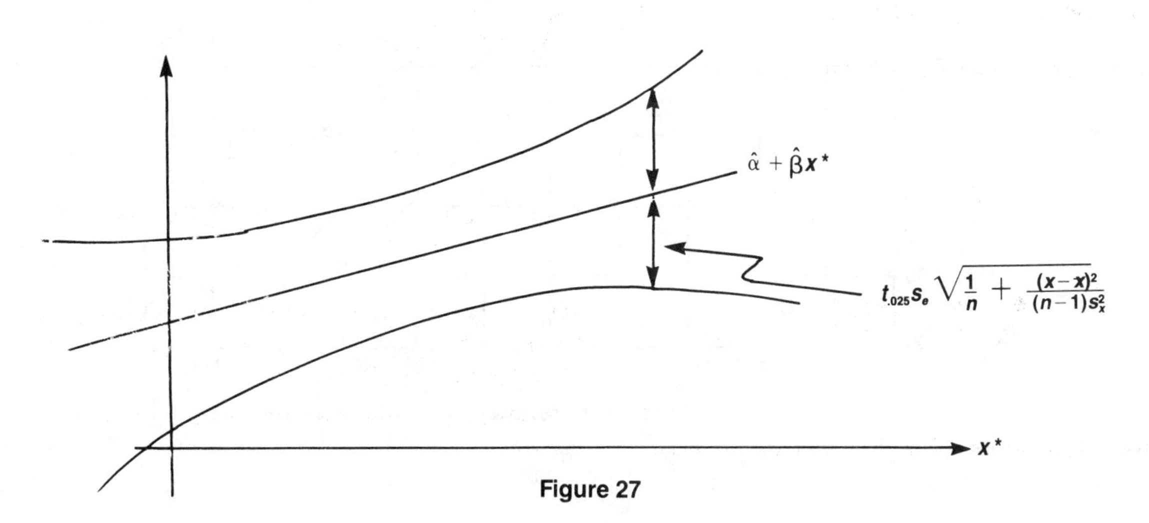

Figure 27

is a "95% confidence" interval for Y at x^*. (The $t_{.025}$ used here is for $n-2$ degrees of freedom.)

If we plot this interval for all x^*, the result is called a *prediction band*. From such a plot we can see (for any x^* we might consider) where 95% of our Y's produced at that x^* will lie. Note that as n grows, the length of the prediction interval does *not* go to zero, since future Y values will be variable.

Example. A survey of agricultural yield against the amount of a pest allowed (the latter can be controlled by crop treatment) has produced the data shown in Example 7.

Example 6: Pest Rate (x_i) and Crop Yield (Y_i) Data

i	1	2	3	4	5	6	7	8	9	10	11	12
x_i	8	6	11	22	14	17	18	24	19	23	26	40
Y_i	59	58	56	53	50	45	43	42	39	38	30	27

Find $\hat{\alpha}$ and $\hat{\beta}$. Plot the regression line with confidence bands and prediction bands (each at the 95 percent confidence level), and with the data points superimposed on the graph. If the pest rate is allowed to go to $x^* = 50$ next year, within what limits can we be 95 percent sure our crop yield will fall?

Solution. Calculations here yield $\hat{\alpha} = 64.247$ and $\hat{\beta} = -1.0130$, so the regression line is estimated as $y = 64.247 - 1.0130x$. The plot is shown in Figure 28.

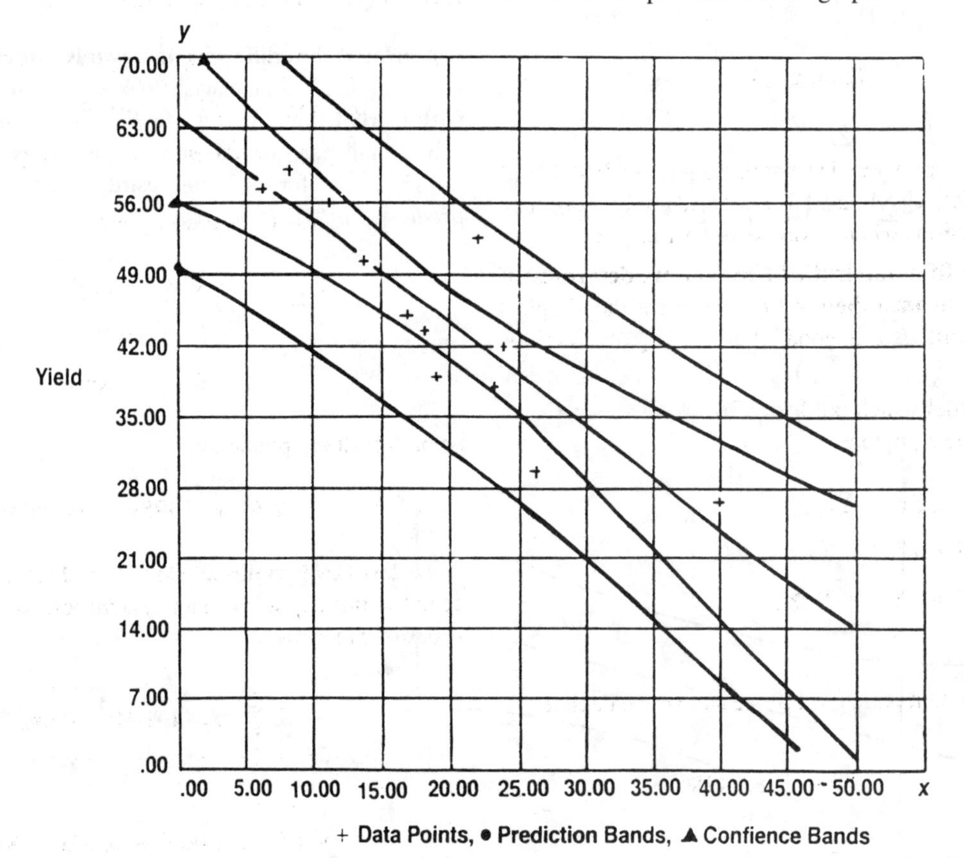

Figure 28

and shows that if we let our pest reach level $x^* = 50$, we can be 95 percent sure our yield will be between 0 and 31.5 units. (The mean yield at $x^* = 50$ is between 1.4 and 26.3 units with 95 percent confidence.)

Note that if one extrapolates outside the range where the experiment was run, $(x^* - \bar{x})^2$ grows large and the interval becomes wide rapidly. (Intuitively, this is due to greater "wobble" of the fitted line as we move away from $\bar{x}$.) Thus, "mis-specification" of the range where the experiment is run is as serious as a large inherent variability (as estimated by s_e^2).

An additional danger of extrapolation is that the response may not be linear (or quadratic, etc.) outside the range where the experiment was run, and we have no way to check that without further experimentation.

For the inverse regression (or calibration) problem, we noted (p. 430) that with the classical solution it is possible to **obtain a confidence interval for x after observing Y,** and will now detail this aspect.

Here if one finds $Y = 35.00$, for example, and wants to put a 95% confidence interval on x (which is now unknown), one proceeds as follows. Draw a horizontal line at height 35.00 on the prediction band graph. Draw vertical lines at the points where this horizontal line intersects the prediction bands. The interval of x-values between the points where these two vertical lines hit the x-axis is our confidence interval for x.

Curvilinear regression. It is not always the case that we will wish to fit $\alpha + \beta x$ to Y. Sometimes we will know some other $f(x)$ is appropriate. The theory for $\alpha + \beta x$ can be easily used to fit *some* **other important functions**, such as **exponential** to fit $\gamma\delta^x$ to Y, fit $\ln(\gamma) + (\ln\delta)x$ to $\ln(Y)$, i.e., fit $\alpha + \beta x$ to z (where $\alpha = \ln(\gamma)$, $\beta = \ln(\delta)$, $z = \ln(Y)$). Similarly one can fit $\dfrac{1}{\alpha+\beta x}$ to Y, and γx^β to Y.

If the functional form is unknown, a polynomial can approximate it well: fit

$$\beta_0 + \beta_1 x + \beta_2 x^2 + \ldots + \beta_p x^p \text{ to } Y$$

(for some p, as small as can be used and still obtain a good fit). The cases $p = 0,1,2$, and sometimes 3, are most often used in practice. Calculations require judgment, matrices, and computers.

Example. In our analytical chemistry example, two additional observations are available (see Example 7):

x_i	12	15
Y_i	.387	.463

Example 7

At these x's, our previous least-squares line predicts Y's of .4067 and .5072, which are off by considerably more than were our predictions when x was between 1 and 10.

How can we fit a quadratic? How can we determine if it gives a significantly better fit? (We want the simplest model which will give a good fit—i.e., a **parsimonious model.**)

To fit the model which predicts Y by $\alpha + \beta x + \gamma x^2$, we need to solve the equations

$$\begin{cases} na + b\Sigma x_i + c\Sigma x_i^2 = \Sigma Y_i \\ a\Sigma x_i + b\Sigma x_i^2 + c\Sigma x_i^3 = \Sigma x_i Y_i \\ a\Sigma x_i^2 + b\Sigma x_i^3 + c\Sigma x_i^4 = \Sigma x_i^2 Y_i. \end{cases}$$

Using *all seven* data pairs (x_i, Y_i), we can now calculate

$$\sum_{i=1}^{7} x_i = 53, \qquad \bar{x} = 7.571428571$$

$$\sum_{i=1}^{7} Y_i = 1.743, \qquad \bar{Y} = .249$$

$$\sum_{i=1}^{7} x_i Y_i = 17.865$$

$$\sum_{i=1}^{7} x_i^2 = 553, \qquad \sum_{i=1}^{7} Y_i^2 = .578377$$

$$\sum_{i=1}^{7} x_i^3 = 6599, \qquad \sum_{i=1}^{7} x_i^2 Y_i = 210.571$$

$$\sum_{i=1}^{7} x_i^4 = 84469$$

The *line* fitted to all 7 points would therefore have

$$\hat{\beta} = \frac{17.865 - \dfrac{(53)(1.743)}{7}}{553 - \dfrac{(53)^2}{7}} = .0307683615$$

$$\hat{\alpha} = .249 - (.03076836)(7.571428571) = .01603955,$$

i.e. the line is $.0160 + .0308x$.

The *quadratic* comes from tedious (and error-prone, because of build-up of inaccuracy due to use of a limited number of significant digits) computations. Modern computing facilities have computer programs capable of solving this (and much more complex problems) accurately, and it is recommended these be used. The system

$$\begin{cases} 7a + 53b + 553c = 1.743 \\ 53a + 553b + 6599c = 17.865 \\ 553a + 6599b + 84469c = 210.571, \end{cases}$$

solved by a pocket calculator, yields

$$c \quad \hat{\gamma} = -.00055$$
$$b \quad \hat{\beta} = .0395$$
$$a \quad \hat{\alpha} = -.0067,$$

so our quadratic prediction is

$$-.0067 + .0395x - .00055x^2.$$

To test the hypothesis that $\gamma = 0$ (that the relation is actually linear, rather than quadratic), we calculate the *residual* variances

$$\hat{\sigma}_1^2 = \frac{\sum_{i=1}^{7} [Y_i - (.0160 + .0308x)]^2}{5}$$

$$\hat{\sigma}_2^2 = \frac{\sum_{i=1}^{7} [Y_i - (-.0067 + .0395x - .00055x^2)]^2}{4}$$

and calculate

$$F = \frac{\hat{\sigma}_1^2 - \hat{\sigma}_2^2}{\hat{\sigma}_2^2}$$

which (under the hypothesis) has an F-distribution with 1 and $7-3=4$ degrees of freedom.

Our calculations here yield the data found in Table 11.

Table 11

Y_i	x_i	$.0160 + .0308x_i$	$-.0067 + .0395x_i - .00055x_i^2$
.032	1	.0468	.03225
.106	3	.1084	.10685
.179	5	.1700	.17705
.243	7	.2316	.24285
.333	10	.3240	.3333
.387	12	.3856	.3881
.463	15	.4780	.4621

Hence

$$\hat{\sigma}_1^2 = .000148744,\ \hat{\sigma}_2^2 = .00000672$$

and

$$F = 21.13.$$

From our Table 6, we arrive at Figure 29,

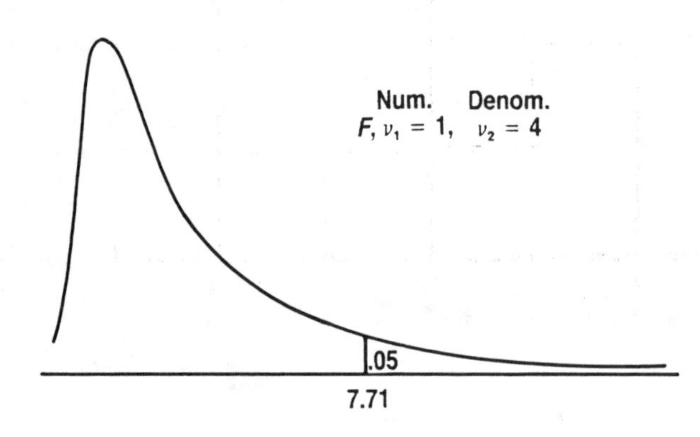

Figure 29

Also, Figure 30.

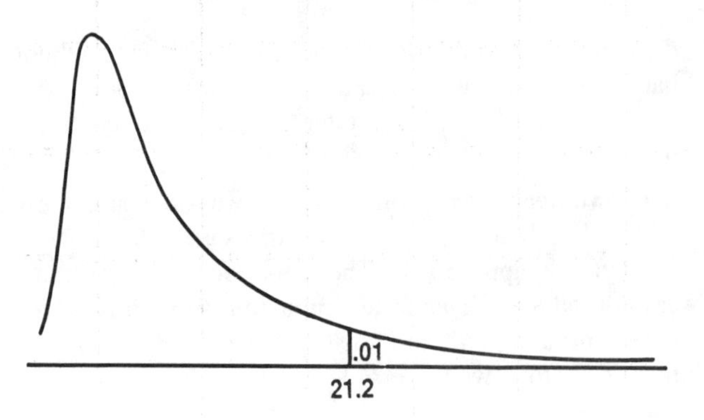

Figure 30

so we *reject* the hypothesis that $\gamma = 0$ at level .05. (At level .01 we almost reject.)

Note that **in regression design of experiments one studies the "best" x's at which to take observations Y. For a line,** it is best (in terms of later estimation of the line) to take $\frac{1}{2}$ the observations at each end of the interval of interest; **for a quadratic** one takes $\frac{1}{4}$ at each end and $\frac{1}{2}$ in the middle. However, this assumes (perhaps from past experimentation) a model valid and known over the range, and does not allow testing of the model.

Multiple regression is useful where several variables (such as amount x_1 of fertilizer and amount x_2 of seed) **are set before Y** (such as turf density) **is observed.** Then, in a simple case, we may model

$$E(Y) = \alpha + \beta x_1 + \gamma x_2.$$

An important problem in this setting is to efficiently experiment to find the (x_1,x_2) pair which maximizes $E(Y)$. We will touch on some aspects of this problem later.

So far, we have assumed that x is fixed while Y is random. Often, however, we may *try* to control x, but are unable to do so exactly (e.g., the goal may be to run a reaction at $x = 270°$F, but it is actually run at a random temperature X with $E(X) = 270°$F due to inherent variability.

1) **Which, if any, of our results still hold if X and Y are random?**
2) **What new wrinkles arise with X and Y both random?**

Example: We may want to measure Y, but instead measure X because the correlation is "high" but it is

- cheaper
- less time consuming (e.g., chemical assay *vs.* waiting a growing season)
- possible (Y may not be available until later)

to measure X.

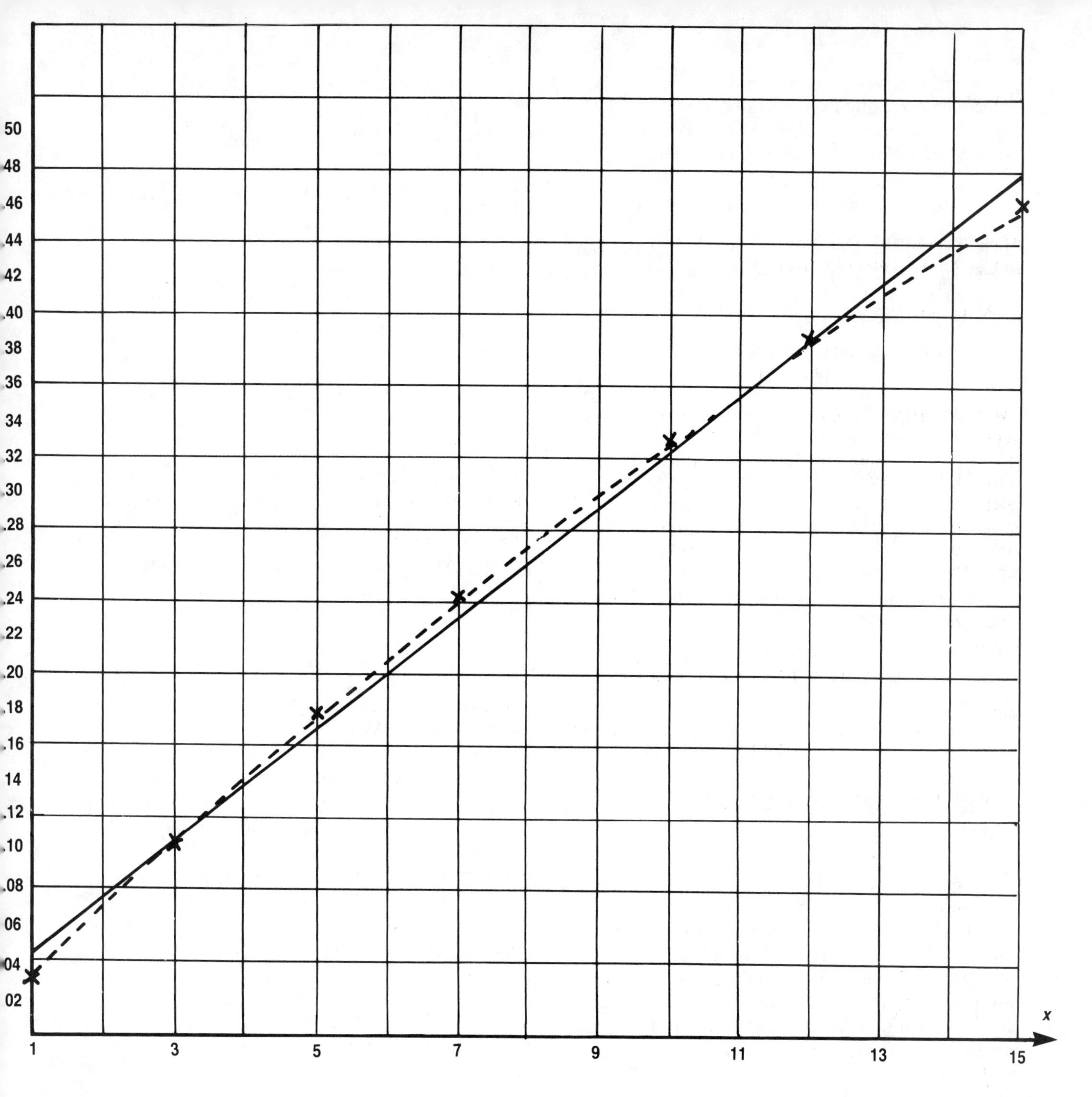

x = Data Point
/ = Least-Squares Line
/ = Least-Squares Quadratic

Figure 31

If X and Y are both random, we call the function $f(X)$ for which

$$E\,(Y - f(X))^2$$

is minimized, the regression curve of Y on X. The α and β for which

$$E\,(Y - (\alpha + \beta X))^2$$

is minimized are called the **regression intercept** and **regression slope** (of Y on X), respectively; the corresponding line is called the **regression line of Y on X,** $y = \alpha + \beta x$. This line has

$$\alpha = E(Y) - \rho(X,Y)\frac{\sigma(Y)}{\sigma(X)}E(X)$$

$$\beta = \rho(X,Y)\frac{\sigma(Y)}{\sigma(X)}$$

where $\sigma(Y) = \sqrt{\text{Var}(Y)}$, $\sigma(X) = \sqrt{\text{Var}(X)}$, and $\rho(X,Y)$ is called the **correlation coefficient** between X and Y. *If* we knew this α and β, which we usually *do not*, our mean squared error of prediction using a line would be

$$E\,(Y - (\alpha+\beta X))^2 = (1 - \rho^2(X,Y))\,\sigma^2\,(Y).$$

It is therefore said that "The square of the correlation coefficient is the proportion of the variance (of Y) accounted for by linear regression on X," and $E(Y - (\alpha+\beta X))^2$ is called the **residual variance.**

Note that the equation of this line is

$$\alpha + \beta X = \{E(Y) - \rho(X,Y)\frac{\sigma(Y)}{\sigma(X)}E(X)\} + \rho\,(X,Y)\frac{\sigma(Y)}{\sigma(X)}X$$

while before (with *fixed x*) we had

$$\hat{\alpha} + \hat{\beta}x = \{\overline{Y} - r_{xY}\frac{s_Y}{s_x}\overline{x}\} + r_{xY}\frac{s_Y}{s_x}x,$$

which has *essentially* replaced unknown quantities by their estimators.

The method of least squares can be used to estimate α and β, and *yields the $\hat{\alpha}$ and $\hat{\beta}$ studied already*. The quantity r_{xY} estimates ρ.

Difference in interpretation for X random and x fixed cases: correlation *vs.* causality (e.g. hospitals and death correlation).

In terms of expectations and variances, the *definition of* $\rho\,(X,Y)$ is

$$\rho\,(X,Y) = \frac{E(XY) - E(X)\,E(Y)}{\sqrt{\text{Var}(X)}\,\sqrt{\text{Var}(Y)}}.$$

We have already worked a lot with $E(X)$, $E(Y)$, $\text{Var}(X)$, and $\text{Var}(Y)$. The *new* quantity involved here is $E(XY)$, which **involves the joint probabilities** of X having a value (say 10) and Y having a value (say 8.5). In fact, if ρ is to measure dependence of Y on X, it must depend on such joint characteristics as "How often is $Y=8.5$ *when* $x=10$?" To illustrate ρ (and how it is a measure of *linear* dependence), let's look at an example.

Example. Suppose X and Y are such that

$X=1$ and $Y=1$	50% of the time
$X=3$ and $Y=9$	25% of the time
$X=4$ and $Y=13$	25% of the time
	100%

Then $E(X) = 1\cdot\frac{1}{2} + 3\cdot\frac{1}{4} + 4\cdot\frac{1}{4} = \frac{9}{4}$

$$E(X^2) = 1^2\cdot\frac{1}{2} + 3^2\cdot\frac{1}{4} + 4^2\cdot\frac{1}{4} = \frac{27}{4}$$

$$\text{Var}(X) = E(X^2) - [E(X)]^2 = \frac{27}{4} - \left(\frac{9}{4}\right)^2 = \frac{27}{16}$$

$$E(Y) = 1\cdot\frac{1}{2} + 9\cdot\frac{1}{4} + 13\cdot\frac{1}{4} = 6$$

$$E(Y^2) = 1^2\cdot\frac{1}{2} + 9^2\cdot\frac{1}{4} + 13^2\cdot\frac{1}{4} = 63$$

$$\text{Var}(Y) = E(Y^2) - [E(Y)]^2 = 63 - 36 = 27$$

$$E(XY) = (1)(1)\cdot\frac{1}{2} + (3)(9)\cdot\frac{1}{4} + (4)(13)\cdot\frac{1}{4} = \frac{81}{4}.$$

So

$$\rho(X,Y) = \frac{\frac{81}{4} - \frac{9}{4}\cdot 6}{\sqrt{\frac{27}{16}}\sqrt{27}} = \frac{81-54}{27} = \frac{27}{27} = 1.$$

Here the mean squared error of prediction is $(1-\rho^2\,(X,Y))\,\sigma^2\,(Y) = (1-1^2)\cdot 27 = 0$: $\alpha + \beta X$ predicts Y *perfectly* (see Example 8). $Y = -3 + 4X$. We can show **for all X,Y, $-1 \leq \rho\,(X,Y) \leq 1$.**

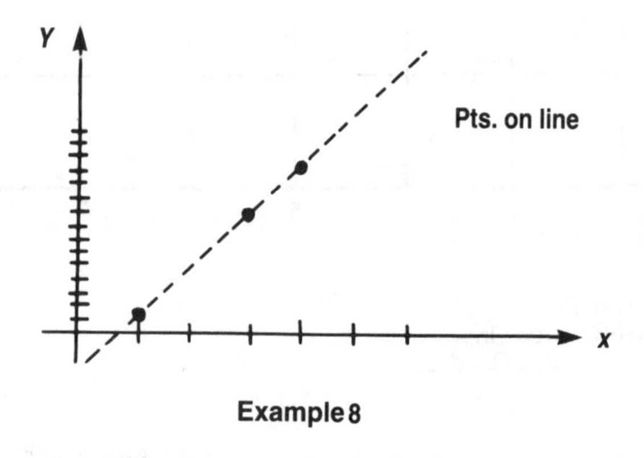

Example 8

RIDGE REGRESSION

As we have seen above (Figure 25), the "usual" estimators of regression coefficients are unbiased and (of all unbiased estimators) have the smallest variance (Gauss-Markov Optimality Theorem). **When multiple linear regression is applied to an existing data set, rather than to data from an experiment which was designed statistically, often problems of "multicollinearity" arise.** In those settings (and perhaps others also), **the technique of "ridge regres-**

sion" due to Hoerl and Kennard in 1970 may be useful in providing estimators of regression coefficients which have a smaller mean squared error (squared bias plus variance) than the "usual" estimators provided by least squares methods. (Hoerl and Kennard showed that in such settings the usual estimators tend to become too large in absolute value, and may even have the wrong sign.) More than 200 papers on this technique have appeared since 1970, and they are cited and summarized in "Ridge Regression—1980, Advances, Algorithms, and Applications" by A. E. Hoerl and R. W. Kennard, *American Journal of Mathematical and Management Sciences*, Vol. 1 (1981), pp. 5–83. The basic ridge regression theory, advances in extending it, and algorithms and their use, are summarized in the following extracts from Hoerl and Kennard (1981).

1. Basic Ridge Regression

The standard model for multiple linear regression will be denoted by

$$y = X\beta + \epsilon \tag{5}$$

where X is nxp and rank p, β is pxl, $E[\epsilon] = 0$, and $E[\epsilon\epsilon'] = \sigma^2 I_n$. **It is assumed that $X'X$ is in the form of a correlation matrix;** one ensures this **by scaling the variables X_i so that each has sum of squares one.** If a distributional assumption is needed, it is assumed that ϵ is multivariate normal. The model (1) can be reduced to a form orthogonal in the factors by using $X = P\Lambda^{1/2}Q'$, the singular value decomposition. The matrix Q is pxp and consists of the orthonormalized eigenvectors of $X'X$. The matrix P consists of the orthonormalized eigenvectors associated with the p nonzero eigenvalues of $X'X$. The eigenvalues of $X'X$ are given by $\Lambda = \text{diag}(\lambda_1, \lambda_2, \ldots, \lambda_p)$.

Let $Z = XQ$ and $\alpha = Q'\beta$. Since $X'X = Q\Lambda Q'$, the transformation $Z = XQ$ is a transformation to principal components when the eigenvalues are ordered so that $\lambda_1 \geq \lambda_2 \ldots \geq \lambda_p$. The matrix Q orients the principal axes with respect to the original axes defined by X. Since $Z = P\Lambda^{1/2}$, the matrix P gives the coordinates with respect to the principal axes normalized so that each column of P has unit length. The singular values, $\sqrt{\lambda_i}$, give the lengths of the principal axes; the length of the ith principal axes is $2\sqrt{\lambda_i}$.

Then (5) becomes

$$y = Z\alpha + \epsilon. \tag{6}$$

The least-squares estimator of α is

$$\hat{\alpha} = (Z'Z)^{-1}Z'Y = \Lambda^{-1}Z'y. \tag{7}$$

The generalized ridge estimator is defined by

$$\hat{\alpha}^*(K) = [\Lambda + K]^{-1}Z'y \tag{8}$$

where $K = \text{diag}(k_1, k_2, \ldots, k_p)$, $k_i \geq 0$. If $K = kI_p$, $k \geq 0$, then the estimator is defined to be the **simple ridge estimator** or just ridge estimator and

$$\hat{\alpha}^*(k) = [\Lambda + kI_p]^{-1}Z'y. \tag{9}$$

From (7) $Z'y = \Lambda\hat{\alpha}$. Hence (8) and (9) can be written as $[\Lambda + K]^{-1}\Lambda\hat{\alpha}$ and $[\lambda + kI_p]^{-1}\Lambda\hat{\alpha}$, respectively, and the **ridge estimators are seen to be "shrunken" least-squares estimators.**

Using the definitions of Z, α, etc., (8) and (9) are readily transformed to β form as follows:

$$\hat{\beta}^*(K) = [X'X + QKQ']^{-1}X'y \tag{10}$$

$$\hat{\beta}^*(k) = [X'X + kI_p]^{-1}X'y. \tag{11}$$

In Hoerl and Kennard (1970a,b) attention was focused on (11); subsequent work has had the same focus. Define the mean square error (*MSE*) of estimation as

$$E[L_1^2] = E[(\hat{\beta}^* - \beta)'(\hat{\beta}^* - \beta)] = E[(\hat{\alpha}^* - \alpha)'(\hat{\alpha}^* - \alpha)] \tag{12}$$

Then **the heart of Hoerl and Kennard (1970a) is that there always exists a $k > 0$ such that the *MSE* for the ridge estimator is smaller than that for least squares.**

2. Extensions of the Theory

Hoerl and Kennard (1970a) stated that the existence theorem relative to k could be extended to the prediction mean square error, namely to

$$\begin{aligned} E[L_2^2] &= E[(\hat{y}^* - E[y])'[(\hat{y}^* - E[y])] \\ &= E[(\hat{\beta}^* - \beta)'X'X(\hat{\beta}^* - \beta)]. \end{aligned} \tag{13}$$

An important generalization of the existence theorems is given in Theobald (1974). In summary the paper states the following: Let B be any non-negative definite matrix of proper dimension, and let $\tilde{\beta}$ be any estimator of β; then consider the *MSE*

$$E[(\tilde{\beta} - \beta)'B(\tilde{\beta} - \beta)]. \tag{14}$$

Theobald showed that a sufficient condition for $\hat{\beta}^*(k)$ to have a smaller *MSE* than the least squares estimator, $\hat{\beta}$, is that $k < 2\sigma^2/(\beta'\beta) = 2\sigma^2/(\alpha'\alpha)$. The result follows from a theorem which states that the following conditions are equivalent: (i) $M_1 - M_2$ is non-negative definite; (ii) $m_1 - m_2 \geq 0$ for all non-negative definite B. Here $\tilde{\beta}_1$ and $\tilde{\beta}_2$ are two estimators of β and

$$\begin{aligned} M_j &= E[(\tilde{\beta}_j - \beta)(\tilde{\beta}_j - \beta)']\; j = 1,2, \\ m_j &= E[(\tilde{\beta}_j - \beta)'B(\tilde{\beta}_j - \beta)]\; j = 1,2. \end{aligned} \tag{15}$$

Another important insight into ridge regression and its

relationship to other regression estimation procedures is given in Hocking, Speed, and Lynn (1976). With the orthogonal form (6) as the base, they define a class of estimators by

$$\hat{\alpha}_b^* = B\hat{\alpha} \tag{16}$$

where $B = \text{diag}(b_1, b_2, \ldots, b_p)$, $0 \leq b \leq 1$. The class defined by B is, therefore, all those estimators obtained by shrinking one or more of the components of $\hat{\alpha}$. The *MSE* for $\hat{\alpha}_b^*$ is

$$E[(\hat{\alpha}_b^* - \alpha)' (\hat{\alpha}_b^* - \alpha)] = \sigma^2 \Sigma b_i^2/\lambda_i + \Sigma \alpha_i (b_i - 1)^2. \tag{17}$$

The dominant position of generalized ridge is exhibited when **the values of the b_i that give minimum *MSE* are determined.** Setting $\vartheta E/\vartheta b_i = 0, i = 1, \ldots, p$ gives

$$b_i = \lambda_i/(\lambda_i + \sigma^2/\alpha_i^2). \tag{18}$$

This is **recognized as the form for generalized ridge.** The matrix B is $\Lambda(\Lambda + K)^{-1}$ where $K = \text{diag}(k_1, k_2, \ldots, k_p)$. Here $k_i = \sigma^2/\alpha_i^2$ which are the values which give minimum *MSE* for generalized ridge. All other shrinkage forms impose restrictions on the b_i and, hence, cannot achieve a guaranteed global minimum for the mean square error functions. Other shrinkage forms include common shrinkage (each component has the same shrinkage multiplier), principal components, fractional rank, and simple ridge.

Simple ridge estimators have the same mathematical form as the posterior mean in a Bayesian analysis—if normality is assumed in (1) and there is *a prior* distribution on the regression coefficients that is normal with zero means, common variance σ^2/k, and zero covariances. Similar statements can be made for generalized ridge estimators.

3. Choice of the Biasing Parameter

In Hoerl and Kennard (1968), **three methods for choosing a value for the biasing parameter, *k*,** were outlined: **(i)** examination of the ridge trace, **(ii)** use of $k = p\hat{\sigma}^2/(\hat{\beta}'\hat{\beta})$ for simple ridge, and **(iii)** iteration on $k_I = \hat{\sigma}^2/\hat{\alpha}_i^2$ for generalized ridge. However, many questions were left unanswered and no statistical properties of any method of choice were given. Subsequently **a number of authors have attacked the problem of a method for the choice of a value of *k*;** they have also attempted to determine some of the statistical properties of their methods.

3.1 Simple Ridge Regression

In simple ridge regression the intractable analytic problems in determining an "optimum" choice for a value for k and the inherently unmanageable distributional problems with reasonable methods of choice have forced all authors to resort to simulation. In this section the results of four primary studies to date are summarized.

$\boldsymbol{k = k_a = p\hat{\sigma}^2/(\hat{\beta}'\hat{\beta}) = p\hat{\sigma}^2/(\hat{\alpha}'\hat{\alpha})}$. Hoerl, Kennard, and Baldwin (1975) published the results of an extensive simulation aimed at determining the properties of the algorithm $k_a = p\hat{\sigma}^2/(\hat{\beta}'\hat{\beta})$ for selecting the value for the biasing parameter k. This simulation demonstrated the existence of at least one algorithm that could be used routinely to ensure that the undesirable properties of least-squares estimates could be overcome. The primary results of the simulation were:

- The use of the ridge estimator with biasing parameter $k_a = p\hat{\sigma}^2/(\hat{\beta}'\hat{\beta})$ has a probability greater than 0.5 of producing estimates with a smaller *MSE* than least squares.
- The probability of a smaller *MSE* using k_a increases as p, the dimension of the factor space, becomes larger.
- For a given size p, the probability of a smaller *MSE* using k_a increases as the spread in the eigenvalues of $X'X$ increases, that is, when $X'X$ is less well-conditioned.

Iteration—$\boldsymbol{k_{at}}$. Hoerl and Kennard (1976) followed the k_a results with the results for an estimator labeled k_{at}. The algorithm for computing k_{at} is based on an iterative procedure that begins with k_a. The rationale for k_{at} is as follows: It was shown in Hoerl and Kennard (1970a) that $E[\hat{\beta}'\hat{\beta}] = \beta'\beta + \sigma^2 \text{Trace}[X'X]^{-1}$. Thus, $\hat{\beta}'\hat{\beta}$ is larger than $\beta'\beta$ and the poorer the conditioning of $X'X$, the larger the difference between $\hat{\beta}'\hat{\beta}$ and $\beta'\beta$. The squared length of the ridge estimator based on k_a, i.e., $\hat{\beta}^*(k_a)$, is going to be smaller than $\hat{\beta}'\hat{\beta}$. It seems reasonable to expect an improvement if the $\hat{\beta}$ in k_a is replaced by $\hat{\beta}^*(k_a)$. Clearly, one could consider iteration using this same strategy.

The algorithm is based on a sequence of estimates of β and values of k as follows: $\hat{\beta}$, $k_{ao} = p\hat{\sigma}^2/(\hat{\beta}'\hat{\beta})$; $\hat{\beta}^*(k_{ao})$, $k_{a1} = p\hat{\sigma}^2/([\hat{\beta}^*(k_{ao})]'[\hat{\beta}^*(k_{ao})])$; $\hat{\beta}^*(k_{a1})$, $k_{a2} = p\hat{\sigma}^2/([\hat{\beta}^*(k_{a1})]'[\hat{\beta}^*(k_{a1})])$; . . .; $\hat{\beta}^*(k_{at})$. The sequence is terminated when **(δ-criterion)**

$$(k_{a,i+1} - k_{a,i})/k_{a,i} \leq \delta = 20T^{1.30}$$

where $T = \text{Trace}(X'X)^{-1}/p$.

The simulation produced the following results:

- A significant reduction in *MSE* can be made using the δ-criterion and making more than one iteration when the criterion calls for it. The improvement in *MSE* becomes greater as the value of T increases, that is, as $X'X$ is less well-conditioned.
- The δ-criterion not only produces a smaller *MSE* than least squares or a single iteration, but it also produces an error distribution with a smaller standard deviation.
- The use of the ridge estimator based on the δ-criterion has a probability greater than 0.5 of producing estimates with a smaller *MSE* than least squares. The fraction of the time that the δ-criterion produces an *MSE* less than least squares is somewhat smaller than

the single iteration estimate (k_a); however, this is more than offset by the smaller *MSE*.

$k = k_b = p\hat{\sigma}^2/\Sigma\hat{\alpha}_i^2\lambda_i = 1/F$. Shortly after the Hoerl and Kennard publication of the k_{at} simulation results, Lawless and Wang (1976) published simulation results based on using estimates where the value of the biasing parameter is computed from $k_b = p\hat{\sigma}^2/\Sigma\hat{\alpha}_i^2 \lambda_i = 1/F$, where λ_i are the eigenvalues of $X'X$ and F is the ratio of the regression mean square to the residual mean square. Lawless and Wang used the same simulation strategy as that used to assess k_a and k_{at}. The results are similar to those obtained for k_{at}, and the conclusions stated there can be stated for k_b.

RIDGM-k_d. Dempster, Schatzoff, and Wermuth (1977) published a simulation study of 56 alternatives to ordinary least-squares estimation in multiple regression. The best of these alternatives (called RIDGM), based on a Bayesian interpretation of the mathematical form of simple ridge, leads to a value of k, say k_d, determined from the solution of the nonlinear equation

$$(1/\hat{\sigma}^2) \sum_{i=1}^{p} \hat{\alpha}_i [(1/k_d) + (1/\lambda_i)]^{-1} - p = 0.$$

The simulation conducted in this study is different from those referenced in the k_a, k_{at}, and k_b descriptions above, so that a direct comparison cannot be made. In a comment to the study, Hoerl did two small simulations that allow a direct comparison of k_a, k_{at}, and k_d. These limited results give an indication that the performance of k_d may be similar to that of k_{at}.

The four algorithms given for selecting a value of k in a particular problem have all been shown to perform appreciably better than least squares across a broad spectrum of regression problems. The three estimators based on k_{at}, k_b, and k_d are all better than the one based on k_a. For k_{at}, k_b, and k_d, more information is needed to permit a definitive recommendation. The estimator based on $k_b = 1/F$ is the simplest to compute; the other two need either a number of iterations or the solution of a nonlinear equation. One has to be cautious about k_b when there is a substantial signal-to-noise ratio, that is, when the F-ratio is overwhelmingly significant or when the number of factors is small; in this case, the performance can be poorer than least squares. The RIDGM estimator shows evidence of the same characteristics. Clearly, there is more work that must be done. Today the computation cost for regression analysis is modest. A reasonable strategy would be to compute at a sufficient number of values of k to obtain a good characterization of the ridge trace; k_{at}, k_b, and k_d should be included among these values. The trace will give an indication of sensitivity.

The computations for ridge regression are relatively straightforward and **can be made with simple modifications to a standard linear regression program.** Some background on the computations is given by Evans (1973), Bolding and Houston (1974), and Gunst (1979). **There are also generally available programs for ridge computations;** some are specific for ridge while others provide subroutines or special functions for more general programs. *Specific ridge programs* are described by Bradley and McGann (1977), Jain, Mahajan, and Bergier (1977), Hoerl (1979), Hui and Jagpal (1979), and Bush (1980). Two widely used commercial programs for statistics are the Biomedical Computer Program (BMDP) from the Health Sciences Computing Facility, University of California, Los Angeles, and the Statistical Analysis System (SAS) from the SAS Institute, Cary, North Carolina. Hill (1975) gives an example of the *use of one program in the* ***BMDP*** *system to compute the regression coefficients for different values of k to enable the construction of a ridge trace*. Carmer and Hsieh (1979), Sinha and Hardy (1979), and Rogers and Hildebrand (1980) all describe **SAS macros** *for ridge regression computations*. Available through the Massachusetts Institute of Technology is TROLL, an interactive, time-shared system with capabilities for statistical research. The facility is connected to a national data-communication network. The *ridge capability* is described in MIT (1975).

3.2 Generalized Ridge Regression

For generalized ridge regression, **it can be shown that the minimum *MSE* is obtained if the values of the p biasing parameters are** σ^2/α_i^2, $i = 1, 2, \ldots, p$. Hence, it seems reasonable to use estimates of the unknowns and use $k_i = \hat{\sigma}^2/\hat{\alpha}_i^2$ to define the ridge estimates. For the same reasons as outlined for k_{at}, one could consider iterating by using the ridge estimates of the α_i, $\hat{\alpha}_i^*(k_i)$, in successive steps. Hemmerle (1975) showed that this iteration method converges to a closed form and a non-iterative explicit solution. Unfortunately, the converged values tend to give too much bias. Therefore, the iterative procedure must be constrained from going to its limit. Hemmerle used a constraint of limiting the increase in the residual sum of squares to a pre-set value, say 20%.

Hocking, Speed, and Lynn (1976) show the potential of generalized ridge regression. However, the work of Hemmerle is the only work to date on an algorithm, and nothing has been done on the statistics of the estimates that result. Clearly, more work can be done. The relationships $k_i = \sigma^2/\alpha_i^2$ show that the amount of bias that can be used is inversely proportional to the size of the regression coefficient. However, the uncertainties associated with estimating individual α_i may be such that the use of a single k that is some kind of average of the individual estimates of the α_i will attain all the reduction in mean square error that is practical. The k_a, k_{at}, and k_b values are of such character.

4. Example

To demonstrate some of the properties of ridge regression and the operation of the algorithms in "Choice of the Biasing Parameters," a small example has been constructed. The **data and some pertinent statistics are given in Table 12.**

TABLE 12: Data for Example.

#	x_1	x_2	x_3	x_4	x_5	y
1	−.423863	−.294275	−.182889	−.285749	−.357062	−12.30606
2	−.391153	−.294275	−.281452	−.299270	−.313518	− 9.82692
3	−.186718	−.172277	−.380015	−.245185	−.117569	− 6.57254
4	−.170363	−.150095	−.215784	−.272228	−.248202	− 6.89471
5	−.158097	−.105732	−.330734	−.123494	−.182885	− 6.15960
6	−.100855	−.105732	−.248598	−.150537	−.204658	− 2.63322
7	−.055879	−.294275	.030664	−.177579	−.161113	− 5.79484
8	−.023169	−.227731	.047091	−.109973	−.161113	− 5.93513
9	−.023169	−.017006	.014237	.011718	.013063	.57523
10	.017718	−.005915	.079946	−.028845	.296100	5.05922
11	.058605	.182628	.096373	.187495	.187240	5.90959
12	.254863	.271354	.244217	.295665	.426733	10.40055
13	.308016	.315717	.277072	.349750	.296100	9.63669
14	.426589	.404444	.359208	.363271	.121924	11.29182
15	.467476	.493170	.490625	.484962	.404961	14.40038

$$X'X = \begin{bmatrix} 1.00000 & & & & \text{symmetric} \\ .92715 & 1.00000 & & & \\ .90679 & .83387 & 1.00000 & & \\ .95248 & .97249 & .90951 & 1.00000 & \\ .85771 & .87045 & .82124 & .88910 & 1.00000 \end{bmatrix}$$

$$(X'X)^{-1} = \begin{bmatrix} 12.64773 & & & & \text{symmetric} \\ -4.01315 & 26.90726 & & & \\ -4.11822 & 8.89978 & 9.40264 & & \\ -4.13492 & -29.48497 & -12.75180 & 47.38467 & \\ -\ .29635 & -1.07297 & -\ .59877 & -2.44586 & 4.85450 \end{bmatrix}$$

$\lambda_1 = 4.57887$ $\lambda_2 = .194057$ $\lambda_3 = .154886$ $\lambda_4 = .0583434$ $\lambda_5 = .0138449$

The data were generated by taking a factor structure from a real data set, and then choosing β_1, β_2, β_3, β_4, and β_5 at random with the constraint that $\beta'\beta = 300$. Then normal error with mean zero and variance one was added to form the observed response values, y.

Figure 30 is a ridge trace for the example, that is, $\hat{\beta}_i^*(k)$ plotted as function of k for each of the five factors. An examination of the trace shows that the coefficients for factors 2 and 4 change fairly rapidly as the value of k moves from zero. From the information given in Table 12, it is seen that the coefficients for factors 2 and 4 have the highest variances at the least-squares value. It is also seen that the coefficients for factors 2 and 4 have the largest covariance. Another observation that can be made is that the sign of the coefficient for factor 4 is incorrect at $k = 0$, the least-squares value. For the example the value of β_4 is known; in a real problem, the context of the data could also tell us that the sign in incorrect. For $k \geq 0.02$ it is seen that the sign for factor 4 is correct and the values for the coefficients are tending to "stabilize," that is, there is little change in the coefficient values as the value of k increases. Hence, it should be possible to obtain estimates of the coefficients with smaller mean square than those of least squares.

The values of k for the four algorithms are:

$$k_a = .0148$$
$$k_b = .0076$$
$$k_{at} = .0245$$
$$k_d = .0200.$$

Since the data were generated with known coefficient values, **the mean square error function** can be calculated. It **is plotted in Figure 31.** The computed values of the coefficients for several values of k together with the true values are tabulated at the top of the figure.

One clearly should not infer that every problem is like this one. However, the ridge trace is a graphical analytical tool that can portray the effects of the factor correlation structure

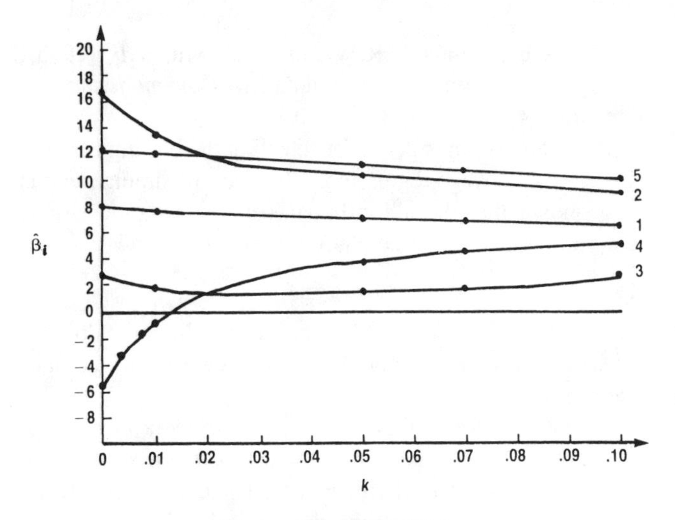

Figure 32: Ridge Trace

on coefficient estimation. The algorithms can lead to estimates with properties better than those of least squares. The specifics were delineated in "Choice of the Biasing Parameter," above.

References

1968

Hoerl, A.E., and Kennard, R.W. (1968). Ridge Regression: Biased Estimation for Nonorthogonal Problems. E. I. du Pont de Nemours & Co., *Engineering Department Report, Accession No. 13183.*

Contains the material of the Hoerl and Kennard (1970a,b) plus the following: (i) comments on the criterion $L_2^2 = (\hat{\beta}^* - \beta)'X'X(\hat{\beta}^* - \beta)$; (ii) Bayesian interpretation of ridge; (iii) the comparison of ridge and estimation of β when β is constrained to a bounded convex set; and (iv) k_a and k_{at} as presented here in "Simple Ridge Regression" as methods of choosing k.

1970

Hoerl, A.E. and Kennard, R.W. (1970a). "Ridge Regression: Biased Estimation of Nonorthogonal Problems." *Technometrics, 12,* 55–67.

In multiple regression it is shown that parameter estimates based on minimum residual sum of squares have a high probability of being unsatisfactory, if not incorrect, if the prediction vectors are not orthogonal. Proposed is an estimation procedure based on adding small, positive quantities to the diagonal of $X'X$. Introduced is the ridge trace, a method for showing in two dimensions the effects of nonorthogonality. It is then shown how to augment $X'X$ to obtain biased estimates with smaller mean square error.

Hoerl, A.E. and Kennard, R.W. (1970b). "Ridge Regression: Applications to Nonorthogonal Problems." *Technometrics, 12,* 69–82.

This paper is an exposition of the use of ridge regression methods. Two examples from the literature are used as a base. Attention is focused on the ridge trace which is a two-dimensional graphical procedure for portraying the complex relationships in multifactor

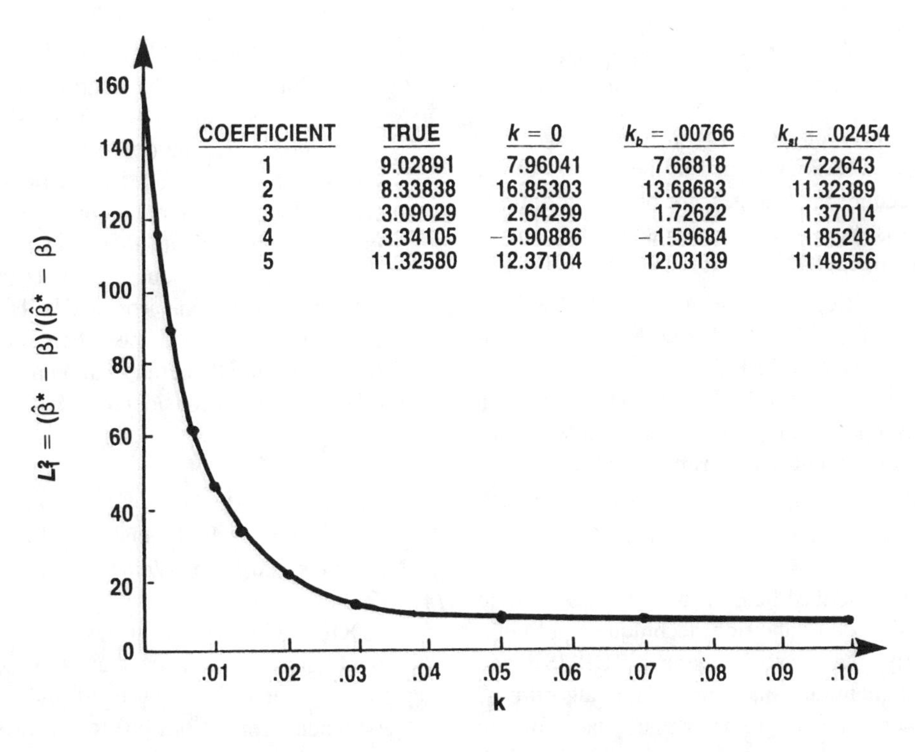

COEFFICIENT	TRUE	$k = 0$	$k_b = .00766$	$k_{at} = .02454$
1	9.02891	7.96041	7.66818	7.22643
2	8.33838	16.85303	13.68683	11.32389
3	3.09029	2.64299	1.72622	1.37014
4	3.34105	−5.90886	−1.59684	1.85248
5	11.32580	12.37104	12.03139	11.49556

Figure 33: Error Curve and Algorithm Solutions.

data. Recommendations are made for obtaining a better regression equation than that given by ordinary least squares estimation.

1973

Evans, D.A. (1973). "Computers in the Teaching of Statistics." *Journal of the Royal Statistical Society, Series A, 136*, 153–190.

Section 3.2 is devoted to ridge regression (pp. 159–162). The computer's ability to evaluate and plot within a few minutes the results for 15 to 20 multiple regressions for, say, 10 variables and 200 observations makes this type of statistical analysis quite feasible for classroom use.

1974

Bolding, J.T. and Houston, S.R. (1974). "FORTRAN Computer Program for Computation of Ridge Regression Coefficients." *Educational and Psychological Measurements, 34*, 151–152.

A FORTRAN program is described which computes ridge regression coefficients.

Theobald, C.M. (1974). "Generalizations of Mean Square Error Applied to Ridge Regression." *Journal of the Royal Statistical Society, Series B, 36,* 103–105.

Considered is a generalization of mean square error to $E[(\tilde{\beta}-\beta)'B(\tilde{\beta}-\beta)]$ where $\tilde{\beta}$ is any estimate of β and B is any nonnegative matrix of proper dimension. It is shown that a sufficient condition for the ridge estimator to have a smaller mean square error than the least squares estimator is that $k \leq 2\sigma^2/(\beta'\beta)$.

1975

Hemmerle, W.J. (1975). "An Explicit Solution for Generalized Ridge Regression." *Technometrics, 17,* 309–314.

The general form of ridge regression proposed by Hoerl and Kennard is examined in the context of the iterative procedure they suggest for obtaining optimal estimators. It is shown that a noniterative, closed-form solution is available for this procedure.

Hill, M.A. (1975). "Ridge Regression Using BMDP2R." *BMD Communications No. 3 (UCLA Health Sciences Computing Facility)*, February 1975, 1–2.

Gives an example of the use of the stepwise regression program to compute the regression coefficients for different values of k in order to construct a ridge trace.

MIT (1975). TEP: Robust and Ridge Regression. *D0070N*, 123 pp. MIT Information Processing Services, Cambridge, Mass. 02139.

The programs described here enable TROLL users to experiment with robust and ridge techniques, individually or jointly. Two large programs ROBUST and RIDGE implement the main regression algorithms. Other, smaller-scale programs prepare data for the main programs and perform related analyses. The documentation is complete with theory and examples.

Hoerl, A.E., Kennard, R.W., and Baldwin, K.F. (1975). "Ridge Regression: Some Simulations." *Communications in Statistics, 4(2)*, 105–123.

An algorithm is given for selecting the biasing parameter, k, in ridge regression. By means of simulation it is shown that the algorithm has a number of desirable properties. (See Section 3.1.)

1976

Hocking, R.R., Speed, F.M., and Lynn, M.J. (1976). "A Class of Biased Estimators in Linear Regression." *Technometrics, 18,* 425–438.

The purpose of this paper is to provide a unified approach to the study of biased estimators in an effort to determine their relative merits. The class of estimators includes the simple and the generalized ridge estimators proposed by Hoerl and Kennard, the principal component estimator with extensions such as that proposed by Marquardt, and the shrunken estimator proposed by Stein. The problem of estimating the biasing parameters is considered and illustrated with two examples.

Hoerl, A.E. and Kennard, R.W. (1976). "Ridge Regression: Iterative Estimation of the Biasing Parameter." *Communications in Statistics—Theory and Methods, A5(1)*, 77–78.

An interative method is given for selecting the biasing parameter, k, in ridge regression. The method produces a distribution of squared errors for the regression coefficients that has a smaller mean and a smaller variance than least squares or the single interation estimate, k_a. (See Section 3.1.)

Lawless, J.F. and Wang, P. (1976). "A Simulation Study of Ridge and Other Regression Estimators." *Communications in Statistics—Theory and Methods, A5(4)*, 307–323.

Considered are a number of estimators of regression coefficients, all of generalized ridge, or "shrinkage," type including principal components. Results of a simulation study indicate that with respect to two commonly used mean square error criteria, two ordinary ridge estimators, one proposed by Hoerl, Kennard, and Baldwin (1975), and the other introduced here, perform substantially better than both least squares and the other estimators considered.

1977

Bradley, C.E. and McGann, A.F. (1977). "RIDGEREG: A Program to Improve the Precision of Regression Estimates for Nonorthogonal Data." *Journal of Marketing Research, 14*, 412–413.

RIDGEREG is written in BASIC. Output includes the standardized covariance matrix, its determinant, and estimates of the standardized and regular regression coefficients for k from 0.0 to 0.5 in steps of 0.1. The program will handle up to 10 factors and 50 observations. (The program listing can be obtained from the authors at University of Wyoming, Laramie, Wyoming 82701.)

Dempster, A.P., Schatzoff, M., and Wermuth, N. (1977). "A Simulation Study of Alternatives to Ordinary Least Squares." *Journal of the American Statistical Association* (with discussion), *72*, 77–106.

Estimated regression coefficients and error in these estimates are computed for 160 artificial data sets drawn from 160 normal linear models structured according to factorial designs. Ordinary multiple regression (OREG) is compared with 56 alternatives which pull some or all estimated regression coefficients some or all the way to zero. Substantial improvements over OREG are exhibited when collinearity effects are present, noncentrality in the original model is small, and selected true regression coefficients are small. Ridge regression emerges as an important tool, while a Bayesian extension of variable selection proves valuable when the true regression coefficients vary widely in importance. (See Section 3.1.)

Comments

Efron, B., and Morris, C.

The use of RIDGM (see Section 3.1) or EBMLE seems to be justified in the aggregate, although the statistician who looks at the 160 individual problems might choose not to use the same rule in all these situations. Other experiments could give opposite conclusions, so the reader's faith in the results of the experiment ultimately depends on how much he believes the authors' data sets typify real world experience.

Hoerl, A.E.

Two simulations which are comparable to those in Hoerl and Kennard (1976) suggest that the authors' algorithm RIDGM is equivalently effective to the iterative algorithm (see Section 3.1).

Allen, D.M.

The commenter gives his own thoughts regarding regression and ill-conditioning in particular. Attention is given to the interpretation of β_j as a partial derivative and some possible pitfalls. It is noted that the dimensionality of a typical regression problem makes a comprehensive simulation study impossible; the authors are commended for their simulation design. It is suggested that the simulation should have been built around α rather than β.

Smith, A.F.M.

Attention is drawn to the relationship between discrete and continuous shrinking methods as given in Leamer and Chamberlain (1976). Criticism is made of comparing estimators on the basis of prediction performance over the estimation data set.

Bingham, C. and Larntz, K.

The point is made that the orientation of β with respect to the relevant coordinate system is important. For example, subset methods shrink the coefficient vector in the direction of planes determined by a subset of the basis vectors. If that is appropriate, they do well. The conclusion is made that the closer the coefficient vector is to the space spanned by the eigenvectors corresponding to the larger eigenvalues, the more improvement ought to be possible over least squares. Using the results of some small simulations reported, it is inferred that it is not clear that ridge methods offer a clear-cut improvement over least squares root, and shrunken estimator. Each of the biased estimators is shown to offer improvement in mean squared error over least squares for a wide range of choices of the parameters of the model. The results of a simulation involving all five estimators indicate that the principal components and latent root estimators perform best overall, but that the ridge regression estimator has the potential of a smaller mean squared error than either of these. The potential for improvement in mean squared error using ridge regression is seen to depend, however, on a better selection of the ridge parameter, k.

Jain, A.K., Mahajan, V., and Bergier, M. (1977). "RRIDGE: A Program for Estimating Parameters in the Presence of Multicollinearity." *Journal of Marketing Research, 14*, 561.

RRIDGE is a FORTRAN IV program that will handle up to 30 factors, 200 observations, and 20 values of k. Output includes the correlation matrix, eigenvalues, and parameters for different values of k. (The program with documentation is available from the authors at School of Management, SUNY at Buffalo, Buffalo, New York 14214.)

1979

Carmer, S.G. and Hsieh, W.T. (1979). "Exploring Biased Regression with SAS." *Proceedings of the Fourth Annual SAS Users Group International Conference,* SAS Institute Inc., Cary, N.C., 27511, 223–228.

This paper reports on a set of SAS76 MACROs with which the user can obtain biased regression analyses using one or more of a number of biased estimators. In brief, the user chooses from among estimators which are either: (i) principal components, shrunken, or ridge; (ii) a combination of principal components and shrunken; or (iii) a combination of principal components and ridge.

Gunst, R.F. (1979). "An Approach to the Programming of Biased Regression Algorithms." *Communications in Statistics—Simulation and Computation, B8(2)*, 151–159.

Since few computer algorithms exist for calculating estimators and ancillary statistics that are needed for biased regression methodologies, users are forced to write their own programs. Brute-force coding of such programs can result in a great waste of computer core and computing time, as well as inefficient and inaccurate computing techniques. This article proposes some guides to more efficient programming by taking advantage of mathematical similarities among several of the more popular biased regression estimators (ridge, principal components, latent root, and shrunken).

Hoerl, A.E. (1979). "A Full Ridge Regression Program." *Department of Mathematical Sciences,* University of Delaware, Newark Del. 19711.

The program is written in FORTRAN IV and provides many features for ridge regression analysis. Included are regression coefficients in correlation and raw units for 32 typical values of k, a ridge trace, residual tables for specified k values, and the specific k values of k_a, k_b, and k_d as described in Section 3.1. (This is a commercial program available from the author.)

Hui, B.S. and Jagpal, H.S. (1979). "RIDGE: An Integrated Ridge Regression Program." *Journal of Marketing Research, XVI,* 571–572.

RIDGE is described as a user-oriented ridge regression program that (i) provides several criteria for selecting k, (ii) gives a plot of the ridge trace, and (iii) allows an examination of the stability of different ridge solutions when the data are perturbed. The standardized and unstandardized parameters, the square of the multiple correlation coefficient, and the residual sum of squares for all ridge solutions are computed. (Available from the first author at Rutgers University, 92 New Street, Newark, New Jersey 07102.)

Sinha, A.N. and Hardy, K.A. (1979). "A SAS Macro for Ridge Analysis of Multivariate General Linear Models." *Proceedings of the Fourth Annual SAS Users Group International Conference,* SAS Institute Inc., Cary, N.C. 27511, 229–233.

This paper reviews the general theory of ridge regression analysis for the multivariate general linear model and presents a SAS macro which enables the user to obtain both ordinary least-squares estimates and ridge estimates for purposes of comparing the two techniques and visualizing the effects of nonorthogonality.

1980

Bush, A.J. (1980). "Ridge: A Program to Perform Ridge Regression Analysis." *Behavior Research Methods and Instrumentation, 12,* 73–74.

A FORTRAN IV program that is comprehensive in its average of the elements of ridge regression analysis. Methods for selecting the biasing parameter k include those of Dempster, Schatzoff, and Wermuth (1977); Hoerl, Kennard, and Baldwin (1975); Kasarda and Shih (1977); Lawless and Wang (1976); and McDonald and Galaraneau (1975). (The program is available from the author at Memphis State University, Memphis, Tennessee 38152.)

Rogers, R. and Hildebrand, E. (1980). "MACRO RIDGREG." *SUGI SASWARE Index. Proceedings of the Fifth Annual SAS Users Group International Conference,* SAS Institute Inc., Cary, N.C. 27511, 490.

Calculates ridge regression coefficients. User specifies the range of values for the biasing parameter, k. Outputs means, variances, correlation matrix, variance inflation factors, and ridge trace. (The program is available from the authors at USDA Forest Service, University of Missouri, Columbia, Missouri 65201.)

EXPERIMENTAL DESIGNS

After an overview of the basic ideas of design, we have covered the statistical principles on which experimentation is based, linear models, and model interpretation and analysis. Even with proper statistical analysis, **statistical design can lead to a saving of much experimentation**. In this section we will cover the basics of design, first noting some examples where it is important.

Example 1. In a certain chemical process, Chemicals A and B are reacted in the presence of a catalyst to form a product, where at least 40% of end product C is desired. The percent of C in the end product is a function of (1) the ratio of A to B in the mix, and (2) the amount of the catalyst used. **How should one experiment in order to find the optimal operating conditions for this process,** i.e. the ratio X_1 and the catalyst amount X_2 which will guarantee at least 40% C at the minimal materials cost?

Example 2. The active ingredient in a certain drug may be subject to deterioration (fully or partially), in which case the drug is ineffective for its intended purpose. If this is found, it is believed that the measured deterioration is a function of: (1) the source of the active ingredient used, of which there are 3; (2) the type of container used, of which there are 2 (clear and dark); (3) the solution in which the ingredient is dissolved, of which there are 2; and (4) the air space in the container, which may be regular or slim. **How can we find which variables (of the four noted) do in fact influence the deterioration, or indeed if any of them do? How can we prevent the effects of other factors, such as production shift or time of day or supplier of raw material, from giving us misleading results** (and, perhaps, how can we allow for a later study of these factors, using the same experimental data)?

Example 3. If we are experimenting with new weed killers to try to find one which is at least as effective as the standard formulation, **how can we take account of the fact that agricultural plots will vary** in the extent of the weeds present if untreated? **How can we evaluate the significance of observed differences and be sure they are not explainable by random variation? How can we guarantee a high probability of actually selecting the best treatment as a result of our experiments?**

The answers to some of these questions are intricate, the answers to some are simple, and the answers to others fall somewhere between these two extremes. However, the answers to all will be clearer if we approach them carefully from **a careful definition, motivation, and development of experimental design and analysis** material. First, let us consider

The Need for Experimental Design (and, in particular, the Randomized Blocks and Latin Squares designs).

Suppose we have four new weed killer formulations, and run an experiment at an experimental field to compare weed density when each of the four is used. As seen from the air, the field where the experiment is run might look like the following grid (Figure 34) when divided into experimental plots.

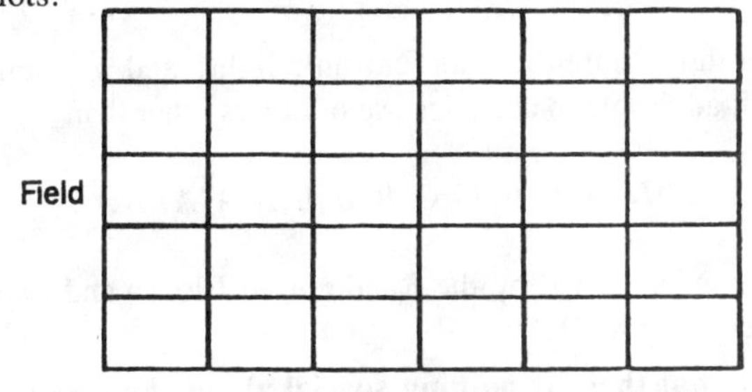

Figure 34

If we choose to apply each formulation to four plots (note: we'll see how to make a rational choice of the number of plots later), we may decide to apply formulations to plots as follows (where "*A*" denotes that formulation number 1 was applied to a plot, "*B*" denotes application of formulation number 2 to a plot, "*C*" denotes formulation 3, and "*D*" denotes formulation 4) (see Figure 35):

A	B	C	D
A	B	C	D
A	B	C	D
A	B	C	D

Figure 35

This is certainly **easy to implement**, and runs little risk of misrecording of data or misapplication of formulations (applying a formulation to the wrong plot, for example) by field personnel. **However, it has a problem**: if in fact the field is not uniform but has a gradient (either of soil quality, or of water table, etc.)

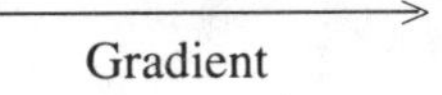

then formulation 4 will uniformly receive the greatest benefit from this gradient, 3 the next greatest, and 1 the least. **The Randomized Blocks Design is used to overcome this problem:** We now consider the field to consist of 4 blocks of 4 plots each (see Figure 36):

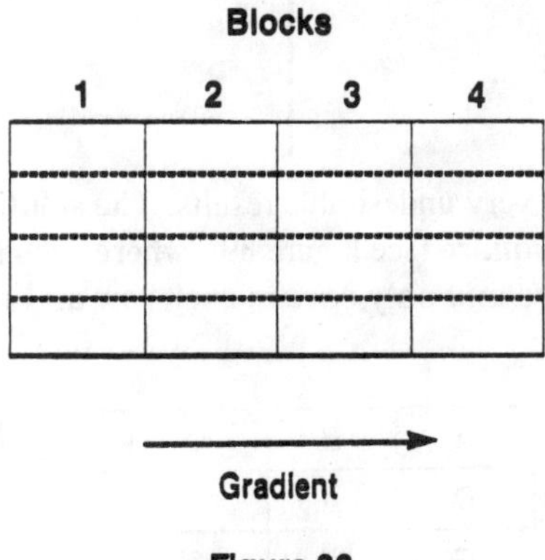

Figure 36

and have *A, B, C, and D* in each and every block, the order within each block being determined by **randomization** (e.g., have slips numbered 1, 2, 3, and 4 in a hat and draw them at random to determine plot order). Now each formulation receives the same detriment or benefit from each level of the gradient, a typical field layout (Figure 37) being

C	D	C	
A	A	B	
B	B	A	
D	C	C	
Hat #1	Hat #2	. . .	

Figure 37

[*Note* that similar considerations apply if 4 unknowns are to be studied with 4 scales (or 4 technicians, or 4 laboratories, etc.).] **However**, the layout (Figure 37) could (by randomization) come out to be

A	A	A	A
C	C	C	C
B	B	B	B
D	D	D	D

Figure 38

in which case a soil or water gradient in the other direction

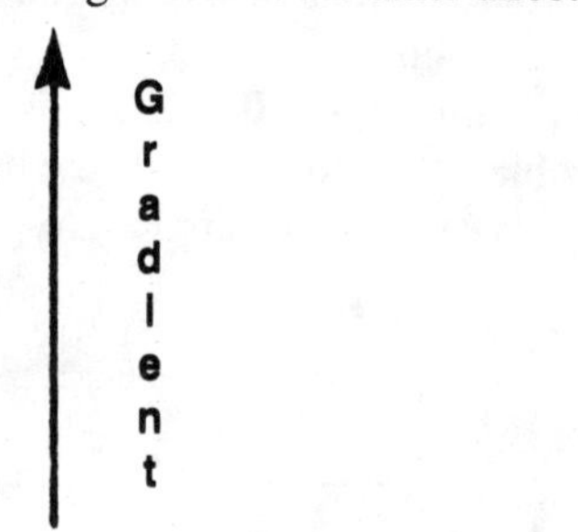

could have very undesirable results. **The solution to this is the Latin Square** (see Figure 39), **where each formulation is used once and only once in each row and each column in the field:**

A	B	C	D
D	A	B	C
C	D	A	B
B	C	D	A

Figure 39

Similar considerations can apply in other contexts. For example, in an experiment to study product from a production line under four different temperature conditions, we may desire to subject 4 units to each condition. If the production line yields 16 consecutive units, we might (where "*A*" denotes the first temperature condition, "*B*" a second, "*C*" a third, and "*D*" the last temperature condition) allocate the 16 as shown in Figure 40.

If there are gradients in the production process, however, for example an increasing percent of one of the key ingredients, this will assure a trend even if temperature has no effect. Randomized blocks could guard against this. (See Figure 41.)

Latin Squares could also guard against this. (See Figure 42.)

Similar examples occur in other contexts (see Figure 43), e.g. when plants are kept in a greenhouse in a row and (e.g.) 4 different formulations are applied to them in order to evaluate formulation effectiveness,

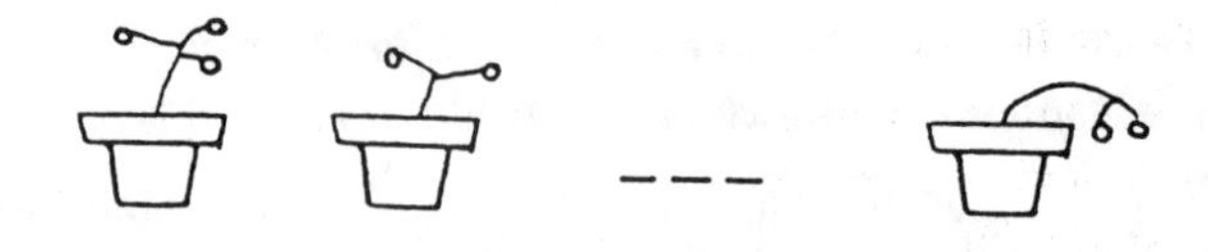

Figure 43

temperature, humidity, air current, light, and gradients make it advisable to consider use of orders other than

D D D D C C C C B B B B A A A A ,

e.g. those suggested by the Randomized Blocks and Latin Squares designs.

Note that there is nothing special about the case of 4 formulations illustrated above, and similar considerations are easily given for other numbers of formulations.

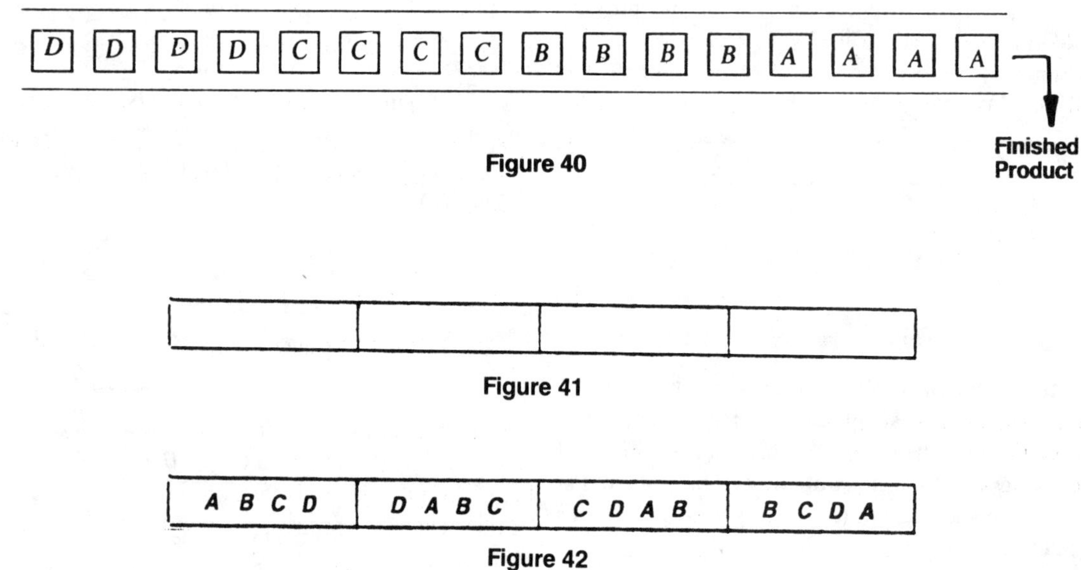

Figure 40

Figure 41

Figure 42

It should now be apparent that **an experiment must be designed (then run, then analyzed)**, *not* run in an undesigned fashion with the hope it may yield informative results. Such results come from recognition, and control via design, of unwanted sources of variability.

Having detailed *the need for experimental design*, we now wish **to compare two common types of experimental designs**

(1) the "one-at-a-time method," and

(2) the complete factorial method

and note what happens to one of these methods when **interaction** (in drugs, synergism) is present.

Suppose we have a production process that has customarily been run at temperature T_1 and pressure P_1, and we wish to study effects of pressure and temperature on yield, in particular temperatures T_1 and T_2 and pressures P_1 and P_2. (See Figure 44.)

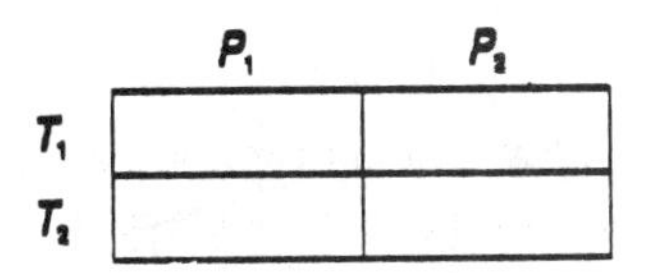

Figure 44

One might then fix T_1 and run the process at P_1 and also at P_2 and observe the yields as shown in Figure 45.

	P_1	P_2
T_1	x	x
T_2		

Figure 45

One may decide to choose the pressure giving the larger yield, and run a 3rd experiment in its column, taking the larger yield there to give the best temperature as shown in Figure 46.

	P_1	P_2
T_1	x	**X**
T_2		x

Figure 46

This method assumes that the effects of temperature and of pressure on yield are *additive*, that is, that there is no *joint* (or *interaction*) *effect* of temperature and pressure together on yield.

For example, if our experiments yield the data shown in

°F	$P_1 = 30$	$P_2 = 40$ psi
$200 = T_1$	2.6	2.3
$300 = T_2$	3.0	4.3

Figure 47

Figure 47, then there is apparent *interaction* because any determination of which pressure is superior (higher yield) depends on the temperature used: P_1 at T_1, but P_2 at T_2.

If we also run an experiment at T_2 with P_1, we then have an instance of the *complete factorial method*, which cannot be misled by interaction in the way that the one-at-a-time method can.

Note that these considerations generalize easily to factors (such as temperature and pressure), which each have several levels or settings as shown in Figure 48, as well as to experiments with more than two factors.

Finally, before going to specific formulas and examples for the basic experimental designs, we should note **the difference between quantitative and qualitative variables,** and the different analyses which are appropriate to each.

Basically, **a one-factor experiment with a quantitative variable involves examining yield (e.g.) as a function of a variable which can be varied continuously,** such as

- pressure,
- temperature,
- queue arrival rate, or
- amount of fertilizer used.

In this case *interpolation* (curve fitting, as discussed in the earlier sections) is possible. For example, one may fit a parabola and solve for the temperature which maximizes the expected yield in the fitted parabola. (See Figure 49.)

On the other hand, **in a one-factor experiment with a qualitative variable, the variable cannot be varied continuously.** For example,

- agricultural variety,
- catalyst, and
- queue discipline

are variables of this type. In such a case, *interpolation is not possible*. Typical questions to be asked here are How much better are the test catalysts than the standard catalyst? and Which catalyst gives the highest mean yield? (See Figure 50.)

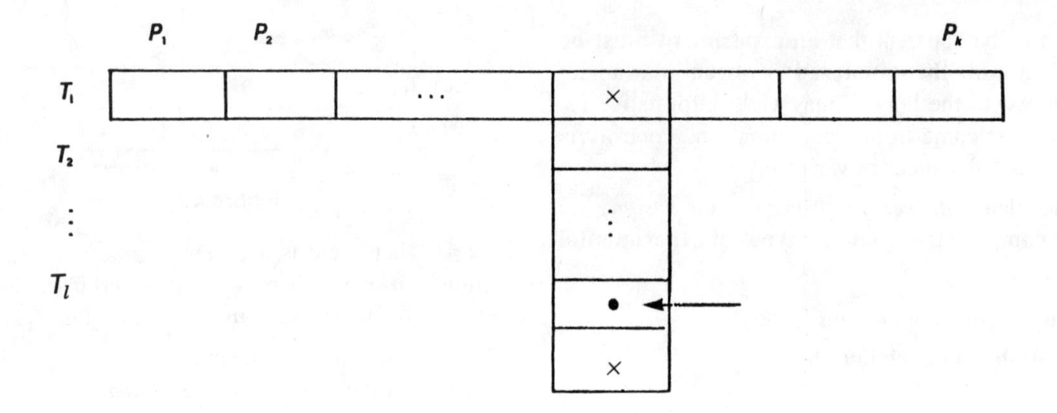

Figure 48

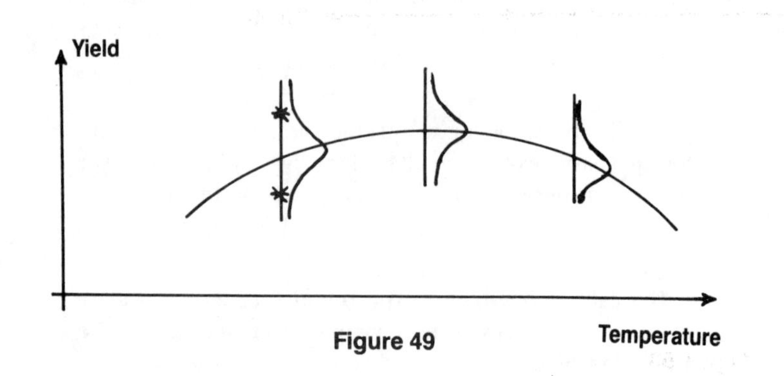

Figure 49

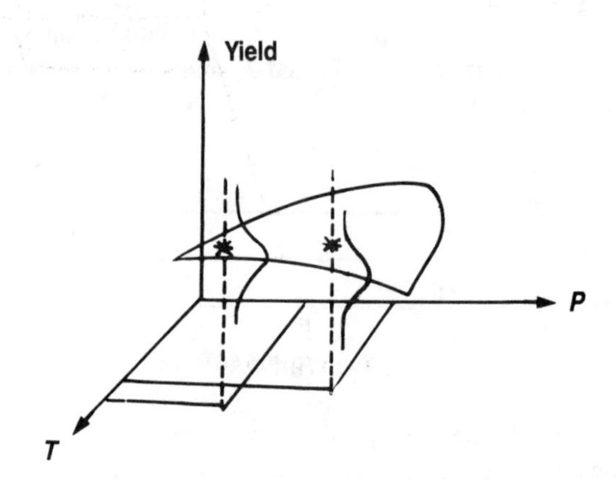

Figure 51

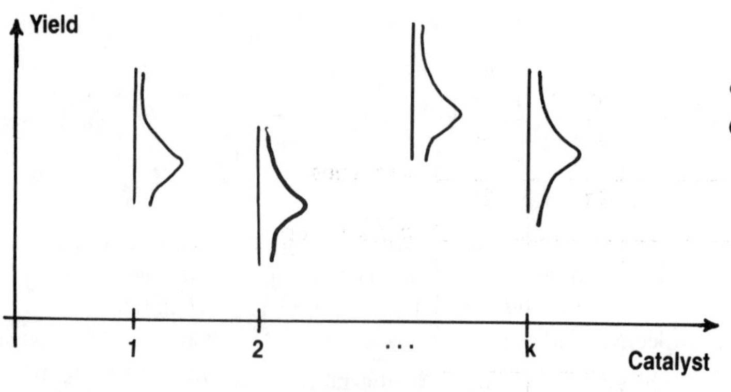

Figure 50

Note that the experimental quantity of interest need not be yield, but could just as well be variability of output, or proportion of defective units, and so on.

Also, similar considerations arise when two (or more) variables affect our yield. We may have:

Quant.–Quant.—Interpolation. (See Figure 51.) Study surface to find P-T for max. yield.

Quant.–Qual.—Interp. in one direction. (See Figure 52.) Find type, P for max. yield.

or

Qual.–Qual.—No interpolation. (See Figure 53.)

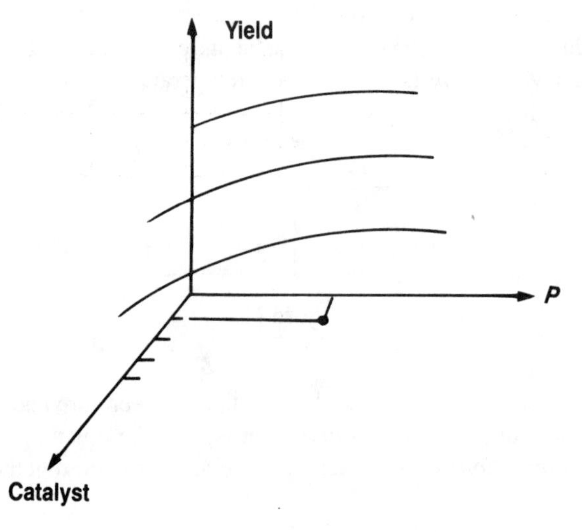

Figure 52

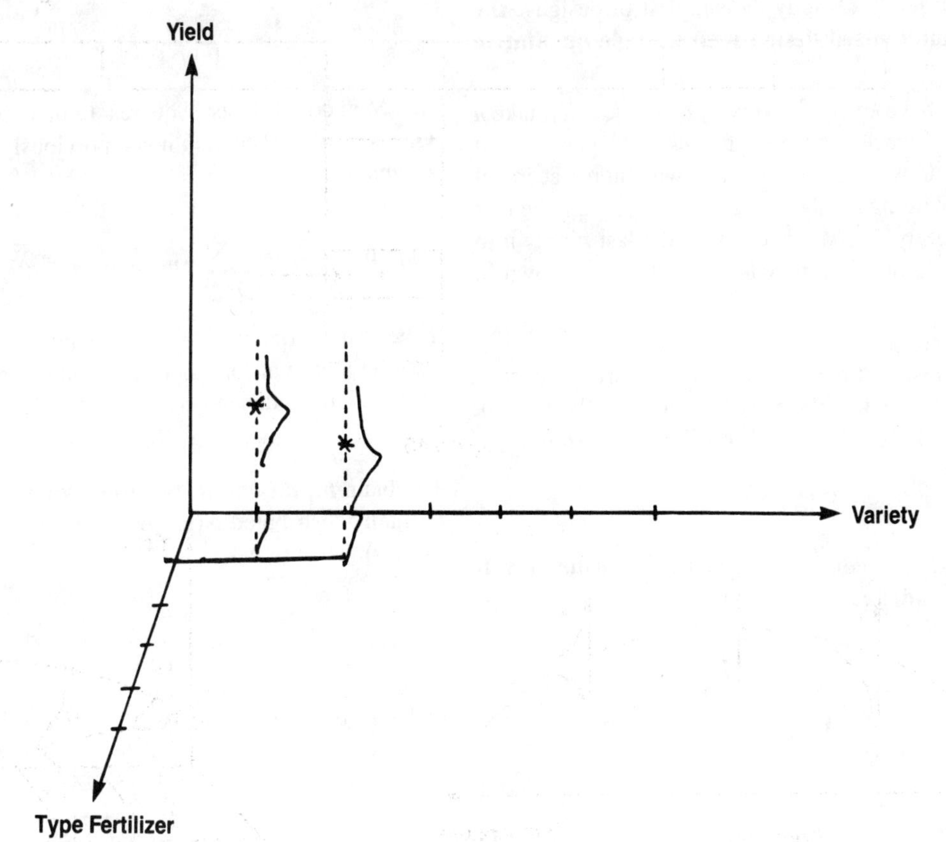

Figure 53

Field	1	2	3	...	$n-1$	n
1	1	2 A	3		$n-1$ A	n
2	$n+1$	$n+2$	$n+3$ A		$2n-1$	$2n$
k		A				nk

k varieties

n obs./variety

Figure 54

We are now ready to study the simplest of designs, **the completely randomized design with a single qualitative factor.**

Suppose we have k varieties (say $A, B, C, \ldots, K$), take n observations on each, and allot the nk field positions at random (e.g., draw numbers at random without replacement from a hat containing numbers $1, 2, \ldots, nk$; assign the first n drawn to variety $A, \ldots,$ and assign the last n drawn to variety K). Then our field may look like the one shown in Figure 54.

Let X_{ij} = the result of the j_{th} application of variety i ($i = 1, 2, \ldots, k$). (*Note* that "variety" could equally well be "catalyst," "scale," and so on.) **We will assume that**

$$E(X_{ij}) = \mu_i$$

(a number not depending on j), and **it is customary to re-write the model as**

$$\begin{aligned} E(X_{ij}) &= \mu_i \\ &= \left(\mu_i - \frac{1}{k}\sum_{i=1}^{k}\mu_i\right) + \left(\frac{1}{k}\sum_{i=1}^{k}\mu_i\right) \\ &= \alpha_i + \mu, \end{aligned}$$

where α_i is called the effect of the ith treatment; note that $\sum_{i=1}^{k}\alpha_i = 0$. **We** will assume that $\text{Var}(X_{ij}) = \sigma^2$, a common variance, and **study the α_i's to obtain:**

a) point estimates,

b) confidence interval estimates,

c) tests of hypotheses (e.g., $H: \alpha_i = \ldots = \alpha_k$), and

d) a decision as to which is largest.

a. If we seek, as estimators of $\mu_1, \ldots, \mu_k$, the $m_1, \ldots, m_k$ which minimize the sum of squared deviations

$$\sum_{i=1}^{k}\sum_{j=1}^{n}(X_{ij} - m_i)^2,$$

the Method of Least Squares (which is justified by the Gauss-Markov Theorem noted previously), we find as *point estimators*

$$\hat{\mu}_1 = \overline{X}_1 = \frac{1}{n}\sum_{j=1}^{n}X_{1j}, \ldots, \hat{\mu}_k = \overline{X}_k = \frac{1}{n}\sum_{j=1}^{n}X_{kj}.$$

Note that we obtained these estimators without any need to assume a normal (or other) distribution for X_{ij}: it is enough that we have independence and $\text{Var}(X_{ij}) = \sigma^2$. (See Figure 53.)

To obtain ***b., c.,* and *d.***, we usually also assume that X_{ij} is normally distributed $N(\mu+\alpha_i, \sigma^2)$ ($i = 1, \ldots, k$; $j = 1, \ldots, n$). Then (e.g.)

$$\overline{X}_1 - \overline{X}_2 \text{ is } N(\alpha_1 - \alpha_2, \sigma^2\frac{2}{n}),$$

and a 95% confidence interval for $\mu_1 - \mu_2$ $(= \alpha_1 - \alpha_2)$ is

$$\overline{X}_1 - \overline{X}_2 \pm t_{.025}\sqrt{\frac{2}{n}}\sqrt{\frac{\sum_{i=1}^{k}\sum_{j=1}^{n}(X_{ij} - \overline{X}_i)^2}{k(n-1)}}.$$

Note that this $t_{.025}$ is found with $k(n-1)$ degrees of freedom. The estimate of σ^2 is reasoned to as follows: We could estimate σ^2 by any one of

$$s_1^2 = \frac{\sum_{j=1}^{n}(X_{1j} - \overline{X}_1)^2}{n-1}, \ldots, s_k^2 = \frac{\sum_{j=1}^{n}(X_{kj} - \overline{X}_k)^2}{n-1},$$

hence also by their average

$$\hat{\sigma}_W^2 = \frac{s_1^2 + \ldots + s_k^2}{k} = \frac{\sum_{i=1}^{k}s_i^2}{k(n-1)}$$

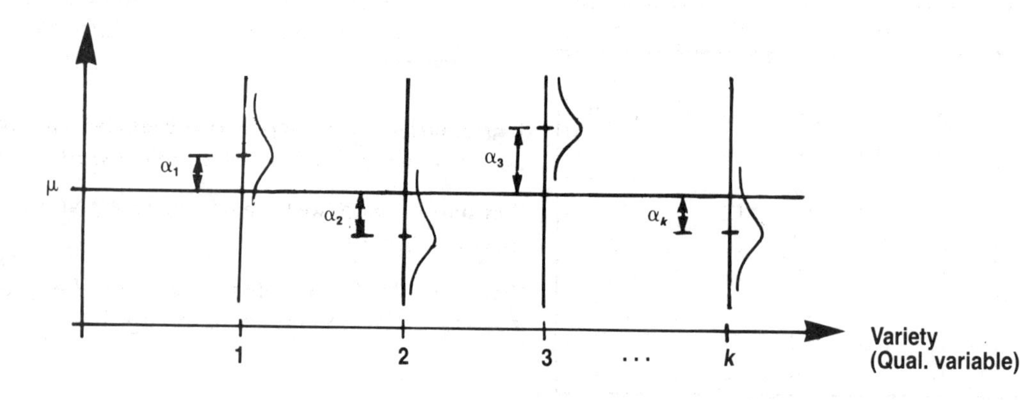

Figure 55

$$= \frac{\sum_{i=1}^{k} \sum_{j=1}^{n} (X_{ij} - \overline{X}_i)^2}{k(n-1)}$$

Within sample variance

Each population yields us $n-1$ degrees of freedom for estimating σ^2, hence we have a total of $k(n-1)$ degrees of freedom.

Motivation for the (ANOVA) test of H: $\alpha_1 = \ldots = \alpha_k\ (=0)$.

If H is true, then the variance of $\overline{X}_1, \ldots, \overline{X}_k$ estimates $\sigma^2/n = \mathrm{Var}(\overline{X}_1) = \ldots \mathrm{Var}(\overline{X}_k)$. This estimate is

$$\frac{\sum_{i=1}^{k} (X_i - X.)^2}{k-1}$$

where $\overline{X}. = (\overline{X}_1 + \ldots + \overline{X}_k)/k$. Thus n times this,

$$\hat{\sigma}_B^2 = n \frac{\sum_{i=1}^{k} (\overline{X}_i - \overline{X}.)^2}{k-1}$$

Between sample variance

estimates σ^2 and is based on $k-1$ degrees of freedom. If H is true, $\hat{\sigma}_W^2$ and $\hat{\sigma}_B^2$ are independent estimates of σ^2, hence

$$F = \frac{\hat{\sigma}_B^2}{\hat{\sigma}_W^2}$$

has an F-distribution with $k-1$ and $k(n-1)$ degrees of freedom *when H is true*. But if H is false, we expect $\hat{\sigma}_B^2 > \sigma_W^2$, **so to test H we reject if $F > F_\alpha$.**

Example. 10 treatments on 4 units each. (See Example 9.)

Treatment	X_{i1}	X_{i2}	X_{i3}	X_{i4}	$\overline{X}_i$
$i=1$	2	0	3	3	2.00
$i=2$	5	5	12	10	8.00
$i=3$	7	10	10	12	9.75
$i=4$	2	10	5	10	6.75
$i=5$	2	10	15	12	9.75
$i=6$	0	10	12	10	8.00
$i=7$	0	1	10	10	5.25
$i=8$	1	3	12	12	7.00
$i=9$	3	5	2	7	4.25
$i=10$	0	0	0	2	.50

$\overline{X}. = 6.125$

Example 9

For this data, we calculate

$$\hat{\sigma}_W^2 = \frac{494.75}{30} = 16.49166667,$$

$$\hat{\sigma}_B^2 = 4\,\frac{87.40625}{9} = 38.84722222,$$

and

$$F = 2.3556.$$

Since, with $k-1 = 9$ and $k(n-1) = 30$ degrees of freedom, $F_{.05} = 2.21$, we reject $H: \alpha_1 = \ldots = \alpha_{10} = 0$ at level .05. A 95% confidence interval for $\mu_1 - \mu_2$ is

$$\overline{X}_1 - \overline{X}_2 \pm t_{.025} \sqrt{\frac{2}{4}} \sqrt{\sigma_W^2} \qquad \hat{\sigma}_W^2$$

i.e.

$$2 - 8 \pm 2.04 \sqrt{\frac{2}{4}} \sqrt{16.49} = -6 \pm 5.86$$

i.e.

$$-11.86 \text{ to } -.14.$$

Thus treatments 1 and 2 are significantly different at the 5% level. Treatment 2 produces from .14% to 11.86% more than does treatment 1.

Note: $F_{.01} = 3.07$, so we do not reject at level .01.

Note that the sample variance of all nk observations has numerator

$$\sum_{i=1}^{k} \sum_{j=1}^{n} (X_{ij} - \overline{X}.)^2,$$

which is called the **total sum of squares**, and denominator $nk-1$. It is easy to prove

Theorem:

$$\text{Total } SS = \Sigma\Sigma (X_{ij} - \overline{X}.)^2 = \underbrace{\Sigma\Sigma (\overline{X}_i - \overline{X}.)^2}_{\substack{\text{Treatment } SS \\ \text{Between-Samples } SS \\ SS\ (\text{Tr})}} + \underbrace{\Sigma\Sigma (X_{ij} - \overline{X}_i)^2}_{\substack{\text{Residual } SS \\ \text{Error } SS \\ SSE}}$$

Customarily, these facts and this analysis are summarized in an analysis of variance table. (See Table 13.)

For our example, we have seen that the ANOVA table is as shown in Table 14.

Note how the $E(MS)$ column makes the F-test intuitively reasonable: the MS ratio on the average is

$$\frac{\sigma^2 + \frac{4}{9}\Sigma\alpha_i^2}{\sigma^2} = 1 + \frac{4}{9}\frac{\Sigma\alpha_i^2}{\sigma^2}.$$

Table 13

Source of Variation	SS	d.f.	MS	$E(MS)$
Treatments	$\Sigma\Sigma\,(\bar{X}_i - \bar{X}_\bullet)^2$	$k-1$	$SS\,(\text{Tr})/k-1$	$\sigma^2 + \dfrac{n\Sigma\alpha_i^2}{k-1}$
Residual	$\Sigma\Sigma\,(X_{ij} - \bar{X}_i)^2$	$k(n-1)$	$SSE/k(n-1)$	σ^2
Total	$\Sigma\Sigma\,(X_{ij} - \bar{X}.)^2$	$nk-1$		

Table 14

Source of Variation	SS	d.f.	MS	$E(MS)$
Treatments	349.625	9	38.85	$\sigma^2 + \frac{4}{9}\Sigma\alpha_i^2$
Residual	494.75	30	16.49	σ^2
Total	844.375	39		

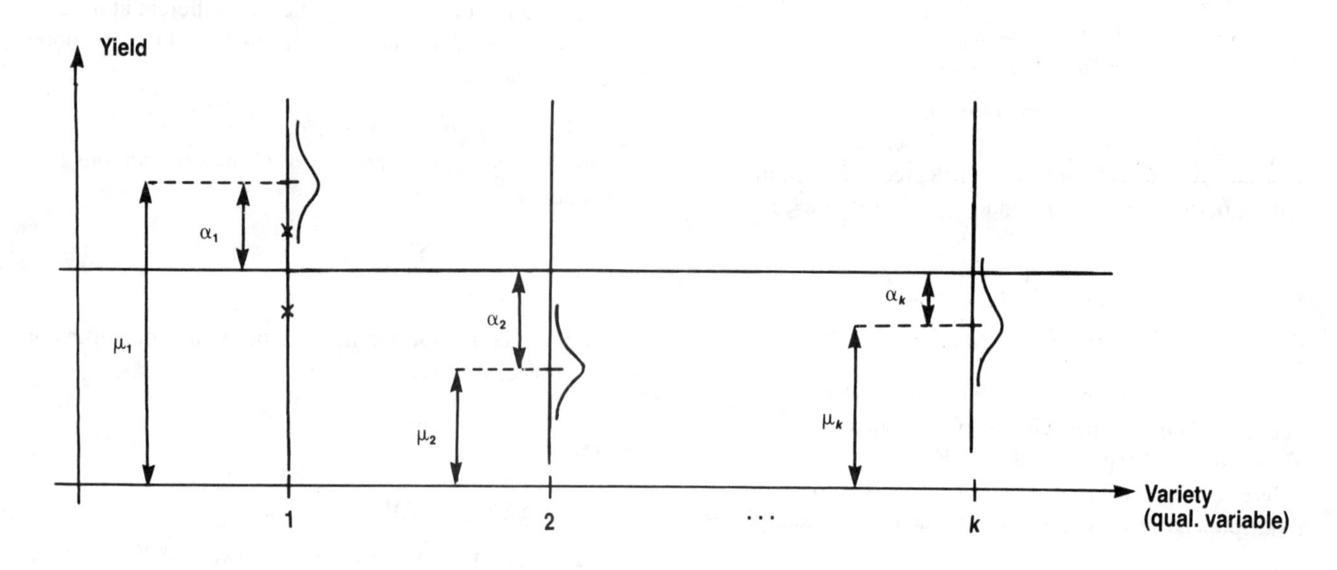

Figure 56

Thus $H : \alpha_1 = \ldots = \alpha_k\ (=0)$ will be easy to detect if $\Sigma\alpha_i^2$ is large or if σ^2 is small. **Thus anything we can do to reduce our experimental error (σ^2) will help us have better power for our test.** *Note* that some short-cut formulas are available for these computations and that the case of n_i observations on treatment i ($i = 1, \ldots, k$) is also easily handled similarly. Since most calculations are done with a statistical package, we omit those formulas here.

Before reviewing the experimental design results for the *completely randomized* design, and going on to the *randomized blocks* design and their comparison, let us try to answer the **question "Why use *variances* to test hypotheses about *means*?"** In fact, we'll cover two answers to this question: one intuitive and designed to have intuitive appeal, the other also being precise and indicating how appropriate test statistics can be obtained in any design of any complexity.

Answer 1. For the completely randomized design, the ANOVA test of $H : \alpha_1 = \ldots = \alpha_k\ (=0)$ rejects H when

$$F = \frac{\hat{\sigma}_B^2}{\hat{\sigma}_W^2} = \frac{\dfrac{n}{k-1}\,\Sigma(\bar{X}_i - \bar{X}_\bullet)^2}{\dfrac{\Sigma\Sigma(X_{ij} - \bar{X}_i)^2}{k(n-1)}}$$

is large. Recall that if $Y_1, \ldots, Y_r$ are independent $N(a,b^2)$ random variables, then we estimate the mean a by $\bar{Y} = \frac{Y_1+\ldots+Y_r}{r}$ and we estimate the variance b^2 by

$$s_Y^2 = \frac{\sum_{i=1}^{r}(Y_i-\bar{Y})^2}{r-1}.$$

Now graphically our setting is as shown in Figure 56.

Observations	Observations	Observations
$X_{11}, X_{12}, \ldots, X_{1n}$	$X_{21}, X_{22}, \ldots, X_{2n}$	$X_{k1}, X_{k2}, \ldots, X_{kn}$
are	are	are
$N(\underbrace{\mu+\alpha_1}_{\mu_1}, \sigma^2)$.	$N(\underbrace{\mu+\alpha_2}_{\mu_2}, \sigma^2)$.	$N(\underbrace{\mu+\alpha_k}_{\mu_k}, \sigma^2)$.
Est. σ^2 by $\frac{\Sigma(X_{1j}-\bar{X}_1)^2}{n-1}$	Est. σ^2 by $\frac{\Sigma(X_{2j}-\bar{X}_2)^2}{n-1}$	. . . Est. σ^2 by $\frac{\Sigma(X_{kj}-\bar{X}_k)^2}{n}$.

and we wish to test $H: \alpha_1 = \ldots = \alpha_k\ (=0)$, the hypothesis that all k varieties have the same mean yield μ. Whether this is true or not, a good estimator of σ^2 is

$$\frac{\frac{\Sigma(X_{1j}-\bar{X}_1)^2}{n-1}+\frac{\Sigma(X_{2j}-\bar{X}_2)^2}{n-1}+\ldots+\frac{\Sigma(X_{kj}-\bar{X}_k)^2}{n-1}}{k} = \hat{\sigma}_W^2.$$

Now $\Sigma\,(\bar{X}_i-\bar{X}.)^2$ almost looks like an estimator of $\Sigma(\mu_i-\mu)^2$, since on the average $\bar{X}_i$ is μ_i, and on the average $\bar{X}.$ is μ. So if

$$\Sigma\,(\bar{X}_i - \bar{X}_\bullet)^2$$

is large, we should reject the hypothesis that $\mu_1 = \ldots = \mu_k = \mu$. That is, reject when

$$\frac{n}{k-1}\Sigma\,(\bar{X}_i - \bar{X}_\bullet)^2 = \hat{\sigma}_B^2$$

is large. But the distribution of $\hat{\sigma}_B^2$ is not one we can use, since it involves the unknown variance σ^2. However

$$F = \frac{\hat{\sigma}_B^2}{\hat{\sigma}_W^2}$$

has a distribution which does not depend on σ^2, namely the distribution in Table 6 (when $H: \alpha_1 = \ldots = \alpha_k$ is true), so we reject when F is "large."

Answer 2. If we wish to estimate $\mu_1, \ldots, \mu_k$, the Method of Least Squares tells us to choose our estimators $\hat{\mu}_1, \ldots, \hat{\mu}_k$ as the quantities $a_1, \ldots, a_k$ which minimize the sum of squared errors

$$\Sigma\Sigma\,(X_{ij} - a_i)^2\,.$$

As one would expect, this is minimized when $a_1 = \bar{X}_1, \ldots, a_k = \bar{X}_k$. So, the sum of squared errors using the best estimators is

$$S_1^2 \equiv \Sigma\Sigma\,(X_{ij} - \bar{X}_i)^2\,.$$

Now if, in fact, $\mu_1 = \ldots = \mu_k$ (which is the hypothesis of interest to us), then we should be able to make the sum of squared errors

$$\Sigma\Sigma\,(X_{ij} - a)^2$$

almost as small as we could before. The best estimator of the common mean turns out to be $\hat{\mu} = \bar{X}.$, so the smallest sum of squares when we restrict ourselves as if $\mu_1 = \ldots = \mu_k$ is

$$S_2^2 \equiv \Sigma\Sigma(X_{ij} - \bar{X}.)^2\,.$$

Thus it is reasonable to reject the hypothesis *if S_2^2 does not* seem to be "almost as small as S_1^2," i.e., if

$$\frac{S_2^2-S_1^2}{S_1^2} \text{ is "large."}$$

However

$$\frac{S_2^2-S_1^2}{S_1^2} = \frac{\Sigma\Sigma(X_{ij}-\bar{X}.)^2 - \Sigma\Sigma(X_{ij}-\bar{X}_i)^2}{\Sigma\Sigma(X_{ij}-\bar{X}_i)^2}$$

$$= \frac{\Sigma\Sigma(\bar{X}_i-\bar{X}.)^2}{\Sigma\Sigma(X_{ij}-\bar{X}_i)^2} = \frac{k-1}{n}\,\frac{1}{k(n-1)}\,\frac{\hat{\sigma}_B^2}{\hat{\sigma}_W^2}$$

Algebraic fact, given on p.453

so this test *is* our ANOVA test given before.

This *method*, of looking at $(S_2^2 - S_1^2)/S_1^2$ and rejecting when it is "large," **is the method used to derive all ANOVA tests for all experimental designs** (randomized block, Latin Square, balanced incomplete block, etc., as well as completely randomized). S_2^2 is the smallest that the sum of squared deviations can be made, while S_1^2 is the smallest that the sum of squared deviations can be made *when* the hypothesis to be tested is in fact true. **This ratio (when multiplied by a ratio of "degrees of freedom") always has an F-distribution when the hypothesis is true, and we reject for large values of the ratio.**

Recall that in **the completely randomized design,** we had k treatments and (to talk in the agricultural field setting, although we can just as well be dealing with scales, technicians, catalysts, laboratories, and so on) we used each on n plots. These n plots were chosen at random from the nk plots available, with n distinct plots being allocated to each treatment (and no plot receiving more than one treatment).

Since $E(x_{ij}) = \mu_i$ (for treatments $i = 1, \ldots, k$ on their $j =$ 1st, 2nd, . . ., nth plots), the average treatment yield is

$$\mu = \frac{\mu_1 + \ldots + \mu_k}{k}$$

and we may express each yield μ_i as μ plus a deviation α_i from μ:

$$E(X_{ij}) = \mu_i = \mu + (\mu_i - \mu)$$
$$= \mu + \alpha_i.$$

Then

$$\alpha_1 + \ldots + \alpha_k = (\mu_1 - \mu) + \ldots + (\mu_k - \mu)$$
$$= \mu_1 + \ldots + \mu_k - k\mu$$
$$= \mu_1 + \ldots + \mu_k - k\frac{\mu_1 + \ldots + \mu_k}{k} = 0.$$

We assume that the X_{ij}'s are independent normal random variables with $\text{Var}(X_{ij}) = \sigma^2$ and wish to obtain:

a.) point estimates of $\alpha_1, \ldots, \alpha_k$ (or, of $\mu_1, \ldots, \mu_k$),

b.) confidence intervals on (e.g.) $\alpha_1 - \alpha_2$ $(= \mu_1 - \mu_2)$,

c.) tests (e.g. of H: $\alpha_1 = \ldots = \alpha_k (= 0)$, i.e. $\mu_1 = \ldots = \mu_k$), and

d.) a decision as to which of the treatments is best (i.e., which has smallest of $\mu_1, \ldots, \mu_k$, in our example).

Solutions. (*Note* that many of the related calculations are very conveniently summarized in the *ANOVA table*.)

a.) From the Method of Least Squares, we find our estimators are

$$\hat{\mu}_1 = \overline{X}_1, \ldots, \hat{\mu}_k = \overline{X}_k \text{ (or } \hat{\alpha}_1 = \overline{X}_1 - \overline{X}_., \ldots, \hat{\alpha}_k = \overline{X}_k - \overline{X}_.).$$

b.) If Y is $N(a,b^2)$, then a 95% confidence interval for the mean is

$$Y \pm t_{.025}\sqrt{(\text{Est. of } b^2)}$$

where $t_{.025}$ is found bearing in mind how many "degrees of freedom" our estimate of b^2 has.

Therefore, since $\overline{X}_1$ is $N(\mu_1, \sigma^2/n)$ and $\overline{X}_2$ is $N(\mu_2, \sigma^2/n)$ and they are independent, $\overline{X}_1 - \overline{X}_2$ is $N(\mu_1 - \mu_2, \frac{\sigma^2}{n} + \frac{\sigma^2}{n}) = N(\mu_1 - \mu_2, 2\frac{\sigma^2}{n})$, and a 95% interval for $\mu_1 - \mu_2$ is

$$\overline{X}_1 - \overline{X}_2 \pm t_{.025}\sqrt{(\text{Est. of } 2\frac{\sigma^2}{n})}$$

i.e.

$$\overline{X}_1 - \overline{X}_2 \pm t_{.025}\sqrt{\frac{2}{n}}\sqrt{(\text{Est. of } \sigma^2)}$$

i.e.

$$\overline{X}_1 - \overline{X}_2 \pm t_{.025}\sqrt{\frac{2}{n}}\sqrt{\frac{\Sigma\Sigma(X_{ij} - \overline{X}_i)^2}{k(n-1)}}$$

and we have $k(n-1)$ "degrees of freedom."

c.) We reject H: $\mu_1 = \ldots = \mu_k$ if

$$F = \frac{SS(\text{Tr})/k-1}{SSE/k(n-1)} > F_\alpha.$$

d.) We select as "best" the treatment which yielded the smallest of the sample means $\overline{X}_1, \ldots, \overline{X}_k$.

In each of these cases (a., b., c., d.), the inference we make improves if the experimental variability σ^2 is decreased or if the sample size n is increased. For example, in case:

a.) The point estimates become less variable;

b.) The 95% confidence intervals become shorter;

c.) The test of level α has a better chance of rejecting H *if* H is false (i.e., is more powerful);

d.) The selection has a higher probability of being correct.

The next design, **the randomized blocks design with a single qualitative factor, has as one of its main purposes the lowering of the basic σ^2 underlying the experiment.** This results in less variable point estimates, shorter confidence intervals, more powerful tests of hypotheses, and higher probability of correctly selecting the best treatment, all without any increase in the sample size.

We now have our field split into n blocks which (we believe) may be more homogeneous within themselves than the overall field is, and each block is split into k plots. In each block we use each treatment on exactly one plot, randomizing the order within the block.

Let X_{ij} = the result of the use of treatment i in block j $(i = 1, \ldots, k; j = 1, \ldots, n)$. **We will assume the additive model**

$$\mathbf{E(X_{ij}) = \lambda_i + \tau_j,}$$

and **it is customary to rewrite the model as**

$$E(X_{ij}) = \lambda_i + \tau_j,$$

$$= \underbrace{(\lambda_i - \frac{1}{k}\sum_{i=1}^{k}\lambda_i)}_{\alpha_i} + \underbrace{(\tau_j - \frac{1}{n}\sum_{j=1}^{n}\tau_j)}_{\beta_j} + \underbrace{(\frac{1}{k}\sum_{i=1}^{k}\lambda_i + \frac{1}{n}\sum_{j=1}^{n}\tau_j)}_{\mu}$$

where α_i is called the effect of the *i*th treatment, β_j is called the effect of the *j*th block, and μ is called the grand mean. Note that $\sum_{i=1}^{k}\alpha_i = \sum_{j=1}^{n}\beta_j = 0$. We will assume that $\mathrm{Var}(X_{ij}) = \sigma^2$ (this is *not* the same σ^2 as in the C.R. design *unless* $\beta_1 = \ldots = \beta_n = 0$; we hope effective blocking has yielded a smaller underlying variance), and again **study the α_i's (and now also the β_j's) to obtain**

a.) point estimates,

b.) confidence interval estimates,

c.) tests of hypotheses (e.g., H: $\alpha_1 = \ldots = \alpha_k$), and

d.) a decision as to which is the largest.

It will be helpful to introduce the notations

$$\bar{X}_{i\cdot} = \frac{1}{n}\sum_{j=1}^{n} X_{ij},\ \bar{X}_{\cdot j} = \frac{1}{k}\sum_{i=1}^{k} X_{ij},\text{ and } \bar{X}_{\cdot\cdot} = \frac{1}{kn}\sum_{i=1}^{k}\sum_{j=1}^{n} X_{ij}.$$

a. Again using the Method of Least Squares, we find the point estimators

$$\hat{\alpha}_i = \bar{X}_{i\cdot} - \bar{X}_{\cdot\cdot},\ \hat{\beta}_j = \bar{X}_{\cdot j} - \bar{X}_{\cdot\cdot},\text{ and } \hat{\mu} = \bar{X}_{\cdot\cdot}.$$

Again, these estimators are obtained under the assumption that the X_{ij}'s are independent in an additive (no interaction) model with variance σ^2.

To obtain *b.*, *c.*, and *d.*, we also assume that X_{ij} is normally distributed $N(\mu + \alpha_i + \beta_j, \sigma^2)$ ($i = 1, \ldots, k$; $j = 1, \ldots, n$). Then (e.g.)

$$\bar{X}_{1\cdot} - \bar{X}_{2\cdot} \text{ is } N(\alpha_1 - \alpha_2, \sigma^2 \frac{2}{n})$$

and a 95% confidence interval for $\alpha_1 - \alpha_2$ is

$$\bar{X}_{1\cdot} - \bar{X}_{2\cdot} \pm t_{.025}\sqrt{\frac{2}{n}}\sqrt{(\text{Est. of } \sigma^2)}\ .$$

The *estimator of* σ^2 can be motivated by the following

Theorem:

Total $SS = \Sigma\Sigma(X_{ij} - \bar{X}_{\cdot\cdot})^2$

$= \Sigma\Sigma\,[(\bar{X}_{i\cdot} - \bar{X}_{\cdot\cdot}) + (\bar{X}_{\cdot j} - \bar{X}_{\cdot\cdot}) + (X_{ij} - \bar{X}_{i\cdot} - \bar{X}_{\cdot j} + \bar{X}_{\cdot\cdot})]^2$

$= \Sigma\Sigma\,(\bar{X}_{i\cdot} - \bar{X}_{\cdot\cdot})^2$ **Treatment SS** **SS (Tr)**

$+ \Sigma\Sigma\,(\bar{X}_{\cdot j} - \bar{X}_{\cdot\cdot})^2$ **Block SS** **SS (Bℓ)**

$+ \Sigma\Sigma\,(X_{ij} - \bar{X}_{i\cdot} - \bar{X}_{\cdot j} + \bar{X}_{\cdot\cdot})^2$ **Residual *SS*** ***SSE*.**

Table 15

Source of Variation	*SS*	d.f.	*MS*	*E(MS)*
Treatments	$\Sigma\Sigma\,(\bar{X}_{i\cdot} - \bar{X}_{\cdot\cdot})^2$	$k-1$	$SS\,(\mathrm{Tr})/(k-1)$	$\sigma^2 + \frac{n\Sigma\alpha_i^2}{k-1}$
Blocks	$\Sigma\Sigma\,(\bar{X}_{\cdot j} - \bar{X}_{\cdot\cdot})^2$	$n-1$	$SS\,(B\ell)/(n-1)$	$\sigma^2 + \frac{k\Sigma\beta_j^2}{n-1}$
Residual	$\Sigma\Sigma\,(X_{ij} - \bar{X}_{i\cdot} - \bar{X}_{\cdot j} + \bar{X}_{\cdot\cdot})^2$	$(n-1)(k-1)$	$SSE/(n-1)(k-1)$	σ^2
Total	$\Sigma\Sigma\,(X_{ij} - \bar{X}_{\cdot\cdot})^2$	$nk-1$		

Table 16

Source of Variation	*SS*	d.f.	*MS*	*E(MS)*
Treatments	349.625	9	38.85	$\sigma^2 + \frac{4}{9}\Sigma\alpha_i^2$
Blocks	269.875	3	89.9583	$\sigma^2 + \frac{10}{3}\Sigma\beta_j^2$
Residual	224.875	27	8.3287	σ^2
Total	844.375	39		

The analysis of variance table corresponding to these and their expected values is as shown in Table 15.

Now it so happens that the data of our previous example (p. 453) *was* taken in blocks, each column of the data table that was previously given representing a block. If we now calculate the analysis of variance table for that data bearing this fact in mind, we find the data given in Table 16.

since (utilizing our previous C.R. calculations to the maximum possible extent) the new items follow from

$$\bar{X}_{.1} = 2.2, \bar{X}_{.2} = 5.4, \bar{X}_{.3} = 8.1, \bar{X}_{.4} = 8.8$$
$$k\Sigma(\bar{X}_{.j} - \bar{X}_{..})^2 = 10(26.9875) = 269.875.$$

Thus our *confidence interval* on $\alpha_1 - \alpha_2$ is

$$\bar{X}_{1.} - \bar{X}_{2.} \pm t_{.025}\sqrt{\frac{2}{n}}\sqrt{8.3287}$$

i.e.

$$2 - 8 \pm 2.052\sqrt{\frac{2}{4}}\sqrt{8.3287} = -6 \pm 4.187$$

i.e.

$$-10.19 \text{ to } -1.81.$$

(Ignoring blocking our confidence interval had length 11.72, while now it has length 8.38; the previous interval was 40% longer!)

Note that **blocking reduced our (estimated) underlying σ^2 from 16.49 to 8.33 . . . about half of our observed variability was in fact due to block differences.**

Relevant hypotheses here are tested as follows:

$$\boldsymbol{H}: \alpha_1 = \ldots = \alpha_k\ (=0) \text{ (No treatment effects)}$$

$$F = \frac{38.85}{8.3287} = 4.665.$$

We compare this with F_α with 9 and 27 d.f. and $F_{.05} = 2.25$, $F_{.01} = 3.15$. Hence we reject H at level .01. (Loss of degrees of freedom increases F_α, but hurt us very little . . . as Table 6 shows, this is true unless we are dealing with very small degrees of freedom. The reduction in underlying variability benefitted us much more!)

$$\boldsymbol{H}: \beta_1 = \ldots = \beta_n\ (=0) \text{ (No block effects)}$$

$$F = \frac{89.9583}{8.3287} = 10.80$$

and, with 3 and 27 d.f., $F_{.05} = 2.95$ while $F_{.01} = 4.60$. Thus the block effects are statistically significant.

Notes

1. Simpler computing formulas are available, but are not our main point. Calculations today are usually done on a large computer, hence ease of hand-computation is not overly important (though accurate calculation is, and good computer programs are designed to calculate accurately).
2. If *interaction* is present, $E(X_{ij}) = \mu_{ij}$, then only point estimates can be obtained unless we can *replicate* (repeat block conditions or split plots in a block); otherwise no degrees of freedom are left for estimation of σ^2.
3. Missing values in the data set require different formulas.
4. Incomplete blocks (block size smaller than number of treatments) designs are also available.
5. The method of obtaining an appropriate test statistic for *any* model is that considered earlier, using S_2^2 and S_1^2.

If we attempt to put **a confidence interval on the difference in the treatment effects of the two most effective treatments,** we note that these are (in this sample) Treatment 1 with $\bar{X}_1 = 2.00$, and Treatment 10 with $\bar{X}_{10} = .50$.

From the *completely randomized* analysis we would find $\alpha_{10} - \alpha_1$ is in

$$.50 - 2.00 \pm 5.86$$

i.e.

$$-7.36 \leq \alpha_{10} - \alpha_1 \leq 4.36;$$

that is, we're 95% sure Treatment 10 allows from 7.36% less, to 4.36% more, than does Treatment 1.

From the *randomized blocks* analysis we would find $\alpha_{10} - \alpha_1$ is in

$$.50 - 2.00 \pm 4.187$$

i.e.

$$-5.69 \leq \alpha_{10} - \alpha_1 \leq 2.69;$$

that is, we're 95% sure Treatment 10 allows from 5.69% less, to 2.69% more, than does Treatment 1.

A Latin square analysis might be able to yield an even smaller variability underlying the experiment (a smaller σ^2), and hence a shorter interval on $\alpha_{10} - \alpha_1$ (the current one does *not* reject the hypothesis that $\alpha_{10} - \alpha_1$). That analysis cannot be done, however, since the experiment was not run in that design.

Next we will discuss **(1) how one can test for interaction in the randomized blocks model, (2) the possible errors in looking at confidence intervals (e.g., on $\alpha_{10} - \alpha_1$, $\alpha_1 - \alpha_9$, and so on) and how multiple-comparisons procedures control these errors, and (3) how the**

probability of correctly selecting the best treatment may be evaluated and controlled and how experimental design and replication help to increase this probability.

Recall that we have seen that *interaction* is one reason for running factorial experiments, and that interaction can and does occur in many experiments. **Since our randomized blocks analysis was based on a no-interactions assumption, $E(X_{ij}) = \mu + \alpha_i + \beta_j$, it would be advisable to check the validity of this assumption.** In the general case with interaction our model would have

$$E(X_{ij}) = \mu + \alpha_i + \beta_j + \eta_{ij},$$

so the hypothesis of interest here is that $\eta_{11} = \eta_{12} = \ldots = \eta_{kk} = 0$. **A popular test, given by Tukey,** computes

$$T = \frac{\{\Sigma\Sigma(X_{ij} - \bar{X}_{i\cdot} - \bar{X}_{\cdot j} + \bar{X}_{\cdot\cdot})(\bar{X}_{i\cdot} - \bar{X}_{\cdot\cdot})(\bar{X}_{\cdot j} - \bar{X}_{\cdot\cdot})\}^2}{\Sigma(\bar{X}_{i\cdot} - \bar{X}_{\cdot\cdot})^2 \, \Sigma(\bar{X}_{\cdot j} - \bar{X}_{\cdot\cdot})^2},$$

$$W^* = (nk - n - k)\frac{T}{SSE - T},$$

notes that when the hypothesis of no interaction is true W^* has an f-distribution with 1 and $nk - n - k$ degrees of freedom, and thus **decides there is interaction if $W^* > F_{.05}$.** (See V. Hegemann and D.E. Johnson, "The power of two tests for nonadditivity," *Journal of the American Statistical Association*, Vol. 71 (1976), pp. 945–948.)

As motivation for Tukey's test statistic for interaction, note that

$$\begin{cases} \bar{X}_{\bullet\bullet} \text{ estimates } \mu \\ \bar{X}_{i\cdot} \text{ estimates } \mu + \alpha_i \\ \bar{X}_{\cdot j} \text{ estimates } \mu + \beta_j \\ X_{ij} \text{ has mean } \mu + \alpha_i + \beta_j + \eta_{ij}, \end{cases}$$

so that

$$X_{ij} - \bar{X}_{i\cdot} - \bar{X}_{\cdot j} + \bar{X}_{\cdot\cdot}$$

estimates

$$(\mu + \alpha_i + \beta_j + \eta_{ij}) - (\mu + \alpha_i) - (\mu + \beta_j) + \mu = \eta_{ij}.$$

Thus, since $\bar{X}_{i\cdot} - \bar{X}_{\cdot\cdot}$ estimates $(\mu + \alpha_i) - \mu = \alpha_i$ and $\bar{X}_{\cdot j} - \bar{X}_{\cdot\cdot}$ estimates $(\mu + \beta_j) - \mu = \beta_j$, we may think of T as estimating

$$\frac{(\Sigma\Sigma\alpha_i\beta_j\eta_{ij})^2}{(\Sigma\alpha_i^2)(\Sigma\beta_j^2)},$$

and it is intuitively reasonable to reject the hypothesis of no interaction (i.e., the hypothesis that $\eta_{11} = \eta_{12} = \ldots = \eta_{kk} = 0$) if this estimate is "large."

For our example, we can calculate

$$T = \frac{\{311.084375\}^2}{(349.625)(269.875)} = 1.025631289$$

$$W^* = 26\,\frac{1.026}{224.875 - 1.026} = .119,$$

while $F_{.05} = 4.23$. We therefore do not reject the hypothesis that there are no interactions in this setting. **(If this test indicated significant interactions, one would often transform the data, e.g. to $Y_{ij} = \log(X_{ij})$,** if possible, **and then perform ANOVA tests.** For items other than tests, e.g. confidence intervals on differences of means, the situation is not handled well by transformations, but one must then go back to the basic experimental goals and interpret them in light of the interaction. In our present example, e.g., interactions between blocks and treatments would indicate very different types of blocks very often, and one would want to examine the field in some detail to see whether all were equally relevant, or whether some were perhaps unsuited to the experiment—did they have greatly different soils, etc.)

Note that a value of .119 is not very small for an F-distribution with 1 and 26 degrees of freedom. While such a distribution will yield a value greater than 4.23 only 5% of the time ($F_{.05} = 4.23$), it will* yield a value less than .004 only 5% of the time as well. Thus even .119 is *not* in the extreme lower tail of this distribution, and the small value should not alarm us.

Our next statistical analysis method, that of **multiple-comparisons, is needed because after our experiment we will either be looking at confidence intervals on differences that we did not know beforehand would be the differences of main interest** (e.g., in the example experiment, we did not know in advance that $\alpha_{10} - \alpha_1$ would be of prime interest) **or we will be looking at many comparisons** (e.g., all new treatments compared each with the standard treatment, in which case we may construct confidence intervals on all of $\alpha_{10} - \alpha_1$, $\alpha_9 - \alpha_1$, $\alpha_8 - \alpha_1$, ..., $\alpha_3 - \alpha_1$, $\alpha_2 - \alpha_1$).

The problem which multiple-comparisons procedures are designed to deal with can be seen from a simple example. Suppose we put a 95% confidence interval on $\alpha_{10} - \alpha_9$ and, independently, another 95% confidence interval on $\alpha_5 - \alpha_4$. Then the probability that *both* intervals are *correct* (i.e., cover the true difference) is only $.95 \times .95 = .9025$. Thus while each interval by itself has a 5% chance of being in error, the chance that at least one of the two is wrong is 10%. With five 95% intervals, the probability that at least one is wrong is .23. The problem is to construct our intervals in such a way that we can be 95% sure that all are right.

*This follows from the fact that if W^* has an F-distribution with 1 and 26 degrees of freedom, then $1/W^*$ has an F-distribution with 26 and 1 degrees of freedom. Hence (interpolating in our table)

$$.05 = P[\frac{1}{W^*} \geq 249.6] = P[\frac{1}{249.6} \geq W^*] = P[.004 \geq W^*].$$

With Scheffé's multiple-comparisons method we are allowed to be interested in any and every linear function of $\alpha_1, \ldots, \alpha_k$, say $a_1\alpha_1 + \ldots + a_k\alpha_k$, for which $a_1 + \ldots + a_k = 0$; such a linear function is called a *contrast*, and examples are $\alpha_{10} - \alpha_1$, $2\alpha_{10} - \alpha_1 - \alpha_2$, and $\alpha_{10} - \frac{\alpha_1 + \ldots + \alpha_9}{9}$. We then estimate $\alpha_1, \ldots, \alpha_{10}$ by $\bar{X}_1, \ldots, \bar{X}_{10}$ in the completely randomized design (or $\bar{X}_{1\cdot}, \ldots, \bar{X}_{10\cdot}$ in the randomized blocks design) and take, as our interval on a $a_1\alpha_1 + \ldots + a_k\alpha_k$,

$$(a_1\bar{X}_1 + \ldots + a_k\bar{X}_k) \pm \sqrt{(k-1)F_{.05}}\sqrt{\sum_{i=1}^{k} a_i^2}\sqrt{(\text{Est. of } \sigma^2)}\ ;$$

here $F_{.05}$ is calculated with $k-1$ and "residual" degrees of freedom (the latter being $k(n-1)$ for the completely randomized design and $(k-1)(n-1)$ for the randomized blocks design). **We are 95% sure that "every contrast is covered by the interval this yields on it," so the probability that "all such intervals are correct" is .95.**

Applying this method to our example, let's find joint intervals for $\alpha_1 - \alpha_2$ and $\alpha_{10} - \alpha_1$.

In the completely randomized design

$$\begin{aligned} &\sqrt{(k-1)F_{.05}}\sqrt{2}\sqrt{(\text{Est. of } \sigma^2} \\ &= \sqrt{(9)(2.21)}\sqrt{2}\sqrt{16.49} \\ &= 25.61, \end{aligned}$$

so our intervals are

$$-31.61 \le \alpha_1 - \alpha_2 \le 19.61 \text{ and}$$
$$-27.11 \le \alpha_{10} - \alpha_1 \le 24.11.$$

We are 95% sure that *both* of these statements hold. (The individual 95% intervals were, recall, $-11.86 \le \alpha_1 - \alpha_2 \le .14$ and $-7.36 \le \alpha_{10} - \alpha_1 \le 4.36$.)

In the randomized blocks design

$$\sqrt{(k-1)F_{.05}}\sqrt{2}\sqrt{(\text{Est. of } \sigma^2} = \sqrt{(9)(2.25)}\sqrt{2}\sqrt{8.33} = 18.37,$$

so our intervals are

$$-24.37 \le \alpha_1 - \alpha_2 \le 12.37 \text{ and}$$
$$-19.87 \le \alpha_{10} - \alpha_1 \le 16.87.$$

We are 95% sure that *both* of these statements hold. (The individual 95% intervals were, recall, $-10.19 \le \alpha_1 - \alpha_2 \le 1.81$ and $-5.69 \le \alpha_{10} - \alpha_1 \le 2.69$.)

If we are *only* interested in confidence intervals on the pairs $\alpha_i - \alpha_j$ $(i, j = 1, \ldots, k)$, then we should not use Scheffé's method, which gives joint (simultaneous) intervals on all contrasts $a_1\alpha_1 + \ldots + a_k\alpha_k$, since *the more items we're allowed to put intervals on*, the wider the intervals will be.

With Tukey's multiple-comparisons method we're allowed to be interested in all pairwise differences $\alpha_i - \alpha_j$ $(i, j = 1, \ldots, k)$. We estimate $\alpha_i - \alpha_j$ by $\bar{X}_i - \bar{X}_j$ (or $\bar{X}_{i\cdot} - \bar{X}_{j\cdot}$ in the randomized blocks design) and take, as our interval on $\alpha_i - \alpha_j$,

$$(\bar{X}_i - \bar{X}_j) \pm \sqrt{q_{.05}}\sqrt{(\text{Est. of } \sigma^2)}\ ;$$

here $q_{.05}$ comes from the tables of the Studentized Range with "$p = k$ and $n_2 =$ d.f. for est. of σ^2."

In our example, in the completely randomized design, this yields

$$\sqrt{q_{.05}}\sqrt{(\text{Est. of } \sigma^2)} = \sqrt{4.83}\sqrt{16.49} = 8.92,$$

so our intervals are

$$-14.92 \le \alpha_1 - \alpha_2 \le 2.92 \text{ and } -10.42 \le \alpha_{10} - \alpha_1 \le 7.42.$$

We are 95% sure that *both* of these statements hold.

In the randomized blocks design,

$$\sqrt{q_{.05}}\sqrt{(\text{Est. of } \sigma^2)} = \sqrt{4.88}\sqrt{8.33} = 6.38,$$

so our intervals are

$$-12.38 \le \alpha_1 - \alpha_2 \le 0.38 \text{ and } -7.88 \le \alpha_{10} - \alpha_1 \le 4.88.$$

We are 95% sure that *both* of these statements (and statements about any other pairs of interest) hold.

If we know before the experiment that we will be interested only in pairs like $\alpha_{10} - \alpha_1, \alpha_9 - \alpha_1, \ldots, \alpha_2 - \alpha_1$ (i.e., all treatments compared with a "standard"), we can shorten the intervals further by using Dunnett's multiple-comparison method. However, if we wish to compare the "best-looking few," we seem forced to use Tukey's method. Scheffé's method seems unsuited to distinct treatments, but is often of interest in other areas.

Related methods are those of Newman-Keuls and of Duncan. These methods increase our risk of finding a hodgepodge of "significant differences" which are not meaningful but rather result from the randomness inherent in the experiment. **These procedures are unnecessary* and confuse the issue for users of statistics, and we will not cover them further.** (For more detail, see p. 302ff of *Statistical Issues*, edited by Roger E. Kirk, Brooks/Cole Pub. Co., Monterey, Calif., 1972.)

We may, if we wish, **interpret our experiment as**

*For shorter intervals, use $\alpha = .10$. Or, use t intervals on pairs.

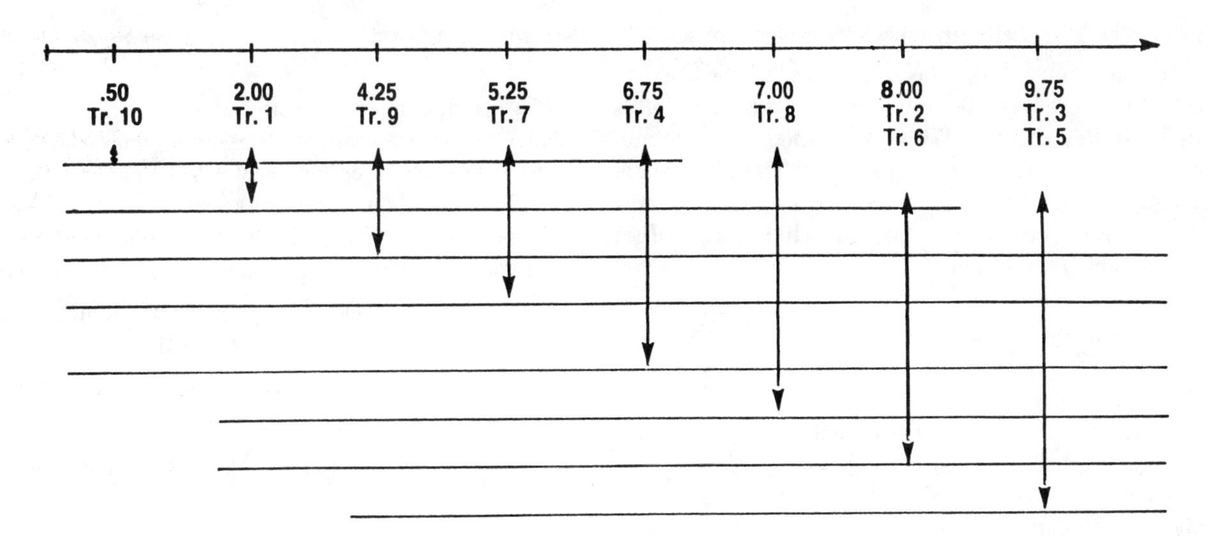

Figure 57

saying that "α_i is (stat.) significantly different from α_j" if $\bar{X}_{i\cdot}$ differs from $\bar{X}_{j\cdot}$ by more than $\sqrt{q_{.05}}\ \sqrt{(\text{Est. of } \sigma^2)}$. E.g., in the example $\sqrt{q_{.05}}\ \sqrt{(\text{Est. of } \sigma^2)} = 6.38$ and graphically (underscoring nonsignificant differences) as shown in Figure 57.

A statistical area we have alluded to, from time to time, is that of *ranking and selection procedures*. While this area is a broad one, it will be clearest if we restrict ourselves to talking about **the problem of selecting the best treatment**. In-depth coverage is given in a separate book.

As we have seen previously, **the goals we have in running an experiment are: sometimes goals which lead us naturally to point estimates** (e.g., "estimate the effect of the *i*th treatment"); **sometimes goals which lead us naturally to testing hypotheses** (e.g., "is there any difference among these *k* treatments?"); **sometimes goals which lead us to interval estimation** (e.g., "within what range does the difference in effect of treatments 1 and 10 lie?"); **and sometimes goals which lead to selection problems** (e.g., "which is the best treatment?").

In each case the problem is statistical, since we only observe effects plus random errors, and we wish to have our procedures satisfy reasonable properties so we are not led either to make incorrect decisions, or to infer more than is warranted from the data.

E.g., in the example experiment, using the randomized blocks design, our estimate of $\alpha_{10}-\alpha_1$ was

$$.50 - 2.00 = -1.50.$$

However, our individual 95% confidence interval on $\alpha_{10}-\alpha_1$ was

$$-5.69 \leq \alpha_{10}-\alpha_1 \leq 2.69$$

which says: we're not even sure that α_{10} and α_1 are different (since $\alpha_{10}-\alpha_1 = 0$, i.e. $\alpha_{10}=\alpha_1$, is possible), but we are sure that

$$\alpha_{10} \leq \alpha_1 + 2.69 \text{ and } \alpha_{10} \geq \alpha_1 - 5.69.$$

This means something quite different to us than if we had obtained either $-3.01 \leq \alpha_{10}-\alpha_1 \leq 0.01$ or $-1.51 \leq \alpha_{10}-\alpha_1 \leq 1.49$ as our interval on $\alpha_{10}-\alpha_1$. All these intervals are centered about an estimate $\alpha_{10}-\alpha_1 = -1.50$, but the narrower ones give us more certitude about the magnitude of possible differences.

Thus if we act as though $\alpha_{10}-\alpha_1 = -1.50$, when in fact we only have reasonable certitude that $-5.69 \leq \alpha_{10}-\alpha_1 \leq 2.69$, we may make erroneous decisions which we will regret later. **The uncertainty present as expressed by the 95% interval about the point estimate of -1.50 should be used to temper decisions based on the point estimate.**

Similarly, if our *goal* is to select the best treatment, we will look at the resulting $\bar{X}_{i\cdot}$ ($i=2, 3, \ldots, 10$) (note that $i=1$ represents the standard treatment) and choose the treatment yielding the *smallest** average (i.e., the smallest of $\bar{X}_{2\cdot}, \bar{X}_{3\cdot}, \ldots, \bar{X}_{10\cdot}$) as being best. But, **is it truly the best, or is it very possible that the inherent variability present in the experiment has made a treatment other than the truly best one look best in our experiment? Can we be 95% sure we have really selected the best treatment when we choose the one which yielded the smallest of $\bar{X}_{2\cdot}, \bar{X}_{3\cdot}, \ldots, \bar{X}_{10\cdot}$?**

(Note that since we found

$\bar{X}_{2\cdot} = 8.00, \bar{X}_{3\cdot} = 9.75, \bar{X}_{4\cdot} = 6.75, \bar{X}_{5\cdot} = 9.75,$

$\bar{X}_{6\cdot} = 8.00, \bar{X}_{7\cdot} = 5.25, \bar{X}_{8\cdot} = 7.00, \bar{X}_{9\cdot} = 4.25$, and $\bar{X}_{10\cdot} = .50$,

*In this example the product is an *undesirable* by-product.

we would be led to select treatment 10 as being the best.)

One way of answering this question as to how sure we can be that treatment 10 is best, **is to use S.S. Gupta's subset selection method. With this method, we select not just one treatment, but a set of several: The smallest set of treatments in which we can be 95% sure that the best treatment is included.** To do this, **we include in the subset all treatments *j* for which**

$$\bar{X}_{j\cdot} \leq \{\text{Smallest of } \bar{X}_{2\cdot}, \ldots, \bar{X}_{10\cdot}\} + D\frac{\sqrt{MSE}}{\sqrt{n}}$$

where n is the sample size on which each of the averages $\bar{X}_{2\cdot}, \ldots, \bar{X}_{10\cdot}$ is based ($n=4$ in our experiment), MSE is our estimate of σ^2 ($MSE = 8.3287$ in our randomized blocks analysis of the data), and D is a constant from tables given by Gupta and Sobel:

***D* for 95% chance of including the best treatment in the selected subset** as shown in Table 17.

Table 17

ν \ k	2	5	10	20	50
15	2.48	3.34	3.78	4.17	4.62
20	2.44	3.25	3.67	4.02	4.44
24	2.42	3.22	3.63	3.97	4.37
30	2.40	3.19	3.59	3.92	4.31
100	2.35	3.09	3.47	3.78	4.14

k = No. of treatments being considered
ν = degrees of freedom for MSE

From S. S. Gupta and M. Sobel, "On a statistic which arises in selection and ranking problems," *Annals of Mathematical Statistics,* Vol. 28 (1957), pp. 957–967.

Since we have $k=9$ and $\nu=27$, our D is approximately 3.60. Hence we include in our selected subset all treatments j for which

$$\bar{X}_{j\cdot} \leq 0.50 + 3.60\frac{\sqrt{8.3287}}{\sqrt{4}}$$

i.e.,

$$\bar{X}_{j\cdot} \leq 5.695.$$

We are therefore able to state that *we are 95% sure that the best treatment* (of the 9 treatments) *is among treatments 7, 9, and 10.*

If we wish comparable certitude about which of these 3 treatments is best, it will be necessary to run further experiments with these three treatments. *If* we choose the "best-looking one" without further experiments, we should not be surprised if (in future experiments run in future years) one of the other treatments from this subset turns out to be better.

Note that if we had used a completely randomized experiment, we would have had in the selected subset all treatments j for which

$$\bar{X}_{j\cdot} \leq 0.50 + 3.51\frac{\sqrt{16.49}}{\sqrt{4}} = 7.63,$$

namely treatments 4, 7, 8, 9, 10. Thus the variance reduction due to blocking allowed us to select a subset of 3 treatments (rather than of 5), which is a very substantial improvement.

For the additional experiment to select the one best of treatments 7, 9, and 10, we use R.E. Bechhofer's indifference-zone selection method. The sample size needed to be 95% sure we really select the best treatment **is approximately**

$$n = -4\ln(.05)\frac{MSE}{(\delta^*)^2} = 12\frac{MSE}{(\delta^*)^2}$$

where δ^* is the "smallest difference worth detecting." Thus, if we desire fine discrimination (δ^* small), it costs us more experimentation than if a coarser discrimination (δ^* larger) will suffice. In terms of the example (where $MSE = 8.3287$), we have the data shown in Example 11.

δ^*	$n = 12\frac{MSE}{(\delta^*)^2}$
1	100
2	25
3	11

Example 11

Thus, if we run 11 more plots on treatments 7, 9, and 10, we will be able to select the best-looking one *and* be 95% sure that it is within $\delta^* = 3$ units of the best.

MULTIPLE COMPARISONS WITH A CONTROL OR STANDARD

As we noted on pages 460 and 461, often one wishes to specify **which treatments are "better" or "worse" when compared with the standard treatment.** Here **"the least significant difference" (LSD) yardstick of Fisher,**

Table 18: Table of t for Two-Sided Comparisons Between p Treatments and a Control for a Joint Confidence Coefficient of $P = 95\%$

	p = *number of treatment means (excluding the control)*													
d.f.	1	2	3	4	5	6	7	8	9	10	11	12	15	[20]
5	2.57	$3.03^{2.3}$	$3.29^{3.6}$	$3.48^{4.6}$	$3.62^{5.4}$	$3.73^{5.9}$	$3.82^{6.4}$	$3.90^{6.8}$	$3.97^{7.2}$	$4.03^{7.5}$	$4.09^{7.8}$	$4.14^{8.0}$	$4.26^{8.7}$	$4.42^{9.4}$
6	2.45	$2.86^{2.1}$	$3.10^{3.4}$	$3.26^{4.3}$	$3.39^{5.0}$	$3.49^{5.6}$	$3.57^{6.0}$	$3.64^{6.4}$	$3.71^{6.8}$	$3.76^{7.1}$	$3.81^{7.4}$	$3.86^{7.6}$	$3.97^{8.2}$	$4.11^{9.0}$
7	2.36	$2.75^{2.0}$	$2.97^{3.2}$	$3.12^{4.1}$	$3.24^{4.8}$	$3.33^{5.3}$	$3.41^{5.7}$	$3.47^{6.1}$	$3.53^{6.5}$	$3.58^{6.7}$	$3.63^{7.0}$	$3.67^{7.2}$	$3.78^{7.8}$	$3.91^{8.6}$
8	2.31	$2.67^{2.0}$	$2.88^{3.1}$	$3.02^{3.9}$	$3.13^{4.5}$	$3.22^{5.1}$	$3.29^{5.5}$	$3.35^{5.9}$	$3.41^{6.2}$	$3.46^{6.5}$	$3.50^{6.7}$	$3.54^{6.9}$	$3.64^{7.5}$	$3.76^{8.2}$
9	2.26	$2.61^{1.9}$	$2.81^{3.0}$	$2.95^{3.8}$	$3.05^{4.4}$	$3.14^{4.9}$	$3.20^{5.3}$	$3.26^{5.6}$	$3.32^{5.9}$	$3.36^{6.2}$	$3.40^{6.5}$	$3.44^{6.7}$	$3.53^{7.2}$	$3.65^{7.9}$
10	2.23	$2.57^{1.8}$	$2.76^{2.9}$	$2.89^{3.6}$	$2.99^{4.2}$	$3.07^{4.7}$	$3.14^{5.1}$	$3.19^{5.4}$	$3.24^{5.7}$	$3.29^{6.0}$	$3.33^{6.2}$	$3.36^{6.5}$	$3.45^{7.0}$	$3.57^{7.7}$
11	2.20	$2.53^{1.8}$	$2.72^{2.8}$	$2.84^{3.5}$	$2.94^{4.1}$	$3.02^{4.6}$	$3.08^{4.9}$	$3.14^{5.3}$	$3.19^{5.6}$	$3.23^{5.8}$	$3.27^{6.1}$	$3.30^{6.3}$	$3.39^{6.8}$	$3.50^{7.5}$
12	2.18	$2.50^{1.7}$	$2.68^{2.7}$	$2.81^{3.4}$	$2.90^{4.0}$	$2.98^{4.4}$	$3.04^{4.8}$	$3.09^{5.1}$	$3.14^{5.4}$	$3.18^{5.7}$	$3.22^{5.9}$	$3.25^{6.1}$	$3.34^{6.6}$	$3.45^{7.3}$
13	2.16	$2.48^{1.7}$	$2.65^{2.7}$	$2.78^{3.4}$	$2.87^{3.9}$	$2.94^{4.3}$	$3.00^{4.7}$	$3.06^{5.0}$	$3.10^{5.3}$	$3.14^{5.5}$	$3.18^{5.8}$	$3.21^{6.0}$	$3.29^{6.5}$	$3.40^{7.1}$
14	2.14	$2.46^{1.7}$	$2.63^{2.6}$	$2.75^{3.3}$	$2.84^{3.8}$	$2.91^{4.2}$	$2.97^{4.5}$	$3.02^{4.9}$	$3.07^{5.2}$	$3.11^{5.4}$	$3.14^{5.6}$	$3.18^{5.8}$	$3.26^{6.3}$	$3.36^{7.0}$
15	2.13	$2.44^{1.7}$	$2.61^{2.6}$	$2.73^{3.2}$	$2.82^{3.8}$	$2.89^{4.2}$	$2.95^{4.5}$	$3.00^{4.8}$	$3.04^{5.1}$	$3.08^{5.3}$	$3.12^{5.5}$	$3.15^{5.7}$	$3.23^{6.2}$	$3.33^{6.8}$
16	2.12	$2.42^{1.6}$	$2.59^{2.5}$	$2.71^{3.2}$	$2.80^{3.7}$	$2.87^{4.1}$	$2.92^{4.4}$	$2.97^{4.7}$	$3.02^{5.0}$	$3.06^{5.2}$	$3.09^{5.4}$	$3.12^{5.6}$	$3.20^{6.1}$	$3.30^{6.7}$
17	2.11	$2.41^{1.6}$	$2.58^{2.5}$	$2.69^{3.1}$	$2.78^{3.6}$	$2.85^{4.0}$	$2.90^{4.4}$	$2.95^{4.7}$	$3.00^{4.9}$	$3.03^{5.1}$	$3.07^{5.3}$	$3.10^{5.5}$	$3.18^{6.0}$	$3.27^{6.6}$
18	2.10	$2.40^{1.6}$	$2.56^{2.5}$	$2.68^{3.1}$	$2.76^{3.6}$	$2.83^{4.0}$	$2.89^{4.3}$	$2.94^{4.6}$	$2.98^{4.8}$	$3.01^{5.1}$	$3.05^{5.3}$	$3.08^{5.4}$	$3.16^{5.9}$	$3.25^{6.5}$
19	2.09	$2.39^{1.6}$	$2.55^{2.5}$	$2.66^{3.1}$	$2.75^{3.5}$	$2.81^{3.9}$	$2.87^{4.2}$	$2.92^{4.5}$	$2.96^{4.8}$	$3.00^{5.0}$	$3.03^{5.2}$	$3.06^{5.4}$	$3.14^{5.8}$	$3.23^{6.4}$
20	2.09	$2.38^{1.6}$	$2.54^{2.4}$	$2.65^{3.0}$	$2.73^{3.5}$	$2.80^{3.9}$	$2.86^{4.2}$	$2.90^{4.5}$	$2.95^{4.7}$	$2.98^{4.9}$	$3.02^{5.1}$	$3.05^{5.3}$	$3.12^{5.7}$	$3.22^{6.3}$
24	2.06	$2.35^{1.5}$	$2.51^{2.3}$	$2.61^{2.9}$	$2.70^{3.4}$	$2.76^{3.7}$	$2.81^{4.0}$	$2.86^{4.3}$	$2.90^{4.5}$	$2.94^{4.7}$	$2.97^{4.9}$	$3.00^{5.1}$	$3.07^{5.5}$	$3.16^{6.0}$
30	2.04	$2.32^{1.5}$	$2.47^{2.3}$	$2.58^{2.8}$	$2.66^{3.2}$	$2.72^{3.6}$	$2.77^{3.9}$	$2.82^{4.1}$	$2.86^{4.3}$	$2.89^{4.5}$	$2.92^{4.7}$	$2.95^{4.8}$	$3.02^{5.2}$	$3.11^{5.8}$
[40]	2.02	$2.29^{1.4}$	$2.44^{2.2}$	$2.54^{2.7}$	$2.62^{3.1}$	$2.68^{3.4}$	$2.73^{3.7}$	$2.77^{3.9}$	$2.81^{4.1}$	$2.85^{4.3}$	$2.87^{4.5}$	$2.90^{4.6}$	$2.97^{5.0}$	[$3.06^{5.5}$]
60	2.00	$2.27^{1.4}$	$2.41^{2.1}$	$2.51^{2.6}$	$2.58^{3.0}$	$2.64^{3.3}$	$2.69^{3.5}$	$2.73^{3.7}$	$2.77^{3.9}$	$2.80^{4.1}$	$2.83^{4.2}$	$2.86^{4.4}$	$2.92^{4.7}$	$3.00^{5.1}$
120	1.98	$2.24^{1.3}$	$2.38^{2.0}$	$2.47^{2.5}$	$2.55^{2.8}$	$2.60^{3.1}$	$2.65^{3.3}$	$2.69^{3.5}$	$2.73^{3.7}$	$2.76^{3.8}$	$2.79^{4.0}$	$2.81^{4.1}$	$2.87^{4.4}$	$2.95^{4.8}$
∞	1.96	$2.21^{1.3}$	$2.35^{1.9}$	$2.44^{2.3}$	$2.51^{2.7}$	$2.57^{2.9}$	$2.61^{3.1}$	$2.65^{3.3}$	$2.69^{3.5}$	$2.72^{3.6}$	$2.74^{3.7}$	$2.77^{3.8}$	$2.83^{4.1}$	$2.91^{4.5}$

The tabular value is the critical value of t appropriate when $p = 0.5$ or $n_c/n_t = 1$. The value shown as a superscript, when multiplied by $(1 - 2p)/(1 - p)$ or $1 - n_t/n_e$, gives the percentage increase required in the critical value of t valid for $.125 < p < .5$ or $n_c/n_t > 1$.

We should note that these tables can also be used for cases with unequal sample sizes, for unequal variances, and (with extrapolation) for $p > 20$.

$$\text{LSD} = t_\nu(a/2)\sqrt{\frac{2}{n}}\sqrt{\text{MS}_{\text{ERROR}}}$$

is widely used, but it controls only the per-comparison error rate (and not the experiment-wise error rate). **For comparisons with a control, a better yardstick is given by Dunnett (multiple comparisons with a control):**

$$d' = tD\,(a/2;\,k,\,\nu)\sqrt{\frac{2}{n}}\sqrt{\text{MS}_{\text{ERROR}}}\,.$$

In a case with $k=21$ treatments at level $\alpha=.05$, with $n=3$ uses of each treatment, and $\nu=40$ (i.e., $(k-1)\times(n-1)$), we find (from tables)

$$tD\,(\alpha/2;\,k,\,\nu) = 3.06.$$

If $s=.441$ is found, then $d' = 1.10$ is the yardstick to be used. The following table, from "New tables for multiple comparisons with a control" by C.W. Dunnett, *Biometrics*, Vol. 20 (1964), pp. 482–491, gives the multiplier tD which is appropriate for setting a yardstick to compare p treatments with 1 standard, and is marked for the above example. (See Table 18.)

We should note that **these tables can also be used for cases with unequal sample sizes, for unequal variances, and (with extrapolation) for $p > 20$.**

In general, the best experiment (i.e., the most informative) is obtained by taking

$$\frac{n_s}{n_t} = \sqrt{p}$$

where n_s is the no. of observations on the standard, n_t is the no. of observations on each treatment, and p is the no. of treatments (excluding the standard). Thus, *if* the main item of interest is comparisons with a standard treatment, we would recommend that (if $n_t = 3$, as before) one should use

$$n_s = n_t\sqrt{p} = 3\sqrt{20} \approx 13.$$

Thus, equal rep. numbers on the treatments is reasonable, but the standard can (with a gain in correct inferences) be allocated a larger sample size. Since, in some of these experiments, variances appear to be greatly unequal, **we will now detail the procedure which should be used to compare treatments with a control with unequal sample sizes and/or variances.**

The procedures will be illustrated on the data in Table 19. Here we have $p = 14$ treatments and 1 control. An ANOVA yields a pooled variance estimate of $s^2 = 1.685$. [Note that, if the data are run in a randomized blocks design, it is in general preferable to enter Blocks as an ANOVA factor, because this will (especially if blocking was effective) generally yield a smaller s^2, and hence more precise comparisons.]

Table 19.

Treatment $= i$	$\overline{X}_i$	s_i^2	n_i
Std.	6.778	.536	18
1	1.857	5.143	7
2	8.000	1.000	5
3	7.000	.500	5
4	7.000	1.000	5
5	8.000	.000	5
6	6.800	.200	5
7	6.200	.700	5
8	6.200	.200	5
9	7.200	.200	5
10	6.800	1.200	5
11	7.400	.300	5
12	8.400	.300	5
13	8.200	1.200	5
14	8.400	.300	5

a. Case of Equal Variances. In this case, **a 95% confidence interval for the difference $\mu_i - \mu_0$** of mean response (between the ith treatment and the standard) **is given by**

$$(\overline{X}_i - \overline{X}_0) \pm |d|_{14,75}^{.05}\left(1 + \left(1 - \frac{n_i}{n_0}\right)\frac{a_{14,75}^{.05}}{100}\right) s\sqrt{\frac{1}{n_i} + \frac{1}{n_0}}$$

where:

$.05 = 1 - .95$ (95% confidence),

$14 = p =$ number of treatments,

$75 =$ number of degrees of freedom for s^2,

$\overline{X}_i =$ mean response for treatment i,

$\overline{X}_0 =$ mean response for standard,

$n_i =$ sample size for treatment i,

$n_0 =$ sample size for standard,

$s = \sqrt{s^2}$, $s^2 =$ error variance,

$|d|_{14,75}^{.05}$ comes from Table III, p. 488 of Dunnett (1964) and above, and

$a_{14,75}^{.05}$ is the superscript on $|d|_{14,75}^{.05}$ in the Dunnett table.

In our case $|d|_{14,75}^{.05}$ is not tabled (the values 14 and 75 fall between those in Table 20, so two-way linear interpolation is used to approximate it as shown in Table 20.

Thus, $|d|_{14,75}^{.05} = 2.89$. Similarly, we find the value for $a_{14,75}^{.05}$ by interpolation on the superscripts as follows as shown in Table 21.

Table 20: 2-Way Interpolation for $|d|^{.05}_{14,75}$

df \ p	12	14	15
60	2.86		2.92
75			
120	2.81		2.87

: from Dunnett's Table II

↓

df \ p	12	14	15
60	2.86		2.92
75	2.8475		2.9075
120	2.81		2.87

$$2.8475 = 2.81 + (2.86-2.81)\frac{75-120}{60-120}$$

$$2.9075 = 2.87 + (2.92-2.87)\frac{75-120}{60-120}$$

↓

df \ p	12	14	15
60	2.86		2.92
75	2.8475	2.89	2.9075
120	2.81		2.87

$$2.89 = 2.8475 + (2.9075-2.8475)\frac{14-12}{15-12}$$

Table 21: Interpolation for $a^{.05}_{14,75}$

df \ p	12	14	15
60	4.4		4.7
75			
120	4.1		4.4

: from Dunnett's Table II superscripts

↓

df \ p	12	14	15
60	4.4		4.7
75	4.325		4.625
120	4.1		4.4

$$4.325 = 4.1 + (4.4-4.1)\frac{75-120}{60-120}$$

$$4.625 = 4.4 + (4.7-4.4)\frac{75-120}{60-120}$$

↓

df \ p	12	14	15
60	4.4		4.7
75	4.325	4.53	4.625
120	4.1		4.4

$$4.53 = 4.325 + (4.625-4.325)\frac{14-12}{15-12}$$

Thus, $a^{.05}_{14,75} = 4.53$, and our interval on $\mu_i - \mu_0$ is given by

$$(\bar{X}_i-\bar{X}_0) \pm (2.89)\left(1+\left(1-\frac{n_i}{n_0}\right)\frac{4.53}{100}\right)\sqrt{1.685}\sqrt{\frac{1}{n_i}+\frac{1}{n_0}},$$

that is

$$(\bar{X}_i-\bar{X}_0) \pm (3.75)\left(1+\left(1-\frac{n_i}{n_0}\right)\frac{4.53}{100}\right)\sqrt{\frac{1}{n_i}+\frac{1}{n_0}}.$$

Performing this calculation for the 14 treatments in Table 19 yields Table 22, where the answer to "Is there a significant difference?" is "Yes" when zero is not in the given interval.

Table 22: Intervals and Significances

Difference	Interval	Sig. Diff.?
$\mu_1 - \mu_0$	(–5.686 , –4.156)	Yes
$\mu_2 - \mu_0$	(.232 , 2.212)	Yes
$\mu_3 - \mu_0$	(– .768 , 1.212)	No
$\mu_4 - \mu_0$	(– .768 , 1.212)	No
$\mu_5 - \mu_0$	(.232 , 2.212)	Yes
$\mu_6 - \mu_0$	(– .968 , 1.012)	No
$\mu_7 - \mu_0$	(–1.568 , .412)	No
$\mu_8 - \mu_0$	(–1.568 , .412)	No
$\mu_9 - \mu_0$	(– .568 , 1.412)	No
$\mu_{10} - \mu_0$	(– .968 , 1.012)	No
$\mu_{11} - \mu_0$	(– .368 , 1.612)	No
$\mu_{12} - \mu_0$	(.632 , 2.612)	Yes
$\mu_{13} - \mu_0$	(.432 , 2.412)	Yes
$\mu_{14} - \mu_0$	(.632 , 2.612)	Yes

This analysis shows treatment 1 to be inferior to the standard, while treatments 2, 5, 12, 13, and 14 are all superior. The mean differences are also quantified by the middle column of Table 22.

b. Case of Unequal Variances. In this case we proceed similarly to the case of Equal Variances, except that **the interval is now**

$$(\bar{X}_i - \bar{X}_0) \pm |d|^{.05}_{14,\nu_i} \left(1 + \left(1 - \frac{n_i}{s_1^2}\frac{s_0^2}{n_0}\right)\frac{a^{.05}_{14,\nu_1}}{100}\right)\sqrt{\frac{s_i^2}{n_i} + \frac{s_0^2}{n_0}}$$

where:

s_i^2 is calculated as per Table 19, and

$$\nu_i = \frac{\frac{s_i^2}{n_i}(n_i - 1) + \frac{s_0^2}{n_0}(n_0 - 1)}{\frac{s_i^2}{n_i} + \frac{s_0^2}{n_0}}.$$

c. Are Variances Equal or Unequal? Tests are available to assess the equality of variances. However, here it appears that, outside of treatment 1 (where data and its entry into the computer data base should be checked), variances will not be assessed unequal by either **Bartlett's test or the F-max test.** (This occurs if s_5^2 is not entered as a zero, but instead is approximated, bearing in mind that two readings of "8" are not exactly equal observations, but are rather two observations each of which is between 7.5^+ and 8.5^-.) It may therefore be reasonable here to delete treatment 1 (which is substantially inferior) and to recalculate the rest of the analysis as in the case of Equal Variances, but now use $p = 13$ and a *new* s^2. This will not result in many more significances (but it may yield one or two more). It is very undesirable to retain the observations made in treatment 1 and go to the case of Unequal Variances, as this is expected to spuriously submerge a number of the significant differences because of the low degrees of freedom attributable to the individual treatments (but their variances are not substantially unequal, so pooling is quite valid there).

Details of the Bartlett and F-max tests can be programmed as, for example, SAS routines. However, automatic use of these tests to choose between analyses in the cases of Equal Variance and Unequal Variances, as described above, is not recommended because (as seen in the example considered) it could lead to the use of a much less powerful (fewer significances) analysis when a more powerful analysis is appropriate. Thus, in any case, both analyses would be better performed, and the **data suggesting that the case of Unequal Variances should be chosen should then be analyzed for outliers, punching errors, etc., before the Case of Unequal Variances is used in preference to using the Case of Equal Variances.**

d. Comparison with the LSD technique. The above analyses (for Equal or Unequal Variance, as appropriate) **give us inferences** in Table 2 **for which we are 95% sure that all inferences made are correct. The LSD technique** will give more significances (and be easier to calculate), but it **does not satisfy this strong statement as to the validity of its inferences.**

In the case of Equal Variances, with LSD one tests the null hypothesis $\mu_0 = \mu_1 = \ldots = \mu_{14}$ via ANOVA. If the test rejects, the intervals computed (and examined for zero-inclusion to test individual differences via LSD) are

$$(\bar{X}_i - \bar{X}_0) \pm t^{.025}_{75}\, s \sqrt{\frac{1}{n_i} + \frac{1}{n_0}}$$

where $t^{.025}_{75}$ comes from tables of Student's t-distribution. In this example, $t^{.025}_{75} = 1.99$. In the case of Unequal Variances, LSD would use the intervals

$$(\bar{X}_i - \bar{X}_0) \pm t^{.025}_{\nu_i} \sqrt{\frac{s_i^2}{n_i} + \frac{s_0^2}{n_0}}$$

with ν_i as in the case of Unequal Variances described above.

e. Comparison with Duncan's Multiple-Range Test (MRT). In this example, it turns out that the treatments found (superior or inferior) are precisely those which Duncan's MRT makes those assertions for. However, **with Duncan's MRT one cannot be assured of the same 95% confidence that all statements will be correct. Also, the MRT gives no intervals.** Thus, using the MRT, one is not in a position to precisely assess whether statistically proven differences are of a size to have practical value, or whether non-proven differences could be large but may have failed to be detected simply because of large underlying variability.

f. Recommendations. For reasons discussed in d. and e., **the Dunnett procedures of a. and b., with careful use of equality of variances tests as in c., are recommended for customary use** in analysis of means in comparison-with-a-standard-type studies.

QUANTITATIVE VARIABLE DESIGNS

Existing Data Sets. It is often desired to analyze existing data sets, e.g. where an output Y at each of a number of points where the variables $X_1, X_2, X_3, \ldots, X_k$ have been measured. **This setting is extremely different than that in which experiments have been run at each of a number of preselected settings of $X_1, \ldots, X_k$,** and **often little can be salvaged from such data even with the most thorough statistical analysis. Some reasons for this are:**

- **the X's may be highly correlated with each other;** hence it may not be possible to separate an effect as being due to, for example, X_1 or X_2
- **The X's may have been manipulated** in order to try to control the "output" Y of the process (some of them perhaps even manipulated in directions which move the output in directions which are not that which is desired), hence giving spurious indications of the directions of effects when analyzed.
- **The X's may cover a very small part of the possible operating range,** so small that any indications of changes in Y attributable to changes in the X's may be overwhelmed by the size of the standard deviation of the process.
- **Other variables which affect the output of the process** (e.g., time of day, atmospheric conditions, operator running the process, etc.) **may not have been held constant**, and may in fact be the real causes of changes observed in the process (while an analysis conducted based only on the X's may erroneously conclude a model based on X's which has no basis in reality).

For these and other reasons, much more information can generally be obtained from a carefully designed experiment than can be obtained from extensively analyzing historical data sets collected in uncontrolled circumstances. The best that one can usually hope for from such a historical data set analysis is **an indication of the most important variables to include in the designed experiment.**

One method useful for examining data sets (in their X's) to see if they are obviously unsuitable for analysis is that of **crossplots**. For example, suppose that an output Y has been measured for each of a number of values of the variables X_1, X_2, X_3, X_4, and X_5. Then one plots a dot (or other symbol) at each coordinate point where a Y was observed, i.e. on each of the 10 pairs of axes

$$\begin{array}{r} X_1 \text{ vs. } X_2, X_1 \text{ vs. } X_3, X_1 \text{ vs. } X_4, X_1 \text{ vs. } X_5 \\ X_2 \text{ vs. } X_3, X_2 \text{ vs. } X_4, X_1 \text{ vs. } X_5 \\ X_3 \text{ vs. } X_4, X_3 \text{ vs. } X_5 \\ X_4 \text{ vs. } X_5. \end{array}$$

If the points on any one or more of these plots do not "cover" the space well, then this is an indication that the data set is not suitable for analysis; however, even if they do cover well, the data may still be unsuitable because these bivariate plots do not account for poor coverage of triples (or quadruples, etc.). For example, if each of $X_1, \ldots, X_5$ had two levels and whenever either X_1 or X_2 was at its larger level then X_3 was also at its larger level, any conclusions drawn from the analysis would be very risky in view of the relationship shown . . . at best one could obtain some (shaky) input for the experimental design process.

VALIDITY AND VARIABILITY OF MEASUREMENT

It is often the case, when variability of results is a problem, that a substantial part of the problem is due to **measurement error** (e.g., in the laboratory analysis of specimens). This is true when samples are submitted to independent laboratories for analysis, and also when the laboratory is within the same company. **Careful study of laboratory error is strongly recommended before any experimental effort is initiated which will be costly or conducted over a long period of time,** particularly when that error has not been subject to such studies (in designed experiments) in the recent past.

Let us now study **how process variability and laboratory error combine into the measurement variability seen by the experimenter.**

If we are interested (for example) in Y at some specified set of operating conditions $X_1, X_2, \ldots,$ we may run m_1 replicate experiments at those conditions. Assuming independent replications, we will have $E(Y)$ at these conditions estimated by $\bar{Y}$, with variance

$$\text{Var}(\bar{Y}) = \frac{\sigma^2}{m_1}.$$

If the **process variability is σ_1^2 and** the **measurement variability is σ_2^2, then,** assuming independence of process and measurements,

$$\sigma^2 = \sigma_1^2 + \sigma_2^2.$$

If each experiment's output is sampled and analyzed m_2 times, independently, this becomes

$$\sigma^2 = \sigma_1^2 + \frac{\sigma_2^2}{m_2}.$$

If, for example, $\sigma = 8$ for Y values, implications of measurement error are then as follows: If $m_1 = 1$, the precision of observed values as affected by m_2 is shown in Table 23.

Table 23

σ_2^2 as a proportion of σ^2	m_2	Std. Dev. (Y reported)
0.00	1	8.00
	2	8.00
	3	8.00
0.50	1	8.00
	2	6.93
	3	6.53
0.75	1	8.00
	2	6.32
	3	5.66
1.00	1	8.00
	2	5.66
	3	4.62

If σ_2^2 is 50% of σ^2, then eliminating σ_2^2 (making it ≈ 0.00) would mean 5 experimental replications would yield as precise an estimate of $E(Y)$ then as 10 replications would before.

There are also implications regarding control of a process.

Index

Author Biographies

Dr. Edward J. Dudewicz is Professor and Chairman, University Statistics Council, at Syracuse University, Syracuse, New York. Dr. Dudewicz has previously authored four books, including *The Handbook of Random Number Generation and Testing* (1981), as well as the forthcoming *Modern Mathematical Statistics* (Wiley). He is a consultant and short-course instructor for both industry and government. His awards include the JACK YOUDEN PRIZE and JACOB WOLFOWITZ PRIZE, as well as designation as a FELLOW of the American Statistical Association, American Society for Quality Control, Institute of Mathematical Statistics, and New York Academy of Sciences. He is an Elected Member of the International Statistical Institute, and has authored over 100 technical publications. He serves as Editor of the *American Journal of Mathematical and Management Sciences* and of *Statistics & Decisions* (West Germany), and has been Invited Lecturer/Visiting Professor at Stanford University, University of Louvain (Belgium), University of London, and in Holland, India, Saudi Arabia, and Australia. His current consulting includes statistics and law, and classification via nuclear magnetic resonance (NMR).

Zaven A. Karian is a Professor and Chairman, Department of Mathematical Sciences at Denison University. He has a B.A. degree from American International College, an M.A. degree from the University of Illinois, and the M.S. and Ph.D. degrees from The Ohio State University. Dr. Karian is frequently invited to give lectures on computer simulation as well as other computer science topics. He is the author of articles on computer simulations and has served as a consultant to many organizations. Dr. Karian has been particularly active in offering workshops and Professional Development Seminars on computer simulation sponsored by academic institutions and by professional organizations.